Guide to Stability Design Criteria for Metal Structures

Guide to Stability Design Criteria for Metal Structures

4th Edition

Edited by Theodore V. Galambos

Professor of Civil Engineering
University of Minnesota

WILEY

A Wiley-Interscience Publication

John Wiley & Sons

New York • Chichester • Brisbane • Toronto • Singapore

***Library of Congress Cataloging in Publication Data*:**

Guide to stability design criteria for metal structures.

"A Wiley-Interscience publication."
Bibliography: p.
Includes indexes.
1. Columns. 2. Girders. I. Galambos, T. V.
(Theodore V.)

TA660.C6G85 1988   624.1'82   87-23212
ISBN 0-471-09737-3

Printed in the United States of America

10 9 8 7 6 5 4 3

Foreword

Since its founding in 1944, the principal objectives of the Structural Stability Research Council (formerly the Column Research Council) have been to foster research on the behavior of compressive components of metal structures and to assist in the development of improved design procedures. The Council provides guidance to practicing engineers and writers of design specifications, codes, and standards in offering both simplified and refined procedures applicable to design and assessing their limitations.

The initial outline of the guide was prepared in 1956 by Lynn S. Beedle and Jonathan Jones. The first edition, published in 1960, was dedicated to the Council's first chairman with these words: "As first Chairman of Column Research Council, Shortridge Hardesty gave freely for twelve years his time, devotion, and material assistance. His mind grasped both the practical problems of engineering application and the fundamental knowledge necessary to research. His influence was a personal inspiration to all who worked in Column Research Council."

There are three new chapters in the fourth edition of the guide. (The first edition had five chapters, the second seven, and the third nineteen, whereas the fourth edition has twenty-one.) Among the subjects that are new in the fourth edition are an introduction to stability theory, a chapter on box girders, and the application of the finite-element method to the solution of stability problems. Fourteen of the nineteen chapters from the third edition are either completely new or have been extensively reworked.

The Council was fortunate in having Bruce G. Johnston as editor of the first three editions of the guide and is indebted to him for the time and effort he has devoted to this work. As a result of Dr. Johnston's guidance, inspiration, and example, the Task Groups of the SSRC have produced the world's foremost collection of ideas and information on the subject of the stability of metal structures in the first three editions of the guide. This present fourth edition is but an evolutionary extension of the pioneering earlier editions.

The concept that the Task Groups of the SSRC would have the responsibility for most individual chapters was retained in the fourth edition. Additional substantial contributions were provided by others, who wrote drafts of chapters or who reviewed manuscripts at various stages of completion of the guide. We sincerely thank all those who had a hand in this effort. (See the end of Chapter 1 for a list of those primarily responsible for each chapter.) Special thanks go to Dr. Bruce G. Johnston for preparing the index for the fourth edition.

The first edition of the guide received special financial support from the Engineering Foundation and the Association of American Railroads. Costs in preparing the second edition were borne jointly by the American Institute of Steel Construction and the Column Research Council. Preparation of the third edition was supported by the National Science Foundation and the American Institute of Steel Construction. The preparation of the fourth edition was supported by a grant from the National Science Foundation. The Council further enjoys the financial support of many organizations, listed in Appendix C, which provide for its continuity and make it possible to sponsor publication of the guide. Through the members of these organizations we also maintain the vital creative interaction between structural engineering practice and research.

Committee on the *Guide*

B. G. Johnston, Chairman
L. S. Beedle
T. V. Galambos, Editor
J. S. B. Iffland
J. Springfield

Minneapolis
November 1987

Contents

Notation and Abbreviations **xiii**

Chapter One Introduction **1**

1.1 The Metal Column 1
1.2 Scope and Summary of the Guide 2
1.3 Mechanical Properties of Structural Metals 3
1.4 Definitions 6
1.5 Postbuckling Behavior 10
1.6 Credits for the Chapters in the Fourth Edition of the SSRC Guide 12

Chapter Two Stability Theory **15**

2.1 Introduction 15
2.2 Bifurcation Buckling 15
2.3 Limit Load Buckling 23

Chapter Three Centrally Loaded Columns **27**

3.1 Introduction 27
3.2 Column Strength 28
3.3 Influence of Residual Stress 33
3.4 Influence of Out-of-Straightness 45
3.5 Influence of End Restraint 48
3.6 Effective Length of Columns 51
3.7 Lateral Bracing Requirements 55
3.8 Strength Criteria for Steel Columns 57
3.9 Aluminum Columns 66
3.10 Development of Column Strength Criteria 81

Chapter Four Plates 89

4.1 Introduction 89
4.2 Local Buckling and Postbuckling Strength of Plates 90
4.3 Interaction Between Plate Elements 117
4.4 Local Buckling and Postbuckling Strength of Stiffened Plates 122
4.5 Buckling of Orthotropic Plates 138
4.6 Laterally Loaded Plates in Compression 146

Chapter Five Laterally Unsupported Beams 155

5.1 Introduction 155
5.2 Elastic Lateral-Torsional Buckling 157
5.3 Inelastic Lateral-Torsional Buckling 170
5.4 Bracing Requirements 171
5.5 Design of Laterally Unsupported Beams 177
5.6 Stability of Horizontally Curved Beams 181

Chapter Six Plate Girders 189

6.1 Introduction 189
6.2 Web Buckling as a Basis for Design 190
6.3 Shear Strength of Plate Girders 192
6.4 Girders with No Intermediate Stiffeners 204
6.5 Bending Strength of Plate Girders 205
6.6 Combined Bending and Shear 208
6.7 Plate Girders with Longitudinal Stiffeners 211
6.8 End Panels 219
6.9 Design of Stiffeners 220
6.10 Panels under Edge Loading 223
6.11 Fatigue 230
6.12 Design Principles and Philosophies 231
6.13 Research Needs 233

Chapter Seven Box Girders 241

7.1 Introduction 241
7.2 Bases of Design 242
7.3 Buckling of Wide Flanges 246
7.4 Bending Strength of Box Girders 263
7.5 Shear Strength of Box Girders 263
7.6 Combined Bending and Shear Strength of Box Girders 266

7.7 Influence of Torsion on Strength of Box Girders 269
7.8 Diaphragms 269
7.9 Unstiffened Diaphragms 272
7.10 Stiffened Diaphragms 277
7.11 Problems Needing Further Research 279

Chapter Eight Beam-Columns **283**

8.1 Introduction 283
8.2 Strength of Beam-Columns 285
8.3 Uniaxial Bending: In-plane Strength 287
8.4 Uniaxial Bending: Lateral-Torsional Buckling 297
8.5 Equivalent Uniform Moment Factor 305
8.6 Biaxial Bending 306
8.7 New Design Approaches 315
8.8 Cyclic Loading of Beam-Columns 317
8.9 Eccentrically Loaded Angle Struts 317

Chapter Nine Tapered Structural Members **329**

9.1 Introduction 329
9.2 Frame Analysis 330
9.3 Stress Analysis 337
9.4 Stability of Tapered Members 341
9.5 Development of Axial Compression Formulas 343
9.6 Development of Bending Formulas 349
9.7 Interaction: Axial Compression and Bending 356
9.8 Summary 357

Chapter Ten Composite Columns

10.1 Introduction 359
10.2 Strength of Composite Cross Sections 363
10.3 Strength of Slender Composite Columns 377
10.4 Shear Strength of Composite Columns 383
10.5 Connections to Composite Columns 383

Chapter Eleven Columns with Lacing, Battens, or Perforated Cover Plates **387**

11.1 Introduction 387
11.2 Shear in Built-Up Columns 387
11.3 Effect of Shear Distortion on Critical Load 389

11.4 Laced Structural Members 390
11.5 Columns with Battens 394
11.6 Stay Plates and Spaced Columns 399
11.7 Columns with Perforated Plates 403
11.8 Design of Lattice Towers and Masts 405

Chapter Twelve Mill Building Columns 409

12.1 Introduction 409
12.2 Effective Length of Stepped Columns 411
12.3 Design Procedure for Stepped Columns 413
12.4 Design Trends and Research Needs 420

Chapter Thirteen Thin-Walled Metal Construction 423

13.1 Introduction 423
13.2 Flexural Members 424
13.3 Compression Members 434
13.4 Diaphragm Action of Thin-Walled Panels 446
13.5 Bracing Requirements 446
13.6 Stainless Steel Structural Members 449
13.7 Aluminum Members 451

Chapter Fourteen Circular Tubes and Shells 461

14.1 Introduction 461
14.2 Description of Buckling Behavior 465
14.3 Unstiffened or Heavy-Ring-Stiffened Cylinders 467
14.4 General Instability of Ring-Stiffened Cylinders 488
14.5 Stringer- or Ring-and-Stringer-Stiffened Cylinder 496
14.6 Effects on Column Buckling 498
14.7 Cylinders Subjected to Combined Loadings 502

Chapter Fifteen Members with Elastic Lateral Restraints 515

15.1 Introduction 515
15.2 Buckling of the Compression Chord 517
15.3 Effect of Secondary Factors on Buckling Load 523
15.4 Top-Chord Stresses Due to Bending of Floorbeams and to Initial Chord Eccentricities 523
15.5 Design Procedures 524
15.6 Plate Girder with Elastically Braced Compression Flange 527
15.7 Guyed Towers 528

Chapter Sixteen Frame Stability **531**

16.1 Introduction 531
16.2 Behavior of Frames 532
16.3 Analytical Models 536
16.4 Factors Influencing Frame Stability 540
16.5 Stability Design Procedures 542
16.6 Stability of Frames Subjected to Dynamic Loads 558
16.7 Special Topics 565
16.8 Concluding Remarks 570

Chapter Seventeen Arches **575**

17.1 In-Plane Stability of Arches, Introduction 575
17.2 In-Plane Linear Stability 577
17.3 In-Plane Nonlinear Elastic Stability 581
17.4 In-Plane Ultimate Load 585
17.5 Design of Arches for In-Plane Stability 592
17.6 Out-of-Plane Stability of Arches, Introduction 593
17.7 Out-of-Plane Buckling of Circular Arches 594
17.8 Out-of-Plane Buckling of Parabolic Arches 597
17.9 Braced Arches and Requirements for Bracing Systems 597
17.10 Ultimate Strength of Steel Arches Subjected to Uniformly Distributed Vertical Loads 599
17.11 Ultimate Strength of Steel Arch Bridges Subjected to Vertical and Lateral Uniform Loads 603

Chapter Eighteen Doubly Curved Shells and Shell-Like Structures **609**

18.1 Introduction 609
18.2 General Considerations 612
18.3 Local Buckling of Spherical Shells 613
18.4 Local Buckling: Spherical Shell-Like Structures 618
18.5 General Buckling: Spherical Shell-Like Structures 619
18.6 Member Buckling 623
18.7 Design Trends and Research Needs 623

Chapter Nineteen Selected Topics in Dynamic Stability **627**

19.1 Introduction 627
19.2 Parametric Resonance 628
19.3 Stability of Impulsively Loaded Columns 634

19.4 Dynamic Snap-Through of Shallow Structures 640
19.5 Flow-Induced Instability 643
19.6 Suddenly Loaded Structures 648

Chapter Twenty Structural Safety **661**

20.1 Introduction 661
20.2 Concepts of Structural Safety 661
20.3 Reliability of Axially Loaded Columns 666
20.4 Conclusions 670

Chapter Twenty-One Finite-Element Analysis of Stability Problems **673**

21.1 Introduction 673
21.2 Formulation of Finite-Element Analysis for Elastic Instability 674
21.3 Computational Procedures for Finite-Element Stability Analysis 678
21.4 Prismatic Members 684
21.5 Framework Instability Analysis 685
21.6 Flat Plates 686
21.7 Shell Buckling 690
21.8 Finite-Element Software 692

Appendix A General References on Structural Stability **699**

Appendix B Technical Memorandums of Structural Stability Research Council **701**

B.1 The Basic Column Formula 701
B.2 Notes on the Compression Testing of Metals 703
B.3 Stub-Column Test Procedure 708
B.4 Procedure for Testing Centrally Loaded Columns 717
B.5 General Principles for the Stability Design of Metal Structures 732
B.6 Determination of Residual Stresses 734
B.7 Tension Testing 744

Appendix C Structural Stability Research Council **751**

Name Index **757**

Subject Index **781**

Notation

A	A coefficient. Area of cross section. Flexural stiffness of arch cross section about x-axis. Arc length in a shell.
A_1	Internal area of cylindrical tube.
A_2	External area of cylindrical tube.
A_b	Area of all battens within one batten spacing.
A_c	Area of compression flange. Area of longitudinal in laced or battened member.
A_d	Area of diagonal in a laced member.
A_e	Area of cross section remaining elastic. Required area of plate-girder end plate.
A_{eff}	Effective area in a thin-walled section such that $P = \sigma_y A_{eff}$.
A_f	Area of flange.
A_g	Gross area of composite column.
A_S	Area of large stiffener plus total area of shell between small stiffeners.
A_s	Area of steel in a composite column. Area of stiffener cross section. Area of small stiffener plus total shell area between stiffeners.
A_{st}	Area of intermediate stiffener.
A_t	Area of tension flange.
A_u	Area of plate-girder bearing stiffener.
A_w	Area of web.
a	Length of side of stiffened plate. Length of perforation in a perforated plate. Torsion bending constant for an I-section. Coefficient defining stable region under dynamic load. Distance between plate-girder stiffeners.
B	A coefficient. Coefficient in postbuckling plate formula. Coefficient to correct for a one-sided stiffener. Flexural stiffness of an arch about y-axis.

$\bar{B}$	Coefficient in postbuckling plate formula.
B_c	Coefficient in design formula for aluminum columns.
B_p	Coefficient in design formula for aluminum-alloy plates.
$B_x(B_y)$	Distributed bending stiffness in a stiffened plate about $x(y)$-axis.
b	Width of rectangular cross section. Width of plate. Width of pony-truss bridge, center-to-center of trusses. Length of short side of a box section, center-to-center of long sides. Length of side of stiffened plate. Transverse distance from edge of a perforation to nearest line of longitudinal fasteners. Coefficient defining stable region under dynamic load.
b_c	Width of compression flange.
b_e	Effective plate width.
b_f	Width of flange of a W shape.
b_t	Width of tension flange.
b_w	Width between centers of flanges in a wide-flange column.
C	A coefficient. Transverse pony-truss bridge frame spring constant, particularly the least one. Torsional rigidity. Coefficient in formula for design of one-sided stiffeners.
C_1, C_2, C_3, C_4	Coefficients for lateral-torsional buckling of a beam.
C_{AA}, C_{BB}, C_{AB}	Coefficients in generalized slope-deflection equations.
C_a	Membrane stiffness of a stiffened cylinder.
C_b	Coefficient for a laterally unsupported bent beam. Bending stiffness of a stiffened cylinder.
$C_{b\gamma}$	Coefficient for a tapered beam.
C_c	Column formula coefficient.
C_m	Coefficient to determine equivalent uniform moment in a beam-column.
C_{req}	Required transverse pony-truss bridge frame spring constant.
C_w	Torsional warping constant.
c	Distance to extreme fiber of beam or column section in bending. Distance center-to-center of perforations in a perforated plate. Distance from middle plane of channel web to centroid of section. Distance from stiffener to assumed location of plastic hinge in plate-girder flange.
D	Flexural rigidity of a plate per unit width. Mean diameter of a cylindrical tube.
D_c	Coefficient in design formula for aluminum columns.
D_p	Coefficient in design formula for aluminum plates.

d	Depth of a section. Diameter of circular cross section. Long side of box section, center-to-center of short sides. Transverse distance between lines of longitudinal fasteners in a perforated plate. Stiffener spacing for a stiffened plate.
d_e	Diameter of elastic portion of a circular cross section.
d_z	Depth of tapered member at z.
E	Modulus of elasticity.
E_r	Reduced modulus.
E_s	Secant modulus.
E_{st}	Strain-hardening modulus (initial).
E_t	Tangent modulus.
e	Distance from centroid of girder cross section to shear center (positive if shear center lies between centroid and compression flange, otherwise negative). Distance from shear center to the middle plane of a channel web. Width of end post in a plate girder. Eccentricity of end load in a beam-column. Initial out-of-roundness of an unstiffened shell.
e_0	Assumed equivalent eccentricity (representing defects, etc.).
F_a	Allowable average compressive stress in axially loaded members.
F_{a0}	Allowable stress for a column having zero slenderness ratio.
F_b	Allowable compressive bending stress.
F_c	Maximum allowable compressive stress on unstiffened element (AISI).
F_{cr}	Critical thrust load applied at ends of an arch.
F_s	Allowable stress for steel in a composite column. Axial force in a transverse plate-girder stiffener.
F_y	Yield point of structural steel.
FS	Factor of safety.
f_a	Average compressive stress due to axial load.
f_b	Compressive stress due to bending moment.
f_v	Average shear stress in a beam web.
f'_c	Compressive strength of concrete.
f''_c	Average compressive strength of concrete at ultimate load.
G	Elastic shear modulus.
G, G_A, G_B	Joint bending stiffness ratio. (Subscripts apply to respective ends of the column.)

G_{eff} Effective shear modulus of diaphragm.
G_{st} Shearing strain-hardening modulus (initial).
g Distance from shear center of girder to point of application of transverse load (positive when load is below shear center, otherwise negative). Equivalent-length factor in a tapered beam. End-panel length of a plate girder.

H Horizontal component of arch thrust. Distributed torsional stiffness in a stiffened plate. Horizontal force at a multistory frame floor level.
H'_i Sway force in multistory frame due to vertical loads.
h Depth of a rectangular cross section. Clear depth of plate girder or beam web between flanges. Depth of pony truss at truss vertical, measured from center of floorbeam to center of top chord. Long side of box section, center-to-center of short sides. Distance between beam- or girder-flange centroids. Rise of arch axis. Effective-length factor in a tapered beam.
h_e Distance to compression-flange centroid from centroid of section.
h_f Distance between flange centroids.
h_i Height of story i in building frame.
h_s, h_w Effective-length factors in tapered beams.
h_t Distance to tension-flange centroid from centroid of section.
h_w Depth of web.

I Moment of inertia of cross section.
I_b Moment of inertia of floorbeam in a pony truss.
I_c Moment of inertia of column cross section. Moment of inertia of compression flange about the y-axis. Moment of inertia of truss vertical in a pony truss. Moment of inertia of longitudinal component of spaced column. Moment of inertia at crown of an arch.
I_E Moment of inertia of large stiffener plus effective width of shell L_E.
I_e Moment of inertia of cross section remaining elastic.
I_{eq} Equivalent uniform moment of inertia for buckling of a nonuniform arch.
I_{FE} Value of I_E that makes the large ring stiffener in a shell equivalent to a bulkhead.
I_g Moment of inertia of girder cross section.
I_0 Polar moment of inertia about shear center. Moment of

	inertia of a segment of a stiffened plate. Moment of inertia at smaller end of a tapered beam.
I_p	Polar moment of inertia of cross section about centroid.
I_s	Moment of inertia of transverse stiffener.
I_t	Moment of inertia of tension flange about y-axis.
I_x, I_y	Moment of inertia of cross section, x and y denoting the coordinate axes.
$(I_y)_{\text{eff}}$	Effective moment of inertia about y-axis.
I_{yc}, I_{yt}	Moment of inertia of compression and tension portions of section about axis parallel to the web, respectively.
J	Torsion constant. Coefficient in AASHTO formula for transverse stiffener design.
$J_{DX}(J_{DY})$	Torsion constant of stiffeners in $X(Y)$ direction.
j	Lateral-torsional buckling constant. Number of panels in a stiffened plate.
K	Effective- or equivalent-length factor. Spring constant. Coefficient in edge-loading analysis of a plate-girder web.
K'	Modified effective-length factor.
$K_{11}, K_{22}, K_{33}, K_{23}$	Coefficients in flexural-torsional buckling.
K_γ	Effective-length factor in a tapered beam.
K_m	Average effective-length factor of all panel length compression chords in a pony truss.
K_s	Elastic support restraint coefficient for a shallow arch. Horizontal elastic support coefficient for an arch.
k	Coefficient of proportionality. Coefficient applied in plate buckling. Elastic foundation modulus. $\sqrt{P/EI}$
k_1, k_2, k_3	Coefficients in design formulas for aluminum-alloy plates.
k_h	Local buckling parameter for box columns.
k_s	Buckling coefficient for a plate in pure shear.
k_w	Local buckling parameter for wide-flange columns.
L	Span of an arch.
L, l	Length of member, particularly a laterally unbraced length.
L_b	Length of shell between bulkheads.
L_c	Unbraced length of a column.
L_e	Effective width of shell acting as part of a stiffener.
L_F	Center-to-center spacing of large shell stiffeners.
L_f	Center-to-center spacing of stiffener rings.
L_g	Unbraced length of a girder.
LF	Load factor.
l	Panel length in a pony-truss bridge.

M	Bending moment.
M', M''	Rotational stiffness of near end of member with far end fixed or hinged, respectively, but with no end translation.
$\bar{M}', \bar{M}''$	Rotational stiffness of near end of member with far end fixed or hinged, respectively, but with near end translationally restrained by a linear spring.
M_{AB}	The moment at $A(B)$ acting on member AB.
M_a, M_b	End moments acting on a beam-column at ends a and b, respectively.
M_c	Critical-bending moment.
M_e	End moment for a framed column.
M_{eq}	Equivalent uniform moment in a beam column.
M_{exp}	Resisting moment by test of plate girder.
M_f	Plate-girder moment contributed by flanges.
$M_{\max}$	Maximum-bending moment.
M_0	Applied-end moment.
M_0, M_{0x}, M_{0y}	Moment in a beam-column without regard to moment caused by deflection.
M_p	Plastic-bending moment.
M_{pc}, M_{pcx}, M_{pcy}	Plastic bending moment modified by axial force.
M_{px}, M_{py}	Plastic-resisting moment about indicated axis.
M_{th}	Theoretical plate-girder resisting moment.
M_u, M_{ux}, M_{uy}	Ultimate bending moment of a plate girder. Ultimate bending moment in the absence of axial load in a laterally unsupported beam-column.
M_{ucx}, M_{ucy}	Maximum moments resisted by a biaxially loaded beam-column in the presence of axial load.
M_w	Warping torsional moment.
M_x, M_y, M_z	Moment about coordinate axes x, y, and z, respectively.
M_y	Yield moment.
m	Width of a perforation in a perforated plate. Mass per unit length of column. Buckling coefficient in an arch. Coefficient in tapered beam analysis. Number of panels in transversely stiffened plate. Coefficient in out-of-plane arch buckling.
m_{pb}, m_{pt}	Plastic resisting moment of bottom and top flanges of a plate girder.
m_{pw}	Plastic moment of web.
N	Nominal axial load. Number of panels in a stiffened plate.
N, n	A factor of safety.

$N_x(N_y)$	Force per unit length in $x(y)$ direction.
n	Number of perforated plates used in a column. Number of parallel planes of battens in a battened column. Number of panels in a pony truss. Number denoting an individual compression member as one of several meeting at a common joint. Coefficient in tapered beam analysis. Number of circumferential lobes in a shell at collapse.
P	Column axial load. Concentrated load on a plate-girder web.
$P(t)$	Variable axial force.
P_1, P_2	Axial compression force in truss member. (Subscripts refer to first and second member, respectively.)
$P_{\phi e}$	Critical load for pure torsional buckling.
P_a	Permissible axial load on composite column.
P_c	Chord stress in a truss at maximum load.
P_{cr}	Critical load.
P_e, P_{ex}, P_{ey}	Euler buckling load. Critical thrust at quarter points for a uniform arch.
P_F	Probability of failure.
P_f	Conservative estimate of failure load in torsional-flexural buckling.
P_k	Critical load in kth mode.
P_{max}	Maximum column load.
P_n	Axial compressive force of nth member.
P_0	Static component of dynamic load.
P_p	Column load at proportional limit.
P_r	Reduced-modulus column load.
P_S	Probability of survival.
P_{TF}	Critical load in combined flexural-torsional buckling.
P_t	Tangent-modulus column load.
P_u	Ultimate load of axially loaded column. Ultimate patch load on plate-girder web.
P_{ue}	Ultimate eccentric load.
$P_{uex}(P_{uey})$	Failure loads for bending about $x(y)$-axis for long column.
P_{ux}	Failure load of long column under axial load constrained to permit bending only about major axis.
P_y	Column axial load at full-yield condition.
p	Difference between uniform loads on two halves of an arch.
p_1	Internal pressure in a cylindrical tube.
p_2	External pressure on a cylindrical tube.

p_B	Critical pressure of small stiffener plus total area of shell between stiffeners.
p_c	Critical external pressure to cause buckling between stiffeners.
p_{crE}	Critical pressure for elastic buckling of a spherical shell-like structure.
p_{crP}	Critical pressure for plastic buckling of a spherical shell-like structure.
p_e	Elastic buckling pressure on a shell.
p_F	Critical pressure in shell buckling.
p_i	Inelastic buckling pressure on a shell.
p_y	Hydrostatic pressure at initial yield in unstiffened shell with initial out-of-roundness.
Q	Transverse shear in centrally loaded column. Form factor for thin-walled section. Shear rigidity of diaphragm. Unit radial load on a shell stiffener ring.
Q_a	Area factor to modify members composed entirely of stiffened elements.
Q_s	Stress factor to modify members composed entirely of unstiffened elements.
R	Mean radius of cylindrical tube. Resistance, or load capacity. Radius of a circular arch. Spherical radius of a shell.
R_1	A specific low value of resistance strength chosen to ensure a safe design.
R_c	Radius to centroidal axis of combined stiffener and effective shell ring.
$R_c(R_s)$	Ratio of critical compressive (shear) stress in combined shear and direct stress to critical compressive (shear) stress in pure compression (shear).
R_d	Radius to centroidal axis of large stiffener plus effective width of shell.
R_m	Mean value of resistance strength.
R_n	Nominal value of resistance strength.
R_0	Outside radius of shell.
r	Radius of gyration of member.
r_c	Radius of gyration of concrete core.
r_e	Radius of gyration of a stiffener-panel combination.
r_f	Radius of gyration of column flange.
r_0	Polar radius of gyration of the cross section about its shear center. Radius of gyration of one chord in a battened column.

r_s	Radius of gyration of steel tube.
r_x	Radius of gyration about the centroidal axis *x-x* (strong axis).
r_y	Radius of gyration about the centroidal axis *y-y* (weak axis).
S	Load, in a generalized sense. Circumferential half length of an arch axis from springline to centerline.
S_1	A specific high value of load chosen to provide a safe design.
S_c	Section modulus for compression.
S_{eff}	Effective section modulus.
S_L	Section modulus of longitudinal stiffener.
S_T	Section modulus of transverse stiffener in a plate girder.
S_t	Section modulus for tension.
S_x	Section modulus about *x-x* axis.
S_{xc}	Section modulus for compression stress about *x-x* axis.
SF_x	Safety factor applied to longitudinal stress.
$s_x(s_y)$	Spacing of stiffeners in $x(y)$ direction.
T	Tensile residual stress designation.
T1, T2, T3, T4	Heat treatment designations for aluminum alloys.
t	Thickness of plate or tubular wall.
t_B	Effective-bending thickness of a shell-like structure.
t_b	Thickness of side b of box-section column.
t_c	Thickness of compression flange.
t_h	Thickness of side h of box-section column.
t_m	Effective-membrane thickness of a shell-like structure.
t_t	Thickness of tension flange.
t_w	Thickness of web plates of box-section beam. Thickness of web.
U	Cube strength of concrete.
u	Displacement in the x direction.
V	Transverse shear force in plate girder.
V_i'	Additional shear at *i*th floor due to sway forces.
V_σ	Shear strength of plate girder due to tension-field action.
V_τ	Shear strength of plate girder due to beam action.
V_{ex}	Experimental value of plate-girder shear strength.
V_p	Plastic shear strength of plate girder.
V_{th}	Theoretical value of plate-girder shear strength.
V_u	Ultimate shear strength of plate girder.

V_w	Shear strength of plate-girder web.
v	Displacement in the y direction.
W	Uniformly distributed total lateral load in a beam-column.
W	Wide-flange shape symbol.
WW	Welded wide-flange shape.
W_{ar}, W_{bm}	Relative magnitudes of total load carried by arch and beam action respectively, in a shallow arch.
W_{cr}	Concentrated critical load at a node in a shell-like structure.
w	Uniform load intensity. Displacement in the z direction. Unit weight of concrete in a composite column. Distributed radial load on arch.
w_c	Critical radial load at arch or ring buckling. Critical load for out-of-plane arch buckling.
$(w_c)_s$	Sinusoidal load for symmetrical buckling of a shallow arch.
$(w_c)_u$	Sinusoidal load for unsymmetrical buckling of a shallow arch.
w_e	Critical uniform buckling load for an arch.
$(w_u)_c$	Unsymmetrical buckling of an arch under uniform load.
$(w_u)_{cs}$, $(w_u)_{cu}$	Uniform load for symmetrical and unsymmetrical buckling, respectively, for a shallow arch.
X_e	Width of rectangular cross section remaining elastic.
X-X, x-x	Coordinate axis.
x	Coordinate axis, particularly a principal axis. A distance.
x_0	Distance between the shear center and the centroid in the direction of the x-axis.
Y_e	Depth of rectangular cross section remaining elastic.
Y-Y, y-y	Coordinate axis.
y	Coordinate axis, particularly a principal axis.
y_c	Distance from centroidal axis x-x to face of tee flange. Height of loaded arch axis at midspan. Distance from neutral axis to compression edge web.
$(y_c)_s$	Height of arch axis at midspan at symmetrical buckling.
$(y_c)_u$	Height of arch axis at midspan at unsymmetrical buckling.
y_0	Distance between the shear center and the centroid in the direction of the y-axis. Portion of plate-girder web in compression.
y_s	Distance from neutral axis to longitudinal stiffener.
y_t	Distance from neutral axis to tension edge of web.
Z	Plastic modulus.
z	Coordinate axis. Distance along the z-axis.

α	Aspect ratio a/b or a/h for stiffened plates. Load ratio P/P_e. Ratio of moments of inertia of adjacent framed columns. Buckling coefficient for uniform arch. Angle subtended by whole span of a circular arch.
α_1	Buckling coefficient for a nonuniform arch.
β	Constant for stiffened plates. Angle of twist of cross section. Buckling parameter in a stiffened arch. Buckling parameter for a shallow arch. Equivalent-length modification factor for unsymmetrical framed columns.
β_2	Coefficient in torsional-flexural buckling.
γ	Buckling parameter for a stiffened plate. Uncertainty factor in nominal load. Taper coefficient in tapered beam analysis.
Δ	A deflection.
Δ_ϵ	An increment of strain.
Δ_σ	An increment of stress.
Δ_{AB}	Deflection of point A relative to point B.
δ	Column deflection caused by bending moment due to an axial load P. Buckling parameter for a stiffened plate. Amplification factor in a beam-column.
δ_0	Maximum initial out-of-straightness of a column. Deflection without regard to moment induced by axial load.
ϵ	Strain. Coefficient in plate-buckling equation.
ϵ_m	Maximum strain.
ϵ_{st}	Strain at initial strain hardening.
ϵ_y	Elastic strain at yield stress.
ζ	First mode damping ratio for an unloaded rod. Coefficient in tapered beam analysis. Exponent in formula for biaxially loaded short beam-column.
η	Ratio of tangent modulus to elastic modulus, E_t/E. Plasticity reduction factor for shell buckling. Exponent in formula for biaxially loaded beam-column. Coefficient in stiffened plate buckling formula. Eccentricity parameter for a composite column.
η_b, η_c	Shear shape factors for longitudinal and batten elements.
Θ	An angle.
θ	Angle of tension-field yield band. A shell geometry parameter. Subtending angle of half-span of a circular arch.
$\theta_{AB}(\theta_{BA})$	Rotation of tangent to elastic curve at $A(B)$ with respect to the line $AB(BA)$.
θ_d	Angle of panel diagonal with flange.
κ	Moment coefficient for lateral-torsional buckling.
λ	Slenderness function $\sqrt{\sigma_y/\sigma_e}$, $\sqrt{\sigma_y/\sigma_c}$. Length parameter in

	a stiffened shell. Load ratio between adjacent framed columns.
λ_e	Equivalent slenderness function.
μ	Shear flexibility parameter in a laced or battened member. Coefficient in tapered beam analysis.
ν	Poisson's ratio.
ξ	Coefficient for buckling for a stiffened plate.
ξ_a, ξ_b	Coefficients in laced and battened columns.
ρ	Ratio of tension bands in a plate-girder web.
σ	Normal stress.
σ_1, σ_2	Normal stress at edges of a plate under nonuniform edge loading.
σ_ϕ	Circumferential stress in a shell.
σ_a, σ_{av}	Average normal stress.
σ_c	Critical stress.
σ_c^*	Critical stress for a plate under normal compression alone, as used in an interaction formula for critical stress under mixed boundary stresses.
σ_{cb}	Maximum nonuniform "pure bending" component of normal stress in a plate at critical load.
σ_{cb}^*	Value of σ_{cb} as used in an interaction formula for combined normal stress and shear stress.
$\sigma_{c(v)}$	Critical stress for a variable cross section.
σ_{cy}	Compressive yield strength, 0.2% offset.
σ_e	Average stress at Euler buckling load. Edge stress in a buckled plate.
σ_{eb}	Elastic buckling stress for a beam.
σ_f	Normal stress in a plate-girder flange.
σ_i	Normal stress in the inelastic range (in a shell).
σ_m	Maximum stress at mid-length of column by the secant formula.
σ_{max}	Maximum combined stress due to column load and bending moment.
σ_n	Transverse normal stress in a plate-girder web.
σ_p	Proportional limit stress.
σ_r	Local residual stress. Residual tension stress in plate-girder web. Radial stress in a shell.
σ_{rc}	Maximum residual compressive stress.
σ_t	Tension-field stress in plate girder.
σ_u	Average stress at failure in thin plate. Ultimate unit strength of a column.
σ_{uy}	Upper yield-point stress.
σ_w	Warping normal stress.

σ_x	Longitudinal stress in a compressed shell.
σ_y	Yield-stress level. Yield strength.
σ_{yf}	Yield-stress in girder flange.
σ_{ym}	Empirical stress level at which flange buckling is likely to occur.
σ_{yr}	Reduced effective hoop yield stress in a shell.
σ_{yw}	Yield stress in plate-girder web.
τ	Shear stress. Time parameter in modal response.
τ_c, τ_{cr}	Critical shear stress in a plate.
τ_c^*	Critical stress in pure shear as used in an interaction formula for critical plate stress under combined shear and normal stress.
τ_{cri}	Inelastic critical buckling stress in web.
τ_u	Shear stress at optimum tension-field angle.
τ_w	Warping shear stress.
τ_y	Shear stress at tension yield in plate girder.
τ_{yw}	Shear yield stress in plate-girder web.
ϕ	Angle of rotation. Uncertainty factor of nominal resistance. Angle of slope of arch axis.
ψ	Parameter used in beam-column formulas. Parameter to convert R_m to R_1.
Ω	Arch buckling reduction factor for unsymmetrical loads.
ω_k	kth natural frequency of unloaded column.

Abbreviations

AA	Aluminum Association
AASHTO	American Association of State Highway and Transportation Officials
AISC	American Institute of Steel Construction
AISE	Association of Iron and Steel Engineers
AISI	American Iron and Steel Institute
AREA	American Railway Engineering Association
ASCE	American Society of Civil Engineers
ASME	American Society of Mechanical Engineers
ASTM	American Society for Testing and Materials
CISC	Canadian Institute of Steel Construction
CRC	Column Research Council
CSA	Canadian Standards Association
ECCS	European Convention for Construction Steelwork
NACA	National Advisory Committee for Aeronautics
SSRC	Structural Stability Research Council
WRC	Welding Research Council

Chapter One

Introduction

1.1 THE METAL COLUMN

Whether a structure be man-made or created by nature, the column is the key element in resisting collapse under gravity loads, in buildings and bridges, in trees and plants. This guide is a summary of modern knowledge on the behavior under load of metal columns. Development of the metal structural column has involved the interrelated development of theory, materials, testing machines, test instruments, design procedures, and design standardization. The history of column theory goes back to the work of the Swiss mathematician Leonard Euler, who in 1744 published his famous column formula. Since then the theoretical developments represented some of the finest achievements in the discipline of applied mechanics. Bruce G. Johnston, editor of the first three editions of this guide, has given a clear review of this history in the paper "Column Buckling Theory: Historic Highlights" (Johnston, 1983).

Since 1944, when the Column Research Council (CRC) was founded, much of the theoretical and practical work related to metal column design was performed under the auspices of this Council. In 1976 the name of the Council was changed to Structural Stability Research Council (SSRC) to reflect the broadened scope of the research. The history of CRC-SSRC—its accomplishments and the personalities involved—has been recounted by Bruce G. Johnston (1981) from the firsthand point of view of an active and creative initial and continuous participant.

Ever since the inception of the Council it has played a leading role in developing rational design criteria based on research not only for metal columns but also for all types of structures and structural elements where stability is a controlling feature of behavior under load. This accumulated knowledge has been disseminated by SSRC in many forms, but the chief vehicles for presenting the sum of it have been the three previous editions of this guide (1960, 93 pages; 1966, 217 pages; 1975, 616 pages). The present fourth edition aims to continue this tradition.

1.2 SCOPE AND SUMMARY OF THE GUIDE

The continued importance and vitality of the research on stability problems is due to technical and economic developments which demand the use of ever-stronger and ever-lighter structures in an increasingly wider range of applications. Such an expansion of usage is made possible by developments in (1) manufacturing, such as metallurgy, cold forming, extruding, welding; (2) theory and understanding of behavior under load; (3) fabrication technology, such as the automated assembly of structural members; (4) computer-aided design; (5) economic competition from nonmetallic materials; and (6) construction efficiency. These developments continually not only change the way in which traditional structures are designed and built, but they also make possible the economic use of material in other areas of application, such as offshore structures, transportation vehicles, and outer-space structures. In all of these applications the demands of higher strength and lighter weight inexorably lead to structures in which a consideration of stability must play a crucial role in design. Increased strength and increased slenderness invariably spell more problems with instability.

The third edition of this guide (published in 1975) was a substantial expansion over the second edition (published in 1966), introducing a number of new chapters which reflected the expanded scope of the Council. This fourth edition does not include major expansions in topics, as can be seen from the comparison of the contents of the two editions presented in Table 1.1. Three new chapters were added: Chapter 2, "Stability Theory," presents a brief primer of the fundamentals; Chapter 7, "Box Girders," deals with the special stability problems of box girders with slender elements; and Chapter 21, "Finite-Element Analysis of Stability Problems," gives a review of this important method of calculating instability problems.

The remaining chapters treat the same topics as those in the third edition. However, much of the material has been rewritten, reflecting the substantial new insights gained in the last decade in theory, computation, and experimentation. This fourth edition especially emphasizes (1) that it is the maximum strength of a structure which is of vital concern to the designer and to the structural specification writer; and (2) that for the economically increasingly important slender structures the traditional idealization of ideal geometry is no longer feasible, so that the practically unavoidable initial geometric imperfections, which are entirely within the acceptable fabrication tolerances, can not longer be ignored. In many

Table 1.1.

Third Edition	Fourth Edition
1 Introduction	1 Introduction*
2 Structural Safety	20 Structural Safety*
3 Centrally Loaded Columns	3 Centrally Loaded Columns*
4 Local Buckling of Plates	4 Plates*
5 Dynamic Load Effects	19 Selected Topics in Dynamic Stability*
6 Laterally Unsupported Beams	5 Laterally Unsupported Beams*
7 Plate Girders	6 Plate Girders*
8 Beam-Columns	8 Beam-Columns*
9 Thin-Walled Metal Construction	13 Thin-Walled Metal Construction*
10 Circular Tubes and Shells	14 Circular Tubes and Shells*
11 Tapered Structural Members	9 Tapered Structural Members
12 Columns with Lacing, Battens, or Perforated Cover Plates	11 Columns with Lacing, Battens, or Perforated Cover Plates
13 Mill Building Columns	12 Mill Building Columns
14 Members with Elastic Lateral Restraints	15 Members with Elastic Lateral Restraints
15 Multi-story Frames	16 Frame Stability*
16 Arches	17 Arches*
17 Stiffened Flat Plates	18 Doubly Curved Shells and Shell-like Structures*
18 Shells and Shell-like Structures	10 Composite Columns*
19 Composite Columns	2 Stability Theory*
	7 Box Girders*
	21 Finite-Element Analysis of Stability Problems*

* New chapter or old chapter with essentially new material.

instances postbuckling behavior and dynamic behavior must be included in the modeling of performance under load. Since the effects of initial imperfections are generally detrimental, the counteracting beneficial effects of small but always present restraints also receive considerable attention in this guide. Simply put, the theoretical models of the structures in the fourth edition more closely represent the actual real structures.

1.3 MECHANICAL PROPERTIES OF STRUCTURAL METALS

A knowledge of the stress-strain relationship during the elastic and initial inelastic ranges of behavior is an essential requisite to compression-member analysis. In the elastic range there are accepted average values of the modulus of elasticity, and test values vary within reasonably small limits. Specified values of the yield point or yield strength (depending on whether the initiation of yielding is a sudden or gradual process) are provided by the various specifications of the American Society for Testing and Materials (ASTM) and by product information from the manufacturers. In this guide the term "yield stress" generally is used to denote either the yield point or yield strength, whichever is applicable.

Stress-Strain Relationships. The initial portions of the typical stress-strain curves for structural steels in compression or tension are shown in Fig. 1.1. The strengths of beams and columns are largely determined by stress-strain characteristics in the range shown. (The complete curves

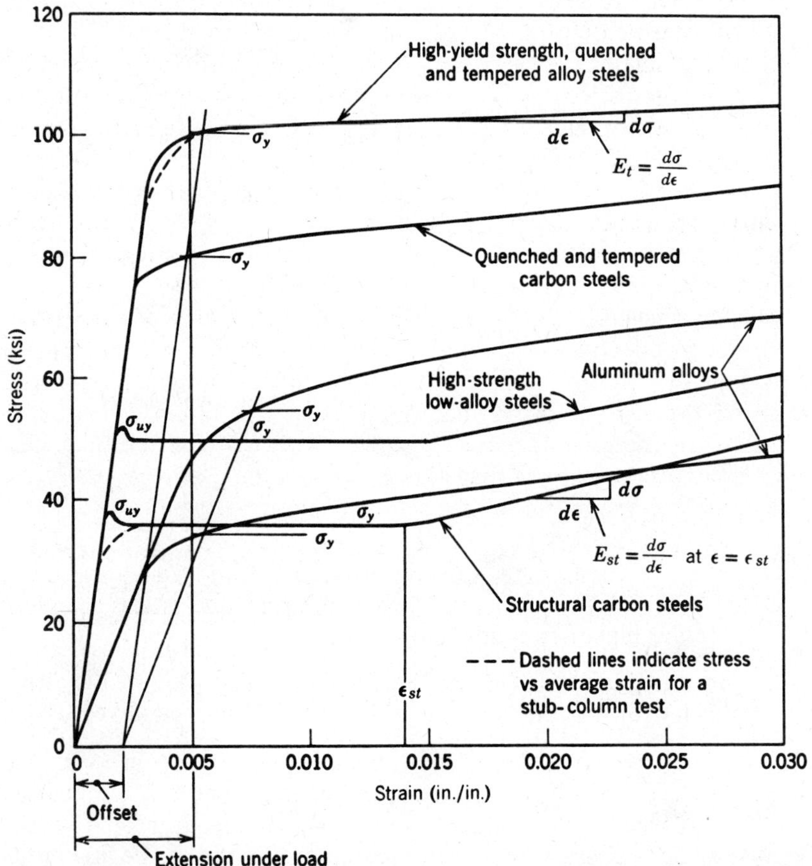

Fig. 1.1 Initial stress-strain relationships for structural steels in compression or tension.

plotted to the same scale as Fig. 1.1 would take up a horizontal space between 20 and 30 times that available on the drawing.)

The structurally significant aspects of a stress-strain curve for carbon or high-strength low-alloy structural steels can be characterized by the following five properties (see Fig. 1.1):

E = modulus of elasticity = slope of stress-strain curve in the elastic range
σ_{uy} = upper yield point (maximum stress prior to yield stress level)
σ_y = yield-stress level (stress at a constant strain rate in the flat portion of the stress-strain curve after initial yield)
ϵ_{st} = strain at initial strain hardening
$E_{st} = (d\sigma/d\epsilon)_{\epsilon=\epsilon_{st}}$ = strain-hardening modulus (initial)

These properties are generally sufficient for calculation of the inelastic strength and plastic deformation of structural steel members.

Structurally significant properties of the aluminum alloys, quenched and tempered steels, cold-worked steels, and stainless steels include the modulus of elasticity, E; the yield strength, preferably determined by the offset method (ASTM Designation A370); and the tangent modulus, $E_t = d\sigma/d\epsilon$, which varies with stress for strains greater than the elastic limit. For all steels and aluminum alloys the maximum tensile (ultimate) strength, based on original area, is also a part of the mill test report, although of no direct relevance to compression-member behavior.

The yield stress of both steel and aluminum alloy varies with temperature, rate of strain, and the surface characteristics of the test specimen, as well as with the testing machine and test method. The yield stress is a function of the rate of strain, becoming lower as the testing speed is lowered. "Zero strain rate" defines a lower limit of the testing speed corresponding to the lowest value of yield-stress level for structural steels. ASTM specifications establish a maximum allowable strain rate. Tests made according to these specifications may be suitable for quality control but indicate yield-stress values as much as 15% greater than those from tests at low rates of strain. The influence of strain rate is less, percentagewise, for the higher-strength steels.

For as-rolled structural steels the yield-stress level in a tension or compression test can be regarded as the level of stress, after initial yield, that is sufficient at a given temperature and rate of strain to develop successively new planes of slip in the portions of the test specimen that remain in the elastic state. After initial yielding has proceeded discontinuously from point to point throughout the specimen, a general strain hardening begins, and the stress rises with further increase in strain. The sharp yield point may disappear with cold work or heat treatment.

The yield-stress level is structurally more significant than the upper yield point, and its existence for relatively large average strains with no appreciable change in stress is taken advantage of in plastic design and inelastic analyses by permitting the assumption that the stress is constant and equal to the yield stress across yielded portions of the cross section.

The plot of average stress versus strain as determined by a stub-column test is somewhat different from that resulting from the test of a small specimen. Residual stresses are one cause of these differences (indicated qualitatively in Fig. 1.1); other factors are the lack of uniformity of yield stress over the cross section and varying degrees of working during the rolling process. Similarly, strain hardening caused by the forming processes in cold-formed members may result in changes in yield stress which tend to shift the curve of average stress versus strain toward higher values of stress and more gradual yield development. Cold-forming effects are particularly pronounced for the stainless steels.

1.4 Definitions

The following list of terms defines their usage in this guide. These terms are supplementary to the list of symbols provided in the notation and include primarily those for which variations in meaning are prevalent in the technical literature.

Beam — A straight or curved structural member, primarily supporting loads applied at right angles to the longitudinal axis. The internal stresses on a transverse cross section may be resolved into one or more of three resultant components: a transverse shear, a bending moment, and a torsional moment.

Beam-column — A beam that also functions to transmit compressive axial force.

Bifurcation — A term relating to the load-deflection behavior of a perfectly straight and perfectly centered compression element at critical load. Bifurcation can occur in the inelastic range only if the pattern of post-yield properties and/or residual stresses is symmetrically disposed so that no bending moment is developed at subcritical loads. At the critical load a member can be in equilibrium in either a straight or slightly deflected configuration, and a bifurcation results at a branch point in the plot of axial load versus lateral deflection from which two alternative load-deflection plots are mathematically valid.

Braced frame — A frame in which the resistance to both lateral load and frame instability is provided by the combined action of floor diaphragms and a structural core, shear walls, and/or a diagonal, K brace, or other auxiliary system of bracing.

Buckle — To kink, wrinkle, bulge, or otherwise lose original shape as a result of elastic or inelastic strain.

Buckled — Descriptive of the final shape after buckling.

Buckling load — The load at which a compressed element, member, or frame collapses in service or buckles in a loading test.

Critical load — The load at which bifurcation (*see* Bifurcation) occurs as determined by a theoretical stability analysis.

Effective length — The equivalent or effective length (KL) which, in the buckling formula for a hinged-end column, results in the same elastic critical load as for the framed member or other compression element under consideration at its theoretical critical load. The use of the effective length concept in the inelastic range implies that the ratio between elastic and inelastic critical loads for an equivalent hinged-end column is the same as the ratio between elastic and inelastic critical loads in the beam, frame, plate, or other structural element for which buckling equivalence has been assumed.

Effective width — A reduced width of plate, slab, or flat segment of a cross section which, assuming uniform stress distribution, leads to the same behavior of a structural member as the actual section of plate and the actual nonuniform stress distribution.

First yield	A limiting stress level above which a permanent set results upon removal of the load.
Initial imperfection	An unavoidable deviation from perfect geometry, for example, initial crookedness of a member, initial out-of-plumb of a story, initial out-of-flatness of a plate, or initial denting or bulging of a shell, which is within the accepted practical tolerance of the particular applicable fabrication technology.
Instability	A condition reached during buckling under increasing load in a compressive member, element, or frame at which the capacity for resistance to additional load is exhausted and continued deformation results in a decrease in load-resisting capacity.
Proportional limit	The load or stress beyond which there is a significant amount of deviation from a prior linear load-deformation or stress-strain relationship. The term is usually used in connection with a tensile or compressive test, and the sensitivity of the strain or deformation measuring device is a determining factor in the evaluation.
Residual stress	The stresses that exist in an unloaded member after it has been formed into a finished product. Such stresses can be caused by cold bending, finishing, straightening, flame cambering, oxygen cutting, welding, cooling after rolling, or quenching during heat treatment.
Restraint	Deviation from the ideal articulated boundary condition or unbraced condition of an element, a member, or a structure.
Stability	The capacity of a compression member or element to remain in position and support load, even if forced slightly out of line or position by an added lateral force. In the elastic range, removal of the added lateral force would result in a re-

turn to the prior loaded position, unless the disturbance causes yielding to commence.

Strain-hardening modulus — For structural steels that have a flat (plastic) region in the stress-strain relationship, the strain-hardening modulus is the initial slope of the stress-strain curve just beyond the terminus of the flat region. It depends on prior strain and thermal history and exhibits a much greater range of variation than does the elastic modulus of the material.

Stub column — A short compression test specimen utilizing the complete cross section, sufficiently long to provide a valid measure of the stress-strain relationship as averaged over the cross section, but short enough so that it will not buckle as a column in the elastic or plastic range.

Tangent modulus — The slope of the stress-strain curve of material in the inelastic range, at any given stress level, as determined by the compression test of a small specimen under controlled conditions. The "effective tangent modulus" (as determined by a stub-column test) is modified by nonhomogeneity of material properties and by residual stresses.

Tangent-modulus load — The critical column load obtained by substituting E_t, the tangent modulus, in place of E in the Euler formula.

Tension-field action — A description of the postbuckling behavior of a plate girder panel under shear force, during which diagonal compressive stresses cause the web to form diagonal waves. The tensile stresses, parallel to the wave troughs, induce compressive stresses in the transverse stiffeners.

Unbraced frame — A frame in which the resistance to lateral load is provided primarily by the bending resistance of the frame members and their connections.

Yield point	The maximum stress recorded in a tensile or compressive test of steel specimen prior to entering the plastic range.
Yield strength	In a tension or compression test, the stress at which there is a specified amount of measured deviation from an extension of the initial linear stress-strain plot, commonly taken as the intersection of the stress-strain curve and a line parallel with the linear portion of the curve but offset by a strain of 0.002.
Yield stress	A general term, denoting either yield strength, yield-stress level, or yield point, as herein defined.
Yield-stress level	For carbon and low-alloy structural steels, the stress immediately beyond the elastic strain range, within which range the strain appears to increase without change in stress. It may be arbitrarily defined as the stress determined at a strain of 0.005.

1.5 POSTBUCKLING BEHAVIOR

Load-deflection relationships in the postbuckling range have an important bearing on the structural design significance of the critical load.

For the idealized "perfect" compression element—one that is perfectly elastic, devoid of imperfection, and within which the load-induced stress is perfectly uniform—three different types of postbuckling behavior are typified by (1) the column, (2) the stiffened plate, and (3) the cylindrical shell. For each of these three cases Fig. 1.2 illustrates with light lines the load-deflection curves beyond the critical load for each "perfect" element, which, for a given situation and given buckling mode, are uniquely determinable by a theoretical analysis. The heavy lines in Fig. 1.2 indicate the theoretical behavior for the same elements when a given degree of imperfection is assumed. The heavy lines approach the light lines as a limit, as the degree of imperfection is assumed to diminish toward zero. The heavy lines are also indicative of what may be expected in a laboratory test.

In the elastic range of behavior of a slender column, (1), the critical load and the maximum load carried by an imperfect column are in reasonable agreement; thus the critical load provides a satisfactory basis for computing the design load. For the stiffened plate, (2), if the added

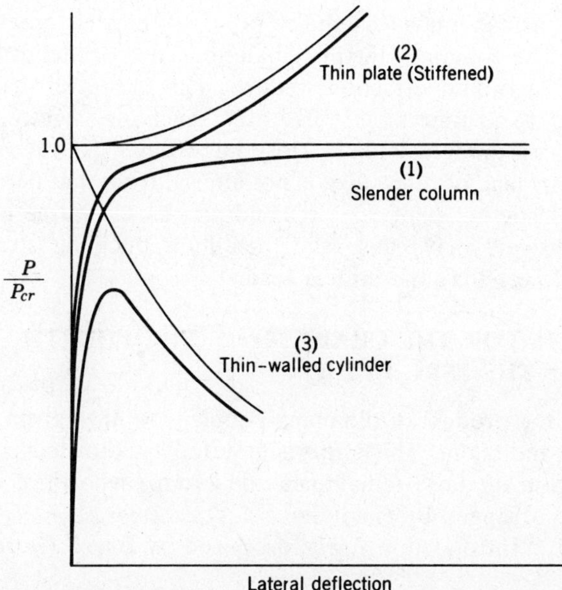

Fig. 1.2 Elastic postbuckling curves for compressed elements.

postbuckling strength is achieved with acceptably small lateral deflections, a greater design load in relation to the critical load might be acceptable. But for the thin-walled cylinder, (3), the maximum load in the real situation is drastically reduced with respect to the critical load; it is to an uncertain degree dependent on the amount of imperfection. Thus the critical load in this case is not a suitable criterion on which to base an allowable design load.

Figure 1.2 depicts only elastic behavior; inelastic behavior of the material may alter the relationships in either of two ways: Buckling may be elastic, as indicated, but the added stress (1) due to bending may cause the combined stress to exceed the elastic stress range, or (2) due either to residual stress or inherent nonlinearity of the stress-strain relationship, the critical and buckling loads may occur in the inelastic range.

In the first case, when the yield point is reached after initial elastic behavior, the curves shown in Fig. 1.2 will simply branch into new paths below those shown. In the second case, the light lines emanating from the critical load bifurcation point will each take on a different initial increment of slope. In the case of the column, the horizontal line indicative of Euler buckling will be replaced by a curved line, initially sloping upward and reaching a maximum (instability point) somewhat greater than the critical load. The column with very small imperfections tends to approach this behavior as the imperfections reduce, with the result that with small imperfections a column under test may reach or slightly exceed the critical (tangent modulus) load. Most important, the relevance of the critical load, or lack of relevance, is not altered from that pertinent to the completely elastic behavior, as discussed previously. Large initial imperfections, however, may cause the maximum buckling strength to be significantly lower than the critical load.

1.6 CREDITS FOR THE CHAPTERS IN THE FOURTH EDITION OF THE SSRC GUIDE

This book is the product of the many people who have given unstintingly of their time and talent. This effort is gratefully acknowledged. Following is a recognition of those individuals and groups who have made major contributions arranged by chapters:

Chapter 1, "Introduction," was produced by Bruce G. Johnston and the editor.

Chapter 2, "Stability Theory," was written by Alex Chajes.

Chapter 3, "Centrally Loaded Columns," was produced by SSRC Task Group 1, Centrally Loaded Columns. The following persons contributed substantial input: Reidar Bjorhovde, Wilfred Chen, John Clark, Bruce Johnston, Lynn Beedle, Cedric Marsh, and Lambert Tall. Based on this input, the editor wrote the chapter.

Chapter 4, "Plates," was produced by SSRC Task Group 13, Thin-Walled Metal Construction. The chapter was written by Shien T. Wang, with substantial assistance from Wei-Wen Yu, Sri Sridharan, Teoman Peköz, and Samuel Errera.

Chapter 5, "Laterally Unsupported Beams," was written by the editor, who received substantial advice from Srilunamu Vinnakota, Nicholas Trahair, Yuhshi Fukumoto, Joachim Lindner, David Nethercot, and Kit Kitipornchai, all of SSRC Task Group 15, Laterally Unsupported Beams.

Chapter 6, "Plate Girders," was produced by SSRC Task Group 16, "Plate Girders," with a substantial contribution by Mohamed Elgaaly, the chairman of the Task Group.

Chapter 7, "Box Girders," was produced by Patrick Dowling and his colleagues at Imperical College in London.

Chapter 8, "Beam-Columns," was written by David Nethercot, with assistance from the members of SSRC Task Group 3, Beam-Columns, and from the editor.

Chapter 9, "Tapered Structural Members," was produced by George Lee.

Chapter 10, "Composite Columns," was written by Richard Furlong, assisted by the members of SSRC Task Group 20, Composite Members and Systems.

Chapter 11, "Columns with Lacing, Battens, or Perforated Cover Plates," written by Fung-Jen Lin and Bruce Johnston.

Chapter 12, "Mill Building Columns," was written by Bruce Johnston.

Chapter 13, "Thin-Walled Metal Construction," is the result of the work of SSRC Task Group 13, with Wei-Wen Yu as editor and special assistance from Samuel Errera, Teoman Pekoz, John Springfield, and Sri Sridharan.

Chapter 14, "Circular Tubes and Shells," is the result of Donald Sherman's work, assisted by Clarence Miller and the members of SSRC Task Group 18, Unstiffened Tubular Members.

Chapter 15, "Members with Elastic Lateral Restraints," was written by Mohamed Elgaaly and Bruce Johnston.

Chapter 16, "Frame Stability," is mainly the work of Jerry Iffland with assistance from Le Wu Lu, Joseph Yura and Franklin Cheng, and SSRC Task Group 4, Frame Stability and Columns as Framed Members.

Chapter 17, "Arches," was written by Walter Austin and Srilunamu Vinnakota, with assistance from S. Kuranishi, T. Yabuki and T. Sakimoto.

Chapter 18, "Doubly Curved Shells and Shell-like Structures," was written by Clarence Miller and edited by Kenneth Buchert and the editor.

Chapter 19, "Selected Topics in Dynamic Stability," was written by the following five persons: D. Krajcinovic, A. E. Sommers, R. Plaut, S. S. Chen, and G. J. Simitses. F. Cheng assisted the editor.

Chapter 20, "Structural Safety," was written by the editor.

Chapter 21, "Finite-Element Analysis of Stability Problems," was written by Richard Gallagher.

Special recognition is due to Bruce Johnston for writing the Index, and to all members of the Executive Committee for reviewing chapters.

References

Johnston, B. G. (1981), "History of Structural Stability Research Council," *ASCE J. Struct. Div.*, Vol. 107, No. ST8, pp. 1529–1550.

Johnston, B. G. (1983), "Column Buckling Theory: Historic Highlights," *ASCE J. Struct. Eng.*, Vol. 109, No. 9, pp. 2086–2096.

Chapter Two

Stability Theory

2.1 INTRODUCTION

The principal subject matter of this guide is the stability of metal structures. This chapter introduces the various types of instability encountered in the other chapters by presenting the solutions to several simple illustrative problems.

Instability is a condition wherein a compression member loses the ability to resist increasing loads and exhibits instead a decrease in load-carrying capacity. In other words, instability occurs at the maximum point on the load-deflection curve.

Problems in instability of compression members can be subdivided into two categories: those associated with the phenomenon called bifurcation of equilibrium, and those in which instability occurs when the system reaches a maximum, or limit, load without any previous bifurcation. In the first case a perfect member, when subjected to increasing load, initially deforms in one mode and then, at a load referred to as the critical load, the deformation suddenly changes into a different pattern. Axially compressed columns, plates, and cylindrical shells experience this type of instability. By comparison, members belonging to the latter category deform in a single mode from the beginning of loading until the maximum load is reached. Shallow arches and spherical caps subjected to uniform external pressure are examples of the second type of instability.

2.2 BIFURCATION BUCKLING

2.2.1 Initially Perfect Systems

The critical load of a compression member, obtained from the linear analysis of an idealized perfect member, does not necessarily coincide with the load at which collapse of a real imperfect member occurs. To determine the failure load of an actual member it is necessary to take initial imperfections into account and to consider the entire nonlinear load-deflection curve of the member. Unfortunately, the process of obtaining such a curve is often too difficult and time consuming to be used in routine engineering design. Instead, the maximum load of a compression member is generally calculated by semiempirical means, that is, using curves fitted to numerically obtained maximum strength curves or combining test results with a qualitative understanding of the nonlinear load-deflection behavior of the imperfect member.

A general understanding of the basic characteristics of the elastic buckling and postbuckling behavior of members that become unstable as a result of bifurcation can be obtained by considering the simple model in Fig. 2.1. The model consists of two rigid bars hinged to one another and to the supports, and restrained laterally by a nonlinear elastic spring. A similar model has been used by many, including Budiansky and Hutchinson (1964) and Hoff (1966).

The restraining force F exerted by the spring on the bars is assumed to be related to the lateral displacement x by

$$F = k_1\epsilon - k_2\epsilon^2 + k_3\epsilon^3 \tag{2.1}$$

where $\epsilon = x/L$. If the model is initially straight, equilibrium in a deformed configuration requires that

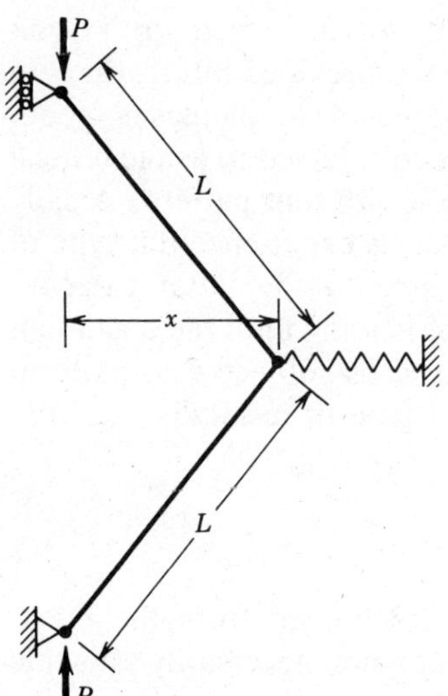

Fig. 2.1 Bifurcation-buckling model.

$$P\epsilon = \frac{F}{2}(1-\epsilon^2)^{1/2} = \tfrac{1}{2}(k_1\epsilon - k_2\epsilon^2 + k_3\epsilon^3)(1-\epsilon^2)^{1/2} \tag{2.2}$$

Letting ϵ become infinitesimally small in Eq. 2.2, we obtain for the critical load

$$P_c = \frac{k_1}{2} \tag{2.3}$$

Based on the work of Koiter (1970), it has been demonstrated that the essential characteristics of the postbuckling behavior of a member can be determined by considering the initial stages of the postbuckling curve, in the vicinity of the critical load. Thus ϵ is assumed to be small but finite, which reduces Eq. 2.2 to

$$P\epsilon = \tfrac{1}{2}(k_1\epsilon - k_2\epsilon^2 + k_3\epsilon^3) \tag{2.4}$$

In view of Eq. 2.3, the foregoing expression can be rewritten in the form

$$P = P_c(1 - a\epsilon + b\epsilon^2) \tag{2.5}$$

where $a = k_2/k_1$ and $b = k_3/k_1$.

Certain structures behave in a symmetric manner; that is, the buckling characteristics are the same regardless of whether the deformation is positive or negative. To simulate the behavior of such structures, we let $a = 0$. Equation 2.5 then reduces to

$$P = P_c(1 + b\epsilon^2) \tag{2.6}$$

The load-deflection curves corresponding to Eq. 2.6 are shown in Fig. 2.2. The type of behavior depicted by these curves is referred to as bifurcation buckling. The member initially deforms in one mode, the prebuckling deformation, and then at the critical load, due to a branch in

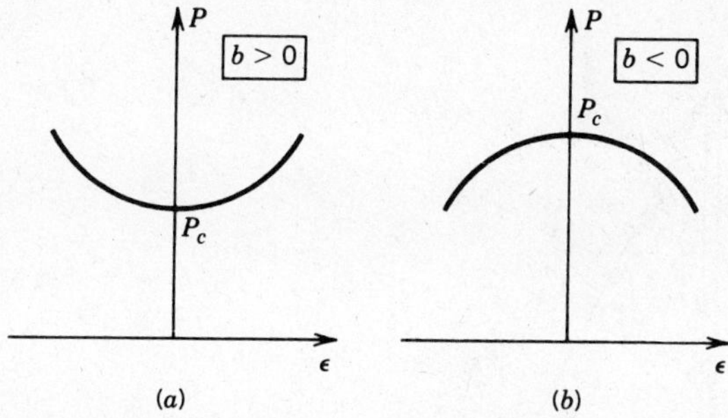

Fig. 2.2 Symmetric buckling of bifurcation model: (*a*) stable postbuckling curve; (*b*) unstable postbuckling curve.

the load-deflection curve, the deformation suddenly changes into a different pattern, the buckling mode. For example, axially loaded columns initially shorten due to axial compression. Then at the critical load the member suddenly begins to bend.

The curve in Fig. 2.2*a* results if $b > 0$ and the curve in Fig. 2.2*b* if $b < 0$. These two cases correspond to models whose springs become either stiffer or more flexible with increasing lateral deflection. In a similar manner, the stiffness of an actual structure may either increase or decrease subsequent to the onset of buckling. In other words, the load required to keep the structure in a deformed configuration may either increase or decrease as the deformation increases in magnitude. If the load that the structure can support subsequent to the onset of buckling increases with increasing deformation, as shown in Fig. 2.2*a*, the structure is said to have a stable postbuckling curve. By comparison, if the load decreases, as indicated in Fig. 2.2*b*, the member has an unstable postbuckling curve.

An axially compressed plate is an example of a structure with a stable postbuckling curve. As the plate buckles, the buckling deformations give rise to tensile membrane stresses which increase the stiffness of the plate and give it the capacity to resist increases in the load. By comparison, the guyed tower in Fig. 2.3 has an unstable postbuckling curve. As the top of the tower deflects laterally, some of the cables are stretched, causing them to push down on the post. As a consequence the external load required to maintain equilibrium decreases with the magnitude of the lateral deflection.

The most notorious example of a structure with an unstable postbuckling curve is the axially compressed cylindrical shell. However, this system does not buckle in a symmetric manner and its behavior is therefore not fully described by Fig. 2.2b.

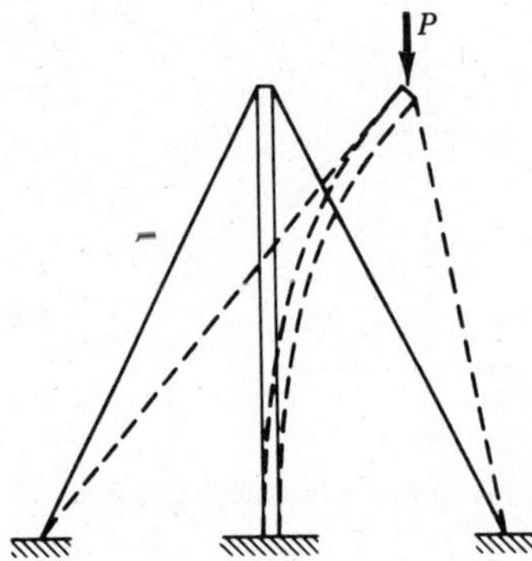

Fig. 2.3 Guyed tower.

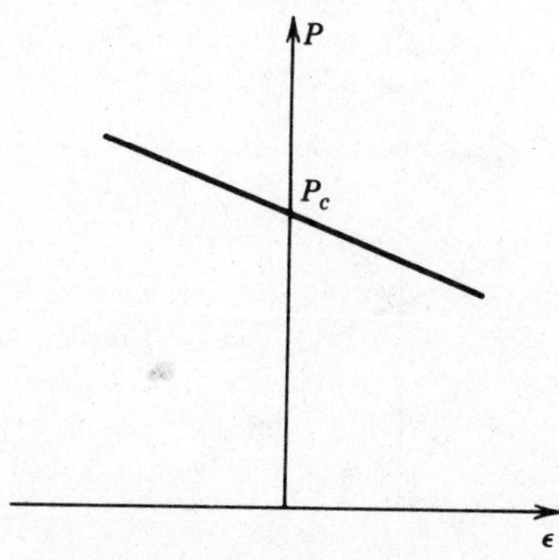

Fig. 2.4 Asymmetric buckling of bifurcation model.

To simulate the behavior of structures that behave in an asymmetric manner, we let $b = 0$ in Eq. 2.5. Thus

$$P = P_c(1 - a\epsilon) \tag{2.7}$$

The load-deflection curve corresponding to Eq. 2.7 is shown in Fig. 2.4. Unlike the symmetric system the unsymmetric one becomes stiffer if it deflects one way and more flexible if it deflects the opposite way.

The simple frame in Fig. 2.5*a* is an example of a structure that has an asymmetric postbuckling curve. If the frame buckles as shown in Fig. 2.5*b*, a secondary tension force V is induced in the vertical member. As a consequence the external load P that the structure can support increases with increasing deformations. By comparison, if the frame buckles as indicated in Fig. 2.5*c*, a secondary compression force is induced in the vertical member and the resistance of the system to applied loads decreases with increasing deformations. The foregoing analytically predicted behavior of the frame in Fig. 2.5 has been verified experimentally by Roorda (1965).

2.2.2 Initially Imperfect Systems

The postbuckling curve of an initially perfect system does not by itself give sufficient information to allow one to determine when failure takes place. To obtain that information, one must also consider the initial imperfections of shape and eccentricities of loading that are present in all real structures.

Assuming that our model has an initial deformation x_0, as indicated in Fig. 2.6, Eq. 2.4 takes the form

$$P(\epsilon + \epsilon_0) = \tfrac{1}{2}(k_1\epsilon - k_2\epsilon^2 + k_3\epsilon^3) \tag{2.8}$$

where $\epsilon_0 = x_0/L$. In view of Eq. 2.3 the relation above can be rewritten as

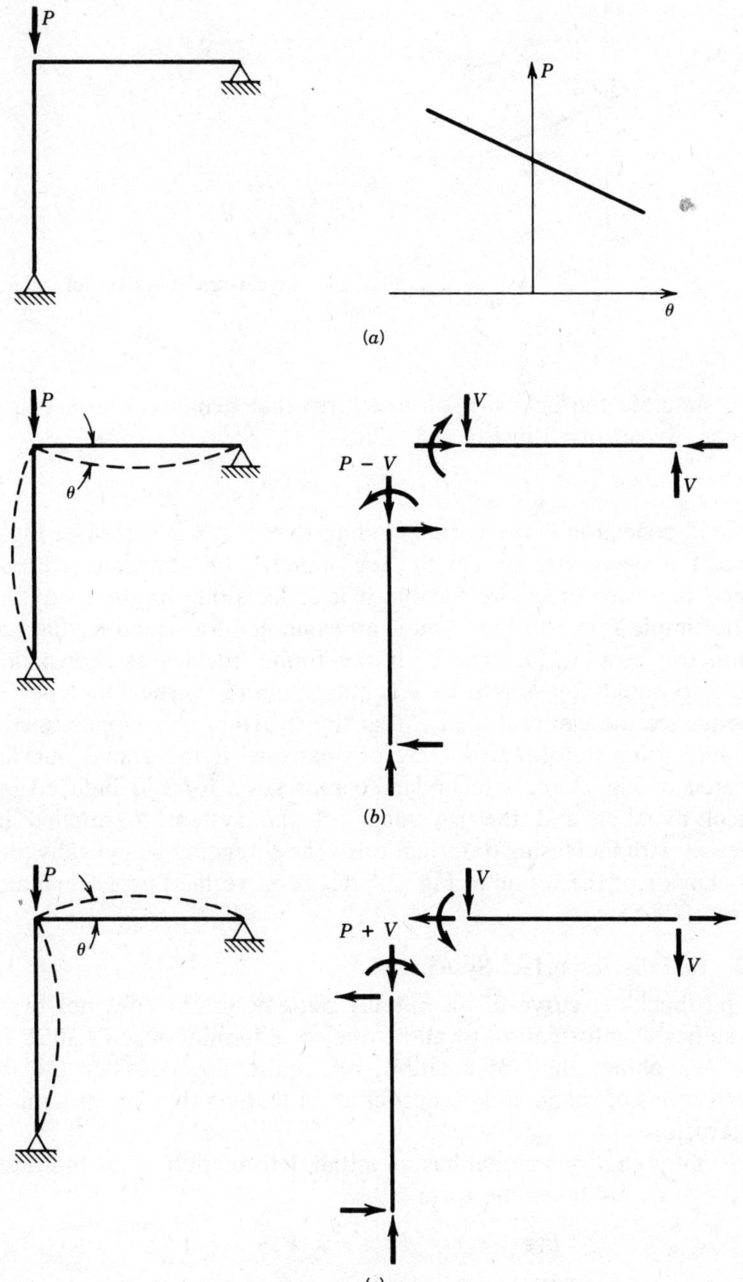

Fig. 2.5 Buckling of L-shaped frame.

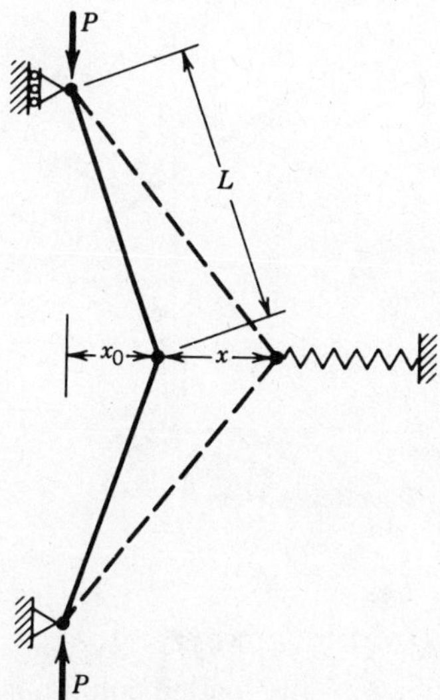

Fig. 2.6 Model of initially imperfect system.

$$P = \frac{P_c(\epsilon - a\epsilon^2 + b\epsilon^3)}{\epsilon + \epsilon_0} \tag{2.9}$$

For symmetric behavior $a = 0$ and

$$P = \frac{P_c(\epsilon + b\epsilon^3)}{\epsilon + \epsilon_0} \tag{2.10}$$

and for asymmetric behavior $b = 0$ and

$$P = \frac{P_c(\epsilon - a\epsilon^2)}{\epsilon + \epsilon_0} \tag{2.11}$$

The load-deflection curves corresponding to Eqs. 2.10 and 2.11 are shown as dashed lines in Fig. 2.7. It is evident from these curves that small initial imperfections do not significantly affect the behavior of the systems with stable postbuckling curves. These members can continue to resist increasing loads above the critical load, and failure takes place only after yielding of the material has occurred.

The amount of postbuckling strength that a system with a stable postbuckling curve possesses depends on two factors: the steepness of the postbuckling curve and the relative magnitude of the critical load and the load at which yielding begins. For example, axially compressed plates possess a relatively steep postbuckling curve and as a consequence often exhibit sizable postbuckling strength. Failure loads three or four times as

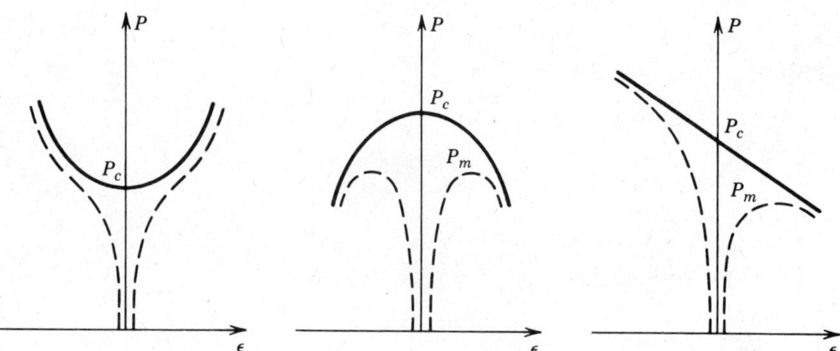

Fig. 2.7 Postbuckling curves of initially imperfect system.

large as the critical load have been obtained (Gerard, 1957). By comparison, the slope of the postbuckling curve of an axially loaded column is extremely small and the failure load of such a member therefore coincides, very nearly, with the critical load.

In addition to possessing a relatively steep postbuckling curve, a system must have a yield load that is considerably in excess of its critical load if the system is to exhibit significant postbuckling strength. A very rough estimate of the postbuckling strength of an axially compressed plate is given by the expression

$$\frac{P_c}{P_f} = \left(\frac{P_c}{P_y}\right)^{1/2} \tag{2.12}$$

where P_c is the critical load, P_f the failure load, and P_y the load when yielding commences. According to Eq. 2.12, a plate possesses significant postbuckling strength when P_c/P_y is considerably smaller than unity. Hence only thin plates can be expected to display sizable postbuckling strength.

Whereas small initial imperfections have only a negligible effect on the behavior of systems with stable postbuckling curves, they have a very marked effect on systems with unstable postbuckling curves. As indicated by the curves in Fig. 2.7, the presence of small initial imperfections will cause systems with unstable postbuckling curves to fail at loads below the critical load. These structures are accordingly referred to as being imperfection sensitive.

By setting $dP/d\epsilon = 0$ for Eqs. 2.10 and 2.11, the following approximations of the maximum load P_m can be obtained. For the symmetric system with $b < 0$,

$$\frac{P_m}{P_c} = 1 - 3\left(-\frac{b}{4}\right)^{1/3} \epsilon_0^{2/3} \tag{2.13}$$

and for the asymmetric system,

$$\frac{P_m}{P_c} = 1 - 2(a\epsilon_0)^{1/2} \tag{2.14}$$

Equations 2.13 and 2.14 indicate that the larger the initial imperfection x_0, and the steeper the postbuckling curve (i.e., the larger a or b), the smaller will be the ratio of P_m to P_c. Axially compressed cylindrical shells that have a very steep postbuckling curve have been found to fail at loads significantly below the critical load (Brush and Almroth, 1975). Using both theory and tests it has been demonstrated that initial imperfections whose magnitude is only 10% of the shell thickness can result in maximum loads whose magnitude is 60% of the critical load (Hutchinson and Koiter, 1970). Conversely, by manufacturing and testing near-perfect shell specimens, failure loads only slightly below the critical load have been obtained (Tennyson, 1964).

In conclusion it is evident that the behavior of real imperfect members can be predicted from the shape of the postbuckling curve for perfect systems. Members with stable postbuckling curves will fail at loads equal to or above the critical load, whereas members with unstable postbuckling curves will fail at loads below the critical load.

2.3 LIMIT-LOAD BUCKLING

Buckling that is associated with a bifurcation of equilibrium is not the only form of instability that can occur. A second type of instability that can take place is illustrated by the model in Fig. 2.8. The model consists of a simple arch, formed by two elastic bars hinged to each other and to

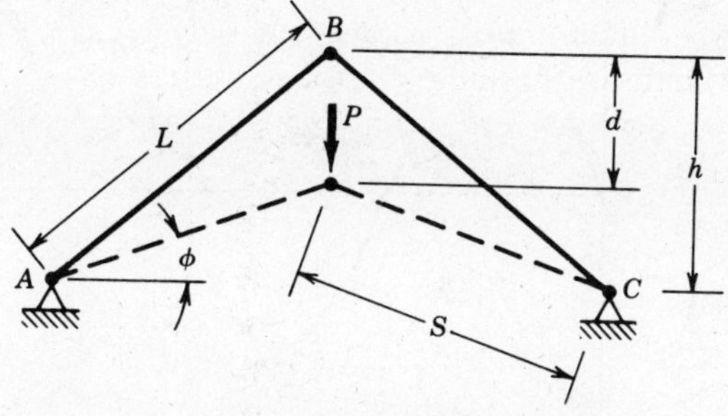

Fig. 2.8 Limit-load buckling model.

the supports. As the load P acting on the model increases, legs AB and BC shorten by an amount Δ, and point B moves down a distance d. The axial force F induced in the bars by the applied load P is equal to

$$F = \frac{P}{2 \sin \phi} = \frac{PS}{2(h-d)} \tag{2.15}$$

and the axial shortening Δ of each bar is given by

$$\Delta = \frac{F}{K} = \frac{PS}{2K(h-d)} \tag{2.16}$$

in which $S = \sqrt{L^2 + d^2 - 2dh}$ is the length of the compressed bars and $K = AE/L$ is the stiffness of the bars. Substitution of $\Delta = L - S$ in Eq. 2.16 leads to

$$L - S = \frac{PS}{2K(h-d)}$$

or

$$L - \sqrt{L^2 + d^2 - 2dh} = \frac{P\sqrt{L^2 + d^2 - 2dh}}{2K(h-d)} \tag{2.17}$$

If the rise h of the arch is assumed to be small compared to L, Eq. 2.17 reduces to

$$P = \frac{Kh^3}{L^2}(2\delta - 3\delta^2 + \delta^3) \tag{2.18}$$

in which $\delta = d/h$.

The load-deflection relation corresponding to Eq. 2.18 is depicted by the solid curve in Fig. 2.9. It is evident that no bifurcation of equilibrium

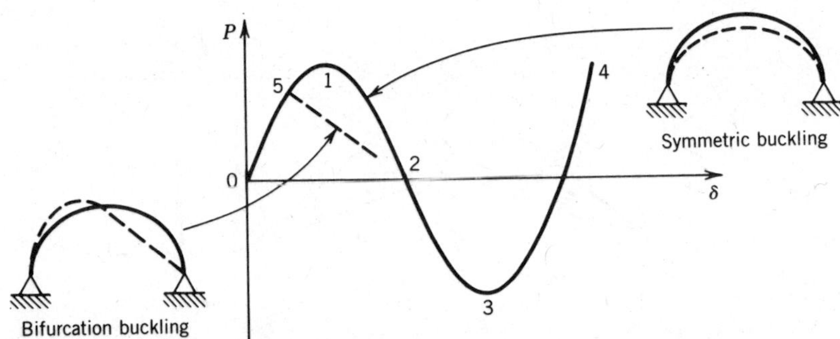

Fig. 2.9 Load-deflection curve of limit-load model.

exists. Instead, the load and deformation increase simultaneously until a maximum or limit load is reached (point 1) beyond which the system becomes unstable.

If the rise h of the model is large enough compared to L, the axial forces in the legs may reach their critical loads, causing the legs to buckle as hinged-hinged columns before the entire system reaches its limit load at point 1. In that case buckling occurs as a result of a bifurcation of equilibrium at point 5 on the curve.

The behavior of arches and spherical shells subject to uniform external pressure is similar to that described by the curves in Fig. 2.9. Arches and spherical caps with a large rise-to-span ratio fail in an asymmetric mode as a result of bifurcation buckling, whereas shallow arches and spherical caps fail in a symmetric mode, due to limit-load buckling.

References

Brush, D. O., and Almroth, B. O. (1975), *Buckling of Bars, Plates and Shells*, McGraw-Hill, New York.

Budiansky, B., and Hutchinson, J. W. (1964), "Dynamics Buckling of Imperfection-Sensitive Structures," *Proc. 11th Int. Cong. Appl. Mech.*, Munich.

Gerard, G. (1957), "Handbook of Structural Stability: Part IV. Failure of Plates and Composite Elements," *NACA Tech. Note No. 3784*, Aug.

Hoff, N. J. (1966), "The Perplexing Behavior of Thin Circular Cylindrical Shells in Axial Compression," *Isr. J. Technol.*, Vol. 4, No. 1.

Hutchinson, J. W., and Koiter, W. T. (1970), "Postbuckling Theory," *Appl. Mech. Rev.* Vol. 23.

Koiter, W. T. (1970), "On the Stability of Elastic Equilibrium," *Tech. Rep. No. AFFDL-TR-70-25*, Air Force Flight Dynamics Laboratory, Wright-Patterson Air Force Base, Ohio, Feb.

Roorda, J. (1965), "Stability of Structures with Small Imperfections," *ASCE J. Eng. Mech. Div.*, Vol. 91, No. EM1, pp. 87–106.

Tennyson, R. C. (1964), "An Experimental Investigation of the Buckling of Circular Cylindrical Shells in Axial Compression Using the Photoelastic Technique," *Rep. No. 102*, Institute of Aerospace Sciences, University of Toronto, Toronto, Nov.

Chapter Three

Centrally Loaded Columns

3.1 INTRODUCTION

The cornerstone of column theory is the Euler column, a mathematically straight, prismatic, pin-ended, perfectly centrally loaded strut that is slender enough to buckle at a stress below the proportional limit of the material. The "buckling load" or "critical load" or "bifurcation load" (see Chapter 1 for a discussion of the significance of these terms) is defined as

$$P_E = \frac{\pi^2 EI}{L^2} \tag{3.1}$$

where EI is the elastic stiffness and L is the length of the column. The "Euler" load P_E is the reference value to which the strength of actual columns is usually compared.

If end conditions other than perfectly frictionless pins can be defined mathematically, the critical load is expressed by

$$P_{EK} = \frac{\pi^2 EI}{(KL)^2} \tag{3.2}$$

where KL is an "effective length" defining a portion of the buckled deflection between points of zero curvature. In other words, KL is the length of an equivalent pinned end column buckling at the same load as the end-restrained column. For example, for columns in which one end of the member is prevented from translating with respect to the other end, K can take on values between 0.5 and 1, depending on the endrestraint.

The isolated column is a theoretical concept; it rarely exists in practice. Usually, a column forms part of a structural frame and its stability is interrelated with the stability of the entire structure. The structure imposes not only axial forces but also end restraints and flexural and torsional forces on the column. This interrelationship is treated elsewhere in many parts of this guide, especially in Chapter 16. This chapter will treat only the isolated column because (1) many structural design situations are idealized such that elements can be thought of as centrally loaded columns (e.g., truss members), and (2) the centrally loaded column is a limiting point in the mathematical space defining the interaction between axial and flexural forces in a member in a structure. Thus an understanding of the strength of individual centrally loaded columns is essential to the development of design criteria for general compression members.

3.2 COLUMN STRENGTH

3.2.1 General Comments

Column strength is characterized by the maximum axial force that can be supported without excessive lateral deformations. Column strength has been studied extensively for several centuries, and many reviews of this rich and varied subject of structural mechanics exist in papers and textbooks. The previous editions of this guide provide an excellent introduction to those wishing to learn more on the subject (see also Chapter 1 for further references on the history of column formulas). Here only the concepts essential to developments in the later portions of this chapter will be briefly explained.

Column strength can be approximated by theoretically considering either (1) a column with mathematically perfect geometry and perfect centroidal loading: "critical load theory," and (2) a column in which the geometry and/or the loading deviate slightly from the perfect: "theory of imperfect columns." The imperfections are practically unavoidable and they represent acceptable construction tolerances which are, as a general rule, not visible to the naked eye, nor can they be quantified precisely beforehand.

The reason for this dual representation (i.e., "perfect" and "imperfect") is that for practical purposes some types of columns (e.g., cold-formed steel and aluminum columns) can be idealized as "perfect," while for other columns (e.g., hot-rolled or welded built-up structural steel columns), it is necessary to consider the effects of the imperfections.

The following discussion will focus on columns failing by instability of the entire member. Columns made from thin plate elements can also fail by local plate or shell instability, and in some cases there is interaction between "overall" and "local" buckling. These topics are treated in greater detail in Chapters 4, 13, and 14. Furthermore, only prismatic columns will be considered. Nonprismatic columns are dealt with in Chapters 9, 11, and 12.

3.2.2 Critical-Load Theory

The strength of a perfectly straight prismatic column with perfect central loading and well-defined end restraints is the Euler load, P_E (Eq. 3.2), as long as the material is still elastic when buckling occurs. When the axial load attains P_E, a stable equilibrium configuration is possible even in the presence of lateral deflection (Fig. 3.1*a*), while the load remains essentially constant (Fig. 3.1*b*, lines *OAB*). Even if an initial deflection, and/or an initial load eccentricity is present, the maximum load will approach the Euler load asymptotically as long as the material remains elastic (curve *C* in Fig. 3.1*b*).

Many practical columns are in a range of slenderness where at buckling portions of the columns are no longer elastic, and thus one of the key assumptions underlying Euler column theory is violated. Essentially, the stiffness of the column is reduced by yielding. This degradation of the stiffness may be the result of a nonlinearity in the material itself (e.g., aluminum, which has a nonlinear stress-strain curve), or it may be due to partial yielding of the cross section at points of compressive residual stress (e.g., steel shapes).

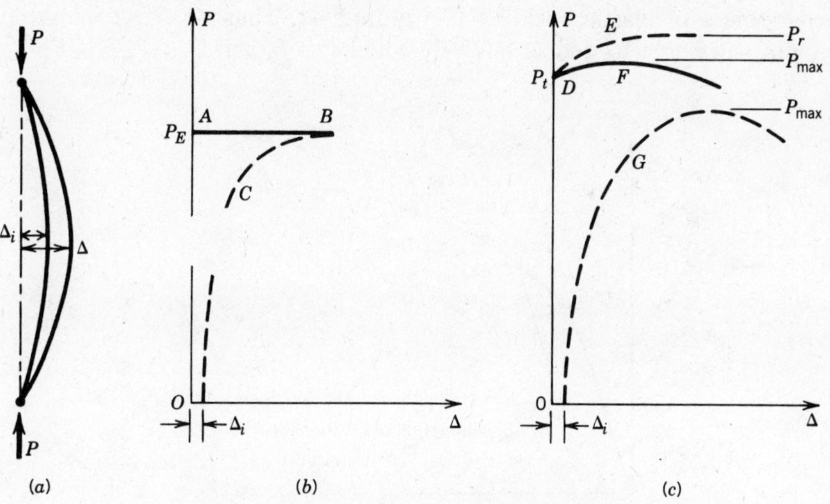

Fig. 3.1 Behavior of perfect and imperfect columns.

The postbuckling behavior of such a column is radically different from the elastic column: Bifurcation buckling occurs at the "tangent modulus" load (point D in Fig. 3.1c),

$$P_t = \frac{\pi^2 E_t I}{L^2} \tag{3.3}$$

but further lateral deflection is possible only if the load increases. If there were no further changes in stiffness due to yielding, the load would asymptotically approach the "reduced modulus" load (E in Fig. 3.1c),

$$P_r = \frac{\pi^2 E_r I}{L^2} \tag{3.4}$$

as the deflection tends to large values. The increase in load is due to the elastic unloading of some fibers in the cross section, which results in an increase in stiffness. The tangent modulus E_t is the slope of the stress-strain curve (Fig. 3.2) when the material is nonlinear, but E_r, and E_t when residual stresses are present, depend also on the shape of the cross section.

Since increased loading beyond the tangent modulus load results in further yielding, stiffness continues to be reduced and the load-deflection curve achieves a peak (P_{max}, point F in Fig. 3.1c) beyond which it falls off.

The improved understanding of the postbuckling behavior of inelastic columns made possible by Shanley (1947) represented the single most significant step in understanding column behavior since Euler's original development of elastic buckling theory in 1744. Thus a perfect inelastic column will begin to deflect laterally when $P = P_t$ and $P_t < P_{max} < P_r$.

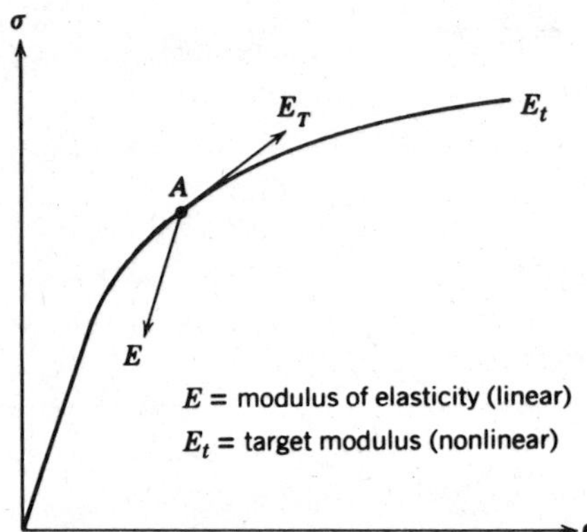

Fig. 3.2 General stress-strain relationship.

3.2.3 Imperfect Columns

Geometric imperfections, in the form of tolerable but unavoidable out-of-straightness of the column and/or eccentricity of the axial load, will introduce bending from the onset of loading, and curve G in Fig. 3.1c characterizes the performance of such a column. Lateral deflection exists from the start of loading, and the maximum load is reached when the internal moment capacity at the critical section is equal to the external moment caused by the product of the load and the deflection. The maximum load is thus a function of the imperfection. For some types of columns the nature of the problem is such that the maximum capacity of the imperfect column is closely approximated by the tangent modulus load of the perfect column, but for many types of columns the imperfections must be included to give a realistic maximum load.

In general, the strength of columns must be determined by including both the imperfections and the material nonlinearity and/or the residual stress effects. The following parts of this chapter will examine in detail how the maximum strength of metal columns can be determined and how this information is used in developing design methods.

3.2.4 Design of Metal Columns as Related to Strength Theories

The determination of the maximum strength of metal columns can be a complicated process, often involving numerical integration, especially when initial imperfections and material nonlinearities or residual stresses must be considered. The structural designers, having to proportion many columns in the course of their working day, cannot be concerned with such lengthy calculations. Simplified column formulas are usually provided for design office practice. These column formulas involve the major parameters of strength, such as the yield point, the length, and the cross-sectional properties, and factors of safety (or load factors) are prescribed to give designs of acceptable safety. Many column formulas have been used throughout the history of structural engineering, and the reader can consult many textbooks, including the third edition of this guide, to learn more about this subject.

While elastic columns can be designed on the basis of the Euler theory, inelastic columns, by necessity of having to account for geometric imperfections and material nonlinearities, are usually designed with empirical column formulas. There are essentially four basic ways by which column design formulas, curves, or charts have been developed:

1. *Empirical formulas based on the results of column tests*. Such formulas are applicable only to the material and the cross section for which the tests were performed. The earliest column formulas (from the

1840s) are of this type. However, some recent papers (e.g., Hall, 1981, and Fukumoto et al., 1983) have utilized the availability of computerized data banks which contain a large proportion of all the column tests reported in the literature. It appears that such formulas may have some difficulty in gaining modern acceptance because they are based entirely on tests. Also, they cannot, in a rational way, account for end restraint.

2. *Formulas based on the yield limit state*. These formulas define the strength of a column as that axial load which will give an elastic stress for an initially imperfect column equal to the yield stress. Such column formulas have a long history, also dating back to the middle of the nineteenth century, and they continue to enjoy popularity to the present (e.g., the British use of the Perry-Robertson formula) (Trahair, 1977). Empirical factors can account for initial imperfections of geometry and loading, but the formulas do not consider the necessarily inelastic basis of column strength, nor can they rationally account for end restraint.

3. *Formulas based on the tangent-modulus theory*. Such formulas can rationally account for the bifurcation load (but not the maximum strength) of perfectly straight columns. If the effects of imperfections are such that they just reduce the maximum strength to the tangent-modulus strength, these formulas have empirical justification. On the other hand, if the perfect column can be thought of as an anchor point in an interaction surface, initial imperfections of geometry and loading can be represented as flexural effects in the interaction equation. The latter approach is used in the European interaction equation (SSRC, ECCS, CRCJ, CMEA, 1982), while the former approach is the basis of the column formulas for cold-formed columns and aluminum columns.

The "CRC—Column Strength Curve," after the acronym of the former name of the Structural Stability Research Council (i.e., Column Research Council), is given by

$$\frac{\sigma_{cr}}{\sigma_y} = 1 - \frac{\lambda^2}{4} \qquad \text{for } \lambda \leqq \sqrt{2} \tag{3.5}$$

and

$$\frac{\sigma_{cr}}{\sigma_y} = \frac{1}{\lambda^2} \qquad \text{for } \lambda \geqq \sqrt{2} \tag{3.6}$$

where

$$\lambda = \frac{KL}{r} \frac{1}{\pi} \sqrt{\frac{\sigma_y}{E}} \tag{3.7}$$

This curve was recommended in the first edition of the guide (1960) and has been used for many steel design specifications in North America and elsewhere. It is based on the average critical stress for small and medium-sized hot-rolled wide-flange sections of mild structural steel, with a symmetrical residual stress distribution typical of such members. The column curves based on the tangent-modulus theory can also accurately account for end restraints (Yura, 1971).

4. *Formulas based on maximum strength*. Modern trends in column design involve column formulas which are a numerical fit of curves obtained from maximum strength analyses of representative geometrically imperfect columns containing residual stresses. The third edition of the guide (1976) presented new column curves based on this principle. Following this, SSRC published Technical Memorandum 5, stating the principle that design of metal structures should be based on the maximum strength, including the effects of geometric imperfections. It was also suggested that the strength of columns might be represented better by more than one column curve, thus introducing the concept of "multiple column curves." The SSRC Curves 1, 2, and 3 are one set of such curves; another example is the set of curves advanced by the European Convention of Constructional Steelwork (ECCS).

3.3 INFLUENCE OF RESIDUAL STRESS

Structural steel shapes and plates contain residual stresses that result primarily from uneven cooling after rolling. Welded built-up members exhibit tensile residual stresses in the immediate vicinity of the welds due to the cooling of the weld metal. These are equal to the yield point of the weld metal, which will normally be somewhat greater than the yield point of the parent metal.

In 1908, in a discussion of the results of column tests at the Watertown Arsenal, residual stresses due to the cooling of hot-rolled steel columns were cited as the probable cause of reduced column strength in the intermediate slenderness range (Howard, 1908). The possible influence of residual stresses on the buckling strength of both rolled members and welded plates in girders was subsequently noted by others (Salmon, 1921; Madsen, 1941). Systematic research on the effect of residual stress on column strength was initiated in the late 1940s under the guidance of Research Committee A of the Column Research Council (Osgood, 1951; Yang et al., 1952; Beedle and Tall, 1960). Recent work in Europe on these effects should also be noted, in order to fully appreciate the magnitude and complexity of the problem (Sfintesco, 1970; Beer and Schultz, 1970).

At the time of the first edition of the guide (1960), the critical-load tangent-modulus curve, based on the effect of typical residual-stress distributions in hot-rolled steel shapes, seemed a proper basis for the determination of allowable column design stresses. Column strength theory could thus be unified, and the tangent-modulus concept, as modified for steel shapes with residual stress, could be extended to all metals. The "CRC Column Strength Curve," based on computed column curves for rolled steel H-shaped members, taken as an approximate average of the strong- and weak-axis buckling curves, served as a basis for the column design provisions of the AISC and CSA Specifications. The second edition of the guide (1966) mentioned the increasing use of columns made of (1) high-strength steels with yield stresses up to 70 ksi, and (2) heat-treated steels with yield stresses up to 100 ksi or more. It noted the importance of initial imperfections as well as of residual stresses in determining the strengths of pinned-end columns made of higher-strength steels.

One of the possible ways of discriminating between different categories of column strength is through the use of the concept of multiple column curves, such as those that were developed through research work at Lehigh University (Bjorhovde and Tall, 1971; Bjorhovde, 1972) and those that have been provided by the studies in Europe (Beer and Schultz, 1970). In addition, large numbers of column tests have also been performed, in some cases on a systematic basis, to provide further assurance of the theoretical results obtained by computer studies. The single largest group of such column tests is probably the more than 1000 tests that were made at a number of European universities and laboratories, as well as a number of tests on heavy shapes at Lehigh University, under the auspices of the European Convention for Constructional Steelworks (Sfintesco, 1970). Over the years, a great many other tests have also been performed, and these have been summarized by Fukumoto et al. (1983).

3.3.1 Hot-Rolled Shapes

The magnitude and distribution of residual stresses in hot-rolled shapes depend on the type of cross section, rolling temperature, cooling conditions, straightening procedures, and metal properties (Beedle and Tall, 1960). Examples of residual-stress distributions resulting from cooling without straightening of wide-flange shapes are shown in Fig. 3.3 (Tall, 1964). For the heavier shapes, residual stresses vary significantly through the thickness. Figure 3.4 shows residual stresses measured (Brozzetti et al., 1970a) in the heaviest (W14 × 730) rolled shape currently produced.

The effect or steel grade on the residual-stress distribution is not as great as the effect of geometry (Tall, 1964). Residual-stress measurements in the flanges of similar shapes made of different steel grades show that the distributions and magnitudes of the residual stress are similar and it is here—in the flanges—that residual stresses have a major effect on the column strength of H-shaped sections.

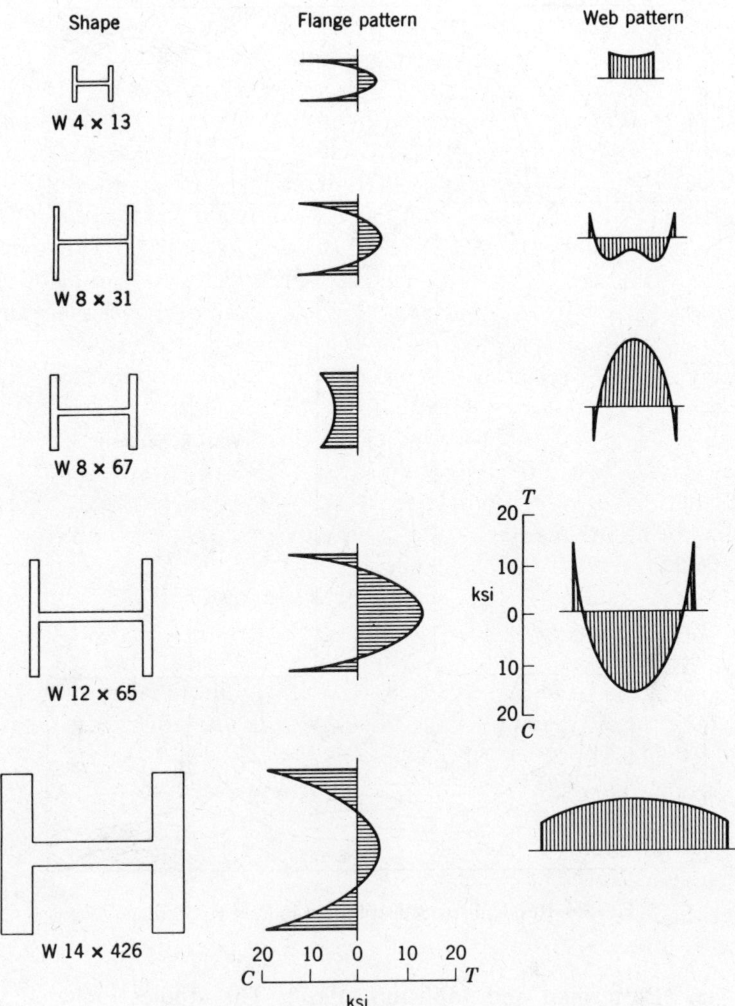

Fig. 3.3 Residual-stress distribution in rolled wide-flange shapes.

Calculated critical-load column curves based on the residual stresses found in the five particular shapes of Fig. 3.3 are plotted in Fig. 3.5 for buckling about the minor axis, along with maximum strength column curves determined by computer analyses and based on a combination of the measured residual stresses and an initial curvature equal to the maximum tolerated by the ASTM A6 Standard.

To interrelate the residual stress and initial curvature effects systematically, extensive column strength analyses were made at the University of

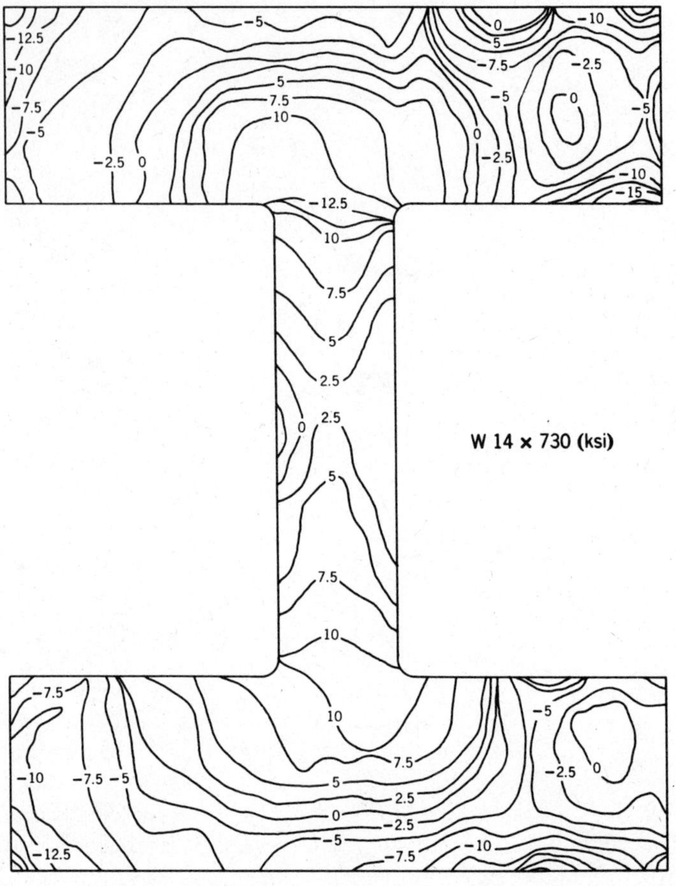

Fig. 3.4 Residual-stress distribution in W14 × 730 shape.

Michigan (Batterman and Johnston, 1967). The studies included yield stresses of 36, 60, and 100 ksi (251, 419, and 699 MPa); maximum compressive residual stresses of 0, 10, and 20 ksi (0, 69, and 138 MPa); five curvatures corresponding to initial midlength out-of-straightness ranging from 0 to $0.004L$; and slenderness ratios ranging from 20 to 240. Emphasis was on buckling about the minor axis, and the results for this condition are presented graphically in the work by Batterman and Johnston (1967), permitting maximum strength evaluation within the range of the parameters cited. On the basis of a maximum residual stress of 13 ksi, which is the scaled average maximum for the five sections shown in Fig. 3.3, together with a yield stress of 36 ksi, the maximum

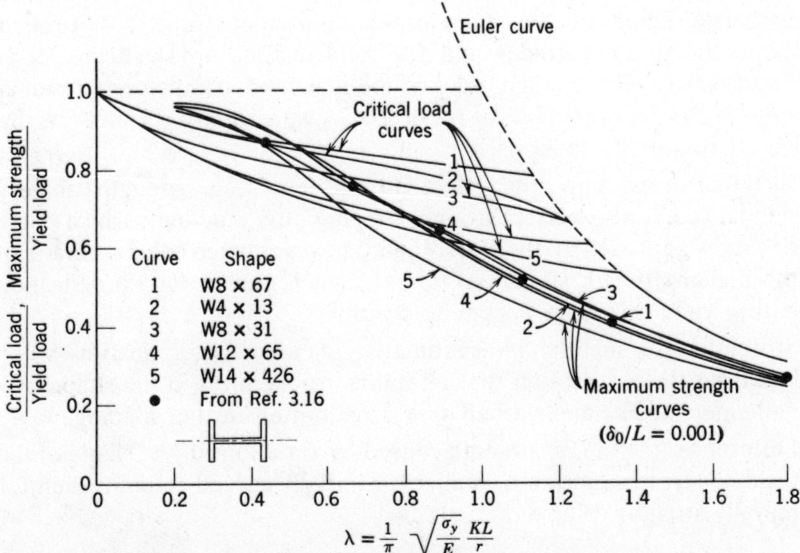

Fig. 3.5 Critical-load curves for straight columns compared with maximum-strength curves for initially curved rolled steel W-shapes.

column strength predicted by Batterman and Johnston is shown by the solid circles on Fig. 3.5. The solid curves are from an analysis neglecting the webs. Although the shapes and residual stress distributions are different, there is good correlation between the two independently developed analysis procedures. These findings are also corroborated by those of a wide-ranging investigation of column strength (Bjorhovde, 1972), which examined the behavior and strength of a large and diverse number of structural shapes. The computational procedure is very accurate but requires knowledge of the residual stresses and the out-of-straightness. In the study performed at Lehigh University (Bjorhovde, 1972), the full range of structural steel grades and shapes was examined, as well as a number of welded build-up box and H shapes.

The results obtained by the studies of Batterman and Johnston (1967) and Bjorhovde (1972) show clearly that:

1. The separate effects of residual stress and initial curvature cannot be added to give a good approximation of the combined effect on the maximum column strength. Thus in some cases and for some slenderness ratios, the combined effect is less than the sum of the parts (intermediate slenderness ratios, low residual stresses). In other cases the combined effect is more than the sum of the parts. The latter

applies to the intermediate slenderness ratio range for heavy hot-rolled shapes in all steel grades and for welded build-up H shapes. It is emphasized that the magnitudes of the maximum compressive residual stresses in a large number of these shapes were at 50% or more of the yield stress of the steel itself.

2. Residual stress has little effect on the maximum strength of very slender columns, either straight or initially crooked, which have strengths approaching the Euler load. However, such columns made of the higher-strength steels can tolerate much greater lateral deflection before yield or before becoming unstable.
3. Strengths are slightly underestimated in a computer analysis when based on the assumption that an initial crookedness in the shape of a half-sine wave remains a half-sine wave during further loading.
4. Differences in column strength caused by variations in the shape of the residual stress pattern are smaller for initially curved columns than for initially straight columns.

3.3.2 Welded Columns

Residual stresses resulting either from welding or from the manufacture of the component plates have a significant influence on the strength of welded H- or box-section columns. The maximum tensile residual stress at a weld or in a narrow zone adjacent to a flame-cut edge is equal to or greater than the yield stress of the plates (Fujita, 1949, Alpsten and Tall, 1970; Alpsten, 1972a; Brozzetti et al., 1970b; Bjorhovde et al., 1972). Welding modifies the prior residual stresses due either to flame cutting or cooling.

Figure 3.6 shows that the strengths of welded columns made of higher-strength steels appear to be influenced relatively less by residual stresses than are the strengths of similar columns made of lower-strength steels (Tall, 1966). It is also evident that the differences in strengths of columns with the maximum permissible initial crookedness are less than the differences in critical loads of initially straight columns (see Fig. 3.5).

As shown in Fig. 3.7, plates with mill-rolled edges (universal mill plates) have compressive residual stresses at the plate edges, whereas flame-cut plates have tensile residual stresses at the edges. In built-up H shapes made of universal mill plates, the welding will increase the compressive stress at the flange tips, thus enlarging the region of compressive residual stress and adversely affecting the column strength. Thus, as illustrated in Fig. 3.8, an H-shape column made from flame-cut plates will have favorable tensile residual stresses at the flange tips. It therefore exhibits greater column strength than that of a column of the same

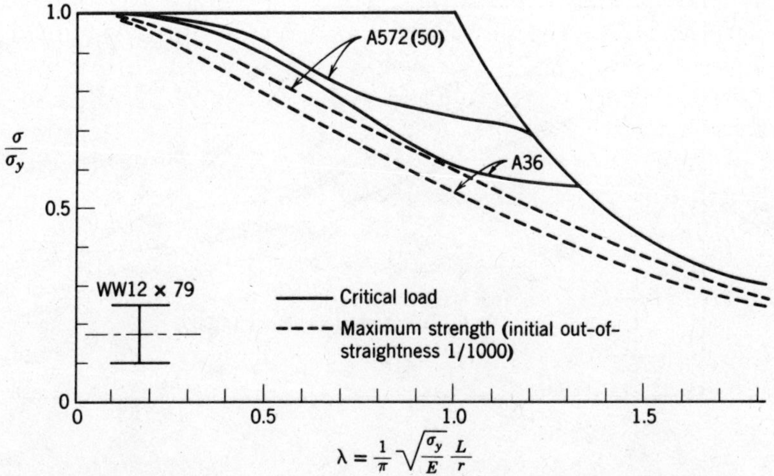

Fig. 3.6 Critical-load curves for welded WW12 × 79 of flame-cut plates compared with maximum-strength curves for initially curved members.

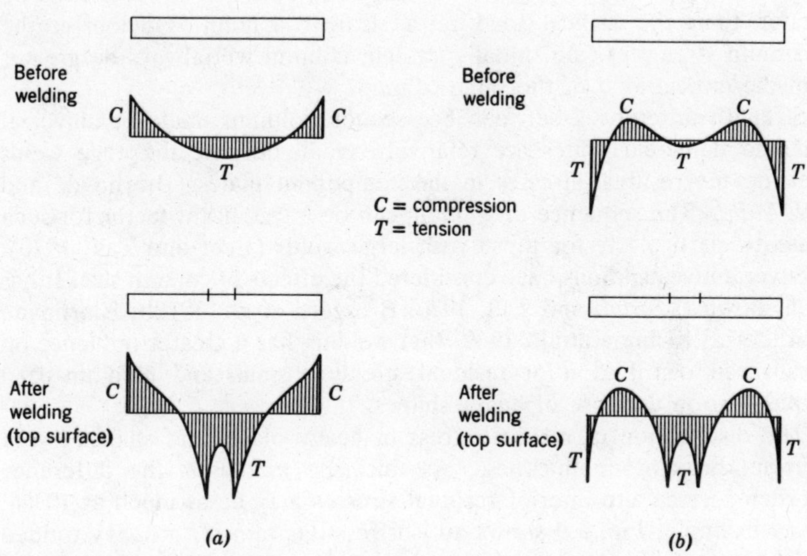

Fig. 3.7 Qualitative comparison of residual stresses in as-received and center-welded universal mill and oxygen-cut plates: (*a*) Universal mill plate; (*b*) oxygen-cut plate.

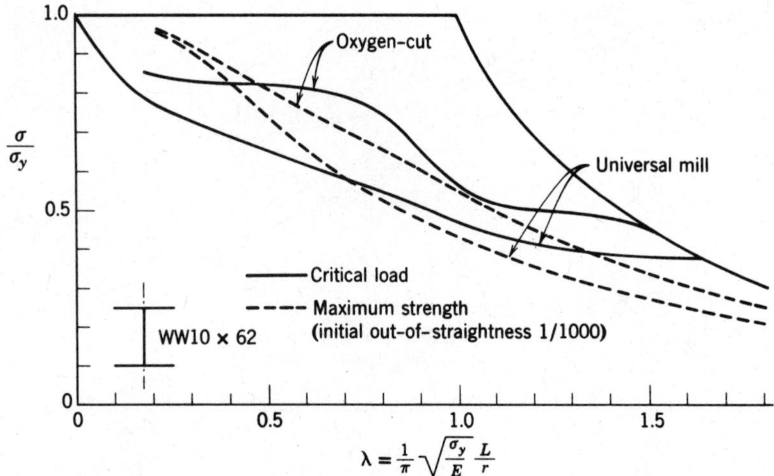

Fig. 3.8 Comparison of column curves for WW10 × 62 (A7 steel) with universal mill versus oxygen-cut plates.

section with flanges consisting of a universal mill plate in which both rolled edges are retained. It is also seen that for short welded columns, the maximum strength of an initially curved column may in some cases be greater than the critical load of a straight column. Obviously, the maximum strength of an initially straight column will always be greater than the critical load of the same column.

Strength differences between box-section columns made of universal mill and flame-cut plates are relatively small, because the edge welds override the residual stresses in the component plates (Bjorhovde and Tall, 1971). The sequence of welding can be a significant factor for such columns, particularly for those with large welds (Beer and Tall, 1970).

Several investigations have considered the effects of column size. It has been shown (Alpsten and Tall, 1970; Brozzetti et al., 1970b; Bjorhovde et al., 1972; Kishima et al., 1969) that welding has a greater influence on the overall distribution of residual stress in small and medium-sized shapes than in the case of heavy shapes.

The distribution of residual stress in heavy plates and shapes is not uniform through the thickness. As thickness increases, the difference between surface and interior residual stresses may be as much as 10 ksi. As an example, Fig. 3.9 shows an isostress diagram for a heavy welded shape made from flame-cut plates. However, it has been found that calculated critical loads and maximum column strengths are only a few percent less when based on the complete residual-stress distributions, as

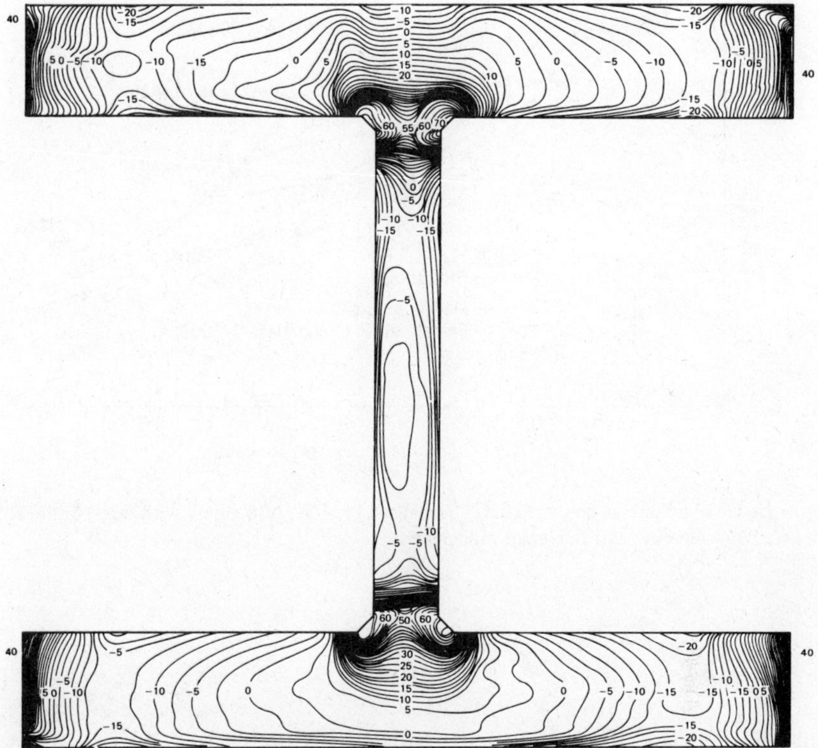

Fig. 3.9 Isostress diagram for WW23 × 681 welded built-up shape (stresses in kips per square inch).

compared with analyses that assume the stress to be constant through the thickness and equal to the surface-measured residual stress.

In general, shapes made from flame-cut plates exhibit higher strength than shapes that are made from universal mill plates. This is demonstrated by the curves in Fig. 3.10. Similarly, flame-cut shapes tend to have strengths that are comparable with those of similar rolled shapes, whereas universal mill shapes tend to be comparatively weaker (Fig. 3.10).

Figure 3.11 compares the strengths of two typical welded columns with flame-cut flange plates, and one being distinctly heavier than the other. It is seen that the heavier shape tends to be relatively stronger than the lighter one. This is even more accentuated for shapes that are welded from universal mill plates; for these the strength of the lighter shape will be significantly lower than the heavy one (Bjorhovde and Tall, 1971; Bjorhovde, 1972). This is the reason the Canadian Standards Association

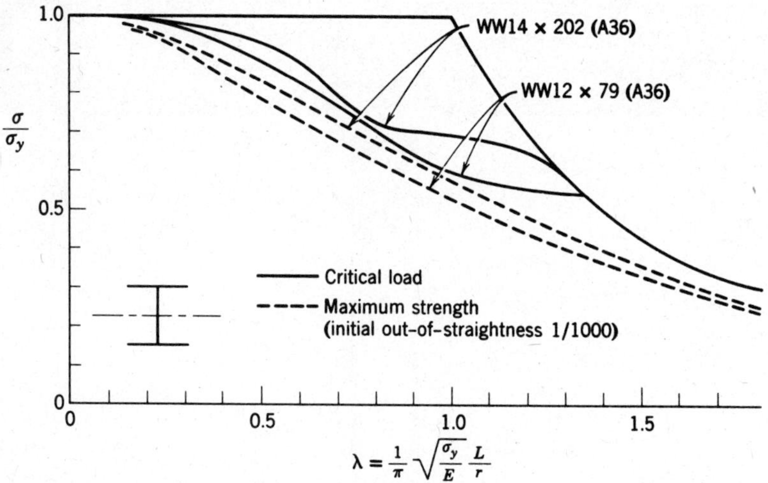

Fig. 3.10 Comparison of column curves for WW24 × 428 (A36 steel) with stress-relieved, oxygen-cut, and universal mill plates.

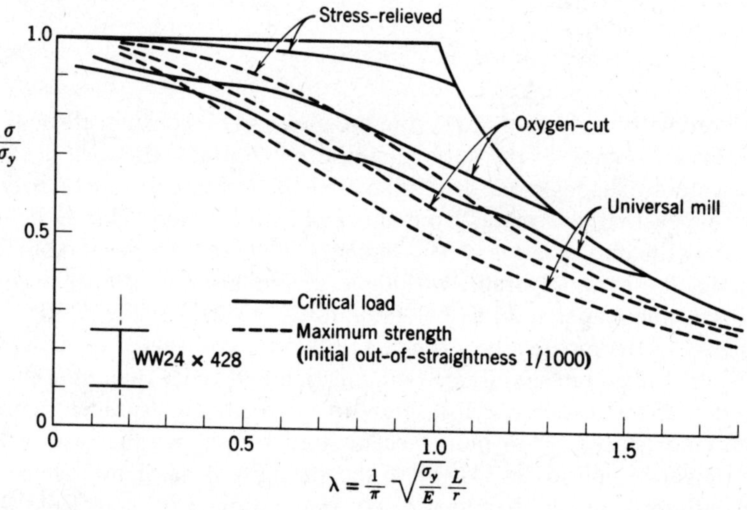

Fig. 3.11 Column curves for heavy and light welded wide-flange shapes.

in its limit states design standard (CSA, 1974) requires that welded built-up shapes can only be made from flame-cut plates.

The sequence of welding and the number of welding passes are factors influencing the distribution of residual stresses. Other welding parameters, such as voltage, speed of welding, and temperature and areas of preheating have less influence (Brozzetti et al., 1970b). Stress-relief annealing of the component plates prior to welding of the shape raises column strength very significantly by reducing the magnitude of the residual stresses, even though it lowers the yield stress of the steel. Figure 3.10 compares the column curves for shapes made from flame-cut and universal mill plates, along with curves for the same shapes made from stress-relieved plates.

3.3.3 Cold-Straightened Columns

Cold straightening of structural sections to meet tolerances for camber and sweep induces a redistribution and reduction of the residual stresses that were caused by earlier rolling and cooling.

In current mill practice, either rolls or gag presses are used to straighten structural shapes (Alpsten and Tall, 1970). In roller straightening (also called "rotorizing"), the shape is passed through a train of rolls that bend the member back and forth with progressively diminishing deformation. This process is used only on small shapes. In gag straightening, concentrated forces are applied locally along the length of the member to bend it to approximate straightness. The process is used on structural shapes of all sizes.

The roller-straightening process redistributes and reduces the initial residual stresses in the flanges, as shown in Fig. 3.12. In gag straightening, moments that approximate the full plastic value, M_p, are produced at the points where the forces are applied, and the cooling residual stresses are therefore redistributed only at or near points of loading. In the usual case of gag straightening, to remove sweep [curvature about the weak $(y-y)$ axis of a wide-flange shape], the change of the residual stress from compression to tension takes place locally at the edges on the side of the flanges where the load is applied. Figure 3.13 shows the residual stresses measured in a W8 × 31 shape (Huber, 1956) after gag straightening about the weak axis.

The strength of a cold-straightened column is, in general, greater than that of the corresponding as-rolled member (Alpsten, 1970; Frey, 1969), because of the improved straightness and the redistribution of residual stress. Roller straightening produces a greater improvement than gag straightening and, according to theoretical analyses and experimental results, it may increase the column strength by as much as 20% when

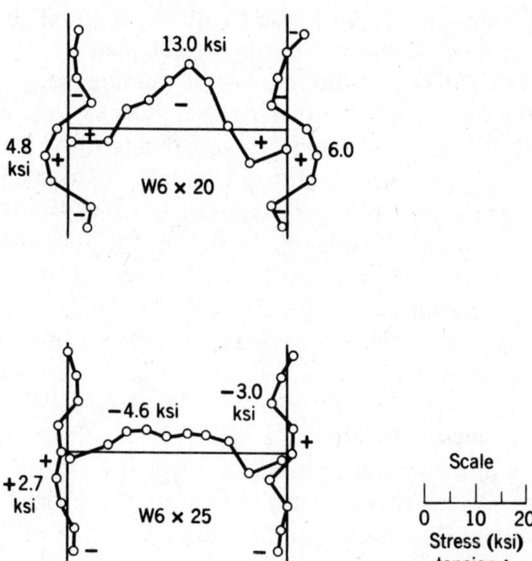

Fig. 3.12 Residual stresses in roller-straightened shapes.

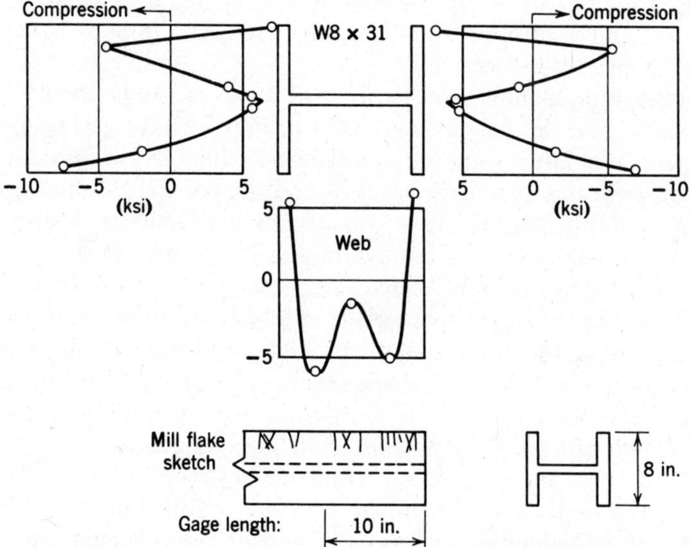

Fig. 3.13 Residual stresses in a gag-straightened shape, bent about the weak axis to remove sweep.

compared at the same slenderness ratio and initial out-of-straightness (Alpsten, 1970, 1972b). However, other investigators (Pavolovic and Stevens, 1981) have found that rotorized columns actually lose some of their strength. This has been attributed to the Bauschinger effect.

The strength and behavior of cold-straightened columns is still only partly understood, and a concentrated research effort should be undertaken to detail all of the individual influences and effects. This is particularly important in view of the fact that almost all hot-rolled wide-flange shapes are straightened in the mill to meet straightness requirements. For tubular shapes the sitation is somewhat different. The final mill process in most cases is cold forming or rolling, producing very small initial out-of-straightnesses, followed in some mills by partial stress relieving.

3.4 INFLUENCE OF OUT-OF-STRAIGHTNESS

The material that has been presented in Section 3.3 has detailed the influence of the residual stresses, which signify one of the primary column strength factors. Another factor is the initial out-of-straightness (also referred to as initial crookedness or initial curvature). The behavior of the initially crooked column has already been examined in the context of critical load theory; in brief, the presence of any geometric imperfection makes the column buckling problem one of a load-deflection type, as opposed to a bifurcation type of stability problem.

For members that fail in the inelastic range of column behavior, the effects of initial curvature are not clear-cut. Some of the characteristics of the behavior and strength of inelastic, initially curved columns have already been discussed in the evaluation of residual-stress influences; the two parameters interact in many ways. Thus the explanation of the ranges of slenderness ratios and column types, for which the combined effect of residual stress and initial crookedness is more than the sum of the parts, emphasizes the complexity of the phenomenon.

The analyses that have been made of the strength of inelastic, initially curved columns have either made use of assumed values and shapes of the initial out-of-straightness, or have used actually measured data. The former is by far the most common, mostly because the measurements that are available for columns are scarce. This applies in particular to the magnitude of the maximum out-of-straightness, normally assumed to occur at midheight of the member, as well as the shape of the bent member. The latter is usually thought to be that of a half-sine wave (Batterman and Johnston, 1967; Bjorhovde and Tall, 1971; Bjorhovde, 1972). The real configuration of the initial out-of-straightness of a column may be very complicated, often expressed as a simultaneous crookedness about both principal axes of the cross section. Systematic measurements have been made in some laboratories in conjunction with testing programs (Bjorhovde, 1972, 1977; Beer and Schultz, 1970; Bjorhovde and Birkemoe, 1979; Fukumoto, 1983), but very few data are available for columns in actual structures (Tomonaga, 1971; Lindner and Gietzelt, 1984).

The magnitude of the initial out-of-straightness is limited by the structural steel delivery specifications, normally expressed as a fraction of the length of the member. Thus wide-flange shapes are required to have less than a maximum initial crookedness of $L/960$ ($\frac{1}{8}$ in. in 10 ft of length), which is usually given as $L/1000$ for convenience. Tubular shapes are required to meet a straightness tolerance of $L/480$, commonly given as $L/500$. The measurements that are available show that most shapes tend to have values toward the maximum permissible (Bjorhovde, 1972), with an average of approximately $L/1500$. This applies to wide-flange shapes; for tubular members the averages tend to be significantly smaller than the specification limitations. For the latter, results (Bjorhovde, 1977; Bjorhovde and Birkemoe, 1979) show that these members exhibit out-of-straightnesses on the order of $L/3000$ to $L/8000$, with an average of approximately $L/6300$. On the whole, it is relatively uncommon to encounter test columns that have out-of-straightnesses larger than the maximum permitted values; in structures, such members are not acceptable in the first place.

In the development of column design criteria such as the SSRC curves (see Section 3.8) and the ECCS curves (Beer and Schultz, 1970), the maximum permissible values of the initial crookedness were utilized. This was done for several reasons, the primary one being that $L/1000$ constituted the upper limit of what is acceptable for actually delivered members, and therefore could be regarded as a conservative measure. It is conceivable that as the amount of data on the values of the curvature increases, other measurements, such as for example the mean, may be used. This was done by Bjorhovde (1972) in conjunction with his development of the original SSRC curves, using the mean of $L/1470$ that was found through statistical evaluations. The resulting multiple column curves are shown in Fig. 3.14, where the curves labeled as 1P, 2P, and 3P have used $L/1470$. For comparison, the SSRC curves have been included in the figure; these have used an initial out-of-straightness of $L/1000$. The mathematical equations for both sets of curves are given in Section 3.8.

Variations in the magnitude of the initial crookedness were considered in the study by Bjorhovde (1972). The strength of the 112 columns that were included in the investigation was examined using maximum initial

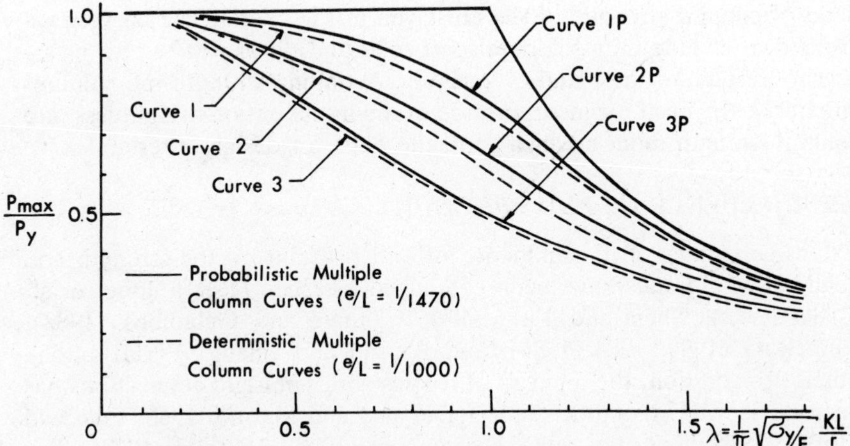

Fig. 3.14 Comparison of multiple column curves developed on the basis of mean out-of-straightness ($= L/1470$) and maximum permissible out-of-straightness ($L/1000$) (Bjorhovde, 1972).

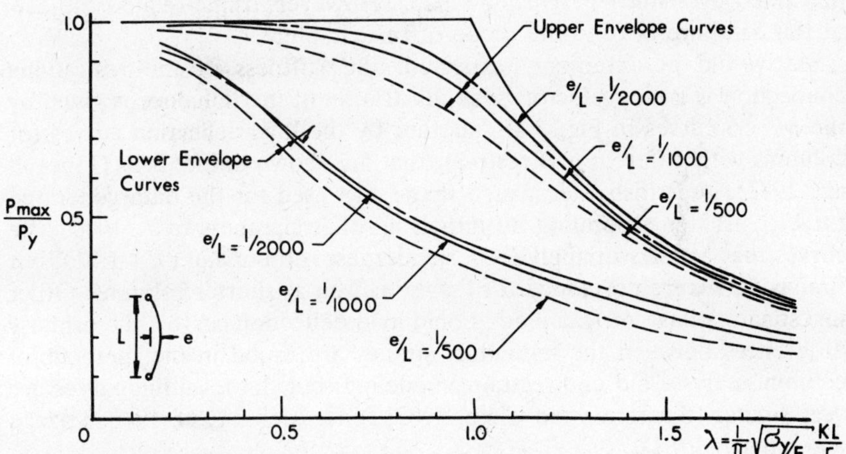

Fig. 3.15 Column curve bands for 112 columns, based on initial out-of-straightnesses of $L/500$, $L/1000$, and $L/2000$ (Bjorhovde, 1972).

out-of-straightnesses of $L/500$, $L/2000$, and $L/10{,}000$; the results for the band of column strength curves are given in Fig. 3.15. (The curves that are shown include only the data for $L/500$ to $L/2000$.)

The results of the studies on the maximum strength of columns emphasize the need for incorporation of the initial out-of-straightness into column strength models which form the basis for design criteria.

3.5 INFLUENCE OF END RESTRAINT

Extensive studies on the influence of end restraint on the strength and behavior of columns have been conducted by Chen (1980), Jones et al. (1980, 1982), Shen and Lu (1983), Chapius and Galambos (1982), Vinnakota (1982, 1983, 1984), and Razzaq and Chang (1981), among others. In addition, the analysis of frames with semirigid connections has been dealt with in several studies (DeFalco and Marino, 1966; Frye and Morris, 1975; Romstad and Subramanian, 1970; Ackroyd, 1979). The column investigations have examined different aspects of restrained member behavior, specifically, determining the influence of:

1. Type of beam-to-column connection.
2. Length of column.
3. Magnitude and distribution of residual stress.
4. Initial out-of-straightness.

The frame analysis studies have focused on evaluations of the drift characteristics of frames with less than fully rigid connections, in part prompted by a study by Disque (1975). However, frame-related subjects of this kind are beyond the scope of this chapter.

As would be intuitively expected, the stiffness of the restraining connection is a major factor. One illustration of the influence is given by the $M - \phi$ curves in Fig. 3.16, another by the load-deflection curves for columns with different end restraint that are shown in Fig. 3.17 (Jones et al., 1982). A British wide-flange shape was used for the data generated for Fig. 3.17 incorporating an initial out-of-straightness of $L/1000$. The curves that are shown apply for a slenderness ratio of 120 ($\lambda = 1.31$), but similar data were developed for longer as well as shorter columns. Other investigators have provided additional load-deflection curves; the primary differences between the individual studies are found in the methods of column analysis and end-restraint modeling, but the resulting curves are very similar (Sugimoto and Chen, 1982; Jones et al., 1980, 1982; Razzaq and Chang, 1981).

Figure 3.17 also includes the load-deflection curve for a pinned-end column. As is evident, the higher the connection restraint, the stiffer will

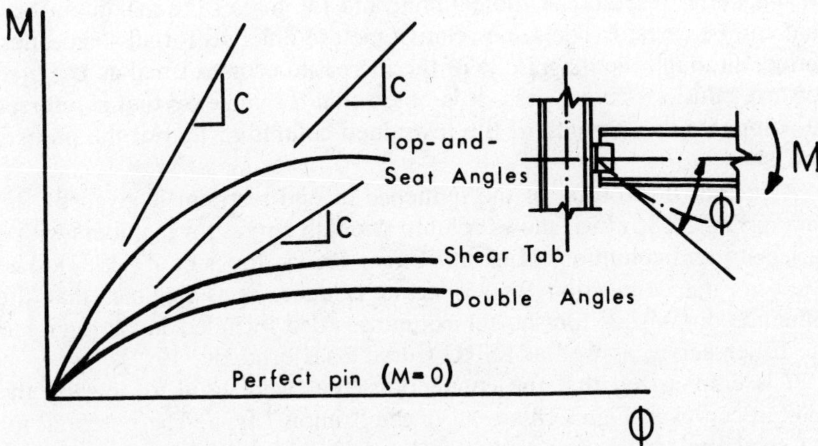

Fig. 3.16 Moment-rotation curves for some typical simple connections (schematic).

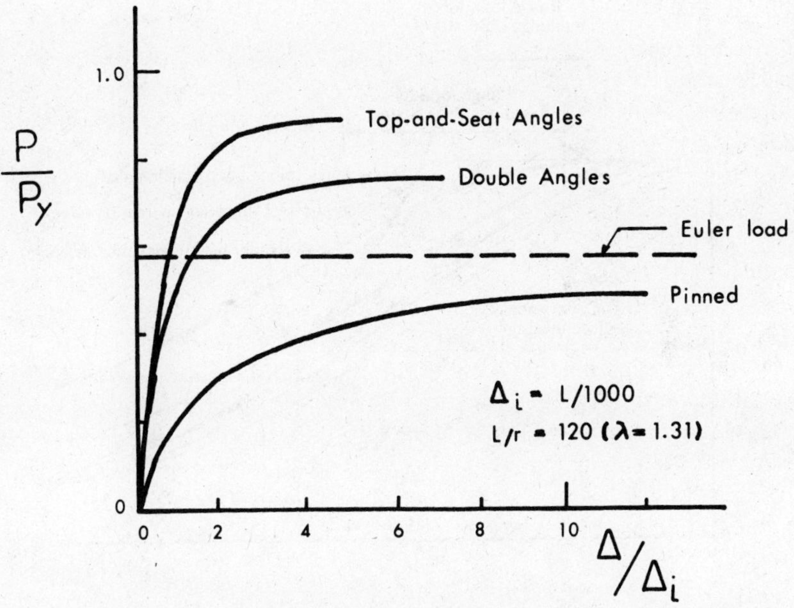

Fig. 3.17 Typical load-deflection curves for columns (Jones et al., 1982).

be the initial response of the column, and the higher the maximum load that can be carried. The same relative picture emerges for all slenderness ratios, although the magnitude of the increase becomes small as L/r goes toward values of 50 and less. It is noted that the increase that is referred to compares the strength of the restrained column to that of the pinned-end one.

A further illustration of the influence of end restraint is given by the data in Fig. 3.18, which shows column strength curves for members with a variety of end conditions (Lui and Chen, 1983a; Jones et al., 1982). The effect of the connection type is again evident, as is the fact that the influence diminishes for shorter columns. Also included in the figure is the Euler curve, as well as SSRC Curve 2 (Bjorhovde, 1972).

It is emphasized that the connections that were used to develop the column curves in Fig. 3.18 are all of the "simple" type. The potential for the structural economies that may be gained by incorporating the end restraint into the column design procedure is clear, although the realistic ranges for the values of λ must be borne in mind. The latter have been delineated in Fig. 3.18 for steels with yield stresses of 36 and 50 ksi. Consequently, the very large column strength increases that have been reported by several researchers are real (Lui and Chen, 1983b; Sugimoto

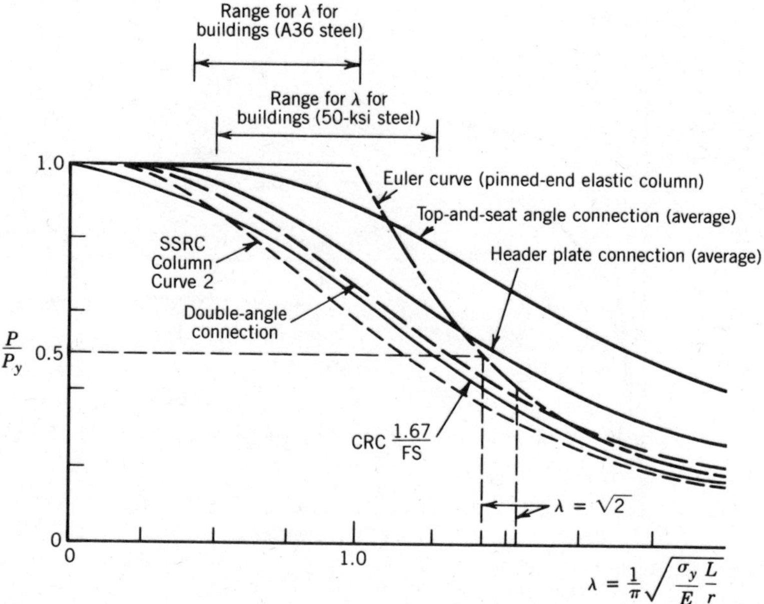

Fig. 3.18 Column curves for members with different types of end restraint.

and Chen, 1982; Jones et al., 1980, 1982; Razzaq and Chang, 1981; Chen, 1980), but they occur for slenderness ratios that are in excess of practical values (Bjorhovde, 1981; Ackroyd and Bjorhovde, 1981). Overall, the end restraint has a clear strength-raising effect.

Using the individual column strength studies as the basis, additional research has recently been completed that demonstrates the application of end restraint to the design of columns in frames (Bjorhovde, 1984). Taking into account actual connection stiffness and the influence of beams, effective length factors for columns in frames have been developed. The method incorporates the use of the well-known alignment charts for the effective length of framed columns, and recognizes that buckling of a column in a frame is influenced by the end-restraint relative stiffness distribution factor, G_r, given as

$$G_r = \frac{\sum (EI/L)_{\text{columns}}}{C^*} \tag{3.8}$$

where C^* is the effective end restraint that is afforded to a column in a beam and column subassemblage, using connections whose initial stiffness is C (= initial slope of the moment rotation curve, as given by the various C values in Fig. 3.16).

The G_r procedure (Bjorhovde, 1984) also incorporates detailed design suggestions, as well as applications of inelastic K-factor principles, as developed by Yura (1971) and expanded by Disque (1973). Practical design examples illustrate the potential for significant structural economies by using the Bjorhovde approach. However, it is emphasized that it is necessary to have the data for the actual end-restraint characteristics of the connections in order to apply these principles. Specifically, the C value must be known.

3.6 EFFECTIVE LENGTH OF COLUMNS

The effective-length factor, K, was discussed briefly when introduced in Eq. 3.2. The coverage of effective column length in this chapter is limited to certain idealized cases and to certain special situations that occur in compression members of trusses. A more general treatment of effective length of columns in continuous and multistory frames is provided in Chapter 16. The effective-length concept has also been applied to members of nonuniform cross section, whereby they are converted to an equivalent pinned-end member with an effective moment of inertia that is referenced to a particular location of the nonuniform member. These approaches are discussed in Chapters 9 and 12.

Figure 3.19 gives theoretical K values for idealized conditions in which

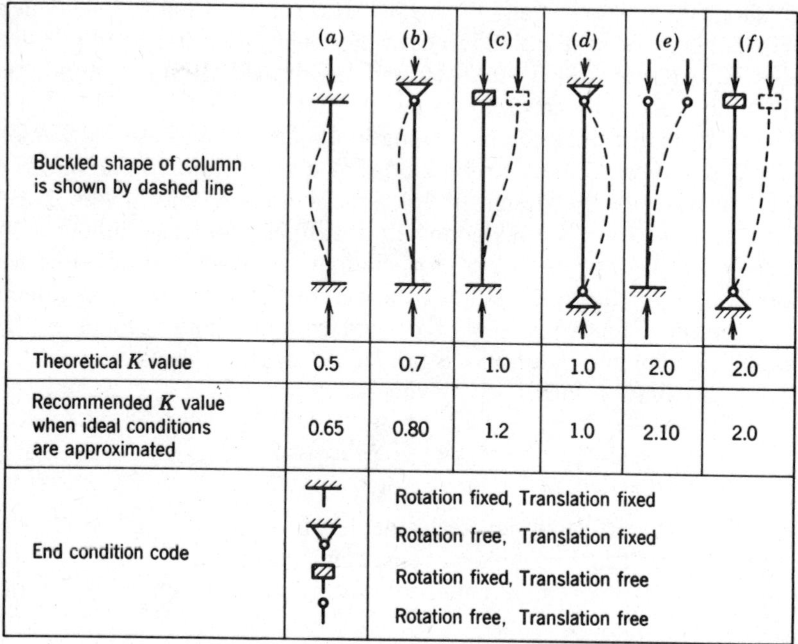

	(a)	(b)	(c)	(d)	(e)	(f)
Buckled shape of column is shown by dashed line						
Theoretical K value	0.5	0.7	1.0	1.0	2.0	2.0
Recommended K value when ideal conditions are approximated	0.65	0.80	1.2	1.0	2.10	2.0
End condition code		Rotation fixed, Translation fixed Rotation free, Translation fixed Rotation fixed, Translation free Rotation free, Translation free				

Fig. 3.19 Effective-length factors K for centrally loaded columns with various end conditions.

the rotational and/or translational restraints at the ends of the column are either fully realized or are nonexistent. At the base, shown fixed under conditions a, b, c, and e, in Fig. 3.19, the condition of full fixity can be approached only when the column is anchored securely to a footing for which the rotation is negligible. Column conditions a, c, and f are approached when the top of the column is integrally framed to a girder many times more rigid than the column. Column condition c is the same as a except that translational restraint is either absent or minimal at the top. Condition f is the same as c except that there is no rotational restraint at the bottom. The recommended design values of K are modifications of the ideal values, taking into account the fact that neither perfect fixity nor perfect flexibility can be attained in practice.

The more general determination of K for a compression member as part of any framework requires the application of methods of indeterminate structural analysis, modified to take account of the effects of axial load and inelastic behavior on the rigidity of the members. Gusset-plate effects can be included; for this case extensive charts for modified slope-deflection equations, and for moment-distribution stiffness and carryover factors, respectively, have been developed (Goldberg, 1954;

Michalos and Louw, 1957). These procedures are not directly applicable to routine design, but they can be used to determine end restraints and resultant modified effective lengths of the component members of a framework.

In triangulated truss frameworks, loads are usually applied only at the joints, producing only axial loads in the members if the joints are hinged. Deflections of the joints are caused by the axial deformations of the members under load and are relatively small. On the other hand, if the joints are welded or heavily bolted or riveted, some secondary bending is induced. The effect of secondary distortions on the buckling strength of truss members is usually small and can be neglected in the buckling analysis.

If every member in a truss were designed to minimum weight and reached maximum stress under the same loading condition, buckling stresses in compression members and yield stresses in tension members would be approached at the same level of live load. On this basis, no restraint would be supplied at the joints, and K would be unity for compression chords and the equivalent lengths would be equal to the full distance between panel points. In a roof truss of nearly constant depth, where a single compression chord of constant cross section is used for the full length of the truss, K may be taken as 0.9. In a continuous truss, K may be taken as 0.85 for the compression chord connecting to the joint where the chord stress changes from compression to tension.

When the magnitude of stress in the compression chord changes at a subpanel point that is not braced normal to the plane on the main truss (Fig. 3.20), the effective-length factor for chord buckling normal to the

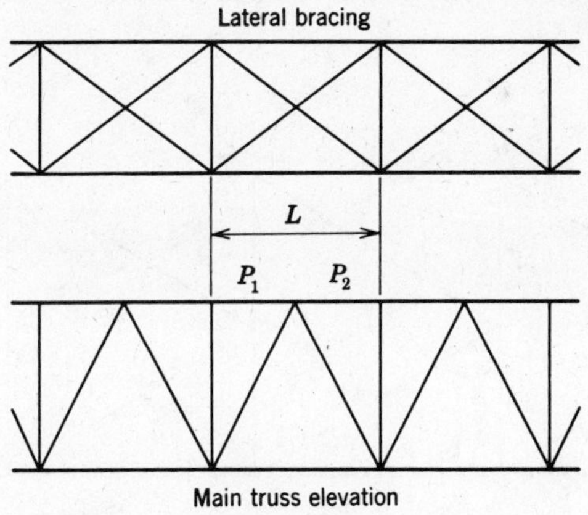

Fig. 3.20

plane of the main truss can be approximated from the two compressive forces P_2 and P_1, as follows:

$$K = 0.75 + 0.25(P_1/P_2) \tag{3.9}$$

where $P_2 < P_1$.

Web members in trusses designed for moving live-load systems may be designed with $K = 0.85$. This is because the position of live load that produces maximum stress in the web member being designed will result in less than maximum stresses in members framing into it, so that rotational restraints will be developed. K should also be taken as unity for web members in a truss designed for a fixed-load system, where maximum stress occurs in all members simultaneously.

The design of vertical web members, U_iL_i, of a K-braced truss (Fig. 3.21) should be based on the length KL. Web-member buckling occurs normal to the plane of the truss, and Eq. 3.9 again applies. Also, P_2 is to be taken as negative in Eq. 3.9 since it is tension. When P_1 and P_2 are numerically equal, Eq. 3.9 yields a value of $K = 0.5$.

For buckling normal to the plane of a main truss, the web compression members should be designed for $K = 1$ unless detailed knowledge of the makeup of the cross frames (perpendicular to the truss) is available. For example, with cross frames of type 1 (Fig. 3.22), it is satisfactory to take $K = 0.8$, and for type 2 it is satisfactory to use $K = 0.7$, provided that the top and bottom lateral bracing systems are adequate to prevent joint translation. When translation of the cross frames is possible, a more exact analysis of web-member stability should be undertaken.

In the case of redundant trusses, there is a reserve strength above the load at initial buckling of any compression member. Masur (1954) has reviewed developments on this subject and established upper and lower bounds for the ultimate load of the buckled members of elastic redundant trusses.

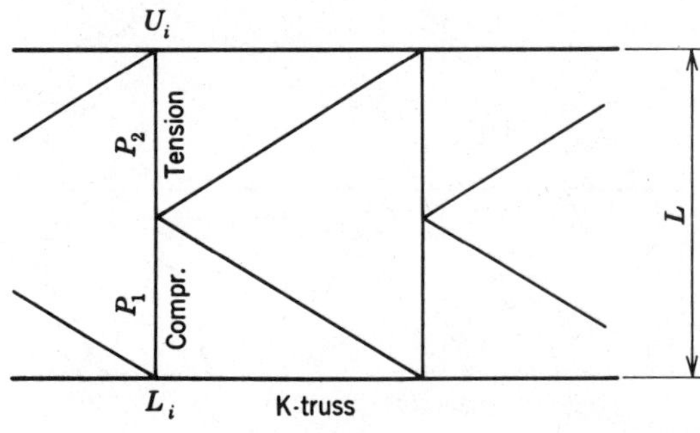

Fig. 3.21

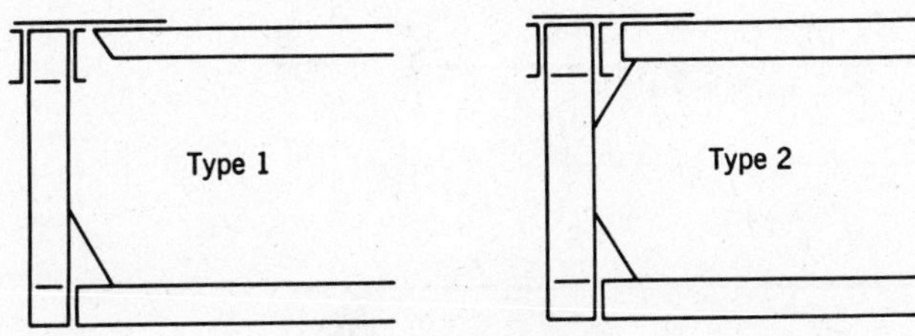

Fig. 3.22

3.7 LATERAL BRACING REQUIREMENTS

Lateral bracing between the ends of columns provides a means of increasing the load-carrying capacity. It is most effective when it is attached at the location along the column where there is a point of contraflexure in one of the higher buckling modes of the column. For example, lateral bracing at the center of a pinned-end column will increase the elastic buckling load by the factor of 4. In other words, the effective length of the column is reduced from the full length to one-half of the length.

Bracing must exhibit both strength and stiffness to perform its intended function. The definitive work on column and beam bracing was that of Winter (Winter, 1960; see also McGuire, 1968, Chap. 3) who demonstrated that both the strength and the stiffness requirements are relatively modest if the bracing is provided at the node points of the higher buckling modes. The required bracing stiffness k_{ideal} is

$$k_{\text{ideal}} = \frac{C\pi^2 E_t I}{L^3} = \frac{CP_{cr}}{L} \tag{3.10}$$

where the coefficient C varies from 2 to 4 (Winter, 1960) and is given in Fig. 3.23. The required bracing strength is

$$F_{\text{bracing}} = \Delta_{\text{initial}}\left(\frac{k_{\text{ideal}}}{1 - k_{\text{ideal}}/k_{\text{actual}}}\right) \tag{3.11}$$

As can be seen from Eq. 3.11, the actual stiffness must be larger than the ideal stiffness. The initial deflection Δ_{initial}, is the misalignment of the brace point with respect to the ideally straight column. If $\Delta_{\text{initial}} = L/500$, and the ratio $k_{\text{ideal}}/k_{\text{actual}} = \frac{1}{2}$, the required bracing force becomes equal to

n	1	2	3	4	∞
F_{bracing}	$0.008P_{cr}$	$0.012P_{cr}$	$0.014P_{cr}$	$0.015P_{cr}$	$0.016P_{cr}$

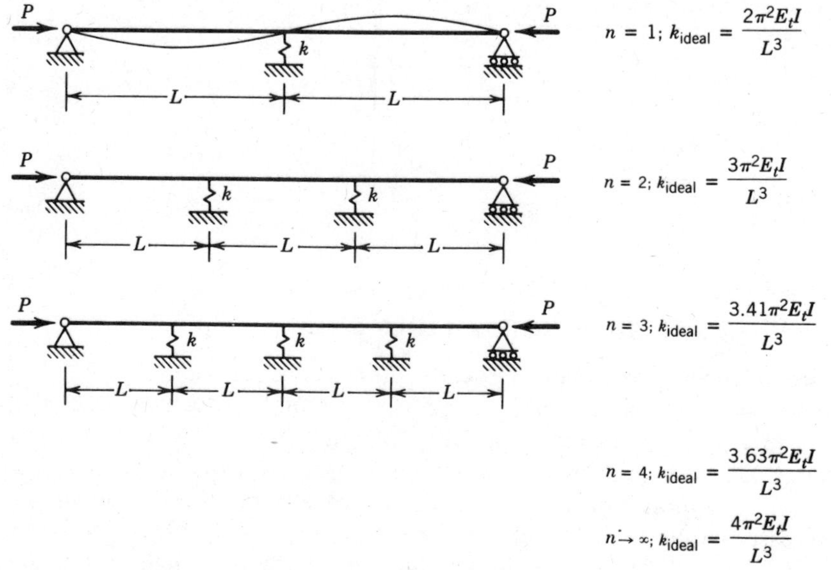

Fig. 3.23 Effect of lateral bracing.

The usual conservative practice is to take as the brace design force two percent (2%) of the force in the column to be braced. If the points of bracing are not evenly spaced (i.e., they are not at the node points of higher buckling nodes), and if the axial loads and the member sizes are different in each braced segment, it is necessary to perform a second-order buckling analysis based on first principles of stability theory to determine the required brace properties. Usually, however, not all the column segments are critical and only one or a few segments will buckle under the factored design loads. In this case the ideal brace stiffness is the maximum value obtained from the following formula, which is evaluated for each pair of adjacent segments:

$$k_{\text{ideal}}^{(i)} = \frac{P_i}{L_i} + \frac{P_{i+1}}{L_{i+1}} \tag{3.12}$$

The required brace force can then be evaluated using Eq. 3.11. The actual bracing stiffness should preferably be at least twice the ideal stiffness, and the axial load in each segment may not exceed the column strength of a pinned end column with a length equal to the corresponding bracing spacing.

There are several additional important points to consider in the design of bracing:

1. Even though the bracing requirements are modest, braces are nevertheless vital parts of the structure and should not be relegated to a negligible role. If the braces are improperly attached, they will be ineffective.
2. Braces must be properly attached to the member to be braced, and their ends must be anchored to rigid supports. Bracing two adjacent columns to each other is useless, since both columns may buckle in the same direction in a lower mode.
3. Bracing must restrain twisting as well as lateral motion, to prevent a lower torsional buckling mode.
4. Bracing systems that restrain multiple parallel columns must be stiff enough to take care of the sum of the axial forces in all the columns. For example, if two lines of bracing ($n = 2$) brace five columns each supporting a load P, the ideal stiffness and the required brace force is five times that needed for one column. A detailed analysis of braced multiple columns and beams is provided by Medland (1979, 1980).

3.8 STRENGTH CRITERIA FOR STEEL COLUMNS

3.8.1 Introduction

The position of the Structural Stability Research Council on the basis for the design of columns is stated in Technical Memorandum No. 5. The following quote from this Memorandum summarizes this position:

> Maximum strength, determined by the evaluation of those effects that influence significantly the maximum load-resisting capacity of a frame, member, or element, is the proper basis for the establishment of strength design criteria.

It is therefore emphasized that the proper column strength model is one that incorporates both the residual stress and initial out-of-straightness.

3.8.2 Multiple Column Curves

Research leading to the use of multiple column curves has been under way for some years. In 1959 the German standard DIN 4114 introduced a special column curve for tubes and used another curve for all the other column types. The later work of the European Convention for Constructional Steelwork (Sfintesco, 1970; Beer and Schultz, 1970; Jacquet, 1970) resulted in recommended design application and code adoption in several countries (Sfintesco, 1976). Research basic to the development of multiple column curves was conducted starting in the 1960s at Lehigh University (Bjorhovde and Tall, 1971) and elsewhere (Birkemoe, 1977a,b,

Bjorhovde, 1977; Bjorhovde and Birkemoe, 1979; Kato, 1977; Sherman, 1976).

The maximum strength of steel columns depends, in addition to the length and the cross-sectional properties and the material properties, F_y and E, on (1) the residual-stress magnitude and distribution, (2) the shape and magnitude of the initial out-of-straightness, and (3) the end restraint. The effects of these three variables were discussed in detail earlier in this chapter. Unless special procedures are utilized in the manufacture of steel columns, such as stress relieving or providing actual pins at each end all three of these effects are present and should be accounted for. The present state of research is such that if the following information is known, accurate calculation of the maximum strength is possible (Bjorhovde, 1972, 1978; Chen and Lui, 1985):

1. *Material properties* (i.e., the yield stress F_y and the modulus of elasticity E); in some cases it is necessary to know the variation of the yield stress across the cross section (welded build-up shapes) or the entire stress-strain curve (cold-formed shapes).
2. *Cross-section dimensions*; in cases of columns with variable sections, the dimensions along the column length need also be known.
3. *Distribution of the residual stresses* in the cross section, including variations through the plate thickness if the shape is a tubular shape or if the plate elements are thick.
4. *The shape and the magnitude of the initial out-of-straightness.*
5. *The moment-rotation relationship of the end restraint.*

Maximum strength may be calculated by postulating suitable but realistic idealisations so that closed-form algebraic expressions may be derived, or one of many available numerical techniques may be used. In numerical calculations it is usually assumed that deformations are small and that plane sections remain plane. The literature on determining the maximum strength of columns is rich and diverse, but the major methods are described in textbooks (see Chen and Atsuta, 1976, and Chen and Hahn, 1985 for example).

Residual stress, initial crookedness, and end restraint are random properties, and complete statistical information is lacking. The method of calculating the maximum strength is also a complicated matter. The question thus arises about what to do in recommending design methods and formulas. An answer to the problem was provided by Bjorhovde (1972), who proceeded as follows.

A computerized maximum strength analysis was performed first on basic data available from carefully constructed column tests performed at

Lehigh University, and it was demonstrated that the method of numerical analysis gave an accurate prediction of the test strengths. Next, a set of 112 column curves was generated for members for which measured residual-stress distributions were available, assuming that the initial crookedness was of a sinusoidal shape having a maximum amplitude of 1/1000 of the column length and that the end restraint was zero. These shapes encompassed the major shapes used for columns, including rolled and welded shapes from light to heavy dimensions. The column curves thus obtained represent essentially the whole spectrum of steel column behavior. The resulting curves are shown in Fig. 3.24.

Bjorhovde then observed that there were definite groupings among the curves and from these three subgroups were identified, each giving a single "average" curve for the subgroup (Bjorhovde and Tall, 1971; Bjorhovde, 1972). The resulting three curves are known as *SSRC Column Strength Curves 1, 2, and 3*, and they are reproduced in Figs. 3.25 through 3.27. These figures contain:

1. The number of column curves used as a basis for the statistical analysis and the width of their scatter band.
2. The calculated $2\frac{1}{2}$ and $97\frac{1}{2}$ percentile lower and upper probability limits for the particular set of curves.
3. The column types to which each of the three curves is related.

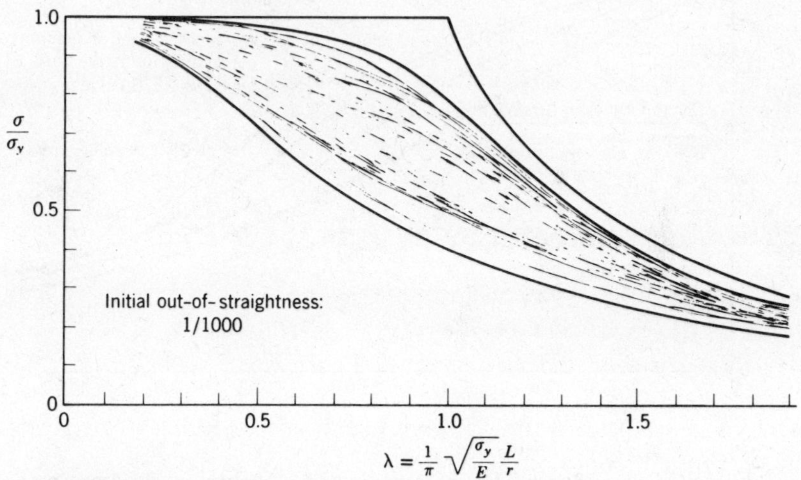

Fig. 3.24 Maximum-strength curves for a number of different column types (Bjorhovde, 1972).

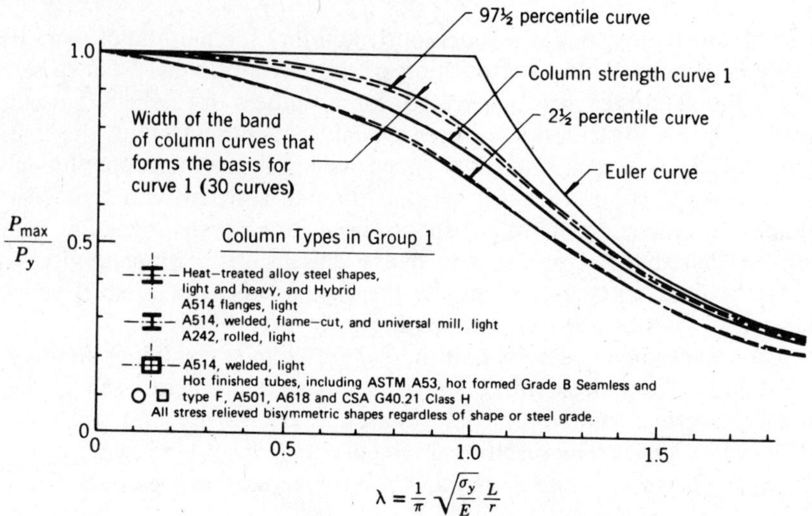

Fig. 3.25 SSRC Column Strength Curve 1 for structural steel (Bjorhovde, 1972). (Based on maximum strength and initial out-of-straightness of $\delta_0 = 0.001L$.)

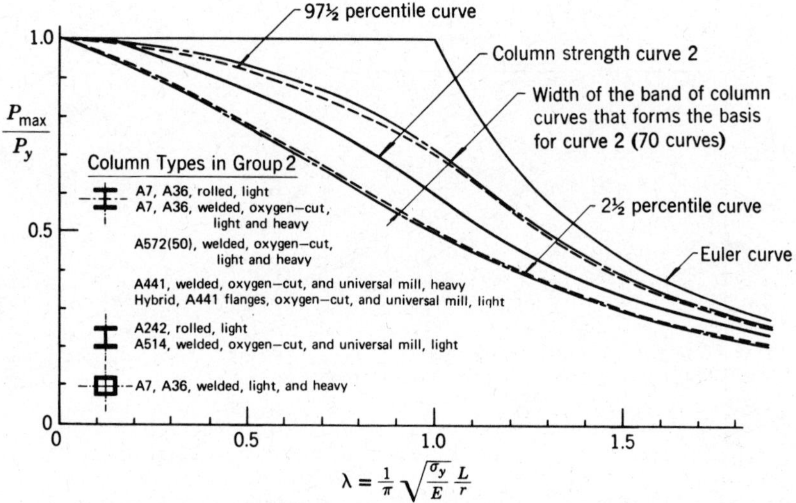

Fig. 3.26 SSRC Column Strength Curve 2 for structural steel (Bjorhovde, 1972). (Based on maximum strength and initial out-of-straightness of $\delta_0 = 0.001L$.)

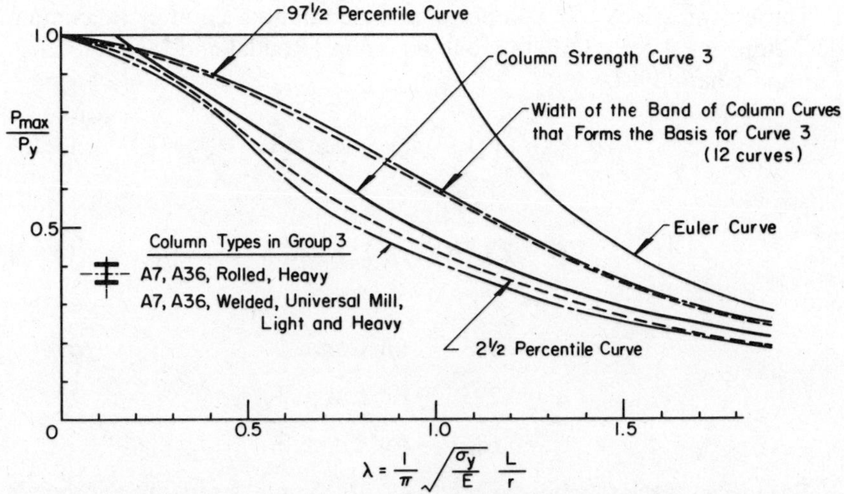

Fig. 3.27 Column Strength Curve 3 for structural steel (Bjorhovde, 1972). (Based on maximum strength and initial out-of-straightness of $\delta_0 = 0.001L$.)

Algebraic representations of the three column strength curves were obained by curve fitting, and the resulting equations are given as Eqs. 3.13 through 3.15.

SSRC Curve 1

$$\left.\begin{array}{lll} 1. & \text{For } 0 \le \lambda \le 0.15 & \sigma_u = \sigma_y \\ 2. & \text{For } 0.15 \le \lambda \le 1.2 & \sigma_u = \sigma_y(0.990 + 0.122\lambda - 0.367\lambda^2) \\ 3. & \text{For } 1.2 \le \lambda \le 1.8 & \sigma_u = \sigma_y(0.051 + 0.801\lambda^{-2}) \\ 4. & \text{For } 1.8 \le \lambda \le 2.8 & \sigma_u = \sigma_y(0.008 + 0.942\lambda^{-2}) \\ 5. & \text{For } \lambda \ge 2.8 & \sigma_u = \sigma_y\lambda^{-2} (= \text{Euler curve}) \end{array}\right\} \tag{3.13}$$

SSRC Curve 2

$$\left.\begin{array}{lll} 1. & \text{For } 0 \le \lambda \le 0.15 & \sigma_u = \sigma_y \\ 2. & \text{For } 0.15 \le \lambda \le 1.0 & \sigma_u = \sigma_y(1.035 - 0.202\lambda - 0.222\lambda^2) \\ 3. & \text{For } 1.0 \le \lambda \le 2.0 & \sigma_u = \sigma_y(-0.111 + 0.636\lambda^{-1} + 0.087\lambda^{-2}) \\ 4. & \text{For } 2.0 \le \lambda \le 3.6 & \sigma_u = \sigma_y(0.009 + 0.877\lambda^{-2}) \\ 5. & \text{For } \lambda \ge 3.6 & \sigma_u = \sigma_y\lambda^{-2} (= \text{Euler curve}) \end{array}\right\} \tag{3.14}$$

SSRC Curve 3

$$\left.\begin{array}{lll} 1. & \text{For } 0 \le \lambda \le 0.15 & \sigma_u = \sigma_y \\ 2. & \text{For } 0.15 \le \lambda \le 0.8 & \sigma_u = \sigma_y(1.093 - 0.622\lambda) \\ 3. & \text{For } 0.8 \le \lambda \le 2.2 & \sigma_u = \sigma_y(-0.128 + 0.707\lambda^{-1} - 0.102\lambda^{-2}) \\ 4. & \text{For } 2.2 \le \lambda \le 5.0 & \sigma_u = \sigma_y(0.008 + 0.792\lambda^{-2}) \\ 5. & \text{For } \lambda \ge 5.0 & \sigma_u = \sigma_y\lambda^{-2} (= \text{Euler curve}) \end{array}\right\} \tag{3.15}$$

These expressions can also be represented quite accurately (maximum deviations −2.1 to +3.6%) by one equation (Rondal and Maquoi, 1979; Lui and Chen, 1984):

$$\sigma_u = \frac{\sigma_y}{2\lambda^2}\left(Q - \sqrt{Q^2 - 4\lambda^2}\right) \le \sigma_y \tag{3.16}$$

where

$$Q = 1 + \alpha(\lambda - 0.15) + \lambda^2 \tag{3.17}$$

and

$$\alpha = 0.103 \qquad \text{for Curve 1}$$
$$\alpha = 0.293 \qquad \text{for Curve 2}$$
$$\alpha = 0.622 \qquad \text{for Curve 3}$$

Based on the probabilistic analyses of column strength, Bjorhovde (1972) also developed multiple column curves where the initial out-of-straightness was equal to its mean value of 1/1470 of the column length (Fig. 3.14). The mathematical equations describing these curves are given as Eqs. 3.18 through 3.20.

SSRC Curve 1P

$$\left.\begin{array}{lll} 1. & \text{For } 0 \le \lambda \le 0.15 & \sigma_u = \sigma_y \\ 2. & \text{For } 0.15 \le \lambda \le 1.2 & \sigma_u = \sigma_y(0.979 + 0.205\lambda - 0.423\lambda^2) \\ 3. & \text{For } 1.2 \le \lambda \le 1.8 & \sigma_u = \sigma_y(0.03 + 0.842\lambda^{-2}) \\ 4. & \text{For } 1.8 \le \lambda \le 2.6 & \sigma_u = \sigma_y(0.018 + 0.881\lambda^{-2}) \\ 5. & \text{For } \lambda \ge 2.6 & \sigma_u = \sigma_y\lambda^{-2}(= \text{Euler curve}) \end{array}\right\} \tag{3.18}$$

SSRC Curve 2P

$$\left.\begin{array}{lll} 1. & \text{For } 0 \le \lambda \le 0.15 & \sigma_u = \sigma_y \\ 2. & \text{For } 0.15 \le \lambda \le 1.0 & \sigma_u = \sigma_y(1.03 - 0.158\lambda - 0.206\lambda^2) \\ 3. & \text{For } 1.0 \le \lambda \le 1.8 & \sigma_u = \sigma_y(-0.193 + 0.803\lambda^{-1} + 0.056\lambda^{-2}) \\ 4. & \text{For } 1.8 \le \lambda \le 3.2 & \sigma_u = \sigma_y(0.018 + 0.815\lambda^{-2}) \\ 5. & \text{For } \lambda \ge 3.2 & \sigma_u = \sigma_y\lambda^{-2} \end{array}\right\} \tag{3.19}$$

SSRC Curve 3P

$$\left.\begin{array}{lll} 1. & \text{For } 0 \le \lambda \le 0.15 & \sigma_y = \sigma_y \\ 2. & \text{For } 0.15 \le \lambda \le 0.8 & \sigma_u = \sigma_y(1.091 - 0.608\lambda) \\ 3. & \text{For } 0.8 \le \lambda \le 2.0 & \sigma_u = \sigma_y(0.021 + 0.385\lambda^{-1} + 0.066\lambda^{-2}) \\ 4. & \text{For } 2.0 \le \lambda \ge 4.5 & \sigma_u = \sigma_y(0.005 + 0.9\lambda^{-2}) \\ 5. & \text{For } \lambda \ge 4.5 & \sigma_u = \sigma_y\lambda^{-2}(= \text{Euler curve}) \end{array}\right\} \tag{3.20}$$

A single expression for these curves has not been developed; however, this can be achieved relatively easily, using the format of Rondal and Maquoi (1979). A single equation that fits Curve 2P has been produced in the development of the AISC Load and Resistance Factor Design (LRFD) specification (see Section 3.8.3).

3.8.3 Design Procedure Alternatives

It was demonstrated in the preceding section that it is possible to develop multiple column curves, into which column types can be grouped for convenience. In developing column design cirteria, the following questions should be considered:

1. What should be the shape and the amplitude of the initial out-of-straightness? As to the shape, there is fairly uniform agreement among researchers that a sinusoidal shape with the maximum amplitude in the center is a conservative and reasonable assumption. The maximum amplitude is, however, a crucial quantity since changes affect the strength, especially in the intermediate slenderness range. Knowledge about initial out-of-straightness is available from measurements of laboratory specimens used for column tests, but there is a paucity of field data. Initial out-of-straightness is a function of the manufacturing process, and some column types, such as manufactured tubes, tend to be very straight. In the development of the SSRC and ECCS multiple column curves the position was taken that an initial amplitude of 1/1000 of the length, essentially the mill tolerance, is a reasonable and conservative value for the basis of developing column curves. One can argue that this conservative approach is justified because the initial out-of-straightness takes care of all other geometric imperfections, such as initial out-of-plumb and the unavoidable eccentricity of the axial load. The out-of-straightness thus may be used to represent the effects of other types of geometric imperfections.

In opposition, it can be argued that all geometric imperfections are small enough so that their effect can be relegated to be accounted for by the factors of safety or the resistance factors. This was the underlying philosophy of the use of the CRC column curve, which has its basis in the tangent-modulus theory, with a factor of safety that depends on the column slenderness ratio. This design approach was entirely sensible when it was initially formulated (in the 1950s), but a convincing mass of research has since shown that the maximum strength can be determined from a knowledge of initial imperfections, and that the economical and safe use of steel columns should take cognizance of the out-of-straightness.

A Task Force of SSRC (1985) has taken an intermediate position, recommending that the basis for the development of design curves for steel columns should be an initial out-of-straightness of 1/1500 of the length. This is close to the average measured in laboratory columns (Bjorhovde, 1972; Fukumoto et al., 1983) and reflects the current position of the SSRC in this matter. For all practical purposes, Eqs. 3.18 through 3.20 represent this condition.

2. Should column curves be based on the tangent-modulus theory for straight columns, and geometric imperfection be accounted for by a specified eccentricity of the load which produces an end moment on the column? Column design would then always be performed with an interaction equation. Such an approach is entirely rational and it seems to be a consideration in the column design criteria for beam-columns as proposed in Europe, where, however, it seems that both initial crookedness and load eccentricity are used for beam-columns, but where intentionally axially loaded columns are designed with consideration of out-of-straightness only. Although this method of column design is a rational alternative, at the present time it does not have a great deal of support within SSRC.

3. Should design be based on the concept of multiple column curves, or should one composite column curve, or formula, be used for the design of all steel columns? The European (ECCS) answer to this question has been to recommend multiple column curves (these are shown in Fig. 3.28), and a number of European codes have already adopted their use. Researchers associated with SSRC have championed and developed the basis for the use of multiple column curves, and the Council favors their use, because column design can be made more efficient and reliability can be made more uniform by their use. As a first step in North America, the Canadian Standards Association (CSA) has adopted the use of SSRC Curve 2 as the basic design curve in CSA Standard S16.1-74 "Steel Structures for Buildings—Limit States Design." In 1980 CSA also adopted SSRC Curve 1 for hollow structural sections, cold formed to final shape and stress relieved. This recommendation was based in part on research on such columns (Bjorhovde and Birkemoe, 1979; Birkemoe, 1977b; Bjorhovde, 1977). The reluctance to adopt multiple column curves is due to the complications in design if more than one basic column curve is used. Another reason is that the assignment of column types to any of the three column strength categories needs further study, pending in part on as yet incomplete or lacking research on the strength of straightened columns, asymmetric and built-up columns, cold-formed tubes, heavy columns, and so on (Bjorhovde, 1980). However, the research and the

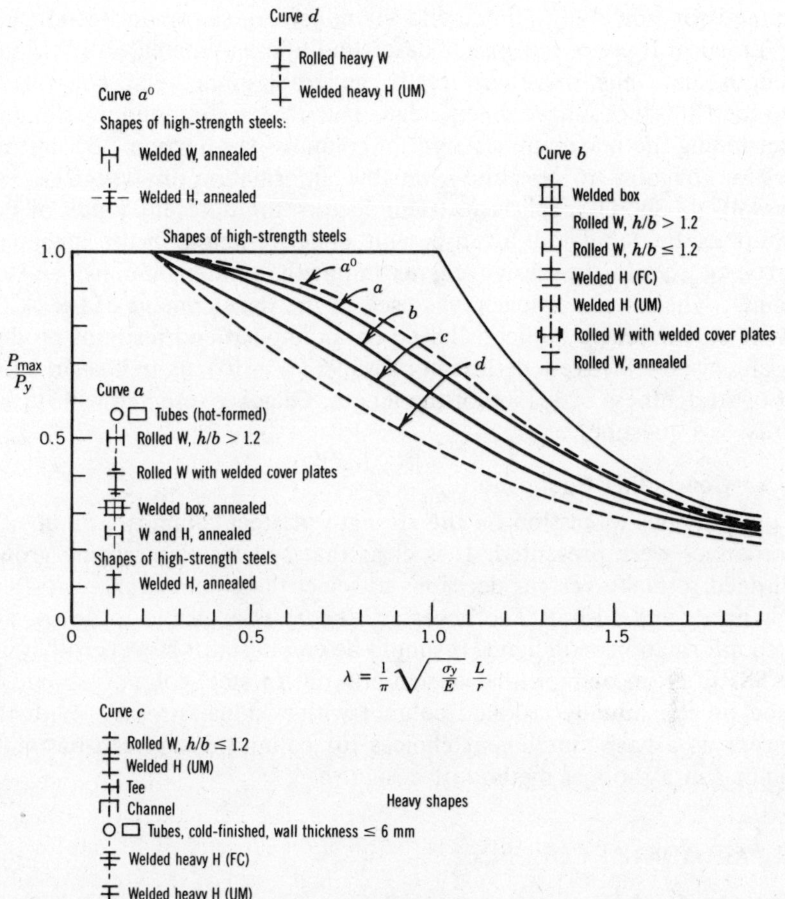

Fig. 3.28 European multiple column curves, recommended by the European Convention for Constructional Steelwork (ECCS). (Based on initial out-of-straightness, $\delta_0 = 0.001L$.)

discussion continue; and in the future these questions will be resolved through the effort of the researchers in SSRC. In the meantime, AISC has adopted a single column curve for its new Load and Resistance Factor Design (LRFD) specification, which is discussed in detail in Chapter 20. This curve is essentially identical to Eqs. 3.19, although a simpler single formula is used.

4. The final question to be asked is: What end restraint should be

assumed for nominally pinned-end columns? As shown in Section 3.5, any practical framing scheme or base condition will increase the column strength and thus there are really no truly pinned-end columns in existence. Methods have been developed to use this end restraint in determining the maximum strength of columns (see Section 3.5), but the question of how to use the available information in design is still unresolved. Should explicit restraint factors for different kinds of end conditions be tabulated for use with effective-length-factor alignment charts, or should the design curves implicitly contain minimal end restraints? The latter approach was used in the development of the AISC LRFD column curve, which is based on an implicit end restraint producing an elastic effective-length factor of 0.96 ($G = 10$), as well as an initial out-of-straightness of 1/1500 of the length. Ongoing research will further clarify this question.

3.8.4 Concluding Comments

In the previous discussion on the strength of steel columns a number of alternatives were presented. It is clear that specification writing groups will need to make various decisions to select the column curve satisfying their needs and wishes. The necessary theory is available to do so, and much information is on hand. It should be emphasized, however, that it is the SSRC's firm opinion that design criteria for steel columns should be based on the initially crooked column with residual stresses. With this concept as a basis, intelligent choices for column design can be made, resulting in a rational method of design.

3.9 ALUMINUM COLUMNS

3.9.1 Material Properties

Alloying elements, heat treatment, and working have a great influence on all the essential properties of aluminum, with the exception of the elastic modulus, which falls within the range 9900 to 10,200 ksi (68,240 to 70,310 MPa) for other than aircraft alloys. In general engineering, a value of 10,000 ksi (68,930 MPa) is used.

Yield strength is more strongly influenced by heat treatment and working than is the ultimate strength. For heat-treated or strain-hardened alloys there is also an increased sharpness of the "knee" between the elastic and plastic ranges, which is significant for columns in the lower range of slenderness ratios. For this reason column formulas for aluminum alloys are divided into two groups, heat-treated and non-heat-treated, which reflect the differing ratio $\sigma_{0.2}/\sigma_{0.1}$. Alloys that are solution heat-treated but not artificially aged, seldom used in engineering, fall in the non-heat-treated group. Guaranteed values for the yield strength, defined by the 0.2% offset, and the ultimate strength are established at levels at which 99% of the material is expected to conform at a 0.95 confidence level. In practice, typical values are some 15% above the guaranteed value; thus the use of the guaranteed value in a design formula which has been formulated on the basis of measured values provides an additional factor to be considered when selecting resistance factors or factors of safety for columns with short and medium slenderness ratios.

Yield strength, σ_y, is the value of the stress in the stress-strain curve at a specified offset from the initial elastic line, usually 0.2% ($\sigma_{0.2}$) and sometimes 0.1% ($\sigma_{0.1}$).

3.9.2 Imperfections

Deviations of real columns from the perfect elastic-plastic model are of two essential forms: nonlinearity of the stress-strain relationships, and geometric imperfections such as eccentricity and initial curvature. Nonlinearity of the stress-strain relationship arises as a natural property of the material, as a consequence of the presence of residual stresses, or as a consequence of the local change in yield stress caused by welding. Residual stresses in aluminum extruded members are small because of the method of production and the straightening of the finished member by stretching. The fact that residual stress effects on column strength of aluminum extruded members are insignificant has been confirmed in European studies (Mazzolani and Frey, 1977). Nonlinearity in unwelded bars is thus attributable only to the material behavior and will vary with the type of alloy and heat treatment. The value of the yield strength does not vary significantly across a profile (Bernard et al., 1973).

Welding introduces residual tensile stresses in the weld bead on the order of the yield strength for the annealed material, and compressive stresses elsewhere. There is also a local reduction in mechanical properties effected by welding that is significant in heat-treated or work-hardened material.

Residual stresses created by cold forming are related to the yield strength in the same manner as in steel; however, longitudinal beads are considered to have little influence on column strength, as a result of either the residual stresses or of any strain hardening. Geometric imperfections fall into two groups: those that are length dependent, such as initial curvature, and those independent of length, such as eccentricity in the profile of the section itself and inaccuracies in the dimensioning of the assembly.

Commercial tolerances on initial curvature allow $L/500$ in some extruded structural members, but in reality $L/1000$ is rarely exceeded. If a bar with such an initial curvature forms two or more bays of the chord of a truss, the final out-of-straightness will be negligible in comparison to the inaccuracies in the assembly. Even in laboratory tests, the initial curvature of the bar as supplied has usually been less than the error in centering the specimen. End moments, due to frame action or eccentricities in joints, will in most cases dominate any moments due to initial curvature. For large assembled columns, such as latticed masts, an initial curvature of $L/1000$ has been found to be representative, and design as a beam-column using this value has been adopted.

3.9.3 Strength of Aluminum-Alloy Columns

Aluminum-alloy column strength has generally been based, in design application, on the tangent-modulus theory—justifiably so because of the generally good agreement with column test results.

Evaluation of the maximum strength of both straight and initially curved columns is practicable by use of the digital computer. A systematic study of the effects of important parameters that affect column strength has been made by Batterman and Johnston (1967) and Hariri (1967).

Initially Straight Columns. Such members were studied by Duberg and Wilder (1959), and Johnston (1963, 1964). In a further paper on the behavior above the critical load (Batterman and Johnston, 1967), the assumed stress-strain curve of the material represented the average of a large number of tests of aluminum alloy 6061-T6. By considering both strong- and weak-axis buckling of an H-type section, the practical range of the shape effect was approximately covered. A section having a depth equal to the width was chosen, with flange thickness approximately one-tenth of the depth, and with a web having a thickness two-thirds that of the flange. The maximum increase in strength above the tangent-modulus load was found to be about 2% for weak-axis bending. This small difference further justifies the use of the tangent-modulus load as a reasonable basis for estimating the strength of initially straight aluminum-alloy columns.

Initially Curved Columns. The maximum strength of initially curved hinged-end aluminum-alloy columns can be evaluated by use of the computer, as described previously (Batterman and Johnston, 1967). For a typical H shape of alloy 6061-T6 with buckling about the strong axis, Fig. 3.29 show typical plots of load versus midheight lateral deflection for an

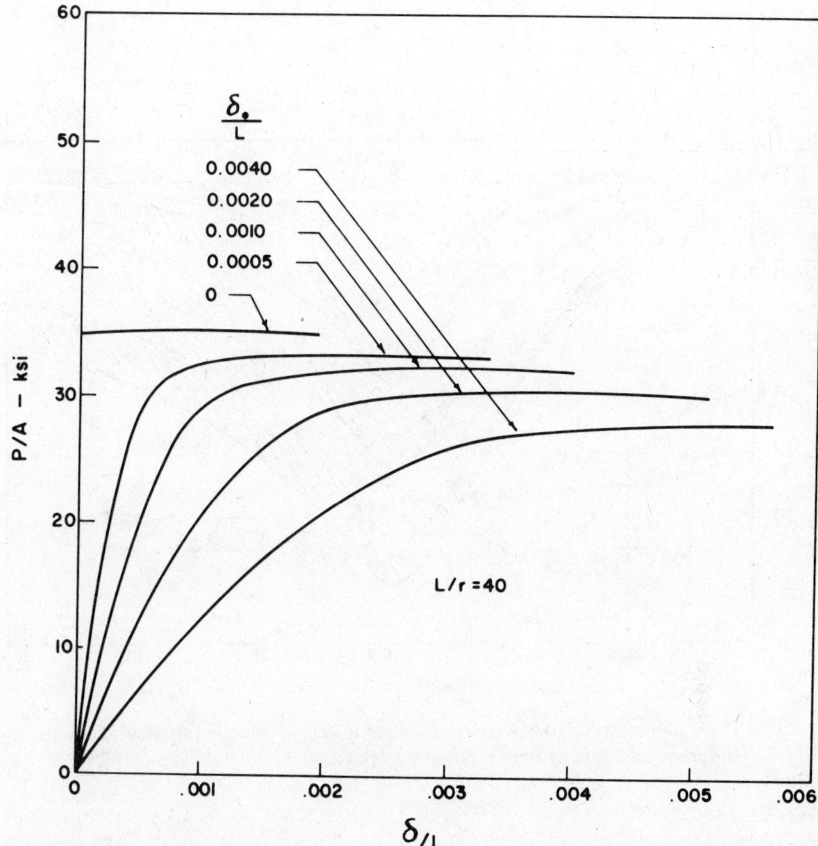

Fig. 3.29 Effect of initial curvature on load-deflection relationships of an aluminum-alloy column (Batterman and Johnston, 1967).

L/r of 40 and for initial midheight out-of-straightness ranging from zero up to $0.004L$. Figure 3.30 illustrates the strength-reduction factor, referenced to the critical load and plotted as a function of L/r for both strong- and weak-axis bending.

The effects of initial curvature are accentuated in unsymmetrical sections, such as the T, as illustrated by the computer-plotted curves in Fig. 3.31 (Hariri, 1967). An initial crookedness of $0.001L$ (with the flange on the convex side) reduced the ultimate strength in comparison with the tangent-modulus load by about 18% at $L/r = 40$, which is about twice the

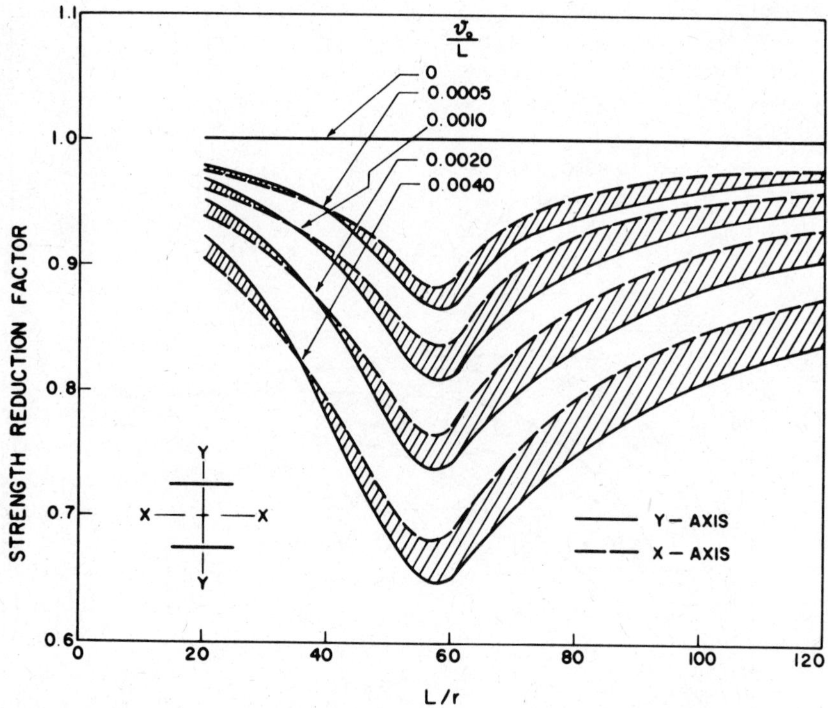

Fig. 3.30 (Maximum strength)/(tangent-modulus load) of aluminum-alloy columns for different amounts of initial curvature (Batterman and Johnston, 1967).

reduction shown in Fig. 3.30 for the doubly symmetric section. Buckling of axially loaded straight members (as confirmed by tests) will occur so as to put the flange of the T on the convex side of the column. The shaded lines at the top of Fig. 3.31 indicate the upper bounds of theoretical strength of a straight column, that is, the inelastic buckling gradients and the reduced-modulus strengths for buckling in either of the two possible directions (Johnston, 1964). Tests also indicated that the effect of end eccentricity was somewhat more deleterious than the effect of the same magnitude of initial out-of-straightness (Hariri, 1967).

The effect of end restraint on initially curved aluminum columns has been studied by Chapuis and Galambos (1982). They conclude, "it is conservative to base a design on K_{el} (the effective length factor determined from elastic buckling analysis) and a column curve derived for pinned crooked columns."

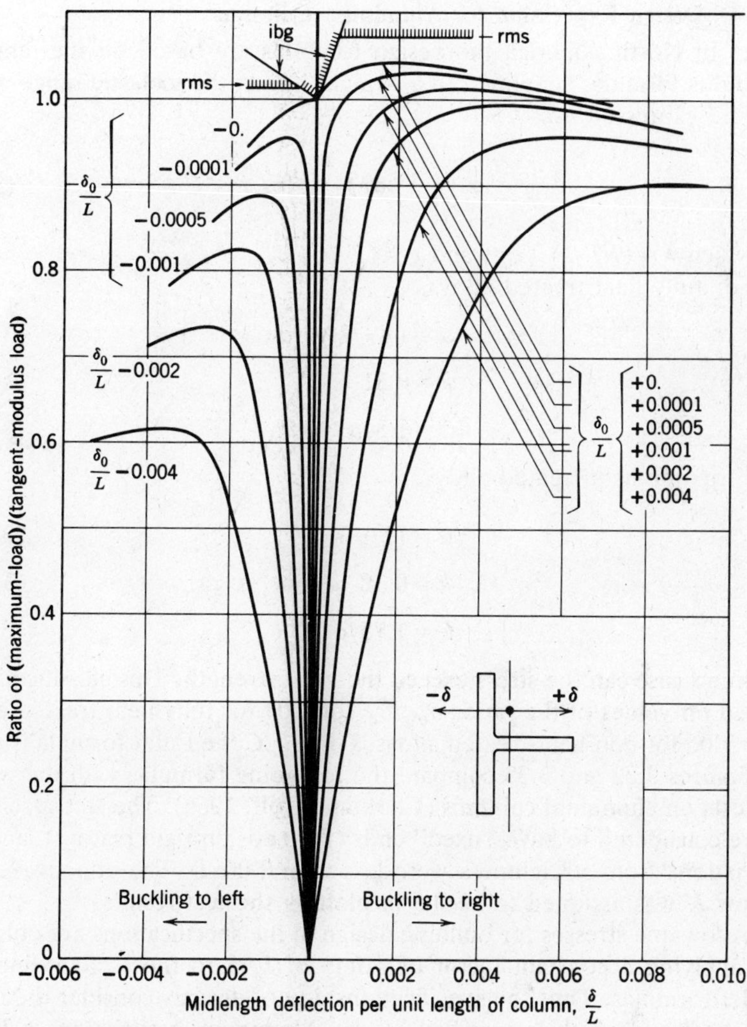

Fig. 3.31 Theoretical behavior of straight and initially curved T-section columns of aluminum alloy (Hariri, 1967).

3.9.4 Design Expression for Aluminum Columns

1. In North America the design formulas are based on the tangent-modulus formula, simplified to a straight line in the inelastic range which can be expressed as

$$\frac{\sigma_c}{\sigma_y} = \alpha^2(1 - k\alpha\lambda) \qquad \text{for } \lambda < C \tag{3.21}$$

in which $\lambda = (KL/r)/(\pi g)$ and $g = (E/\sigma_y)^{1/2}$.

For fully heat-treated alloys,

$$\alpha = (1 + 2/g)^{1/2} \tag{3.22a}$$

$$k = 0.31 \tag{3.22b}$$

$$C = 1.3/\alpha \tag{3.22c}$$

and for non-heat-treated alloys,

$$\alpha = (1 + 3/g)^{1/2} \tag{3.23a}$$

$$k = 0.38 \tag{3.23b}$$

$$C = 1.75/\alpha \tag{3.23c}$$

In no case can the stress exceed the yield strength. This classification is based on values of the ratios $\sigma_{0.2}/\sigma_{0.1}$ of 1.04 for fully heat-treated alloys and 1.06 for non-heat-treated alloys. For $\lambda > C$ the Euler formula is used.

Figures 3.32 and 3.33 compare the foregoing formulas with the results of tests on aluminum columns (Clark and Rolf, 1966). The test specimens were considered to have "fixed" ends (flat ends on rigid platens), and the deviations from straightness were less than $0.001L$. The effective-length factor K was assumed to be 0.5 in plotting the test results.

Allowable stresses for building design in the specifications are obtained by applying a constant factor of safety of 1.95 to the straight-line and Euler formulas. Thus the specifications do not directly consider the initial crookedness, which is specified by the Aluminum Association as $L/480$ for standard structural shapes but as $L/960$ for most other extruded shapes. Batterman and Johnston (1967) showed that a small initial crookedness can appreciably reduce the factor of safety, especially in the transition region between elastic and inelastic buckling. Figure 3.34 illustrates the results of calculations for columns of 6061-T6 alloy with $\delta_0 = 0.001L$. In a discussion of the specification, Hartmann and Clark (1963) noted that the effects of small amounts of initial crookedness or eccentricity may be offset by the use of conservative values of the equivalent-length factor as a basis for the specification formulas. To

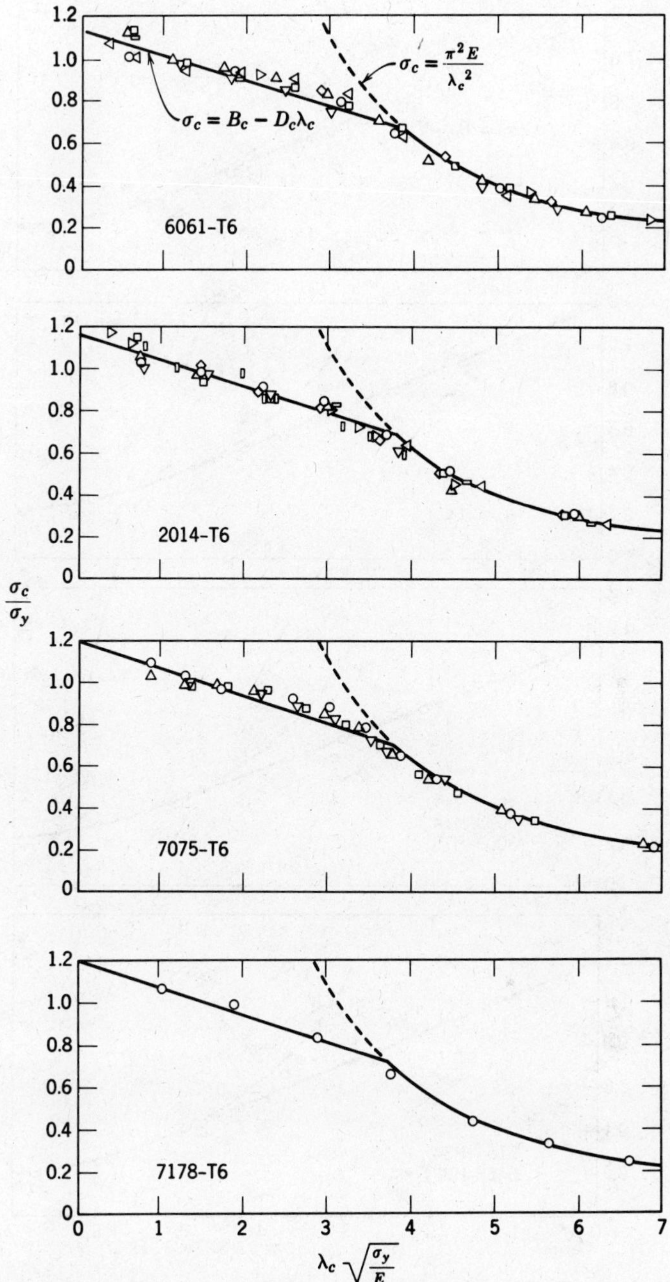

Fig. 3.32 Column strength of aluminum alloys (artificially aged) (Clark and Rolf, 1966).

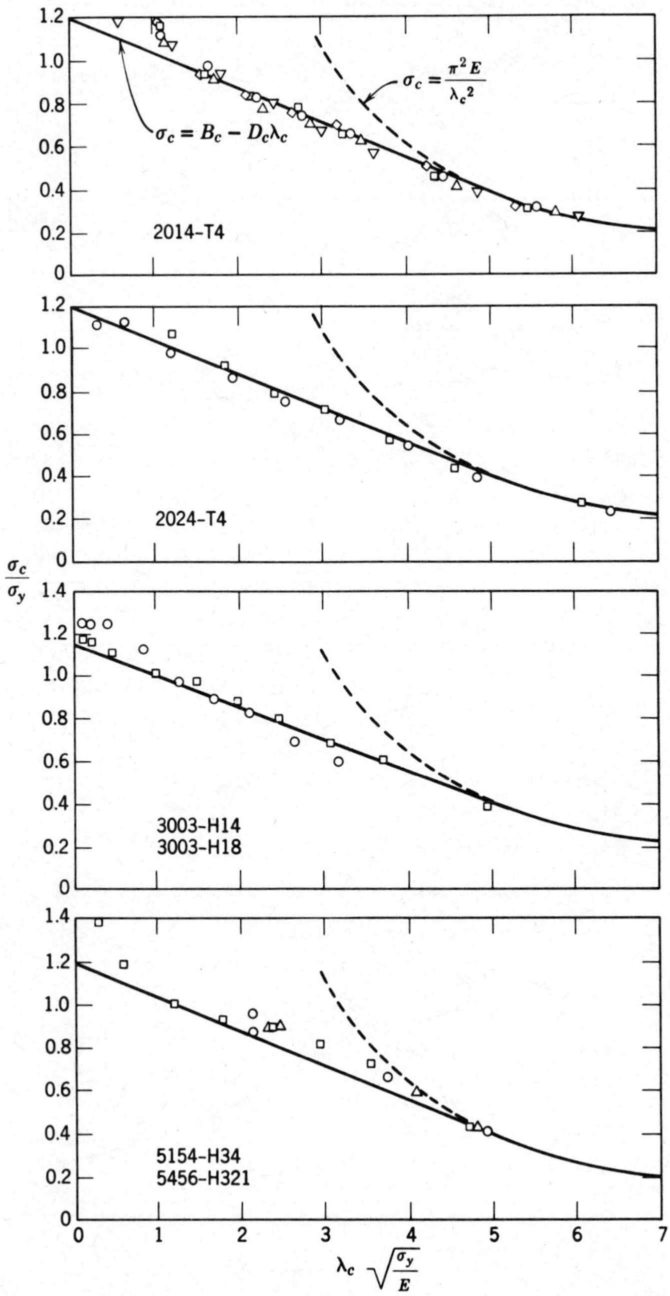

Fig. 3.33 Column strength of aluminum alloys (not artificially aged) (Clark and Rolf, 1966).

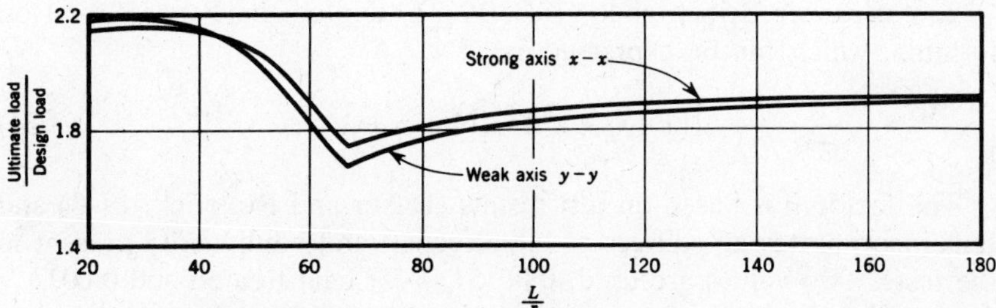

Fig. 3.34 Design load factor for wide-flange shapes of aluminum-alloy columns with $\delta_0 = 0.001L$ (Batterman and Johnston, 1967).

illustrate this, Batterman and Johnston (1967) considered the hypothetical case of a column with typical material properties and ends restrained such that throughout loading to maximum strength, the column segment between inflection points is always $0.9L$. For bending about the weak $(y-y)$ axis and an initial crookedness of $0.001L$, the theoretical safety factor is plotted in Fig. 3.35 both for this case and for the case of pinned ends. Thus for a relatively slight amount of end restraint, the safety factor varies from values slightly above 2.2 in the short column range to a minimum of 1.97 at $L/r = 67$. It then increases to slightly above 2.2 at $L/r = 120$. The effect of the initial crookedness in reducing the safety factor from 1.95 to 1.67 is thus offset if the column is restrained at the ends such that an actual K value of 0.9 is produced.

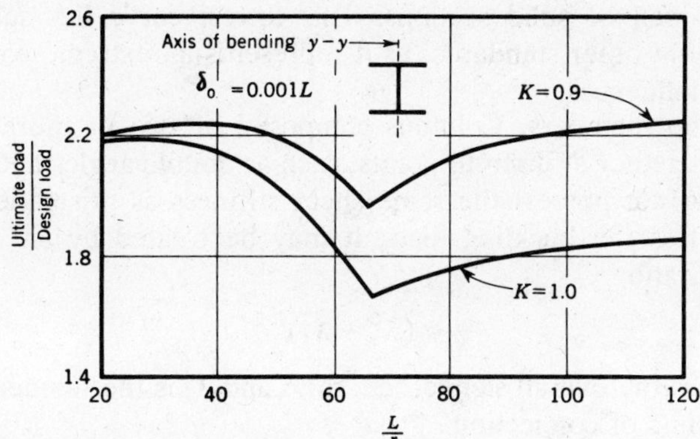

Fig. 3.35 Comparison of safety factors for columns with and without end restraint (Batterman and Johnston, 1967).

2. An earlier British code (ISE, 1962) adopted the Perry-Robertson formula, which can be expressed as

$$\frac{\sigma_c}{\sigma_y} = \frac{1}{2\lambda^2}\{(1+\eta+\lambda^2) - [(1+\eta+\lambda^2)^2 - 4\lambda^2]^{1/2}\} \tag{3.24}$$

The factor η is based on test results (Baker and Roderick, 1948) and thus incorporates all sources or imperfection and nonlinearity present in the tests. Two values are used, $0.0015L/r$ for heat-treated and $0.003L/r$ for non-heat-treated alloys. In a later code (BSI, 1969) straight-line formulas or the type used in North America were adopted.

3. In Europe a program of research (Mazzolani and Frey, 1980; Bernard et al., 1973) provided the background information on imperfections, residual stresses, and the influence of welds, which lead to the preparation of the ECCS recommendations (ECCS, 1978). The Ramberg-Osgood formula was used to model the stress-strain relationship in the form

$$\epsilon = \sigma/E + 0.002(\sigma/\sigma_{0.2})^n \tag{3.25}$$

Steinhardt's suggestion (1971) that n is equal to the value of the yield strength in kN/cm^2 gives close agreement with test results (n is a coefficient that reflects the shape of the stress-strain curve).

Three curves are proposed using values of n of 20, 15, and 10 to represent the various alloy types. Included in the computer evaluation were initial imperfections of both curvature and eccentricity, as well as unsymmetrical sections.

The curves are shown in Fig. 3.36. The uppermost curve is for heat-treated, symmetrical, open or solid sections, with the lowest curve for all hollow sections and all non-heat-treated sections other than symmetric open or solid sections. This lowest curve lies substantially below those in other standards, as it represents an extreme combination of adverse influences.

4. Built-up members. Columns composed of two or more elements connected together at discrete points, such as double angles and battened channels, do not possess the same shear stiffness as prismatic sections. This influences the buckling load. It may be treated by an equivalent slenderness ratio:

$$\lambda = (\lambda_1^2 + \lambda_2^2)^{1/2} \tag{3.26}$$

in which λ_1 is the overall slenderness ratio and λ_2 is the slenderness ratio between points of connection.

In many standards the influence is ignored, although for a double-angle column with stitch bolts at third points, the reduction in capacity due to shear flexibility often exceeds 20% and can be 50% if λ_1 and λ_2 are

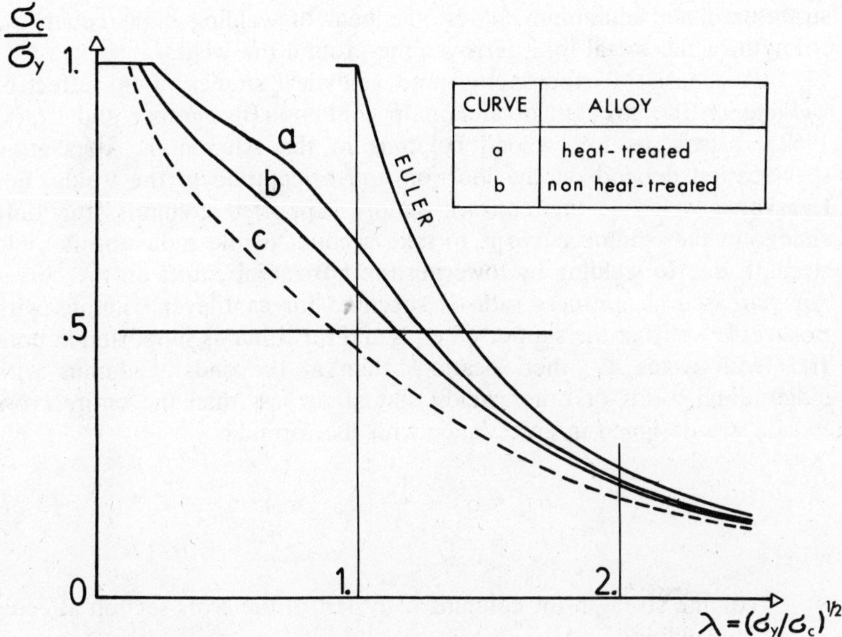

Fig. 3.36 Basic column buckling curves (ECCS aluminum structures).

equal, as some standards permit. Double angles with b/t ratios in excess of 12 should also be examined for torsional buckling, and in particular for combined torsion-flexure buckling.

3.9.5 Assemblies of Interconnected Bars

For cross-braced diagonals, one in tension, the other in compression, the compression bar is effectively supported at its center and will buckle at a load associated with an effecive length of $L/2$. Other combinations of bars common in latticed transmission towers are dealt with in the ECCS Manual (1978).

It is assumed that the sum of the loads carried by a group of bars so interconnected that they must buckle together is equal to the sum of the capacities of the bars, independently of the distribution of the loads between the bars; tension loads are treated as opposite sign to compression loads. This is valid if the mean stress in no bar exceeds the elastic limit, which is common for the diagonals in transmission towers.

3.9.6 Welded Columns

Welding may affect the strength of aluminum columns by introducing out-of-straightness and residual stresses. In addition, in heat-treated or strain-hardened aluminum alloys, the heat of welding may reduce the strength of the metal in a narrow zone around the weld.

On the basis of experimental and analytical studies of the effect of welding on the strength of aluminum columns (Brungraber and Clark, 1962), provisions for welded columns in the Aluminum Association specification depend on the location and orientation of the welds. For transverse welds at the ends of simply supported columns, the only change in the column curve is to take account of the reduction in yield strength due to welding by lowering the horizontal cutoff on the curve. An increased slenderness ratio is specified for cantilever columns with transverse welds at the supported ends and for columns supported at both ends with welds at other locations than at the ends. Columns with longitudinal welds or other welds that affect less than the entire cross section are designed in accordance with the formula

$$\sigma_{pw} = \sigma_n - \frac{A_w}{A}(\sigma_n - \sigma_w) \tag{3.27}$$

where

σ_{pw} = column strength for columns with part of the cross section affected by welding
σ_n = column strength for the same column if there were no welds
σ_w = column strength for the same column if the whole cross section were heat-affected
A_w = area of heat-affected zone
A = total area

The heat-affected zone is specified to be a width of 1 in. (25 mm) on either side of the center of a butt weld or the heel of a fillet weld.

In Canadian practice (CSA, 1980) the term A_w/A in Eq. 3.27 is replaced by I_w/I, in which I_w is the moment of inertia of the heat-affected zone about the axis of bending and I is the total moment of inertia. This method takes account of the position of the weld in the cross section.

Research undertaken at the initiative of the ECCS Committee No. T2 (Mazzolani, 1974; Gatto et al., 1979) has provided the basic information on the distribution of residual stresses and mechanical properties across longitudinally welded I and box sections and on the geometric imperfections created. Modeling the stress-strain relationship of the different zones with the Ramberg-Osgood formula, computer analyses were used to predict the maximum capacity of welded columns which incorporated all the applicable imperfections. A program of column tests provided supporting data.

A simple treatment has been proposed, which takes the form of a reduction factor applied to the standard buckling curves. This reduction factor, n, is related to the reduced area (Brungraber and Clark, 1962) as follows: For heat-treated alloys:

$$0<\lambda<1: \qquad n=\frac{A_r}{A}-\left(\frac{A_r}{A}-0.85\right)\lambda \tag{3.28a}$$

$$1<\lambda<3: \qquad n=0.775+0.075\lambda \tag{3.28b}$$

in which

A = gross area

$$A_r=[\sigma_y(A-A_w)+\sigma_{yw}A_w]/\sigma_y \tag{3.29}$$

σ_y = yield strength of parent metal
σ_{yw} = yield strength in heat-affected zones
A_w = area of heat-affected zones, taken as 20 mm each side of the welds

$$\lambda=(\sigma_y/\sigma_c)^{1/2} \tag{3.30}$$

and for nontreated alloys:

$$0<\lambda<1: \qquad n=1-0.2\lambda \tag{3.31a}$$

$$1<\lambda<3: \qquad n=0.7+0.1\lambda \tag{3.31b}$$

3.9.7 Postbuckling Behavior

With the growing use of highly redundant latticed structures, design methods aimed at taking advantage of the postbuckling behavior of compression members have received considerable attention. Van den Broek (1948) described how the "elastic-plastic" characteristic of long compression diagonals in steel transmission towers was used as early as 1907. The extension of this idea to chord members, however, is restricted by the loss of "ductility" as the slenderness ratio decreases. For critical stresses higher than half the yield strength, the failure is essentially brittle.

A number of models have been proposed for the postbuckling regime. Exact load-shortening relationships can be obtained by computer analysis or by direct testing. To include them in analysis for general use is seldom justified.

An elastic analysis for an ideal single column, bowing until the extreme fiber yields, gives the load-shortening relationships of Fig. 3.37. Actual behavior is typified by the curved line.

To give a lower-bound methods suited for computer analysis, Marsh

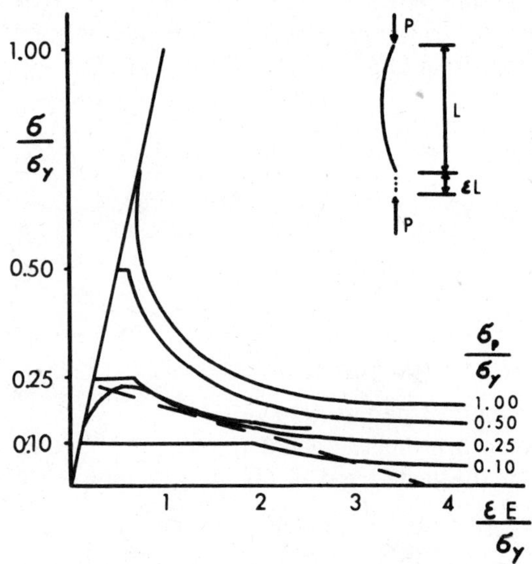

Fig. 3.37 Load-shortening relationship for axially loaded struts.

and Silva (1977) proposed the use of straight unloading lines passing under the actual unloading lines (a typical line is shown dashed in Fig. 3.37). These straight lines can be introduced into the analysis by the expedient of giving the member an appropriate negative area.

A representative theoretical analysis by Supple (1978) showed how the failure load is influenced by the slope of the unloading line. Extensive experimental work and proposals for a design treatment by Schmidt and Morgan (1974) have not been encouraging. Model space trusses, supported on all edges, with uniform chords have failed in tests at loads below those given by an elastic analysis. This was attributed to the influence of misfit on the distribution of the bar loads and the brittle failure of the most heavily loaded bar. The overall mechanism of failure followed the predicted pattern, but no useful postbuckling capability was shown to exist when chord buckling governed. Analytical studies, based on representative load-shortening relationships and incorporating misfit and eccentricity have been conducted. In general it may be stated that capacity is not greatly influenced by consideration of the postbuckling behavior for those cases where the main chords buckle first.

When space trusses are supported on columns widely spaced in both directions, an optimum design requires heavier chords along the column lines, with zones of lighter chords spanning between them. It is possible to arrange the proportions such that slender chords buckle first, allowing load to be transferred in a ductile manner. The buckling of a single slender chord does not limit the capacity, as the axial strain is controlled by the elastic shortening of the unbuckled heavier chords (Marsh and Silva, 1977).

3.10 DEVELOPMENT OF COLUMN STRENGTH CRITERIA

When detailed strength and performance data are not available for a specific column shape, it is possible for individual researchers or code-writing bodies to develop column curves of types that are similar to those that have been presented in this guide. The following gives a brief outline of the assumptions that should be used, the type of data that are required, and the computational technique that is suitable for these types of problems.

Data That Are Needed. The following data are needed for the computation of the maximum strength of columns:

1. Type of material and its material properties (i.e., yield stress, yield strain, modulus of elasticity).
2. Distribution of the residual stresses in the cross section, including variation through the thickness, if the shape is large or is tubular.
3. Variation of the yield stress throughout the cross section. This is in most cases needed only for welded built-up shapes and cold-formed shapes, where the yield stress at a weld or a cold-formed corner, for example, may differ significantly from the nominal properties.
4. In case the material is of a type or grade that exhibits nonlinear stress-strain characteristics (e.g., stainless steel), a complete, typical stress-strain curve is required.
5. Maximum value of the initial out-of-straightness.

Assumptions for the Analysis. The following assumptions are normally conservative in nature, with the result that computed column strengths are usually somewhat less than those obtained in actual tests:

1. Material is linearly elastic, perfectly plastic.
2. The initial and all subsequent deflection shapes of the column can be described by a half sine-wave.
3. The residual stresses are constant in an element of the cross section along the full length of the column.
4. Sections that are originally plane remain so for the range of deflections that is suitable for column studies.
5. Yielded fibers in the cross section will unload elastically.

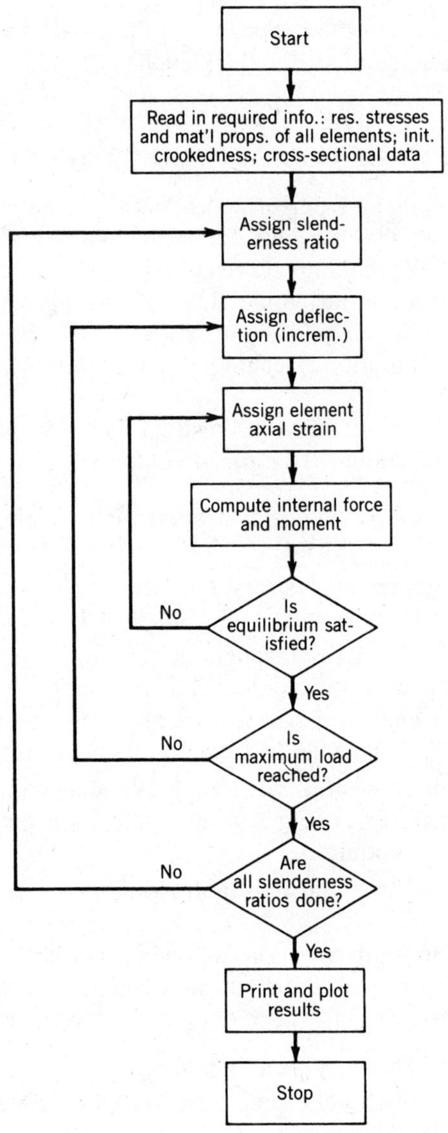

Fig. 3.38 Flowchart for maximum column strength computation.

6. The yield stress may vary across the width and through the thickness of the component plates of the cross section but does not vary along the length of the column.
7. In line with assumption 2, only stresses and strains at midheight of the column are considered in the analysis.

It should be pointed out that if detailed yield stress and other material data are not available for the elements in the cross section, the results of a stub column test can be used. If this is not available, tension test results for various parts of the shape can be used; the properties to utilize in the computations should then be based on a weighted average.

Computational Technique. The solution of the problem of the maximum strength of a column needs the availability of a high-speed computer. A detailed procedure for the computation of the maximum strength has been given by Bjorhovde (1972) and Batterman and Johnston (1967). The computation is based on an incremental solution procedure, whereby equilibrium is established for every load and deflection level. This requires iteration over the cross section to determine when individual fibers yield or unload. Both force and moment equilibrium are enforced. The computations are carried to a level where the column cannot take any additional load when an additional amount of deflection is imposed; this constitutes the maximum strength level. It is recommended that the incrementation procedure be based on deflection increments rather than load, due to the convergence problems that may be encountered as the maximum load is approached when load increments are used. The procedure above leads to the development of the load-deflection curve for the columns with a given slenderness ratio or length. To obtain the complete column curve, it must be repeated for a range of lengths. As an illustration of the basic steps in the column strength computations, Fig. 3.38 gives a flowchart that indicates the necessary major parts of the solution.

References

Ackroyd, M. H. (1979), "Nonlinear Inelastic Stability of Flexibly Connected Plane Steel Frames," Ph.D. dissertation, University of Colorado, Boulder.

Ackroyd, M. H., and Bjorhovde, R. (1981), "Effect of Semi-rigid Connections on Steel Column Strength" (discussion), *J. Constr. Steel Res. London*, Vol. 1, No. 3, pp. 48–51.

Alpsten, G. A. (1970), "Residual Stresses and Strength of Cold-Straightened Wide Flange Shapes," *Jernkontorets Ann.*, Vol. 154 (in Swedish).

Alpsten, G. A. (1972a), "Prediction of Thermal Residual Stresses in Hot Rolled Plates and Shapes," *9th Cong. IABSE, Final Rep.*, May, pp. 1–13.

Alpsten, G. A. (1972b), "Residual Stresses Yield Stress and Column Strength of Hot-Rolled and Roller-Straightened Steel shapes," *Coll. Column Strength*, Paris.

Alpsten, G. A., and Tall, L. (1970), "Residual Stresses in Heavy Welded Shapes," *Weld. J.*, Vol. 49.

Baker, J. F., and Roderick, J. W. (1948), "The Strength of Light Alloy Struts, *Res. Rep. No. 3*, Aluminum Development Association, London.

Batterman, R. H., and Johnston, B. G. (1967), "Behavior and Maximum Strength of Metal Columns," *ASCE J. Struct. Div.*, Vol. 93, No. ST2, pp. 205–230.

Beddle, L. S., and Tall, L. (1960), "Basic Column Strength," *ASCE J. Struct. Div.*, Vol. 86, No. ST7, pp. 139–173.

Beer, H., and Schultz, G. (1970), "Theoretical Basis for the European Column Curves," *Constr. Met.*, No. 3, p. 58.

Beer, G., and Tall, L. (1970), "The Strength of Heavy Welded Box Columns," Fritz Eng. Lab. Rep. No. 337.27, Lehigh University, Bethlehem, Pa., December.

Bernard, A., Frey, F., Janss, J., and Massonnet, C. (1973), "Research on the Buckling Behavior of Aluminum Columns," *IABSE Mem.*, Vol. 33-1, pp. 1–33 (in French).

Birkemoe, P. C. (1977a), "Column Behavior of Heat-Treated Cold-Formed Hollow Structural Shapes," *Stability of Structures under Static and Dynamic Loads*, Proc. 2nd Int. Coll., Washington, D.C.

Birkemoe, P. C. (1977b), "Development of Column Curves for H.S.S.," *Int. Symp. Hollow Struct. Sec.*, CIDECT, Toronto.

Bjorhovde, R. (1972), Deterministic and Probabilistic Approaches to the Strength of Steel Columns, Ph.D. dissertation, Lehigh University, Bethlehem, Pa., May.

Bjorhovde, R. (1977), "Strength and Behavior of Cold-Formed H.S.S. Columns," *Struct. Eng. Rep. No. 65*, University of Alberta, Edmonton, December.

Bjorhovde, R. (1978), "The Safety of Steel Columns," *ASCE J. Struct. Div.*, Vol. 104, No. ST3, pp. 463–477.

Bjorhovde, R. (1980), "Research Needs in Stability of Metal Structures," *ASCE J. Struct. Div.*, Vol. 106, No. ST12, pp. 2425–2442.

Bjorhovde, R. (1981), "End Restraint and Column Stability" (discussion), *ASCE J. Struct. Div.*, Vol. 107, No. ST8, pp. 1696–1700.

Bjorhovde, R. (1984), "Effect of End Restraint on Column Strength: Practical Applications Eng., *J. Am. Inst. Steel Constr.*, Vol. 21, No. 1, pp. 1–13.

Bjorhovde, R., and Birkemoe, P. O. (1979), "Limit States Design of H.S.S. Columns," *Can J. Civ. Eng.*, Vol. 8, No. 2, pp. 276–291.

Bjorhovde, R., and Tall, L. (1971), "Maximum Column Strength and the Multiple Column Curve Concept," *Rep. No. 337.29*, Lehigh University, Fritz Eng. Lab., Bethlehem, Pa., October.

Bjorhovde, R., Brozzetti, J., Alpsten, G. A., and Tall, L. (1972), "Residual Stresses in Thick Welded Plates," *Weld J.*, Vol. 51.

British Standards Institution (1969), *The Structural Use of Aluminum*, British Standard Code of Practice CP118, BSI, London.

Brozzetti, J., Alpsten, G. A., and Tall, L. (1970a), "Residual Stresses in a Heavy Rolled Shape 14WF730," *Fritz Eng. Lab. Rep. No. 337.1*, Lehigh University, Bethlehem, Pa., January.

Brozzetti, J., Alpsten, G. A., and Tall, L. (1970b), "Welding Parameters Thick Plates and Column Strength," *Fritz Eng. Lab. Rep. No. 337.21*, Lehigh University, Bethlehem, Pa., February.

Brungraber, R. J., and Clark, J. W. (1962), "Strength of Welded Aluminum Columns," *Trans. Am. Soc. Civ. Eng.*, Vol. 127, Part II, pp. 202–226.

Canadian Standards Association (1974), *Steel Structures for Buildings—Limit States Design*, CSA Standard S16.1-74, CSA, Rexdale, Ontario.

Chapuis, J., and Galambos, T. V. (1982), "Restrained Crooked Aluminum Columns," *ASCE J. Struct. Div.*, Vol. 108, No. ST3, pp. 511–524.

Chen, W. F. (1980), "End Restraint and Column Stability," *ASCE J. Struct. Div.*, Vol. 105, No. ST11, pp. 2279–2295.

Chen, W. F., and Atsuta, T. (1976), *Theory of Beam-Columns*, Vol. 1, McGraw-Hill, New York.

Chen, W. F., and Han, D. C. (1985), *Tubular Members in Offshore Structures*, Pitman, London.

Chen, W. F., and Lui, E. M. (1985), "Stability Design Criteria for Steel Members and Frames in the United States," *J. Constr. Steel Res.*, Vol. 5, No. 1, pp. 51–94.

Clark, J. W., and Rolf, R. L. (1966), "Buckling of Aluminum Columns, Plates and Beams," *ASCE J. Struct. Div.*, Vol. 92, No. ST3, pp. 17–38.

DeFalco, F., and Marino, F. J. (1966), "Column Stability in Type 2 Construction," *Eng J. Am. Inst. Steel Constr.*, Vol. 3, No. 2, pp. 67–71.

Disque, R. O. (1973), "Inelastic K-Factor in Column Design," *Eng. J. Am. Inst. Steel Constr.*, Vol. 10, No. 2, pp. 33–35.

Disque, R. O. (1975), "Directional Moment Connections—A Proposed Design Method for Unbraced Steel Frames," *Eng. J. Am. Inst. Steel Constr.*, Vol. 12, No. 1, pp. 14–18.

Duberg, J. E., and Wilder, T. W. (1950), "Column Behavior in the Plastic Strength Range," *J. Aeronaut. Sci.*, Vol. 17, No. 6, p. 323.

European Convention for Constructional Steelwork (1978), *European Recommendation for Aluminum Structures*, ECCS Committee T2, Brussels.

Frey, F. (1969), "Effect of the Cold-Bending of H-Shaped Rolled Steel Sections on Column Strength," *IABSE Mem.*, Vol. 29I, pp. 101–124 (in French).

Frye, M. J., and Morris, G. A. (1975), "Analysis of Flexibly Connected Steel Frames," *Can. J. Civ. Eng.*, Vol. 2, No. 3, pp. 280–291.

Fukumoto, Y., Nethercot, D. A., and Galambos, T. V. (1983), "Experimental Data for the Buckling of Steel Structures—NDSS Stability of Metal Structures," *3rd. Int. Colloq. SSRC*, Toronto, May, pp. 609–630.

Gatto, F., Mazzonlani, F. M., and Morri, D. (1979), "Experimental Analysis of Residual Stresses and of Mechanical Properties in Welded Profiles of Al-Si-Mg Alloy," *Ital. Mach. Equip.*, Vol. 2, March.

Goldberg, J. E. (1954), "Stiffness Charts for Gusseted Members under Axial Load," *Trans. Am. Soc. Civ. Eng.*, Vol. 119, p. 43.

Hall, D. H. (1981), "Proposed Steel Column Design Criteria," *ASCE J. Strut. Div.*, Vol. 107, No. ST4, pp. 649–670.

Hariri, R. (1967), Post Buckling Behavior of Tee-Shaped Aluminum Columns, Ph.D. dissertation, University of Michigan, June.

Hartmann, E. C., and Clark, J. W. (1963), "The U.S. Code," *Proc. Symp. Alum. Struct. Eng.*, London, June 11–12, The Aluminum Federation.

Howard, J. E. (1908), "Some Results of the Tests of Steel Columns in Progress at the Watertown Arsenal," *Proc. Am. Soc. Test. Mater.*, Vol. 8, p. 336.

Huber, A. W. (1956), The Influence of Residual Stress on the Instability of Columns, Ph.D. Dissertation, Lehigh University, Bethlehem, Pa., May.

Institution of Structural Engineers (1962), Report on the Structural Use of Aluminum, ISE, London.

Jacquet, J. (1970), "Column Tests and Analysis of Their Results," *Constr. Met.*, No. 3, pp. 13–36.

Johnston, B. G. (1963), "Buckling Behavior above the Tangent Modulus Load," *Trans. Am. Soc. Civ. Eng.*, Vol. 128, Part I, pp. 819–848.

Johnston, B. G. (1964), "Inelastic Buckling Gradient," *ASCE J. Eng. Mech. Div.*, Vol. 90, No. EM5, pp. 31–48.

Jones, S. W., Kirby, P. A., and Nethercot, D. A. (1980), "Effect of Semi-rigid Connections on Steel Column Strength," *J. Const. Steel Res. London*, Vol. 1, No. 1, pp. 35–46.

Jones, S. W., Kirby, P. A., and Nethercot, D. A. (1982), "Columns with Semi-rigid Joints," *ASCE J. Struct. Div.*, Vol. 108, No. ST2, pp. 361–372.

Kato, B. (1977), "Column Curves for Cold-Formed and Welded Tubular Members," *2nd Int. Colloq. Stab. Steel Struct.*, Liege, Belgium, pp. 53–60.

Kishima, V., Alpsten, G. A., and Tall, L. (1969), "Residual Stresses in Welded Shapes of Flame-Cut Plates in ASTM A572(50) Steel," *Fritz Eng. Lab. Rep. No. 321.2*, Lehigh University, Bethlehem, Pa., June.

Lindner, J., and Gietzelt, R. (1984), "Imperfecktionsannahmen für Stützenschiefstellungen," *Stahlbau*, Vol. 53, No. 4, pp. 97–101.

Lui, E. M., and Chen, W. F. (1983a), "Strength of H-Columns with Small End-Restraints," *Struct. Engrs.*, Vol. 61B, No. 1, Part B, London, pp. 17–26.

Lui, E. M., and Chen, W. F. (1983b), "End Restraint and Column Design Using LRFD," *Eng. J. Am. Inst. Steel Constr.*, Vol. 20, No. 1, pp. 29–39.

Lui, E. M., and Chen, W. F. (1984), "Simplified Approach to the Analysis and Design of Columns with Imperfections," *Eng. J. Am. Inst. Steel Constr.*, Vol. 21, No. 2, pp. 99–117.

Madsen, I. (1941), "Report of Crane Girder Tests," *Iron Steel Eng.*, Vol. 18, No. 11, p. 47.

Marsh, C., and Silva, N. F. (1977), "The M-Dec Space Truss in Brazil," *ASCE Conf., Preprint 2928*, San Francisco.

Masur, E. F. (1954), Lower and Upper Bounds to the Ultimate Loads of Buckled Trusses," *Q. Appl Math.*, Vol. 11, No. 4, p. 385.

Mazzolani, F. M. (1974), "Structural Imperfections in Welded Aluminum Assemblies," *Rev. Alum.*, July–August.

Mazzolani, F. M., and Frey, F. (1977), "Buckling Behavior of Aluminum Alloy Extruded Members," *2nd Int. Colloq. Stab. Steel Struct.*, Liege, Belgium, pp. 85–94.

Mazzolani, F.M., and Frey, F. (1980), "The Bases of the European Recommendations for the Design of Aluminum Alloy Structures," *Alluminio*, No. 2.

McGuire, W. (1968), *Steel Structures*, Prentice-Hall, Englewood Cliffs, N.J.

Medland, I. C. (1979), "Flexural-Torsional Buckling of Interbraced Columns," *Eng. Struct.*, Vol. 1, No. 3, pp. 131–138.

Medland, I. C. (1980), "Buckling of Interbraced Beam Systems," *Eng. Struct.*, Vol. 2, No. 2, pp. 90–96.

Michalos, J., and Louw, J. M. (1957), "Properties for Numerical Analysis of Gusseted Frameworks," *Proc. Am Railw. Eng. Assoc.*, Vol. 58, p. 1.

Osgood, W. R. (1951), "The Effect of Residual Stress in Column Strength," *Proc. 1st U.S. Natl. Cong. Appl. Mech.*, June, p. 415.

Pavolovic, M. N., and Stevens, L. K. (1981), "Effect of Prior Flexural Prestraint on the Stability of Structural Steel Columns," *Eng. Struct.*, Vol. 3, No. 2, pp. 66–70.

Razzaq, Z., and Chang, J. G. (1981), "Partially Restrained Imperfect Columns," *Proc. Conf. Joints Struct. Steelwork*, Teeside, England, April, Pentech Press Ltd., Chichester, West Sussex, England.

Romstad, K. M., and Subramanian, C. V. (1970), "Analysis of Frames with Partial Connection Rigidity," *ASCE J. Struct. Div.*, Vol. 96, No. ST11, pp. 2283–2300.

Rondal, J., and Maquoi, R. (1979), "Single Equation for SSRC Column Strength Curves," *ASCE J. Struct. Div.*, Vol. 105, No. ST1, pp. 247–250.

Salmon, E. H. (1921), *Columns*, Oxford Technical Publishers, London.

Schmidt, L. C., and Morgan, P. R. (1974), "Structural Behavior of Regular Space Trusses," *Trans Inst. Eng., Aust.*, Vol. CE16, No. 1, pp.

Sfintesco, D. (1970), "Experimental Basis for the European Column Curves," *Constr. Met.*, No. 3, p. 5.

Sfintesco, D., ed. (1976), *ECCS Manual on the Stability of Steel Structures*, 2nd ed., European Convention for Constructional Steelwork, Brussels.

Shanley, F. R. (1947), "Inelastic Column Theory,"*J. Aeronaut. Sci.*, Vol. 14, No. 5, p. 261.

Shen, Z.-Y., and Lu, L.-W. (1983), "Analysis of Initially Crooked, End Restrained Columns," *J. Constr. Steel Res.*, Vol. 3, No. 1, pp. 10–18.

Sherman, D. R. (1976), *Tentative Criteria for Structural Applications of Steel Tubes*, American Iron and Steel Institute, Washington, D.C.

SSRC, ECCS, CRCJ, CMEA (1982), *Stability of Metal Structures—A World View*, American Institute of Steel Construction, Chicago.

Steinhardt, G. (1971), "Aluminum in Engineered Construction," *Aluminum*, No. 47 (in German).

Sugimoto, H., and Chen, W. F. (1982), "Small End Restraint Effects in Strength of H-Columns," *ASCE J. Struct. Div.*, Vol. 108, No. ST3, pp. 661–681.

Supple, W. J. (1978), "A Plastic Collapse Model for Space Trusses," in *Stability Problems in Engineering Structures and Components* (eds. T. H. Richards and P. Stanley), Applied Science Publishers, Barking, Essex, England.

Tall, L. (1964), "Recent Developments in the Study of Column Behavior," *J. Inst. Eng. Aust.*, Vol. 38, No. 12, pp.

Tall, L. (1966), "Welded Built-up Columns," *Fritz Eng. Lab. Rep. No.* 249.29, Lehigh University, Bethlehem, Pa., April.

Tomonaga, K. (1971), "Actually Measured Errors in Fabrication of Kasumigaseki Building," *Proc. 3rd Reg. Conf. Plann. Des. Tall Build.*, Tokyo, September.

Trahair, N. S. (1977), *The Behavior and Design of Steel Structures*, Chapman & Hall, London.

Van den Broek, J. A. (1948), *Theory of Limit Design*, Wiley, New York.

Vinnakota, S. (1982), "Planar Strength of Restrained Beam-Columns," *ASCE J. Struct. Div.*, Vol. 108, No. ST11, Nov., pp. 2496–2516.

Vinnakota, S. (1983), "Planar Strength of Directionally and Rotationally Restrained Steel Columns," *Proc. 3rd Int. Coll. Stab. Met. Struct.*, Toronto, May, pp. 315–325.

Vinnakota, S. (1984), "Closure," *ASCE J. Struct. Div.*, Vol. 110, No. 2, pp. 430–435.

Winter, G. (1960), "Lateral Bracing of Columns and Beams," *Trans. Am. Soc. Civ. Eng.*, Vol. 125, p. 807.

Yang, H., Beedle, L. S., and Johnston, B. G. (1952), "Residual Stress and the Yield Strength of Steel Beams," *Weld J., Res. Suppl.*, Vol. 31, pp. 224–225.

Yura, J. A. (1971), "The Effective Length of Columns in Unbraced Frames," *Eng. J. Am. Inst. Steel Constr.*, Vol. 8, No. 2, pp. 37–42.

Chapter Four

Plates

4.1 INTRODUCTION

Figure 4.1 shows a number of cross-sectional shapes for metal compression or flexural members. Except for the hollow cylinder, (*a*), all of the members are composed of connected elements which, for purposes of analysis and design, can be treated as flat plates. When a plate element is subjected to direct compression, bending, shear, or a combination of these stresses in its plane, theoretical critical loads may be evaluated indicating that the plate may buckle locally before the member as a whole becomes unstable or before the yield stress of the material is reached. Such behavior is characterized by distortion of the cross section of the member. The almost inevitable presence of initial out-of-planeness may result in a gradual growth of cross-sectional distortion with no sudden discontinuity in real behavior at the theoretical critical load.

The theoretical critical load for a plate is not necessarily a satisfactory basis for design, since the ultimate strength can be much greater than the critical buckling load. For example, a plate loaded in uniaxial compression, with both longitudinal edges supported, will undergo stress redistribution as well as develop transverse tensile membrane stresses after buckling that provide postbuckling support. Thus additional load may often by applied without structural damage. Initial imperfections in such a plate may cause bending to begin below the buckling load, yet the plate, unlike an initially imperfect column, may sustain loads greater than the theoretical buckling load.

This chapter considers local buckling theory and postbuckling behavior for plates with or without stiffeners which are used in open sections made up of plate elements and rectangular tubes. Interaction between plate elements in a section is discussed. Interaction between local plate buckling and overall member buckling is treated in Chapter 13. Buckling of

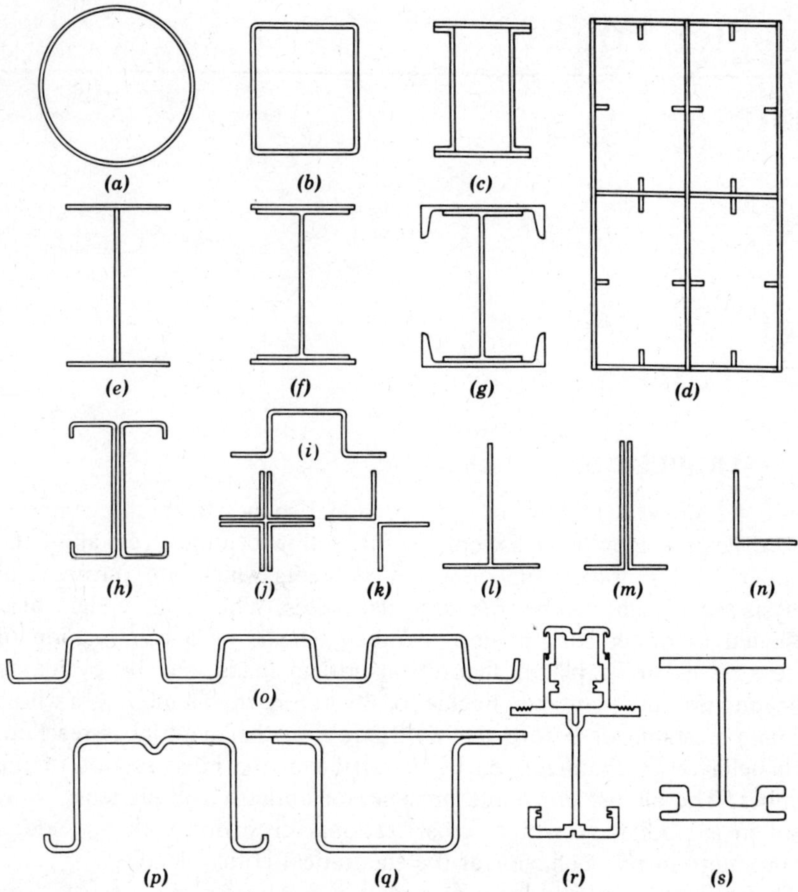

Fig. 4.1 Compression or flexural members.

hollow cylinders, with or without stiffeners, is treated in Chapter 14. Design applications of thin-walled metal construction are covered in Chapter 13, and plate buckling and postbuckling problems in relation to plate girder and box girder design are discussed in Chapters 6 and 7.

4.2 LOCAL BUCKLING AND POSTBUCKLING STRENGTH OF PLATES

An examination of the buckling behavior of a single plate supported along its edges is an essential preliminary step toward the understanding of local buckling behavior of plate assemblies. The buckling stresses are obtained from the concept of bifurcation of an initially perfect structure. In practice the response of the structure is continuous, due to the inevitable presence of initial imperfections. Thus the critical stress must be viewed as a useful index to the behavior, and plates can continue to carry additional loads well after initial buckling. Postbuckling resistance in plates is due to the redistribution of axial compressive stresses and, to a lesser extent, to the membrane tension and shear that accompany the out-of-plane bending of the plate in both the longitudinal and transverse directions. The longitudinal stresses tend to concentrate in the vicinity of the longitudinally supported edges, which are the stiffest parts of the buckled plate. As a result, yielding begins along these edges, which limits the load-carrying capacity.

4.2.1 Uniaxial Uniform Compression

Buckling Strength

Long rectangular plates. In 1891, Bryan (1891) presented the analysis of the elastic critical stress for a rectangular plate simply supported along all edges and subjected to a uniform longitudinal compressive stress. The elastic critical stress of a long plate segment is determined by the plate width-to-thickness ratio b/t, by the restraint conditions along the longitudinal boundaries, and by the elastic material properties. It is expressed as

$$\sigma_c = k \frac{\pi^2 E}{12(1 - \nu^2)(b/t)^2} \tag{4.1}$$

in which k is a "buckling coefficient" determined by a theoretical critical-load analysis. It is a function of plate geometry and boundary conditions such as those shown in Fig. 4.2.

When the member cross section is composed of various connected elements (see Fig. 4.1) a lower bound of the critical stress can be determined by assuming, for each plate element, a simple support condition for each edge attached to another plate element, or a free condition for any edge not so attached. The smallest value of the critical stress found for any of the elements is a lower bound of the critical stress for the cross section. This stress will be conservative because the element providing the lower bound will be restrained by the more stable adjoining plate elements if longitudinal edge joints provide effective continuity.

More complete information on k factors as influenced by the interaction between plate components can be found in a number of references (Stowell et al., 1952; Gerard and Becker, 1957/1958); Timoshenko and Gere, 1961; Bleich, 1952).

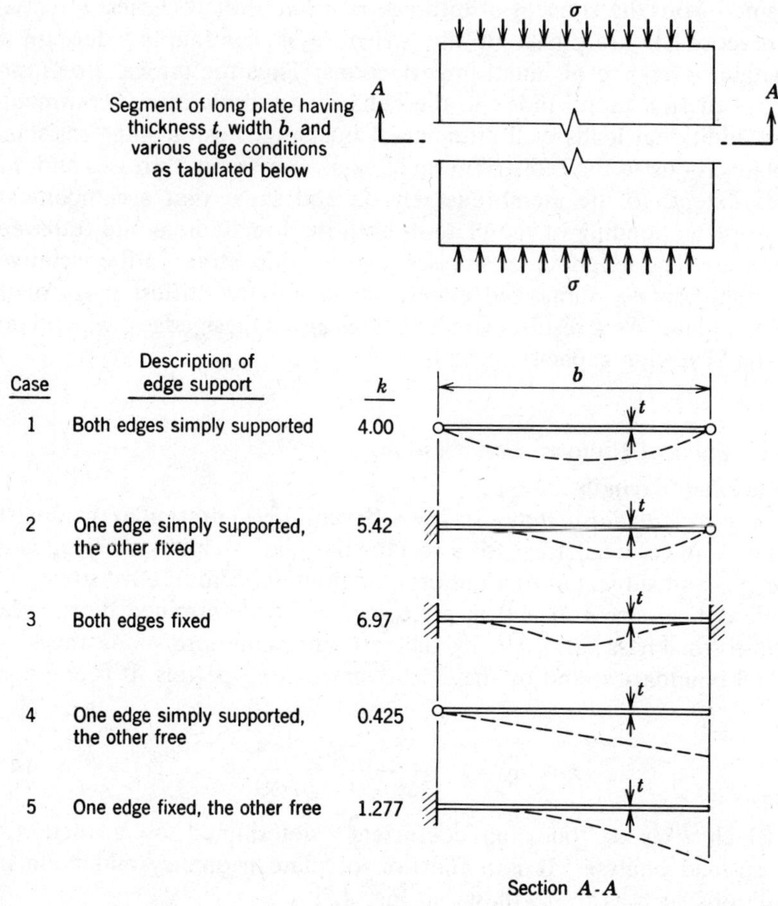

Fig. 4.2 Coefficients k for Eq. 4.1.

Short plates. When a plate element is relatively short in the direction of the compressive stress (i.e., $a/b \ll 1$), the critical stress may be conservatively estimated by assuming that a unit width of plate behaves like a column.

Inelastic buckling. Bleich (1952) generalized the expression for the critical stress of a flat plate under uniform compressive stress in either the elastic or inelastic range in the following manner:

$$\sigma_c = k \frac{\pi^2 E \sqrt{\eta}}{12(1-\nu^2)(b/t)^2} \tag{4.2}$$

in which $\eta = E_t/E$.

This modification of Eq. 4.1 to adapt it to a stress higher than the proportional limit is a conservative approximation to the solution of a complex problem that involves a continuous updating of the constitutive relations depending on the axial stress carried (Stowell, 1948; Bijlaard, 1949, 1950).

Postbuckling Strength. Local buckling causes a loss of stiffness and a redistribution of stresses. Uniform edge compression in the longitudinal direction results in a nonuniform stress distribution after buckling (Fig. 4.3); and the buckled plate derives almost all of its stiffness from the longitudinal edge supports.

Elastic postbuckling stiffness is measured in terms of the apparent modulus of elasticity, E^* (the ratio of the average stress carried by the plate to the average strain) (see Fig. 4.4). The values of E^* for long plates ($a \gg b$) for some typical longitudinal edge conditions are given by Allen and Bulson (1980). The values given below are sufficiently accurate up to twice the critical stress.

- *Simply supported longitudinal edges*. Sides straight but free to move laterally:

$$E^* = 0.5E \tag{4.3a}$$

Sides free to move:

$$E^* = 0.408E \tag{4.3b}$$

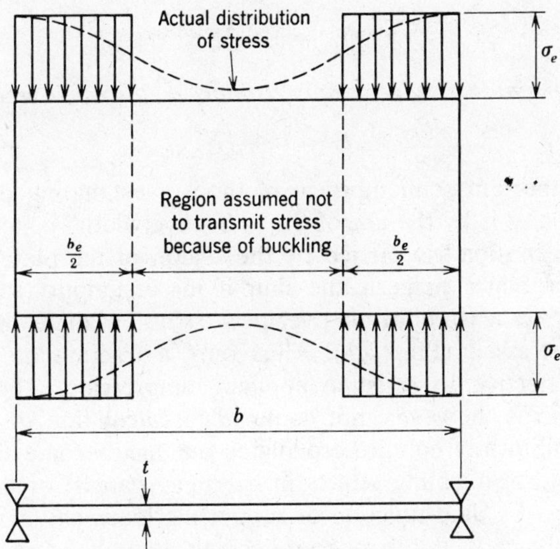

Fig. 4.3 Definition of effective width.

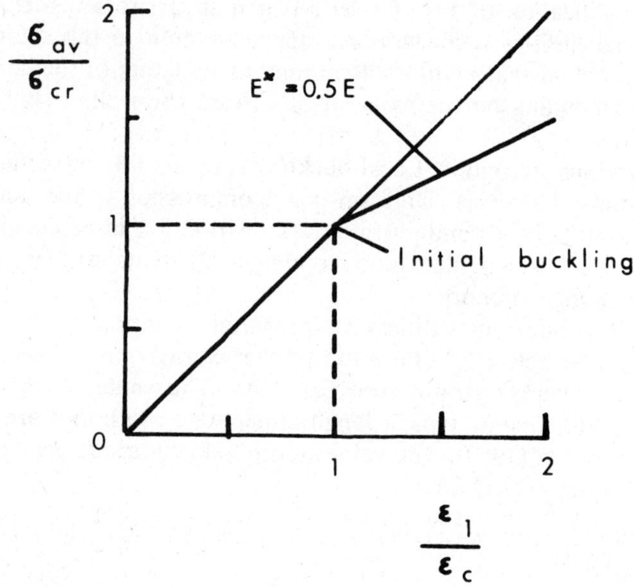

Fig. 4.4 Postbuckling stiffness of plates having simply supported edges (Allen and Bulson 1980).

- *Clamped longitudinal edges*. Sides straight but free to move laterally:

$$E^* = 0.497E \tag{4.3c}$$

- *One longitudinal edge simply supported, the other free*

$$E^* = 0.444E \tag{4.3d}$$

A very important semiempirical method of estimating the maximum strength of plates is by the use of the "effective width" concept. The fact that much of the load is carried by the region of the plate in the close vicinity of the edges suggests the simplifying assumption that the maximum edge stress acts uniformly over two "strips" of plate and the central region is unstressed (Fig. 4.3). Thus only a fraction of the width is considered effective in resisting applied compression. The concept of effective width is, however, not confined to calculation of postbuckling strength of uniformly compressed plates but has become the means of allowing for local buckling effects in columns, panels, or flexural members that have the dual function of supporting loads and acting as walls, partitions, bulkheads, floors, or roof decking. In a plate structure, the use of the effective width leads to an effective cross section consisting of portions of members meeting along a junction. It is near these functions that the plates will begin to yield preceding failure.

The effective-width concept has been used in design specifications for many years. Specifications of the AISI (1986), the Aluminum Association (1982), and the AISC (1978) all permit the use of an effective width in the design of members having plate elements with b/t ratios greater than the limits for full effectiveness.

The effective-width concept seems to have had its origin in the design of ship plating (Murray, 1946). It had been found that longitudinal bending moments in ships caused greater deflections than those calculated using section properties based on the gross area of the longitudinal members. Deflections could be calculated more accurately by considering only a strip of plate over each stiffener having a width of 40 or 50 plate thicknesses as effective in acting with the stiffeners in resisting longitudinal bending.

The advent of all-metal aircraft construction provided another opportunity for the use of the effective-width concept, since it was advantageous to consider some of the metal skin adjacent to stiffeners as being part of the stiffener in calculating the strength of aircraft components. Cold-formed members used in steel buildings also provide useful applications of stiffened-sheet construction. A discussion of the effective-width concept as applied to cold-formed steel design has been prepared by Winter (1983).

Tests by Schuman and Back (1930) of plates supported in V-notches along their unloaded edges demonstrated that, for plates of the same thickness, increasing the plate width beyond a certain value did not increase the ultimate load that the plate could support. It was observed that wider plates acted as though narrow side portions or "effective load-carrying areas" took most of the load. Newell (1930) and others were prompted by these tests to develop expressions for the ultimate strength of such plates. The first to use the effective-width concept in handling this problem was von Kármán (1932). He derived an approximate formula for the effective width of simply supported plates, and, in an appendix to his paper, Sechler and Donnell derived another formula based on slightly different assumptions. Subsequently, many other effective-width formulas have been derived, some empirical, based on approximate analyses, and some based on the large-deflection plate-bending theory, employing varying degrees of rigor.

For plates under uniform compression, stiffened along both edges parallel to the direction of the applied compression stress, von Kármán (1932) developed the following approximate formula for effective width, based on the assumption that two strips along the sides, each on the verge of buckling, carry the entire load:

$$b_e = \left[\frac{\pi}{\sqrt{3(1-\nu^2)}}\sqrt{\frac{E}{\sigma_e}}\right] t \tag{4.4}$$

Combining Eqs. 4.4 and 4.1, for $k = 4$ (simple edge supports), the formula suggested by Ramberg et al. (1939) is obtained (see Fig. 4.3 for notation):

$$\frac{b_e}{b} = \sqrt{\frac{\sigma_c}{\sigma_e}} \tag{4.5}$$

From Fig. 4.3, the average stress is

$$\sigma_{av} = \frac{b_e}{b}\,\sigma_e \tag{4.6}$$

Substituting Eq. 4.5 into Eq. 4.6 with the edge stress equal to the yield stress ($\sigma_e = \sigma_y$) gives us

$$\sigma_{av} = \sqrt{\sigma_c \sigma_y} \tag{4.7}$$

As a result of many tests and studies of postbuckling strength, Winter (1947) and Winter et al. (1950) suggested the formula for effective width that was adopted in the 1946 through 1962 editions of the AISI Specifications for light-gage cold-formed steel:

$$\frac{b_e}{t} = 1.9\sqrt{\frac{E}{\sigma_e}}\left(1 - 0.475\sqrt{\frac{E}{\sigma_e}}\,\frac{t}{b}\right) \tag{4.8}$$

or, alternatively, in the form of Eq. 4.5,

$$\frac{b_e}{b} = \sqrt{\frac{\sigma_c}{\sigma_e}}\left(1 - 0.25\sqrt{\frac{\sigma_c}{\sigma_e}}\right) \tag{4.9}$$

Equations 4.8 and 4.9 are basically the same as Eqs. 4.4 and 4.5, respectively, but include a correction coefficient determined from tests and reflecting the total effect of various imperfections, including initial deviations from planeness. Equation 4.8 was found to be satisfactory also for austenitic stainless steel in the annealed and flattened condition (Johnson and Winter, 1966) and for quarter- and half-hard type 301 stainless steel (Wang, 1969, 1975).

Introducing the coefficient $B = (b/t)\sqrt{\sigma_e/E}$, Winter's formula, Eq. 4.8, can be written as

$$\frac{b_e}{b} = \frac{1.90}{B} - \frac{0.90}{B^2} \tag{4.10}$$

A formula proposed by Conley et al. (1963) is nearly the same as that proposed by Winter and can be expressed as

$$\frac{b_e}{b} = \frac{1.82}{B} - \frac{0.82}{B^2} \tag{4.11}$$

A useful form of Eq. 4.10 or 4.11 is obtained by introducing the material yield strength into the dimensionless parameter B. If $\bar{B}$ is defined as

$$\bar{B} = B\sqrt{\frac{\sigma_y}{\sigma_e}} = \frac{b}{t}\sqrt{\frac{\sigma_y}{E}} \tag{4.12}$$

and if σ_{av}/σ_e is substituted for b_e/b, and both sides of the equation are multiplied by σ_e/σ_y, Eq. 4.11 can be written

$$\frac{\sigma_{av}}{\sigma_y} = \frac{1.82}{\bar{B}}\sqrt{\frac{\sigma_e}{\sigma_y}} - \frac{0.82}{\bar{B}^2} \tag{4.13}$$

By introducing discrete values of $\bar{B}$ into Eq. 4.13, the relationships shown in Fig. 4.5 between σ_{av}/σ_y and σ_e/σ_y for $\bar{B}$ values greater than 1.0,

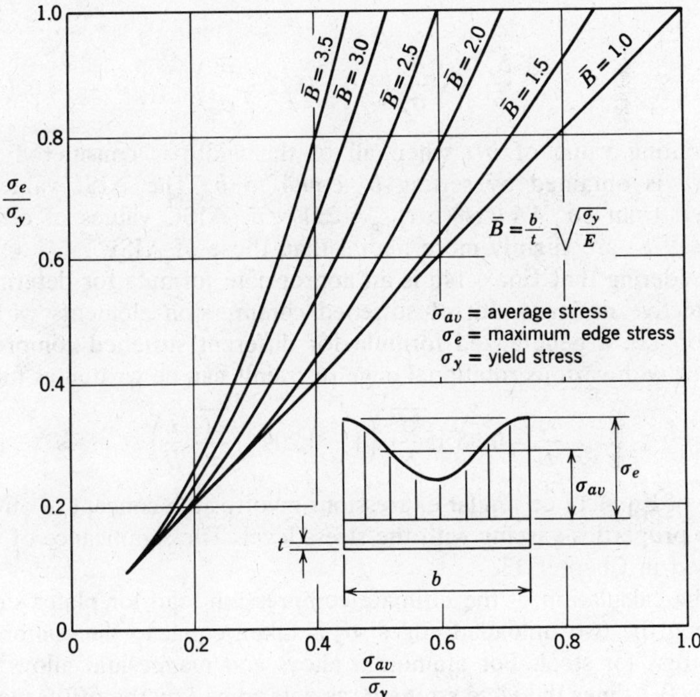

Fig. 4.5 Chart for determining σ_e/σ_y.

were determined. It was assumed that there would be no loss of plate effectiveness for values of $\bar{B} \leqslant 1.0$ and thus the straight line from (0, 0) to (1.0, 1.0) was drawn for $\bar{B} = 1.0$. Lines of constant $\bar{B}$ when plotted fully are tangent to the $\bar{B} = 1.0$ line, and only their upper portions are shown. Thus for any given strength level of plate steel, a relationship between average stress after buckling and the maximum or edge stress of the plate panel is established as a function of the actual b/t ratio. This relationship is valid for stiffened plates in which the ratio of stiffener cross-sectional area to plate-panel cross-sectional area is small. If the cross section of a structural member includes a buckled plate, the effective-width approach should be used in computing deflections, in determining the location of the neutral axis, or in other calculations where the effective moment of inertia or radius of gyration of the member is important.

In the 1968 and later editions of the AISI Specification for cold-formed steel members, the coefficients in Eqs. 4.8 and 4.9 were reduced slightly, giving the following expressions for effective width:

$$\frac{b_e}{t} = 1.9 \sqrt{\frac{E}{\sigma_e}} \left(1.0 - 0.415 \sqrt{\frac{E}{\sigma_e}} \frac{t}{b}\right) \tag{4.14a}$$

or

$$\frac{b_e}{b} = \sqrt{\frac{\sigma_c}{\sigma_e}} \left(1.0 - 0.22 \sqrt{\frac{\sigma_c}{\sigma_e}}\right) \tag{4.14b}$$

The limiting value of b/t when all of the width is considered to be effective is obtained by setting b_e equal to b. The AISI value thus obtained from eq. 4.14 is $(b/t)_{\text{lim}} = 221/\sqrt{\sigma}$. AISC values of effective width (1978) are slightly more liberal than those of AISI.

Considering that Eq. 4.14a is an appropriate formula for determining the effective design width of stiffened compression elements with a k value of 4.0, a generalized formula for different stiffened compression elements with various rotational edge restraints can be written as follows:

$$\frac{b_e}{t} = 0.95 \sqrt{\frac{kE}{\sigma_e}} \left(1 - 0.209 \sqrt{\frac{kE}{\sigma_e}} \frac{t}{b}\right) \tag{4.15}$$

Use of Eq. 4.14 or similar expressions involves the concept of effective section properties varying with the stress level. The significance of this is discussed in Chapter 13.

In the calculation of the ultimate compression load for plates supported along the two unloaded edges, σ_e is taken equal to the compressive yield stress for steel. For aluminum alloys and magnesium alloys, σ_e is taken as 0.7 times the yield strength, as determined by the offset method. However, if the buckling stress σ_c exceeds 0.7 times the yield strength, the load capacity as determined by inelastic plate-buckling analysis may be taken as $bt\sigma_c$ in which σ_c is determined by Eq. 4.2 and the effective width need not be calculated. The use of the ultimate compressive buckling load in specifications for aluminum structures is discussed later in this section.

Equation 4.14b can be used to determine a nondimensional ultimate-strength curve for steel plates in the postbuckling range. The average stress on the plate at ultimate load, σ_{av}, is the ultimate load divided by the total area. From Eq. 4.14b,

$$\sigma_{av} = \frac{P_{ult}}{bt} = \sqrt{\sigma_c \sigma_y}\left(1.0 - 0.22\sqrt{\frac{\sigma_c}{\sigma_y}}\right) \tag{4.16}$$

In Fig. 4.6 the average stress at ultimate load, by Eq. 4.16, is compared with the uniform-edge compression stresses to cause buckling. A method for predicting the strength of simply supported plates, taking into account initial out-of-flatness, is given by Abdel Sayed (1969) and Dawson and Walker (1972). Out-of-flatness, residual stress, and strain hardening are considered by Dwight and Ratcliffe (1968).

For a plate supported along only one longitudinal edge, the effective width has been experimentally determined by Winter as

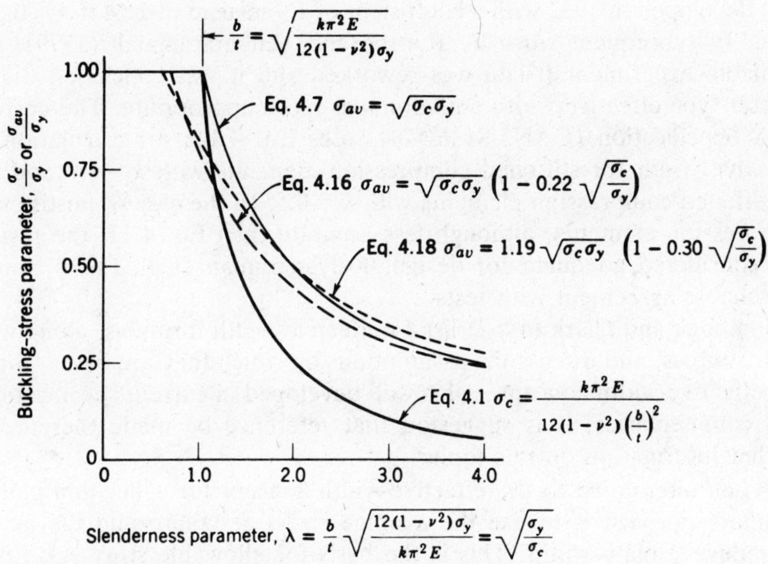

Fig. 4.6 Nondimensional buckling curves for plates under uniform edge compression (adapted from Brockenbrough and Johnston, 1974).

$$\frac{b_c}{b} = 1.19\sqrt{\frac{\sigma_c}{\sigma_e}}\left(1 - 0.30\sqrt{\frac{\sigma_c}{\sigma_e}}\right) \qquad (4.17)$$

This equation has also been confirmed by an analysis carried out by Kalyanaraman (Kalyanaraman and Pekoz, 1978; Kalyanaraman et al., 1977). The average stress at ultimate load can then be expressed as

$$\sigma_{av} = 1.19\sqrt{\sigma_c \sigma_y}\left(1 - 0.30\sqrt{\frac{\sigma_c}{\sigma_y}}\right) \qquad (4.18)$$

Equation 4.18 is also shown in Fig. 4.6. This curve actually falls above and to the right of Eq. 4.16, but it must be remembered that in the elastic range σ_c for a plate supported on both longitudinal edges is about eight times as large as that for the same plate supported along only one longitudinal edge.

After a statistical study of the available test data, Lind et al. (1976) suggest a compact expression for the effective width at ultimate load in the form

$$\frac{b_e}{t} = 1.64\sqrt{\frac{E}{\sigma_y}} \qquad (4.19)$$

and this formula has been incorporated into the Canadian specifications (CSA, 1974). This can be compared to von Kármán's expression (Eq. 4.4) developed in 1932 with a coefficient of 1.9 instead of 1.64 if $\nu = 0.3$ is used. In subsequent work by Roorda and Venkataramaiah (1979), the available experimental data was reworked and it was concluded that a Winter-type effective-width equation was more appropriate. The current CSA Specification (CAN3-S136-M84) uses Eq. 4.14a for calculation of effective width for stiffened compression elements with $k = 4.0$ and for unstiffened compression elements with $k = 0.5$. In the case of unstiffened compression elements, although less accurate than Eq. 4.17, the results are considered adequate for design. Kalyanaraman et al. (1977) found reasonable agreement with tests.

Jombock and Clark (1962) list 14 effective-width formulas, along with their sources, and discuss the assumptions on which they are based. Since the effective-width concept is also well developed in current specifications and commentaries, it is suggested that reference be made thereto for further information on this topic.

As an alternative to the effective-width concept for wide, thin plates, another approach is to use the average stress at failure and the actual (unreduced) plate width. This is the basis for allowable stresses on thin sections in the Aluminum Association Specifications (1982). In applying these specifications the designer does not, in general, calculate an effective width but uses instead an allowable stress that has been derived by applying a factor of safety to the average stress at failure for plate elements. For plates that buckle in the inelastic stress range, the average stress at failure is considered to be the same as the local-buckling stress, since plates of these proportions have little postbuckling strengh (Jombock and Clark, 1968). Inelastic local-buckling strength for aluminum plates is represented in the specifications by the following straight-line approximation to Eq. 4.1 (Clark and Rolf, 1966):

$$\sigma_c = B_p - D_p k_1 \frac{b}{t} \tag{4.20}$$

where

$$B_p = \sigma_y \left[1 + \frac{(\sigma_y)^{1/3}}{k_2} \right]$$

$$D_p = \frac{(B_p)^{3/2}}{k_3 (E)^{1/2}}$$

$$k_1 = \sqrt{\frac{12(1 - \nu^2)}{k}}$$

For aluminum alloys that are artificially aged (temper designations beginning with T5, T6, T7, T8, or T9), $k_2 = 11.4\,\text{ksi}^{1/3}$ ($78.6\,\text{MPa}^{1/3}$) and $k_3 = 10$.

For other aluminum alloys (temper designations beginning with 0, H, T1, T2, T3, or T4), $k_2 = 7.6\,\text{ksi}^{1/3}$ ($52.4\,\text{MPa}^{1/3}$) and $k_3 = 10\sqrt{2/3}$.

Equation 4.20 has been shown to agree well with the results of tests on aluminum plate elements (Jombock and Clark, 1968; Clark and Rolf, 1966).

For plates that buckle elastically, the average stress at failure is represented for purposes of the aluminum specifications as

$$\sigma_{av} = \sqrt{\sigma_e \sigma_c} \tag{4.21}$$

Equation 4.21 corresponds to Eq. 4.7. Jombock and Clark (1968) demonstrated that the edge stress at failure σ_e for aluminum plates could be represented by a function of the intercept B_p in Eq. 4.20. This results in a simple relationship between the ultimate-strength curves corresponding to elastic and inelastic buckling. Generally, this edge stress at failure for aluminum alloys is about $0.7\sigma_y$.

The formulas used in the Aluminum Association Specifications (1982) are illustrated in Fig. 4.7. Comparisons with test results were published by Jombock and Clark (1968). The use in these specifications of the

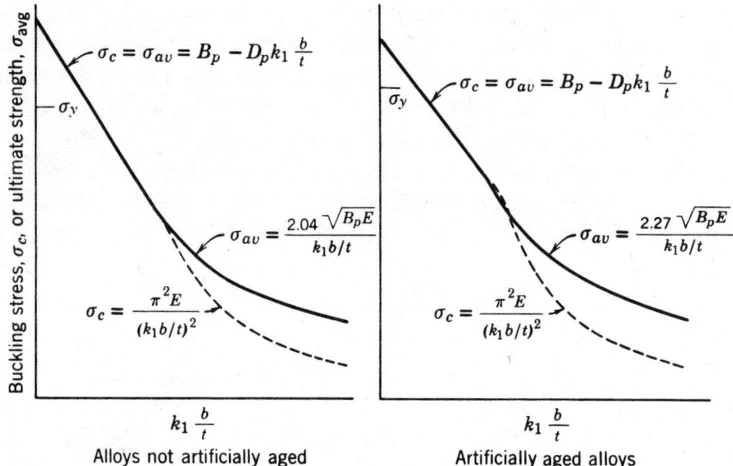

Fig. 4.7 Equations for buckling stress and ultimate strength of plates used in Aluminum Association Specifications (1982).

average stress at failure for thin sections results in some simplification, since the designer does not have to calculate an effective width. However, it sacrifices some of the flexibility of the effective-width approach and does not give as accurate a picture of the true physical behavior. For example, the average stress-at-failure method does not treat the change in moment of inertia of a member when its compression elements are in the post-buckling range, and hence does not lend itself to calculation of deflections. Therefore, the Aluminum Association Specifications include an effective-width formula to be used in calculating deflections in the postbuckling range.

Certain box-girder-bridge structural failures in the early 1970s have interjected a warning note to designers of plate structures that many aspects of plate behavior need further investigation. For example, the concepts of effective width and average stress at failure consider plate strains only up to the maximum plate capacity—that is, the ultimate load as characterized by reaching the yield stress at the edge. Research has shown that inelastic strains beyond the point may lead to a sudden and substantial reduction of the plate's load-carrying capacity (Dwight, 1971; Dwight and Moxham, 1969).

4.2.2 Compression and Bending

Buckling Strength. When compression plus bending loads are applied to a structural member, plate elements of the member can be subjected to in-plane stresses which vary along the loaded edges of the plate, from a maximum compressive stress, σ_1, to a minimum stress, σ_2, as shown in Fig. 4.8. For this situation, elastic critical plate stresses are dependent on the edge-support conditions and the ratio of bending stress to uniform compression stress. Values of k_c that can be substituted for k in Eq. 4.1 are tabulated in Fig. 4.8 for several cases. For intermediate stress ratios (σ_{cb}/σ_c), values of k_c can be estimated by linear interpolation. For plates with a free edge the k_c values vary slightly with Poisson's ratio. In the inelastic range, an estimate of the buckling stress can be obtained by using k_c for k in Eq. 4.2. Diagrams for buckling coefficients for rectangular plates under combined bending and compressive stresses in two perpendicular directions are given by Yoshizuka and Naruoka (1971).

Loading	Ratio of Bending Stress to Uniform Compression Stress σ_{cb}/σ_c	Minimum Buckling Coefficient, $^*K_c$					
				Top Edge Free		Bottom Edge Free	
		Unloaded Edges Simply Supported	Unloaded Edges Fixed	Bottom Edge Simply Supported	Bottom Edge Fixed	Top Edge Simply Supported	Top Edge Fixed
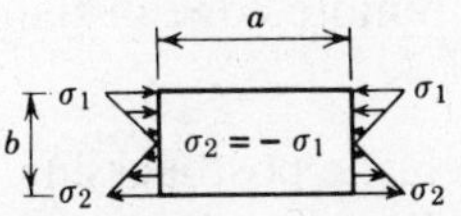 $\sigma_2 = -\sigma_1$	∞ (pure bending)	23.9	39.6	0.85	2.15		
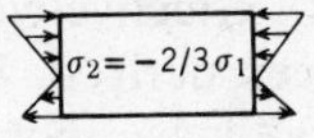 $\sigma_2 = -2/3\sigma_1$	5.00	15.7					
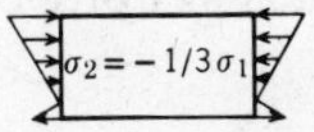 $\sigma_2 = -1/3\sigma_1$	2.00	11.0					
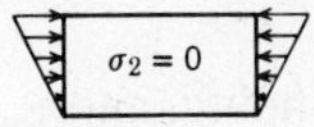 $\sigma_2 = 0$	1.00	7.8	13.6	0.57	1.61	1.70	5.93
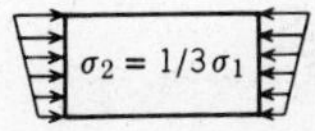 $\sigma_2 = 1/3\sigma_1$	0.50	5.8					
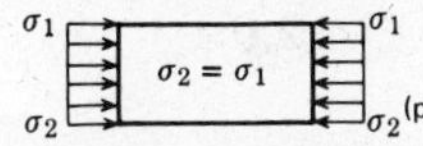 $\sigma_2 = \sigma_1$	0.0 (pure compression)	4.0	6.97	0.42	1.33	0.42	1.33

*Values given are based on plates having loaded edges simply supported and are conservative for plates having loaded edges fixed.

Fig. 4.8 Buckling coefficients for long flat plates under compression and bending (Brockenbrough and Johnston, 1974; Bijlaard, 1957).

Postbuckling Strength. Plate elements and channel sections subjected to prescribed loading and displacement eccentricity have been studied by Rhodes and Harvey (1971, 1976). These solutions are based on the assumption of a fixed buckling mode and are thus, in principle, restricted in their applicability to the immediate vicinity of bifurcation, but for the case of single plates have been shown to be in reasonable agreement with experimental results up to three times the critical load.

In the case of simply supported plates under eccentric loading P, the failure loads for plates with various loading eccentricities can be accurately predicted by a simple expression of the form (Rhodes and Harvey, 1977).

$$\bar{P}_{\text{ult}} = \frac{pb}{\pi^2 D} = \frac{\bar{\sigma}_y + 11.4}{b(e/b) + 0.85} \tag{4.22}$$

where

$D = Et^3/12(1 - \nu^2)$

$\bar{\sigma}_y = \sigma_y(b^2 t/\pi^2 D)$

e = distance from the point of load application to the remote edge of the plate

A study has been carried out by Usami (1982) on elastic postbuckling of plates in compression and bending using compatibility equations and the energy method. Based on the numerical results, effective-width formulas in combined compression and bending have been derived. It is shown that Winter's formula (Eq. 4.14) forms a good lower bound. Based on Usami's results, the following empirical formula was proposed to predict the postbuckling strength, σ:

$$\frac{\sigma}{\sigma_e} = \sqrt{\frac{\sigma_c}{\sigma_e}}\left[(1 + 0.1\xi_0) - (0.22 + 0.05\xi_0)\sqrt{\frac{\sigma_c}{\sigma_e}}\right] \tag{4.23a}$$

in which $\xi_0 = 0$ for pure compression and 2.0 for pure bending, and ξ_0 can be calculated from

$$\begin{aligned} k &= \frac{8.4}{2.1 - \xi_0} && \text{for } 0 \leqq \xi_0 \leqq 1.0 \\ k &= 10\xi_0^2 - 13.736\xi_0 + 11.372 && \text{for } 1.0 \leqq \xi_0 \leqq 2.0 \end{aligned} \tag{4.23b}$$

for different geometric boundary conditions. This equation reduces to Winter's equation with $\xi_0 = 0$. If the plate yield stress σ_y is substituted for σ_e, σ becomes the ultimate stress of the plate.

Based on an experimental study, LaBoube and Yu (1982) have found that the postbuckling strength of beam web elements subjected to pure bending stress is a function of web slenderness ratio, the bending stress ratio of the web, the width-to-thickness ratio of the compression flange, and the yield point of the steel. The effective depth equations for beam webs have been developed through a statistical analysis to predict the ultimate bending capacity of cold-formed steel beams when used in conjunction with the assumed bending stress distribution.

The Aluminum Association Specifications (1982) treat postbuckling strength of webs in bending by means of an average stress approach similar to that used for plates in compression. This approach is compared with test results by Jombock and Clark (1968).

4.2.3 Shear

Buckling Strength. When a plate is subjected to edge shear stresses as shown in Fig. 4.9, it is said to be in a state of "pure shear." Tension and compression stresses exist in the plate, equal in magnitude to the shear stress and inclined at 45°. The destabilizing influence of compressive stresses is resisted by tensile stresses in the perpendicular direction. Unlike the case of edge compression, the buckling mode is composed of a combination of several waveforms and this is part of the difficulty in the buckling analysis for shear.

The critical shear stresses can be obtained by substituting τ_c and k_s for σ_c and k in Eq. 4.1, in which k_s is the buckling coefficient for shear buckling stress. Critical-stress coefficients k_s for plates subjected to pure shear have been evaluated for three conditions of edge support. In Fig. 4.9 these are plotted with the side b, as used in Eq. 4.1, always assumed to be shorter than side a. Thus α is always greater than 1 and by plotting k_s in terms of $1/\alpha$, the complete range of k_s can be shown and the magnitude of k_s remains manageable for small values of α. However, for application to plate-girder design it is convenient to define b (or h in plate-girder applications) as the vertical dimension of the plate-girder web for a horizontal girder. Then α may be greater or less than unity and empirical formulas for k_s together with source data are as follows:

Plate simply supported on four edges. Solutions developed by Timoshenko (1910), Bergmann and Reissner (1932), and Seydel (1933) are approximated by Eqs. 4.24a and 4.24b, in which $\alpha = a/b$:

$$k_s = 4.00 + \frac{5.34}{\alpha^2} \quad \text{for } \alpha \lesssim 1 \tag{4.24a}$$

$$k_s = 5.34 + \frac{4.00}{\alpha^2} \quad \text{for } \alpha \geq 1 \tag{4.24b}$$

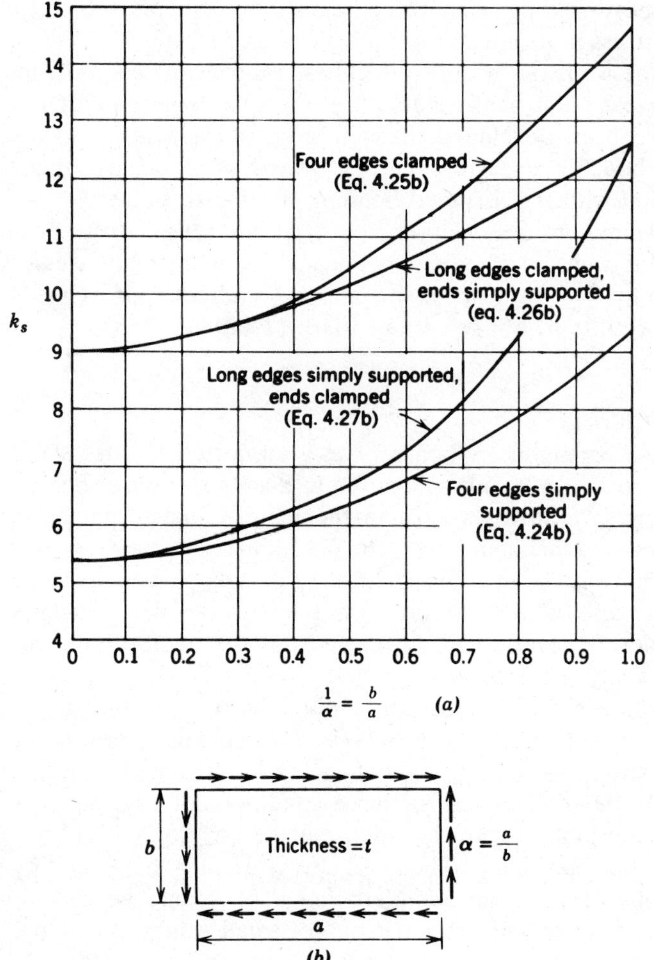

Fig. 4.9 Buckling coefficients for plates in pure shear. (Side b is the short side.)

Plate clamped on four edges. In 1924, Southwell and Skan obtained $k_s = 8.98$ for the case of the infinitely long rectangular plate with clamped edges. For the finite-length rectangular plate with clamped edges, Moheit (1939) obtained

$$k_s = 5.6 + \frac{8.98}{\alpha^2} \qquad \text{for } \alpha \lesssim 1 \tag{4.25a}$$

$$k_s = 8.98 + \frac{5.6}{\alpha^2} \qquad \text{for } \alpha \geq 1 \tag{4.25b}$$

Plate clamped on two opposite edges and simply supported on the other two edges. A solution for this problem has been given by Iguchi (1938) for the general case, and by Leggett (1941) for the case of the square plate. Cook and Rockey (1963) later obtained solutions considering the antisymmetric buckling mode which was not considered by Iguchi. The expressions below were obtained by fitting a polynomial equation to the Cook and Rockey results as shown in Fig. 2.36 of the book by Bulson (1970).

Long edges clamped:

$$k_s = \frac{8.98}{\alpha^2} + 5.61 - 1.99\alpha \qquad \text{for } \alpha \lesssim 1 \tag{4.26a}$$

$$k_s = 8.98 + \frac{5.61}{\alpha^2} - \frac{1.99}{\alpha^3} \qquad \text{for } \alpha \geq 1 \tag{4.26b}$$

Short edges clamped:

$$k_s = \frac{5.34}{\alpha^2} + \frac{2.31}{\alpha} - 3.44 + 8.39\alpha \qquad \text{for } \alpha \lesssim 1 \tag{4.27a}$$

$$k_s = 5.34 + \frac{2.31}{\alpha} - \frac{3.44}{\alpha^2} + \frac{8.39}{\alpha^3} \qquad \text{for } \alpha \geq 1 \tag{4.27b}$$

Curves for $\alpha \geq 1$ are plotted in Fig. 4.9. If the predicted critical stress in shear is greater than the proportional limit of the material, the buckling will be inelastic.

Postbuckling Strength. The initial mode of buckling in pure shear, which takes the form of a half wave in the tension direction and at least one full wave in the compression direction (Fig. 4.10*a*), undergoes a change in the advanced postbuckling range, eventually taking the form of a family of diagonal folds (Fig. 4.10*b*). These folds carry significant tensile stresses developed in the postbuckling range and the displacement pattern is called a "tension field."

The maximum shear load that can be applied before failure occurs due to a breakdown of the material in the tension field, and it is influenced by the rigidity of the edge members supporting the plate. This problem is dealt with in greater detail in Chapter 6. For a plate with infinitely stiff edge members, the maximum shear strength can be estimated by the formula (Allen and Bulson, 1980)

$$\bar{V}_u = \tau_c bt + \tfrac{1}{2}\sigma_{ty} bt \tag{4.28}$$

where

$$\sigma_{ty} = \sqrt{\sigma_y^2 - 0.75\tau_c^2} - 1.5\tau_c \qquad \text{provided that } \tau_c \ll \sigma_y$$

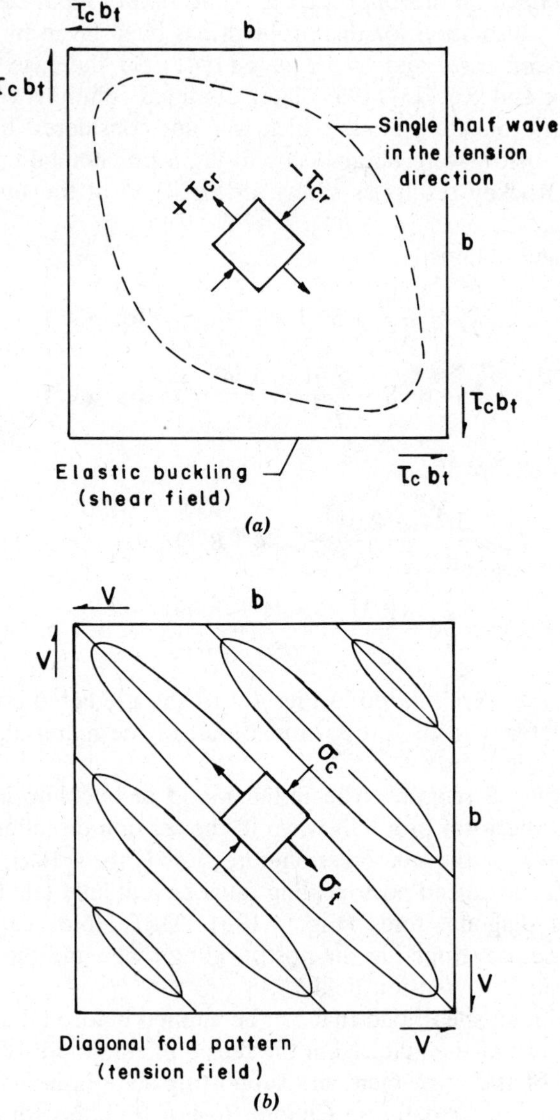

Fig. 4.10 Shear and tension fields (square plate).

4.2.4 Combined Stresses

Shear Combined with Direct Stress. The case of shear combined with longitudinal compression, with all sides simply supported, was treated by Iguchi (1938). His results are approximated by the following interaction equation, also shown graphically in Fig. 4.11:

$$\frac{\sigma_c}{\sigma_c^*} + \left(\frac{\tau_c}{\tau_c^*}\right)^2 = 1 \tag{4.29}$$

where σ_c^* and τ_c^* denote critical stress, respectively, under compression or shear alone.

Equation 4.29, shown in Fig. 4.11, is for ratios of a/b greater than unity. Batdorf and associates (Batdorf and Houbolt, 1946; Batdorf and

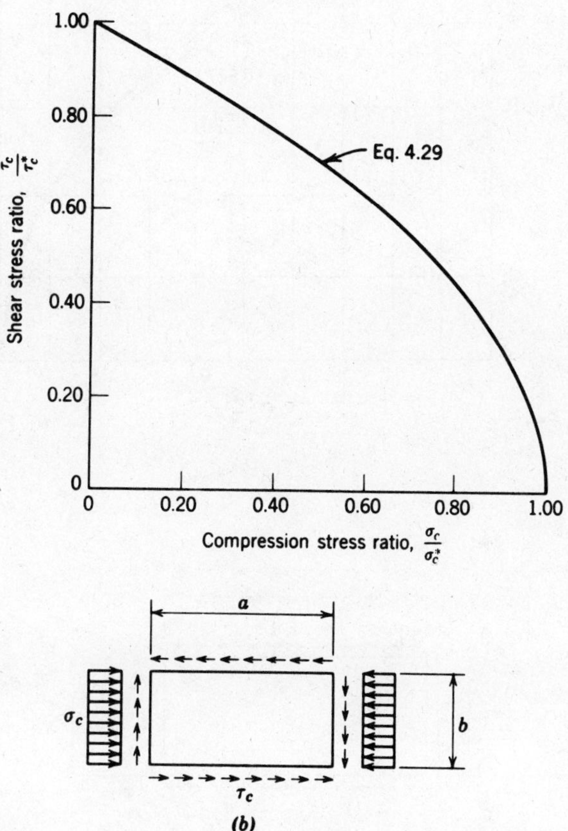

Fig. 4.11 Interaction curve for buckling of flat plates under shear and uniform compression.

Stein, 1947) have shown that when the loaded side b is more than twice as long as a, Eq. 4.29 becomes overly conservative. This situation is the exception in actual practice, and Eq. 4.29 may be accepted for engineering design purposes.

The work of Stowell (1949) and Peters (1954) may be referred to with regard to buckling in the inelastic range under combined compressive and shear stress for loads applied in constant ratio. Peters found that a

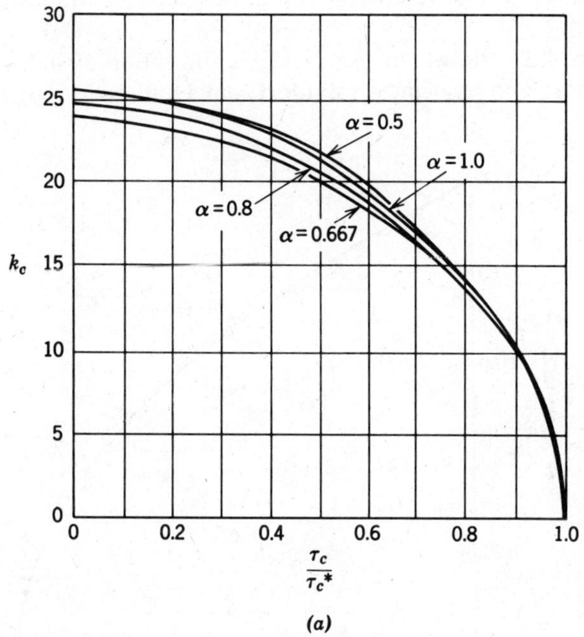

(a)

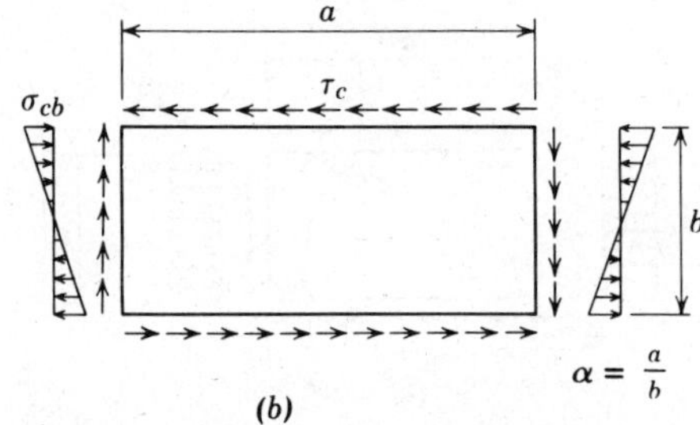

(b)

Fig. 4.12 Buckling coefficients for plates in combined bending and shear.

circular stress-ratio interaction formula as expressed by Eq. 4.30 was conservative and agreed better with test results than Eq. 4.29.

$$\left(\frac{\sigma_c}{\sigma_c^*}\right)^2 + \left(\frac{\tau_c}{\tau_c^*}\right)^2 = 1 \tag{4.30}$$

Shear Combined with Bending. For a plate simply supported on four sides, under combined bending and pure shear, Timoshenko (1934) obtained a reduced k_c as a function of τ_c/τ_c^* for values of $\alpha = 0.5$, 0.8, and 1.0, where τ_c is the actual shearing stress and τ_c^* is the buckling stress for pure shear. This problem was also solved by Stein (1936) and Way (1936), whose results for four values of α are plotted in Fig. 4.12.

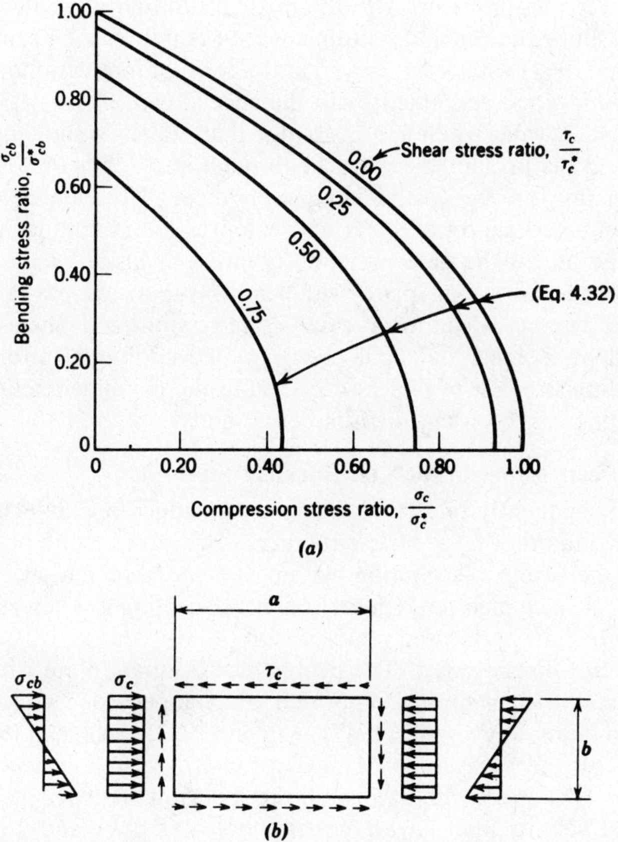

Fig. 4.13 Interaction curve for buckling of flat plates under shear, compression, and bending.

Chwalla (1936a) suggested the following approximate interaction formula, which agrees well with the graphs of Fig. 4.12.

$$\left(\frac{\sigma_{cb}}{\sigma^*_{cb}}\right)^2 + \left(\frac{\tau_c}{\tau^*_c}\right)^2 = 1 \tag{4.31}$$

For a plate simply supported on four sides, under combined bending and direct stress at the ends (of dimension b), combined with shear, an approximate evaluation of the critical combined load is obtained by use of a three-part interaction formula, Eq. 4.32 (Gerard and Becker, 1957/1958).

$$\left(\frac{\sigma_c}{\sigma^*_c}\right) + \left(\frac{\sigma_{cb}}{\sigma^*_{cb}}\right)^2 + \left(\frac{\tau_c}{\tau^*_c}\right)^2 = 1 \tag{4.32}$$

The foregoing problem, with the further addition of vertical compressive force along the top and bottom edges of length a, has been treated by McKenzie (1964) with results given in the form of interaction graphs. The results are in good agreement with the special case of Eq. 4.32. Interaction Eq. 4.32, valid when a/b is greater than unity, is shown graphically in Fig. 4.13, as presented in Brockenbrough and Johnston (1976).

Information on the postbuckling strength of plate elements subjected to the combined action of shear and compression is limited. A semiempirical method for the determination of stress levels at which permanent buckles occur in a long plate with simply supported edges under the combined action of uniform axial compression and shear has been suggested by Zender and Hall (1960). Some additional information on postbuckling strength of plates subjected to the combined action of shear and bending can be found in Chapters 6 and 7.

4.2.5 Effects of Perforation on Buckling Strength

Designers frequently find it necessary to introduce openings in the webs of girders and other large plate structures. The introduction of an opening changes the stress distribution within the member and will, in many instances, also change the mode of failure. Buckling is a key aspect in the behavior of thin perforated plates.

Plates in Compression. The problem of a square plate with a central hole, having either simply supported or clamped-edge conditions, has been studied by Levy et al. (1947), Kumai (1952), Schlack (1964), Fujita et al. (1970), and Kawai and Ohtsubo (1968).

Figure 4.14 gives the ratios of the buckling coefficients of simply supported square plates with circular holes (k_{pc}/k) and square holes (k_{ps}/k) to the coefficient of unperforated plates. For a given size of perforation, the reduction in the critical load for a plate having a square

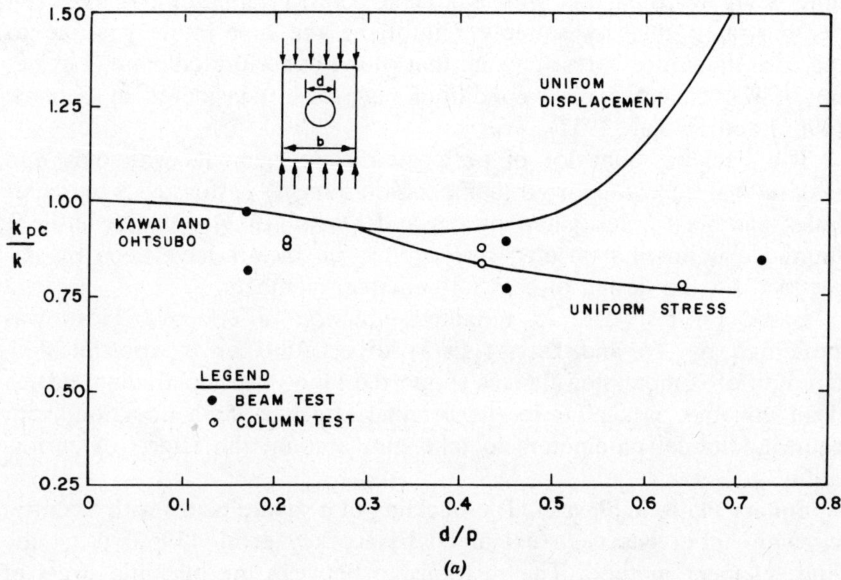

(a)

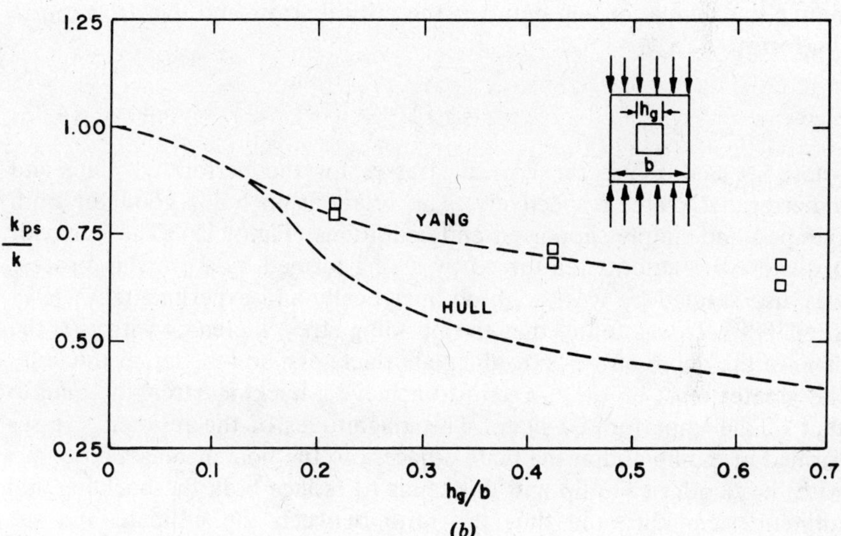

(b)

Fig. 4.14 Buckling of plates with holes: (*a*) effect of d/b ratio on buckling coefficient and comparison of test data with theoretical buckling coefficients (Yu and Davis, 1973); (*b*) effect of h_1/b ratio on buckling coefficient and comparison of test data with theoretical buckling coefficients (Yu and Davis, 1973).

hole is greater than that with a circular hole (Yang, 1969). It has been demonstrated that by suitably reinforcing the hole, it is possible to increase the critical stress beyond that of the unperforated plate (Levy et al., 1947). Clamped edge conditions have been considered by Schlack (1964) and Kumai (1952).

The buckling behavior of perforated plate elements with only one longitudinal edge supported (compression flanges) perforated by circular holes has been investigated by Yu and Davis (1973). On the basis of limited experimental results, stress reduction factors have been recommended for the design of cold-formed steel members.

Based on test data, a modified equation of effective width was presented by Yu and Davis (1973) to account for the postbuckling strength of compression flanges supported along two longitudinal edges. That equation is similar to the original effective-width equation, with some additional parameters to take into account the effects of perforation.

Square Plates in Shear. The buckling of a square plates with a central circular cutout has been examined by Rockey et al. (1969) using the finite-element method. The relationship between the buckling stress of the plate and the relative size of the hole (d/b) was obtained for both simply supported and clamped-edge conditions. Rockey's work suggests a simple linear relationship between the critical stress and the d/b ratio in the form

$$\tau_{cp} = \tau_c\left(1 - \frac{d}{b}\right) \tag{4.33}$$

where τ_{cp} and τ_c are the critical stresses for the perforated plate and unperforated plate, respectively. The relationship holds good for both clamped and simply supported end conditions (Fig. 4.15). The behavior of plates with cutouts reinforced by a ring formed by a pressing process was also studied by Rockey, both analytically and experimentally (Rockey, 1980). It was found that the buckling stress increases with t_r/t, the ratio of the depth of the lip to the plate thickness, and the larger the hole, the greater must be the t_r/t ratio to achieve a buckling strength equal to that of the unperforated plate. The magnitudes of the residual stresses formed in the lip and in the plate adjacent to the hole increase, however, with the depth of the lip and this tends to reduce both the buckling and ultimate strengths; and thus the ratio between the ultimate and the buckling loads is not significantly affected as t_r/t is increased beyond a certain value.

Shear buckling of square perforated plates was also investigated by

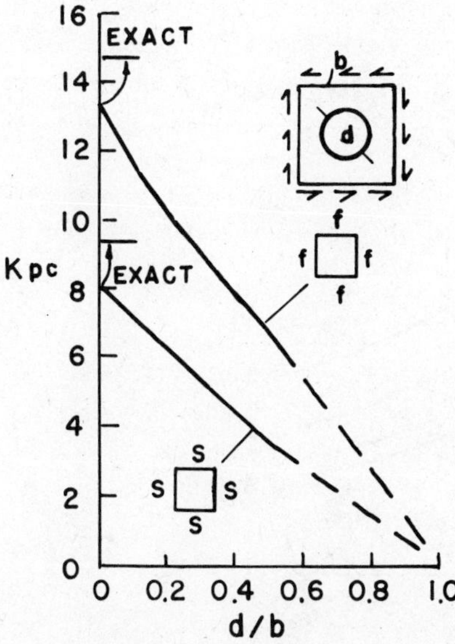

Fig. 4.15 Simply supported and clamped plates with hole under shear loading (Rockey et al., 1969).

Grosskurth et al. (1976) using the finite-element approach. They considered the case of uniform shear deformation instead of uniform shear stress and obtained critical stresses that were in closer agreement with, although higher than, the experimental values.

Shear Webs with Multiple Holes. The problem of a long shear web with holes has received some attention in the literature. Based on the available data, Michael (1960) has suggested semiempirical expressions for the critical stress in terms of d/a and a/b (notation indicated in Fig. 4.16). These are graphically plotted in Fig. 4.16 and are applicable for the web fixed along the top and bottom edges.

Webs with Holes under Combined Loading. Redwood and Uenoya (1979) have investigated the problem of webs with holes subjected to combined bending and shear. Using the finite-element approach for the solution of the plane stress problem and a Rayleigh-Ritz procedure for the buckling analysis, they studied the problem of shear webs with aspect ratios from 1.5 to 2.5 with circular or rectangular holes. They suggested an interaction formula for τ_c and σ_{cb} (critical values of the maximum shear and bending stresses, respectively) in the form

$$\left(\frac{\tau_c}{\tau_{cp}}\right)^2 + \left(\frac{\sigma_{cb}}{\sigma_{cbp}}\right)^2 = 1.0 \tag{4.34}$$

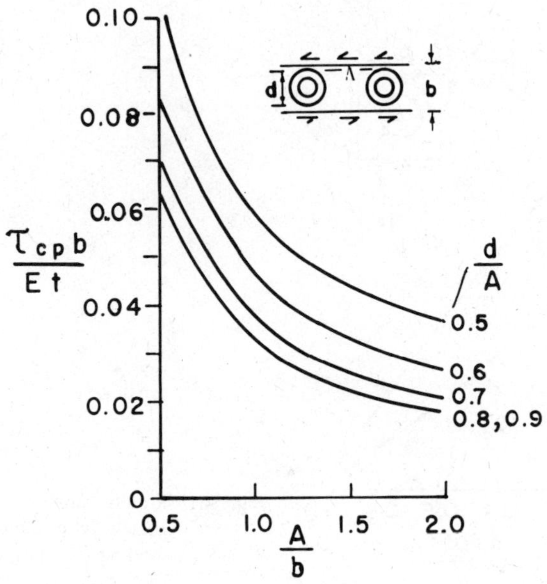

Fig. 4.16 Critical shear stress for webs with holes (Michael, 1960).

in which τ_{cp} and σ_{cbp} are the pure shear and pure bending critical stresses of the plate with the hole. These, in turn, can be expressed in terms of the corresponding critical stresses of plates without holes (τ_c^*, σ_{cb}^*) and the relative sizes of the holes with respect to the plate dimensions. With the notation indicated in Fig. 4.17, the expressions for plates with rectangular holes take the form

$$\sigma_{cbp} = \left[1.02 - 0.04\left(\frac{a}{H}\right)\right]\sigma_{cb}^* \qquad \text{but} \le \sigma_{cb}^* \tag{4.35a}$$

$$\tau_{cp} = \left[1.24 - 1.16\left(\frac{2H}{h}\right) - 0.17\left(\frac{a}{H}\right)\right]\tau_c^* \qquad \text{but} \le \tau_c^* \tag{4.35b}$$

and for circular holes,

$$\sigma_{cbp} = \sigma_{cb}^* \tag{4.36a}$$

$$\tau_{cp} = \left[1.15 - 1.05\left(\frac{2R}{h}\right)\right]\tau_c^* \qquad \text{but} \le \tau_c^* \tag{4.36b}$$

where R is the radius of the hole.

The values of σ_{cb}^* and τ_c^* can be obtained from a knowledge of the aspect ratio and boundary conditions of the plate. For example, for a

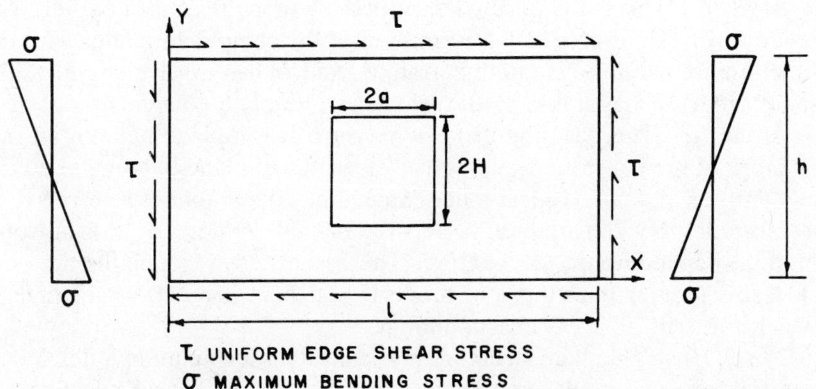

Fig. 4.17 Perforated rectangular plate under combined action of shear and bending.

simply supported plate with an aspect ratio of 2, these stresses are given with sufficient accuracy by the following expressions:

$$\sigma^*_{cb} = 23.90\sigma^*_c \tag{4.37a}$$

$$\tau^*_c = 6.59\sigma^*_c \tag{4.37b}$$

in which

$$\sigma^*_c = \frac{\pi^2 E}{12(1-\nu^2)}\left(\frac{t}{h}\right)^2$$

4.3 INTERACTION BETWEEN PLATE ELEMENTS

In the preceding section, attention has been confined to the behavior of a single plate element supported along one or both of its longitudinal edges. The structural sections employed in practice (Fig. 4.1) are composed of plate elements arranged in a variety of configurations. It is clear that the behavior of an assembly of plates would be governed by an interaction between the plate components. In this section the mechanics of such an interaction and its implication in design are discussed briefly.

4.3.1 Buckling Modes of a Plate Assembly

Unlike a single plate element supported along the unloaded edges, a plate assembly can buckle in one of several possible modes. For the case of axial compression, the buckling mode can take one of the following forms:

Mode I. This is the purely local buckling mode discussed earlier. The mode involves out-of-plane deformation of the component plates with the junctions remaining essentially straight, and it has a wavelength of the same order of magnitude as the widths of the plate elements.

Mode II. The buckling process may involve in-plane bending of one or more of the constituent plates as well as out-of-plane bending of all the elements as in a purely local mode. Such a buckling mode is referred to as a "stiffener buckling mode," "local torsional mode," or "orthotropic mode," depending on the context. The associated wavelengths are considerably greater than those of mode I, but there is a half-wavelength at which the critical stress is a minimum.

Mode III. The plate structure may buckle as a column in a flexural or flexural-torsional mode with or without interaction of local buckling.

Attention in this chapter will primarily be on the mode I type of buckling. Column behavior (mode III) and interaction between local and overall modes of buckling are treated in Chapter 13 and elsewhere in the guide.

4.3.2 Buckling of a Plate Assembly

A prismatic plate structure is often viewed simply as consisting of "stiffened" and "unstiffened" plate elements. The former are plate elements supported on both of their longitudinal edges by virtue of their connection to adjacent elements, while the latter are those supported only along one of their longitudinal edges. Thus the critical local buckling stress of a plate assembly may be taken as the smallest of the critical stresses of the plate elements, each treated as simply supported along its junctions with other plates. However, such a calculation must be used with caution for the following reasons:

1. The results can be unduly conservative when the plate structure consists of elements with widely varying slendernesses. This is the result of neglecting the rotational restraints at the functions.
2. The results are inapplicable unless it is ensured that all the plate elements buckle "locally" (i.e., the junctions remain essentially straight). If, on the other hand, mode II or III type of buckling is critical, the result of such a simplified calculation would be on the unsafe side.

The intervention of stiffener buckling (mode II) is usually averted in practice by designing "out" the stiffener buckling mode by the provision of stiffeners (edge or intermediate) of adequate rigidity. This is appropriate because of the limited postbuckling resistance associated with the mode II type of buckling. A design approach for edge and intermediate stiffeners has been proposed by Desmond et al. (1981).

Optimal design of the cross section would be one that made all the component plates equally stiff with due regard to their condition (i.e., whether they were "stiffened" or "unstiffened." This would also ensure that all the elements participate equally in the local buckling process, thus making it realistic to use design concepts such as the effective width.

In plastic design of steel structures, it is necessary that the moment capacity of the member not be impaired by local buckling until the required rotation is achieved. This can be achieved by limiting the width-to-thickness ratios of elements that are vulnerable to local buckling in the inelastic range. Such limiting width-to-thickness ratios have been

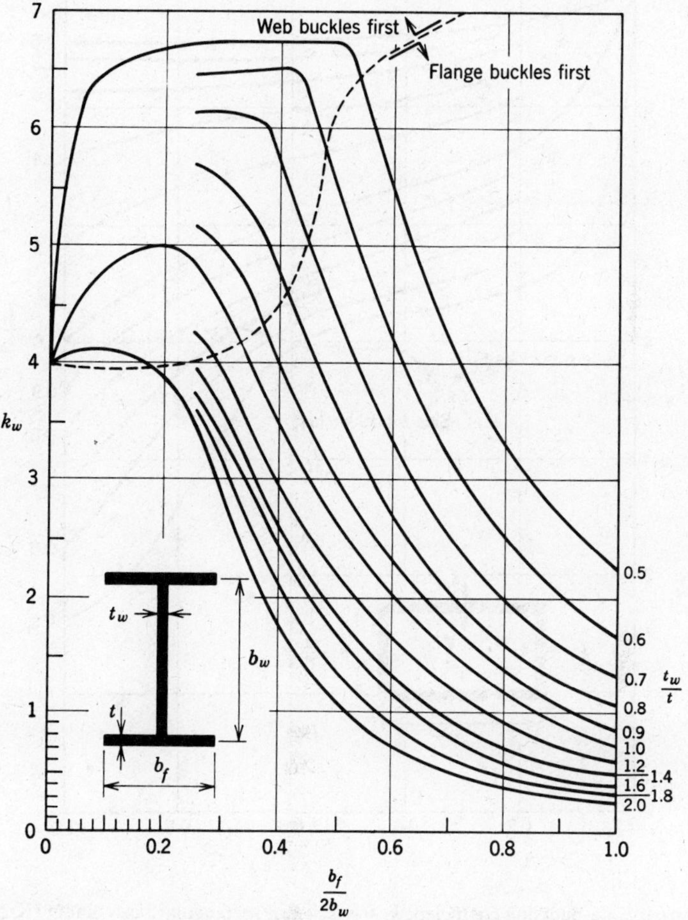

Fig. 4.18 (*a*) Plate-buckling coefficient k_w for wide-flange columns (Kroll et al., 1943).

proposed for flanges of I-beams by Lay (1965). Such provisions are also available in design specifications (AISC, 1978).

Figure 4.18 gives the variations of the local buckling coefficients k_w for a wide-flange I-section, a box section and a Z- or channel section, respectively, with respect to the geometrical properties of the member. Each of these charts is divided into two portions by a dashed line running across it (Kroll et al., 1943). In each portion, buckling of one of the elements dominates over the other, but the proportions exactly on the dashed line represent the most structurally efficient configuration in that

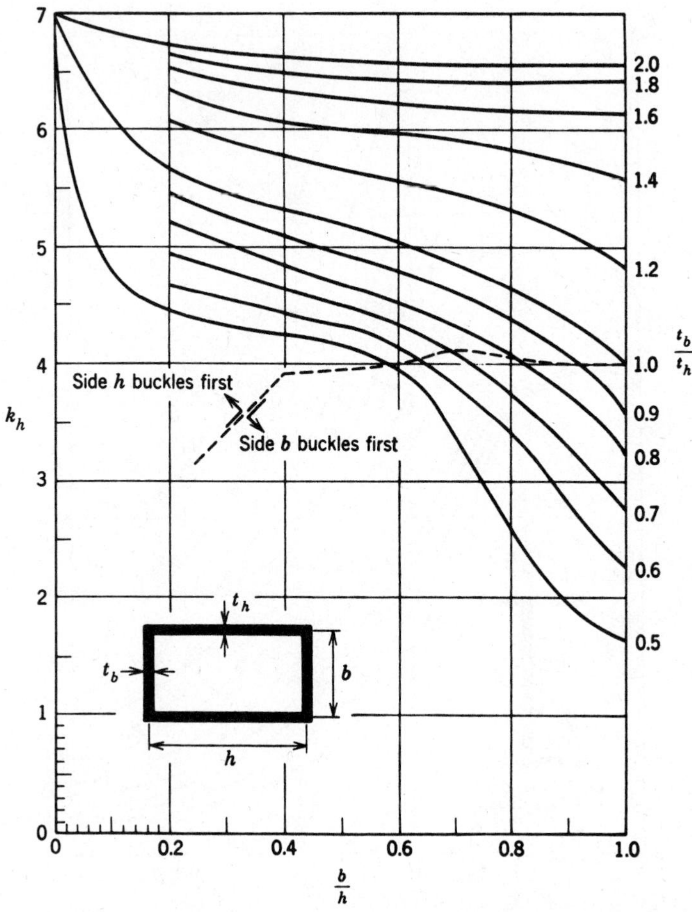

Fig. 4.18 (*b*) Plate-buckling coefficient k_h for side h of rectangular box column (Kroll et al., 1943).

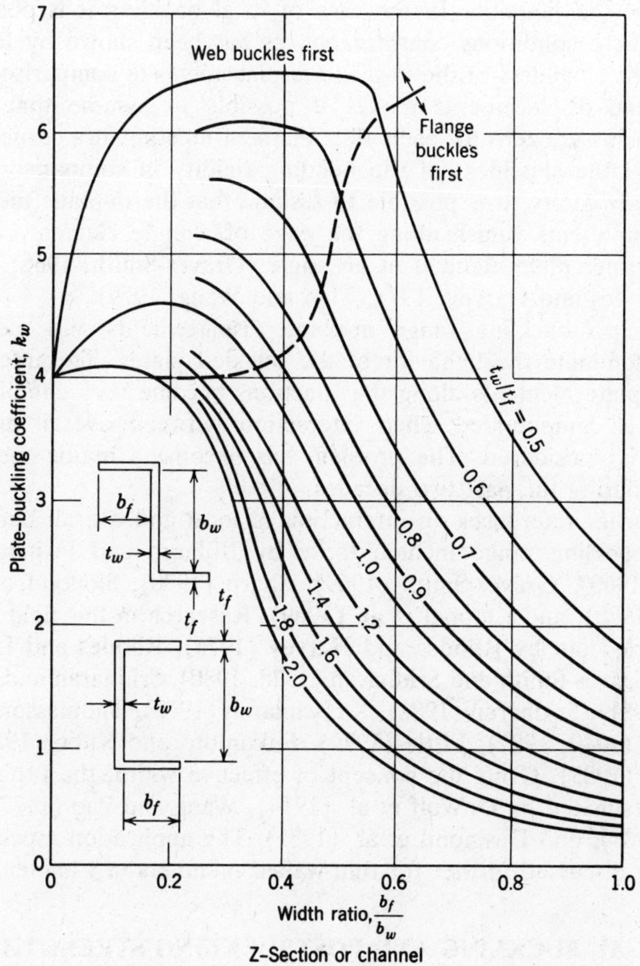

Fig. 4.18 (*c*) Plate-buckling coefficient k_w for Z-sections and channels (Kroll et al., 1943).

there occurs a complete participation of all plates in the local-buckling process. Additional charts and related information may be found in the references such as the Japanese *Handbook of Structural Stability* (CRCJ, 1971).

4.3.3 Postbuckling of a Plate Assembly

Interaction between the elements of a plate assembly is inescapable because of the equilibrium and compatibility conditions that must be satisfied at the junction. In the case of local buckling it is possible to simplify these conditions considerably, as has been shown by Benthem (1959). The smallness of the in-plane displacements in comparison to the out-of-plane displacements makes it possible to assume that normal displacements are zero for each plate element meeting at a corner. Also, because of the smallness of the bending rigidity in comparison to the extensional rigidity, it is possible to assume that the in-plane membrane stress components vanish along the edge of a plate element, where it meets another plate element at an angle (Graves-Smith, 1968; Walker, 1964; Rhodes and Harvey, 1971; Tien and Wang, 1979).

In the postbuckling range in-plane displacements and membrane stresses dominate the behavior of the buckled plates. The interactions between plate elements along the junctions become very complex. The problem is compounded when interaction between overall and local buckling is considered. The problem has become a major subject for research during the past two decades.

The earlier references on interaction of local and overall buckling in the postbuckling range include those by Bijlaard and Fisher (1952), Cherry (1960), Graves-Smith (1969), Sharp (1970), Škaloud and Zornerova (1970), and Klöppel et al. (1969). Research in this field has also been carried out by Rhodes and Harvey (1976), Rhodes and Loughlan (1980), Graves-Smith and Sridharan (1978, 1980), Sridharan and Graves-Smith (1981), Sridharan (1982), Kaliyamaran (1978), Thomasson (1978), Hancock (1980, 1981), Little (1979), Fukumoto and Kubo (1982), and Mulligan (1983). Using the concept of effective width, the problem has been investigated by DeWolf et al. (1974), Wang and Pao (1977), Wang et al. (1981), and Desmond et al. (1981). The application aspect of this subject is discussed further for thin-walled members in Chapter 13.

4.4 LOCAL BUCKLING AND POSTBUCKLING STRENGTH OF STIFFENED PLATES

This section deals with the buckling and ultimate strength of stiffened flat plates under various combinations of loadings. The behavior of the stiffened plate as a unit is emphasized rather than the stability of its individual elements. However, in design, the structural properties of all of the components must be considered.

Stiffened plates can fail through instability in essentially two different ways. In one, overall buckling, the stiffeners buckle along with the plate, and in the other, local buckling, the stiffeners form nodal lines and the plate panels buckle between the stiffeners. In either case the stiffness of the combination may be such that initial buckling takes place at fairly low stress levels. Nevertheless, a significant amount of postbuckling strength may remain in the stiffened plate, provided that proper attention is given to the design and fabrication of the structural details. A great deal of information on this subject can be found in the *Handbook of Structural Stability* (CRCJ, 1971) and in a book by Troitsky (1976).

4.4.1 Uniaxial Compression

Buckling Strength. It is often economical to increase the compressive strength of a plate element by introducing longitudinal and/or transverse stiffeners (Fig. 4.19). In the following paragraphs methods are presented

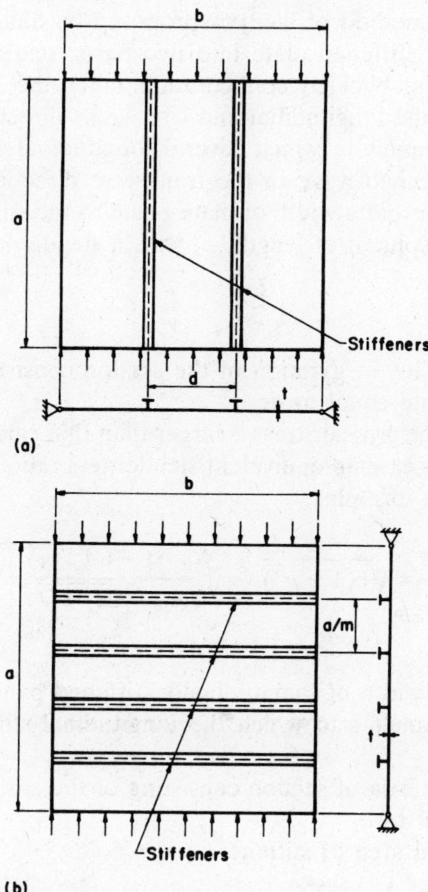

Fig. 4.19 Stiffened plate panels: (*a*) panel with longitudinal stiffeners; (*b*) panel with transverse stiffeners.

for determining the compressive strength of stiffened plate panels. The edges of the plate are assumed to be simply supported in all cases, and it is also assumed that individual elements of the panel and stiffeners are not subject to instability.

Longitudinal stiffeners. Timoshenko and Gere (1961), Bleich and Ramsey (1951), and Seide and Stein (1949) have presented charts and tables for determining the critical stress of plates simply supported on all edges and having one, two, or three equally spaced longitudinal stiffeners parallel to the direction of the applied compressive load. The solutions of Seide and Stein are also useful for other numbers of equally spaced stiffeners. In all of these solutions the stiffeners are assumed to have zero torsional rigidity.

A conservative method of analysis proposed by Sharp (1966) divides the analysis of the stiffened plate into two parts: one applying to short panels in which the buckled configuration takes the form of a single half-wave in both the longitudinal and transverse directions and another applying to long panels in which several longitudinal waves may occur along with a single half-wave in the transverse direction. In very short panels, the stiffener and a width of plate equal to the stiffener spacing, d, are analyzed as a column of length, a, with a slenderness ratio,

$$\left(\frac{L}{r}\right)_{eq} = \frac{a}{r_e} \tag{4.38}$$

where r_e is the radius of gyration of the section consisting of a stiffener plus a width of plate equal to d.

In long panels the critical stress is larger than that calculated by the use of Eq. 4.38. In this case an equivalent slenderness ratio is defined for use in column strength formulas,

$$\frac{L}{r_{eq}} = \sqrt{6(1-\nu^2)}\,\frac{b}{t}\,\sqrt{\frac{1+(A_s/bt)}{1+\sqrt{(EI_e/bD)+1}}} \tag{4.39}$$

where

$b = Nd$ = overall width of longitudinally stiffened panel

N = number of panels into which the longitudinal stiffeners divide the plate

I_e = moment of inertia of section consisting of the stiffener plus a width of plate equal to d

A_s = cross-sectional area of stiffener

$$D = \frac{Et^3}{12(1-\nu^2)}$$

The smaller of the values from Eqs. 4.38 and 4.39 is used in the analysis, and it is assumed that the plate is fully effective over the panel width d. For greater values of d, buckling of the stiffeners and of the plate between the stiffeners would need consideration.

A flat aluminum sheet with multiple longitudinal stiffeners, or a formed stiffened sheet, subjected to a uniform longitudinal compression (Sherbourne et al., 1971) will buckle into waves of length ψb, in which b is the plate width and $\psi = 1.8(I_x/t^3)^{1/4}$, where I_x is the moment of inertia in the strong direction and t is the plate thickness.

For a formed aluminum stiffened sheet this becomes $\psi = 1.8(\rho r_x/t)^{1/2}$, in which ρ is the ratio of the developed sheet width to the net width, and r_x is the radius of gyration in the strong direction.

If the spacing, a, of transverse supports is less than ψb, the elastic critical stress is given approximately by

$$\sigma_c = \pi^2 E\left[1 + \left(\frac{a}{\psi b}\right)^4\right] \Big/ \left(\frac{a}{r_x}\right)^2 \tag{4.40a}$$

If the spacing exceeds ψb, then

$$\sigma_c = 2\pi^2 E \Big/ \left(\frac{\psi b}{r_x}\right)^2 \tag{4.40b}$$

The Canadian standard (CSA, 1983), using this method, reduces the design procedure to determining the equivalent slenderness ratios, which are, respectively, for the two cases,

$$\lambda = \frac{a}{r_x} \Big/ \left[1 + \left(\frac{a}{\psi b}\right)^4\right]^{1/2} \tag{4.41a}$$

$$\lambda = 0.7\psi b / r_x \tag{4.41b}$$

Transverse stiffeners. The required size of transverse stiffeners for plates loaded in uniaxial compression has been defined by Timoshenko and Gere (1961) for one, two, or three equally spaced stiffeners, and by Klitchieff (1949) for any number of stiffeners. The stiffeners as sized provide a nodal line for the buckled plate and thus prohibit overall buckling of the stiffened panel. The strength of the stiffened panel would be limited to the buckling strength of the plate between stiffeners. These authors also give formulas for calculating the buckling strength for smaller stiffeners. The required minimum value of γ given by Klitchieff is

$$\gamma = \frac{(4m^2 - 1)[(m^2 - 1)^2 - 2(m^2 + 1)\beta^2 + \beta^4]}{2m[5m^2 + 1 - \beta^2]\alpha^3} \tag{4.42}$$

where

$$\beta = \frac{\alpha^2}{m} \qquad \alpha = \frac{a}{b} \qquad \gamma = \frac{EI_s}{bD}$$

and m = number of panels
$m - 1$ = number of stiffeners
EI_s = flexural rigidity of one transverse stiffener

An approximate analysis that errs on the conservative side but gives estimates of the required stiffness of transverse stiffeners for plates, either with or without longitudinal stiffeners, may be developed from a consideration of the buckling of columns with elastic supports.

Timoshenko and Gere (1961) show that the required spring constant, K, of the elastic supports for a column, for the supports to behave as if absolutely rigid, is given by

$$K = \frac{mP}{Ca} \tag{4.43}$$

where

$$P = \frac{m^2\pi^2(EI)_c}{a^2}$$

and

C = constant depending on m, which decreases from 0.5 for $m = 2$ to 0.25 for infinitely large m
$(EI)_c$ = flexural stiffness of column
m = number of spans
a = total length of column

In the case of a transversely stiffened plate (Fig. 4.19*b*), a longitudinal strip is assumed to act as a column which is elastically restrained by the transverse stiffeners. Assuming also that the loading from the strip to the stiffener is proportional to the deflection of the stiffener, the spring constant for each column support can be estimated. For a deflection shape of a half-sine wave the spring constant is

$$K = \frac{\pi^4(EI)_s}{b^4} \tag{4.44}$$

Equating Eqs. 4.43 and 4.44 and inserting the value given for P results in the following:

$$\frac{(EI)_s}{b(EI)_c} = \frac{m^3}{\pi^2 C(a/b)^3} \tag{4.45}$$

In the case of panels without longitudinal stiffeners $(EI)_c = D$ and the left side of Eq. 4.45 is γ.

Values of C are tabulated by Timoshenko and Gere (1961) for $m \leq 11$. As shown by Fig. 4.20, these values are given approximately by

$$C = 0.25 + \frac{2}{m^3} \tag{4.46}$$

In Table 4.1 values calculated using Eq. 4.45 are compared to corresponding values tabulated by Timoshenko and Gere (1961) for one, two, and three stiffeners and to those calculated by Eq. 4.42 for 10 stiffeners. Equation 4.45 is always conservative and is highly accurate for cases in which several stiffeners subdivide the panel.

Longitudinal and transverse stiffeners. For a combination of longitudinal and transverse stiffeners, Gerard and Becker (1957/1958) give figures showing the minimum value of γ as a function of α for various combinations of equally stiff longitudinal and transverse stiffeners. The same procedure used to establish Eq. 4.45 can also be applied to this

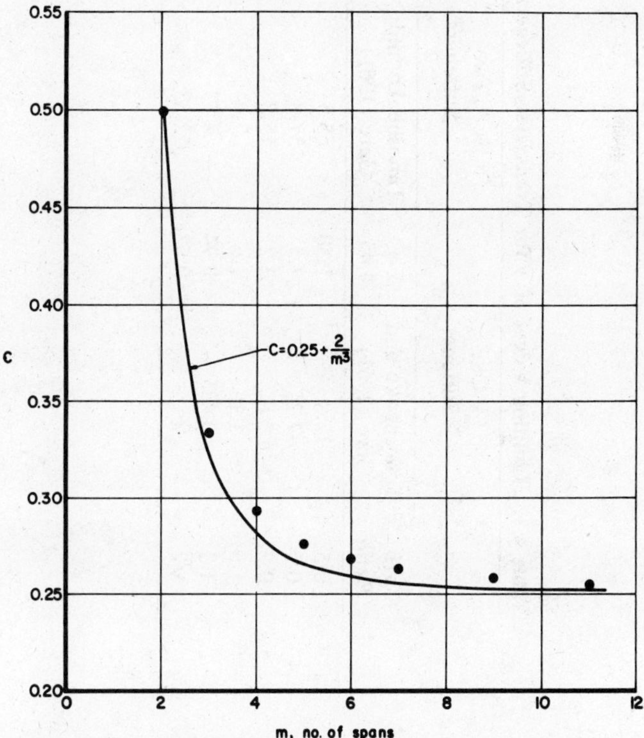

Fig. 4.20 Buckling of columns with elastic supports.

Table 4.1. Limiting Values of γ for Transverse Stiffeners

	One Stiffener		Two Stiffeners		Three Stiffeners		Ten Stiffeners	
a/b Ratio	Timoshenko and Gere (1961)	Eq. 4.45	Timoshenko and Gere (1961)	Eq. 4.45	Timoshenko and Gere (1961)	Eq. 4.45	Eq. 4.42	Eq. 4.45
0.5	12.8	1.30	65.5	65.5	177.0	177.0	4170	4220
0.6	7.25	7.5	37.8	38.0	102.0	122.0	2420	2440
0.8	2.82	3.2	15.8	16.0	43.1	43.2	1020	1030
1.0	1.19	1.6	7.94	8.20	21.9	22.1	522	528
1.2	0.435	0.94	4.43	4.73	12.6	12.8	301	304
$\sqrt{2}$	0	0.57	2.53	2.90	7.44	7.85	185	187

case. The difference in the development is that the spring constant, K, of the support is dependent on the number of longitudinal stiffeners as illustrated below. The flexural stiffness $(EI)_c$ in these cases is equal to the average stiffness per unit width of the plate-longitudinal stiffener combination.

Number and Spacing of Longitudinal Stiffeners	Spring Constant, K	$(EI)_s/b(EI)_c = \gamma$
One centrally located	$\dfrac{48(EI)_s}{b^3}$	$\dfrac{0.206m^3}{C(a/b)^3}$
Two equally spaced	$\dfrac{162}{5}\dfrac{(EI)_s}{b^3}$	$\dfrac{0.152m^3}{C(a/b)^3}$
Four equally spaced	$\dfrac{18.6(EI)_s}{b^3}$	$\dfrac{0.133m^3}{C(a/b)^3}$
Infinite number equally spaced	$\dfrac{\pi^4(EI)_s}{b^3}$	$\dfrac{m^3}{\pi^2C(a/b)^3} = \dfrac{0.1013m^3}{C(a/b)^3}$

The required size of transverse stiffeners in a panel also containing longitudinal stiffeners is thus approximately

$$\frac{(EI)_s}{b(EI)_c} = \frac{m^3}{\pi^2 C(a/b)^3}\left(1 + \frac{1}{N-1}\right) \tag{4.47}$$

This formula yields essentially the same values as does the formula presented for aluminum panels in the Alcoa Handbook (1966). With this size of transverse stiffener the strength of the panel is limited to the buckling strength of the longitudinally stiffened panel between transverse stiffeners.

Stiffener type. The methods of analysis described above are directly applicable to open-section stiffeners having negligible torsional stiffness and are conservative when applied to stiffeners with appreciable torsional stiffness. The influence of torsional stiffness on overall panel buckling has been studied by Kusuda (1959) for the case of one longitudinal or one transverse stiffener. Stiffeners with large torsional rigidity also provide partial or complete fixity of the edges of subpanel plating, thereby increasing their critical stresses.

It has been shown by Lind (1973) and Fukumoto et al. (1977) that the stiffener type affects the buckling mode as well as the ultimate carrying capacity of the stiffened plate. It has also been shown by Tvergaard (1973), and Fok et al. (1977) that local imperfections of stiffeners influence significantly the overall buckling behavior of stiffened plate panels.

Postbuckling Strength. Buckling of a stiffened-plate panel may occur by primary instability with a half wavelength which is on the order of the panel length, or by local instability with a half-wavelength which is on the order of the width of plate elements of the plate and stiffeners. As plate panel length increases, the ultimate stress decreases, until at large slenderness ratio the panel fails in flexure as a long column. The ultimate strength in the long-column range can be predicted by the Euler column formula, where the radius of gyration is computed for the combined section of the stiffener and the effective width of the plating. At an intermediate slenderness ratio there is a transition in the mode of failure from the purely local mode to one dominated by overall panel failure. In the transition zone the panel fails through a combination of the primary buckling and flexural modes and may involve stiffener twisting.

It is well known that plates supported at their edges are often able to sustain compressive load far in excess of their buckling loads. The margin between the buckling load and the ultimate load in plates, known as the postbuckling strength, depends on whether the critical stress is reached below or above the proportional limit of the material. If the buckling stress is well below the proportional limit, the ultimate load may be many times greater than the buckling strength, depending on the aspect ratio, a/b, of the plate element. As the buckling stress approaches the yield strength of the material, the reserve strength in the postbuckling range approaches zero.

Initial buckling modes of stiffened flat panels vary with the slenderness ratio of the panel and with the type of construction—monolithic or built-up. In the following, as a basis for evaluating postbuckling strength, a brief discussion is given regarding predictions of initial buckling stresses and ultimate strength.

Local instability mode. The initial buckled form has one transverse sinusoidal half-wave, with perhaps a number of longitudinal half-waves. As the compressive load is increased, the central portion of the transverse half-wave becomes flattened, and, as shown in Fig. 4.21 the transverse deformation is no longer a simple sinusoidal curve. The well-known effective-width approach to the analysis of the postbuckling strength of a flat plate is based on the stress distribution associated with this buckled form, as has been discussed in detail in Section 4.2.

Another change of buckled form is possible when a rectangular flat plate, simply supported on all sides, is subjected to uniform edge compression and is free to expand laterally. A dynamic snap from one buckled form to another may occur by a sudden change of wavelength of

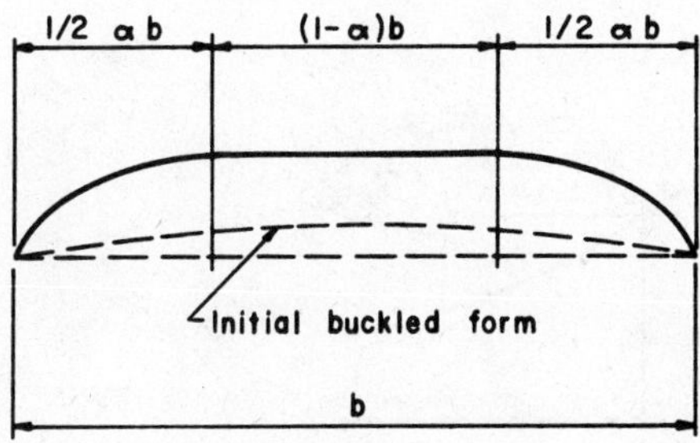

Fig. 4.21 Initial and final buckled shapes.

buckles along the direction of compression. The behavior of the flat plate in this sense is analogous to the elastic postbuckling of a column supported laterally by a nonlinear elastic medium (Koiter, 1963; Tsien, 1947; Stein, 1959). A column supported laterally by a finite number of nonlinear elastic restraints buckles initially into m sinusoidal half-waves over its length; then subsequently the buckled form may become unstable, and the column may snap into n half-waves, with n being greater than m. The exact analysis of transition between the two modes of buckling is not known at present. Sherbourne et al. (1971) determined the terminal wavelength for flat plates at the ultimate capacity.

Failure strength of very short stiffened panels. For a short stiffened plate with the slenderness ratio smaller than 20, Gerard and Becker (1957/1958) note that the failure stress is independent of the panel length. The average stress at failure in this slenderness range is known as the crippling, crushing, or local-failure stress, and will be represented by $\bar{\sigma}_f$. Gerard and Becker (1957/1958) present a method for determining the crippling stress of short longitudinally stiffened panels. However, for most cases, the yield stress of the material can be considered the failure stress for short stiffened panels.

Ultimate strength of intermediate and long stiffened panels. Gerard and Becker (1957/1958) describe a method to predict the buckling failure of stiffened panels based on a curve analogous to the Johnson parabola shown in Fig. 4.22. At stresses lower than the local buckling stress σ_c and the proportional limit σ_{pl}, the Euler column equation is used. In the transition range between $L/r = 20$ and the long column, a parabola of the following form is used:

$$\bar{\sigma}_c = \bar{\sigma}_f\left[1 - \frac{\sigma_c}{\sigma_e}\left(1 - \frac{\sigma_c}{\bar{\sigma}_f}\right)\left(\frac{\sigma_{20}^{1/2} - \sigma_e^{1/2}}{\sigma_{20}^{1/2} - \sigma_c^{1/2}}\right)^2\right] \tag{4.48}$$

where

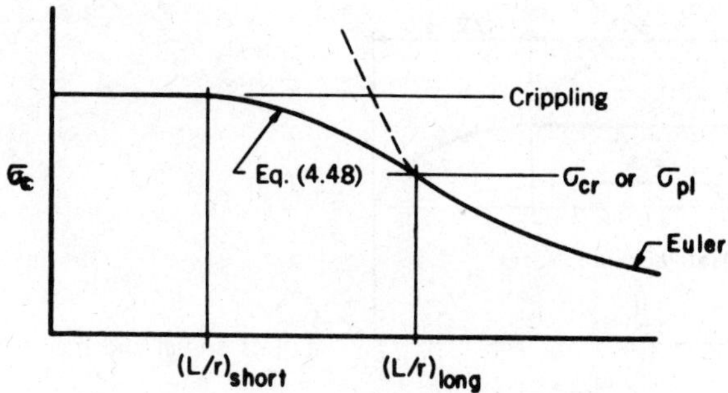

Fig. 4.22 Column curve for stiffened panels.

$\bar{\sigma}_c$ = failure stress
$\sigma_e = \pi^2 E/(L/r)^2$ = Euler stress for the panel
$\bar{\sigma}_f$ = failure stress for a short stiffened panel
σ_{20} = Euler stress evaluated at $L/r = 20$

Many direct reading column charts have been prepared for panel ultimate strength. The type of plot is shown in Fig. 4.23. Gerard and Becker (1957/1958) have provided references and examples of this work.

In the determination of panel strength it is necessary to estimate the effective column length of the plate-stiffener combinations as well as the effective width of plating that acts in conjunction with the stiffener. When

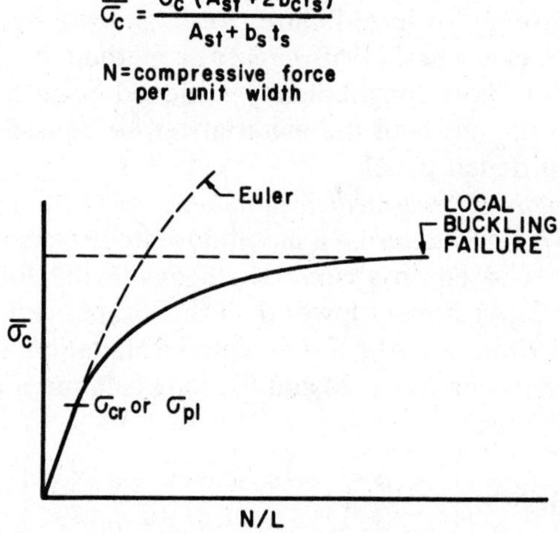

Fig. 4.23 Column chart by Gerard and Becker (1957/1958).

the critical stress for the individual panel of plating between stiffeners is greater than the critical stress for the stiffened panel, the plating may be assumed completely effective, and the effective column length of the panel is determined by the end conditions. When the critical stress for the individual plating panel is significantly less than that for the stiffened panel as a unit, the ultimate strength of the stiffened plate is considered to be the lesser of (1) the load that causes the stress at the juncture of plate and stiffener to reach the yield strength of the material, or (2) the column strength of the stiffener in conjunction with an effective width of plating that is less than the actual width of plating between stiffeners.

The application of the first criterion for ultimate strength assumes that the stiffener is stiff enough to allow the plate-edge stress to reach the yield stress before the stiffener buckles as a column and that the stress distribution across the buckled plate can be determined, while the second criterion assumes that the effective column length of the stiffener and the effective width of the associated plating can be properly determined and that residual stresses and distortions due to fabrication are properly accounted for with regard to column buckling. The effective-width concept is treated in Section 4.2.

To provide a better understanding of the behavior of stiffened plates, a theoretical study was carried out by Wittrick (1968). The collapse of box-girder bridges in the early 1970s precipitated a great interest in research on various aspects of box girders, especially the interactive buckling of an assembly of plates and the ultimate strength of a stiffened plate. The research efforts in many ways were centered around the Merrison Committee (1973). Murray (1975) reported on analysis and design procedure for the collapse load for the stiffened plates. Crisfield (1975) presented a finite-element formulation for the large-deflection elastic-plastic full-range analysis of stiffened plating. A simple approach for the design of stiffened steel compression flanges was proposed by Dwight and Little (1976).

A theoretical and experimental study on the inelastic buckling strength of stiffened plates was reported by Fukumoto et al. (1977). Residual stresses were considered and the stiffened plates treated had relatively low width-to-thickness ratios and with relatively rigid stiffeners. It was found that partial yielding in the flat stiffeners considerably reduced the buckling strength of a stiffened plate. In the case of plates stiffened by tee-type stiffeners, the strength reduction due to partial yielding in the stiffeners was much less pronounced. This is probably due to the fact that the tee stiffeners were more rigid and less susceptible to initial imperfections than were the flat stiffeners. Murray (1973) has indicated that based on experimenal results there is often little margin of strength above the load at which yielding first occurs in a stiffener. This phenomenon can lead to a triggering effect which results in a sudden failure of the stiffened plate. Such a viewpoint is also shared by Horne and Narayanan (1977).

Horne and Narayanan (1975, 1977) have proposed a design method for the prediction of collapse loads of stiffened plates subjected to axial compression. They have compared the results obtained from their method, methods proposed by other researchers, and experiments. The results are all close to the observed strength. Horne and Narayanan's method is based on the British Perry-Robertson column formula and consists of analyzing the stiffened plates as a series of isolated columns comprising a stiffener and the associated effective width of plating. The criterion for plate failure is the attainment of yielding stress at the plate stiffener boundary. The criterion for the stiffener initiated failure is by yielding or by instability of the stiffener.

Little (1976) and Elsharkawi and Walker (1980) have studied the effects of continuity of longitudinal stiffeners on the failure mode and strength of stiffened plates consisting of several bays between cross frames. Little has indicated that there is a tendency for longitudinal continuity to be strengthened where failure occurs in the plate, and weakened where failure occurs in the stiffener. A simplified design method to account for the effects of continuity has been proposed by Elsharkawi and Walker (1980).

Desmond et al. (1981a,b) have presented an experimental study of edge-stiffened compression flanges and intermediately stiffened compression flanges. The compression elements are either adequately stiffened, partially stiffened, or unstiffened. A stiffener requirement that provides the minimum stiffener stiffness to support these compression flanges adequately is presented. The effective-width approach derived from the experiments appears to give satisfactory predictions for the ultimate strength of the members tested. Nguyen and Yu (1982) have reported a study on longitudinally reinforced cold-formed steel beam webs. Based on the experimental results, an effective-width procedure is proposed to predict the ultimate strength of such members subjected to bending. Additional information on this subject matter is presented in Chapters 6 and 7.

4.4.2 Compression and Shear

Buckling Strength

Initial buckling. Analytical and experimental results on the buckling behavior of stiffened plates under combined compression and shear are relatively scarce. Recourse is therefore usually made to data for unstiffened plates supplemented with whatever data are available for the type of longitudinally stiffened construction considered to be most important in practice, that is, that shown in Fig. 4.24. The case of unstiffened rectangular plates under combined compressive and shear stresses has been presented in Section 4.2.4, which showed that a simple parabolic relationship of the form of Eq. 4.49 was satisfactory for engineering purposes for all ranges of elastic restraint from free rotation to complete fixity. The relationship takes the form

$$R_c + R_s^2 = 1 \tag{4.49}$$

where

R_c = ratio of compressive stress when buckling occurs in combined shear and direct stress to compressive stress when buckling occurs in pure compression

R_s = ratio of shear stress when buckling occurs in combined shear and direct stress to shear stress when buckling occurs in pure shear

Johnson (1957) treated the problem of long plates with one and two stiffeners acted on by axial compression and shearing stresses. The stiffeners were assumed to have bending stiffness only and the resulting interaction curves show discontinuities which reflect the mode into which the plate buckles. When the bending stiffness ratio of the stiffener to that of the plate (EI/bD) is low, the buckle goes through the stiffener. However, when the stiffness ratio is increased so that nodal lines occur along the stiffener lengths, buckling takes place without deflection along the stiffeners. In this case another interaction relation exists between

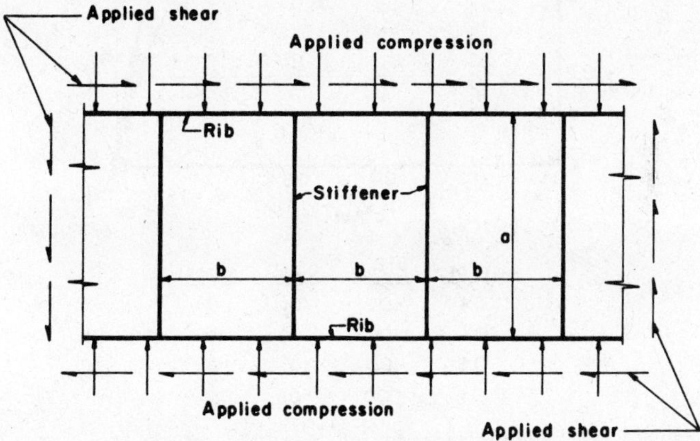

Fig. 4.24 Longitudinally stiffened plate under combined compression and shear stresses.

compressive and shearing stresses. Another somewhat more limited study of the interaction relationship of infinitely wide stiffened plates under compression and shearing stresses was reported by Harris and Pifko (1969). In this study the finite-element method was used. The stiffeners were assumed to have both bending and torsional stiffness, and the grid refinement used was judged to be adequate to ensure accurate results. The results shown in Fig. 4.25 were compared to the parabolic expression

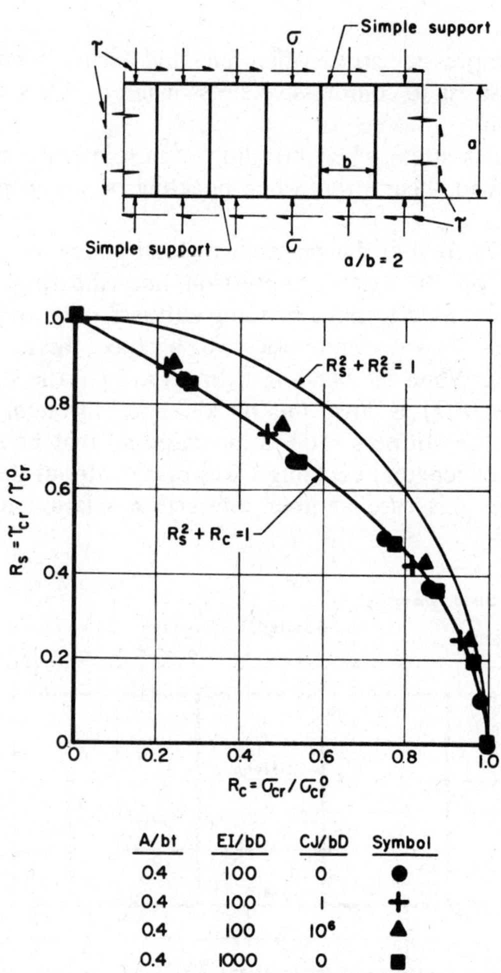

A/bt	EI/bD	CJ/bD	Symbol
0.4	100	0	●
0.4	100	1	+
0.4	100	10^6	▲
0.4	1000	0	■

Fig. 4.25 Analytical interaction relations for infinitely wide stiffened plate under combined compression and shearing stresses.

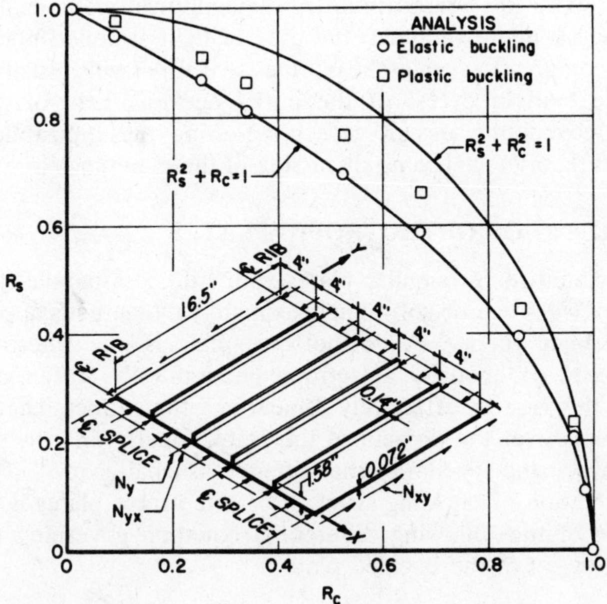

Fig. 4.26 Analytically predicted elastic and inelastic interaction curves for an integrally stiffened panel.

(Eq. 4.49). Except for the case of assumed large torsional stiffness ratio ($GJ/bD = 10^6$) the analytical points follow the parabolic relationship very well.

Buckling in the inelastic range. Analytic prediction of the interaction relationship for stiffened panels which buckle in the inelastic range under combined compression and shear are practically nonexistent in the literature. One such computation reported by Harris and Pifko (1969) was made for an integrally stiffened panel made of aluminum 2024-T351 and having the dimensions shown in Fig. 4.26. The predicted interaction curves are shown in this figure for both the elastic case, which agrees very well with the parabolic relationship (Eq. 4.49), and the inelastic-buckling case. Because of the limited nature of the data, no general relationship for the inelastic-buckling case can be derived. However, it should be noted from Fig. 4.26 that the circular relationship lies above the analytical curve of the inelastic-buckling case.

Postbuckling Strength. Postbuckling behavior is complex, and exact treatment of the problem has not yet been achieved. Because of the importance of predicting the strength of stiffened panels loaded beyond initial buckling in the aircraft industry semiempirical methods have been in existence for many years. In the case of light stiffened plates under combined compression and shear, the structure will usually take a considerable load in excess of the initial buckling load of the plate. Chapter 6 covers the design aspects of using these postbuckling or tension-field theories; hence no discussion of these methods is made here.

4.5 BUCKLING OF ORTHOTROPIC PLATES

Problems related to rectangular plates with stiffeners parallel to one or both pairs of sides can be solved approximately by methods applicable to orthotropic plate theory. An orthotropic plate is one whose material properties are orthogonally anisotropic; a uniformly stiffened plate is reduced to this case by effectively "smearing" the stiffness characteristis of its stiffeners over the domain of the plate. Clearly, the theory is best applicable when the spacing of the stiffeners is small.

The calculation of buckling strength of orthotropic plates is based on the solution of the following differential equation governing the small deflection $w(x, y)$ of the buckled plate:

$$D_1 \frac{\partial^4 w}{\partial x^4} + 2D_3 \frac{\partial^4 w}{\partial x^2 \partial y^2} + D_2 \frac{\partial^4 w}{\partial y^4} + N_x \frac{\partial^2 w}{\partial x^2} + N_y \frac{\partial^2 w}{\partial y^2} + 2N_{xy} \frac{\partial^2 w}{\partial x \partial y} = 0 \tag{4.50}$$

where

$$D_1 = \frac{(EI)_x}{1 - \nu_x \nu_y}$$

$$D_2 = \frac{(EI)_y}{1 - \nu_x \nu_y}$$

$$D_3 = \frac{1}{2} (\nu_y D_1 + \nu_x D_2) + 2(GI)_{xy}$$

in which N_x, N_y, and N_{xy} are in-plane forces per unit width (Fig. 4.27), $(EI)_x$ and $(EI)_y$ are flexural stiffnesses, per unit width, of beam strips in the x and y directions, respectively; ν_x and ν_y are flexural Poisson ratios; and $2(GI)_{xy}$ is a measure of torsional stiffness.

Theoretical buckling data for several cases of rectangular plates with supported edges ($w = 0$), under uniform in-plane loadings, N_x, N_y, and N_{xy}, applied singly or in certain combinations, are presented. The following are some of the additional notations that will be employed: a and b are the lengths of plate in x and y directions, respectively (see Fig. 4.27); m(integer) is the number of buckles or half-waves in the x direction

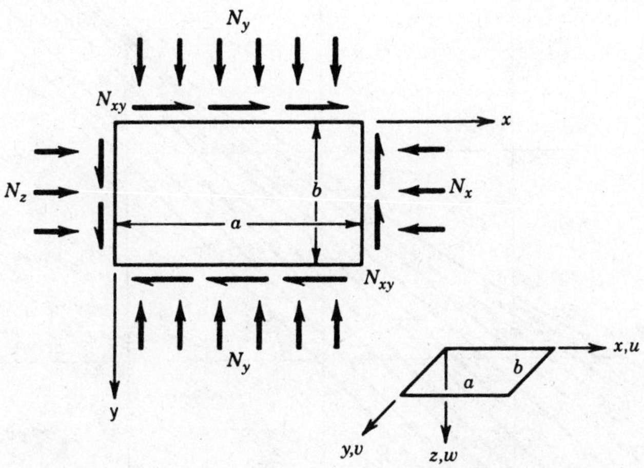

Fig. 4.27 Plate subjected to axial and shear stress.

when the buckle pattern is sinusoidal in that direction; and n(integer) is the number of buckles or half-waves in the y direction when the buckle pattern is sinusoidal in the y direction.

Case I. Uniaxial Compression in x Direction, Loaded Edges Simply Supported, Unloaded Edges Elastically Restrained against Rotation. In this case, referring to Fig. 4.27, N_x is the only loading, the edges $x = 0, a$ are simply supported, and the edges $y = 0, b$ are each elastically restrained against rotation by a restraining medium whose stiffness (moment per unit length per radian of rotation) is K. The quantity Kb/D_2, to be denoted by ϵ, will be used as a dimensionless measure of this stiffness. An exact solution for this case leads to the following formula defining the value of N_x that can sustain a buckle pattern containing m sinusoidal half-waves in the x direction:

$$\frac{N_x b^2}{\pi^2 D_3} = \left(\frac{b}{a/m}\right)^2 \frac{D_1}{D_3} + 2 + f_1\left(\epsilon, \left(\frac{a/m}{b}\right)^2 \frac{D_2}{D_3}\right) \qquad (4.51)$$

where f_1 is the function plotted in Fig. 4.28. If $(a/mb)^2(D_2/D_3) > 0.4$, Eq. 4.51 can be very closely approximated by the formula

$$\frac{N_x b^2}{\pi^2 D_3} = \left(\frac{b}{a/m}\right)^2 \frac{D_1}{D_3} + 2 + f_2(\epsilon) + \left(\frac{a/m}{b}\right)^2 \frac{D_2}{D_3} f_3(\epsilon) \qquad (4.52)$$

where f_2 and f_3 are the functions plotted in Fig. 4.29. The buckling load is the smallest N_x obtained by substituting different integer values of m ($m = 1, 2, 3, \ldots$) into Eq. 4.51 or, if it applies, Eq. 4.52. In performing

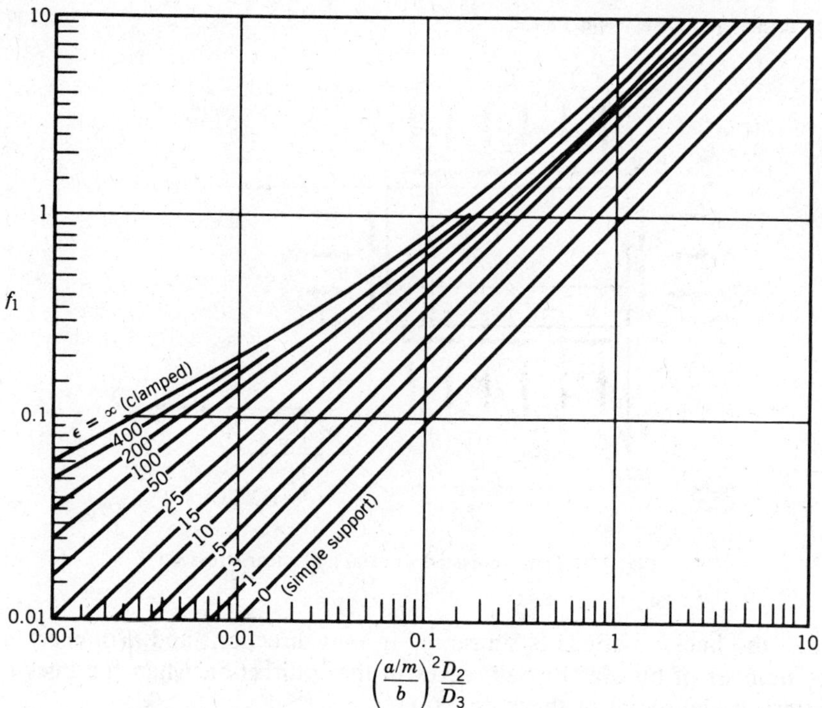

Fig. 4.28 Function f_1 in Eq. 4.51.

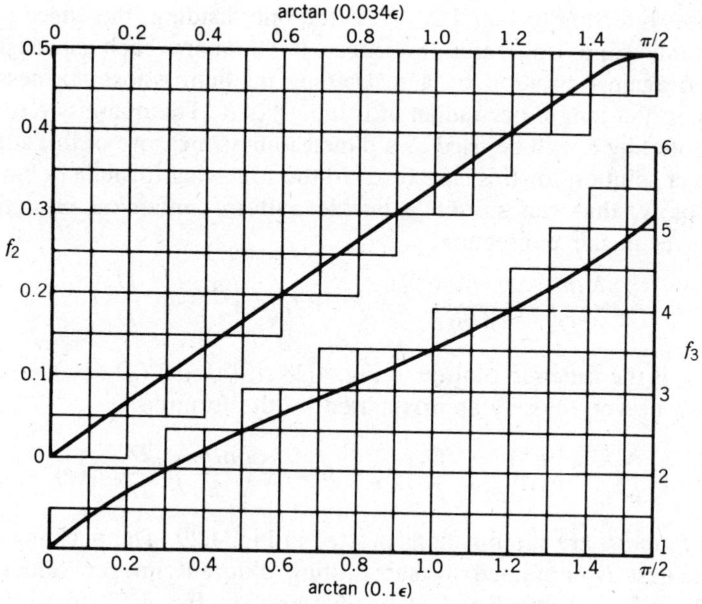

Fig. 4.29 Functions f_2 and f_3 in Eq. 4.52.

this minimization, one should take into account the fact that for most practical restraining media the stiffness K is not fixed but is a function of the half-wavelength a/m of the edge rotation. K may also depend on the axial load in the restraining medium, therefore on N_x, necessitating a trial-and-error calculation to determine N_x for any selected m. Negative values of K are physically possible but are excluded from consideration in Figs. 4.28 and 4.29. Unequal restraints along the edges $y = 0$ and b can be handled approximately by first assuming the $y = 0$ constraint to be present at both edges, then the $y = b$ constraint, and averaging the two values of N_x thus obtained.

Case II. Biaxial Compression, All Edges Simply Supported. When all four edges are simply supported, combinations of N_x and N_y that can sustain a buckle pattern of m sinusoidal half-waves in the x direction and n sinusoidal half-waves in the y direction are defined exactly by the interaction equation

$$\frac{1}{n^2}\frac{N_x b^2}{\pi^2 D_3} + \frac{1}{m^2}\frac{N_y a^2}{\pi^2 D_3} = \left(\frac{b/n}{a/m}\right)^2 \frac{D_1}{D_3} + 2 + \left(\frac{a/m}{b/n}\right)^2 \frac{D_2}{D_3} \tag{4.53}$$

In using Eq. 4.53, one must substitute different combinations of m and n until that combination is found which minimizes N_x for a given N_y, N_y for a given N_x, or N_x and N_y simultaneously for a given ratio between them. It can be shown (Libove, 1983) that these minima will occur with at least one of the two integers equal to unity. Therefore, only the combinations $m = 1$; $n = 1, 2, 3, \ldots$ and $n = 1$; $m = 2, 3, \ldots$ need to be tried. The condition

$$\frac{N_y b^2}{\pi^2 D_2} < 1 \tag{4.54}$$

is sufficient to ensure that $n = 1$ will govern. In that case, with N_y regarded as given, Wittrick (1952) has shown that the N_x required for buckling is defined by

$$\frac{N_x b^2}{\pi^2 D_3} = 2 + (k-2)\left[\frac{D_1 D_2}{D_3^2}\left(1 - \frac{N_y b^2}{\pi^2 D_2}\right)\right]^{1/2} \tag{4.55}$$

where k is the function plotted as curve (a) in Fig. 4.30. The case of the infinitely long plate ($a/b = \infty$) is of interest as the limiting case of a very long plate. For that case it can be deduced from Eq. 4.53 that if $N_x b^2/\pi^2 D_3 \leq 2$, the longitudinal compression N_x is too small to sustain a multilobed buckle pattern, and the plate will buckle in a cylindrical mode (i.e., as a wide plate-column of length b) when $N_y b^2/\pi^2 D_2 = 1$. On the other hand, if $N_x b^2/\pi^2 D_3 > 2$, the buckle pattern will be sinusoidally lobed in the x direction with a half-wave length-to-width ratio of

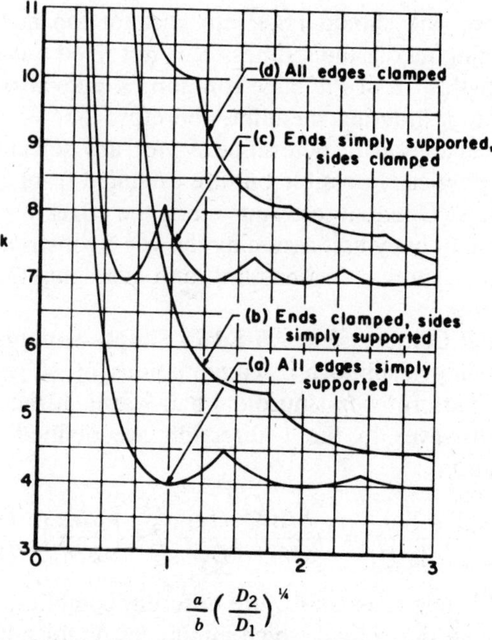

Fig. 4.30 Buckling coefficients for orthotropic plates.

$$\frac{a/m}{b} = \left[\frac{D_2}{D_1}\left(1 - \frac{N_y b^2}{\pi^2 D_2}\right)\right]^{-1/4} \tag{4.56}$$

and the buckling will occur when N_x and N_y satisfy Eq. 4.55 with k set equal to 4. The interaction curve of Fig. 4.31 summarizes the results just given for the case $a/b = \infty$.

Case III. Compression in *x* Direction Applied to Clamped Edges, Limited Compression in *y* Direction Applied to Simply Supported Edges. Referring to Fig. 4.27, and considering the case in which the edges $x = 0, a$ are clamped, the edges $y = 0, b$ are simply supported, N_{xy} is absent, and N_y satisfies the inequality (Eq. 4.54). Then, with N_y regarded as given, the value of N_x required to cause buckling is defined by Eq. 4.55 with k taken from curve (b) of Fig. 4.30 (Wittrick, 1952).

Case IV. Uniaxial Compression, Loaded Edges Simply Supported, Unloaded Edges Clamped. Here N_x is the only loading, the edges $x = 0$ and a are simply supported, and the other two edges are clamped. The exact solution for this case is contained in the subcase $\epsilon = \infty$ of case I. An approximate solution is given by Wittrick (1952) in the form

$$\frac{N_x b^2}{\pi^2 \sqrt{D_1 D_2}} = k - c\left(1 - \frac{D_3}{\sqrt{D_1 D_2}}\right) \tag{4.57}$$

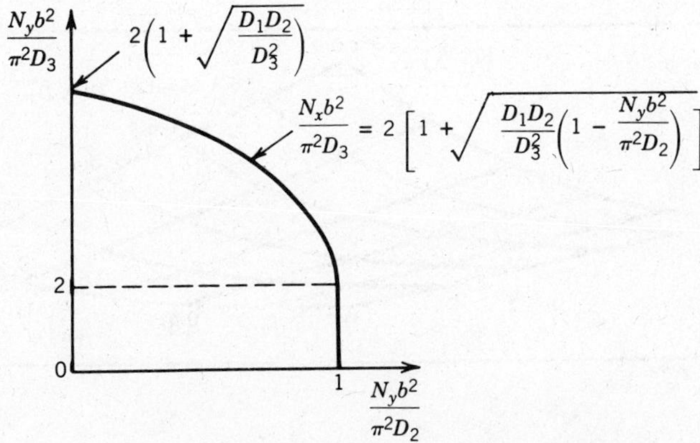

Fig. 4.31 Interaction curve for orthotropic plates.

where $c = 2.4$ and k is taken from curve (c) of Fig. 4.30. Equation 4.57 is virtually equivalent to Eq. 4.52 and must therefore be subject to the same restriction, [i.e., $(a/mb)^2(D_2/D_3) > 0.4$].

Case V. Uniaxial Compression, All Edges Clamped. Wittrick (1952) gives an approximate solution for this case in the form of Eq. 4.57 with $c = 2.46$ and k taken from curve (d) of Fig. 4.30.

Case VI. Shear, Various Boundary Conditions. Theoretical data for the shear flow N_{xy} required to cause buckling of rectangular orthotropic plates have been collected by Johns (1971). Three of his graphs are reproduced in Fig. 4.32. They apply, respectively, to the boundary conditions of (a) all edges simply supported; (b) edges $y = 0$ and $y = b$ simply supported, the other two edges clamped; and (c) all edges clamped. In Fig. 4.32, k_s stands for $N_{xy}b^2/\pi^2 D_1^{1/4} D_2^{3/4}$.

To make use of the buckling data presented above, one must of course, know the values of the elastic constants appearing in Eq. 4.50. These constants are best determined experimentally, by tests such as those described by Libove and Batdorf (1948) and Becker and Gerard (1963). If the plates are of a simple enough construction, the constants can also be evaluated theoretically. For example, for a sheet of thickness t and shear modulus G orthogonally stiffened by x-wise stiffeners of torsional stiffness C_1 spaced a distance b_1 apart, and y-wise stiffeners of torsional stiffness C_2 spaced a distance a_1 apart, $(EI)_x$ may be taken as the flexural stiffness of the composite beam consisting of one x-wise stiffener and its associated width b_1 of sheet, divided by b_1, while $(EI)_y$ is computed in an analogous manner using a y-wise stiffener and its associated width a_1 of sheet. For such plates, ν_x and ν_y may usually be taken as zero with little error, and then

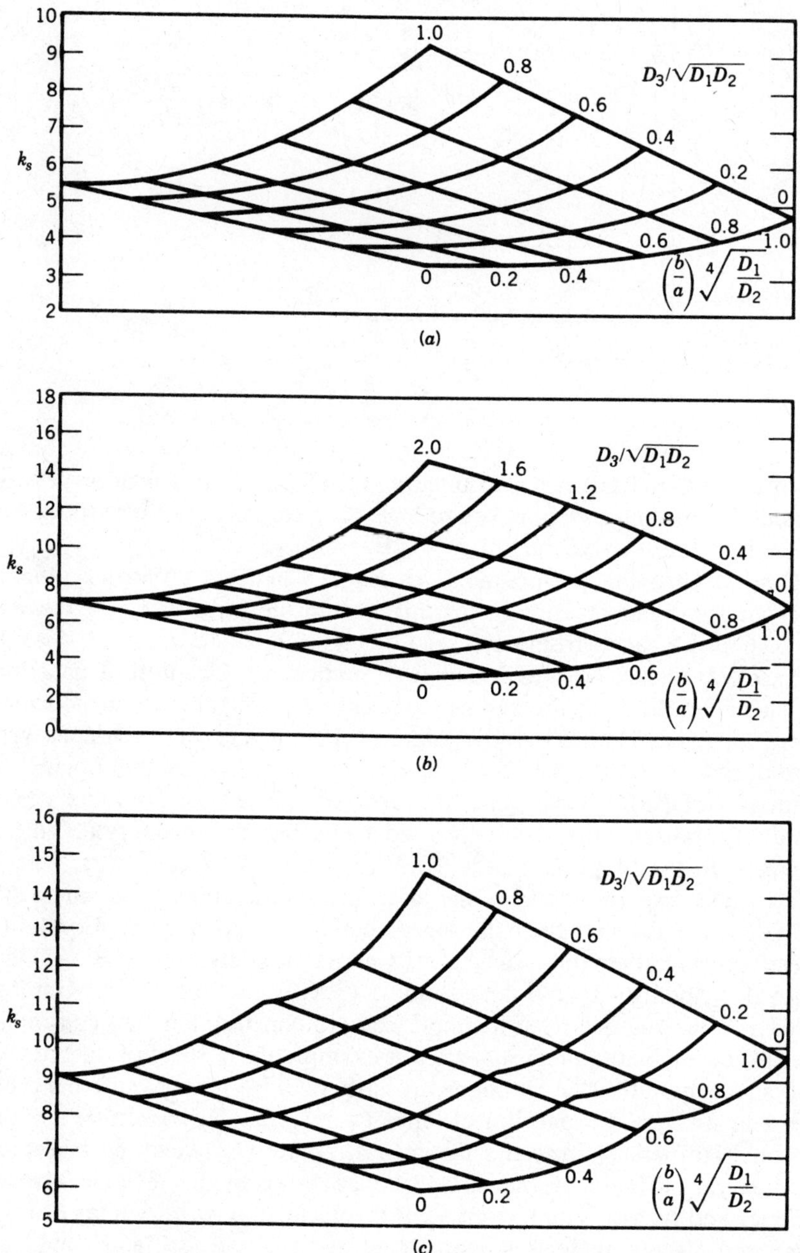

Fig. 4.32 Shear buckling coefficients for orthotropic plates. [Adapted from Johns (1971), printed with the permission of Her Majesty's Stationery Office.]

$$D_3 = 2(GI)_{xy} = \frac{Gt^3}{6} + \frac{1}{2}\left(\frac{C_1}{b_1} + \frac{C_2}{a_1}\right) \tag{4.58}$$

(The factor $\frac{1}{2}$ in this formula is sometimes erroneously omitted.) Plates with integral waffle-like stiffening may also be modified as orthotropic plates, provided that the ribbings are so oriented as to create axes of elastic symmetry parallel to the plate edges; formulas for estimating the elastic constants of such plates are derived by Dow et al. (1954). Corrugated-core sandwich plates with the corrugations parallel to the x or y axis can similarly be treated as orthotropic plates, and formulas for their elastic constants are developed by Libove and Hubka (1951); however, for such sandwich plates, deflections due to transverse shear, neglected in the present discussion, may sometimes be important. It is common practice to treat a corrugated plate along also as an orthotropic plate, and the appropriate elastic constants when the profile of the plate is sinusoidal are discussed by Lau (1981). However, there are indications (Perel and Libove, 1978) that modeling the corrugated plate as an orthoropic plate may lead to an underestimate of its shear buckling strength.

The orthotropic plate model has an additional shortcoming when applied to stiffened plates, namely its neglect of any coupling between in-plane forces and out-of-plane deflections. That is, underlying Eq. 4.50 is the tacit assumption that there exists a reference plane in which the forces N_x, N_y, and N_{xy} can be applied without producing any curvatures or twist. In the case of a sheet with identical stiffening on both sides, there does of course exist such a plane—it is the middle surface of the sheet. If the stiffening is one-sided, however, it is usually not possible to find a reference plane that will eliminate completely the coupling between in-plane forces and out-of-plane deflections. It has been shown (Whitney and Leissa, 1969; Jones, 1975) that such coupling, occurring in the context of composite laminated plates, can have a marked effect on the buckling loads. Therefore, it is very likely that in metal plates with one-sided stiffening it can also have a marked effect on the buckling loads. A thorough investigation of this effect, using an appropriately generalized orthotropic plate theory, would be a worthwhile subject for future research.

Finally, orthotropic plate theory is incapable of modeling local buckling, that is, buckling in which the buckle wavelengths are of the same order as the stiffener spacings or the widths of the plate elements of which the stiffeners are composed. Wittrick and Horsington (1984) have developed more refined approaches that can account for local buckling and modes of buckling in which local and overall deformations appear in conjunction. Their methods are applicable to plates with unidirectional stiffening possssing certain boundary conditions and subjeced to combinations of shear and biaxial compression.

4.6 LATERALLY LOADED PLATES IN COMPRESSION

Plates with stiffeners in the direction of the axial load that are also subjected to a distributed lateral load are commonly encountered as the bottom plates in ships. A study based on a large-deflection postbuckling theory (Supple, 1980) shows that lateral pressure produces effects similar to initial geometric imperfections with a long buckling wavelength. It is shown that a sufficiently high pressure induces stable postbuckling in the long-wave mode.

The stiffened plating is supported on heavy transverse structural members which may be assumed to be rigid. The panels have width-to-length ratios of about 2.5 to 4.0, and there is essentially no interaction in the transverse direction and thus the panel undergoes cylindrical bending. It behaves as if it were a wide beam-column under axial and lateral loads. It is convenient to isolate one longitudinal stiffener together with a plate of the width equal to the spacing of the stiffeners, b, and to consider that all other stiffeners behave in a similar manner. Such a beam-column is shown in Fig. 4.33.

Ultimate-strength analysis, that is, the determination of the maximum values of the lateral and axial loads that can be sustained by the member, requires a consideration of:

1. The yield strength of plate and stiffener materials.
2. Large-deflection theory.
3. Partial plastification of the section.
4. Postbuckling and post-ultimate-strength behavior of the plate.

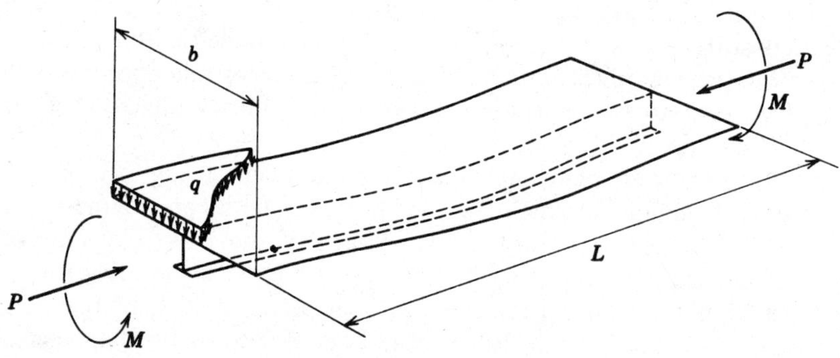

Fig. 4.33 Idealized beam-column.

The complexity of the problem, in particular, consideration of items 2, 3, and 4 suggests that numerical methods be used. Some solutions have been obtained for standard wide-flange shapes by Ketter (1962), and Lu and Kamalvand (1968) in the form of interaction curves between the axial and lateral loads. Unfortunately, these results cannot be used for plating since items 1 and 4 were not considered. Another complication is that the wide-flange sections are doubly symmetrical, whereas the section shown in Fig. 4.33 is singly symmetrical with a very large top flange.

This means, for example, that although it was possible to nondimensionalize the results for wide-flange sections in terms of L/r and make them applicable with a very small error to any wide-flange section, the unsymmetrical section of the stiffened panel must be treated as a special case for every combination of relative proportions.

Nomographs for the ultimate strengths of stiffened plates with a yield point of 47 ksi and the ranges of geometrical and loading parameters commonly encountered in ship structures are given by Vojta and Ostapenko (1967/1968).

References

Abdel-Sayed, G. (1969), "Effective Width of Thin Plates in Compression," *ASCE J. Struct. Div.*, Vol. 95, No. ST10, pp. 2183–2204.

Aluminum Company of America (1960), *Alcoa Structural Handbook* (also *Handbook of Design Stresses for Aluminum*), Alcoa, Pitsburgh, Pa.

Allen, H. G., and Bulson, P. S. (1980), *Background to Buckling*, McGraw-Hill, Maidenhead, Berkshire, England.

Aluminum Association (1982), *Specifications for Aluminum Structures*, AA, Washington, D.C.

American Institute of Steel Construction (1978), *Specification for the Design, Fabrication and Erection of Structural Steel for Buildings*, AISC, Chicago.

American Iron and Steel Institute (1986), *Specification for the Design of Cold-Formed Steel Structural Members*, AISI, Washington, D.C.

Batdorf, S. B., and Houbolt, J. C. (1946), "Critical Combinations of Shear and Transverse Direct Stress for an Infinitely Long Flat Plate with Edges Elastically Restrained Against Rotation," *NACA Rep. No. 847.*

Batdorf, S. B., and Stein, M. (1947), "Critical Combinations of Shear and Direct Stress for Simply Supported Rectangular Flat Plates," *NACA Tech. Note No. 1223.*

Becker, H., and Gerard, G. (1963), "Measurement of Torsional Rigidity of Stiffened Plates," *NASA Tech. Note No. D-2007.*

Benthem, J. P. (1959), "The Reduction in Stiffness of Combinations of Rectangular Plates in Compression after Exceeding the Buckling Load," *NLL-TR S-539.*

Bergman, S., and Reissner, H. (1932), "Über die Knickung von rechteckigen Platten bei Schubbeanspruchung," *Z. Flugtech Motor-luftschiffahrt*, Vol. 23, p. 6.

Bijlaard, P. P. (1949), "Theory and Tests on Plastic Stability of Plates and Shells," *J. Aeronaut. Sci.*, Vol. 16, pp. 529–541.

Bijlaard, P. P. (1950), "In the Plastic Buckling of Plates," *J. Aeronaut. Sci.*, Vol. 17, p. 493.

Bijlaard, P. P. (1957), "Buckling of Plates under Nonhomogeneous Stress," *ASCE J. Eng. Mech. Div.*, Vol. 83, No. EM3, Proc. Paper 1293.

Bijlaard, P. P., and Fisher, G. P. (1952), "Column Strength of H Sections and Square Tubes, in Post Buckling Range of Component Plates," *NACA Tech. Note No. 2994.*

Bleich, F. (1952), *Buckling Strength of Metal Structures*, McGraw-Hill, New York.

Bleich, F., and Ramsey, L. B. (1951), *A Design Manual on the Buckling Strength of Metal Structures*, Society of Naval Architects.

Brockenbrough, R. L., and Johnston, B. G. (1974), *USS Steel Design Manual*, 2nd ed., U.S. Steel Corporation, Pittsburgh, Pa.

Bryan, G. H. (1891), "On the Stability of a Plane Plate under Thrusts in Its Own Plane, with Applications to the 'Buckling' of the Sides of a Ship," *Proc. Lond. Math. Soc.*, Vol. 22.

Bulson, P. S. (1970), *Stability of Flat Plates*, American Elsevier, New York.

Canadian Standards Association (1974, 1984), *Cold Formed Steel Structural Members*, CSA Standard S136, CSA, Rexdale, Ontario.

Canadian Standards Association (1983), *Strength Design in Aluminum*, CAN3-S157-M83, CSA, Rexdale, Ontario.

Cherry, S. (1960), "The Stability of Beams with Buckled Compression Flanges," *Struct. Eng.*, Vol. 38, No. 9.

Chwalla, E. (1936a), Beitrag zur Stabilitätstheorie des Stegbleches vollwandiger Träger," *Stahlbau*, Vol. 9, p. 81.

Chwalla, E. (1936b), "Die Bemessung der waagerecht ausgesteiften Stegbleche vollwandiger Träger," *Prelim. Pub. IABSE, 2nd Cong.*, Berlin-Munich, 957.

Clark, J. W., and Rolf, R. L. (1966), "Buckling of Aluminum Columns, Plates and Beams," *ASCE J. Struct. Div.*, Vol. 92, No. ST3, pp. 17–38.

Column Research Committee of Japan (1971), *Handbook of Structural Stability*, Corona Publishing Co., Tokyo.

Conley, W. F., Becker, L. A., and Allnutt, R. B. (1963), "Buckling and Ultimate Strength of Plating Loaded in Edge Compression: Progress Report 2. Unstiffened Panels," *David Taylor Model Basin Rep. No. 1682.*

Cook, I. T., and Rockey, K. C. (1963), "Shear Buckling of Rectangular Plates with Mixed Boundary Conditions," *Aeronaut. Q.*, Vol. 14.

Crisfield, M. A. (1975), "Full Range Analysis of Steel Plates and Stiffened Plating under Uniaxial Compression," *Proc. Inst. Civ. Eng.*, Part 2, pp. 1249–1255.

Dawson, R. G., and Walker, A. C. (1972), "Post-Buckling of Geometrically Imperfect Plates," *ASCE J. Struct. Div.*, Vol. 98, No. ST1, pp. 75–94.

Desmond, T. P., Peköz, T., and Winter, G. (1981a), "Edge Stiffeners for Thin-Walled Members," *ASCE J. Struct. Div.*, Vol. 107, No. ST2, Proc. Pap. 16056, pp. 329–354.

Desmond, T. P., Peköz, T., and Winter, G. (1981b), "Intermediate Stiffeners for Thin Walled Members," *ASCE J. Struct. Div.*, Vol. 107, No. ST4, Proc. Pap. 16194, pp. 627–648.

DeWolf, J. T., Peköz, T., and Winter, G. (1974), "Local and Overall Buckling of Cold Formed Members," *ASCE J. Struct. Div.*, Vol. 100, No. ST10, pp. 2017–2036.

Dow, N. F., Libove, C., and Hubka, R. E. (1954), "Formulas for the Elastic Constants of Plates with Integral Waffle-Like Stiffening," *NACA Rep. No. 1195.*

Dwight, J. B. (1971), "Collapse of Steel Compression Panels," *Proc. Conf. Dev. Bridge Des. Constr.*, Crosby Lockwood, London.

Dwight, J. B., and Little, G. H. (1976), "Stiffened Steel Compression Flanges, a Simpler Approach," *Struct. Eng.*, Vol. 54, No. 12, pp. 501–509.

Dwight, J. B., and Moxham, K. E. (1969), "Welded Steel Plates in Compression," *Struct. Eng.*, Vol. 47, No. 2.

Dwight, J. B., and Ratcliffe, A. T. (1968), "The Strength of Thin Plates in Compression," in *Thin-Walled Steel Strucures* (ed. K. C. Rockey and H. V. Hill), Crosby Lockwood, London.

Elsharkawi, K., and Walker, A. C. (1980), "Buckling of Multibay Stiffened Plate Panels," *ASCE J. Struct. Div.*, Vol. 106, No. ST8, pp. 1695–1716.

Fok, W. C., Walker, A. C., and Rhodes, J. (1977), "Buckling of Locally Imperfect Stiffeners in Plates," *ASCE J. Eng. Mech. Div.*, Vol. 103, No. EM5, pp. 895–911.

Fujita, Y., Yoshida, K., and Arai, H. (1970), "Instability of Plates with Holes," 3rd Report, *J. Soc. Nav. Archit. Jpn.*, Vol. 127, pp. 161–169.

Fukumoto, Y., and Kubo, M. (1982), "Buckling in Steel U Shaped Beams," *ASCE J. Struct. Div.*, Vol. 108, No. ST5, pp. 1174–1190.

Fukumoto, Y., Usami, T., and Yamaguchi, K. (1977), "Inelastic Buckling Strength of Stiffened Plates in Compression," *IABSE Period.* 3/1977, *IABSE Proc.* p-8/77, pp. 1–15.

Gerard, G., and Becker, H. (1957/1958), "Handbook of Structural Stability," six parts, *NACA Tech. Notes Nos. 3781–3786.*

Graves-Smith, T. R. (1968), "The Postbuckled Behavior of Thin Walled Columns," *8th Cong. IABSE*, New York, pp. 311–320.

Graves-Smith, T. R. (1969), "The Ultimate Strength of Locally Buckled Columns of Arbitrary Length," in *Thin-Walled Steel Structures* (ed. K.C. Rockey and H.V. Hill), Crosby Lockwood, London.

Graves-Smith, T. R., and Sridharan, S. (1978), "A Finite Strip Method for the Buckling of Plate Structures under Arbitrary Loading," *Int. J. Mech. Sci.*, Vol. 20, pp. 685–693.

Graves-Smith, T. R., and Sridharan, S. (1980), "The Local Collapse of Elastic Thin-Walled Columns," *J. Struct. Mech.*, Vol. 13, No. 4, pp. 471–489.

Grosskurth, J. F., White, R. N., and Gallagher, R. H. (1976), "Shear Buckling of Square Perforated Plates," *ASCE J. Eng. Mech. Div.*, Vol. 102, No. EM6, Proc. Pap. 12641, pp. 1025–1040.

Hancock, G. J. (1980), "Design Methods for Thin-Walled Box Columns," *Rep. No. 359*, University of Sydney, Sydney.

Hancock, G.J. (1981), "Interaction Buckling of I-Section Columns," *ASCE J. Struct. Div.*, Vol. 107, No. ST1, pp. 165–181.

Harris, H. G., and Pifko, A. B. (1969), "Elastic-Plastic Buckling of Stiffened Rectangular Plates," *Proc. Symp. Appl. Finite Elem. Methods Div. Eng.*, Vanderbilt University, November.

Horne, M. R., and Narayanan, R. (1975), "An Approximate Method for the Design of Stiffened Steel Compression Panels," *Proc. Inst. Civ. Eng.*, Part 2, Vol. 59, pp. 501–514.

Horne, M. R., and Narayanan, R. (1977), "Design of Axially Loaded Stiffened Plates," *ASCE J. Struct. Div.*, Vol. 103, No. ST11, pp. 2243–2257.

Iguchi, S. (1938), "Die Knickung der rechteckigen Platte durch Schubkräfte," *Ing. Arch.*, Vol. 9, p. 1.

Johns, D. J. (1971), "Shear Buckling of Isotropic and Orthotropic Plates: A Review," *Tech. Rep. R & M No. 3677*, British Aeronautical Research Council.

Johnson, A. E., Jr. (1957), "Charts Relating the Compressive and Shear Buckling Stress of Longitudinally Supported Plates to the Effective Deflectional Stiffness," *NACA Tech. Note No. 4188.*

Johnson, A. L., and Winter, G. (1966), "Behavior of Stainless Steel Columns and Beams," *ASCE J. Struct. Div.*, Vol. 92, No. ST5, p. 97.

Jombock, J. R., and Clark, J. W. (1962), "Postbuckling Behavior of Flat Plates," *Trans. Am. Soc. Civ. Eng.*, Vol. 127, Part I, pp. 227–240.

Jombock, J. R., and Clark, J. W. (1968), "Bending Strength of Aluminum Formed Sheet Members," *ASCE J. Struct. Div.*, Vol. 94, No. ST2, pp. 511–528.

Jones, R. M. (1975), *Mechanics of Composite Matrials*, Scripta, Silver Spring, Md., pp. 264–267.

Kalyanaraman, V., and Pekoz, T. (1978), "Analytical Study of Unstiffened Elements," *ASCE J. Struct. Div.*, Vol. 104, No. ST9, pp. 1507–1524.

Kalyanaraman, V., Pekoz, T., and Winter, G. (1977), "Unstiffened Compression Elements," *ASCE J. Struct. Div.*, Vol. 103, No. ST9, pp. 1833–1848.

Kawai, T., and Ohtsubo, H. (1968), "A Method of Solution for the Complicated Buckling Problems of Elastic Plates with Combined use of Raleigh-Ritz Procedure in the Finite Element Method," *Proc. 2nd Air Force Conf. Matrix Methods Struct. Mech.*, October.

Ketter, R. L. (1962), "Further Studies on the Strength of Beam-Columns," *Trans. Am. Soc. Civ. Eng.*, Vol. 127, No. 2, p. 224.

Klitchieff, J. M. (1949), "On the Stability of Plates Reinforced by Ribs," *J. Appl. Mech.*, Vol. 16, No. 1.

Klöppel, K., Schmied, R., and Schubert, J. (1969), "Ultimate Strength of Thin Walled Box Shaped Columns under Concentric and Eccentric Loads—Post Buckling Behavior of Plate Elements and Large Deformations Are Considered," *Stahlbau*, Vol. 38, pp. 9–19 and 73–83.

Koiter, W. T. (1963), "Introduction to the Post-Buckling Behavior of Flat Plates," *Int. Colloq. Comportement Postcritiques Plaques Utilisée Constr. Metall.*, Institute of Civil Engineering, Liege, Belgium.

Kroll, W. D., Fisher, G. P., and Heimerl, G. J. (1943), "Charts for the Calculation of the Critical Stress for Local Instability of Columns with I, Z, Channel and Rectangular Tube Sections," *NACA Wartime Rep. No. L-429.*

Kumai, T. (1952), "Elastic Stability of a Square Plate with a Central Circular Hole under Edge Thrust," *Rep. Res. Inst. Appl. Mech. Kyusyn Univ.*, Vol. 1, No. 2, p. 1.

Kusuda, T. (1959), "Buckling of Stiffened Panels in Elastic and Strain-Hardening Range," *Report No. 39*, Transportation Technical Research Institute (Unyn-Gijutsu Kenkyrijo Mejiro, Toshima-Ku), Tokyo.

LaBoube, R. A., and Yu, W. W. (1982), "Bending Strength of Webs of Cold-Formed Steel Beams," *ASCE J. Struct. Div.*, Vol. 108, No. ST7, pp. 1589–1604.

Lau, J. H. (1981), "Stiffness of Corrugated Plate," *ASCE J. Eng. Mech. Div.*, Vol. 17, No. EM1, pp. 271–275.

Lay, M. G. (1965), "Flange Local Buckling in Wide-Flange Shapes," *ASCE J. Struct. Div.*, Vol. 91, No. ST6, pp. 49–66.

Legget, D. M. A. (1941), "The Buckling of a Square Panel in Shear When One Pair of Opposite Edges Is Clamped and the Other Pair Simply Supported," *Tech. Rep. R & M No. 1991*, Aeronautical Research Council.

Levy, S., Wolley, R. M., and Knoll, W. D. (1947), "Instability of Simply Supported Square Plate with Reinforced Circular Hole in Edge Compression," *J. Res. Natl. Bur. Stand.*, Vol. 39, pp. 571–577.

Libove, C. (1983), "Buckle Pattern of Biaxially Compressed Simply Supported Orthotropic Rectangular Plates," *J. Compos. Mater.*, Vol. 17, No. 1, pp. 45–48.

Libove, C., and Batdorf, S. B. (1948), "A General Small-Deflection Theory for Flat Sandwich Plates," *NACA Rep. No. 899*.

Libove, C., and Hubka, R. E. (1951), "Elastic Constants for Corrugated-Core Sandwich Plates," *NACA Tech. Note No. 2289*.

Lind, N. C. (1973), "Buckling of Longitudinally Stiffened Sheets," *ASCE J. Struct. Div.*, Vol. 99, No. ST7, pp. 1686–1691.

Lind, N. C., Ravindra, M. K., and Schorn, G. (1976), "Empirical Effective Width Formula," *ASCE J. Struct. Div.*, Vol. 102, No. ST9, pp. 1741–1957.

Little, G. H. (1976), "Stiffened Steel Compression Panels—Theoretical Failure Analysis," *Struct. Eng.*, Vol. 54, No. 12, pp. 489–500.

Little, G. H. (1979), "The Strength of Square Steel Box-Columns—Design Curves and Their Theoretical Basis," *Struct. Eng.*, Vol. 57.

Lu, L. W., and Kamalvand, H. (1968), "Ultimate Strength of Laterally Loaded Columns," *ASCE J. Struct. Div.*, Vol. 94, No. ST6, pp. 1505–1524.

McKenzie, K. I. (1964), "The Buckling of Rectangular Plate under Combined Biaxial Compression, Bending and Shear," *Aeronaut. Q.*, Vol. 15, No. 3, pp. 239–246.

Merrison Committee (1973), "Inquiry into the Basis of Design and Method of Erection of Steel Box Girder Bridges (Interior Design and Workmanship Rules)," London.

Michael, M. E. (1960), *J. R. Aeronaut. Soc.*, Vol. 64, p. 268.

Moheit, W. (1939), "Schubbeulung rechteckiger Platten mit eingespannten Rändern," Thesis, Technische Hochschule Darmstadt, Leipzig.

Murray, J. M. (1946), "Pietzker's Effective Breadth of Flange Reexamined," *Engineering*, Vol. 161, p. 364.

Murray, N. W. (1973), "Buckling of Stiffened Panel Loaded Axially and in Bending," *Struct. Eng.*, Vol. 51, No. 8, pp. 285–301.

Murray, N. W. (1975), "Analysis and Design of Stiffened Plates for Collapse Load," *Struct. Eng.*, Vol. 53, No. 3, pp. 153–158.

Newell, J. S. (1930), "The Strength of Aluminum Alloy Sheets," *Airway Age*, Vol. 11, p. 1420; Vol. 12, p. 1548.

Nguyen, R. P., and Yu, W. W. (1982), "Longitudinally Reinforced Cold-Formed Steel Beam Webs," *ASCE J. Struct. Div.*, Vol. 108, No. ST11, pp. 2423–2442.

Perel, D., and Libove, C. (1978), "Elastic Buckling of Infinitely Long Trapezoidally Corrugated Plates in Shear," *Trans. ASME J. Appl. Mech.*, Vol. 45, pp. 579–582.

Peters, R. W. (1954), "Buckling of Long Square Tubes in Combined Compression and Torsion and Comparison with Flat-Plate Buckling Theories," *NACA Tech. Note No. 3184*.

Ramberg, W., McPherson, A. E., and Levy, S. (1939), "Experiments on Study of Deformation and of Effective Width in Axially Loaded Sheet-Stringer Panels," *NACA Tech. Note No. 684*.

Redwood, R. G., and Uenoya, M. (1979), "Critical Loads for Webs with Holes," *ASCE J. Struct. Div.*, Vol. 105, No. ST10, pp. 2053–2068.

Rhodes, J., and Harvey, J. M. (1971), "Effects of Eccentricity of Load or Compression on the Buckling and Postbuckling Behavior of Flat Plates," *Int. J. Mech. Sci.*, Vol. 13, pp. 867–879.

Rhodes, J., and Harvey, J. M. (1976), "Plain Channel Section Struts in Compression and Bending beyond the Local Buckling Load," *Int. J. Mech. Sci.*, Vol. 8, pp. 511–519.

Rhodes, J., and Harvey, J. M. (1977), "Examination of Plate Post Buckling Behavior," *ASCE J. Eng. Mech. Div.*, Vol. 103, No. EM3, pp. 461–480.

Rhodes, J., and Loughlan, J. (1980), "Simple Design Analysis of Lipped Simple Columns," *Proc. 5th Int. Spec. Conf. on Cold-Formed Steel Struct.*, St. Louis.

Rockey, K. C. (1980), "The Buckling and Post-Buckling Behavior of Shear Panels Which Have a Central Circular Cut Out," in *Thin Walled Strucures* (ed. J. Rhodes and A. C. Walker), Granada, London.

Rockey, K. C., Anderson, R. G., and Cheung, Y. K. (1969), "The Behavior of Square Shear Webs Having a Circlar Hole," in *Thin-Walled Steel Structures* (ed. K. C. Rockey and H. V. Hill), Crosby Lockwood, London.

Roorda, J., and Venkataramaiah, K. R. (1979), "Effective Width of Stiffened Cold-Formed Steel Plates," *Can J. Civ. Eng.*, Vol. 6, No. 3.

Schlack, A. L. (1964), "Elastic Stability of Pierced Square Plates," *Exp. Mech.*, p. 167.

Schuman, L., and Back, G. (1930), "Strength of Rectangular Flat Plates under Edge Compression," *NACA Tech. Rep. No. 356.*

Seide, P., and Stein, M. (1949), "Compressive Buckling of Simply Supported Plates with Longitudinal Stiffeners," *NACA Tech. Note No. 1825.*

Seydel, E. (1933), "Über das Ausbeulen von rechteckigen isotropen oder orthogonalaniso-tropen Platten bei Schubbeanspruchung," *Ing. Arch.*, Vol. 4, p. 169.

Sharp, M. L. (1966), "Longitudinal Stiffeners for Compression Members," *ASCE J. Struct. Div.*, Vol. 92, No. ST5, pp. 187–212.

Sharp, M. L. (1970), "Strength of Beams or Columns with Buckled Elements," *ASCE J. Struct. Div.*, Vol. 96, No. ST5, pp. 1011–1015.

Sherbourne, A. N., Marsh, C., and Lian, C. Y. (1971), "Stiffened Plates in Uniaxial Compression," IABSE Pub., Vol. 31-I, Zurich, pp. 145–178.

Škaloud, M., and Zornerova, M. (1970), "Experimental Investigation into the Interaction of the Buckling of Compressed Thin-Walled Columns with the Buckling of their Plate Elements," Acta Tecnica CSAV, No. 4.

Southwell, R. V., and Skan, S. (1924), "On the Stability under Shearing Force of a Flat Elastic Strip," *Proc. R. Soc.*, Vol. A105.

Sridharan, S. (1982), "A Semi-analytical Method for the Post-Local-Torsional Buckling Analysis of Prismatic Plate Structures," *Int. J. Numer. Methods Eng.*, Vol. 18, pp. 1685–1697.

Sridharan, S., and Graves-Smith, T. R. (1981), "Postbuckling Analysis with Finite Strips," *ASCE J. Eng. Mech. Div.*, Vol. 107, No, EM5, pp. 869–888.

Stein, O. (1936), "Stabilität ebener Rechteckbleche unter Biegung and Schub," *Bauingenieur*, Vol. 17, p. 308.

Stein, M. (1959), "The Phenomenon of Change in Buckle Pattern in Elastic Structures," *NACA Tech. Note No. R-39.*

Stowell, E. Z. (1948), "A Unified Theory of Plastic Buckling of Columns and Plates," *NACA Tech. Note No. 1556.*

Stowell, E. Z. (1949), "Plastic Buckling of a Long Flat Plate under Combined Shear and Longitudinal Compression," *NACA Tech. Note No. 1990.*

Stowell, E. Z., Heimerl, G. J., Libove, C., and Lundquist, E. E. (1952), "Buckling Stresses for Flat Plates and Sections," *Trans. Am. Soc. Civ. Eng.*, Vol. 117, pp. 545–578.

Supple, W. J. (1980), "Buckling of Plates Under Axial Load and Lateral Pressures," in *Thin Walled Structures* (ed. J. Rhodes and A. C. Walker), Granada, London.

Thomasson, P. -O. (1978), "Thin-Walled C-Shaped Panels in Axial Compression" *Document D.I. 1978*, Swedish Council for Building Research,

Tien, Y. L., and Wang, S. T. (1979), "Local Buckling of Beams under Stress Gradient," *ASCE J. Struct. Div.*, Vol. 105, No. ST8, pp. 1571–1588.

Timoshenko, S. (1910), "Einige Stabilitätsprobleme der Elastizitätstheorie," *Z. Math. Phys.*, Vol. 58, pp. 337.

Timoshenko, S. (1934), "Stability of the Webs of Plate Girders," *Engineering*, Vol. 138, p. 207.

Timoshenko, S. P., and Gere, J. M. (1961), *Theory of Elastic Stability*, 2nd ed., McGraw-Hill, New York.

Troitsky, D. S. C. (1976), *Stiffened Plates, Bending, Stability and Vibrations*, Elsevier, New York.

Tsien, H. S. (1942), "Buckling of a Column with Nonlinear Lateral Supports," *J. Aeronaut. Sci.*, Vol. 9, No. 4.

Tvergaard, V. (1973), "Imperfection-Sensitivity of a Wide Integrally Stiffened Panel under Compression," *Int. J. Solids Struct.*, Vol. 9, pp. 177–192.

Usami, T. (1982), "Post Buckling of Plates in Compression and Bending," *ASCE J. Struct. Div.*, Vol. 108, No. ST3, pp. 591–609.

Usami, T., and Fukumoto, Y. (1982), "Local and Overall Buckling of Welded Box Columns," *ASCE J. Struct. Div.*, Vol. 108, No. ST3, pp. 525–542.

Vojta, J. F., and Ostapenko, A. (1967), "Ultimate Strength Design Curves for Longitudinally Stiffened Plate Panels with Large *b/t*," *Fritz Eng. Lab. Rep. No. 248.19*, Lehigh University, Bethlehem, Pa.

von Kármán, T., Sechler, E. E., and Donnell, L. H. (1932), "Strength of Thin Plates in Compression," *Trans. Am. Soc. Mech. Eng.*, Vol. 54, No. APM-54-5, p. 53.

Wang, S. T. (1969), "Cold-Rolled Austenitic Stainless Steel: Material Properties and Structural Performance," *Cornell Univ. Dep. Struct. Eng. Rep.*, No. 334.

Wang, S. T., and Pao, H. Y. (1981), "Torsional-Flexural Buckling of Locally Buckled Columns," *Int. J. Comput. Struct.*, Vol. 11.

Wang, S. T., Errera, S. J. and Winter, G. (1975), "Behavior of Cold Rolled Stainless Steel Members," *ASCE J. Struct. Div.*, Vol. 101, No. ST11, pp. 2337–2357.

Wang, S. T., Yost, M. I., and Tien, Y. L. (1977), "Lateral Buckling of Locally Buckled Beams Using Finite Element Techniques," *Int. J. Comput. Struct.*, Vol. 7.

Walker, A. C. (1964), "Thin Walled Structural Forms under Eccentric Compressive Load Actions," Doctoral thesis, University of Glasgow.

Way, S. (1936), "Stability of Rectangular Plates under Shear and Bending Forces," *J. Appl. Mech. ASME.*

Webb, S. E., and Dowling, P. J. (1980), "Large Deflection Elasto-plastic Behavior of Discretely Stiffened Plates," *Proc. Inst. Civ. Eng.*, Part 2, Vol. 69, pp. 375–401.

Whitney, J. M., and Leissa, A. W. (1969), "Analysis of Heterogeneous Anisotropic Plates," *Trans. ASME J. Appl. Mech.*, Vol. 36, pp. 261–266.

Winter, G. (1947), "Strength of Thin Steel Compression Flanges," *Trans. Am. Soc. Civ. Eng.*, Vol. 112, p. 527.

Winter, G. (1983), "Commentary on the 1968 Edition of the Specification for the Design of Steel Structural Members," *Cold-Formed Steel Design Manual*, American Iron and Steel Institute, Washington, D.C.

Winter, G., Lansing, W., and McCalley, R. B. (1950), "Four Papers on the Performance of Thin Walled Steel Structures," *Eng. Exp. Stu. Rep. No. 33,* Cornell University, pp. 27–32, 51–57.

Wittrick, W. H. (1952), "Correlation between Some Stability Problems for Orthotropic and Isotropic Plates under Bi-axial and Uni-axial Direct Stress," *Aeronaut. Q.*, Vol. 4, pp. 83–99.

Wittrick, W. H. (1968), "A Unified Approach to the Initial Buckling of Stiffened Panels in Compression," *Aeronaut. Q.*, Vol. 19, p. 265.

Wittrick, W. H., and Horsington, R. W. (1984), "Buckling and Vibration of Composite Folded Plate Structures of Finite Length in Combined Shear and Compression," *Proc. R. Soc. London*, Vol. A392, pp. 107–144.

Yamaki, N. (1959/1960), "Postbuckling Behavior of Rectangular Plates with Small Initial Curvature Loaded in Edge Compression," *J. Appl. Mech. ASME*, Vol. 26, pp. 407–414; Vol. 27, pp. 335–342.

Yang, H. T. Y. (1969), "A Finite Element Formulation for Stability Analysis of Doubly Curved Thin Shell Structures," Doctoral thesis, Cornell University, Ithaca, N.Y.

Yoshizuka, J., and Narmoka, M. (1971), "Buckling Coefficient of Simply Supported Rectangular Plates under Combined Bending and Compressive Stresses in Two Perpendicular Directions," *Stahlbau*, Vol. 40, p. 217.

Yu, W. W., and Davis, C. S. (1973), "Cold Formed Steel Members with Perforated Elements," *ASCE J. Struct. Div.*, Vol. 99, No. ST10, p. 2061.

Zender, W., and Hall, J. B., Jr. (1960), "Combinations of Shear Compressive Thermal and Compressive Load Stresses for the Onset of Permanent Buckles in Plates," *NASA Tech. Note No. D-384.*

Chapter Five

Laterally Unsupported Beams

5.1 INTRODUCTION

Beams, girders, joists, and trusses subjected to flexure have much greater strength and stiffness in the plane in which the loads are applied (major principal axis) than in the plane of the minor principal axis. Unless these members are properly braced against lateral deflection and twisting, they are subject to failure by lateral-torsional buckling prior to the attainment of their full in-plane capacity. They are especially prone to this type of buckling during the construction phase, where braces are either absent or different in type from the permanent ones.

Lateral-torsional buckling is a limit state of structural usefulness where the deformation of a beam changes from predominantly in-plane deflection to a combination of lateral deflection and twisting while the load capacity remains first constant, before dropping off due to large deflections and yielding. Lateral-torsional buckling can be avoided by properly spaced and designed lateral bracing, or by using cross sections which are torsionally stiff, such as box-shaped sections or open-section beam groups connected intermittently by triangulated lacing or by diaphragms.

The principal variable affecting lateral-torsional buckling strength is the distance between lateral braces. Other variables are: the type and position of the loads, the restraints at the ends and at intermediate locations, the type of cross sections, continuity at supports, the presence or absence of stiffening devices that restrain warping at critical locations (Ojalvo, 1977), the material properties, the magnitude and distribution of the residual stresses, prestressing forces, initial imperfections of geometry and loading, discontinuities in the cross section (e.g., change of section, holes, and copes), cross-sectional distortion, and interaction between local and overall buckling.

The analytical aspects of determining the lateral-torsional buckling strength are quite complex, and closed-form solutions exist only for the most simple cases. However, research in this field has been conducted intensively since the mid-nineteenth century, and current research interest continues at a high level. The literature in the field is extensive. A review of the early work was given by Johnston (1976), while Lee (1960), Trahair (1977b), and Nethercot (1983) provide reviews of the modern development. The theory and the design applications are covered extensively in textbooks in various languages. Some of the English-language texts are those of Bleich (1952), Timoshenko and Gere (1961), Vlasov (1961), Trahair (1977), Brush and Almroth (1975), Chajes (1974), Allen and Bulson (1980), and Galambos (1968).

Lateral-torsional buckling behavior is presented graphically in a plot relating strength (critical moment) and unbraced length. The solid curve in Fig. 5.1 illustrates schematically the variation of the critical load when a bifurcation-type instability occurs for a perfectly straight beam. The dashed curve represents the case when initial imperfections are present. There are three ranges of behavior: (1) elastic buckling, which governs for long beams (of importance during construction); (2) inelastic buckling, when instability occurs after some portions of the beam have yielded; and (3) plastic behavior, where the unbraced length is short enough so that buckling occurs after the plastic moment is reached. The last two ranges are of importance for the completed structure.

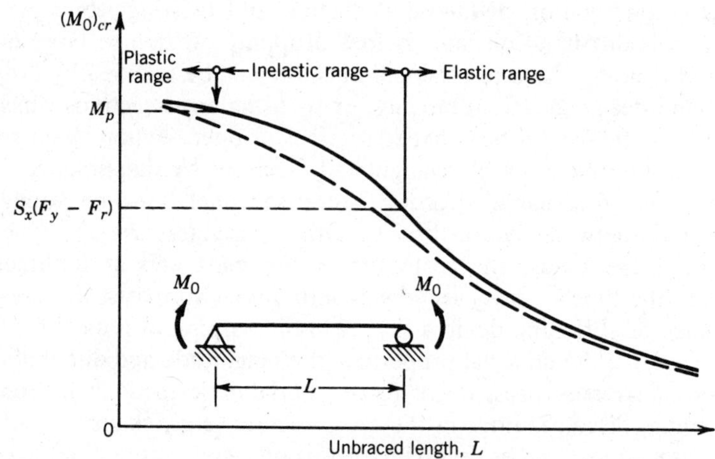

Fig. 5.1 Beam buckling curves.

5.2 ELASTIC LATERAL-TORSIONAL BUCKLING

5.2.1 Simply Supported Doubly Symmetric Beams of Constant Sections

Uniform Moment. This case is the simplest configuration, and buckling analysis leads to a closed-form solution (Timoshenko and Gere, 1961). The ends of the beam are prevented from lateral deflection ($u = 0$) and twisting ($\phi = 0$), but they are free to rotate laterally ($u'' = 0$) and the end cross section is free to warp ($\phi'' = 0$). The critical buckling moment is given by Eq. 5.1:

$$M_{0cr} = (\pi/L)\sqrt{EI_yGJ}\sqrt{1 + W^2} \tag{5.1}$$

where

$$W = \frac{\pi}{L}\sqrt{\frac{EC_w}{GJ}} \tag{5.2}$$

L is the span length, E and G are the elastic and shear moduli, respectively, and I_y, J, and C_w are, respectively, the minor axis moment of inertia, the St.-Venant torsion constant, and the warping constant. The second square root in Eq. 5.1 represents the effect of warping torsional stiffness. The ratio $E/G = 0.385$ for steel and 0.376 for aluminum. The cross-sectional constants I_y, J, and C_w are listed in handbooks or they can be derived [see, e.g., Bleich (1952) or Galambos (1968)] for doubly symmetric open sections. For rectangular solid or boxed beams, W may be taken as zero.

Nonuniform Bending. If the end moments are unequal (Fig. 5.2), numerical or approximate solutions are required to obtain the buckling load. The most extensive early work for this case was performed by Massonet (1947), Salvadori (1955, 1956), and Horne (1954), and their results have been verified many times by other researchers using various numerical techniques. Salvadori (1955) found that a simple modifier to Eq. 5.1 can account for the effect of moment gradient.

$$M_{cr} = C_b M_{0cr} \tag{5.3}$$

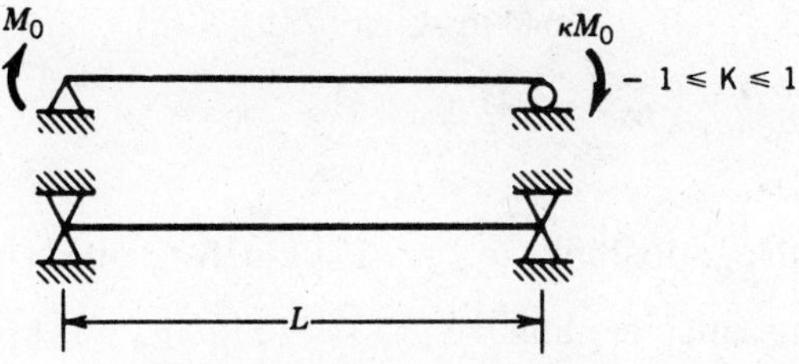

Fig. 5.2 Nonuniform moment.

where M_{0cr} is obtained from Eq. 5.1 and C_b is the "equivalent uniform moment factor." Various lower-bound formulas have been proposed for C_b, but the most commonly accepted are the following:

$$C_b = 1.75 + 1.05\kappa + 0.3\kappa^2 \leq 2.56 \tag{5.4a}$$

$$1/C_b = 0.6 + 0.4\kappa \leq 0.4 \tag{5.4b}$$

The moment ratio κ (Fig. 5.2) is defined as positive when the end moments cause double curvature deflection. The value of C_b is mildly dependent on W [Eq. 5.2; also Nethercot (1983)], especially when κ approaches +1; however, either Eq. 5.4a or 5.4b is accurate enough for most purposes.

When transverse loads are applied between the ends of the beam (Fig. 5.3), without intermediate lateral braces, it is important to consider the location of the applied load with respect to the shear center. The value of C_b may be computed from the following equations (Nethercot, 1983; Nethercot and Rockey, 1972):

$$\begin{aligned} &\text{Top-flange loading:} C_b = A/B \\ &\text{Load at shear center:} C_b = A \\ &\text{Bottom-flange loading:} C_b = AB \end{aligned} \tag{5.5}$$

where A and B are taken from Fig. 5.3. Comparison with accurate numerical solutions indicates that the critical buckling loads computed by Eqs. 5.5 are within 5% of the true solution.

The effect of load location can be appreciated from the following example for a simply supported beam under uniform load, using a W24 × 55 steel beam of 576 in. (14.63 m) length:

$$E = 29{,}000 \text{ ksi } (200{,}000 \text{ MPa}) \qquad G/E = 0.385$$

$$I_y = 29.1 \text{ in.}^4 \ (12.1 \times 10^6 \text{ mm}^4); \quad J = 1.18 \text{ in.}^4 \ (0.49 \times 10^6 \text{ mm}^4)$$

$$C_w = 3870 \text{ in.}^6 \ (1.015 \times 10^{12} \text{ mm}^6)$$

$$W = 0.503 \qquad M_{0cr} = 644 \text{ in.-kips } (731 \text{ kN-m})$$

$$A = 1.13,\ B = 1.230 \qquad (\text{Fig. 5.3})$$

$$w_{cr} = 8M_{cr}/L^2$$

$$\text{Top flange loading, } w_{cr} = 0.171 \text{ kip/ft } (2.50 \text{ kN/m})$$

$$\text{Shear center loading, } w_{cr} = 0.211 \text{ kip/ft } (3.08 \text{ kN/m})$$

$$\text{Bottom flange loading, } w_{cr} = 0.259 \text{ kip/ft } (3.78 \text{ kN/m})$$

Loading	BMD	M	A	B
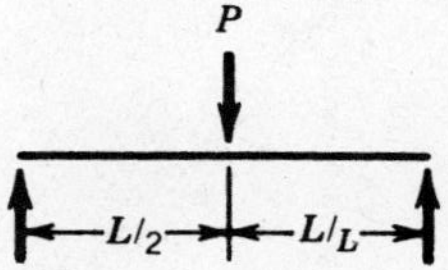 P; $L/2$, L/L	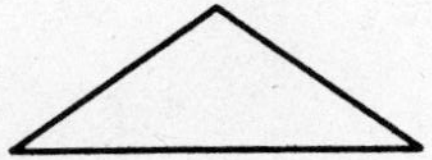	$\frac{PL}{4}$	1.35	$1 - 0.180W^2 + 0.649W$
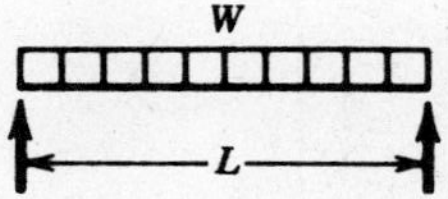 W; L		$\frac{wL^2}{8}$	1.12	$1 - 0.154W^2 + 0.535W$
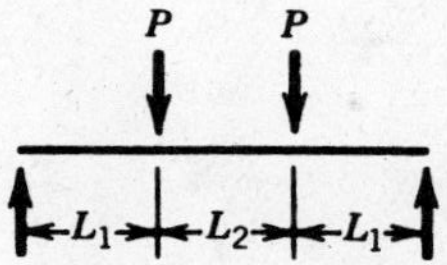 P, P; L_1, L_2, L_1		PL_1	$1 + \left(\frac{L_1}{2L_1 + L_2}\right)^2$	$1 - 0.465W^2 + 1.636W$

Fig. 5.3 Coefficients for transverse loads.

5.2.2 End-Restrained Doubly Symmetric Beams of Constant Section

End restraint has a pronounced effect on the elastic lateral-torsional buckling strength of beams. Special cases of idealized end conditions are shown in Fig. 5.4. Nethercot and Rockey (1972) have presented the following method of solution for the two loading cases of Fig. 5.5.

$$M_{cr} = CM_{0cr} \tag{5.6}$$

where $C = A/B$ for top-flange loading, $C = A$ for loading through the shear center, and $C = AB$ for bottom-flange loading, and A and B are determined by the equations in Fig. 5.5.

In restraint condition II of Fig. 5.4 the end is free to rotate laterally but is prevented from warping. Such a condition can be effectively achieved by welding boxed stiffeners ("tube-type warp restraints," Fig.

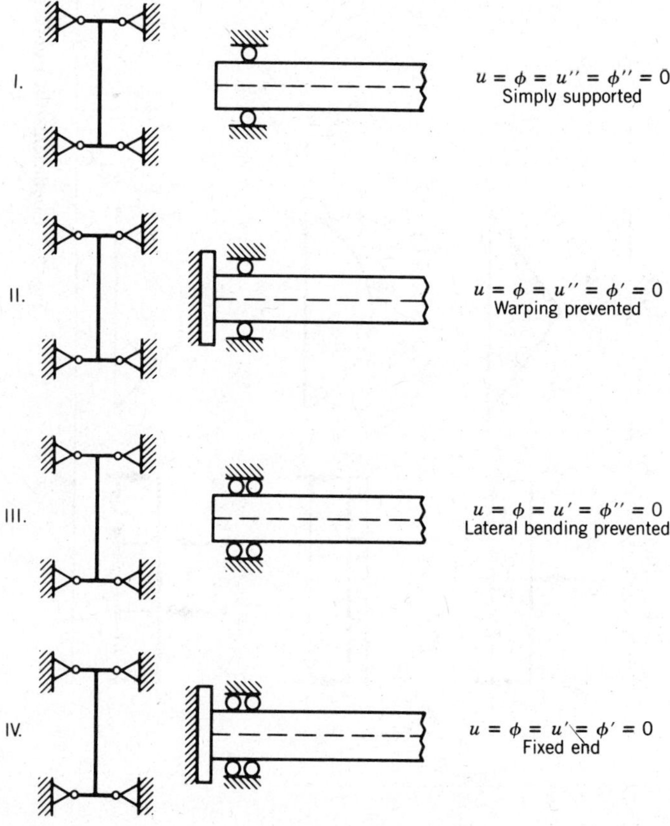

Fig. 5.4 Idealized end restraints.

Loading	Restraint	A	B
	I	1.35	$1-0.180W^2+0.649W$
	II	$1.43+0.485W^2+0.463W$	$1-0.317W^2+0.619W$
	III	$2.0-0.074W^2+0.304W$	$1-0.207W^2+1.047W$
	IV	$1.916-0.424W^2+1.851W$	$1-0.466W^2+0.923W$
	V	$2.95-1.143W^2+4.070W$	1
	I	1.13	$1-0.154W^2+0.535W$
	II	$1.2+0.416W^2+0.402W$	$1-0.225W^2+0.571W$
	III	$1.9-0.120W^2+0.006W$	$1-0.100W^2+0.806W$
	IV	$1.643-0.405W^2+1.771W$	$1-0.339W^2+0.625W$
	V	$2.093-0.947W^2+3.117W$	$1.073+0.044W$

Case V: Lateral support at center. Restraint: Equal at both ends.

Fig. 5.5 Restraint categories.

5.6; Ojalvo, 1977) at or near the end support. Other warp-restraining stiffeners have been also recommended and their effectiveness has been analyzed (Ojalvo and Chambers, 1977; Vacharajittiphan and Trahair, 1974; Heins and Potocko, 1979; Szewczak et al., 1983). The effect of end plates was studied by Lindner and Gietzelt (1983). The effect of warp prevention is illustrated for the W24 × 55 steel beam of ($W = 0.503$; $M_{0cr} = 644$ in.-kips; $A = 1.507$; $B = 1.230$) 576 in. length under uniformly distributed load:

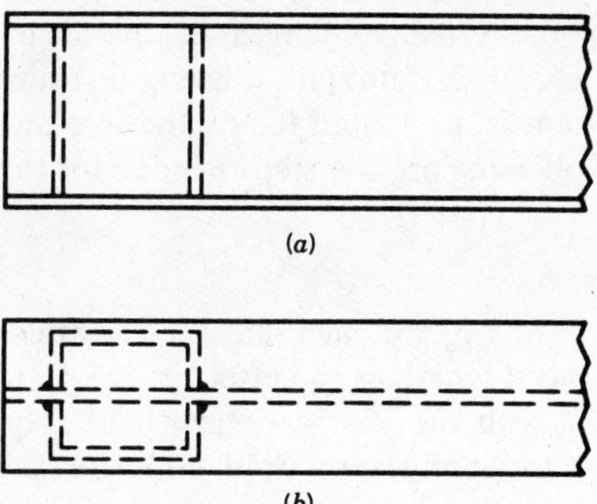

Fig. 5.6 Tube-type warp restraints.

Top-flange loading, $w_{cr} = 0.228$ kip/ft (3.33 kN/m)

Shear center loading, $w_{cr} = 0.281$ kip/ft (4.10 kN/m)

Bottom-flange loading, $w_{cr} = 0.346$ kip/ft (5.05 kN/m)

When the initial elastic buckling load for the case of warping prevented is compared to the case of free warping, as computed previously, it can be seen that the former has a 33% higher buckling load. For other lengths and other sections this ratio is different, of course, but the point is made that the buckling load can be increased substantially with a modest expenditure.

Bracing will contribute significantly to increasing the strength of the beam. In simply supported and continuous beams, lateral and torsional bracing is provided not only at the supports but also at intermediate points, and lateral buckling analysis is performed on each unbraced segment, using Eq. 5.3 along with the equivalent uniform moment factor C_b (Eq. 5.4a or 5.4b). However, it is necessary that the bracing points prevent both lateral and torsional movement of the braced point (Lindner, 1985). The load corresponding to the smaller critical moment is then a conservative lower bound to the elastic buckling load because each unbraced segment is assumed to be laterally and torsionally simply supported. Accounting for the end restraint of the adjacent segments on the critical segment can substantially increase the buckling load (Nethercot, 1983). Considerable research was performed on analyzing the effects of continuity (Vacharajittiphan and Trahair, 1975; Nethercot, 1973a; Powell and Klingner, 1970; Trahair, 1969; Salvadori, 1955; Hartmann, 1967), and Trahair (1977a,b) has recommended the following simple and conservative method, which is based on the analogy that the buckling behavior of continuous beams is the same as the behavior of end-restrained columns. Thus the nomographs for nonsway columns (Johnston, 1976) can be used to obtain the effective length of the beam segment. This method assumes that lateral restraint and warping restraint are identical, and that in-plane restraint is accounted for by the in-plane bending moment diagram (BMD). Following are the steps needed for the analysis:

1. Compute the in-plane BMD (Fig. 5.7*a*).
2. Determine C_b (Eq. 5.4) and M_{cr} (Eq. 5.3) for each unbraced segment in the beam, using the actual unbraced length as the effective length in Eq. 5.1, and identify the segment with the lowest critical load. The critical loads for buckling assuming simply supported ends for the weakest segment and the two adjacent segments are P_m, P_{rL}, and P_{rR}, respectively (Fig. 5.7*b*).

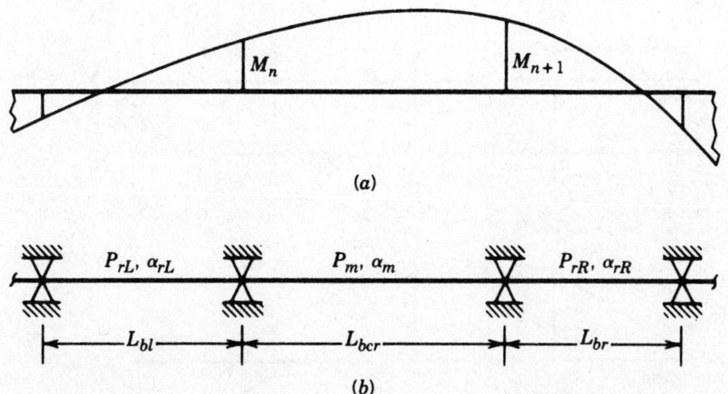

Fig. 5.7 Computation of the effect of end restraint. (*a*) In-plane BMD; (*b*) braced segments.

3. Compute the stiffness ratios for the three segments as follows: for the critical segment

$$\alpha_m = 2EI_y / L_{bcr}$$

for each adjacent segment

$$\alpha_r = n(EI_y / L_b)[1 - (P_m / P_r)]$$

where

$n = 2$ if the far end of the adjacen segment is continuous
$n = 3$ if it is pinned
$n = 4$ if it is fixed

4. Determine the stiffness ratios $G = \alpha_m / \alpha_r$ and obtain the effective length factor K from the nonsway restrained column nomographs.
5. Compute the critical moment and the buckling load of the critical segment from the equation

$$M_{cr} = \frac{C_b \pi \sqrt{EI_y GJ}}{KL} \sqrt{1 + \frac{\pi^2 EC_w}{GJ(KL)^2}} \tag{5.7}$$

The method is illustrated by the example of Fig. 5.8. When the end restraint of the critical segment is accounted for, the buckling load is increased by 34%.

When the lateral and the warping end restraints are unequal, the following general approximate method may be used.

$$M_{cr} = \frac{C_b \pi \sqrt{EI_y GJ}}{K_y L} \sqrt{1 + \frac{\pi^2 EC_w}{GJ(K_z L)^2}} \tag{5.8}$$

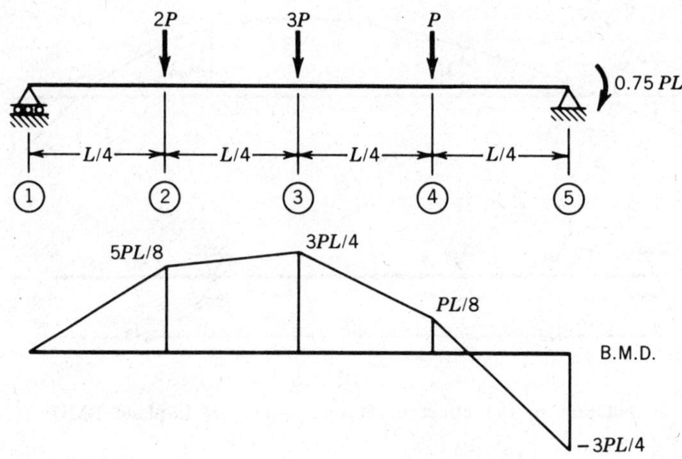

Segment	$M_{\max}$	C_b (Eq. 5.4a)	M_{cr} (Eq. 5.3) [in.-kips (kN-m)]	P_{cr} [kips (kN)]
1-2	$0.625PL$	1.75	8411 (950)	22.43 (99.8)
2-3	$0.75PL$	1.083	5205 (588)	11.57 (51.5)
3-4	$0.75PL$	1.583	7608 (860)	16.91 (72.2)
4-5	$0.75PL$	1.933	9291 (1050)	20.65 (91.9)

$L = 600$ in. (15.24 m); W24 × 55; $I_y = 29.1$ in.^{4} (12.1×10^6 mm^4)
$J = 1.18$ in.^{4} (0.491×10^6 mm^4); $C_w = 3870$ in.^{6} (1.039×10^{12} mm^6). Critical segment: 2–3.

(*a*)

$$\alpha_{12} = \frac{3EI_y}{L/4}\left(1 - \frac{11.57}{22.43}\right) = 1.453\left(\frac{4EI_y}{L}\right)$$

$$\alpha_{23} = \frac{2EI_y}{L/4} = 2\left(\frac{4EI_y}{L}\right)$$

$$\alpha_{34} = \frac{2EI_y}{L/4}\left(1 - \frac{11.57}{16.91}\right) = 0.632\left(\frac{4EI_y}{L}\right)$$

$$G_2 = 2/1.453 = 1.4$$

$$G_3 = 2/0.632 = 3.2$$

From nonsway nomograph: $K = 0.85$.

$$M_{cr} = 6990 \text{ in.-kips } (790 \text{ kN-m})$$

$$P_{cr} = 15.53 \text{ kips } (69.1 \text{ kN})$$

(*b*)

Fig. 5.8 Example of end restraint.

For example, for end condition II in Fig. 5.4, $K_y = 1.0$ ($u'' = 0$) and $K_z = 0.5$ ($\phi' = 0$). When one end is "pinned" and the other "fixed," then $K_z = 0.7$. More exact values of K_z and K_y for many end-restraint combinations are given by Vlasov (1961). Solutions obtained by numerical methods for special cases of end restraint are tabulated in many papers (e.g., Austin et al., 1957; Clark and Hill, 1960). Approximate solutions with coefficients are also given by Clark (1960).

5.2.3 Monosymmetric Beams

The methods and formulas in Section 5.2.2 apply to doubly symmetric beams of uniform section, for example, wide-flange shapes, solid rectangular bars, and rectangular tubes (for the latter sections, $C_w = 0$), and for channels where bending about the major axis is in the plane of the shear center (Fig. 5.9). For monosymmetric beams where bending is in the plane of symmetry (Fig. 5.10), the shear center S and the centroid C do

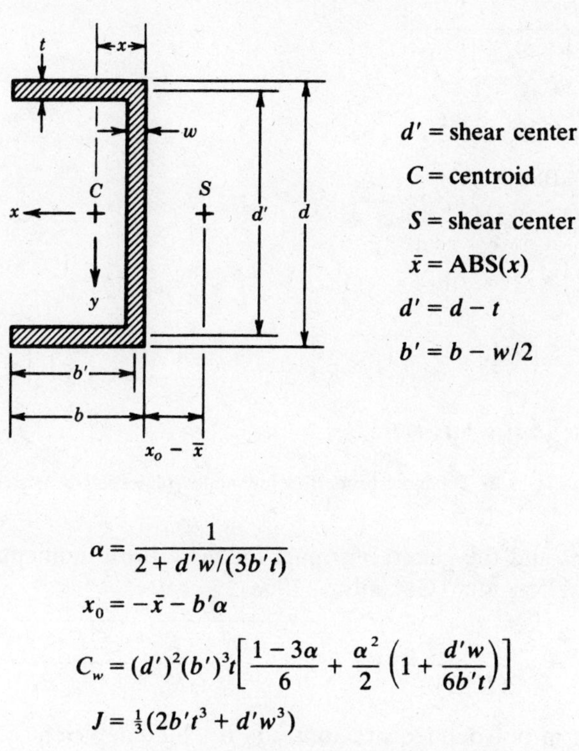

$$\alpha = \frac{1}{2 + d'w/(3b't)}$$

$$x_0 = -\bar{x} - b'\alpha$$

$$C_w = (d')^2(b')^3 t\left[\frac{1-3\alpha}{6} + \frac{\alpha^2}{2}\left(1 + \frac{d'w}{6b't}\right)\right]$$

$$J = \tfrac{1}{3}(2b't^3 + d'w^3)$$

Fig. 5.9 Torsional properties for channels.

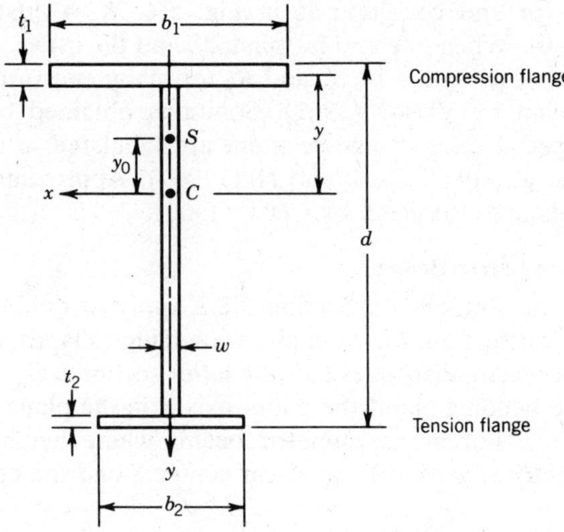

$$\alpha = \frac{1}{1 + (b_1/b_2)^3(t_1/t_2)}$$

$$d' = d - (t_1 + t_2)/2$$

$$C_w = (d')^2 b_1^3 t_1 \alpha / 12$$

$$\bar{y} = \mathrm{ABS}(y)$$

$$y_0 = -\bar{y} - \alpha d'$$

$$\beta_x = \frac{1}{I_x}\left\{(d' - \bar{y})\left[\frac{b_2^3 t_2}{12} + b_2 t_2 (d' - \bar{y})^2 + \frac{w}{4}(d' - \bar{y})^3\right]\right.$$

$$\left. - \bar{y}\left(\frac{b_1^3 t_1}{12} + b_1 t_1 \bar{y}^2\right) + \frac{w}{4}\left[\left(h - \bar{y} - \frac{t_2}{2}\right)^4 - \left(\bar{y} - \frac{t_1}{2}\right)^4\right]\right\} - 2y_0$$

$$J = \frac{1}{3}(b_1 t_1^3 + b_2 t_2^3 + d' w^3)$$

Fig. 5.10 Torsional properties for monosymmetric I-beams.

not coincide, and the general formula for the critical moment is given by the following formula (Galambos, 1968):

$$M_{cr} = \frac{C_b \pi^2 E I_y \beta_x}{2(K_y L)^2}\left\{1 + \sqrt{1 + \frac{4}{\beta_x^2}\left[\frac{C_w K_y^2}{I_y K_z^2} + \frac{GJ(K_y L)^2}{\pi^2 E I_y}\right]}\right\} \tag{5.9}$$

The only term not defined previously is β_x, the coefficient of monosymmetry. The general expression for β_x is

$$\beta_x = \frac{1}{I_x} \int_A y(x^2 + y^2)\, dA - 2y_0 \tag{5.10}$$

where I_x is the major-axis moment of inertia, y_0 is the shear center distance (Fig. 5.10), which is negative if the larger flange is in compression, x and y are centroidal coordinates, and integration is over the whole sectional area A. The value of $\beta_x = 0$ for doubly symmetric shapes, such as an I shape, and for a channel. The formula for β_x for a general I-shaped monosymmetric beam is given in Fig. 5.10. For a T-shaped beam (tee) or a double-angle beam, $C_w = 0$ and β_x is determined from the formula in Fig. 5.10, with either $b_2 = t_2 = w$ or $b_1 = t_1 = w$, depending on whether the flange or the stem is in compression.

For practical purposes β_x of the section shown in Fig. 5.10 can be approximated by (Kitipornchai and Trahair, 1980; for an exact expression, see Nethercot, 1983, p. 25)

$$\beta_x = 0.9d'\left(\frac{2I_{yc}}{I_y} - 1\right)\left[1 - \left(\frac{I_y}{I_x}\right)^2\right] \tag{5.11}$$

where d' is the distance between the centers of areas of the two flanges, I_{yc} the minor-axis moment of inertia of the compression flange, and I_y the minor-axis moment of inertia of the whole cross section. For a tee, $I_{yc} = I_y$ if the flange is in compression and $I_{yc} = 0$ if the stem is in compression. In cases when $K_y = K_z = K$, Eq. 5.9 can be written as (Trahair, 1977a)

$$M_{cr} = \frac{\pi C_b}{KL}\left[\sqrt{EI_y GJ}\left(B_1 + \sqrt{1 + B_2 + B_1^2}\right)\right] \tag{5.12}$$

where

$$B_1 = \frac{\pi \beta_x}{2(KL)} \sqrt{\frac{EI_y}{GJ}} \tag{5.13a}$$

$$B_2 = \frac{\pi^2 EC_w}{(KL)^2 GJ} \tag{5.13b}$$

for tee-shaped sections, $B_2 = 0$.

Charts, tables, and approximate formulas for a vareity of end-restraint conditions and transverse loadings applied at, below, and above the shear center are presented by Clark and Hill (1960) and Anderson and Trahair (1972), and for tapered doubly symmetric and nonsymmetric beams by Nethercot (1973b), Kitipornchai and Trahair (1972, 1975a) and Kitipornchai and Dux (1985). The lateral-torsional buckling of tapered beams is discussed more fully in Chapter 9.

5.2.4 Cantilever Beams

Solutions in the form of tables, charts, and approximate formulas for the lateral-torsional buckling load of cantilevers are available for a variety of loading conditions, including transverse loads applied at the level of either flange (Timoshenko and Gere, 1961; Clark and Hill, 1960; Nethercot, 1973c, 1983; Massey and McGuire, 1971; Trahair, 1977a,b; Nethercot and Rockey, 1972). Despite the fact that the upper flange of the

Restraint Conditions		Effective length	
At root	At tip	Top flange loading	All other cases
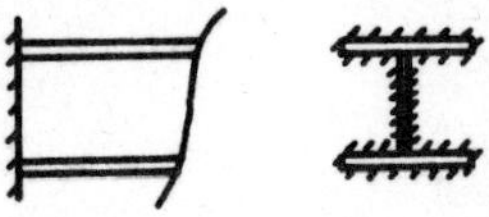		1.4 L	0.8 L
	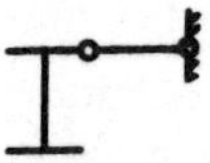	1.4 L	0.7 L
	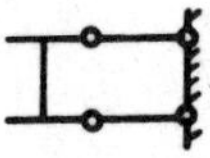	0.6 L	0.6 L
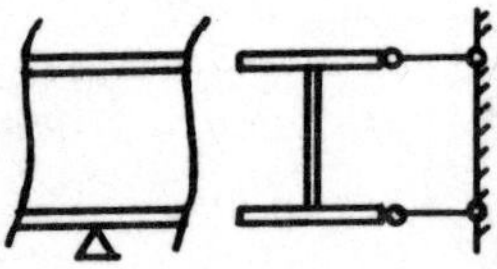		2.5 L	1.0 L
	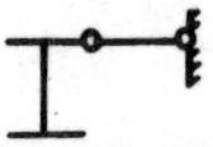	2.5 L	0.9 L
	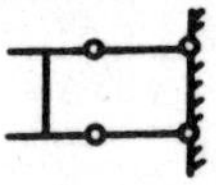	1.5 L	0.8 L
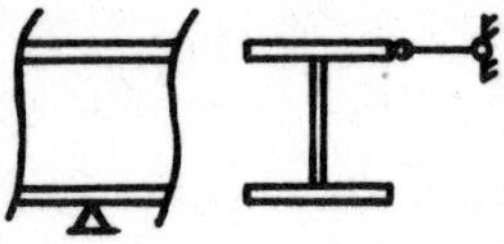		7.5 L	3.0 L
	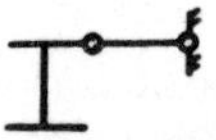	7.5 L	2.7 L
	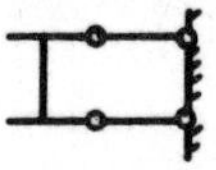	4.5 L	2.4 L

Fig. 5.11 Effective length factors for cantilevers.

cantilever is in tension, it is the flange that deforms more in the buckled shape, and lateral stability is further reduced as the point of load application is raised relative to the shear center. While in a careful analysis recourse can be had to the published solutions, some of which are cited above, Nethercot (1983) has shown that for most applications the simple effective-length method is satisfactory:

$$M_{cr} = \frac{\pi}{KL} \sqrt{EI_y GJ} \sqrt{1 + \frac{\pi^2 EC_w}{(KL)^2 GL}} \tag{5.14}$$

where the effective-length factors for various restraint conditions at the tip and at the root of the cantilever are given in Fig. 5.11. Both end load and uniformly distributed load have been considered.

5.2.5 Summary of Elastic Lateral-Torsional Buckling

The preceding sections of this chapter have presented formulas for the solution of various cases of loading, beam geometry, and end restraint. These formulas are sufficient for developing design criteria in strucural specifications and for solving most of the usual and unusual cases encounered in practice. Since most of the cases of lateral-torsional buckling are not amenable to closed-form solution, the original solutions are numerical, and many papers were cited to enable the user of this guide to obtain the answers which were presented in the form of charts, tables, and special-purpose approximate expressions. Recourse can also be had to solutions that can be adapted to programmable calculators and microcomputers. These methods are based on various energy methods (e.g., Timoshenko and Gere, 1961; Bleich, 1952; Salvadori, 1955; Vlasov, 1961), finite-difference methods (Galambos, 1968; Vinnakota, 1977b), finite-integral methods (Brown and Trahair, 1968) and finite-element methods (Barsoum and Gallagher, 1970; Powell and Klingner, 1970). As computers are increasingly becoming part of the structural engineer's workstation, these computerized numerical methods will be used more and more in lieu of the approximate equations. Thus it is vital that structural engineers possess a thorough understanding of the basic aspects of the theory of lateral-torsional buckling as it is explained in textbooks on stability theory.

The lateral stability of trusses was investigated by Horne (1960), and approximate solutions and a lower-bound design equation for light factory-made truses ("steel joists") are given by Hribar and Laughlin (1968) and Galambos (1970). Lateral buckling of the unsupported compression chord of trusses is discussed in Chapter 15.

The previous discussion in this chapter considered solutions for the elastic lateral-torsional buckling of ideally perfect beams under perfect in-plane loading, using linear buckling theory. This theory has been quite adequate for the development of design rules and it has been substantiated extensively by tests. However, there have also been a number of investigations which have considered the effects of initial geometric imperfections (Vinnakota, 1977a; Nethercot, 1983), prebuckling curvature (Trahair and Woolcock, 1973), and postbuckling strength (Masur and Milbrandt, 1957; Woolcock and Trahair, 1974, 1976). A challenge to part of the theory of monosymmetric beams, the so-called "Wagner hypothesis," has been presented by Ojalvo (1981). Resolution of this challenge awaits experimental confirmation. It should be noted that end connections such as partial end plates (Lindner, 1985) and coped ends (Yura and Cheng, 1985) can influence the buckling load significantly.

The linear buckling theory assumes that the original shape of the cross section does not distort and that local buckling (Chapter 4) and lateral-torsional buckling occur independently. If the plate elements of the cross section are relatively thin, as is the case of light-gage cold-formed beams (Chapter 13), there is a possibility of combined cross-sectional distortion and lateral-torsional and/or local buckling. This effect was investigated by Goldberg et al. (1964), Rajasekaran and Murray (1973), Plank and Wittrick (1974), Akay and Johnson (1977) and Hancock (1978). Lateral-torsional buckling of rolled or welded I-beams is not normally significantly affected by distortion (Hancock, 1978).

5.3 INELASTIC LATERAL-TORSIONAL BUCKLING

Buckling in the elastic range (Fig. 5.1) is important for relatively slender beams, and in practical terms this means that it is more useful when considering the strength of beams during the construction phase. However, an understanding of elastic buckling is also important for another reason: It serves as the model of the analysis of inelastic beams, which buckle laterally and torsionally after some portions of the beam have exceeded the yield strain. Most of the research on inelastic lateral-torsional buckling has used the tangent-modulus theory of inelastic buckling by investigating the stability of an equivalent elastic beam that has its stiffness properties (minor-axis bending stiffness, St.-Venant torsional stiffness, and warping torsional stiffness) reduced due to in-plane flexure prior to buckling. Since stability is checked for the stiffnesses just prior to buckling (tangent modulus), this analysis gives theoretically a lower bound to the buckling load. However, this is masked by the presence of unavoidable initial geometric imperfections, and for practical purposes tests have shown that the tangent-modulus approach is a satisfactory model. The pattern of yielding prior to buckling is affected by the magnitude and the distribution of residual stresses, and so in effect even a doubly symmetric beam will act as a monosymmetric beam in the inelastic range.

Because the extent of yielding varies from section to section in a beam under nonuniform flexure, the stiffness and monosymmetry parameters are not uniform along the length of the beam, and for continuous beams and frames, in-plane plastic force redistribution plays a significant role. By necessity, then, inelastic analysis is always performed by a numerical method.

Research on inelastic lateral-torsional buckling of beams has been performed vigorously since the 1940s, and research continues in the 1980s, especially on problems related to continuous systems. The first paper on the subject was authored by Neal (1950), and the residual stress effect was first considered by Galambos (1963). The combined effects of residual stresses and geometrical imperfections on inelastic lateral-torsional buckling of beams were considered by Lindner (1974), Vinnakota (1977a), and Celigoj (1979). A large number of analytical and experimental studies have been performed on the inelastic buckling of beams, and this is not the place to present a full discussion. Such a review has been given by Trahair (1977a,b, 1983), who presents a detailed recapitulation of the theory, the assumptions, the methods of analysis, experimental verifications, and tabulated and graphical results. All tests reported in the literature have been analyzed statistically and reported by Fukumoto (Fukumoto and Kubo, 1977; Fukumoto et al., 1980; Fukumoto and 1981).

Another aspect of the problem of lateral-torsional buckling of beams is the determination of the maximum permissible unbraced length so that the plastic moment can be attained and maintained for a sufficiently long rotation capacity so as to permit the development of a plastic mechanism (Lay and Galambos, 1965, 1967).

The research on inelastic lateral-torsional buckling is complex, involving extensive numerical calculations and carefully conducted expensive experiments. The significance of this work to the user of this guide lies in the fact that it provides the basis for the development of design rules.

5.4 BRACING REQUIREMENTS

It is evident that the single most important element in preventing the lateral-torsional buckling of beams is the spacing, stiffness, and strength of the bracing. For bracing to be fully effective, it must prevent both twisting and lateral deflection of the cross section of the beam. For example, it is not adequate simply to bolt the bottom flange of a simple beam to the support. The top flange must also be prevented from lateral motion (Trahair, 1977a).

The spacing of lateral bracing determines the buckling load (see the equations for the various conditions discussed above in this chapter). Lateral bracing usually exists in the final structure through the medium of the slab or the roof, which is supported directly on the beam, or through secondary members framing into the beam. In regions of negative moment where only the tension flange is braced, it is necessary to provide lateral bracing for the compression flange. Lateral bracing available to the final structure is generally more than adequate to prevent lateral and torsional displacement at the points of support. However, during construction the amount of bracing provided is usually very scant, and therefore it is necessary to determine its required minimum stiffness and strength.

One conservative method of dealing with this problem has been recommended by Winter (1960): Assume that the compressed portion of the beam is isolated and behaves as a restrained initially imperfect axially loaded column under a load equal to the area of the compressed portion times the maximum compressive stress. Winter (1960) then gives simple rules for determining the required strength and stiffness of the bracing, whether this bracing is continuously applied or acts at discreet locations. [For further elaboration, see also McGuire (1968).]

The example problem in Fig. 5.12 shows the application of this method when the compression flanges of two parallel beams are braced at the third points of the span. The calculation involves the determination of the required stiffness and the strength of the braces. From this example it is evident that (1) the bracing area required is but a small portion of the area of the beam to be braced; (2) the determination of the required strength involves an assumption of initial deflections; (3) the bracing requirements depend on the number of parallel beams to be braced; and (4) the ends of the bracing must be firmly anchored. Further studies of a series of interbraced beams are reported by Medland (1980).

The strength and stiffness requirements of intermittent or continuous bracing have been considered by many researchers (e.g., Mutton and Trahair, 1973; Peköz, 1983; Nethercot and Trahair, 1975, among many others), and this work is summarized by Nethercot and Trahair (1984). Design rules and simple and conservative bracing criteria are given by Gedies (1983). The bracing requirements when the bracing provides both torsional and translational restraints at any point in the cross section can be expressed explicitly as far as the stiffness requirements are concerned, but recourse must be made to results obtained from finite element analysis when determining the required strength (Nethercot and Trahair,

1984). When the tension flange of a simply supported symmetric wide-flange beam is restrained by infinitely stiff continuous lateral restraint, the elastic critical moment is

$$M_{0cr} = \frac{GJ}{d'} + \frac{\pi^2}{L^2}\left(EC_w + \frac{EI_y d'^2}{4}\right) \tag{5.15}$$

where d' is the distance between the centers of area of the flanges.

In-plane loading

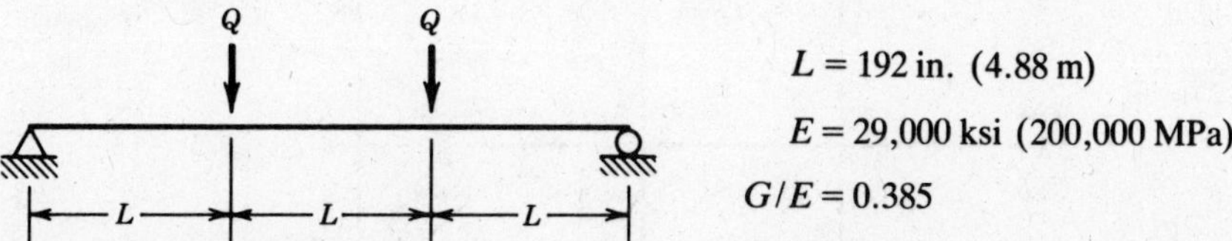

W24 × 55; $A = 16.2\text{ in.}^2$ $(10.45 \times 10^3\text{ mm}^2)$; $J = 1.18\text{ in.}^4$ $(0.491 \times 10^6\text{ mm}^4)$
$I_y = 29.1\text{ in.}^4$ $(12.1 \times 10^6\text{ mm}^6)$; $S_x = 114\text{ in.}^3$ $(1.87 \times 10^6\text{ mm}^3)$; $C_w = 3870\text{ in.}^6$ $(1.039 \times 10^{12}\text{ mm}^6)$.

Lateral-torsional buckling analysis

Lateral braces at supports and under the loads. Critical segment: center span.

$$M_{0cr} = \frac{\pi}{L}\sqrt{EI_y GJ}\sqrt{1 + \frac{\pi^2 ECw}{GJL^2}} = 3125\text{ in.-kips } (353\text{ kN-m})$$

Adjacent span:

$$M_{cr} = C_b M_{0cr} = 1.75 M_{0cr}$$

$$\alpha_m = 2EI_y/L$$

$$\alpha_r = \frac{3EI_y}{L}\left(1 - \frac{M_{cr}}{M_{0cr}}\right) = \frac{3EI_y}{L}\left(1 - \frac{1}{1.75}\right)$$

$$G = \frac{\alpha_m}{\alpha_r} = \frac{\frac{2}{3}}{1 - 1/1.75} = 1.56$$

$$K = 0.825 \quad \text{(nonsway nomograph)}$$

(*a*)

Fig. 5.12 Example of brace design.

$$M_{0cr} = \frac{\pi}{KL}\sqrt{EI_yGJ}\sqrt{1+\frac{\pi^2ECw}{GJ(KL)^2}} = 4362 \text{ in.-kip}^4 \text{ (493 kN-m)}$$

$$f_{cr} = M_{0cr}/S_x = 38.3 \text{ ksi (264 MPa)}$$

$$P_{cr} = f_{cr}A/2 = 310 \text{ kips (1379 kN)}$$

Bracing system (*two beams*)

Compression flange bracing. Bracing must be able to support compression:

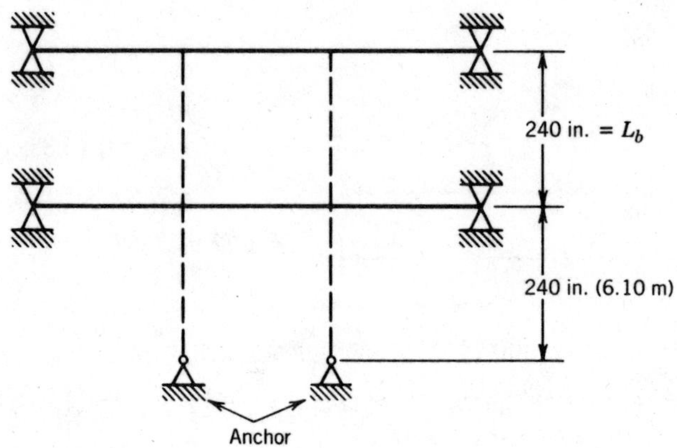

Bracing design:

$L_b/r_b \leq 200$

Use 2L $4 \times 3 \times \frac{1}{4}$; $r_b = 1.28$ in. (32.5 mm); $A = 3.38$ in.^{2} (2.18×10^3 mm^2)

$L_b/r_b = 240/1.28 = 187.5$; $F_a = 4.2$ ksi (29.0 MPa).

Axial stiffness: $k_{\text{act}} = \dfrac{A_bE}{L_b} = 408$ kips/in. (71.5 kN/mm)

k_{act} = actual (available) bracing stiffness

(*b*)

Fig. 5.12 (*Continued*)

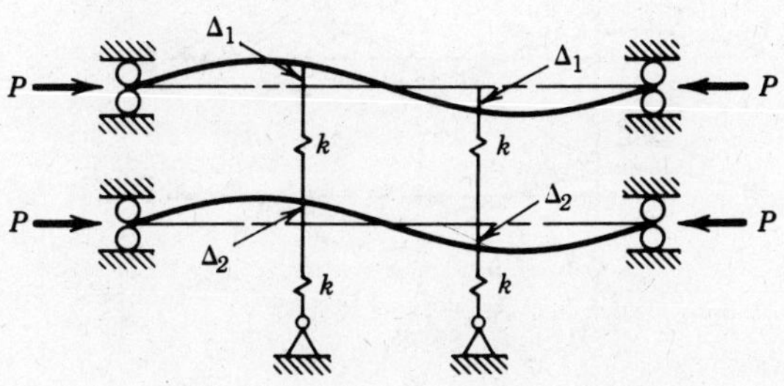

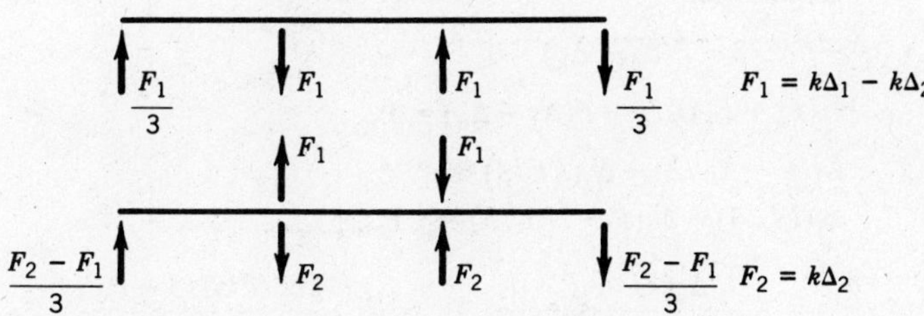

Equilibrium about left interior brace point, assuming zero beam moment:

$$P\Delta_1 - F_1L/3 = 0$$

$$P\Delta_2 - (F_2 - F_1)L/3 = 0$$

$$\left.\begin{aligned} &\Delta_1(P - kL/3) + \Delta_2(kL/3) = 0 \\ &\Delta_1(kL/3) + \Delta_2(P - 2kL/3) = 0 \end{aligned}\right\}$$

$$\text{Determinant} = 0 \rightarrow P_{cr} = kL\left(\frac{3 - \sqrt{5}}{6}\right) = 0.1273kL$$

(*c*)

Fig. 5.12 (*Continued*)

Strength analysis

Assume initial deflection is same as buckled shape.

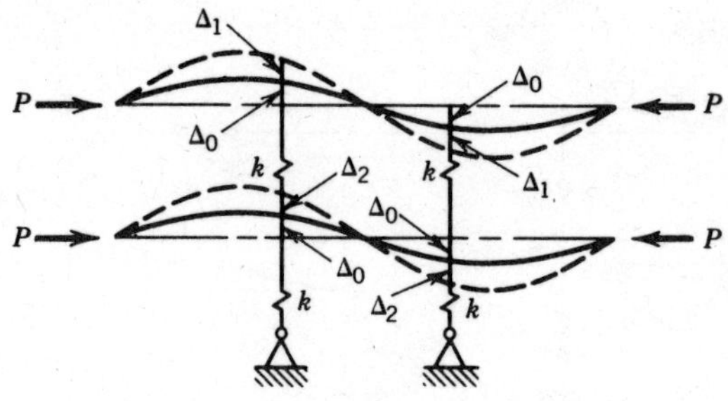

Equilibrium:

$$-F_1L/3 + P(\Delta_1 + \Delta_0) = 0$$

$$-(F_2 - F_1)L/3 + P(\Delta_2 + \Delta_0) = 0$$

$$\left.\begin{array}{l}\Delta_1(P - kL/3) + \Delta_2(kL/3) = -P\Delta_0 \\ \Delta_1(kL/3) + \Delta_2(P - 2kL/3) = -P\Delta_0\end{array}\right\}$$

$$\Delta_1 = \frac{P\Delta_0(kL - P)}{P^2 - PkL + k^2L^2/9}; \qquad \Delta_2 = \frac{P\Delta_0(2kL/3 - P)}{P^2 - PkL + k^2L^2/9}$$

Force in brace:

$$F_1 = k_{\text{act}}(\Delta_1 - \Delta_2)$$

$$F_2 = k_{\text{act}}\Delta_2$$

(*d*)

$P_{cr} = 310$ kips $= 0.1273k_{id}L$

$k_{id} =$ required (ideal) bracing stiffness necessary to reach P_{cr}

$k_{id} = 310/(0.1273 \times 192) = 12.7$ kips/in. (2.22 kN/mm)

$\Delta_0 = L/480 = 0.40$ in. (10.2 mm)

$F_{1\text{req}} = 2.01$ kips (8.94 kN); $\quad F_{2\text{req}} = 3.99$ kip (17.7 kN)

$$\frac{F_{\text{req}}}{P_{cr}} \times 100 = 1.3\%$$

Stress in brace: 3.99/3.38 = 1.2 ksi (8.3 MPa) < 4.2 ksi (29.0 MPa) O.K.

(*e*)

Fig. 5.12 (*Continued*)

Although it appears that the calculation of bracing forces is complex, it should be realized that usual minimum bracing sizes are generally quite adequate. The important point to realize is that bracing must be present at the right places along the length and on the cross section, that the bracing must be adequately attached to the beam, and that the ends of the bracing must be adequately anchored.

5.5 DESIGN OF LATERALLY UNSUPPORTED BEAMS

Design against lateral-torsional buckling is an important aspect of the design process because of the sudden and possibly catastrophic nature of failure. This is especially so during construction, as pointed out before. Lateral-torsional buckling is a complex problem, depending on many parameters which are not well defined at the time of design. In detailed analysis, such as the investigation of failures, it is necessary to perform a careful computation involving all the parameters that can be estimated. In design calculations which are governed by structural codes (specifications), it is usually adequate to provide simple rules based on conservative assumptions regarding the end restraints. Most design specifications assume simple end conditions ($u = u'' = \phi = \phi'' = 0$) of each braced segment, and a variable moment based on the moment diagram as determined by planar structural analysis. The critical elastic moment is thus determined by Eq. 5.3 with M_{0cr} computed by Eq. 5.1. Inelastic buckling is usually determined by specifying some type of empirical curve which gives essentially the elastic solution for long beams, and which terminates at $M_{cr} = M_p$, the plastic moment, for short beams.

Various approximations have been used in specifications to simplify the calculation of M_{0cr} (Eq. 5.1), and these have been described in numerous textbooks on structural steel design. Following is a discussion of the methods of treating lateral-torsional buckling in modern design specifications (SSRC, ECCS, CRCJ, CMEA, 1982).

The basis of these modern methods of design is the proper determination of the appropriate elastic critical load, based on the methods discussed earlier in this chapter for conservative end restraints. The resulting critical moment is M_E. The following empirical methods are used to determine the buckling load in the inelastic range:

1. It is assumed that beams and columns act similarly in the inelastic range, so the lateral-torsional buckling strength is determined by the appropriate column formula for an equivalent slenderness parameter

$$\lambda_{eq} = \sqrt{\frac{M_p}{M_E}} \tag{5.16}$$

For example, the multiple column curves (see Chapter 3) may be expressed as

$$\frac{F_{cr}}{F_y} = \frac{Y - \sqrt{Y^2 - 4\lambda^2}}{2\lambda^2} \leq 1.0 \tag{5.17}$$

where

$$Y = 1 + \alpha(\lambda - 0.15) + \lambda^2 \qquad \text{and} \qquad \lambda = \lambda_{eq} \tag{5.18}$$

and $\alpha = 0.103$, 0.293, and 0.662, respectively, for SSRC Column Categories 1, 2, and 3 (Chaper 3). Other column curve equations could be used, of course. Depending on whether or not the column curve includes initial imperfections, this also reflects on the beam-curve.

This approach is used in many of the lateral-torsional buckling specifications around the world and is the basis of all current U.S. specifications.

2. The West Europeans [European Committee on Constructional Steelwork (ECCS)] use an empirical beam formula of the form

$$\frac{M_{cr}}{M_p} = \left(\frac{1}{1 + \lambda_{eq}^{2n}}\right)^{1/n} \tag{5.19}$$

where n is a coefficient that equals 2.5 to 1.5, depending on the criteria set by the various national code-writing bodies in western Europe (SSRC, ECCS, CRCJ, CMEA, 1982).

3. The proposed AISC Load and Resistance Factor Design (LRFD) (Yura et al., 1978) specification uses a straight-line transition from the elastic buckling curve at $M_{cr} = M_r$, $L = L_r$ to $M_{cr} = M_p$, $L = L_p$, where

$$M_r = S_x(F_y - F_r) \tag{5.20}$$

(S_x is the elastic section modulus, F_y the yield stress, and F_r the compressive residual stress); L_r is determined from Eq. 5.1 for $M_{0cr} = M_r$, and $C_b = 1.0$; and

$$L_p = 1.762\sqrt{E/F_y}\, r_y \tag{5.21}$$

The lateral-torsional buckling strength is then determined for simply supported beams under uniform moment by the following formulas:

$$M_{cr} = M_p \qquad \text{for } L \leq L_p \tag{5.22a}$$

$$M_{cr} = C_b\left[M_p - (M_p - M_r)\frac{L - L_p}{L_r - L_p}\right] \leq M_p \qquad \text{for } L_p \leq L \leq L_r \tag{5.22b}$$

$$M_{cr} = M_E \qquad \text{for } L \geq L_r \tag{5.22c}$$

Simply supported beam under uniform moment

W24 × 55; $L = 150$ in. (3.81 m); $C_b = 1.00$

$F_y = 36$ ksi (248 MPa); $F_r = 10$ ksi (69 MPa); $E = 29{,}000$ ksi (200,000 MPa)

$G/E = 0.385$

$I_y = 29.1$ in.^{4} $(12.1 \times 10^6$ mm$^4)$; $S_x = 114$ in.^{3} $(1.87 \times 10^6$ mm$^3)$; $Z_x = 134$ in.^{3} $(2.19 \times 10^6$ mm$^3)$

$r_y = 1.34$ in. (34 mm); $J = 1.18$ in.^{4} $(0.491 \times 10^6$ mm$^4)$; $C_w = 3870$ in.^{6} $(1.039 \times 10^{12}$ mm$^6)$

$$M_{0cr} = M_E = \frac{\pi}{L}\sqrt{EI_yGJ}\sqrt{1 + \frac{\pi^2 ECw}{GJL^2}} = 4806 \text{ in.-kips (543 kN-m)}$$

$$M_p = F_y Z_x = 4824 \text{ in.-kips (545 kN-m)};\ \lambda = \sqrt{M_p/M_E} = 1.00$$

(1) *Equivalent column method*

SSRC column curve no. 2: $\alpha = 0.293$

$$Y = 1 + \alpha(\lambda - 0.15) + \lambda^2 = 2.253$$

$$\frac{M_{cr}}{M_p} = \frac{Y - \sqrt{Y^2 - 4\lambda^2}}{2\lambda^2} = 0.609$$

$M_{cr} = 2938$ in.-kips (332 kN-m)

(2) *European beam formula*, $n = 2.5$

$$\frac{M_{cr}}{M_p} = \left(\frac{1}{1 + \lambda^{2n}}\right)^{1/n} = 0.756$$

$M_{cr} = 3656$ in.-kips (413 kN-m)

(3) *AISC LRFD method*

$M_r = S_x(F_y - F_r) = 2964$ in.-kips (335 kN-m)

$$M_r = \frac{\pi}{L_r}\sqrt{EI_yGJ}\sqrt{1 + \frac{\pi^2 EC_w}{GJL_r^2}}$$

Solve for $L_r = 198.1$ in. (5.03 m) $> L = 150$ in. (3.81 m)

$$L_p = \frac{300 r_y}{\sqrt{F_y}} = 67.0 \text{ in. (1.70 m)} < L = 150 \text{ in. (3.81 m)}$$

$$M_{cr} = C_b[M_p - (M_p - M_r)(L - L_p)/(L_r - L_p)] \le M_p$$

$M_{cr} = 3646$ in.-kips (412 kN-m)

Fig. 5.13 Example of design methods.

The application of the three methods is shown by the example given in Fig. 5.13, and the three approaches are compared to each other and to analytically obtained results (Kitipornchai and Trahair, 1975b; Vinnakota, 1977a) in Fig. 5.14. The information in Fig. 5.14 pertains to a W10 × 22 beam ($E = 29{,}000$ ksi; $G/E = 0.385$, $F_y = 43.5$ ksi, and $F_r = 12.6$ ksi). When the beam is subject to uniform moment, the ECCS and the LRFD method give essentially the same result; in nonuniform bending the ECCS approach is more conservative than the LRFD method. Of the three methods the LRFD approach is closest to the prediction of inelastic buckling analyses (Yura et al., 1978).

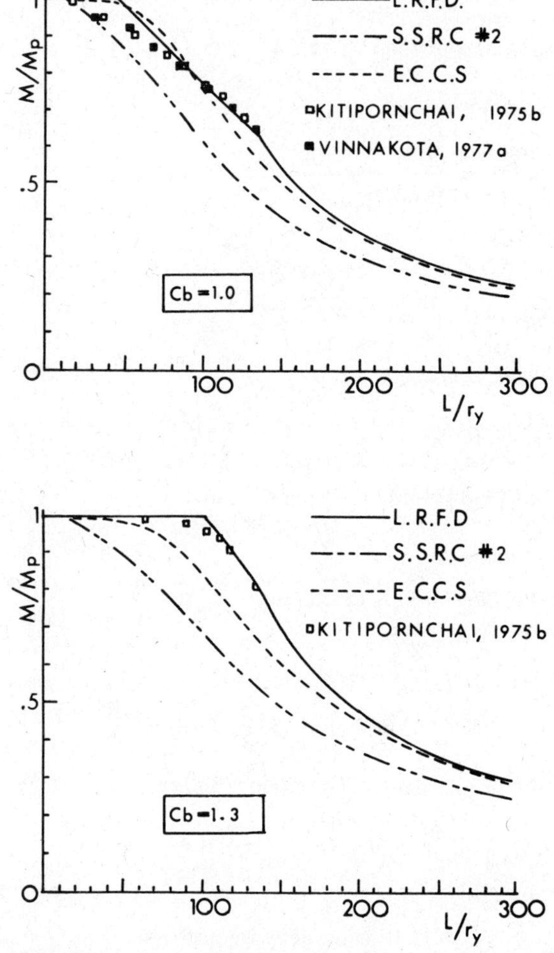

Fig. 5.14 Comparison of design methods and analytical solutions.

Further discussion of design methods and of other variations of the empirical approach are given by Nethercot and Trahair (1983) and Galambos (1983). The experimental statistical data for inelastic beam tests have been given by Fukumoto and co-workers (Fukumoto and Kubo, 1977; Fukumoto et al., 1980; Fukumoto and Itoh, 1981), who also examine the ECCS beam equation (Eq. 5.19) in the light of these tests. Galambos (1983) also uses the Fukutomo data to arrive at resistance factors for beam design according to the AISC LRFD beam strength formulas (Eqs. 5.22).

5.6 STABILITY OF HORIZONTALLY CURVED BEAMS

Curved beams are used extensively in highway bridges, as either box beams or I-shaped beams. In their final constructed stage the multiple I-shaped beams are stabilized by the concrete deck, which is connected to each beam by shear connectors. However, the individual beams are very flexible during construction, so it is necessary to consider their lateral instability.

There is an extensive literature on the first-order analysis of horizontally curved beams; see, for example, Vlasov (1961), Dabrowski (1964), McManus et al. (1969), and Heins (1975). However, there are relatively few references on the subject of second-order or stability analysis.

A horizontally curved I-shaped girder formed in a way that the curvature is in the planes of the flanges and loaded by forces causing flexure in the plane of the web will deflect vertically and horizontally from the beginning of loading, and it will also twist. These deformations will increase nonlinearly due to second-order effects as the load is increased, until the deflections become large or until yielding occurs. Thus there is no buckling in the classical sense, but failure is ultimately due to large deformations and plastification. However, a buckling-type solution has been performed by considering a deviated equilibrium configuration from an already deformed shape. This solution represents the case where the deformations tend to infinity, and the corresponding critical load cannot be practically attained. Such solutions were presented by Culver and McManus (1971), Yoo and Pfeiffer (1983), and Nakai and Kotoguchi (1983) for elastic buckling. Fukumoto and Nishida (1981) give a multiplying factor by which the elastic lateral-torsional buckling load of a simply supported straight beam of the same length is to be multiplied to obtain an "equivalent buckling load" for a curved beam, as follows:

$$MF = \sqrt{1 - \frac{\alpha^2}{\pi^2}} \tag{5.23}$$

where α is the angle (in radians) between the end tangents of the curved beam. Unfortunately, the three solutions (i.e., Culver, Yoo, and Fukumoto and Nishida) do not give the same results, especially for beams with large curvatures. However, the divergence is substantial only for cases outside the usual dimensions of curved bridges, and furthermore, the buckling condition is unattainable, as noted above.

The ultimate strength of horizontally curved girders was investigated by Culver and McManus (1971), Fukumoto and Nishida (1981), and Yoshida and Maegawa (1983, 1984). Experiments on curved girders were performed by Culver and McManus (1971) and Fukumoto and Nishida (1981). Culver and McManus (1971) used their experimental results and a simplified analysis to develop conservative multiplying factors by which the critical load of a straight beam is to be multiplied to account for curvature. These reduction factors are contained in the 1985 AASHTO Specification for Highway Bridges, and they are:

$$RF = \rho_b \rho_w \leq 1 \tag{5.24}$$

where for

$$0.297 \leq \frac{b_f}{2t_f}\sqrt{\frac{F_y}{E}} \leq 0.409 \text{ (noncompact shape)}$$

$$\rho_b = \frac{1}{1 + (L/b_f)(L/R)} \tag{5.25}$$

$$\rho_w = \frac{1}{1 + (f_w/f_b)(1 - L/b_f/75)} \tag{5.26}$$

$$\rho_w = \frac{0.95 + (L/b_f)/(30 + 8000(0.1 - L/R)^2)}{1 - 0.6(f_w/f_b)} \tag{5.27}$$

For

$$\frac{b_f}{2t_f}\sqrt{\frac{F_y}{E}} < 0.297 \text{ (compact shape)}$$

$$\rho_b = \frac{1}{1 + [(L/R) - 0.01]^2 (L/b_f)(1 + L/6b_f)} \tag{5.28}$$

$$\rho_w = 0.95 + 18\left(0.1 - \frac{L}{R}\right)^2 - \frac{(f_w/f_b)[0.3 - 0.1(L/R)(L/b_f)]}{\rho_b(F_{bs}/F_y)} \tag{5.29}$$

In these equations

L = length of the unsupported compression flange between cross diaphragms

R = radius of horizontal curvature of the beam

b_f = total flange width of I shape
f_b = first-order flexural stress in flange (taken to be positive)
f_w = first-order warping normal stress at the flange tip, positive when f_w on the flange tip closest to the center of curvature is in compression (if f_w is positive, Eq. 5.26 is to be used; otherwise, the smaller value of ρ_w from Eqs. 5.26 and 5.27 shall be used)
F_{bs} = allowable bending stress for straight beams

These reduction factors are conservative. The warping stress ratio f_w/f_b may not exceed ± 0.5, L/b_f may not exceed 25, nor may $(b_f/2t_f)\sqrt{F_y/E}$ exceed 0.409.

Fukumoto and Nishida (1981) and Yoshida and Maegawa (1983, 1984) obtained the inelastic ultimate strength of horizontally curved simply supported beams by numerical analysis, considering second-order effects, yielding and residual stresses. Numerical results are presented in curve form for various I-shaped cross sections, for a number of residual stress patterns and for several loading cases (end moments, distributed loads, central concentrated load). Excellent agreement is found between the tests and predictions. The presented data are, unfortunately, not directly suitable for use by designers. Fukumoto and Nishida (1981) give an explicit conservative estimate of the ultimate strength of curved beams with simple supports, suitable for the mildly curved beams used for highway bridges ($L/R < 0.2$). The ultimate moment M_u is obtained from the following equation:

$$\left(\frac{M_u}{M_p}\right)^4 \lambda^4 - \left\{\left[1 + \frac{P_E(d - t_f)}{2M_p}\,\frac{L^2}{2Rb_f}\right]\lambda^4 + 1\right\}\left(\frac{M_u}{M_p}\right)^2 - \frac{L^2}{2Rb_f}\,\frac{M_u}{M_p} + 1 = 0 \tag{5.30}$$

where

$$\lambda = \sqrt{\frac{M_p}{M_E}} \tag{5.31}$$

$$P_E = \frac{\pi^2 EI_y}{L^2} \tag{5.32}$$

and

M_p = plastic moment
M_E = elastic lateral-torsional buckling moment of straight beam
d = depth of beam

Until further studies provide more general design approximations, the AASHTO multiplication factors or Fukumoto's equation (Eq. 5.30) can be used to provide conservative solutions.

References

Akay, H. V., and Johnson, C. P. (1977), "Local and Lateral Buckling of Beams and Frames," *ASCE J. Struct. Div.*, Vol. 103, No. ST9, pp. 1821–1832.

Allen, H. G., and Bulson, P. S. (1980), *Background to Buckling*, McGraw-Hill, Maidenhead, Berkshire, England.

Anderson, J. M., and Trahair, N. S. (1972), "Stability of Monosymmetric Beams and Cantilevers," *ASCE J. Struct. Div.*, Vol. 98, No. ST1, pp. 269–286.

Austin, W. J., Yegian, S., and Tung, T. P. (1957), "Lateral Buckling of Elastically End-Restrained Beams," *Trans. Am. Soc. Civ. Eng.*, Vol. 122, pp. 374–409.

Barsoum, R. S., and Gallagher, R. H. (1970), "Finite Element Analysis of Torsional and Flexural-Torsional Stability Problems," *Int. J. Numer. Methods Eng.*, Vol. 2, pp. 335–352.

Bleich, F. (1952), *Buckling Strength of Metal Structures*, McGraw-Hill, New York.

Brown, P. T., and Trahair, N. S. (1968), "Finite Integral Solutions of Differential Equations," *Civ. Eng. Trans. Inst. Eng. Aust.*, Vol. CE10, No. 2.

Brush, D. O., and Almroth, B. O. (1975), *Buckling of Bars, Plates and Shells*, McGraw-Hill, New York.

Celigoj, C. (1979), "Influence of Initial Deformations on the Carrying-Capacity of Beams," *Stahlbau*, Vol. 48, pp. 69–75, 117–121 (in German).

Chajes, A. (1974), *Principles of Structural Stability Theory*, Prentice-Hall, Englewood Cliffs, N.J.

Clark, J. W., and Hill, H. N. (1960), "Lateral Buckling of Beams," *ASCE J. Struct. Div.*, Vol. 86, No. ST7, pp. 175–196.

Culver, C., and McManus, P. (1971), Instability of Curved Girders—Lateral Buckling of Curved Plate Girders, Carnegie-Mellon University, Dept. of Civ. Eng., September.

Dabrowski, R. (1964), "Analysis of Curved Beams of Thin-Walled Open Cross Section," *Stahlbau*, Vol. 12, pp. 364–372 (in German).

Fukumoto, Y., and Itoh, Y. (1981), "Statistical Study of Experiments on Welded Beams," *ASCE J. Struct. Div.*, Vol. 107, No. ST1, pp. 89–104.

Fukumoto, Y., and Kubo, M. (1977), "An Experimental Review of Lateral Buckling of Beams and Girders," *Proc. Int. Coll. Stab. Struct. Under Static Dyn. Loads*, SSRC, Washington D.C., March.

Fukumoto, Y., and Nishida, S. (1981), "Analysis of Curved Beams of Thin-Walled Open Cross Section," *ASCE J. Eng. Mech. Div.*, Vol. 107, No. EM2, pp. 367–385.

Fukumoto, Y., Itoh, Y., and Kubo, M. (1980), "Strength Variation of Laterally Unsupported Beams," *ASCE J. Struct. Div.*, Vol. 106, No. ST1, pp. 165–182.

Galambos, T. V. (1963), "Inelastic Lateral Buckling of Beams," *ASCE J. Struct. Div.*, Vol. 89, No. ST5, pp. 217–244.

Galambos, T. V. (1968), *Structural Members and Frames*, Prentice-Hall, Englewood Cliffs, N.J.

Galambos, T. V. (1970), "Spacing of Bridging for open-Web Steel Joists," *Tech. Dig.*, No. 2, Steel Joist Institute.

Galambos, T. V. (1983), "A World View of Beam Stability Research and Practice," *Proc. SSRC Annu. Meet.*, Toronto, May.

Gedies, R. W. (1983), "Beam Buckling Tests with Various Brace Stiffnesses," M.S. thesis, University of Texas at Austin.

Goldberg, J. E., Bogdanoff, J. L., and Glantz, W. D. (1964), "Lateral and Torsional Buckling of Thin-Walled Beams," *Proc. IABSE*, Vol. 24, pp. 92–100.

Hancock, G. J. (1978), "Local Distortional and Lateral Buckling of I-Beams," *ASCE J. Struct. Div.*, Vol. 104, No. ST11, pp. 1787–1800.

Hartmann, A. J. (1967), "Elastic Lateral Buckling of Continuous Beams," *ASCE J. Struct. Div.*, Vol. 93, No. ST4, pp. 11–28.

Heins, C. P. (1975), *Bending and Torsional Design in Structural Members*, Lexington Books, Lexington, Mass.

Heins, C. P., and Potocko, R. A. (1979), "Torsional Stiffening of I-Girder Webs," *ASCE J. Struct. Div.*, Vol. 105, No. ST8, pp. 1689–1700.

Horne, M. R. (1954), "The Flexural-Torsional Buckling of Members of Symm. I-Sect. under Comb. Thrust and Unequal Terminal Moments," *Q. J. Mech. Appl. Math.*, Vol. 7, Part 4.

Horne, M. R. (1960), "The Elastic Stability of Trusses," *Struct. Eng.*, Vol. 38.

Hribar, J. A., and Laughlin, W. P. (1968), "Lateral Stability of Welded Light Trusses," *ASCE J. Struct. Div.*, Vol. 94, No. ST3, pp. 809–832.

Johnston, B. G., ed. (1976), *Guide to Stability Design Criteria for Metal Structures* (SSRC), Wiley, New York.

Kirby, P. A., and Nethercot, D. A. (1979), *Design for Structural Stability*, Wiley, New York.

Kitipornchai, S., and Dux, P. F. (1985), "Buckling and Bracing of Elastic Beams and Cantilevers," in *Handbook of Civil Engineering*, Technomic Publishing Co., Lancaster, Pa.

Kitipornchai, S., and Trahair, N. S. (1972), "Elastic Stability of Tapered I-Beams," *ASCE J. Struct. Div.*, Vol. 98, No. ST3, pp. 713–728.

Kitipornchai, S., and Trahair, N. S. (1975a), "Elastic Stability of Tapered Monosymmetric Beams," *ASCE J. Struct. Div.*, Vol. 101, No. ST8, pp. 1661–1678.

Kitipornchai, S., and Trahair, N. S. (1975b), "Inelastic Buckling of Simply Supported Steel I-Beams," *ASCE J. Struct. Div.*, Vol. 101, No. ST7, pp. 1333–1348.

Kitipornchai, S., and Trahair, N. S. (1980), "Buckling Properties of Monosymmetric I-Beams," *ASCE J. Struct. Div.*, Vol. 106, No. ST5, pp. 941–958.

Lay, M. G., and Galambos, T. V. (1965), "Inelastic Steel Beams under Uniform Moment," *ASCE J. Struct. Div.*, Vol. 91, No. ST6, pp. 67–94.

Lay, M. G., and Galambos, T. V. (1967), "Inelastic Beams under Moment Gradient," *ASCE J. Struct. Div.*, Vol. 93, No. ST1, pp. 381–400.

Lee, G. C. (1960), "A Survey of the Literature on the Lateral Instability of Beams," *Weld. Res. Counc. Bull.*, No. 63, August.

Lindner, J. (1974), "Influence of Residual Stresses on the Load-Carrying Capacity of I-Beams," *Stahlbau*, Vol. 43, pp. 39–45, 86–91 (in German).

Lindner, J. (1985), "Influence of Parameters of Lateral Bracing on the Load-Carrying Capac. of Beams," *Proc. SSRC Annu. Tech. Sess.*, Cleveland, April.

Lindner, J., and Gietzelt, R. (1983), "Influence of End Plates on the Ultimate Load of Laterally Unsupported Beams," in *Instability and Plastic Collapse of Steel Strucures* (*M. R. Horne Conf.*), Manchester, Granada, London, pp. 538–546.

Massey, C., and McGuire, P. J. (1971), "Lateral Stability of Nonuniform Cantilevers," *ASCE J. Eng. Mech. Div.*, Vol. 93, No. EM3, pp. 673–686.

Massonnet, C. (1947), "Buckling of Thin-Walled Bars with open Cross Section," *Hommage Fac. Sc. Appl. Univ. Liege* (ed. G. Thone) (in French).

Masur, E. F., and Milbrandt, K. P. (1957), "Collapse Strength of Redundant Beams after Lateral Buckling," *J. Appl. Mech. ASME*, Vol. 24, No. 2, pp.

McGuire, W. (1968), *Steel Structures*, Prentice-Hall, Englewood Cliffs, N.J.

McManus, P. F., Nasir, G., and Culver, C. (1969), "Horizontally Curved Girders—State of the Art," *ASCE J. Struct. Div.*, Vol. 95, No. ST5, pp. 853–870.

Medland, I. C. (1980), "Buckling of Interbraced Beam-Systems," *Eng. Struct.*, Vol. 2, pp. 90–96.

Mutton, B. R., and Trahair, N. S. (1973), "Stiffness Requirements for Lateral Bracing," *ASCE J. Struct. Div.*, Vol. 99, No. ST10, pp. 2167–2182.

Nakai, H., and Kotoguchi, H. (1983), "A Study on Lateral Buckling Strength and Design Aid for Horizontally Curved I-Girder Bridges," *Proc. Jpn. Soc. Civ. Eng.*, No. 339, pp. 195–204.

Neal, B. G. (1950), "The Lateral Instability of Yielded Mild Steel Beams of Rectangular Cross Section," *Phil. Trans. R. Soc. London*, Vol. A242.

Nethercot, D. A. (1973a), "Buckling of Laterally and Torsionally Restrained Beams," *ASCE J. Eng. Mech. Div.*, Vol. 99, No. EM4, pp. 773–792.

Nethercot, D. A. (1973b), "Lateral Buckling of Tapered I-Beams," *IABSE Pub.*, Vol. 33-II, pp. 173–192.

Nethercot, D. A. (1973c), "The Effective Lengths of Cantilevers as Governed by Lateral Buckling," *Struct. Eng.*, Vol. 51, No. 5.

Nethercot, D. A. (1983), "Elastic Lateral Buckling of Beams," in *Beams and Beam Columns—Stability in Strength* (ed. R. Narayanan), Applied Science Publishers, Barking, Essex, England.

Nethercot, D. A., and Rockey, K. C. (1972), "A Unified Approach to the Elastic Lateral Buckling of Beams," *Eng. J. Am. Inst. Steel Constr.*, Vol. 9, No. 3, pp. 96–107.

Nethercot, D. A., and Trahair, N. S. (1975), "Design of Diaphragm-Braced I-Beams," *ASCE J. Struct. Div.*, Vol. 101, No. ST10, pp. 2045–2062.

Nethercot, D. A., and Trahair, N. S. (1983), "Design of Laterally Unsupported Beams," in *Beams and Beam-Columns—Stability and Strength* (ed. R. Narayanan), Applied Science Publishers, Barking, Essex, England.

Nethercot, D. A., and Trahair, N. S. (1984), *Developments in Thin-Walled Structures*, Vol. 2, Bracing of Thin-Walled Structures, Applied Science Publishers, Barking, Essex, England.

Ojalvo, M. (1981), "Wagner Hypothesis in Beam and Column Theory," *ASCE J. Eng. Mech. Div.*, Vol. 107, No. EM4, pp. 649–658.

Ojalvo, M., and Chambers, R. S. (1977), "Effects of Warping Restraints on I-Beam Buckling," *ASCE J. Struct. Div.*, Vol. 103, No. ST12, pp. 2351–2360.

Peköz, T. (1983), "Diaphragm-Braced Thin-Walled Channel and Z-Section Beams," in *Beams and Beam-Columns—Stability and Strength* (ed. R. Narayanan), Applied Science Publishers, Barking, Essex, England.

Plank, R. J., and Wittrick, W. H. (1974), "Buckling under Combined Loading of Thin Flat-Walled Structures by a Complex Finite Strip Method," *Int. J. Numer. Methods Eng.*, Vol. 8, No. 2.

Powell, G., and Klingner, R. (1970), "Elastic Lateral Buckling of Steel Beams," *ASCE J. Struct. Div.*, Vol. 96, No. ST9, pp. 1919–1932.

Rajasekaran, S., and Murray, D. W. (1973), "Coupled Local Buckling in Wide-Flange Beam-Columns," *ASCE J. Struct. Div.*, Vol. 99, No. ST6, pp. 1003–1024.

Salvadori, M. G. (1955), "Lateral Buckling of I-Beams," *Trans. Am. Soc. Civ. Eng.*, Vol. 120, p. 1165.

Salvadori, M. G. (1956), "Lateral Buckling of Eccentrically Loaded I-Columns," *Trans. Am. Soc. Civ. Eng.*, Vol. 121, p. 1163.

SSRC, ECCS, CRCJ, CMEA (1982), *Stability of Metal Structures—A World View*, American Institute of Steel Construction, Chicago.

Szewczak, R. M., Smith, E. A., and DeWolf, J. T. (1983), "Beams with Torsional Stiffeners," *ASCE J. Struct. Div.*, Vol. 109, No. ST7, pp. 1635–1647.

Timoshenko, S. P., and Gere, J. M. (1961), *Theory of Elastic Stability*, McGraw-Hill, New York.

Trahair, N. S. (1969), "Elastic Stability of Continuous Beams," *ASCE J. Struct. Div.*, Vol. 95, No. ST6, pp. 1295–1312.

Trahair, N. S. (1977a), "Lateral Buckling of Beams and Beam-Columns," in *Theory of Beam-Columns*, Vol. 2, W. F. Chen and T. Atsuta, McGraw-Hill, New York, Chap. 3.

Trahair, N. S. (1977b), *The Behaviour and Design of Steel Structures*, Chapman & Hall, London.

Trahair, N. S. (1983), "Inelastic Lateral Buckling of Beams," *Beams and Beam-Columns—Stability and Strength* (ed. R. Narayanan), Applied Science Publishers, Barking, Essex, England.

Trahair, N. S., and Woolcock, S. T. (1973), "Effect of Major Axis Curvature on I-Beam Stability," *ASCE J. Eng. Mech. Div.*, Vol. 99, No. EM1, pp. 85–98.

Vacharajittiphan, P., and Trahair, N. S. (1974), "Warping and Distortion of I-Section Joints," *ASCE J. Struct. Div.*, Vol. 100, No. ST3, pp. 547–564.

Vacharajittiphan, P., and Trahair, N. S. (1975), "Analysis of Lateral Buckling in Plane Frames," *ASCE J. Struct. Div.*, Vol. 101, No. ST7, pp. 1497–1516.

Vinnakota, S. (1977a), "Inelastic Stability of Laterally Unsupported Beams," *Comput. Struct.*, Vol. 7, No. 3.

Vinnakota, S. (1977b), "Finite Difference Methods for Plastic Beam-Columns," in *Theory of Beam-Columns*, Vol. 2, W. F. Chen and T. Atsuta, McGraw-Hill, New York, Chap. 10.

Vlasov, V. Z. (1961), *Thin-Walled Elastic Beams*, Israel Program for Scientific Translations, Jerusalem.

Winter, G. (1960), "Lateral Bracing of Columns and Beams," *Trans. Am. Soc. Civ. Eng.*, Vol. 125, Part I, pp. 808–845.

Woolcock, S. T., and Trahair, N. S. (1974), "Post-Buckling of Determinate I-Beams," *ASCE J. Eng. Mech. Div.*, Vol. 100, No. EM2, pp. 151–172.

Woolcock, S. T., and Trahair, N. S. (1976), "Post Buckling of Redundant I-Beams," *ASCE J. Eng. Mech. Div.*, Vol. 102, No. EM2, pp. 293–312.

Yoo, C., and Pfeiffer, P. A. (1983), "Elastic Stability of Curved Members," *ASCE J. Struct. Eng.*, Vol. 109, No. 12, pp. 2922–2940.

Yoshida, H., and Maegawa, K. (1983), "Ultimate Strength Analysis of Curved I-Beams," *ASCE J. Eng. Mech.*, Vol. 109, No. 1, pp. 192–214.

Yoshida, H., and Maegawa, K. (1984), "Lateral Instability of I-Beams with Imperfections," *ASCE J. Struct. Eng.*, Vol. 110, No. 8, pp. 1875–1892.

Yura, J. A., and Cheng, J. (1985), "Lateral and Local Instability of Coped Beams," *Proc. SSRC Tech. Sess.*, Cleveland, April.

Yura, J. A., Galambos, T. V., and Ravindra, M. K. (1978), "The Bending Resistance of Steel Beams," *ASCE J. Struct. Div.*, Vol. 104, No. ST9, pp. 1355–1370.

Chapter Six

Plate Girders

6.1 INTRODUCTION

This chapter deals with the buckling and strength of plate girders. Curved and box girders are not covered. Considerations of buckling involve not only lateral-torsional buckling and local buckling of the flange, as in the case of beams, but buckling of the web as well. The web is often reinforced with transverse stiffeners, and occasionally with longitudinal stiffeners, to increase its resistance to buckling, and design involves finding a combination of plate thickness and stiffener spacing that will be economical in material and fabrication.

The strength of a centrally loaded column is usually only slightly larger than the theoretical buckling load, but the strength of a stiffened plate may be much larger than its buckling load. Nevertheless, buckling was accepted as a basis for design of plate-girder webs almost exclusively until the early 1960s. This was due principally to the fact that formulas for predicting buckling are relatively simple and have been known for many years, while suitable analyses of postbuckling strength are relatively new. However, the postbuckling strength was acknowledged in most specifications by using smaller factors of safety for web buckling than for yielding or failure of other elements.

The source of the postbuckling strength of stiffened plate-girder webs in shear was explained by Wilson (1886). He had observed in his early experience (25 years before) that railroad plate-girder bridges with webs $\frac{3}{16}$ in. thick and stiffeners at intervals of 5 ft were "bearing up well under use." He discovered:

> By means of a paper model with a very thin flexible web, that when stiffeners were properly introduced the web no longer resisted by compression, but by tension, the stiffeners taking up the duty of compressive resistance, like the posts of a Pratt truss, and dividing the girder into panels equivalent to those of an open truss, the web in each panel acting as an inclined tie.

Wilson says that using this theory he obtained "results that quite agreed with practical examples," but he does not explain his analysis.

Wagner (1931) developed a diagonal-tension theory of web shear. Wagner's work was extended by Kuhn (1956) for applications in aircraft design. Extensive studies, both analytical and experimental, were made in the late 1950s by Basler and Thürlimann on the postbuckling behavior of web panels in bending as well as in shear (Basler, 1963a,b; Basler and Thürliman, 1960a,b, 1963; Basler et al., 1960; Yen and Basler, 1962). Practical procedures were developed and have been adopted in many specifications. Widespread interest in the subject has resulted in a number of modifications to the Basler-Thürlimann approach to achieve better correlation between theory and tests.

Two approaches to the design of plate-girder webs are used: (1) design based on buckling as a limiting condition, with a relatively low factor of safety to allow for postbuckling strength, and (2) design based on yielding or ultimate strength as a limiting condition, with the same factor of safety as for yielding or ultimate strength of other structural members.

6.2 WEB BUCKLING AS A BASIS FOR DESIGN

Where buckling is taken as the basis for design of plate-girder webs the maximum stress in the web, computed by conventional beam theory, should not exceed the buckling stress divided by a factor of safety. A number of formulas and extensive charts and tables have been developed for the buckling analysis of stiffened and unstiffened plates. Buckling of rectangular plate panels is discussed in Chapter 4.

The geometric parameters that determine buckling of plate-girder webs are the web thickness t, the web depth h between flanges, and the spacing a of transverse stiffeners. Four limiting values of web slenderness must be established:

1. A limiting value of h/t to control flexural buckling of a web with no longitudinal stiffener.
2. A limiting value of h/t to control flexural buckling of a web with a longitudinal stiffener.
3. A limiting value of h/t to control shear buckling of a web with no transverse stiffeners.
4. A limiting value of a/t to control shear buckling of a web with transverse stiffeners.

These limits are chosen to give a small factor of safety against buckling. It is usually assumed that the web panel is simply supported at the flanges and stiffeners. For example, if a factor of safety of 1.25 with respect to bend buckling of a web with no longitudinal stiffener is desired, the limiting value of h/t is found from Eq. 4.1 as

$$1.25f_b = \frac{23.9\pi^2 E}{12(1-\nu^2)(h/t)^2} \tag{6.1}$$

where f_b is the service-load bending stress and $k = 23.9$ is from Fig. 4.8. Equation 6.1 gives

$$\frac{h}{t} = \frac{23{,}000}{\sqrt{f_{b,\mathrm{psi}}}} = 4.2\sqrt{\frac{E}{f_b}} \tag{6.2}$$

This is the limiting value of the allowable stress design method of the 1984 AASHTO specifications. The limit can also be expressed in terms of F_y, as in the 1979 AREA formula

$$\frac{h}{t} = \frac{32{,}500}{\sqrt{F_{y,\mathrm{psi}}}} = 6.04\sqrt{\frac{E}{F_y}} \tag{6.3}$$

Equations 6.2 and 6.3 give almost the same results for a web whose bending stress f_b equals the allowable value $0.55F_y$. Limiting values of h/t for cases 2 and 3 can be derived similarly. However, the derivation of formulas for spacing of transverse stiffeners (case 4) is less simple because of the dependence of k_s on the aspect ratio $\alpha = a/h$ (Gaylord and Gaylord, 1972). AREA neglects this dependence and arrives at a simple formula of the same form as Eqs. 6.2 and 6.3.

$$\frac{a}{t} = \frac{10{,}500}{\sqrt{f_{v,\mathrm{psi}}}} \tag{6.4a}$$

where f_v is the average calculated service-load shear stress in psi. AASHTO, by assuming that $k_s = 5(1 + 1/\alpha^2)$ for all values of α, limits the allowable shear stress from web buckling in a web panel to

$$f_v = \frac{7 \times 10^7(1 + 1/\alpha^2)}{(h/t)^2} \le \tfrac{1}{3}F_y \tag{6.4b}$$

Equation 6.4b can be rewritten in the following form:

$$\frac{a}{t} = \frac{8370}{\sqrt{f_v - [8370/(h/t)]^2}} \tag{6.4c}$$

Equations 6.4b and 6.4c contain the term of web depth-to-thickness ratio, h/t, and these equations should therefore provide more accurate results than Eq. 6.4.

The procedures given above for control of web slenderness are generally conservative and, in some cases, extravagant, because they neglect the postbuckling strength of the web.

6.3 SHEAR STRENGTH OF PLATE GIRDERS

In evaluating the behavior of a plate girder subjected to shear it is assumed that the web is plane and the material is elastic-plastic. Such a web buckles at a stress that can be predicted theoretically. The load at this stage corresponds to the beam-action strength of the girder. Subsequent to buckling the stress distribution in the web changes and considerable postbuckling strength may be realized because of the diagonal tension that develops. This is called the tension-field action. Even without transverse stiffeners a plate girder can develop a shear stress at the ultimate load which is several times the shear-buckling stress. Figure 6.1 shows the general distribution of the tension field that develops in a plate girder with transverse stiffeners. This stress distribution has been verified experimentally (Basler et al., 1960; Clark and Sharp, 1971; Steinhardt and Schröter, 1971).

The tension field in the girder with stiffeners is anchored by the flanges and stiffeners. The resulting lateral load on the flanges is illustrated in Fig. 6.2, and it is clear that this causes the flanges to bend inward. Therefore, the nature of the tension field is influenced by the bending

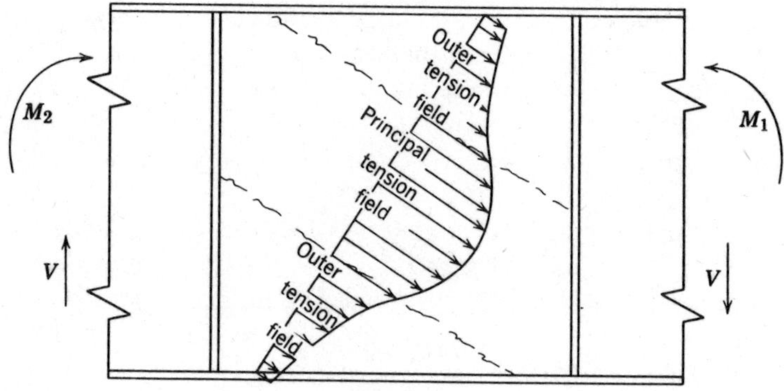

Fig. 6.1 Tension-field action.

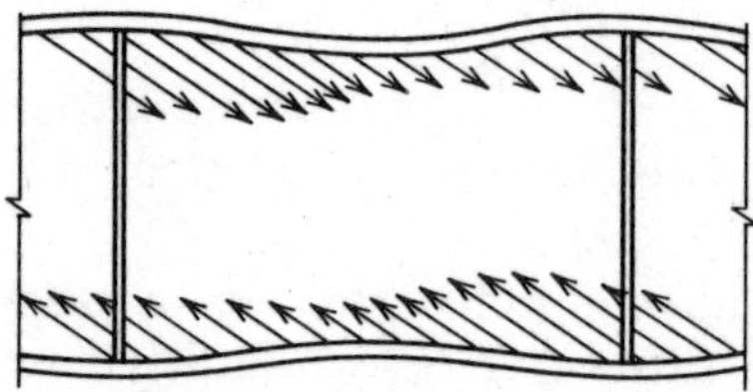

Fig. 6.2 Flange resistance.

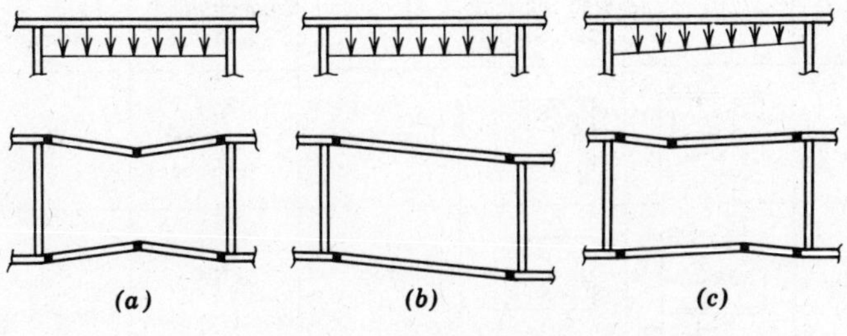

Fig. 6.3 Frame action.

stiffness of the flanges. For example, if the stiffness of the flanges is large relative to the web, the tension field may be uniform over the entire panel. With continued increase in load the tensile membrane stress combines with the shear-buckling stress to cause yielding of the web, and failure of the panel occurs upon formation of a mechanism involving a yielded zone in the web and plastic hinges in the flanges. Figure 6.3 shows three possible failure modes, involving (1) a beam mechanism in each flange, (2) a panel mechanism, and (3) a combined mechanism. The additional shear associated with the formation of a failure mechanism involving plastic hinges in the flanges is called the frame action (Cescotto et al., 1981).

Wagner (1931) used a complete, uniform tension field to determine the strength of a panel in pure shear. The flanges are assumed to be rigid and the web very thin. This model corresponds to that shown in Fig. 6.3*b* except that there are no plastic hinges, since the flanges are infinitely stiff. The Wagner analysis has been found to be quite satisfactory for aircraft structures.

Basler and Thürlimann (Basler, 1963a) were the first to formulate a successful model for plate girders of the type used in civil engineering structures. They assume that the flanges are too flexible to support a lateral loading from the tension field, so that the yield band shown in Table 6.1 determines the shear strength. The inclination and width of the yield band are defined by the angle θ, which is chosen so as to maximize the shear strength. The shear stress τ_u for the optimal value of θ was found to be*

$$\tau_u = \tau_{cr} + \tfrac{1}{2}\sigma_t \sin \theta_d \tag{6.5}$$

where

* Basler gives the result in terms of the aspect ratio α instead of θ_d.

Table 6.1

Investigator	Mechanism	Web Buckling Edge Support	Unequal Flanges	Longitudinal Stiffener	Shear and Moment
Basler (1963-a)	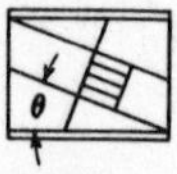 θ	S S S S	Immaterial	Yes, Cooper (1965)	Yes
Takeuchi (1964)	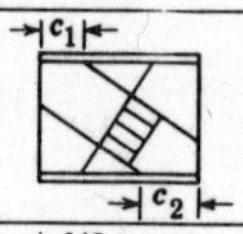 c_1, c_2	S S S S	Yes	No	No
Fujii (1968, 1971)	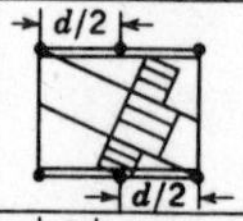 $d/2$, $d/2$	F S S F	Yes	Yes	Yes
Komatsu (1971)	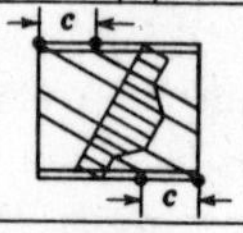 c, c	F S S F	No	Yes, at mid-depth	No
Chem and Ostapenko (1969)		F S S F	Yes	Yes	Yes
Porter et al. (1975)	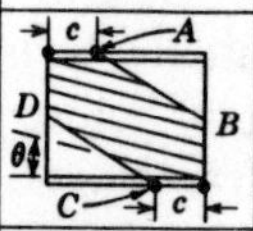 c, A, D, B, θ, C, c	S S S S	Yes	Yes	Yes
Hoglund (1971-a, b)	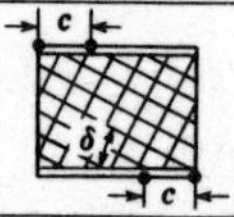 c, δ, c	S S S S	No	No	Yes
Herzog (1974-a, b)	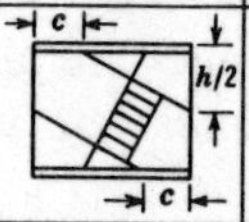 c, $h/2$, c	Web buckling component neglected	Yes, in evaluating c	Yes	Yes
Sharp and Clark (1971)		$F/2$ S S $F/2$	No	No	No
Steinhardt and Schroter (1971)		S S S S	Yes	Yes	Yes

τ_{cr} = shear-buckling stress
σ_t = tension-field stress
θ_d = angle of panel diagonal with flange

Combining the beam shear τ_{cr} and the postbuckling tension σ_t and substituting the results in the Mises yield condition gives

$$\sigma_t = -\tfrac{3}{2}\tau_{cr}\sin 2\theta + \sqrt{\sigma_{yw}^2 + (\tfrac{9}{4}\sin^2 2\theta - 3)\tau_{cr}^2} \tag{6.6}$$

The maximum value of τ_u is then found by substituting σ_t from Eq. 6.6 into Eq. 6.5.

To obtain a simpler solution Basler assumes that σ_t and τ_{cr} are additive, as they would be if σ_t acted at 45°, and uses the resulting combination of principal stresses in a linear approximation of the Mises yield condition. This gives

$$\sigma_t = \sigma_{yw}\left(1 - \frac{\tau_{cr}}{\tau_{yw}}\right) \tag{6.7}$$

where σ_{yw} and τ_{yw} are the web yield stresses in tension and shear, respectively. Values of τ_u by this approximation are less than those by Eq. 6.6, but an investigation of girders with a wide range of proportions showed that the maximum difference was less than 10%. Substituting σ_t from Eq. 6.7 into Eq. 6.5 gives

$$\tau_u = \tau_{cr} + \tfrac{1}{2}\sigma_{yw}\left(1 - \frac{\tau_{cr}}{\tau_{yw}}\right)\sin\theta_d \tag{6.8}$$

Basler assumes that inelastic buckling will occur if τ_{cr} exceeds $0.8\tau_{yw}$ and takes the inelastic buckling stress τ_{cri} to be

$$\tau_{cri} = \sqrt{0.8\tau_{cr}\tau_{yw}} \qquad 0.8\tau_{yw} \lesssim \tau_{cr} \lesssim 1.25\tau_{yw} \tag{6.9}$$

This value is to be substituted for τ_{cr} in Eq. 6.8.

It was shown first by Gaylord (1963) and later by Fujii (1968a) and Selberg (1973) that Basler's formula gives the shear strength for a complete tension field instead of the limited band of Table 6.1. The correct formula for the limited band is

$$\tau_u = \tau_{cr} + \sigma_{yw}\left(1 - \frac{\tau_{cr}}{\tau_{yw}}\right)\frac{\sin\theta_d}{2 + \cos\theta_d} \tag{6.10}$$

Therefore, Basler's formula overestimates the shear strength of a girder whose flanges are incapable of supporting lateral load from a tension field.

Many variations of the postbuckling tension field have been developed since the Basler-Thürlimann solution was published. The principal characteristics of most of these are shown in Table 6.1 and are discussed in the following paragraphs. The table shows the tension field, the positions of the plastic hinges if they are involved in the solution, the edge conditions assumed in computing the shear-buckling stress, and other features of the solutions. In all cases except the Fujii and Herzog models the shear-buckling strength is added to the vertical component of the tension field to give the contribution of the web to the shear strength of the girder panel.

Takeuchi (1964) appears to have been the first to make an allowance for the effect of flange stiffness on the yield zone in the web. He located the boundaries of the tension field at the distances c_1 and c_2 from diagonally opposite corners of the panel (Table 6.1). These distances were assumed proportional to the respective flange stiffnesses I_{f1} and I_{f2} and were chosen to maximize the shear strength. However, shear strengths determined in this way were not in good agreement with test results (Konishi et al., 1965). Lew and Toprac (1968) used this model in their investigation of hybrid plate girders and determined c_1 by a formula established to give agreement with available test results.

Fujii (1968a,b) assumes a tension field encompassing the whole panel, together with beam mechanisms in each flange with the interior hinge at midpanel (Table 6.1). The web compression in the direction perpendicular to the principal tension is assumed equal to the compression in that direction at the initiation of buckling. Tresca's yield condition is then used to determine the magnitude of the tension. If the flanges can resist the tributary web stress with the web in the yield condition, the web yields uniformly over the panel, but if they cannot, there is a central band of yielding with a smaller tension equal to that which the flange can support in the outer triangular portions. Inelastic shear buckling of the web is assumed to begin at $\tau_{cr} = 0.5\tau_{yw}$, and to vary parabolically from that point to τ_{yw} at $h/t = 0$. The theory was extended to include unsymmetrical girders (Fujii, 1971).

Komatsu (1971) gives formulas for four modes of failure. Failure in the first mode occurs in the manner shown in Table 6.1, where the inner band yields under the combined action of the buckling stress and the postbuckling tension field, while the smaller tension in the outer bands is the value that can be supported by the girder flange as a beam mechanism with the interior hinge at the distance c determined by an empirical formula based on tests. The inclination of the yield band is determined so as to maximize the shear, as in Basler's solution, but the optimum inclination must be determined by trial. In the second mode, which is a limiting case of the first mode, the interior hinge develops at midpanel, and the web yields uniformly throughout the panel. In the third mode of failure the flanges are assumed to remain elastic while allowing complete yielding of the web. An optimum value of the tension-field inclination must also be found by trial for this case. The fourth case is a limiting case in which a Wagner field develops along with a panel mechanism of the flanges.

Chern and Ostapenko (1969) proposed the tension field shown in Table 6.1, where the principal band is determined by yielding, taking into account the stress that exists at buckling. A panel mechanism is assumed to develop in the flanges. The resulting ultimate shear strength V_u is given as

$$V_u = \tau_{cr} A_w + \tfrac{1}{2}\sigma_t A_w[\sin 2\theta - (1-\rho)\alpha + (1-\rho)\alpha \cos 2\theta] + \frac{2}{a}(m_{pb} + m_{pt}) \qquad (6.11)$$

where

ρ = ratio of outer-band tension to inner-band tension
m_{pb}, m_{pt} = plastic resisting moment of bottom and top flange
α = panel aspect ratio

Equation 6.6 is used to determine σ_t. In computing m_p the flange is assumed to act with an effective width of web b_e given by

$$b_e = 12.5t\left(0.8 - \frac{\tau_{cr}}{\tau_{yw}}\right) \qquad \frac{\tau_{cr}}{\tau_y} \lessapprox 0.8 \qquad (6.12)$$

Two categories of shear buckling are used to determine τ_{cr} in Eq. 6.6 (1) elastic buckling, assuming the web panel fixed at the flanges and simply supported at the stiffeners, and (2) inelastic buckling if τ_{cr} for elastic buckling exceeds $0.5\tau_{yw}$. The authors differentiate V_u with respect to θ to develop a formula for the optimum value of θ, which must be solved by trial.

The tension field of Porter et al. (1975) shown in Table 6.1 consists of a single band and is a development of one suggested earlier by Rockey and Skaloud (1972) in which the tension band was taken in the direction of the panel diagonal. The tensile membrane stress together with the buckling stress causes yielding, and failure occurs when hinges form in the flanges to produce a combined mechanism that includes the yield zone $ABCD$. The vertical component of the tension field is added to the shear at buckling and combined with the frame action shear. The resulting ultimate shear strength is

$$V_u = \tau_{cr} A_w + \sigma_t A_w\left(\frac{2c}{h} + \cot\theta - \cot\theta_d\right)\sin^2\theta + \frac{4m_p}{c} \qquad (6.13)$$

The coordinate c of the plastic hinge is

$$c = \frac{2}{\sin\theta}\sqrt{\frac{m_p}{\sigma_t t}} \qquad 0 \le c \le a \tag{6.14}$$

where m_p is the plastic moment of resistance of the flange. The flange is assumed to act with an effective width of web b_e given by

$$b_e = 30t\left(1 - 2\frac{\tau_{cr}}{\tau_{yw}}\right) \qquad \frac{\tau_{cr}}{\tau_{yw}} \lessapprox 0.5 \tag{6.15}$$

Formulas to reduce m_p for the effect of the flange axial force are given. The authors suggest Eq. 6.6 for determining σ_t. The elastic shear-buckling stress is calculated with the four edges of the panel simply supported, but if this value exceeds $0.8\tau_{yw}$ inelastic buckling is assumed to occur, with τ_{cri} given by

$$\frac{\tau_{cri}}{\tau_{yw}} = 1 - 0.16\frac{\tau_{yw}}{\tau_{cr}} \tag{6.16}$$

The maximum value of V_u must be found by trial. However, for a given panel θ is the only independent variable in Eq. 6.13, and the maximum is not difficult to determine, particularly since the optimal value of θ usually lies between $\theta_d/2$ and 45° and τ_u is not sensitive to small changes from the optimum of θ. An assumption of $\theta = 0.667\theta_d$ will give either a very close approximation or an underestimation of the failure load.

If the flanges cannot develop moment, then $m_p = 0$ and Eq. 6.14 yields $c = 0$. Substituting this value in Eq. 6.13 and maximizing τ_u by differentiating with respect to θ the true Basler solution (Eq. 6.10) results. It is shown by Porter et al. (1975) that Eq. 6.13 includes several other existing solutions as special cases. A procedure for evaluating the effect on V_u of the reduction in m_p due to flange axial forces is also given.

Höglund (1971a, b, 1973) has developed a theory for girders without intermediate transverse stiffeners which was later extended to girders with intermediate stiffeners. He used the system of bars shown in the figure in Table 6.1 as a model of the web. The compression bars in this system are perpendicular to the tension bars. When the angle δ between the tension bars and the flanges is decreased, the shear-buckling load for the system is increased. If the load is uniformly distributed, δ is varied along the girder because the shear varies. Calculated stresses in the bar system are in good agreement with stresses measured in test girders. The shear strength V_u is given by

$$V_u = V_w + \frac{4m_p}{c}\left(1 - \frac{M}{M_f}\right)^2 \qquad M \le M_f \tag{6.17}$$

where

Table 6.2. Values of V_w in Eq. 6.17[a]

$\sqrt{\tau_y/\tau_{cr}}$		
End Stiffeners Rigid[b]	End Stiffeners Nonrigid	V_w
Less than 0.8	Less than 0.8	$\tau_y ht$
0.8–2.75	0.8–1.25	$\tau_y ht \dfrac{1.8}{1+\sqrt{\tau_y/\tau_{cr}}}$
Over 2.75	—	$\tau_y ht \dfrac{1.32}{\sqrt{\tau_y/\tau_{cr}}}$
Over 2.75	Over 1.25	$\tau_y ht/\sqrt{\tau_y/\tau_{cr}}$

[a] τ_{cr} to be computed from Eq. 4.1 with k from Eqs. 4.24.
[b] See Eqs. 6.18 and Fig. 6.4.

V_w = shear strength of web (Table 6.2)
$c = a(0.25 + m_p/m_{pw})$
m_{pw} = plastic moment of web $(= \sigma_{yw}th^2/4)$
M = moment in panel
M_f = flange moment $= \sigma_{yf}A_f h$

The term involving M/M_f accounts for the reduction in the flange moment m_p and will depend on a mean value of M, but Höglund suggests that the largest value be used to simplify the problem.

It will be noted that V_w depends on the nature of the end stiffeners. End stiffeners can be considered rigid if Eqs. 6.18 are satisfied (see also Fig. 6.4).

$$
\begin{aligned}
e &\geqq 0.18h \\
\frac{3000t}{\sqrt{\tau}} &> g > 0.18h \\
A_e &\geqq 0.1ht \\
A_u &\geqq \frac{V_u}{\sigma_y} - 12t^2 \\
A_{st} &\geqq \frac{16m_p}{a\sigma_{y(st)}}
\end{aligned}
\tag{6.18}
$$

In the second of these equations τ is in kilograms per square centimeter and t in centimeters. The term $3000t/\sqrt{\tau}$ is identical with the right side of Eq. 6.4a. A_{st} is the intermediate-stiffener area required to develop the tension field in the adjacent panel.

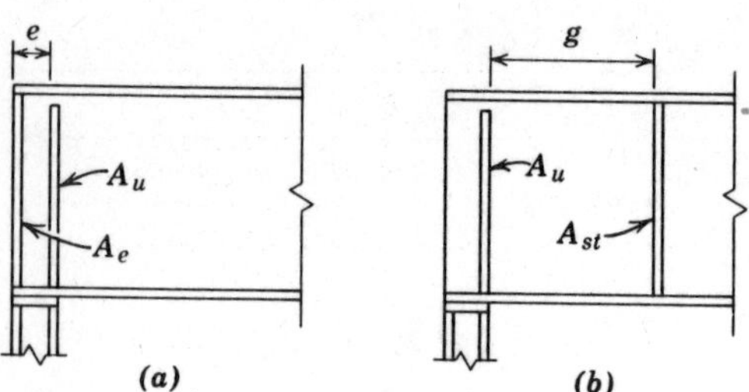

Fig. 6.4 End stiffeners.

Herzog (1974a,b) takes the boundary of the tension field from midheight of the panel at the stiffeners to the plastic hinges in the flanges (Table 6.1). The distance c is based on an average, relative flange stiffness to allow for unequal flanges, and a chart, developed from a study of various test results, particularly those reported by Rockey and Skaloud (1968), is given to determine it. He gives three relatively simple formulas for the ultimate shear, one for each of three ranges of panel aspect ratio, together with a coefficient by which the results are to be multiplied to account for a longitudinal stiffener.

Clark and Sharp (1971) and Sharp and Clark (1971) proposed a tension field for thin-webbed aluminum girders which consists of a Basler field on which is superimposed a complete tension field inclined at 45° (Table 6.1). The flanges are assumed to be elastic beams continuous over the stiffeners and subjected to a uniform load from the 45° field. The shear strength is the sum of the vertical components of the two tension fields and the shear at buckling. Basler's approximation of the von Mises yield condition is used to determine the combination of web stresses that cause yielding. In general, their procedure gives conservative results when compared with tests to ultimate load because it is based on general yielding of the web. However, it has been noted that aluminum girders generally do not develop plastic hinges in the flanges, and failure usually occurs by flange or stiffener buckling or, in riveted and bolted girders, by cracks developing at web holes because of the diagonal tension.

Steinhardt and Schröder (1971) have also suggested a tension field for aluminum girders (Table 6.1). The tension-field band is in the direction of the panel diagonal and its boundaries insersect the midpanel points of the flanges. The tension-field loading on the flange is assumed to vary sinusoidally with a maximum value at the stiffeners. Assuming flange bending to be elastic the corresponding tension-field distribution shown in the figure is derived. The resultant shear is found by adding the vertical component of the tension field to the shear at buckling. The theory was extended to girders of unsymmetrical sections.

Tables 6.3A and 6.3B give comparisons of predicted shear strengths according to Basler, Fujii, Ostapenko, Höglund, Rockey, and Komatsu with results of tests from a number of sources. In computing the values an axial-force reduction in the flange plastic moment is taken into account by Höglund and Rockey but not by Fujii, Ostapenko, and Komatsu. Furthermore, in the five cases in Table 6.3A when M/M_f exceeds unity Höglund has reduced his pure-shear values by Basler's shear-moment interaction formula. Two points should be kept in mind in evaluating the results in these tables: (1) definitions of ultimate strength differ among investigators, and (2) even girders that are identical in design and fabrication can differ considerably in ultimate strength. For example, the ultimate strengths of TG1 and TG1′ in Table 6.3B were 15,450 and 11,850 kg, respectively, yet these were tests on presumably identical halves of a single girder.

Table 6.3A. Shear Strength of Plate Girders

Source	Test Number	$\frac{a}{h}$	$\frac{h}{t}$	V_{ex}/V_u				$\frac{M}{M_f}$
				Basler[a]	Fujii[b]	Ostapenko[a]	Höglund[c]	
Okumura et al.	G1	2.61	55	0.85	1.06	0.87	0.89	0.82
	G2	2.61	55	0.87	1.08	0.88	0.93	1.01
	G3	2.63	70	0.97	1.01	0.88	0.99	1.25
	G5	2.68	70	1.05	1.09	0.94	1.02	0.87
	G6	1.25	70	1.06	1.22	0.96	0.86	0.45
	G7	2.68	70	1.07	1.09	0.94	1.02	0.87
	G9	2.78	90	1.20	0.95	0.90	1.06	1.00
Nishino and Okumura	G1	2.67	60	1.04	0.94	0.98	1.03	0.97
	G2	2.67	60	0.98	1.11	0.93	0.98	1.25
	G3	2.67	77	1.00	0.98	0.86	0.96	1.09
Basler et al.	G6-T1	1.5	259	1.04	1.08	0.96	1.06	0.48
	G6-T2	0.75	259	0.95	0.97	0.94	1.06	0.62
	G6-T3	0.5	259	0.98	1.00	0.93	1.04	0.74
	G7-T1	1.0	255	0.98	1.05	0.96	1.12	0.59
	G7-T2	1.0	255	1.02	1.09	1.00	1.16	0.61
Cooper et al.	H1-T1	3.0	127	1.33	0.96	1.00	1.01	1.00
	H1-T2	1.5	127	1.08	0.92	0.97	1.06	0.60
Konishi et al.	B	1.0	267	0.81	1.02	0.79	1.18	0.53
Okumura and Nishino	G1-1	3.0	182	1.21	1.07	0.99	0.97	1.00
	G1-2	1.5	182	1.03	1.04	0.91	1.08	0.87
	G2-1	3.0	144	1.34	0.96	1.02	1.01	1.16
	G2-2	1.5	144	1.17	1.00	0.96	1.14	0.93
Mean value				1.05	1.03	0.94	1.03	
Standard deviation				0.14	0.07	0.05	0.08	

[a] From Chern and Ostapenko (1969).
[b] From Fujii (1971).
[c] From Höglund (1973).

Table 6.3B. Shear Strength of Plate Girders

Source	Test Number	$\frac{a}{h}$	$\frac{h}{t}$	V_{ex}/V_u Rockey[a]	Höglund[b]	Komatsu[c]	$\frac{M}{M_f}$
Rockey and Škaloud	TG 14	1.0	316	1.02	1.29	—	0.70
	TG 15	1.0	316	1.00	1.29	—	0.53
	TG 16	1.0	316	1.07	1.11	—	0.38
	TG 17	1.0	316	0.99	1.12	—	0.34
	TG 18	1.0	316	0.92	1.15	—	0.33
	TG 19	1.0	316	0.92	1.16	—	0.32
Basler et al.	G6-T1	1.5	259	1.02	1.06	1.03	0.48
	G6-T2	0.75	259	1.14	1.06	0.97	0.62
	G6-T3	0.5	259	1.12	1.04	0.99	0.74
	G7-T1	1.0	255	1.08	1.12	0.96	0.59
	G7-T2	1.0	255	1.04	1.16	1.00	0.61
Škaloud	TG1[d]	1.0	400	—	1.27	1.10	0.65
	TG1′	1.0	400	—	0.97	0.84	0.50
	TG2	1.0	400	—	1.17	0.96	0.28
	TG2′	1.0	400	—	1.01	0.83	0.24
	TG3	1.0	400	—	1.13	1.00	0.20
	TG3′	1.0	400	—	1.12	1.00	0.20
	TG4	1.0	400	—	1.14	1.02	0.19
	TG4′	1.0	400	—	1.08	0.97	0.18
	TG5	1.0	400	—	1.11	0.99	0.15
	TG5′	1.0	400	—	1.08	0.96	0.14
Mean value				1.03	1.13	0.97	
Standard deviation				0.07	0.08	0.07	

[a] From Porter et al. (1975).
[b] From Höglund (1973).
[c] From Komatsu (1971).
[d] TG1, TG1′, and so on, are twin girders. End stiffeners nonrigid according to Höglund (Eqs. 6.18).

The results in Tables 6.3A and 6.3B are summarized in Table 6.4. Basler's formula gives the largest range of values; however, the others do not differ from Basler significantly. Chern and Ostapenko's formulas tend to overestimate shear strength more than the others; this is due in part to the neglect of shear-moment interaction. All but one of the first 10 girders in Table 6.3A were subjected to fairly large moment, and when they were compared by Chern and Ostapenko (1970b) with the Chern-Ostapenko shear-moment interaction formulas the following values of V_{ex}/V_u were obtained: 0.90, 0.92, 0.92, 0.97, 0.97, 0.95, 1.00, 0.96, and 0.91. The average of these 10 values is 0.95; the 10 values in Table 6.3A average 0.91.

Höglund's formulas give the most conservative results, particularly for slender webs (Table 6.3B). This conservatism is explained in part by the fact that he bases the axial-force reduction in flange plastic moment (Eq.

Table 6.4. Comparison of Test Results in Tables 6.3A and B

Investigator	Mean	Standard Deviation	Range of V_{ex}/V_u	Ratio Highest/Lowest
Basler	1.05	0.14	1.33–0.84	1.64
Fujii	1.03	0.07	1.22–0.92	1.33
Chern and Ostapenko	0.94	0.05	1.02–0.79	1.29
Höglund (Table 6.3A)	1.03	0.08	1.18–0.86	1.37
Höglund (Table 6.3B)	1.13	0.08	1.29–1.01	1.28
Rockey	1.03	0.07	1.14–0.92	1.24
Komatsu	0.97	0.07	1.10–0.83	1.32

6.17) on the largest moment in the panel. Rockey uses an average flange stress in computing this effect, and Basler found better correlation in predictions of shear-moment interaction when he used a reduced moment (Basler, 1963b).

Herzog presents in graphical form a comparison of shear strengths by his formulas with results of 96 tests by others, including all of those in Tables 6.3A and B. The mean value of the ratio V_{ex}/V_u is 1.036. However, the standard deviation is 0.16, and in one test (not in Tables 6.3) V_{ex}/V_u is 1.41 while Höglund's formula gives 1.04.

The simplest formulas are Basler's Höglund's, and Herzog's, and of these Höglund's gives the most consistent results except for slender webs ($h/t > 300$). Fujii's formulas and Komatsu's also give good results, but they are more complicated. The formulas by Rockey, Evans, and Porter give good results but require the optimum inclination of the tension field to be determined by trial. Chern and Ostapenko's formulas are complicated and also require a trial determination of the optimum inclination of the tension field.

Yonezawa et al. (1978) proposed an ultimate shear strength theory for plate girders with webs diagonally stiffened between vertical stiffeners. It is assumed that the ultimate shear strength consists of the sum of contributions from three sources: the beam shear force V_{cr} taken by the diagonally stiffened web, the shear force V_t taken by the tension field in the web, and the shear force V_s taken by the diagonal stiffener. Buckling coefficients, obtained from a finite-difference solution of the differential equation for a buckled web with a diagonal stiffener under shear, are presented for the computation of the beam shear force. Coefficients for panels with both compression and tension type stiffeners with either fixed or pinned plate boundaries are given. Expressions for the tension field contribution are based on the work of Rockey et al. The shear force taken by the diagonal stiffener is determined by taking the vertical component of the force acting on the stiffener. Tests on two plate girders showed good agreement between theoretical and experimental ultimate loads. Studies on diagonally stiffened web were also performed at Liege (Jetteur, 1984).

In the case of tapered plate girders, the axial forces in the inclined flanges have vertical components. These components of force may either increase or decrease the shear force carried by the web depending on the direction of taper and on the direction of the applied shear. Shear buckling of simply supported plates of variable depth have been investigated by finite-element analysis and charts for the buckling coefficient k have been prepared by Elgaaly (1973). Two investigations have been reported on the ultimate shear strength of tapered web girders:

Falby and Lee (1976) have proposed a method for estimating ultimate shear strength based on the Basler model. Their method is limited to small tapers and does not accoun for the effect on shear of the axial load in the inclined flange.

Davies and Mandal (1979) extended Rockey's tension-field model to the case of tapered web girders. Their theory is not limited to small tapers and takes into account the influence of the axial force in the sloping flange. Good agreement is obtained between theoretical predictions and experimental collapse loads. The theory is applicable to loading conditions where the girder is loaded within the tip (the intersection point of the flanges). However, it appears that modifications must be made to the theory in order to deal with the more common loading case associated with continuous plate girders with tapered webs at the supports.

6.4 GIRDERS WITH NO INTERMEDIATE STIFFENERS

Plate girders with bearing stiffeners at the supports but with no intermediate stiffeners except for bearing stiffeners at heavy concentrated loads are of practical interest. The postbuckling strength of such girders can be significant. The shear stress at ultimate load in three tests with web slendernesses h/t of 210, 210, and 300 (B1, K1, and B4 of Table 6.5) were 3.69, 4.00, and 4.68 times the theoretical shear-buckling stress assuming panel boundaries simply supported. Höglund's formula (Eq. 6.17), which was developed originally for such girders, is in good agreement with test results (Table 6.5).

The formulas by Ostapenko and Chern are in good agreement with tests on girders with panel aspect ratios of 5.5 if the tension-field contribution is assumed to be zero (Table 6.5); in other words, the ultimate shear is the sum of the critical shear, assuming the web fixed at the flanges, and the flange plastic-hinge contribution.

Table 6.5. Shear Strength of Girders with Long Panels

				V_{ex}/V_{th}	
Source	Test No.	$\frac{a}{h}$	$\frac{h}{t}$	Höglund[a]	Chern and Ostapenko[b]
Carskaddan (1968)	C-AC2	5.5	143	0.99	1.02
	C-AC3	5.5	71	0.97	0.94
	C-AC4	5.5	102	1.04	1.02
	C-AC5	5.5	103	0.96	0.96
	C-AHI	5.5	69	1.03	1.00
Höglund (1971b)	B1[c]	[d]	210	1.14	—
	K1	[d]	210	0.94	—
	B4	[d]	300	1.05	—

[a] From Höglund (1973).
[b] From Chern and Ostapenko (1969).
[c] End stiffeners nonrigid according to Höglund (Eqs. 6.18).
[d] No intermediate transverse stiffeners.

6.5 BENDING STRENGTH OF PLATE GIRDERS

A plate girder subjected primarily to bending moment usually fails by lateral-torsional buckling, local buckling of the compression flange, or yielding of one or both flanges. Buckling of the compression flange into the web (vertical buckling) has been observed in many tests, and the following limiting value of the web slenderness h/t to preclude this mode of failure has been developed by Basler and Thürlimann (1963):

$$\frac{h}{t} \leqq \frac{0.68E}{\sqrt{\sigma_y(\sigma_y + \sigma_r)}} \sqrt{\frac{A_w}{A_f}} \tag{6.19}$$

where A_w is the area of web, A_t the area of one flange, and σ_r the residual tension that must be overcome to achieve uniform yield in compression. In tests in which it has been observed, vertical buckling has occurred only after general yielding of the compression flange in the panel. Therefore, Eq. 6.19 may be too conservative, or even unnecessary, for girders of practical proportions. However, web slenderness must be limited to facilitate fabrication and to avoid fatigue cracking under repeated loads due to out-of-plane web flexing.

As in the case of shear, buckling of the web due to bending does not exhaust the panel capacity. However, the distribution of the bending stress changes in the postbuckling range and the web becomes less efficient. The solution to this problem by most investigators is based on the assumption that a portion of the web becomes ineffective. Basler and Thürlimann assume a linear distribution of stress on the effective cross

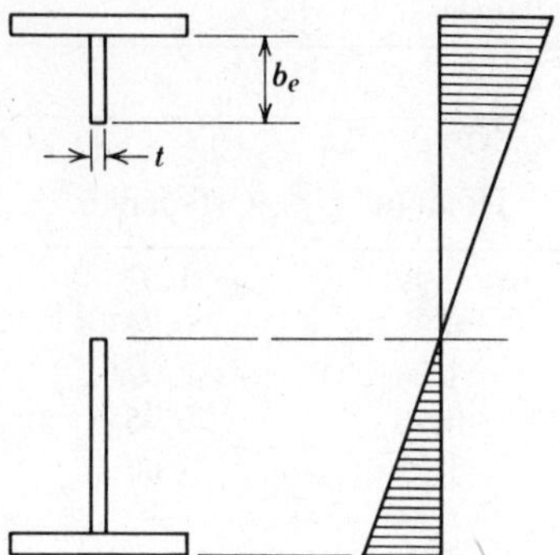

Fig. 6.5 Bending stresses.

section shown in Fig. 6.5, with the ultimate moment being reached when the extreme fiber compression reaches yield stress, or a critical stress if some form of buckling controls. The effective width b_e is assumed to be $30t$ for a web with $h/t = 360$. This is the limiting slenderness according to Eq. 6.19 with $A_w/A_f = 0.5$, $\sigma_y = 33$ ksi, and $\sigma_r = 16.5$ ksi. Bending strength is then assumed to increase linearly from the value for a girder whose web can reach yield stress in bending without buckling. This assumption gives

$$\frac{M_u}{M_y} = 1 - C\left[\frac{h}{t} - \left(\frac{h}{t}\right)_y\right] \tag{6.20}$$

where C is a constant and $(h/t)_y$ is the web slenderness that permits yielding in bending without buckling. In particular, the following formula was suggested:

$$\frac{M_u}{M_y} = 1 - 0.0005\,\frac{A_w}{A_f}\left(\frac{h}{t} - 5.7\sqrt{\frac{E}{\sigma_y}}\right) \tag{6.21}$$

The value of $(h/t)_y$ in this formula is somewhat larger than the theoretical value for hinged-edged panels and was chosen to give the limiting slenderness 170 which was prescribed at that time, for steel with $\sigma_y =$ 33 ksi, by the AISC Specification. Equation 6.21 has been found to be in good agreement with test results (Basler and Thürlimann, 1963; Maeda, 1971; Cooper, 1971) with girders with $h/t = 388$, 444, and 751.

Höglund (1973) assumes the effective width b_e in Fig. 6.5 to be $0.76t\sqrt{E/\sigma_{yf}}$ and considers an additional strip of web of $1.64t\sqrt{E/\sigma_{yf}}$, immediately above the neutral axis to be effective. The following formula for the effective section modulus was derived:

$$S_{\text{eff}} = S\left[1 - 0.15\,\frac{A_w}{A_f}\left(1 - 4.8\,\frac{t}{h}\sqrt{\frac{E}{\sigma_{yf}}}\right)\right] \qquad \frac{h}{t} \leqq 4.8\sqrt{\frac{E}{\sigma_{yf}}} \tag{6.22}$$

where

S_{eff} = effective section modulus
S = section modulus of unreduced cross section

This equation was found to give a slightly larger reduction than Eq. 6.21, and was in good agreement with 11 tests ($0.96 < M_{\text{exp}}/M_{th} < 1.04$). For hybrid girders S_{eff} should be decreased by the amount ΔS given by

$$\Delta S = \frac{h^2 t}{12}\left(2 + \frac{\sigma_{yw}}{\sigma_{yf}}\right)\left(1 - \frac{\sigma_{yw}}{\sigma_{yf}}\right)^2 \qquad \frac{\sigma_{yw}}{\sigma_{yf}} \leqq 1 \tag{6.23}$$

Fujii's formulas (1968a, 1971) for the ultimate moment are more complicated and are restricted to laterally supported girders. They involve the parameter $(t/h)\sqrt{E/\sigma_{yf}}$ of the Höglund formula and the web bend-buckling stress. A multiplying coefficient for hybrid girders is given. A comparison of predicted values with results of tests on 10 nonhybrid girders gave $0.94 < M_{\text{exp}}/M_{th} < 1.11$.

Chern and Ostapenko (1970a) have developed formulas for hybrid girders with unequal flanges. The ultimate moment is the sum of the web-buckling moment based on an effective width similar to Basler's. The postbuckling moment is determined by yielding of the tension flange or by lateral or local buckling of the compression flange. A comparison of predicted values with results of tests gave $0.95 < M_{\text{exp}}/M_{th} < 1.15$ for 14 nonhybrid girders and $0.86 < M_{\text{exp}}/M_{th} < 1.13$ for 10 hybrid girders. In a later report (Ostapenko et al., 1971) the following modification of Basler's formula was proposed:

$$M_u = \frac{I}{y_c}\,\sigma_c\left\{1 - \frac{I_w}{I} + \frac{\sigma_{yw}}{\sigma_c}\left[\frac{I_w}{I} - 0.002\,\frac{y_c t}{A_c}\left(\frac{y_c}{t} - 2.85\sqrt{\frac{E}{\sigma_{yw}}}\right)\right]\right\} \tag{6.24}$$

where

I = moment of inertia of cross section
I_w = moment of inertia of web about centroidal axis of cross section
y_c = distance from neutral axis to compression edge of web
A_c = area of compression flange
σ_c = compression-flange buckling stress

$$\frac{y_c}{t} \gtrless 2.85\sqrt{\frac{E}{\sigma_{yw}}}$$

$$\sigma_{yf} \gtrless \sigma_{yw}$$

This equation has been found to give good correlation with test results. The ultimate moment in terms of the tension flange is

$$M_u = \frac{I\sigma_{yt}}{y_t}\left[1 - \frac{I_w}{I}\left(1 - \frac{\sigma_{yw}}{\sigma_{yt}}\right)\right] \tag{6.25}$$

where y_t is the distance from the neutral axis to the tension edge of the web.

Herzog (1973, 1974b) gives a formula for the general case of the unsymmetrical hybrid girder with one or more longitudinal stiffeners, which includes simple reduction coefficients for vertical, local, and lateral buckling of the compression flange. The ultimate moment is given by

$$M_u = \sigma_{yc}A_c\left(y_c + \frac{t_c}{2}\right) + \sigma_{yt}A_t\left(y_t + \frac{t_t}{2}\right) + \sigma_{ys}A_s y_s + \frac{\sigma_{yw}t_w}{6}\left[(1+\phi)y_c^2 + 2y_t^2\right] \tag{6.26}$$

where

$$\phi = \frac{\sigma_{ys}\sum A_s}{\sigma_{yw}A_w}$$

In Eq. 6.26, *c*, *t*, *s*, and *w* signify compression flange, tension flange, longitudinal stiffener, and web, respectively, and *y* and *t* are defined in Fig. 6.6. The distance y_c to the compression edge of the web is

$$y_c = \frac{2}{3+\phi}\left[h - 2\left(\frac{\sigma_{yc}A_c + \sigma_{ys}A_s - \sigma_{yt}A_t}{\sigma_{yw}t_w}\right)\right] \tag{6.27}$$

A comparison of predicted values by Eq. 6.26 with results of tests on 23 girders without longitudinal stiffeners gave $0.91 < M_{\mathrm{exp}}/M_{th} < 1.19$, with a mean value 1.005 and standard deviation 0.095. Comparisons with tests on 26 longitudinally stiffened girders gave $0.90 < M_{\mathrm{exp}}/M_{th} < 1.16$, with a mean value 1.036 and standard deviation 0.105.

6.6 COMBINED BENDING AND SHEAR

Assuming the shear in a girder to be carried only by the web, as in Basler's solution, shear is a maximum when the web is yielded uniformly, or, if shear buckling occurs at a smaller stress, when it has a fully developed tension field. These values are independent of the bending moment in the panel as long as the moment is less than $M_f = \sigma_{yf} A_f h$, which is the moment that can be carried by the flanges alone (AB in Fig. 6.6*a*). Any larger moment must be resisted in part by the web, which reduces the shear, until the shear capacity finally becomes zero for a panel in pure bending (BC in Fig. 6.6*a*). If the flange contribution to shear is taken into account, as in the more recent theories of shear strength, AB in Fig. 6.6*a* is not correct, because the flange axial force from the moment in the panel reduces the flange plastic moment m_p on which the flange contribution is based. This condition is represented by a line $A'B$, which can be defined, for example, by Höglund's formula (Eq. 6.17).

Basler's interaction diagram (1963b) is shown in Fig. 6.6*b*. The segment BC, for which a formula is given, corresponds to BC in Fig. 6.6*a* with $M_u = M_p$, but is assumed to be invalid for thin-webbed girders when M exceeds M_y. The segment BC can be taken to be a straight line. Correlation with test results was good with M taken at $h/2$ from the high-moment end of the panel or at midpanel if $a < h$.

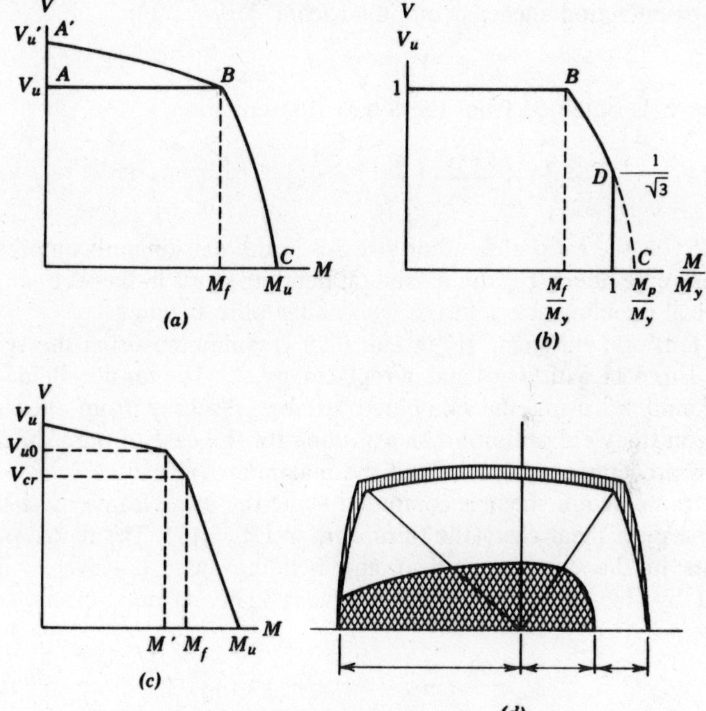

Fig. 6.6 Shear-moment interaction diagrams.

Herzog (1974a, b) assumes a trilinear diagram similar to Basler's. Fujii's interaction diagram is shown in Fig. 6.6*c*, where the additional point (M'_f, V_{uo}) is for the case of a flange with no bending resistance (Fujii, 1971).

Chern and Ostapenko (1970b) assume the ultimate capacity will be dictated by failure of the web, instability of the compression flange, or yielding of the tension flange. The complete interaction behavior is represented schematically by the interaction curve of Fig. 6.6*d*. Curve Q_2–Q_1–Q_4 represents failure of the web, curve Q_2–Q_3 represents buckling of the compression flange, and curve Q_4–Q_5 represents yielding of the tension flange.

When web failure controls (region Q_2–Q_1–Q_4 of Fig. 6.6*d*) the total shear resistance V_{wc} is

$$V_{wc} = V_{crc} + V_{ec} + V_{fc} \tag{6.28}$$

in which V_{crc} is the beam-action shear, V_{ec} the tension-field shear, and V_{fc} the frame-action shear. The subscript c indicates that these shears are associated with combined stresses in the girder.

The beam-action shear is computed from

$$V_{crc} = A_w \tau_c \tag{6.29}$$

in which τ_c is obtained from the interaction equation

$$\left(\frac{\tau_c}{\tau_{cr}}\right)^2 + \frac{1+C}{2}\left(\frac{\sigma_{bc}}{\sigma_{cp}}\right) + \frac{1-C}{2}\left(\frac{\sigma_{bc}}{\sigma_{cp}}\right)^2 = 1 \tag{6.30}$$

in which C is the ratio of bending stresses in the tension and compression at the extreme fibers, σ_{bc} the maximum bending stress in the web, and σ_{cp} the critical buckling stress in the web under pure bending.

The tension-field shear V_{ec} in Eq. 6.28 is computed using the second term in Eq. 6.11 with $\rho = \frac{1}{2}$ and σ_t replaced by σ_{tc}. The tension-field stress σ_{tc} is found by using the combined stresses resulting from shear and bending in the yield criterion. As was done for the case of pure shear, an iteration procedure is used to find the maximum value of V_{wc}.

The frame-action shear is computed using the same frame mechanism as for the pure shear case (the third term in Eq. 6.11). The flange plastic moments in the combined shear and bending case, however, will be affected by the axial forces in the flanges. The resulting frame-action shear is

$$V_{fc} = \frac{1}{a}(m_{cl} + m_{cr} + m_{tl} + m_{tr}) \tag{6.31}$$

in which m_{cl} and m_{cr} are the plastic moments in the compression flange at the left and right sides of the panel, and m_u and m_{tr} are the corresponding plastic moments in the tension flange. All of these moments are modified for the effect of axial force in the flange.

The other conditions for which failure occurs are depicted by curves Q_2–Q_3 and Q_4–Q_5 of Fig. 6.6*d*. Curve Q_2–Q_3 represents the case where buckling of the compression flange controls, in which case a completely developed tension field will not have formed. In the region represented by curve Q_4–Q_5, the tension flange starts yielding before the web plate reaches its ultimate shear strength. Yielding will penetrate into the cross section and the plastic strength of the girder panel will be the ultimate capacity.

The model postulated by Rockey et al. (Rockey, 1971a,b; Rockey and Skaloud, 1972; Rockey et al., 1973; Porter et al., 1975) for predicting the strength of girders without longitudinal stiffeners under bending and shear includes three additional factors not included in the pure bending and pure shear models. These factors are (1) the reduction in the shear buckling stress of the web due to the presence of bending stress, (2) the influence of the in-plane bending stress on the value of the diagonal tension-field stress at failure, and (3) the reduction of the magnitude of the plastic modulus of the flanges resulting from the axial compressive and tensile strength.

The buckling stress reduction is handled by using the interaction equation

$$\left(\frac{\sigma_{bc}}{\sigma_{cp}}\right)^2 + \left(\frac{\tau_c}{\tau_{cr}}\right)^2 = 1 \tag{6.32}$$

to determine the critical shear stress τ_c under combined bending and shear. Note that Eq. 6.32 is for the case of a symmetric cross section, which is a special case of Eq. 6.30 with $C = -1$.

After the panel has buckled the tension-field shear is computed by using a postbuckling shear force modified to include the effect of combined bending and shear. This involves modifying the tension field stress σ_1 to include both bending and shear stresses in the yield criterion and it also involves using an interaction equation to find the flange plastic moments m_f when axial stress resulting from bending acts in the flanges. The equations previously given for pure shear are then used with an iteration approach to determine the strength of the girder panel.

6.7 PLATE GIRDERS WITH LONGITUDINAL STIFFENERS

Longitudinal stiffeners (Fig. 6.7) can greatly increase the bending strength of plate girders. This additional strength can be attributed to

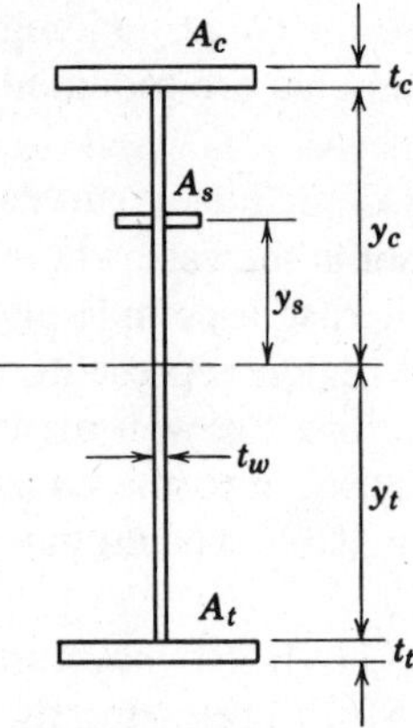

Fig. 6.7 Plate girders with longitudinal stiffeners.

control of the lateral deflection of the web which increases the flexural stress the web can carry and also improves the bending resistance of the flange due to greater web restraint. Rockey and Leggett (1962) have determined that the optimum location for a longitudinal stiffener used to increase the flexural buckling resistance of a panel is 0.22 times the web depth from the compression flange if the web is assumed to be fixed at the flanges and simply supported at all four edges. Accordingly, 0.20 of the depth has been adopted nearly universally by design specifications as the accepted location for a longitudinal stiffener. Where a longitudinal stiffener at 0.20 of the web depth from the compression flange is provided, the value of k in the plate-buckling formula is increased from 23.9 to 129, which means that the elastic critical bending stress is more than five times as large as for a girder with no longitudinal stiffener. Since an unstiffened web of mild steel with a slenderness of about 170 develops yield-stress moment without buckling, a stiffened web can do the same with a slenderness $170\sqrt{123.9/23.9}$, or about 400. Tests show that an adequately proportioned longitudinal stiffener at $0.2h$ from the compression flange eliminates the bend-buckling loss in girders with web slendernesses as large as 450, so that the ultimate moment as determined by compression-flange buckling strength is attained (Cooper, 1967). Girders with larger slenderness are likely to require two or more longitudinal stiffeners to eliminate the web bend-buckling loss. Of course, the increase in bending strength of a longitudinally stiffened thin-web girder is usually small because the web contribution to bending strength is small. However, longitudinal stiffeners can be important in a girder subjected to repeated loads because they reduce or eliminate the transverse bending of the web, which increases resistance to fatigue cracking at the web-to-flange juncture and allows more slender webs to be used (Yen and Mueller, 1966).

The optimum location of a longitudinal stiffener that is used to increase resistance to shear buckling is at middepth. In this case the two subpanels buckle simultaneously and the increase in critical stress can be substantial. For example, $k = 9.34$ and 6.34 for a square panel and the corresponding subpanel, respectively, and the slenderness ratio h/t of the square panel is twice that of the subpanel. Therefore, the elastic shear-buckling stress for the subpanel is 2.7 times as large as for the square panel. Of course, in a web with a longitudinal stiffener not at middepth the larger subpanel buckles first, and at a smaller critical stress than for the stiffener at middepth.

The postbuckling shear strength of longitudinally stiffened girders has been evaluated in two ways: Cooper (1967) assumes that each subpanel develops its own tension field after buckling, while Porter et al. (1975) assume that only one tension field is developed between the flanges and transverse stiffeners even if longitudinal stiffeners are used.

If each subpanel develops its own tension field as suggested by Cooper, the tension field shears for a girder stiffened as shown in Fig. 6.8, are

$$V_{p1} = \frac{\sigma_{t1} b_1 t}{2\sqrt{1 + \alpha_1^2}} \qquad V_{p2} = \frac{\sigma_{t2} b_2 t}{2\sqrt{1 + \alpha_2^2}} \tag{6.33}$$

in which b_1 and b_2 are the depths of the subpanels, σ_{t1} and σ_{t2} the diagonal tension stresses in the subpanels, and α_1 and α_2 the aspect ratios a/b_1 and a/b_2, respectively. Cooper goes on to develop the total shear strength after employing a modified form of the von Mises yield condition to evaluate the diagonal tension stresses. Chern and Ostapenko (1971) extended Cooper's model to include frame action of the flanges and of

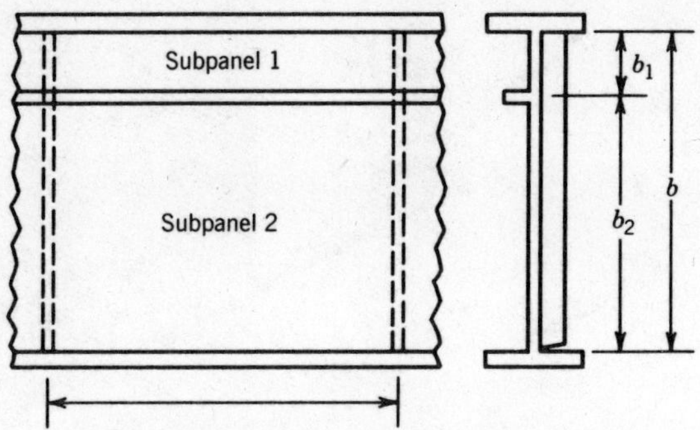

Fig. 6.8 Typical longitudinally stiffened panel.

the longitudinal stiffener. The model used by Rockey et al. (1974) was suggested after it was observed from tests that an overall tension field develops in the web.

Figure 6.9 shows a test girder at failure. The inclined arrows painted on the web show the predicted angle of the tension field and are in good agreement with the axes of the buckles in both subpanels. The predicted positions of the flange plastic hinges, also painted on the girder, are in good agreement with the actual positions. The longitudinal stiffeners were at $h/4$ from the compression flange in the right-hand panel. Failure was in the right-hand panel, but it is clear that the other panel was not far behind. To obtain the tension-field shear resistance of a longitudinally stiffened girder the same approach is used for an unstiffened girder, but in computing the tension field stress associated with failure the critical beam-action shear corresponding to buckling of the largest subpanel is used in the yield criterion.

Where longitudinal stiffeners are used in the panel the shear strength under combined shear and bending is treated by Chern and Ostapenko and by Rockey et al. in much the same way as if the panel were under pure shear. However, modifications to the buckling stresses, tension-field

Fig. 6.9 Longitudinally stiffened girder at ultimate load in shear (Rockey et al., 1974).

stresses, and plastic moments in the flanges to account for the combined stress effects are included. As noted previously, Chern and Ostapenko include the contribution from the longitudinal stiffener in computing the frame-action strength and include the modification of the stiffener's plastic moment capacity to account for the combined stress effects in this case.

Interaction equations to compute the critical buckling stress for each subpanel are presented by Chern and Ostapenko. These equations are similar to Eq. 6.30. The subpanel tension-field stresses are then computed using the state of stress in each subpanel with the yield criterion. Finally, an iterative approach is used to find the shear strength.

Rockey et al. compute the buckling stress in the subpanels by an interaction approach applied to each subpanel stress condition as shown in Fig. 6.10. The state of stress in a subpanel is therefore a shear stress τ_c, a pure bending stress σ_{bc}, and an axial stress σ_{cc}. The buckling condition is

$$\frac{\sigma_{cc}}{\sigma_{crc}} + \left(\frac{\sigma_{bc}}{\sigma_{cp}}\right)^2 + \left(\frac{\tau_c}{\tau_{cr}}\right)^2 = 1 \tag{6.34}$$

The postbuckling strength of the panel is determined as if no longitudinal stiffeners were present.

Combined shear and bending of longitudinally stiffened girders with unequal flanges has been investigated by Chern and Ostapenko (1971). Their analysis for shear is based on their tension-field distribution for girders without longitudinal stiffeners with each subpanel treated independently as in Cooper's analysis, and taking into account the plastic moments of the flanges and stiffener and the axial-force reduction in flange plastic moment. The girder strength is determined by shear until the shear-strength curve intersects the curve for bending strength (Fig.

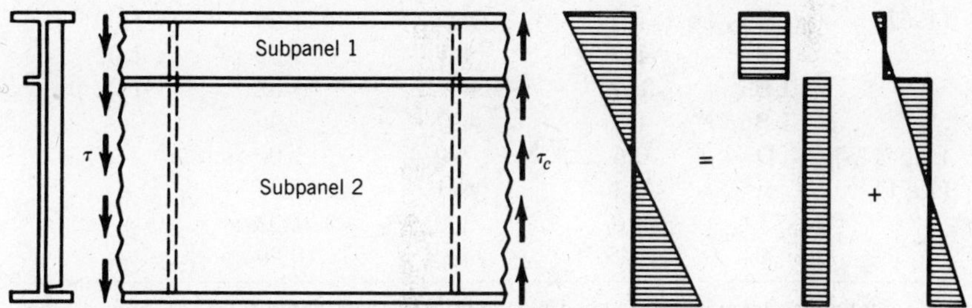

Fig. 6.10 Stress distribution in panels of plate girder under shear and bending.

6.6*d*). The analysis involves some lengthy iterations so that it is not suitable for hand computation.

Of the three analyses discussed in this chapter Cooper's is the most conservative and the easiest to use. Rockey's method requires trial-and-error to determine the optimum inclination of the tension field and, for each trial, an iteration to determine the axial-force reduction in flange plastic moment. Finally, Chern and Ostapenko's solution involves so many iterative steps as to be manageable only by digital computer.

A large amount of experimental data is available from tests on transversely stiffened girders with and without longitudinal stiffeners. Many of these data can be found in the previously cited references.

The various theories for predicting the strength of plate girders in bending agree well with the experimental evidence. For example, Table 6.6 compares the theories of Cooper (1965) and Chern and Ostapenko (1970b) with experimental data for several girders with longitudinal stiffeners. As can be seen, there is a good correlation between theory and experimental results. For the nine bending test results compared with Cooper's theory, the average ratio of test ultimate load to predicted ultimate load is 0.99. For four bending tests the Chern and Ostapenko ratio is 1.01.

Tables 6.7 through 6.9 give results of tests on longitudinally stiffened girders subjected to high shear. These results are compared with Cooper's

Table 6.6. Bending Tests on Longitudinally Stiffened Girders[a]

				Experimental/ Theoretical	
Source of Data	Test Number	Aspect Ratio	Web Depth-to-Thickness Ratio	Cooper (1965)	Ostapenko and Chern (1970)
Massonnet (1962)	LB2	1.0	447	0.99	0.99
	LB3	1.0	447	1.00	1.00
	LB4	1.5	447	1.02	1.02
	LB5	0.75	447	1.02	1.02
	LB6	1.0	407	0.96	—
Dubas (1971)	D	0.6	299	1.00	—
	E	0.4	401	0.94	—
	3	0.6	300	1.02	—
	4	0.45	400	0.99	—

[a] All longitudinal stiffeners were located at one-fifth the web depth from the compression flange.

Table 6.7. Shear Test Results Compared with Cooper (1967) Theory

Test Number	Aspect Ratio	Web Depth-to-Thickness Ratio	Location of Longitudinal Stiffener from Compression Flange[a]	Experimental ÷ Theoretical Load
LS1-T2	1.0	256	0.33b	1.10
LS2-T1	1.0	275	0.33b	1.06
LS3-T1	1.5	276	0.33b	1.10
LS3-T2	1.5	276	0.33b	1.17
LS3-T3	0.75	276	0.33b	1.07
LS4-T1	1.0	260	0.2b	1.00
LS4-T2	1.0	260	0.5b	1.18

[a] b = web depth (Fig. 7.10).

Table 6.8. Comparison of Ostapenko and Chern (1970) Theory with Various Test Results

Source of Data	Test Number	Aspect Ratio	Web Depth-to-Thickness Ratio	Location of Longitudinal Stiffener from Compression Flange[a]	Experimental ÷ Theoretical Load
Cooper (1967)	LS1-T2	1.0	256	0.33b	1.00
	LS2-T1	1.0	275	0.33b	0.94
	LS3-T1	1.5	276	0.33b	1.01
	LS3-T2	1.5	276	0.33b	1.00
	LS3-T3	0.75	276	0.33b	0.93
	LS4-T1	1.0	260	0.2b	1.08
Skaloud (1971)	UG5.1	1.77	400	0.26b	1.06
	UG5.2	1.15	400	0.26b	1.06
	UG5.3	1.46	400	0.26b	0.97
	UG5.4	1.77	264	0.26b	1.00
	UG5.5	0.83	264	0.26b	0.96
	UG5.6	1.77	264	0.26b	0.88
Porter et al. (1975)	F11-T1	1.39	365	0.2b	1.03
	F11-T2	1.20	365	0.2b	1.14
	F11-T3	1.00	365	0.2b	0.94

[a] b, web depth (Fig. 6.10).

Table 6.9. Comparison of Test Results with Theory of Rockey et al. (1974)

Girder Number	Aspect Ratio	Web Depth-to-Thickness Ratio	Location of longitudinal Stiffener from Compression Flange[a]	Experimental ÷ Theoretical Load
SH1	1	387	$0.50b$	0.94
SH1	1	387	$0.33b$	0.99
SH2	1	387	$0.20b$	1.02
SH2	1	387	$0.25b$	1.02
SH2R	1	387	$0.20b$	0.93
SH2R	1	387	$0.25b$	0.93
SH4	1.33	400	$0.25b$	1.09
SH4	1.33	400	$0.33b$	1.07
SH5	1	370	$0.25b$	0.96
SH5	1	370	$0.33b$	0.98
SH7	1.67	295	$0.33b$	1.03
SH7	1.67	295	$0.25b$	1.03
SH8	1.67	295	$0.50b$	0.92
SH8	1.67	295	$0.33b$	1.05
SH9	1	375	$0.50b$	0.99
SH9	1	375	$0.33b$	1.04

Table 6.10. Shear Strength of Longitudinally Stiffened Girders (Chern and Ostapenko, 1971; D'Apice et al., 1966)

Test Number Cooper (1967)	Aspect Ratio	Web Depth-to-Thickness Ratio	Location of Longitudinal Stiffener from Compression Flange[a]	Experimental ÷ Theoretical		
				Cooper (1967)	Rockey (1974)	Ostapenko and Chern (1971)
LS1-T2	1	256	$0.33b$	1.10	1.05	1.00
LS2-T1	1	275	$0.33b$	1.06	0.95	0.94
LS3-T1	1.5	276	$0.33b$	1.10	0.88	1.01
LS3-T2	1.5	276	$0.33b$	1.17	1.05	1.00
LS3-T3	0.75	276	$0.33b$	1.07	0.93	0.93
LS4-T1	1	260	$0.20b$	1.00	0.99	1.08
LS4-T2	1	260	$0.50b$	1.18	0.95	—

theory (1967) in Table 6.7, Ostapenko and Chern's theory (1970) in Table 6.8, and the theory of Rockey et al. (1974) in Table 6.9. In Table 6.10 the aforementioned theories are compared for the series of tests reported by Cooper.

As can be seen, Cooper's theory conservatively estimates the shear strength consistent with his assumptions. The Chern–Ostapenko and Rockey theories bracket the experimental data and are in good agreement with these tests. It is interesting to note that the average ratio of experimental to theoretical shears equals 1.0 for the Chern–Ostapenko theory based on the 15 tests reported. The standard deviation for these tests is 0.065. A similar comparison obtained using the theory of Rockey et al. gives an average of 1.0 with a standard deviation of 0.051 for 16 tests. More recent work on longitudinally stiffened plate girders design is reported by Maquoi et al. (1983).

6.8 END PANELS

The tension field in a plate-girder panel is resisted by the flanges and by the adjacent panels and transverse stiffeners. Since the panels adjacent to an interior panel are tension-field designed, they can be counted on to furnish the necessary support. However, an end panel does not have such support and must be designed as a beam-shear panel unless the end stiffeners are designed to resist the bending effect of a tributary tension field. Basler (1963a) assumed that an end panel designed for beam shear can support a tension field in the adjacent interior panel, and this assumption has been generally accepted. This means that the end-panel stiffener spacing can be based on the shear-buckling stress as discussed by Škaloud (1962). For example, the length of the end panel according to the AISC Specification is determined by a formula which gives a factor-of-safety of 1.65 with respect to shear buckling. If the end panel is designed for tension-field action an end post must be provided. A possible end post consists of the bearing stiffener and an end plate (Fig. 6.4*a*). According to Basler such an end post can be designed as a flexural member consisting of the stiffener and end plate and the portion of the web between, supported at the top and bottom flanges and subjected to the horizontal component of the tension field distributed uniformly over the depth. The required area A_e of the end plate, based on ultimate load considerations, is given by

$$A_e = \frac{(\tau - \tau_{cr})hA_w}{8e\sigma_y} \tag{6.35}$$

where e is the distance between bearing stiffener and end plate (Fig. 6.4).

The bearing stiffener itself is designed to support the end reaction. According to tests reported by Schueller and Ostapenko (1970), web shear may control the design of the end post.

6.9 DESIGN OF STIFFENERS

6.9.1 Transverse Stiffeners

Transverse stiffeners must be stiff enough to preserve the straight boundaries that are assumed in computing shear buckling of plate-girder webs. Stein and Fralich (1950) developed a solution for this problem for an infinitely long web with simply supported edges and equally spaced stiffeners. Results were in fair agreement with tests on 20 specimens. Using their numerical data, Bleich developed a formula for the moment of inertia I of the stiffener which can be put in the following form:

$$I = 2.5ht^3\left(\frac{h}{a} - 0.7\,\frac{a}{h}\right) \qquad a \lessapprox h \tag{6.36}$$

The AASHTO formula for load-factor design is

$$I = Jat^3 \tag{6.37a}$$

in which

$$J = 2.5\left(\frac{h}{a}\right)^2 - 2 \gtrapprox 0.5 \tag{6.37b}$$

This is the same as Eq. 6.36 except that the coefficient of a/h in the second term in parentheses is 0.8 instead of 0.7. The AISC formula is $I = (h/50)^4$, and since it depends on h alone, it appears to be inadequately related to web-buckling parameters.

In girders with longitudinal stiffeners the transverse stiffener must also support the longitudinal stiffener as it forces a horizontal node in the bend-buckling configuration of the web. According to an analysis by Cooper (1967), the required section modulus S_T of the transverse stiffener is given conservatively by $S_T = S_L h/a$, where S_L is the section modulus of the longitudinal stiffener. This requirement is reduced considerably in the AASHTO specification, where $S_T = S_L h/3a$.

Transverse stiffeners in girders that rely on a tension field must also be designed for their role in the development of the diagonal tension. In this situation they are compression members, and so must be checked for local buckling. Furthermore, they must have cross-sectional area adequate for the axial force F that develops. The value of F_s in Basler's solution is

$$F_s = \tfrac{1}{2}\sigma_t at(1 - \cos\theta_d) \tag{6.38a}$$

Substituting the value of σ_t from Eq. 6.7 gives

$$F_s = \tfrac{1}{2}\sigma_{yw} at\left(1 - \frac{\tau_{cr}}{\tau_{yw}}\right)(1 - \cos\theta_d) \tag{6.38b}$$

The AISC formula for the cross-sectional area A_s of stiffeners symmetrical about the plane of the web is derived from Eq. 6.38b by $A_s = F_s/\sigma_{ys}$. If stiffeners are one-sided A_s is multiplied by a factor to correct for eccentricity. The AASHTO formula, given here in the notation of this chapter, is a modification of Eq. 6.38b.

$$A_s = \left[0.15Bht(1 - C)\frac{V}{V_u} - 18t^2\right]\frac{\sigma_{yw}}{\sigma_{ys}}B \tag{6.39}$$

where

$$C = \frac{18{,}000}{\sqrt{\sigma_{y,psi}}}\frac{t}{h}\frac{1}{\cos\theta_d} - 0.3 \le 1$$

In this formula C replaces τ_{cr}/τ_y in Eq. 6.38b, B is a factor to correct for one-sided stiffeners, and the term $18t^2$ is the area of a portion of the web assumed to act with the stiffener. The reduction in required area effected by V/V_u, where V is the design ultimate-load shear and V_u the shear strength of the panel, is also permitted by the AISC specification.

If longitudinally stiffened girders are assumed to develop independent tension fields in their subpanels, as Cooper assumes, then Eq. 6.39 can be used to determine transverse stiffener area if h is taken to be the depth of the deeper subpanel. (The stiffener area requirement by this equation will always be larger for the deeper subpanel.) On the other hand, if a panel of a longitudinally stiffened girder develops the same tension field as it would without the longitudinal stiffener, as Rockey's work appears to show, then transverse-stiffener area requirements are the same for both.

In addition to the effects of axial force from the tension field, British Standard BS 5400 Part 3 (1982) accounts for the destabilizing forces arising from the action of stresses in the plane of the girder web on the initial displacements of the transverse stiffeners. This destabilizing effect is assessed by assuming that adjacent transverse stiffeners have initial displacements which alternate on each side of the girder web in a "sawtooth" progression. Horne (1980) has shown that the magnification of displacements, and the consequential bending stresses in the stiffeners, by the longitudinal and shear stress in the plate panels are equivalent to the effect produced by an imaginary axial force in the stiffeners. An approximate formula for a strut with an initial imperfection of length/750 is utilized for the design of the transverse stiffeners.

6.9.2 Longitudinal Stiffeners

A longitudinal stiffener must be stiff enough to maintain a node in a buckled web and must resist axial compression because of its location in the compression zone of the web. Therefore, both moment of inertia and cross-sectional area enter into a determination of its size. There have been a number of analytical investigations of this problem. Results are usually expressed in terms of three parameters: (1) $\gamma^* = EI/hD$, where I is the moment of inertia of the stiffener and $D = Et^3/12(1-\nu^2)$ (2) $\delta = A_s/ht$, in which A_s is the area of the stiffener, and (3) the panel aspect ratio a/h. Empirical formulas involving these parameters for various positions of the stiffener have been developed from numerical data, and charts (Klöppel and Scheer, 1960) are available.

The AASHTO formula for moment of inertia of a longitudinal stiffener at $h/5$ from the compression flange is

$$I_s = ht^3(2.4\alpha^2 - 0.13)$$

It will be noted that this equation does not contain δ. However, it is in fairly good agreement with the values of γ^* by Dubas (1948) for $\delta = 0.10$ in the range $0.5 \le \alpha \le 1.6$, and since γ^* decreases with decrease in δ, it is a reasonable upper bound for girders of practical proportions. It is important to note that these values of γ^* are derived from linear-buckling analysis and give a stiffness that guarantees only the critical load.

In calculating γ^* a portion of the web should be considered as part of the cross section. Strain measurements reported by Massonnet (1962) showed a mean effective width of $20t$. Tests show that the theoretical values must be increased considerably to develop the ultimate strength. Massonnet found that they should be multiplied by a factor ranging from 3 for a longitudinal stiffener at middepth to 7 for one at $h/5$ from the compression flange. This conclusion has been verified by a number of other investigators (Rockey, 1971a,b; Dubas, 1971). Further extensions were proposed by Maquoi et al. (1983). It is apparent, however, that there is a considerable range in the empirical multiplication factors. Difficulties may arise in practice in deciding on an appropriate factor to be used for a particular combination of applied stress and stiffener geometry.

An alternative approach to the design of a longitudinal web stiffener is to treat it as a column. This approach has been adopted in BS 5400: Part 3 (1982). Axial loading of the strut includes the longitudinal stress due to girder bending moment plus an axial load equivalent to the destabilizing effect of shear stress and in-plane stress transverse to the stiffener. For the design of longitudinal stiffeners an effective width of web plate of $16t$ on each side of the stiffener is considered.

6.10 PANELS UNDER EDGE LOADING

Girders often support loads on the top flange which produce compression on the edge of the web. They may be distributed over large distances, in some cases the length of the girder, or over relatively small distances, in which case they can be taken as concentrated loads. Bearing stiffeners may be needed for concentrated loads, but in some cases the web can carry them unaided.

A number of studies of buckling of plate elements under edge loading have been made (Girkmann, 1936; Zetlin, 1955; Wilkesmann, 1960; Klöppel and Wagemann, 1964; Warkenthin, 1965; Kawana and Yamakoshi, 1965; Rockey and Bagchi, 1970; Bossert and Ostapenko, 1967; Ostapenko et al., 1968; Khan and Walker, 1972; Khan et al., 1977). Basler (1961) presented a simplified analysis, neglecting the effect of bending and shear, which was adopted in the 1963 AISC Specification.

The ultimate strength of a web under edge loading may exceed the buckling load by a considerable margin, and several experimental investigations to determine ultimate strengths have been made. Bossert and Ostapenko (1967) and Ostapenko et al. (1968) used a finite-difference analysis to determine the web-buckling stress for a panel subjected to bending and an edge load distributed uniformly over the panel. The web was assumed to be fixed at the flanges and simply supported at the stiffeners. Ten tests were made on three plate-girder specimens with aspect ratios varying from 0.8 to 1.6 and ratios of extreme-fiber bending stress to edge stress from 0 to 5. The resulting ultimate strengths are shown in Fig. 6.11. The horizontal line through the ordinate 3 gives a lower bound on the experimental strenghs. The following formula for the allowable edge stress $\sigma_{e(\mathrm{all})}$ gives a factor of safety of 2 on the lower-bound ultimate edge stress:

$$\sigma_{e(\mathrm{all})} = \frac{1}{\sqrt{a/h}} \frac{24{,}000}{(h/t)^2} \frac{K}{\sqrt{F_y}} \tag{6.40}$$

where K is a coefficient given in Fig. 6.12.

Rockey (1972), Elgaaly (1975, 1977), Elgaaly and Rockey (1973), and Bagchi and Rockey (1975) used the finite-element method to determine the buckling load of a plate simply supported on all four edges and subjected to a partial edge loading (patch loading). Results were also obtained for the cases of a panel with patch load and shear, and patch

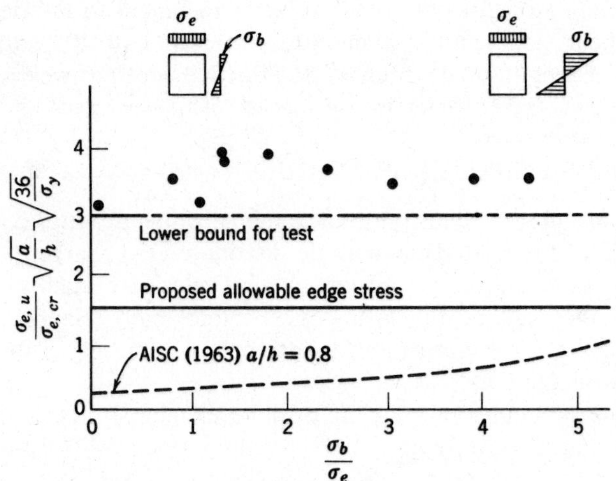

Fig. 6.11 Web plates under edge loading.

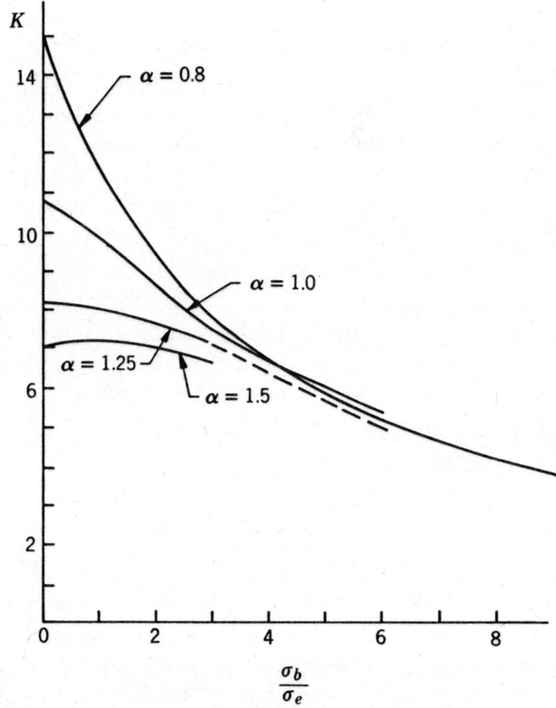

Fig. 6.12 Values of K in Eq. 6.40.

load and moment. Values of the coefficient k in the buckling formula (Eq. 4.1) were determined. Tests were conducted on 28 inverted-trough sheet-steel floor elements to determine both the buckling load and the ultimate load. Variables in the tests were the ratio of loaded length to panel length c/a, the panel aspect ratio a/h, and the web slenderness h/t. The panel was subjected to patch load in 22 tests and to patch load and in-plane moment in six tests. A linear relationship was found to exist between P_u/P_{cr} and both c/a and h/t. For the case $c/a = 0.2$, P_{cu}/P_{cr} was found to be smaller for the smaller of the two aspect ratios tests, which contradicts Eq. 6.40.

Results of patch-load tests on 20 trough-section beams of 15-ft span are reported by Elgaaly and Rockey (1973). The objective was to determine the interaction between edge loading and pure in-plane bending. The ratio c/a was 0.2 for all tests. It was found that the moment did not significantly reduce patch-loading strength until it exceeded 50% of the bending strength of the panel.

Elgaaly (1975) conducted tests on 18 trough-section beams of about 2 ft span, to determine the interaction between the ultimate strength under edge loading and shear. The ratio c/a was 0.2 for all tests and the panel slenderness ratio was 325 for 12 specimens and 200 for the remaining six. The results from the tests show that the presence of shear will reduce the ultimate load-carrying capacity of the web under edge loading, and an approximate relationship for this reduction was established. It was shown also that the postbuckling strength increases with the increase in the panel slenderness ratio. Furthermore, panels subjected to the combination of edge loading and shear exhibit a higher ultimate strength-to-buckling strength ratio than panels subjected to edge loading only.

The effect of flange thickness on web capacity under direct in-plane loading was also examined (Elgaaly, 1977) and results of patch-load tests on five welded girders are reported. The web dimensions were kept identical in all five girders (aspect ratio = 1 and slenderness ratio = 250). The c/a ratio was 0.2 for all tests. The ratios from these tests are compared with the AISC requirements in Fig. 6.13.

A theoretical study (Rockey, 1979) was carried out to determine the influence of a longitudinal stiffener on the ultimate capacity of a concentrated edge load. The study was restricted to the consideration of the buckling of a square plate panel reinforced by a single longitudinal stiffener at the one-fifth depth position. It was shown that the presence of the stiffener significantly increased the edge load. Values of buckling coefficients relating buckling resistance of the plate and the flexural rigidity of the longitudinal stiffener are provided.

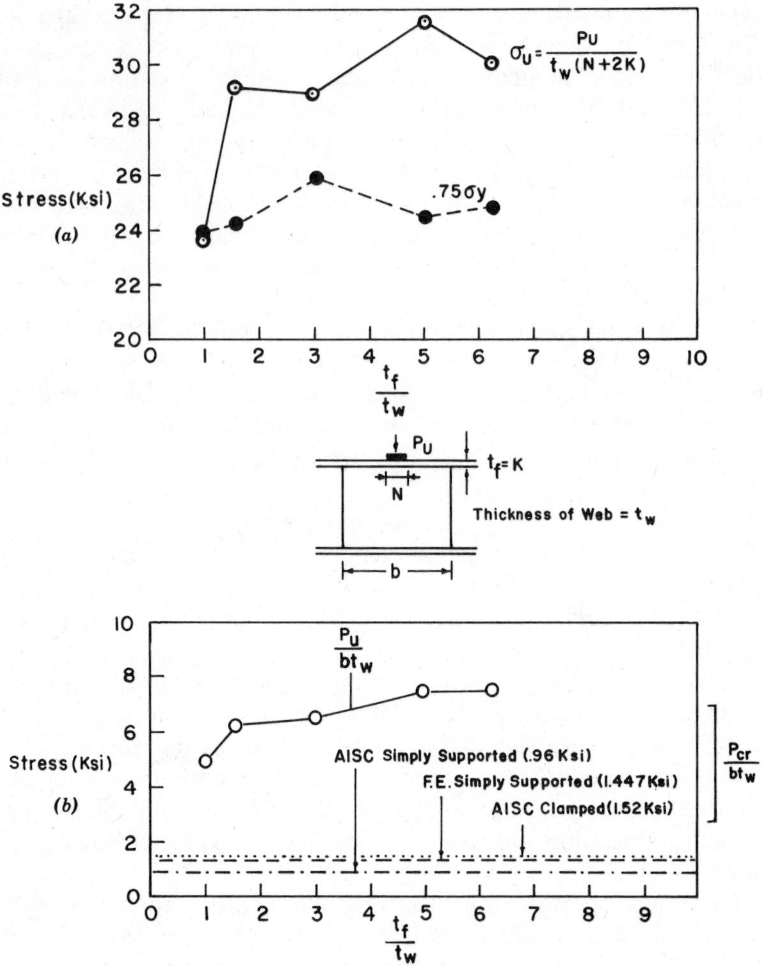

Fig. 6.13 Test results versus AISC requirements (Elgaaly, 1977).

The research described above is applicable to either flangeless panels or to beams with very thin flanges. However, in the case of plate girders and I-beams with reasonably thick flanges, it can be expected that the flanges will make a significant contribution to resisting local loading. On the basis of a preliminary series of tests Granholm (1960) and Bergfelt and Hövik (1968) concluded that web thickness was the most important parameter affecting the ultimate concentrated edge load, and proposed an empirical formula $P_u = 8.5t_w^2$ for mild steel (P_u is obtained in kN if t_w is given in millimeters). These tests had thin flanges and the failure mode was local yielding of the webs under the concentrated load. Further tests by Bergfelt (1967) on specimens with higher ratios of flange-to-web thickness showed that the failure load is also influenced by the bending stiffness of the flange and that failure involved a combination of yielding and buckling of the web. The ultimate load for the "weak" flange case where web yielding controls may be predicted by

$$P_u = 13\eta t_{fi} t_w \sigma_y \tag{6.41}$$

where t_{fi} is an equivalent flange thickness given by $t_{fi} = t_f\,(b/25t_f)^{1/4}$ to account for b/t values other than 25, and η is a parameter dependent on t_{fi}/t_w as follows:

t_{fi}/t_w	0.5	1.0	1.5	2.0
η	0.55	0.65	0.85	2.00

The theoretical basis for Eq. 6.41 is the treatment of the flange as a beam on an elastic foundation with plasticity accounted for by adjustments to the spring constants. Use of the η values, tabulated above, limits the ultimate load to a value associated with the start of rapidly increasing vertical deformation of the flange. For the "strong" flange case the following empirical formula shows good correlation with test results:

$$P_u = 0.6t_w^2(E\sigma_y)^{1/2}(1 + 0.4t_{fi}/t_w) \tag{6.42}$$

The ultimate concentrated load as given by Eqs. 6.41 and 6.42 is shown in Fig. 6.14. Where $t_{fi}/t_w = 2$ for mild steel both equations predict

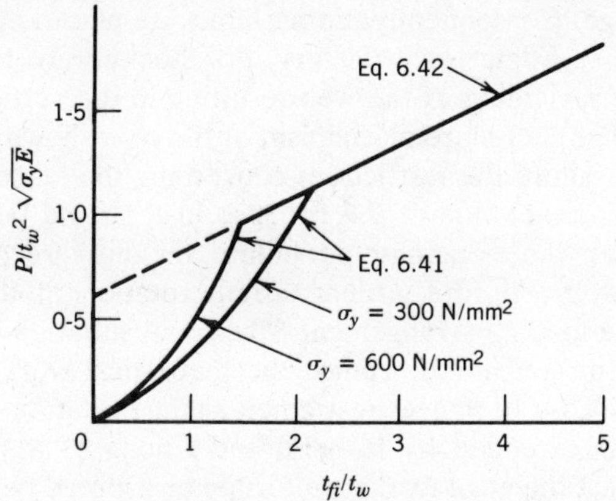

Fig. 6.14 Ultimate concentrated loads from Eqs. 6.41 and 6.42.

a failure load in close agreement with Granholm's original formula. Bergfelt's formulas are applicable when the load is concentrated as a sharp edge load. Where the load is distributed along a web length c the ultimate load given by Eqs. 6.41 and 6.42 should be multiplied by a factor $f(c)$ given by

$$f_{(c)} = \frac{\gamma}{1 - e^{-\gamma} \cos \gamma} \leq 1.3 \qquad (6.43)$$

where $\gamma = c/2L$ and $L = 6.7t_f$.

The limitation of 1.3 on the factor $f(c)$ is imposed to reflect the observation that when the load is distributed along the web, buckling will occur before yielding and thus the distribution length has a small influence on the ultimate load. For combined bending moment and concentrated load it was observed that a reasonable lower bound to test results is obtained when the value of P_u from Eqs. 6.41 and 6.42 is multiplied by the following correction factor proposed by Djubek and Škaloud (1976):

$$[1 - (\sigma_b/\sigma_y)^2]^{1/2} \qquad (6.44)$$

Roberts and Rockey (1979) developed a method based on the upper bound theorem of plastic collapse for predicting the ultimate value of a concentrated edge load. The assumed failure mechanism, involving plastic hinges in the flange and web, is shown in Fig. 6.15*a*. Equating external and internal work gives an expression from which the ultimate load P_u may be determined if the web hinge location α is known. The latter is chosen empirically to agree with test results. The solution presented by Roberts and Rockey (1979) contains an error in the expression which was pointed out and corrected by Chatterjee (1980). The mechanism solution overestimates the experimental collapse load when the loaded length c becomes too large. Consequently, upper limits are placed on c so that the correlation with test data is satisfactory. For stocky webs failure may be initiated by direct yielding of the web resulting in the vertical descent of the applied load and a failure mechanism of the type shown in Fig. 6.15*b*. For this type of failure the restrictions concerning the loaded length c do not apply. The actual value of the collapse load should be taken as the smaller of either the bending-type failure or the direct compression failure of the web. For the girders tested, rotation of the flange was prevented by the loading arrangement—the usual situation in most of the tests with concentrated loads. Hence the theoretical work in general is applicable to the case of flanges restrained against rotation. The mechanism solution was extended by Roberts and Chong (1981) to a loading condition in which the edge load is uniformly distributed between vertical stiffeners as indicated in Fig. 6.15*c*.

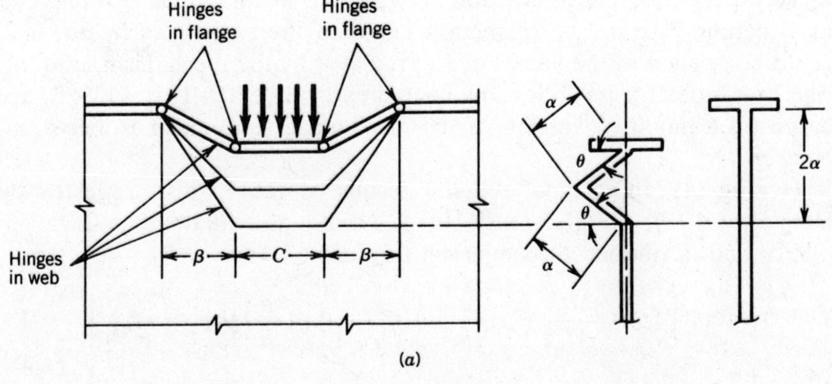

(a)

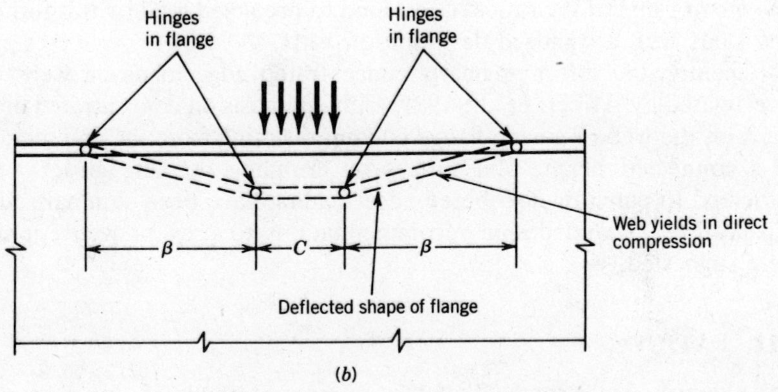

(b)

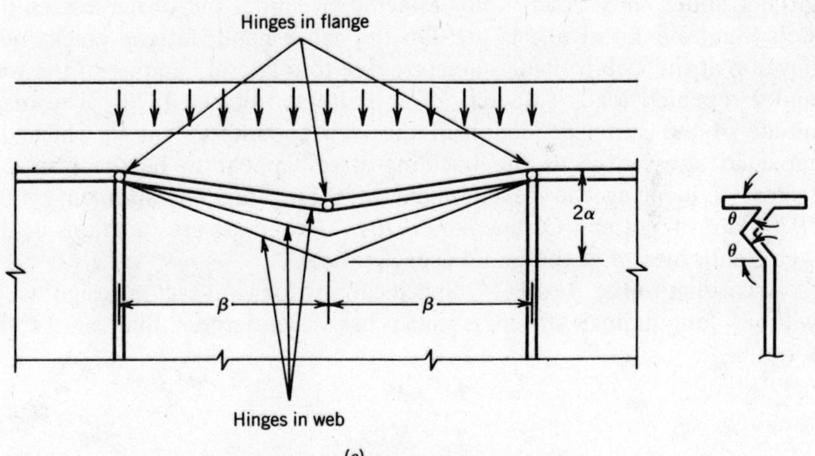

(c)

Fig. 6.15 Failure mechanism models.

To account for the interaction between coexisting global bending and local bending stresses, a correction factor of the type given in Eq. 6.44 could be applied to the values of P_u predicted by the mechanism solution. The provisions for patch loading contained in BS 5400; Part 3 (1982) are based on a modified version of the mechanism solution of Roberts and Rockey.

Herzog (1974b, c) analyzed the results of the 72 tests reported by Bergfelt (1971), Bergfelt and Hövik (1968) and Škaloud and Novak (1972) and developed the empirical formula

$$P-100t^2\left[1.2+\frac{5I_f h}{AI_w t}\left(1+\frac{c}{h}\right)^2\left(0.85+0.01\frac{a}{h}\right)\right]\left[1-\left(\frac{\sigma_b}{\sigma_y}\right)^2\right]^{1/8} \tag{6.45}$$

The mean value of the ratios of test load to predicted load by this formula was 1.001 with a standard deviation of 0.141.

Recently, the effect of nearly concentrated edge loads on webs was investigated by Aribert et al. (1981) with emphasis on concentrated direct loads on the web of a wide-flange column resulting from the end moment of a connected beam. The studies on crippling of plate girder webs, subjected to patch or distributed edge loading have been summarized by Elgaaly (1983), and design recommendations to prevent web crippling were suggested.

6.11 FATIGUE

Investigations have shown that the ultimate strength of thin-web plate girders under static load is not affected by initial out-of-flatness of the web (Shelestenko et al., 1970). On the other hand, fatigue cracks may develop at the web-to-flange juncture due to a lateral bending of the web under repeated loads (Maeda, 1971; Patterson et al., 1970). The magnitude of the initial deflection of the web and the extent to which the repeated stress exceeds the buckling stress appear to be the principal factors influencing the development of these cracks (Patterson et al., 1970; Parsanejad and Ostapenko, 1970). The factors are functions of the web slenderness and the panel aspect ratio.

According to the AASHTO Specifications for load-factor design, webs without longitudinal stiffeners must have slenderness that satisfy the formula

$$\frac{h}{t} \lesssim \frac{36{,}500}{\sqrt{F_{y,\mathrm{psi}}}} \tag{6.46}$$

If the girder has unequal flanges, h in this equation is replaced by $2y_0$, where y_0 is the portion of the web in compression. This limit may be somewhat conservative (Chern, 1970).

6.12 DESIGN PRINCIPLES AND PHILOSOPHIES

Despite the general recognition that the classical buckling theory is an inadequate guide to the prediction of the strength of plates panels in shear or compression, it must also be realized that comprehensive ultimate strength models are not yet fully developed. A situation exists, therefore, where some countries have opted to make use of ultimate strength theory when justified, while others prefer to rely in varying degrees on the use of the classical linear theory of plate buckling. Consequently, current design criteria for plate girders vary widely depending on the philosophy of design. It is interesting to compare the design provisions for plate girders adopted in various countries.

The current ASIC Specification (1978) and the allowable stress and load factor design methods of the AASHTO 1977 Specification recognize the contribution of tension-field action. Design for shear is then based on simplified expressions of Basler's formulas. AASHTO allows the use of a single longitudinal stiffener located at one-fifth of the depth from the compression flange, its principal function being the control of lateral web deflections. This reduction of out-of-plate deflections both improves the fatigue life of the web and permits the use of thinner web plates. The existing AASHTO provisions for transverse and longitudinal stiffeners prescribe minimum rigidities based on elastic buckling theory. These provisions are intended for moderate-span plate-girder bridges and are not adequate for long-span plate girders requiring very deep webs. For such cases, the problems requiring consideration include the design of deep webs with multiple longitudinal stiffeners, the development of new design methods for stiffener proportioning, the establishment of a procedure to account for axial load shedding from web to compression flange, and the development of a general ultimate strength method for transversely and longitudinally stiffened webs. Some of these problems have been investigated in a recent study carried out under the sponsorship of the Federal Highway Administration (FHWA, 1980).

In British Standard BS 5400: Part 3 (1982) girders are classified into two categories—those without longitudinal web stiffeners and those with such stiffeners. For the former, experimental data were thought to be sufficiently complete to justify the use of tension-field theory and an adaptation of the Rockey model is used for the ultimate shear strength. For the latter, it was thought that ultimate strength criteria had not been sufficiently established for longitudinal web (or box girder flange) stiffeners to cope with the associated high shear deformations. Therefore, the design of webs in girders containing longitudinal stiffeners is based on web panel behavior which stops short of the tension-field mechanism. Individual plate panels are subject to checks for yielding and buckling. The buckling check uses coefficients based on a large-deflection, elastic-plastic numerical study of initially imperfect plates under various stress patterns (Harding and Hobbs, 1979). Stiffener design is based on a strength criterion which attempts to assess the function of the stiffeners in controlling the stability of the plate panels. Such tests are currently being conducted at the University of Texas at Austin.

Since comprehensive ultimate strength models for plate girders are not yet available, the recommendations of the European Convention for Constructional Steelwork (ECCS) contain "provisional rules" (ECCS, 1976) based on the linear theory of plate buckling. The rules attempt to derive the ultimate strength from the critical buckling load by the introduction of suitable correction factors c^*, which vary with the type of stress distribution. A value larger than unity is used where a large postbuckling reserve exists, while c^* values less than unity are used where imperfections and residual stresses lower the critical buckling load. To ensure that the postbuckling strength of the plate panels is realized, the stiffeners must have an effective rigidity m times the minimum rigidity γ^* of the linear buckling theory. The provisional rules suggest multiplication factors $m = 4$ for open-section stiffeners and 2.5 for closed-section stiffeners.

In Germany the rules for design of members in compression are contained in DIN 4114, 1952–53 edition, and in the 1973 Interim Provisions to this specification. Proposed new provisions for the design of plates in compression are described by Scheer and Nolke (1976) and are expected to supersede some sections of DIN 4114. The viewpoint of the German committee is similar to that of the ECCS Commission; that is, while considerable progress has been made in developing ultimate design methods, they are not comprehensive and simple enough for immediate practical use. Consequently, while the proposed new German rules continue to be based on the classical buckling theory, they reevaluate safety factors and the interaction of local and overall buckling. On the question of required stiffener rigidity there is a difference of philosophy between ECCS recommendations and the proposed German rules. The former group advocates the use of multiplication factors to ensure that stiffeners remain rigid in the postbuckling domain. The latter group are of the opinion that it is unnecessary and uneconomic to increase the minimum stiffener rigidity beyond that required of the linear buckling theory as long as a degree of safety greater than or equal to that specified can be achieved.

6.13 RESEARCH NEEDS

1. *Combined Shear, Moment, and Axial Force*. Plate girders in cable-stayed girder bridges, self-anchored suspension bridges, arches, and rigid frames are subjected to axial compression or tension in addition to shear and moment. Some of the analyses discussed in this chapter can be extended to cover this situation, but there is little or no experimental research on the ultimate strength of plate-girder panels under combined shear, moment, and axaial force. Consideration of multiple longitudinal stiffeners would be an important aspect of such research.
2. *Longitudinal Stiffeners*. Design procedures for longitudinally stiffened girders should be studied with a view to simplifying the ultimate-strength analyses that show good correlation with test results. These investigations should cover multiple longitudinal stiffeners.
3. *Stiffeners*. Current methods for analysis and design of stiffeners are probably adequate, but further investigations, particularly of longitudinal stiffeners, would be worthwhile.
4. *Panels with Variable Depth*. More work is required to develop general design procedures for the ultimate strength of panels with variable depth.
5. *Loads on Flanges*. Although much progress has been made on the effect of concentrated edge loads on panels and girder flanges, more work is required to produce a comprehensive ultimate-strength design method which accounts for all factors which may influence the magnitude of the ultimate load.
6. *Webs with Holes*. The buckling strength of a thin square plate with a central circular hole, clamped at its edges and subjected to pure shear, were determined analytically and experimentally by Grosskurth et al. (1976). Stability of webs containing circular or rectangular middepth holes was investigated by Redwood and Venoya (1979). The postbuckling strength of webs with holes needs to be investigated. Narayanan and Der-Avanessian (1985) cite seven references on this topic and refer to over 70 tests performed at the University of Cardiff.
7. *Fatigue*. Better knowledge of the fatigue behavior of slender web plates is needed to take full advantage of ultimate-strength design of plate-girder bridges and crane girders.
8. *Corrugated Webs*. Vertically corrugated webs may be an economical alternative to thin webs. There is no theoretical or experimental information on the ultimate strength of such panels.
9. *Composite Girders*. The assumptions on which the analyses for shear and bending of plate-girder panels are based can be expected to apply

to composite girders, but there are no experimental results to confirm this assumption. The large stiffness and low tensile strength of a concrete slab, compared to a steel flange, may result in some differences in the panel behavior. Therefore, tests on thin-webbed composite girders are needed. Such tests are currently being conducted at the University of Texas at Austin.

References

Aribert, J. M., Lachal, A., and Elnawawy, O. (1981), "Modelisation elasto-plastique de la resistance d'un profile en compression locale," *Constr. Met.*, No. 2.

Bagchi, D. K., and Rockey, K. C. (1975), "Postbuckling Behaviour of Web Plate under Partial Edge Loading," *Proc. 3rd Int. Spec. Cold-Formed Steel Struct.*, University of Missouri at Rolla, November 24–25, Vol. 1, pp. 351–384.

Basler, K. (1961), "New Provisions for Plate Girder Designs," *Proc. Natl. Eng. Conf.*, AISC.

Basler, K. (1963a), "Strength of Plate Girders in Shear," *Trans Am. Soc. Civ. Eng.*, Vol. 128, Part II, p. 683.

Basler, K. (1963b), "Strength of Plate Girders under Combined Bending and Shear," *Trans. Am. Soc. Civ. Eng.*, Vol. 128, Part II, p. 720.

Basler, K., and Thürlimann, B. (1960a), "Carrying Capacity of Plate Girders," *IABSE 6th Cong., Prelim. Pub.* Vol. 16.

Basler, K., and Thürlimann, B. (1960b), "Buckling Tests on Plate Girders," *IABSE 6th Cong., Prelim. Pub.* Vol. 17.

Basler, K., and Thürlimann, B. (1963), "Strength of Plate Girders in Bending," *Trans. Am. Soc. Civ. Eng.*, Vol. 128, Part II, p. 655.

Basler, K., Yen, B. T., Mueller, J. A., and Thürlimann, B. (1960), "Web Buckling Tests on Welded Plate Girders," *Weld. Res. Counc. Bull.*, No. 64, September.

Bergfelt, A. (1971), "Studies and Tests on Slender Plate Girders without Stiffeners, *IABSE Coll. Des. Plate Box Girders Ultimate Strength*, London.

Bergfelt, A. (1976), "The Behaviour and Design of Slender Webs under Partial Edge Loading," *Int. Conf. Steel Plated Struct.*, Imperial College, London.

Bergfelt, A., and Hövik, J. (1968), "Thin-Walled Deep Plate Girders under Static Load," *IABSE 8th Cong., Final Rep.*, New York, 1968.

Bossert, T. W., and Ostapenko, A. (1967), "Buckling and Ultimate Loads for Plate Girder Web Plates Under Edge Loading," *Fritz Eng. Lab. Rep. No. 319.1*, Lehigh University, Bethlehem, Pa., June.

British Standards Institution (1982), *Steel, Concrete and Composite Bridges*, British Standard BS 5400: Part 3. *Code of Practice for Design of Steel Bridges*, BSI, London.

Carskaddan, P. S. (1968), "Shear Buckling of Unstiffened Hybrid Beams," *ASCE J. Struct. Div.*, Vol. 94, No. ST8, pp. 1965–1992.

Cescotto, S., Maquoi, R., and Massonnet, C. (1981), "Sur Ordinateur du comportement a la ruine des putres a ame pleine cisailles on fleches," *Constr. Met.*, No. 2, pp. 27–40.

Chatterjee, S. (1980), "Design of Webs and Stiffeners in Plate and Box Girders," *Int. Conf. Des. Steel Bridges*, Cardiff.

Chern, C., and Ostapenko, A. (1969), "Ultimate Strength of Plate Girder under Shear," *Fritz Eng. Lab. Rep. No. 328.7*, Lehigh University, Aug.

Chern, C., and Ostapenko, A. (1970a), "Bending Strength of Unsymmetrical Plate Girders," *Fritz Eng. Lab. Rep. No. 328.8*, Lehigh University, Sept.

Chern, C., and Ostapenko, A. (1970b), "Unsymmetrical Plate Girders under Shear and Moment," *Fritz Eng. Lab. Rep. No. 328.9*, Lehigh University, Bethlehem, Pa., October.

Chern, C., and Ostapenko, A. (1971), "Strength of Longitudinally Stiffened Plate Girders under Combined Loads," *IABSE Coll. Des. Plate Box Girders Ultimate Strength*, London.

Clark, J. W., and Sharp, M. L. (1971), "Limit Design of Aluminum Shear Web," *IABSE Coll. Des. Plate Box Girders Ultimate Strength*, London.

Cooper, P. B. (1965), "Bending and Shear Strength of Longitudinally Stiffened Plate Girders," *Fritz Eng. Lab. Rep. No. 304.6*, Lehigh University, Bethlehem, Pa., September.

Cooper, P. B. (1967), "Strength of Longitudinally Stiffened Plate Girders," *ASCE J. Struct. Div.*, Vol. 93, No. ST2, pp. 419–452.

Cooper, P. B. (1971), "The Ultimate Bending Moment for Plate Girders," *IABSE Coll. Des. Plate Box Girders Ultimate Strength*, London.

D'apice, M. A., Fielding, D. J., and Cooper, P. B. (1966), "Static Tests on Longitudinally Stiffened Plate Girders," *Weld. Res. Counc. Bull., No. 117*, October.

Davies, G., and Mandal, S. N. (1979), "The Collapse Behaviour of Tapered Plate Girders Loaded within the Tip," *Proc. Inst. Civ. Eng.*, Vol. 67, Part 2.

Djubek, J., and Škaloud, M. (1976), "Postbuckled Behaviour of Web Plates in the New Edition of Czechoslovak Design Specifications," *Int. Conf. Steel Plated Struct.*, Imperial College, London.

Dubas, C. (1948), "A Contribution to the Buckling of Stiffened Plates," *IABSE 3rd Cong., Prelim. Pub.*, Liege.

Dubas, P. (1971), "Tests on Postcritical Behavior of Stiffened Box Girders," *IABSE Coll. Des. Plate Box Girders Ultimate Strength*, London.

Elgaaly, M. A. (1973), "Buckling of Tapered Plates under Pure Shear," *Proc. Conf. Steel Struct.*, Timisaora, Romania, October.

Elgaaly, M. (1975), "Failure of Thin-Walled Members under Patch Loading and Shear," *Proc. 3rd Int. Spec. Conf. Cold-Formed Steel Struct.*, University of Missouri at Rolla, Nov. 24–25, Vol. 1, pp. 357–381.

Elgaaly, M. (1977), "Effect of Flange Thickness on Web Capacity under Direct in-Plane Loading," *Proc. Annu. Tech. Sess. SSRC*, May.

Elgaaly, M. (1983), "Web Design under Compressive Edge Loads," *Eng. J. Am. Inst. Steel Constr.*, Vol. 20, No. 4, pp. 153–171.

Elgaaly, M. A., and Rockey, K. C. (1973), "Ultimate Strength of Thin-Walled Members under Patch Loading and Bending," *Proc. 2nd Spec. Conf. Cold-Formed Steel Struct.*, University of Missouri at Rolla, October.

European Convention for Constructional Steelwork (1976), "Conventional Design Rules Based on Linear Buckling Theory," *Proc. 2nd Int. Colloq. Stab., Introd. Rep.*, ECCS, Liège.

Falby, W. E., and Lee, G. C. (1976), "Tension-Field Design of Tapered Webs," *Am. Inst. Steel Constr.*, Vol. 13, No. 1., p. 11.

Federal Highway Administration (1980), "Proposed Design Specifications for Steel Box Girder Bridges," *Rep. No. FHWA-TS-80-205*, U.S. Department of Transportation, Federal Highway Administration.

Fujii, T. (1968a), "On an Improved Theory for Dr. Basler's Theory," *IABSE 8th Cong., Final Rep.*, New York.

Fujii, T. (1968b), "On Ultimate Strength of Plate Girders," *Jpn. Shipbuild. Mar. Eng.*, May, pp.

Fujii, T. (1971), "A Comparison between Theoretical Values and Experimental Results for the Ultimate Shear Strength of Plate Girders," *IABSE Coll. Des. Plate Box Girders Ultimate Strength*, London.

Fujii, T., Fukumoto, Y., Nishino, F., and Okumura, T. (1971), "Research Works on Ultimate Strength of Plate Girders and Japanese Provisions on Plate Girder Design," *IABSE Coll. Des. Plate Box Girders Ultimate Strength*, London.

Gaylord, E. H. (1963), "Discussion of K. Basler 'Strength of Plate Girders in Shear'," *Trans. Am. Soc. Civ. Eng.*, Vol. 128, Part II, p. 712.

Gaylord, E. H., and Gaylord, C. N. (1972), *Design of Steel Structures*, 2nd ed., McGraw-Hill, New York.

Girkmann, K. (1936), "Die Stabilität der Stegbleche vollwandiger Träger bei Berücksichtigung örtlicher Lastangriffe," *IABSE 3rd Cong. Final Rep.*, Berlin.

Granholm, C. A. (1960), "Tests on Girders with Thin Web Plates," *Rapport 202*, Institutionen for Byggnadsteknik, Chalmers Tekniska Hogskola, Goteborg (in Swedish).

Grosskurth, J. F., White, R. N., and Gallagher, R. H. (1976), "Shear buckling of Square Perforated Plates," *ASCE J. Eng. Mech. Div.*, Vol. 102, No. EM6, pp. 1025–1040.

Harding, J. E., and Hobbs, R. E. (1979), "The Ultimate Behaviour of Box Girder Web Panels," *Struct. Eng.*, Vol. 57B, No. 9.

Herzog, M. (1973), "Die Traglast versteifter, dünnwändiger Blechträger unter reiner Biegung nach Versuchen," *Bauingenieur*, September.

Herzog, M. (1974a), "Die Traglast unversteifter und versteifter, dünnwändiger Blechträger unter reinem Schub und Schub mit Biegung nach Versuchen," *Bauingenieur*, October.

Herzog, M. (1974b), "Ultimate Strength of Plate Girders from Tests," *ASCE J. Strut. Div.*, Vol. 100, No. ST5, pp. 849–864.

Herzog, M. (1974c), "Die Krüppellast sehr dünner Vollwandträgerstege nach Versuchen," *Stahlbau*, Vol. 43, pp. 26–28.

Höglund, T. (1971a), "Behavior and Load-Carrying Capacity of Thin-Plate I Girders" *R. Inst. Technol. Bull.*, No. 93, Stockholm (in Swedish).

Höglund, T. (1971b), "Simply Supported Thin Plate I-Girders without Web Stiffeners Subjected to Distributed Transverse Load," *IABSE Coll. Des. Plate Box Girders Ultimate Strength*, London.

Höglund, T. (1973), "Design of Thin-Plate I Girders in Shear and Bending," *R. Inst. Technol. Bull.*, No. 94, Stockholm.

Horne, M. R. (1980), "Basic Concepts in the Design of Webs," *Int. Conf. Des. Steel Bridges*, Cardiff.

Jetteur, P. (1984), "Contribution a la solution de problème particuliers d'instabilité dans la grand poutres metallique," *Coll. Publ. Fac. Sci. Appl.*, Liege, No. 94.

Kawana, K., and Yamakoshi, M. (1965), "On the Buckling of Simply Supported Rectangular Plates Under Uniform Compression and Bending," *J. Soc. Nav. Archit. West Jpn.*, Vol. 29, pp.

Khan, M. Z., and Walker, A. C. (1972), "Buckling of Plates Subjected to Localized Edge Loading," *Struct. Eng.*, Vol. 50, No. 6, pp. 225–232.

Khan, M. Z., Johns, K. C., and Hayman, B. (1977), "Buckling of Plates with Partially Loaded Edges," *ASCE J. Struct. Div.*, Vol. 103, No. ST3, pp. 547–558.

Klöppel, K., and Scheer, J. S. (1960), *Beulwerte ausgesteifter Rechteckplatten*, Wilhelm Ernst, West Berlin.

Klöppel, K., and Wagemann, C. H. (1964), "Beulen eines Bleches unter einseitiger Gleichstreckenlast," *Stahlbau*, Vol. 33, pp. 216–220.

Komatsu, S. (1971), "Ultimate Strength of Stiffened Plate Girders Subjected to Shear," *IABSE Coll. Des. Plate Box Girders Ultimate Strength*, London.

Konishi, I., et al. (1965), "Theories and Experiments on the Load Carrying Capacity of Plate Girders," *Rep. Res. Comm. Bridges Steel Frames Weld. Kansai Dist. Jpn.*, July (in Japanese).

Kuhn, P. (1956), *Stresses in Aircraft and Shell Structures*, McGraw-Hill, New York.

Lew, H. S., and Toprac, A. A. (1968), "Static Strength of Hybrid Plate Girders," *SFRL Tech. Rep.*, University of Texas, p. 550–11, January.

Maeda, Y. (1971), "Ultimate Static Strength and Fatigue Behavior of Longitudinally Stiffened Plate Girders in Bending," *IABSE Coll. Des. Plate Box Girders Ultimate Strength*, London.

Maquoi, R., Jetteur, P., Massonnet, C., and Škaloud, M. (1983), "Calcule des ames et semelles radies des ponts en acier," *Constr. Met.*, No. 4, pp. 15–28.

Massonnet, C. (1962), "Stability Considerations in the Design of Steel Plate Girders," *Trans. Am. Soc. Civ. Eng.*, Vol. 127, Part II, pp. 420–447.

Narayanan, R., Der-Avanessian, N. G.-V. (1985), "Design of Slender Webs Having Rectangular Holes," *ASCE J. Struct. Eng.*, Vol. III, No. 4, pp. 777–787.

Ostapenko, A., and Chern, C. (1970), "Strength of Longitudinally Stiffened Plate Girders," *Fritz Eng. Lab. Rep. No. 328.10*, Lehigh University, Bethelehem, Pa., December.

Ostapenko, A., Yen, B. T., and Beedle, L. S. (1968), "Research on Plate Girders at Lehigh University," *IABSE 8th Cong., Final Rep.*

Ostapenko, A., Chern, C., and Parsanejad, C. (1971), "Ultimate Strength Design of Plate Girders," *Fritz Eng. Lab. Rep. No. 328.12*, Lehigh University, (Bethlehem, Pa., January.).

Parsanejad, S., and Ostapenko, A. (1970), "On the Fatigue Strength of Unsymmetrical Steel Plate Girders," *Weld. Res. Counc. No. 156*, November.

Patterson, P. J., Corrado, J. A., Huang, J. S., and Yen, B. T. (1970), "Fatigue and Static Tests of Two Welded Plate Girders," *Weld. Res. Counc. Bull., No. 155*, October.

Porter, D. M., Rockey, K. C., and Evans, H. R. (1975), "The Collapse Behavior of Plate Girders Loaded in Shear," *Struct. Eng.*, Vol. 53, No. 8, pp. 313–325.

Redwood, R. L., and Venoya, M. (1979), "Critical Loads for Webs with Holes," *ASCE J. Struct. Div.*, Vol. 105, No. ST10, pp. 2053–2068.

Roberts, T. M., and Chong, C. K. (1981), "Collapse of Plate Girders under Edge Loading," *ASCE J. Struct. Div.*, Vol. 107, No. ST8, pp. 1503–1510.

Roberts, T. M., and Rockey, K. C. (1979), "A Mechanism Solution for Predicting the

Collapse Loads of Slender Plate Girders When Subjected to in-Plane Patch Loading," *Proc. Inst. Civ. Eng.*, Vol. 67, Part 2.

Rockey, K. C. (1971a), "An Ultimate Load Method for the Design of Plate Girders," *IABSE Coll. Des. Plate Box Girders Ultimate Strength*, London.

Rockey, K. C. (1971b), "An Ultimate Load Method of Design for Plate Girders," *Proc. Conf. Dev. Bridge Des. Construct.*, Crosby Lockwood, London.

Rockey, K. C. (1971c), "Free Discussion," *IABSE Coll. Des. Plate Box Girders Ultimate Strength, London*, p. 323.

Rockey, K. C., and Bagchi, D. K. (1970), "Buckling of Plate Girder Webs under Partial Edge Loadings," *Int. J. Mech. Sci.*, Vol. 12, Pergamon Press, Oxford.

Rockey, K. C., and Leggett, D. M. A. (1962), "The Buckling of a Plate Girder Web under Pure Bending When Reinforced by a Single Longitudinal Stiffener," *Proc. Inst. Civ. Eng.*, Vol. 21.

Rockey, K. C., and Škaloud, M. (1968), "Influence of Flange Stiffness upon the Load Carrying Capacity of Webs in Shear," *IABSE 8th Cong., Final Rep.*, New York.

Rockey, K. C., and Škaloud, M. (1972), "The Ultimate Load Behavior of Plate Girders Loaded in Shear," *Struct. Eng.*, Vol. 50, No. 1.

Rockey, K. C., Elgaaly, M., and Bagchi, D. K. (1972), "Failure of Thin-Walled Members under Patch Loading," *ASCE J. Struct. Div.*, Vol. 98, No. ST12, pp. 2739–2752.

Rockey, K. D., Evans, H. R., and Porter, D. M. (1973), "Ultimate Load Capacity of Stiffened Webs Subjected to Shear and Bending," *Proc. Conf. Steel Box Girders*, Institution of Civil Engineers, London.

Rockey, K. C., Evans, H. R., and Porter, D. M. (1974), "The Ultimate Strength Behavior of Longitudinally Stiffened Reinforced Plate Girders," *Symp. Nonlinear Tech. Behav. Struct. Anal.*, Transport and Road Research Laboratory, Crowthorne, December.

Scheer, J., and Nolke, H. (1976), "The Background to the Future German Plate Buckling Design Rules," *Int. Conf. Steel Plated Struct.*, Imperial College, London.

Schueller, W., and Ostapenko, A. (1970), "Tests on a Transversely Stiffened and on a Longitudinally Stiffened Unsymmetrical Plate Girder," *Weld. Res. Counc. Bull. No. 156*, November.

Selberg, A. (1973), "On the Shear Capacity of Girder Webs," *University of Trondheim Report.*

Sharp, M. L., and Clark, J. W. (1971), "Thin Aluminum Shear Webs," *ASCE J. Struct. Div.*, Vol. 97, No. ST4, pp. 1021–1038.

Shelestanko, L. P., Dushnitsky, V. M., and Borovikov, V. (1970), "Investigation of the Influence of Limited Web Deformation on the Ultimate Strength of Welded Pate Girders," *Res. Steel Compos. Superstruct. Bridges*, No. 76, Transport, Moscow (in Russian).

Skaloud, M. (1962), "Design of Web Plates of Steel Girders with Regard to the Postbuckling Behavior (Analytical Solution), *Struct. Eng.*, Vol. 40, No. 12, p. 409.

Skaloud, M. (1971), "Ultimate Load and Failure Mechanism of Thin Webs in Shear," *IABSE Coll. Des. Plate Box Girders Ultimate Strength*, London.

Škaloud, M., and Novak, P. (1972), "Postbuckled Behavior and Incremental Collapse of Webs Subjected to Concentrated Load," *IABSE 9th Cong., Prelim. Pub.*, Amsterdam.

Stein, M., and Fralich, R. W. (1950), "Critical Shear Stress of Infinitely Long Simply Supported Plate with Transverse Stiffeners," *J. Aeronaut. Sci.*, Vol. 17.

Steinhardt, O., and Schröter, W. (1971), "Postcritical Behavior of Aluminum Plate Girders with Transverse Stiffeners," *IABSE Coll. Des. Plate Box Girders Ultimate Strength*, London.

Takeuchi, T. (1964), "Investigation of the Load Carrying Capacity of Plate Girders," M.S. thesis, University of Kyoto (in Japanese).

Wagner, H. (1931), "Flat Sheet Metal Girder with Very Thin Metal Web," *NACA Tech. Nemo. Nos. 604, 605, 606.*

Warkenthin, W. (1965), "Zur Beulsicherheit querbelasteter Stegblechfelder," *Stahlbau*, Vol. 34, p. 28.

White, R. N., and Cottingham, W. (1962), "Stability of Plates under Partial Edge Loading," *ASCE J. Eng. Mech. Div.*, Vol. 88, No. EM5, pp. 67–86.

Wilkesmann, F. W. (1960), "Stegblechbeulung bei Längsrandbelastung," *Stahlbau*, Vol. 29.

Wilson, J. M. (1886), "On Specifications for Strength of Iron Bridges," *Trans. Am. Soc. Civ. Eng.*, Vol. 15, Part I, pp. 401–403, 489–490.

Yen, B. T., and Basler, K. (1962), "Static Carrying Capacity of Steel Plate Girders," *Highw. Res. Board Proc.*, Vol. 41.

Yen, B. T., and Mueller, J. A. (1966), "Fatigue Tests of Large-Size Welded Plate Girders," *Weld. Res. Counc. Bull., No. 118*, November.

Yonezawa, H., Miakami, I., Dogaki, M., and Uno, H. (1978), "Shear Strength of Plate Girders with Diagonally Stiffened Webs, *Trans Jpn. Soc. Civ. Eng.*, Vol. 10.

Zetlin, L. (1955), "Elastic Stability of Plates under Edge Loading," *Proc. Am. Soc. Civ. Eng.*, Vol. 81.

Chapter Seven

Box Girders

7.1 INTRODUCTION

Box girders are used extensively for bridges (Fig. 7.1), heavy industrial buildings, offshore platforms, and other structures where large loads are frequently encountered. They are employed to best advantage when use is made of their considerable torsional stiffness. Although box girders may have a variety of cross-sectional shapes (Fig. 7.2) ranging from a deep narrow box to a wide shallow box (perhaps with many webs, some sloping) those occurring most frequently have flanges which are wider and more slender than in plate girder construction. The webs, on the other hand, may be of comparable slendernesses to those of plate girders. However, it is the extensive use of slender plate construction for flanges as well as webs which makes stability considerations so important in the case of box girders.

In this chapter aspects of the stability of box girders and their components, which are additional to those described in Chapter 6 on plate girders, will be discussed. Most of these additional problems are a consequence of the use of wide flanges, that is, the influences of shear lag and lateral loading on the buckling of stiffened compression plates. Others relate to the need to provide diaphragms within a box, not only to retain the box shape but also to distribute forces to the support bearings. In the case of box girders, as opposed to plate girders, the bearings may not be positioned directly beneath and in line with the webs. Much of the information on web stability provided already in the context of plate girders is applicable to box girders, but there are important differences which are discussed. Some of the unresolved problems relating to box girders are also outlined.

Fig. 7.1 Box girder bridges.

7.2 BASES OF DESIGN

Up until the end of the 1960s the design of box girders was not codified in detail in any country. Although a wide variety of approaches was used for the design of box girder bridges, the basis generally adopted was to use a factor of safety on the buckling stress, more often than not the *critical elastic buckling stress*, of the component being designed. These factors of safety varied from country to country and were intended to cover not only the as-built condition but also conditions met during construction. In some cases they also reflected an appreciation of the postbuckling reserve of plates, with lower factors of safety being used than would be the case where no such reserve was anticipated.

Three major collapses which occurred during the erection of box girder bridges, at Milford Haven (1969), West Gate (1970), and Koblenz (1971) (Fig. 7.3), caused the whole basis of the design of box girders to be reviewed, (ICE, 1973; "Inquiry," 1973) not only in Europe but also in the United States and, since then, worldwide. The drafting of new rules in the wake of these tragedies coincided with the trend toward limit-state

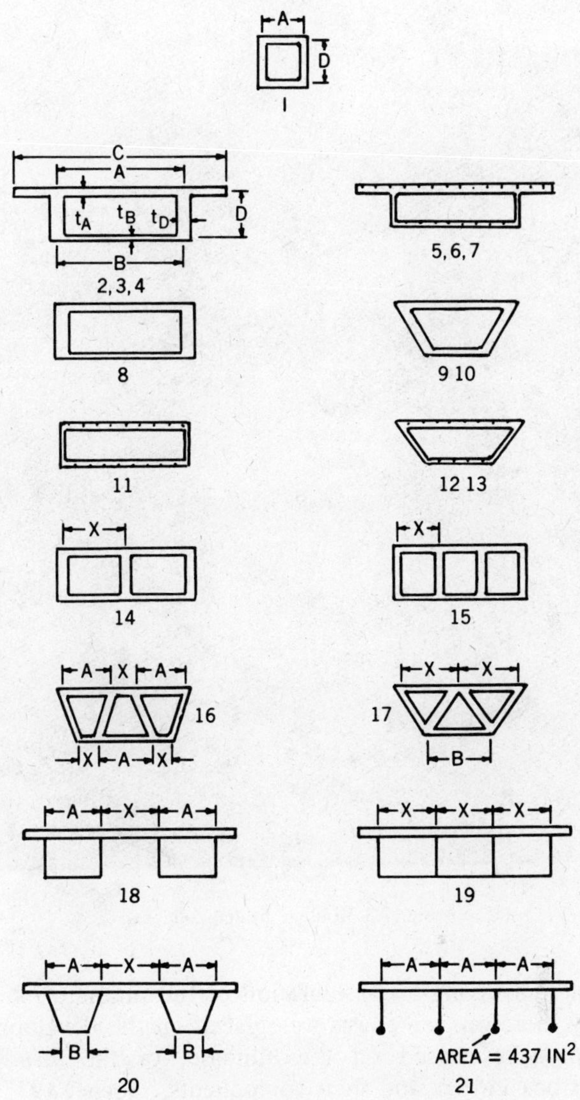

Fig. 7.2 Bridge cross sections.

Fig. 7.3 Koblenz bridge collapse.

format codes involving a consideration of the ultimate limit state. Not surprisingly therefore, emphasis was placed on the development of new design approaches based on the ultimate, or the *inelastic buckling strength*, of box girders and their components (Horne, 1977).

Influence of Initial Geometric Imperfections and Residual Stresses. Much attention was paid to the role of initial imperfections, both geometric and residual stresses, during the investigations of the strength of box girders and their components (Dowling et al., 1977b). Extensive measurements were made of residual stresses produced during box girder fabrication by using both destructive, but mainly nondestructive, methods. These were carried out in model boxes fabricated using techniques representative of full-scale techniques and also in actual box girder bridges under construction. Numerous surveys of the distribution of initial distortions in model and full-scale boxes were also made. Concurrent with these measurements fundamental studies on the prediction of residual stresses, and on their effects, together with the effects of geometric imperfections, were carried out in many places.

The weakening effect of both types of initial imperfection, separately and together, is now well understood and has been incorporated into the various design methods produced to predict the inelastic buckling strength of plated structures. The weakening effect of these imperfections is most pronounced in the range of intermediate slendernesses, that is, those slendernesses at which the critical buckling stress and the yield stress are roughly equal (Frieze et al., 1975). The knockdown in strength is most marked for plates of intermediate slenderness subjected to compressive stressing and is least pronounced for shear-loaded cases, being practically negligible for rectangular plates in pure shear.

Strength curves for plate panels used for design have normally been selected with allowances made for practical combinations of residual stresses and initial imperfections. Ultimate load design methods for stiffeners or stiffened plates also normally incorporate the effects of imperfections in their expressions. Typically, a level of compressive residual stress of the order of 10% of the yield stress, with an initial geometric imperfection which is related to fabrication tolerances, may be adopted by code drafters. In other cases a geometric imperfection is chosen which is considered on the basis of theoretical or experimental evidence to incorporate the combined weakening effect of residual stress and practical initial distortions. It has been shown that provided that the magnitude of the initial distortion chosen is big enough (and this is often of the same order as specified fabrication tolerances), the additional weakening effect of even the largest residual compressive stresses is small for practical plates.

The option of specifying levels of residual stresses to be calculated by designers and measured by fabricators is not a realistic one in terms of the economics of normal fabricated box girders and is avoided in specifications. As a result of the efforts made to reconsider the basis of design for box girders, new rules were produced in the United Kingdom, the United States, and West Germany, among other countries. The first to be fully codified were the British rules; those of the United States are still being developed at the time of writing. For this reason the basis of the design approach contained in BS 5400: Part 3, Design of Steel Bridges (1983) is summarized here in more detail than is devoted to the other rules.

7.3 BUCKLING OF WIDE FLANGES

Box girder flanges are normally stiffened in both the longitudinal and transverse directions (Fig. 7.4). The unstiffened flanges of narrow box girders can be treated as plates, as described in Chapter 4, by using reduced effective widths to account for the effects of buckling. Possible modes of buckling occurring in an orthogonally stiffened plate are illustrated in Fig. 7.5. These include overall buckling of the stiffened flange, buckling of the longitudinally stiffened panels between transverse elements, various forms of local plate and stiffener buckling, and various combinations of these modes.

The linear elastic buckling of stiffened flanges with one or more stiffeners is covered in Chapter 4 and formed the basis of earlier design

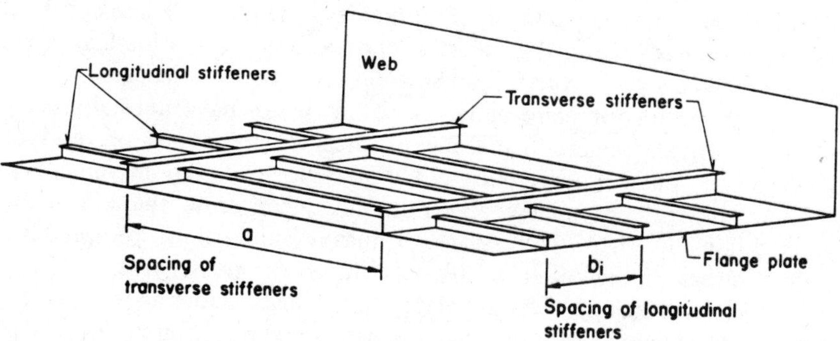

Fig. 7.4 Sample stiffened panel (compression flange).

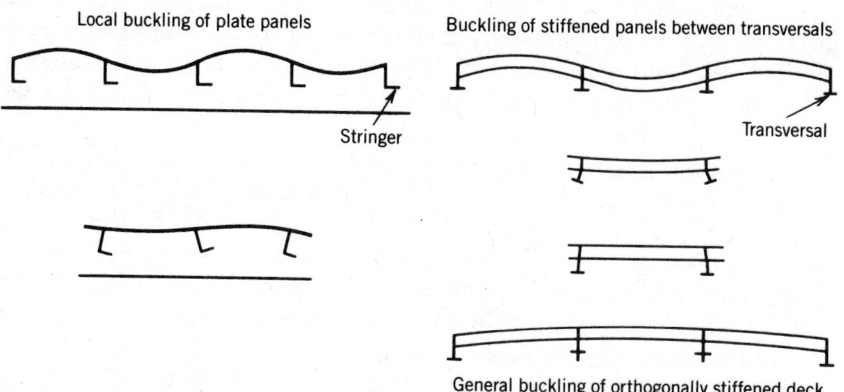

Fig. 7.5 Various forms of local buckling of stringers and transversals.

methods. Indeed, many codes still in use today are based on these methods together with the use of safety factors determined from test results. These are reviewed by Rockey and Evans (1980). In this chapter emphasis will be placed on methods designed to predict the ultimate strength of such flanges, as these are being used to replace the elastic methods. It should be recognized, however, that even now the new rules for stiffened compression flanges are not entirely based on ultimate strength considerations, as pointed out by Dowling (Rockey and Evans, 1980).

In considering the inelastic buckling of wide stiffened flanges, three basic approaches can be adopted. In increasing in order of complexity, the flanges can be treated as:

1. A series of disconnected struts.
2. An orthotropic plate.
3. A discretely stiffened plate.

For most stiffened flanges the "strut" approach is sufficiently accurate and is suitable for design purposes. When the flange stiffening is lighter than normal, advantage can be taken of the postbuckling reserve of the stiffened plate and many methods have been produced based on this type of modeling, notably those of Massonnet and his co-workers (Jetteur, 1983; Maquoi and Massonnet, 1971). However, the flange geometry rarely justifies such an approach in practice and the strut approach will, in any case, produce safe designs. The discretely stiffened plate approach is of interest mainly as an aid to understanding the flange behavior rather than as a design aid, although in the case of a flange stiffened by only one or two stiffeners, it can be used, in the absence of simpler approaches, to proportion compression flanges. Each of these methods is described in turn below.

7.3.1 The Strut Approach

The basis of this approach is to treat a plate stiffened by several equally spaced longitudinal stiffeners as a series of unconnected compression members or struts, each of which consists of a stiffener acting together with an associated width of plate that represents the plate between stiffeners (Fig. 7.6). Where transverse stiffeners are present they are designed to be sufficiently stiff to ensure that they provide nodal lines acting as simple rotationally free supports to the ends of the longitudinal struts. Thus the equivalent buckling length of the longitudinal stiffeners is

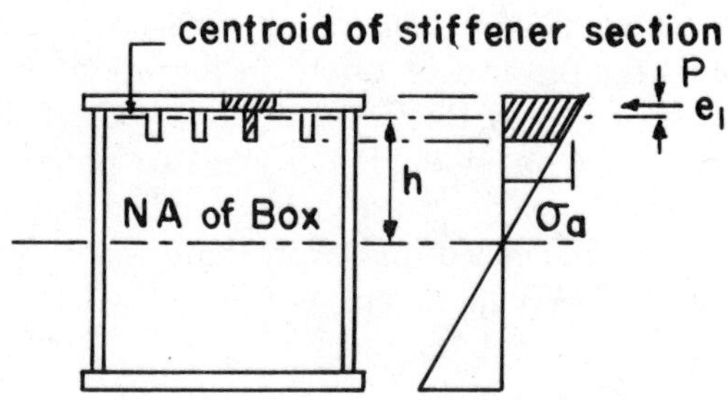

Fig. 7.6 Curvature of box girder.

effectively the distance between transverse elements. Such an approach was suggested by Ostapenko and Vojta (1967) and Dowling and his co-workers, Moolani and Chaterjee (Dowling and Chatterjee, 1977; Dowling et al., 1977a; Chatterjee, 1978; Dwight and Little, 1976). Apart from the overall buckling of the longitudinal stiffeners between transverse stiffeners, allowance is also made for reduction in effectiveness due to buckling of the plate between stiffeners.

Where the longitudinal stiffeners are open sections such as flat bars, tees, or angles, limitations are placed on their cross-sectional geometry to ensure that local buckling of the stiffeners does not precede the attainment of the ultimate strength of the flange. In the case of closed sections such as troughs or vees, used in orthotropic steel deck construction, the limitations on cross-sectional geometries are adjusted to allow for bending in the plane of the stiffener walls and to ensure that economical use of relatively thin walled sections is not precluded.

Limitations of stiffener cross sections are based on controlling the applied stresses under ultimate load to values which are fractions of the elastic critical buckling stresses. In BS 5400 the safety factor used is 2.25, which was derived as a suitable value for flat stiffeners but has been applied to all other types of stiffening. The limitations imposed in the proposed U.S. code (Wolchuk and Mayrbourl, 1980) are somewhat different and correspond to the requirement that the torsional buckling stress should be greater than the yield stress, with relaxations to allow for "thinning out" the stiffeners in low-stress zones. These two sets of criteria lead to the following restrictions:

Draft U.S. rules

$$\text{For } f_{\max} \geq 0.5F_y: \qquad C_S \leq \frac{0.40}{\sqrt{F_y/E}} \tag{7.1}$$

$$\text{For } f_{max} \leq 0.5F_y: \qquad C_S \leq \frac{0.65}{\sqrt{F_y/E}} \tag{7.2}$$

where

f_{max} = maximum factored compression stress
C_S = effective slenderness coefficient

$$= \frac{\mathrm{d}}{1.5\mathrm{t}_0} + \frac{\mathrm{w}}{12\mathrm{t}} \qquad \text{for } \textit{flats} \tag{7.3a}$$

$$= \frac{d}{1.35t_0 + 0.56r_y} + \frac{w}{12t} \qquad \text{for } \textit{tees} \text{ or } \textit{angles} \tag{7.3b}$$

For any outstand of a stiffener,

$$\frac{b'}{t'} \leq \frac{0.48}{\sqrt{F_y/E}} \tag{7.4}$$

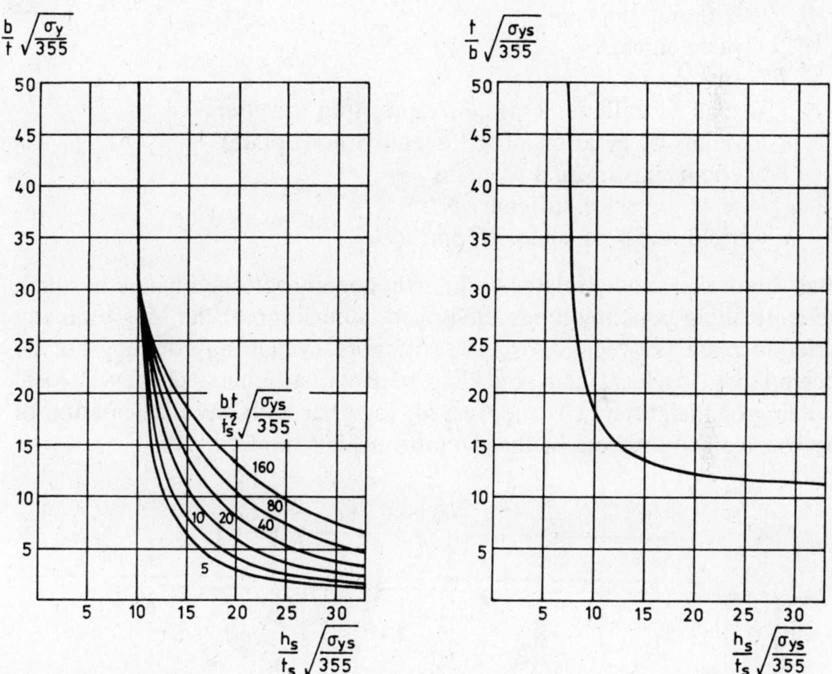

Fig. 7.7(*a*) Limiting slenderness for flat stiffeners; (*b*) limiting slenderness for angle stiffeners.

BS 5400.
Flats:

$$\frac{h_s}{t_s}\sqrt{\frac{\sigma_{ys}}{355}} \leq 10 \tag{7.5}$$

or $\leq$ a higher value obtained from Fig. 7.7*a* when $b/t\sqrt{\sigma_y/355} \leq 30$.
Angles:

$$b_s \leq h_s$$

$$\frac{b_s}{t_s}\sqrt{\frac{\sigma_{ys}}{355}} \leq 11 \tag{7.6}$$

$$\frac{h_s}{t_s}\sqrt{\frac{\sigma_{ys}}{355}} \leq 7$$

or $\leq$ a higher value obtained from Fig. 7.7*b* when $(l_s/b_s)\sqrt{\sigma_{ys}/355} \leq 50$. In these formulas (see Fig. 7.8),

d, h_s = stiffener depth
b_s = width of angle
t_0, t', t_s = stiffener thickness
t = plate thickness
w, b = spacing of stiffeners
l_s = span of stiffener between supporting members
r_y = radius of gyration of stiffener (without plate) about axis normal to plate
σ_{ys} = yield stress of stiffener, N/mm^2
σ_y = yield stress of plate, N/mm^2

Thus the design model relates to an orthogonally stiffened flange in which the controlling buckling mode envisaged is buckling of the longitudinally stiffened plate between transverse stiffeners, which may or may not be accompanied by local plate buckling between stiffeners. However, local buckling of the stiffener is suppressed, together with any participation of the transverse members in the overall buckled mode.

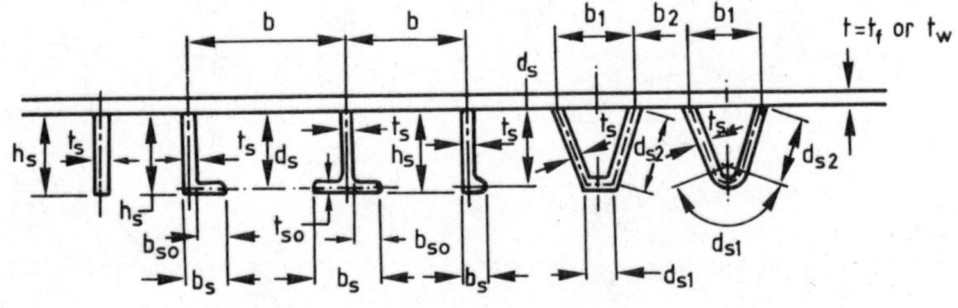

Fig. 7.8 Flange and web stiffeners.

In accounting for the buckling of the repeating combined stiffener/plate "strut" allowance can be made for the effect of initial distortions and residual stresses caused by welding the stiffeners to the plate. The extensive amount of data now available on plate strength can be used to account for interstiffener plate buckling using either an effective width or effective stress approach. Normally, a simple effective width approach is used to account for both the reduced strength and stiffness of the compressed plate. Although there are approximations involved in selecting just one width to account for stiffness as well as strength, the gain in accuracy which could be obtained using a more complex approach is not considered to be sufficient for design.

The effective width used in BS 5400 is derived from the results of a parametric study and uses a strength-based effective width which accounts for practical levels of initial imperfections and compressive residual stresses. Figure 7.9 gives curves from BS 5400 which may be used to calculate the width of plate k_{cb} considered to act effectively with a stiffener. Knowing the cross-sectional properties of the strut, the calculation of the buckling strength of the strut can be obtained from a column buckling formula of the Perry Robertson type which relates to a pin-ended, initially crooked, axially loaded column,

$$\frac{\sigma_{su}}{\sigma_y} = \frac{1}{2}\left\{\left[1+(\eta+1)\frac{\sigma_E}{\sigma_y}\right] - \sqrt{\left[1+(\eta+1)\frac{\sigma_E}{\sigma_y}\right]^2 - \frac{4\sigma_E}{\sigma_y}}\right\} \tag{7.7}$$

where

σ_{su} = limiting applied axial stress on effective strut section
σ_E = Euler stress of effective strut
σ'_y = available yield stress of compressive extreme fiber $= \sigma_{ys}$ when checking stiffener; $= \sigma_{ye}$ (i.e., an effective yield stress allowing for the presence of shear) when checking flange
$\eta = \Delta y / r_{se}^2$
Δ = maximum initial imperfection
y = distance from centroid to compressive extreme fiber
r_{se} = radius of gyration of effective section about axis parallel to plate

As the cross section is asymmetric the possibility of failure in the two directions (one causing compressive failure of the plate, the other causing compressive yielding of the stiffener tip) must be checked. The magnitude of the initial crookedness in these two directions may well be different, although in BS 5400 the bow in the direction causing compression in the stiffener tip is taken to be as large as that in the opposite direction. This allows for the fact that compressive residual stresses

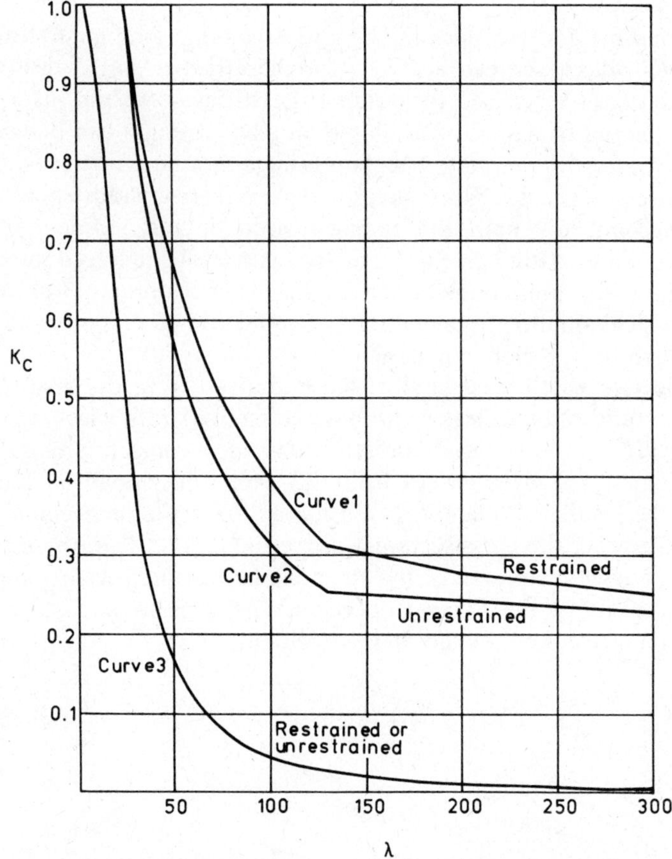

Note: The value of K_c to be used is the higher of the values obtained using either:

(a) curve 1 or 2 as relevant, with

$$\lambda = \frac{b}{t}\sqrt{\frac{\sigma_y}{355}} \quad \text{or}$$

(b) curve 3 with $\lambda = \frac{a}{t}\sqrt{\frac{\sigma_y}{355}}$

where

a is the panel dimension in the direction of stress considered
b is the panel dimension normal to the direction of stress.

Item (a) will always give the higher value for K_c where $a/b \geq 0.5$.

For $a/b < 0.5$, items (a) or (b) may give a higher value.

Fig. 7.9 Coefficient K_c for plate panels under direct compression.

caused by welding may be present at the stiffener tip, which are otherwise unaccounted for in the design procedure. The effect of end eccentricity of loading over the stiffener/plate strut is represented by a term in Δ given by $r_{se}^2/2y_{Bs}$. Here y_{Bs} is the distance from the centroid of the effective stiffener section to the neutral axis of the cross section of the complete beam.

The design of the stiffened flange is checked to ensure that

$$\frac{\sigma_a + 2.5\tau_1 k_{s1}}{k_{l1}\sigma_{ye}} \leq 1 \tag{7.8}$$

$$\frac{\sigma_a + 2.5\tau_1 k_{s2}}{k_{l2}\sigma_{ye}} \leq 1 \tag{7.9}$$

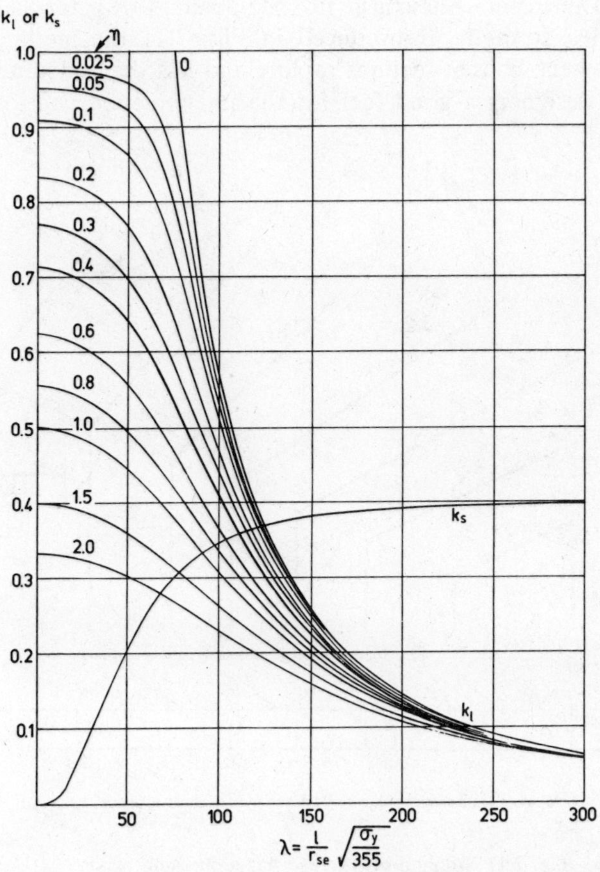

Fig. 7.10 Parameters for the design of longitudinal flange stiffeners.

using the curves plotted in Fig. 7.10. The first of these expressions checks for failure by yielding of the stiffener tip and the second checks for buckling or yielding of the plate panel. If the longitudinal stress varies along the length of the stiffener, then σ_a in the expressions above is taken at a point $0.4l$ from the end where the stress is greater. In these formulas

σ_a = longitudinal stress at centroid of effective stiffener section
τ_1 = in-plane shear stress in flange plate due to torsion
σ_{ys} = yield stress of stiffener material
σ_{ye} = effective yield stress of flange plate material allowing for presence of shear
l = spacing of cross beams

A method proposed by Wolchuk and Mayrbourl (1980) in the proposed AASHTO Code for box girders uses a format based on a German procedure (Deutscher Ausschuss für Stahlbau, 1978) for calculating stiffened flange strength. Reproduced in Fig. 7.11 the method allows designers to check chosen sections rapidly and has the added advantage that it gives designers a good feel for the phenomenon controlling the

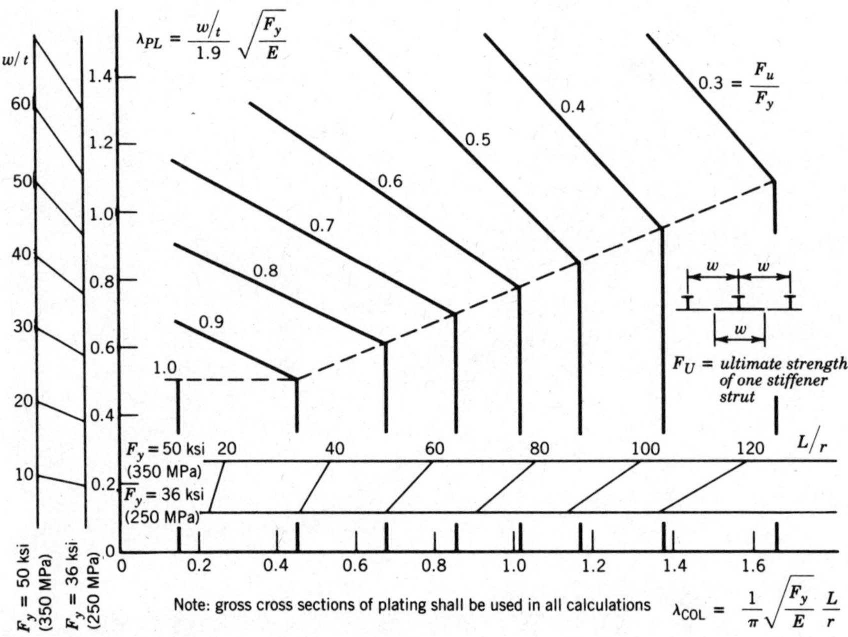

Fig. 7.11 Strength of stiffened flange in compression.

strength (i.e., plate or stiffener buckling). The ultimate flange strength is computed as $P_u = F_u A_f$ or $F'_u A_f$, whichever is less, where

A_f = cross-sectional area of stiffened flange
F_u = ultimate strength of stiffener strut (stiffener with full width of associated plating)
F'_u = modified ultimate strength of stiffener strut under combined compression and shear
$= F_u \quad \text{for } f_v \leq 0.175 F_y$
$= 1.05 F_u \sqrt{1 - 3f_v^2/F_y^2} \quad f_v \geq 0.175 F_y$
f_v = governing shear stress in flange

Design methods based primarily on the results of tests on stiffened panels have more recently been suggested by Japanese authors and were reviewed in the U.S.–Japan Seminar on Inelastic Instability of Steel Structures and Structural Elements held in Tokyo (Fujita and Galambos, 1981). One such proposed strength curve is given by

$$\frac{\sigma_u}{\sigma_y} = 1.24 - 0.54\bar{\lambda} \tag{7.10}$$

where

$$\bar{\lambda} = \frac{b}{t}\sqrt{\frac{\sigma_y}{E}\,\frac{12(1-\nu^2)}{\pi^2 k}} \tag{7.11}$$

and

$k = 4n^2$
n = number of plate panels

It should be noted that this strength curve is chosen to agree with the *mean* of the test results obtained in Japan as shown in Fig. 7.12.

In the case of stiffened flanges not subjected to normal loading, transverse stiffeners are designed to provide sufficient stiffness to ensure that they act as supports which are effectively rigid against movements normal to the plane of the flange plate but offer little resistance to rotation of the longitudinal stiffeners at their junctions. Thus they ensure that the effective buckling length of the latter does not exceed their span between adjacent transverse members. Where flanges form the deck of a box girder bridge which has to carry lateral loading the cross girders are primarily proportioned to resist the lateral loading and will normally have more than sufficient rigidity to constrain the stiffened panels in the manner envisaged.

The stiffness requirement for transverse members stipulated within BS 5400: Part 3 is based on limited experimental observations and provides a

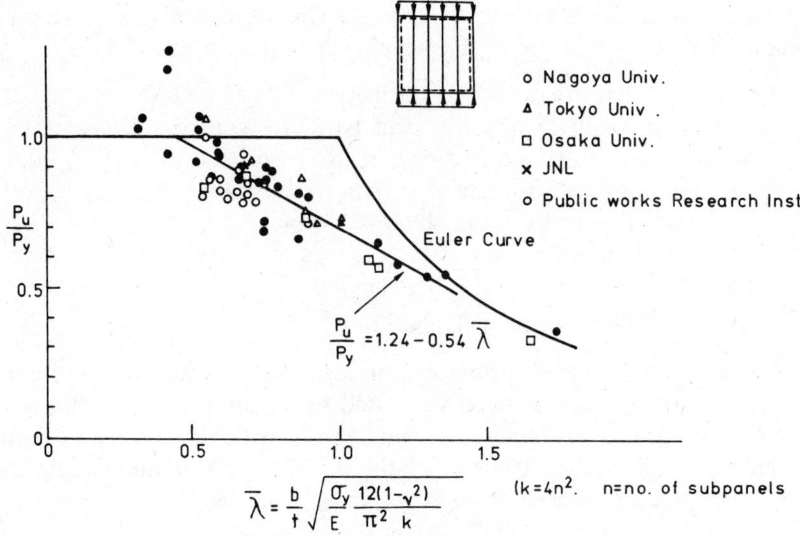

Fig. 7.12 Stiffened plate test results and proposed strength curve.

factor of safety of 3 against overall linear elastic buckling. The approximate expression produced for the minimum second moment of area required of a transverse stiffener and its effective width of attached flange is

$$I_{be} = \frac{9\sigma_f^2 B^4 a A_f^2}{16KE^2 I_f} \tag{7.12}$$

where

σ_f = longitudinal compressive stress in flange
B = spacing of webs of main beam
a = cross-beam spacing
A_f = cross-sectional area per unit width of the flange
I_f = second moment of area per unit width of the flange
K = 24 for segments between *interior* webs of main beams and is determined from Fig. 7.13 in other cases

In addition, the transverse member must be designed to carry any locally applied lateral loading as well as the normal component of in-plane loading in the flange caused by lack of alignment of the transverse members. Two simplified loading cases, selected to cover this latter effect, are stipulated as:

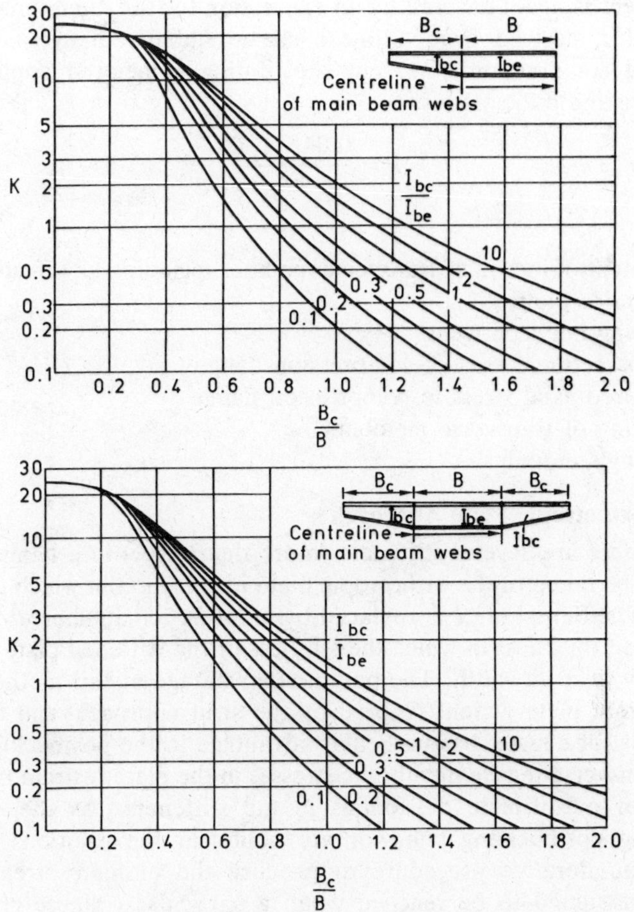

Fig. 7.13 Buckling coefficient K for transverse members.

1. A uniformly distributed load per unit width of $\sigma_f A_f/200$.
2. A concentrated load at a cantilever tip of $\sigma_f A_{sc}/160$, where A_{sc} is the cross-sectional area of the longitudinal stiffening member at the cantilever tip (σ_f and A_{sc} are in SI units).

In the proposed U.S. code the destabilizing effect of flange geometric imperfections is taken into account by assuming the transverse stiffeners or cross-frame members to be loaded by a uniformly distributed load of 1% of the average factored compressive force in the flange. The rigidity requirement is again covered by an expression for the minimum moment of inertia I_t derived from a linear elastic stability analysis of spring supported compression bars, but substituting a factored applied axial stress f for the critical stress.

$$I_t = \frac{0.04b^3 A_f f}{Ea} \tag{7.13}$$

where

I_t = moment of inertia of transverse member including an effective width of flange plate
b = spacing between webs
A_f = cross-sectional area of compression flange
f = factored axial stress in compression flange
a = spacing of transverse members
E = Young's modulus

7.3.2 Orthotropic Plate Approach

Where there are several stiffeners (more than three) in a flange, advantage can be taken of the orthotropic plate idealization in which the actual discretely stiffened plate is replaced by an *ortho*gonally aniso*tropic* plate of constant thickness in which the stiffness of the stiffened plate is spread uniformly over its width. The potential advantage of this method is that the inherent plate action, ignored by the strut approach, can be mobilized. This, of course, has particular advantage in the postbuckling range when transverse tensile membrane stresses in the plate restrain the rate of growth of out-of-plate deflections in the stiffeners. As the equations describing postbuckling behavior are nonlinear the solutions generally involve an iterative procedure to produce the ultimate strength. The latter is assumed to be reached when a particular collapse criterion is satisfied. Collapse criteria suggested include the onset of yield at the flange/web junction, the mean stress along that edge reaching yield, yield at a stiffener tip at the center of the flange, or in-plane yield in the plate in the same region. These methods have been described in some detail by Massonnet and his co-workers Maquoi (Maquoi and Massonnet, 1971) and Jetteur (1983) and are reviewed in the book *Behaviour and Design of Steel Plated Structures*, issued by the ECCS (Dubas and Gehri, 1986). They suffer from the disadvantage that an iterative procedure is needed for their solution, although in some cases this has been greatly aided by the provision of design charts.

7.3.3 Discretely Stiffened Plate Approach

Analytical studies have been made of stiffened panels in which account has been taken of the discrete nature of the stiffening and which incorporate nonlinear geometrical and material effects. Both finite-difference and finite-element numerical formulations have been used for this purpose. The results, although giving insight into the interactive buckling of plate and stiffeners, have not been used to produce a general design method, although the analytical techniques can be used to analyze any particular design. Škaloud and Novotny (1965) gives the background to some available analyses.

Influence of Shear Lag on Flange Buckling. Shear lag arises because the in-plane *shear* straining of a flange causes those parts most remote from a web to *lag* behind those in the vicinity of the web (Fig. 7.14). It is most marked in beams whose flanges are relatively wide compared to their length and therefore is of greater importance in box girder than in plate girder construction. It results in a nonuniform distribution of longitudinal stresses across the flange width, with bigger stresses occurring near flange/web junctions and smaller ones in the regions farthest removed from such junctions. The addition of stiffeners to a wide flange results in the nonuniformity becoming even more pronounced, even though the maximum stresses may be reduced. A systematic study of the factors influencing shear lag in box girder flanges has been reported by Moffatt and Dowling (1975, 1976) and forms the basis of the relevant rules in the proposed U.S. design specification (Fig. 7.15) and in the new

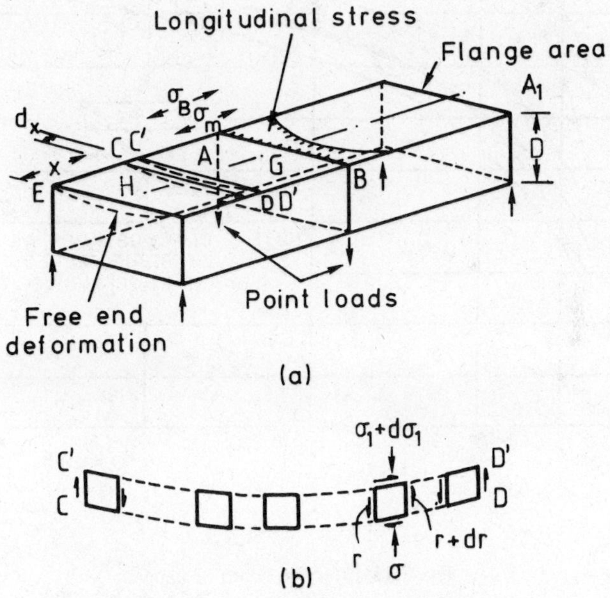

Fig. 7.14 Definition of shear lag.

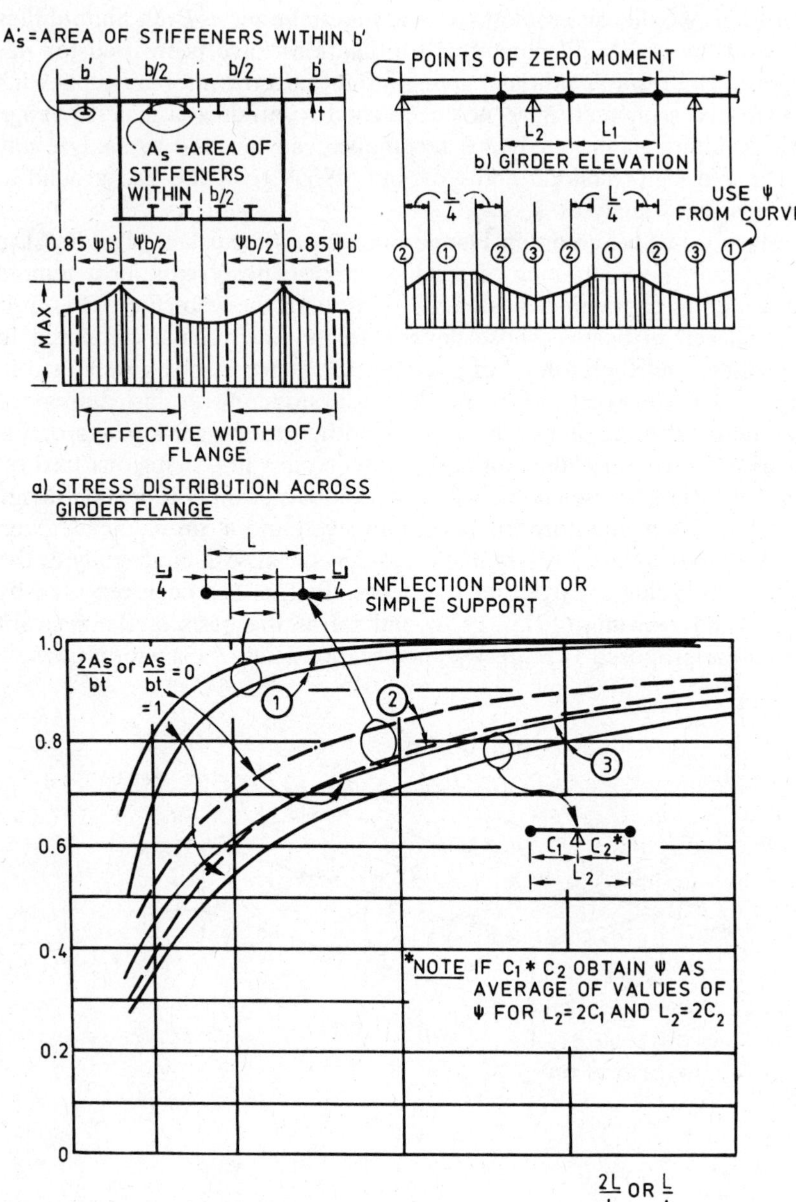

Fig. 7.15 Effective width of flange.

British code. Similar studies have been reported by Abdel Sayed (1969) and others.

It is the nonuniform distribution of in-plane stresses in a box girder flange which is of interest from the point of view of structural stability. This form of distribution can increase or decrease the average stress, causing earlier buckling of the flange compared with the uniformly compressed case, depending on the degree of stiffening in the plate. On the other hand, it is clear that concentration of the applied stresses near the longitudinally compressed edges of the plate will encourage an earlier onset of yield at the edges than would uniform stressing across the flange width. The interaction of these two effects is complex and, indeed, is complicated further by the capacity of the stiffeners and plate panels to redistribute load. Tests (Dowling et al., 1977a; Frieze and Dowling, 1979) have shown, however, that for most practical cases shear lag can be ignored in calculating the ultimate compressive strength of stiffened or unstiffened flanges. This conclusion has been supported by the numerical studies of Lamas and Dowling (1980), Burgan and Dowling (1985), and Jetteur et al. (1984). Thus a flange may normally be considered to be loaded uniformly across its width. Only in the case of flanges with particularly large aspect ratios, or particularly slender edge panels or stiffeners, is it necessary to consider the flange stability in greater detail.

The limits given in BS 5400; Part 3 for neglecting shear lag in the calculation of the strength of unstiffened flanges is for flanges where the maximum longitudinal stress is more than 1.67 times the mean longitudinal stress. For cases in which this limit is exceeded the serviceability limit state has to be checked by including the effects of shear lag in the calculation of applied stresses.

Wolchuk, in the proposed U.S. code, suggests that shear lag may be ignored if the peak stress does not exceed the average stress by more than 20% and a uniform stress distribution is assumed for the calculation of the ultimate flange strength. Where these limits are exceeded it is suggested that the flange capacity at the web be increased to accommodate an additional force, computed as the load-factored stress is excess of 120% of the average stress multiplied by the flange area affected by the excess.

Influence of Lateral Loads. Lateral loads can be of interest in the context of bridges where the deck constitutes the top flange of the box girder. Where the box in question is a marine structure, such as a ship or an offshore platform, the lateral loading may be hydrostatic. In such cases the question arising is: What is the effect of the lateral loading on the stability of the stiffened plate acting as a compression flange to a box girder?

Plates such as those used in bridge decks to resist traffic loading are normally stocky and designed primarily to limit deflections under the lateral loading to very small acceptable values ($\ll t$). The presence of in-plane stresses due to participation of the plate as part of the box girder flange magnifies the deflections and bending stresses by a factor of approximately $1/(1 - \sigma_a/\sigma_{cr})$. In the same way allowance can be made for the magnification of stresses in a compressed stiffened deck carrying lateral loading by also increasing the stresses due to the moments, when appropriate, by this amplification factor.

In BS 5400: Part 3 stiffened compression flanges used to carry local wheel loading, such as steel bridge decks, may be designed for the ultimate limit state ignoring the presence of the local loads. This is based on experimental evidence of a relatively limited nature. However, this is one of the few cases for which it is necessary to carry out a serviceability check using elastic analysis. This check is done to ensure that yielding under working load is prevented. The stresses in both the deck plate and the stiffeners are checked, including all elastic effects, such as shear lag, torsion, and in-plane stressing of the plate due to local bending of the stiffened plate under wheel loading. In calculating the stiffener stresses a distinction is made between the zones of hogging moment in the longitudinal stiffener over a transverse member and the sagging moment occurring between transverse members. Whereas in the latter case the full amplified stresses due to in-plane loading and local bending are combined, in the former only the in-plane stresses are amplified and added to the local bending stresses occurring at the longitudinal to transverse stiffener intersection.

The approximate expressions used to check the stiffener design are as follows:

1. Over transverse members

$$\frac{\sigma_a + 2.5\tau_1 k_{s1}}{k_{l1}\sigma_{ys}} + \frac{\sigma_{f0}}{\sigma_{ys}} \leq 1 \tag{7.14}$$

2. Between transverse members

$$\frac{\sigma_a + 2.5\tau_1 k_{s2}}{k_{l2}\sigma_{ye}} = \frac{\sigma_{fz}}{\sigma_{ye}} \leq 1 \tag{7.15}$$

where

f_0 = stress due to local bending at point on stiffener farthest from flange plate

σ_{fz} = stress at midplane of flange plate due to local bending

and the other symbols are as defined earlier in this section.

The proposed U.S. code calls attention to the need to consider the effects of axial compression in the deck, combined with the effects of bending under lateral loading, when carrying out the design of orthotropic decks on the elastic basis but offers no guidance on how this may be done.

7.4 BENDING STRENGTH OF BOX GIRDERS

A box girder subjected primarily to bending moment will normally fail by buckling of the compression flange. Unlike plate girders, lateral torsional buckling rarely governs for practical box girders. Methods described in Section 7.3 can be used to calculate the compression flange strength.

In the simplest case the bending strength of box girders can be assumed to be contributed by the flanges alone and the moment of resistance can be calculated on the basis of the design force multiplied by the distance separating the flange centroids. Such an approach is allowed within BS 5400: Part 3 and greatly simplifies design.

To calculate the strength of the box girder, allowing for the contribution of the web, a linear distribution of stress over the depth of the cross section may be assumed. Effects of buckling in the web can be represented by using effective widths of web adjacent to the stiffeners as outlined in Chapter 4. The British Code BS 5400 also allows the use of the effective thickness approach for girders, which has the advantage that no recalculation of the position of the neutral axis is needed as is the case with the effective width approach. Cooper's expression for bending strength (1967) is used to give the effective thickness for unstiffened webs, and the full thickness is used for webs stiffened by effective longitudinal stiffeners. Failure is deemed to have occurred when the extreme fiber flange stress reaches the calculated ultimate stress of the compression flange σ_{lc}, or the yield stress of the tension flange σ_{yt}, whichever is the critical criterion. Thus in BS 5400

$$M_D = Z_{xc}\sigma_{lc} \quad \text{or} \quad Z_{xt}\sigma_{yt} \tag{7.16}$$

where Z_{xc} and Z_{xt} are the elastic moduli of the effective section for the extreme compression and tension fibers, respectively.

7.5 SHEAR STRENGTH OF BOX GIRDERS

The key difference between plate and box girders which may influence the shear strength of the webs is the use of relatively thin flanges in box girders at the boundaries of the webs. Caution is needed in applying available tension field models, derived and verified in the context of plate girder webs, to the design of the webs of box girders. Of major concern is the relatively small amount of support against in-plane movement which may be afforded to the web by the thin flange of a box girder, compared with the restraint offered by the thicker and narrower flange of a corresponding plate girder. In the latter case the out-of-plane bending rigidity and in-plane extensional rigidity of the flange to resist movement perpendicular to and parallel to the flange/web junction, respectively, is more effectively mobilized than in the case of thin flanged box girders.

7.5.1 Box Girders without Longitudinal Stiffeners

In BS 5400: Part 3 (BSI, 1983) the tension-field model of Rockey and his co-workers (Porter et al., 1975) has been modified for application to both plate and box girders without longitudinal stiffeners, (see Chapter 6). Thus for box girders with unstiffened flanges but with transverse stiffeners in the webs (or flanges)—a form of construction associated with relatively small boxes—advantage can be taken of postbuckling strength using a tension-field model. Limited use is made of the plastic frame mechanism action in the tension field to keep shear deformations within the limits for the whole girder. To do this the maximum possible shear capacity is limited to the shear yield capacity of the web alone. In the case of a box with thin unstiffened flanges, the width of flange taken as effective is $10t_f\sqrt{355/\sigma_{yf}}$ each side of the web, when the flange projects beyond the web. Where no such projection occurs the effect of frame action is neglected in the calculation of tension-field capacity (Harding and Dowling, 1981).

Wolchuk and Mayrbourl (1980) suggest, in applying the tension-field model to the transversely stiffened webs of box girders, the solution of Basler (1961), which is based on the assumption of negligible flange bending rigidity. This corresponds roughly to neglecting the frame action in the Rockey solution and reflects the caution needed for box girders, where the flanges are generally more slender than in plate girders. Thus the proposed U.S. code suggests that

$$V_u = V_B + V_T \tag{7.17}$$

with

$$V_B = Dt_w F_{vcr} \tag{7.18}$$

$$V_T = \frac{Dt_w F_T}{2(\sqrt{1+\alpha^2}+\alpha)} \tag{7.19}$$

where

D = depth of web between flanges, measured along web
d_0 = transverse stiffener separation

$\alpha = d_0/D$
t_w = web thickness
F_{vcr} = critical buckling shear stress
F_T = tension-field stress

7.5.2 Box Girders with Longitudinal Stiffeners

In the case of box girders with longitudinal stiffeners very little experimental evidence is available to underpin the application of tension-field theories to web design. Code drafters are therefore doubly cautious on account of the unknown interaction between thin longitudinally stiffened webs and thin box girder flanges. Conservative approaches to design have been suggested which still, nonetheless, take advantage of postbuckling reserve, albeit to a lesser extent than might be possible with a full plastic tension-field treatment.

In the British codified method (BSI, 1983) the stiffened web is checked on a panel-by-panel basis. The design procedure consists of calculating the longitudinal stresses using simple bending theory and gross areas. Shear forces are assumed to be distributed uniformly down a cross section. Each panel is then checked for yielding under combined compression, bending, and shear using the interaction formula

$$\left(\frac{\sigma_1 + 0.77\sigma_b}{\sigma_{yw}}\right)^2 + 3\left(\frac{\tau}{\sigma_{yw}}\right)^2 \leq 1 \tag{7.20}$$

and for buckling using

$$\frac{\sigma_1}{\sigma_{yw}K_1} + \left(\frac{\sigma_b}{\sigma_{yw}K_b}\right)^2 + 3\left(\frac{\tau}{\sigma_{yw}K_q}\right)^2 \leq 1 \tag{7.21}$$

where

σ_1 = mean longitudinal stress
σ_b = maximum longitudinal stress due to in-plane bending
τ = average shear stress
σ_{yw} = yield stress of web material
K_1, K_b, K_q = coefficients for ultimate plate strength

In checking yielding, any proportion of the longitudinal stresses σ_1 and σ_b, up to 60% maximum in a panel, can be shed to the flanges while maintaining overall equilibrium. In checking stability, up to 60% of these stresses can be shed from the restrained inner panels, but none can be shed from outer panels, which are considered to be unrestrained.

The proposed U.S. code uses the same approach as for webs stiffened transversely only, except that F_{vcr} is now calculated for each subpanel bounded by longitudinal and transverse stiffening and assumes the lowest value.

7.6 COMBINED BENDING AND SHEAR STRENGTH OF BOX GIRDERS

7.6.1 Box Girders without Longitudinal Stiffeners

If the bending and shear strengths have been calculated without any contributions from the web and flanges, respectively, there is evidence to suggest that girders can safely resist these magnitudes of moment M_R and shear V_R acting simultaneously, so no interaction needs to be considered. For box girders with flanges and webs unstiffened longitudinally this approach provides a simple conservative estimate of combined bending and shear strength.

However, more often webs will have been considered to make some contribution to the bending strength M_D and, in the case where webs have been designed to take advantage of postbuckling strength, flanges may have been considered to contribute to the shear strength V_D through their framing action. BS 5400 proposes the use of an interactive diagram (Fig. 7.16) which can be used for box girders with no longitudinal stiffeners.

The proposed U.S. (Wolchak and Mayrbourl, 1980) specification for transversely stiffened webs of box girders uses an interaction equation to calculate the critical buckling stress for combined shear, bending, and compression, in terms of ratios of the individual stress components to their critical buckling values when acting alone. Thus

$$\left(\frac{F_{vcr}}{F^0_{vcr}}\right)^2 + \left(\frac{F_{bcr}}{F^0_{bcr}}\right)^2 + \frac{F_{ccr}}{F^0_{ccr}} = 1 \qquad (7.22)$$

However, the stress components are interdependent and may be related as follows:

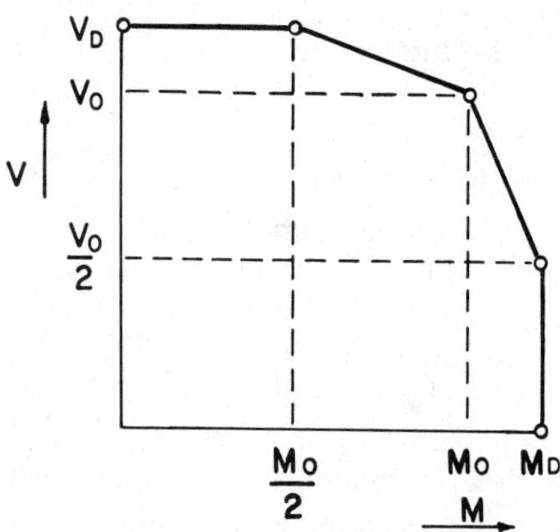

Fig. 7.16 Interaction between shear force and bending moment.

$$F_{bcr} = \frac{1-R}{2} \mu F_{vcr}$$

$$F_{ccr} = \frac{1+R}{2} \mu F_{vcr}$$

where

$$R = \frac{f_{2w}}{f_{1w}}$$

$$\mu = \frac{f_{1w}}{f_v}$$

f_{1w} = governing axial compressive stress at longitudinal edge of web panel
f_{2w} = axial stress at opposite edge of panel
f_v = governing shear stress = V/D_{tw}

These stresses are illustrated in Fig. 7.17. F°_{vcr}, F°_{bcr}, F°_{ccr} are the buckling stresses computed by assuming only shear, bending, or compressive stresses, respectively, were present.

An additional flange force ΔF is added to the flange forces computed in accordance with elastic analysis. It includes that portion which must be transferred to the flanges from the webs after buckling, as well as a

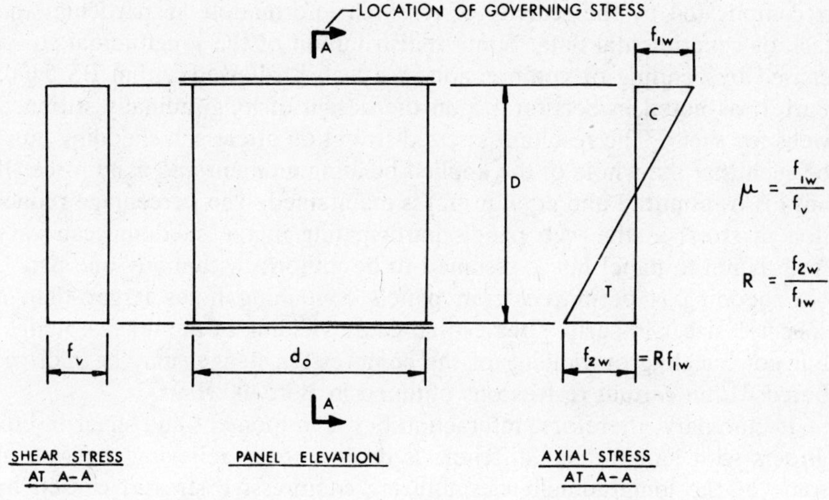

Fig. 7.17 Definition of μ and R for unstiffened and transversely stiffened webs.

portion related to the use of tension-field action for the webs. These are for compression flanges:

$$\Delta F_1 = \left(1 - \frac{\sum V_B}{V_M}\right)[(f_{1R} - f_1)A_{fc} + \tfrac{1}{2}V_M \cot(\tfrac{1}{2}\theta_d)] \tag{7.23}$$

and for tension flanges:

$$\Delta F_2 = \left(1 - \frac{\sum V_B}{V_M}\right)[(f_{2R} - f_2)A_{ft} - \tfrac{1}{2}V_M \cot(\tfrac{1}{2}\theta d)] \tag{7.24}$$

where

V_M = factored shear force acting coincident with maximum moment
$\Sigma V_B = \Sigma Dt_w F_{vcr}$ = sum of buckling shear capacities of all webs at cross section considered
f_1, f_2 = stress in compression and tension flange, respectively, assuming fully participating webs
f_{1R}, f_{2R} = stresses assuming reduced moment of inertia of cross section
A_{fc}, A_{ft} = compression and tension flange area, respectively
θ_d = angle of inclination of web panel diagonal to the horizontal

7.6.2 Box Girders with Longitudinal Stiffeners

For box girders with longitudinally stiffened flanges and webs the sitation is complicated by the scarcity of research information, in particular the lack of experimental data. Some redistribution of the longitudinal stress caused by bending or compression in a web is allowed within BS 5400: Part 3, as noted in Section 7.5 on the design of longitudinally stiffened webs for shear. The resultant stress distribution after such shedding must be such that the whole of the applied bending moment and axial force (if any) is transmitted and equilibrium is maintained. The percentage reduction in stress in the web panels participating in the shedding can vary from panel to panel but is assumed to be uniform within any one panel. No shedding is permitted from panels containing holes larger than a specified size. Similarly, stresses that cause yielding of the tension flange, but not buckling or yielding of the compression flange, may be redistributed within certain restrictions outlined in BS 5400: Part 3.

In summary, therefore, interaction between moment and shear in box girders with longitudinal stiffeners is dealt with by relieving the web of some of the longitudinally destabilizing compressive stresses caused by bending, and distributing the load to the compression flange while maintaining overall equilibrium of the cross section. In the proposed U.S. code combined bending and shear are treated in the same way for webs with and without longitudinal stiffeners. However, with longitudinal stiffening the formulas are applied to each subpanel in turn, rather than considering the overall web depth.

7.7 INFLUENCE OF TORSION ON STRENGTH OF BOX GIRDERS

The level of torsional stress induced in the webs and flanges of practical box girders does not of itself normally constitute an instability problem, although tests on small thin-walled boxes subjected to torsion have shown that local instability of box corners can limit the strength in such cases. Allowance for the effects of the shears induced by torsion can be made in the design of the flanges by reducing the effective yield stress in the plate due to the presence of combined stresses, including torsional shear stresses, in the flange plate. In the design of the webs, additional shear stresses caused by torsion may be added to those associated with bending when calculating the total stresses applied to a web.

7.8 DIAPHRAGMS

Two types of diaphragms are encountered within box girders: intermediate diaphragms and load-bearing diaphragms. The former limit cross-sectional deformation, while the latter are provided at points of support to give load paths for vertical loads (web shears) and horizontal loads (flange shears) through to the support bearings, and to prevent buckling of the webs in the vicinity of these large concentrated loads. Intermediate diaphragms will not be covered here. Load-bearing diaphragms, on the other hand, need careful consideration in relation to buckling, just as support stiffeners do in the case of plate girders.

Most diaphragms are stiffened at the bearing locations with bearing stiffeners. These are often accompanied by short-length stub stiffeners immediately above the bearings, the role of which is to stiffen the diaphragm plate in the vicinity of the bearings so that localized redistribution of stress concentrations can occur through yielding of the plate. In longer diaphragms (Fig. 7.18), secondary stiffening can be provided to stabilize the diaphragm plate as an alternative to the use of very thick plate.

Transverse secondary stiffening may also be necessary to resist any transverse compression caused by the diaphragm's behavior as a deep beam, an action that is accentuated by the use of sloping webs. As the aim of any sensible design must be to eliminate the possibility of diaphragm buckling limiting the strength of the box girder it makes little sense to attempt to economize on diaphragm design and a conservative design approach is usually adopted.

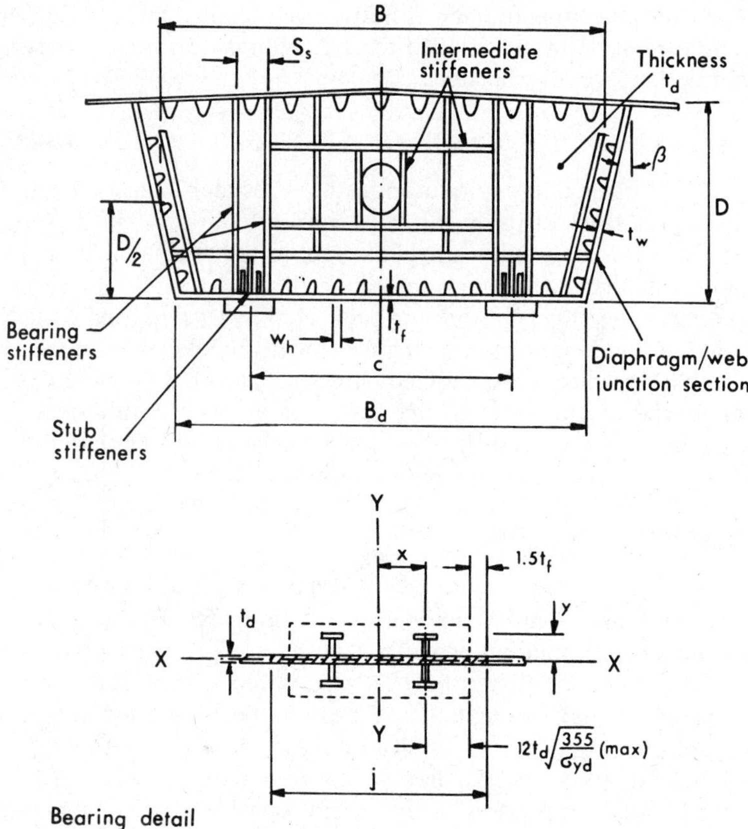

Fig. 7.18 Geometric notation for diaphragms.

Diaphragm design is treated in detail in BS 5400: Part 3. Only single-cell boxes, without steeply sloping webs and with diaphragms normal to the girder axis, are considered. Two approaches to design are contained within that specification, one for simple diaphragms and the other for diaphragms of more complicated geometries. In each case the design procedure involves three main features. These are:

1. Limitations of diaphragm geometry.
2. Analysis of diaphragm stresses.
3. Design checks on diaphragm yielding and buckling.

The designer can choose to use diaphragms of simple layout, referred to as "unstifened" diaphragms, even though they normally have full-height stiffeners at the bearing locations as well as a diaphragm plate (Fig. 7.19). The stiffeners, however, can be omitted in smaller boxes. For larger boxes the designer may opt for "stiffened" diaphragms with arrangements of bearing, stub, and secondary stiffening of the type referred to above.

The design method for the unstiffened diaphragms places greater restrictions on geometry and provides simple formulas for stress analysis,

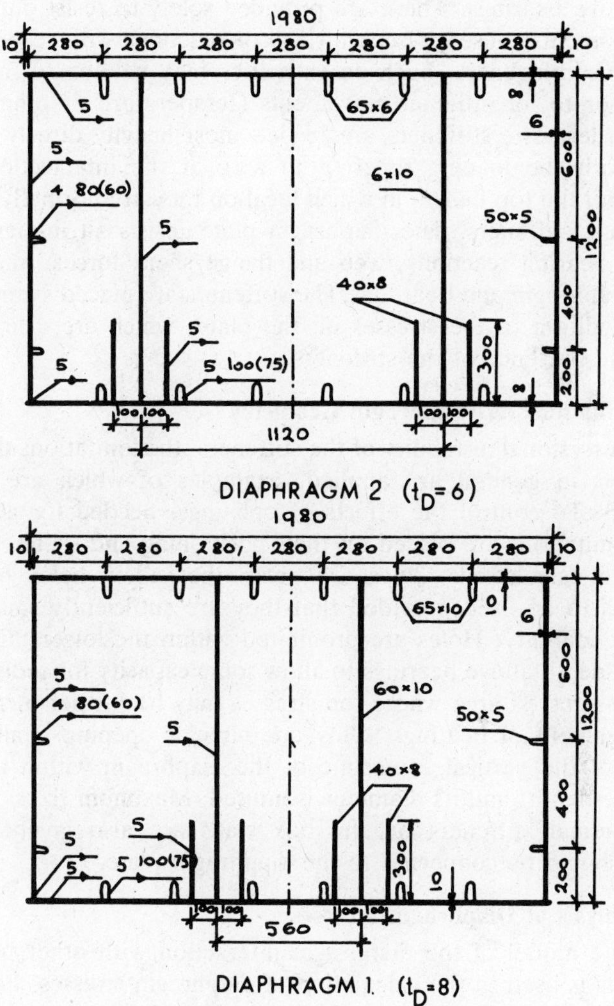

Fig. 7.19 Details of load-bearing diaphragms.

as well as simple yielding and buckling expressions for use in the design checks. Stiffened diaphragms, while still subject to limitations of geometry, albeit less restrictive ones, normally require detailed computer analysis of stresses and detailed stiffener and plate panel failure checks.

7.9 UNSTIFFENED DIAPHRAGMS

Vertical stiffeners, which must be the full length of the diaphragm and positioned symmetrically on both sides of the diaphragm, are placed directly above bearings. These are provided solely to resist out-of-plane bending moments caused by eccentricity of reactions with respect to the diaphragm midthickness. Such eccentricities may be due to fabrication tolerances or to longitudinal movements (temperature or otherwise) of the box girder. The stiffeners are loaded most heavily direcly over the bearings with the loading tapering to zero at the intersection of the stiffener with the top flange, at which location they are normally attached to the flange stiffeners. The diaphragm plate resists all in-plane forces caused by bearing reactions, web and flange shear forces, and friction between diaphragm and bearings. The stiffeners are placed symmetrically to avoid addition to the stresses on the plate, which are calculated by ignoring the presence of the stiffeners.

7.9.1 Limitations on Diaphragm Geometry

To prevent torsional instability of the stiffeners, the limitations that relate to stiffeners in general are applied, examples of which are given in Section 7.3. To control the effects of openings, needed for access and services, limitations are placed on their positioning and sizing. To avoid complicated calculations, rules are given that allow holes of certain proportions to be used provided that they are sufficiently small not to affect plate stability. Holes are prohibited within the lower third of the diaphragm depth above bearings to allow some capacity for redistribution in a highly stressed area where the stresses may be further increased by the misalignment of bearings. Only one circular opening is allowed on each side of the vertical centerline of the diaphragm within the upper third of the depth, and its diameter is limited. Maximum sizes of cutouts for longitudinal stiffeners on the box walls are also given, and the stiffeners should be connected to the diaphragm plate.

7.9.2 Analysis of Diaphragm Stresses

A simplified model of the diaphragm interaction with other box girder components is used in the calculation of diaphragm stresses. Portions of flange are considered to act with the isolated diaphragm, which responds as a deep beam loaded by edge forces and supported by the bearings. The effective widths to be used in the calculations are based on plane stress considerations (shear lag) alone and so may only be used when the transverse stiffness of the compression flange is not significantly reduced by transverse stresses, that is, provided that the calculated stresses do not exceed:

1. $\frac{1}{4}$ of flange longitudinal compressive strength.
2. $\frac{1}{2}(t_f/b)^2E$, where t_f is the flange plate thickness and b is the spacing of longitudinal stiffeners.

Otherwise, a reduced width of flange on either side must be used. It will be noted that no portion of the webs is included. This leads to a conservative estimate of the stresses and greatly simplifies what is in reality a complicated problem due to the high level of both shear and bending stresses, which may coexist in the web at that location if the diaphragm is at an internal support.

For simple diaphragms the shear flows are simplified as shown in Fig. 7.20. Advantage has been taken of the ability of both plate panels and welds connecting the diaphragm and webs to redistribute shears applied in the nonuniform manner predicted by elastic theory. This has enabled the shear flows to be taken as uniform. By also including the effects of inclined webs, the resulting horizontal reference stress is calculated using

$$\sigma_{R2} = \left[\left(\frac{K_d \sum R_v}{2} + \frac{T}{B}\right)x_R + Q_{fv}\frac{l_f}{2}\right]\frac{1}{Z_e} + \frac{\sum R_v \tan\beta}{2A_e} \qquad (7.25)$$

where

$K_d = 2$ usually, and allows for boundary shears
B = average width of diaphragm

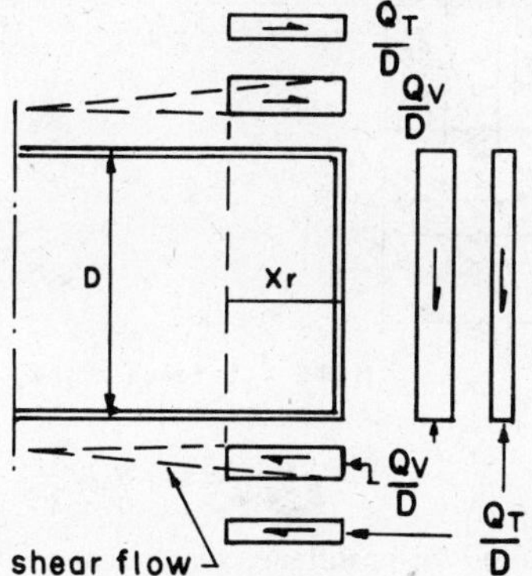

Fig. 7.20 Shear flows assumed to derive K_D.

$\sum R_v$ = total vertical force transmitted to bearings
Q_{fv} = vertical force transmitted to diaphragm by a change of flange shape
l_f = horizontal distance from reference point to nearest edge of bottom flange
T = torque transmitted to diaphragm
Z_e, A_e = effective section modulus and area, respectively, of diaphragm and flanges at the vertical section through the reference point

Other symbols are shown in Fig. 7.21.

The vertical stress is calculated using

$$\sigma_{R1} = \frac{R_v(1 + 4e/t_d)}{(j - \sum w_h)t_d} \tag{7.26}$$

for twin symmetrical bearings, where

t_d = diaphragm thickness
j = effective width of contact of bearing pad allowing for load dispersion through flange (Fig. 7.21)

With a single central bearing the vertical stress has an additional component, given by $0.77(T_b j/2I_{yd})$. Because maximum stressing in the vicinity

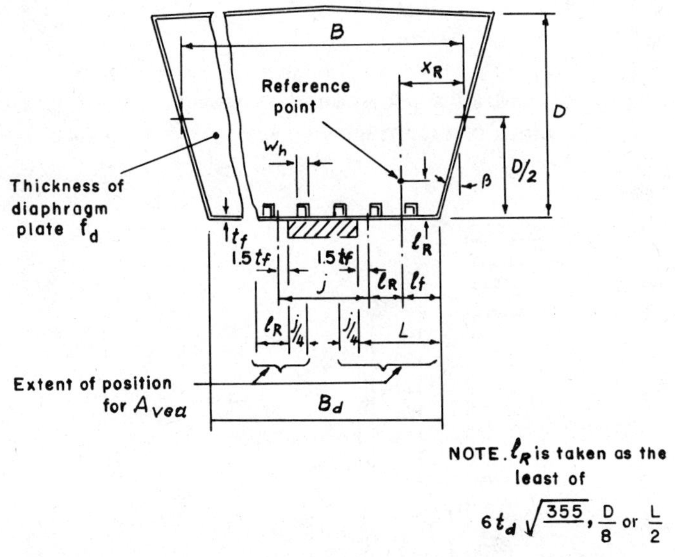

Fig. 7.21 Reference point and notation for unstiffened diaphragms.

of the bearings is very localized, a factor of 0.77 on yield is used here in the yield stress check to make some allowance for plastic redistribution. In these formulas

R_v = total vertical load transmitted to one bearing
T_b = torsional reaction at single central bearing
$\sum w_h$ = sum of widths of cutouts for stiffeners immediately above the flange and within width j
I_{yd} = second moment of area of width j of diaphragm plate, excluding cutouts
e = eccentricity of bearing reaction along span

The shear stress is given by

$$\tau_R = \left(\frac{\sum R_v}{2} + Q_{fv} + \frac{T}{2B}\right)\frac{1}{A_{vea}} + \frac{Q_h}{A_{he}} \tag{7.27}$$

where

Q_h = shear force due to transverse horizontal loads transmitted from top flange
A_{vea} = minimum net area of vertical cross section of diaphragm plating
A_{he} = net area of horizontal cross section through reference point

Thus it is possible to calculate the combined stress at the reference point (Fig. 7.21).

7.9.3 Design Checks on Diaphragm Yielding and Buckling

The bearing stiffener is checked for yielding but not for buckling, as the plate itself is designed to take all in-plane forces from the webs and flanges. However, in the check for yielding the axial stresses above the bearings caused by the reactions at those locations are added to those caused by the out-of-plane moments. The effective area of contact of plate and stiffeners to be used in calculating these stresses is shown in Fig. 7.18.

The diaphragm plate is checked for both yielding and buckling. Use is made of approximate expressions for the elastic critical buckling of rectangular and trapezoidal plates loaded along their edges in checking the buckling strength of the diaphragm plate. Rotational restraint provided by flanges and webs to the diaphragm at their intersections is conservatively neglected. The empirical expressions allow for slope of webs, spacing and width of bearings, panel aspect ratio, and influence of top flange loading in coefficients K_1 to K_4.

The complete buckling check is

$$\sum R_v + \frac{T}{l_b} \leq \frac{0.7KEt_d^3}{D} \tag{7.28}$$

where

$K = K_1K_2K_3K_4$
$K_1 = 3.4 + 2.2D/B_d$
$K_2 = 0.4 + j/2B_d$ for single central bearing
$= 0.4 + (c - j/3)/B_d$ for twin bearings
$K_3 = 1 - \beta/100$

$$K_4 = 1 - \frac{fP_d}{\sum R_v + T/l_b}\left(\frac{2B}{B_d} - 1\right)$$

c = distance between centers of bearings
$l_b = jK$ for single central bearings
$= c$ for twin bearings

$P_d = W_d \sum P/K_5$

W_d = uniformly distributed load applied to top of diaphragm
P = any local load applied to top of diaphragm

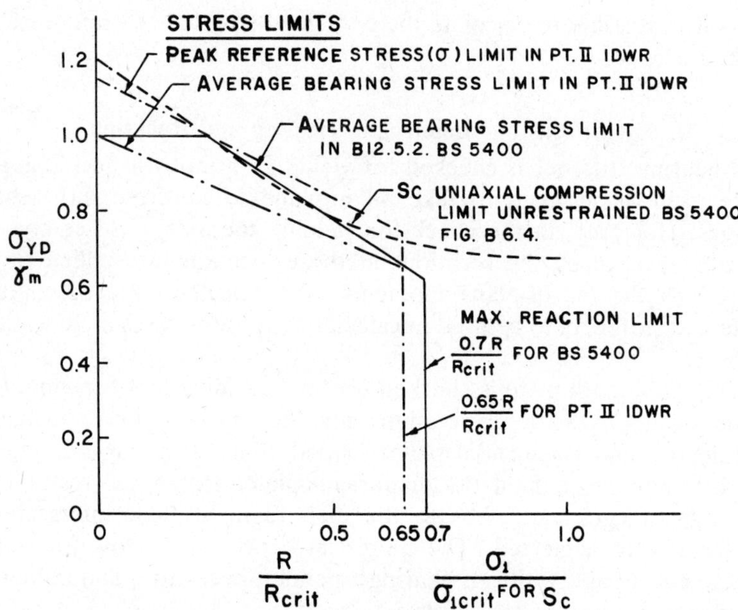

Fig. 7.22 Stress limits in simple plate diaphragms.

$K_5 = 0.4 + w/2B - B_d$
w = width of load P with allowance for dispersal through flange
$f = 0.55$ for $D/B \leq 0.7$
$= 0.86$ for $D/B \geq 1.5$
with linear interpolation for intermediate values

Other symbols are dimensions defined in Figs. 7.18 and 7.20.

The reasoning behind the buckling check above is that the reactions applied to the diaphragm should not exceed 70% of the reactions causing critical buckling. This is based on the results of numerical parametric studies which suggest that provided the average stress above the bearing is contained, no further yielding due to buckling will occur in other parts until the diaphragm is loaded beyond 70% of the critical load. The stress in the vicinity of the bearings may attain average yield stress when the plate is loaded well below critical but is reduced as the stress approaches the cutoff level of 70% of critical load (see Fig. 7.22).

7.10 STIFFENED DIAPHRAGMS

The basis for the design of stiffened diaphragms differs from that of the "simple" unstiffened diaphragms in that the bearing stiffeners are designed not only to resist any out-of-plane moments but also to act as load-bearing stiffeners, in conjunction with narrow associated widths of diaphragm plate, to transmit vertical loads to the bearings. The plate panels are then designed for shear and transverse stress only. Secondary stiffeners are used to stabilize the plate by subdividing it into appropriately sized subpanels. Boundary stiffeners may also be provided at the web/diaphragm junctions to help resist tension field forces in the webs or to transmit reactions from cross girders or floor beams, with or without cantilevers, into the diaphragm.

7.10.1 Limitations on Diaphragm Geometry

The limitations on stiffeners to control stiffener buckling are the same as those for unstiffened diaphragms. Load-bearing stiffeners are not required to be symmetrical and can be welded to a single side, although they are still required to be full height. Unstiffened openings such as might be used in unstiffened diaphragms may be replaced by larger holes framed by stiffeners designed to resist destabilizing forces.

7.10.2 Analysis of Diaphragm Stresses

A rational analysis such as finite-element analysis is often used to determine stresses in a stiffened diaphragm. Such an analysis would model the presence of stiffeners and openings with sufficient accuracy for design purposes. Alternatively, use can be made of the simplified analysis to determine the in-plane transverse stresses in the diaphragm plate. Secondary stiffeners are ignored in calculating diaphragm properties. Stresses must be calculated at the corners of each plate panel and any in-plane bending stresses due to Vierendeel action around large openings must be calculated separately and added to the other in-plane stresses. Load-bearing stiffeners, including a width of plating no more than 12 times the diaphragm thickness, are designed to carry linearly varying axial stresses compatible with the assumption of uniform shear flows and out-of-plane moments that decrease linearly with height.

7.10.3 Design Checks on Diaphragm Yielding and Buckling

Plate panels are checked for yield at all critical parts, ignoring vertical stresses. The panels are checked for buckling using ultimate strength interaction formulas originally devised for use in the design of longitudinally stiffened webs. Panels in the vicinity of the bearings are proportioned so that overall yield precedes buckling. Other panels are designed for buckling.

Load-bearing stiffeners are designed as struts, as in the case of longitudinally stiffened panels. A yield check is made for all sections, with some overstress, $1.33\sigma_{ys}$, permitted at the points of contact with bearings. Buckling checks are confined to the middle third of the height, where destabilizing effects are greatest in stiffeners of constant section. The destabilizing effects of transverse plate stresses are accounted for by means of an additional fictitious load in the stiffener. This load is taken as

$$P_{se} = \frac{\sigma_q l_s^2 t_d k_s}{a_{\max}} \qquad (7.29)$$

where

σ_q = horizontal stress in the middle third of length l_s

l_s = length of stiffener between points of effective restraint

$a_{\max}$ = spacing of vertical stiffeners

and k_s is found from Fig. 7.9 using $l = l_s$.

Intermediate stiffeners are assumed to be free of axial stresses and applied moments and are proportioned to resist the destabilizing effect of transverse and shear stresses in the case of vertical stiffeners and shear stress alone in the case of horizontal stiffeners. A similar approach is used for stiffeners framing large holes, in which case the destabilizing terms are calculated assuming the hole to be absent.

7.11 PROBLEMS NEEDING FURTHER RESEARCH

Despite the enormous amount of research carried out on box girders in the 1970s there are still many aspects of the stability of box girders needing further attention, some of which are summarized below.

Flange buckling

- Design rules are needed for flanges stiffened by one or two stiffeners. The strut approach may not be appropriate in such cases, as it neglects the transverse stiffness of the plate and is a poor model for a single stiffener.
- The limitations on cross-sectional shape to prevent local buckling of open and closed stiffeners need to be defined more accurately.
- There is a need for a simpler approach than those available for the buckling of lightly stiffened flanges which allows the postbuckling reserve of the stiffened flange to be taken into account.
- The limitations on redistribution of stresses caused by shear lag in the flange need to be defined more clearly for both simply and continuously supported box girders.
- A more rational ultimate load method for the design of laterally loaded flanges needs to be evolved.

Web buckling

- Simple tension field design methods need to be developed for application to box girders which can include longitudinal stiffeners and the interaction among shear, bending, and axial stresses in the web.
- The application of corner stiffening to boxes as an aid to stabilizing the web and corner could be usefully investigated.

Diaphragms

- Simplification of the design approach to unstiffened diaphragms should be sought based on further research.
- Methods for the design of diaphragms should be produced to cover the cases of skew diaphragms and twin-walled diaphragms.
- The effects of interactions among diaphragm, flange, and web need more consideration

Behavior of boxes

- The strength of longitudinally stiffened boxes under combined bending and shear needs closer attention in the future.

- The strength of stiffened boxes under combined bending, shear, and torsion also needs attention.

References

Abdel-Sayed, G. (1969), "Effective Width of Steel Deck-Plate in Bridges," *ASCE J. Struct. Div.*, Vol. 95, p. 1459.

Basler, K. (1961), "Strength of Plate Girders in Shear," *ASCE J. Struct. Div.*, Vol. 87, p. 151.

British Standards Institution (1983) *Steel, Concrete and Composite Bridges*," British Standard BS 5400: Part 3. *Code of Practice for Design of Steel Bridges*, BSI, London.

Burgan, B. A., and Dowling, P. J. (1985), "The Collapse Behaviour of Box Girder Compression Flanges—Numerical Modelling of Experimental Results," *CESLIC Rep. BG 83*, Imperial College, University of London.

Chatterjee, S. (1978), "Ultimate Load Analysis and Design of Stiffened Plates in Compression," Ph.D thesis, Imperial College, University of London.

Cooper, P. B. (1967), "Strength of Longitudinally Stiffened Plate Girders," *ASCE J. Struct. Div.*, Vol. 93, p. 419.

Deutscher Ausschuss für Stahlbau (1978), "Beulsicherheitsnachweise für Platten," *DASt Richtlinie 12*, Deutscher Ausschuss für Stahlbau, Cologne, West Germany.

Dowling, P. J., and Chatterjee, S. (1977), "Design of Box Girder Compression Flanges," *2nd Int. Colloq. Stab.*, European Convention for Constructional Steelwork, Brussels, p. 153.

Dowling, P. J., Harding, J. E., and Frieze, P. A., eds. (1977a) "Steel Plated Structures," *Proc. Intl. Conf. Imperial College*, 1976, Crosby Lockwood, London.

Dowling, P. J., Frieze, P. A., and Harding, J. E. (1977b), "Imperfection Sensitivity of Steel Plates under Complex Edge Loading," *2nd Int. Colloq. Stab. Steel Structures*, Liege, p. 305.

Dubas, P., and Gehri, E., eds. (1986), *Behaviour and Design of Steel Plated Structures*, ECCS Technical Committee 8.3, Publ. No. 44, European Convention for Constructional Steelwork, Brussels.

Dwight, J. B., and Little, G. H. (1976), "Stiffened Steel Compression Flanges—A Simpler Approach," *Struct. Eng.*, Vol. 54, p. 501.

Einarsson, B., Crisfield, M. A., and Dowling, P. J. (1982), "Collapse of Stiffened Box Girder Diaphragms," *J. Constr. Steel Res.*, Vol. 2.

Frieze, P. A., Dowling, P. J., and Hobbs, R. E. (1975), "Parametric Study of Plates in Compression," *CESLIC Rep. BG 39*, Imperial College, University of London.

Frieze, P. A., and Dowling, P. J. (1979), "Testing of a Wide Girder with Slender Compression Flange Stiffeners under Pronounced Shear Lag Conditions," *CESLIC Rep. BG 49*, Imperial College, University of London.

Fujita, Y., and Galambos, T. V., eds. (1981), "Inelastic Instability of Steel Structures and Structural Elements," *U.S.-Jpn. Joint Sem.*, Tokyo.

Harding, J. E., and Dowling, P. J. (1981), "The Basis of the Proposed New Design Rules for the Strength of Web Plates and Other Panels Subject to Complex Edge Loading," in

Stability Problems in Engineering Structures & Components (ed. T. H. Richards and P. Stanley), Applied Science Publishers, Barking, Essex, England, p. 355.

Horne, M. R. (1977), "Structural Action in Steel Box Girders," *CIRIA Guide 3*, Construction Industry Research and Information Association, London.

"Inquiry into the Basis of Design and Method of Erection of Steel Box Girder Bridges", (1973) *Report of the Committee and Appendices*, H.M. Stationary Office, London.

Institute of Civil Engineers (1973), "Steel Box Girder Bridges," *Proc. Int. Conf.*, Institute of Civil Engineers, London.

Jetteur, P. (1983), "A New Design Method for Stiffened Compression Flanges of Box Girders," in *Thin-Walled Structures* (ed. J. Rhodes and A. C. Walker), Granada, London, p. 189.

Jetteur, P., et al. (1984), "Interaction of Shear Lag with Plate Buckling in Longitudinally Stiffened Compression Flanges," *Acta Technica CSAV*, No. 3, p. 376.

Lamas, A. R. G., and Dowling, P. J. (1980), "Effect of Shear Lag on the Inelastic Buckling Behaviour of Thin-Walled Structures," in *Thin-Walled Structures* (ed. J. Rhodes and A. C. Walker), Granada, London, p. 100.

Maquoi, R., and Massonnet, C. (1971), "Théorie non-lineaire de la resistance postcritique des grandes poutres en caisson raidies," *Mem. AIPC*, Vol. 31-II, p. 91.

Moffatt, K. R., and Dowling, P. J. (1975), "Shear Lag in Steel Box Girder Bridges," *Struct. Eng.*, Vol. 53, p. 439.

Moffatt, K. R., and Dowling, P. J. (1976), "Discussion," *Struct. Eng.*, Vol. 54, p. 285.

Ostapenko, A., and Vojta, J. F. (1967), "Ultimate Strength Design of Longitudinally Stiffened Plate Panels with Large b/t," *Fritz Eng. Lab. Rep. No. 248.18*, Lehigh University, Bethlehem, Pa., August.

Porter, D. M., Rockey, K. C., and Evans, H. R. (1975), "The Collapse Behaviour of Plate Girders in Shear," *Struct. Eng.*, Vol. 53, p. 313.

Rockey, K. C., and Evans, H. R., eds. (1980), "Design of Steel Bridges," Proc. Int. Conf., at University College, Cardiff, Granada, London.

Škaloud, M., and Novotny, R. (1965), "Üerberkritisches Verhalten einer anfänglich gekrümmten gleichformig gedrüchten, in der Mitte mit einer langsrippe versteiften Platte," *Acta Technica CSAV*, p. 210.

Wolchuk, R., and Mayrbourl, R.M. (1980), "Proposed Design Specification for Steel Box Girder Bridges," *Rep. No. FHWA-TS 80-205*, U.S. Department of Transportation, Federal Highway Administration, Washington, D.C.

Chapter Eight

Beam-Columns

8.1 INTRODUCTION

Beam-columns are defined as members subjected to a combination of axial force and bending moment. They therefore provide a link between the column under pure axial load discussed in Chapter 3 and the beam loaded only by moments, which was the subject of Chapter 5. Indeed, a case can be made for considering all members in frame structures as beam-columns, with columns and beams being the special cases that result when one load component becomes negligibly small. The bending moments in beam-columns may be generated as a result of transverse loading acting between the member's ends, from loading on adjacent members in rigidly framed structures or by the eccentricity of reactions and nominal axial forces in simply framed structures. When considering the behavior of beam-columns in rigidly framed structures it is normally necessary to consider also the influence of the surrounding members, which leads to the study of subassemblages and complete frames of Chapter 16. In the present chapter the subject is treated principally in terms of the response of an isolated member to a known system of end forces and moments.

Depending on the exact manner in which a beam-column is loaded and supported, its response may be categorized in a number of different ways. Perhaps the most fundamental feature is the presence (or absence) of a bracing system which is capable of preventing translation of one end relative to the other. Problems involving sway are more appropriately considered in the context of overall frame behavior. For nonsway beam-columns the three problem classes illustrated in Fig. 8.1 may be identified:

1. The thrust is applied with an eccentricity about the minor axis (or if the eccentricity is about the major axis, then the column is prevented

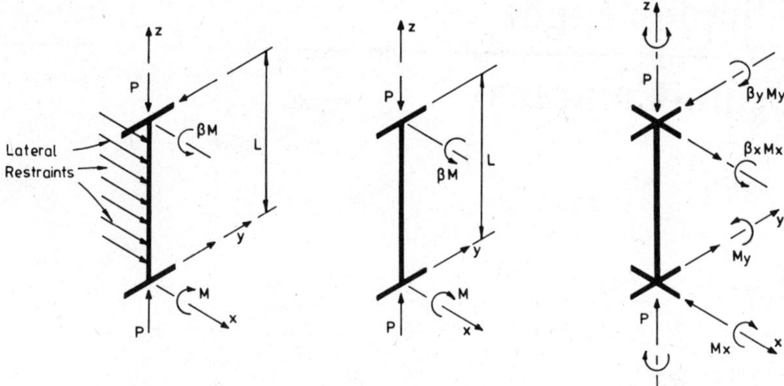

Fig. 8.1 Classes of beam-column behavior.

from deflecting out of this plane by properly designed bracing), in which case the member will collapse by excessive deformation in this plane.

2. The thrust is applied with an eccentricity about the major axis, in which case the member will collapse by deflecting about the minor axis and twisting (i.e., similar to lateral-torsional beam buckling).
3. The thrust is applied with an eccentricity about both axes, in which case the member will collapse by combined bending and twisting.

Thus case 1 represents an interaction between column buckling and simple uniaxial beam bending, case 2 represents an interaction between column buckling and beam buckling, and case 3 represents the interaction of column buckling and biaxial beam bending. Clearly, case 3 is the most general case, with the others being more limited versions.

The analysis of various aspects of these problems has formed the subject of a very large number of research investigations. Initially these were confined to the elastic range, but the increasing availability of computers has meant that by including inelastic material behavior in conjunction with such features as residual stresses and initial deformations realistic maximum strength analyses may now be conducted. The development of the theory of beam-columns has recently been summarized by both Massonnet (1976) and Chen and Atsuta (1977). A more general review of the most significant developments prior to 1976 is available in the previous edition of the guide (Johnston, 1976), while reviews devoted specifically to the biaxial problem have been provided by Chen and Santathadaporn (1968) and by Chen (1977a, 1981). In view of the availability of these recent comprehensive summaries of the general subject area of beam-columns, no general historical review will be presented herein. Rather, the chapter will concentrate on a critical evaluation of the various design and analysis approaches that are currently available for each type of beam-column problem. In this way it is hoped that the material will be of direct use to designers needing to work beyond the limitations of codes of practice, to specification writing bodies engaged in the task of updating design codes, and to researchers concerned with advancing the understanding of particular aspects of the topic. Readers requiring a historical perspective on the subject are advised to consult the reviews referenced above, in which they will find more than 300 relevant papers listed.

8.2 STRENGTH OF BEAM-COLUMNS

The load-carrying capacity of a beam-column depends on several factors, which may conveniently be arranged under the three headings "load-related," "member-related," and "imperfection-related." The first of these is readily appreciated if the strength of a beam-column subject to any combination of axial load P, major-axis moment M_x, and minor-axis moment M_y is displayed on a three-dimensional interaction diagram of the type shown as Fig. 8.2. Clearly, any point located on an axis will correspond to loading of one type only, while the line or surface joining these end points will define strength under two or tree load components, respectively.

In constructing diagrams of the form of Fig. 8.2, use is made of the member's strength as a column P_u and as a beam M_{ux} and M_{uy} in fixing the end points. Procedures for determining these quantities have previously been discussed in Chapters 3 (columns) and 5 (beams). These make use of several properties of the member (e.g., geometrical proportions, material strength, unbraced length, end support conditions, etc.). These same properties will also have some effect on the exact shape of the interaction curves or surfaces (i.e., their degree of concavity or convexity).

The various "imperfections" (e.g., lack of straightness in either plane, residual stresses, variation of material strength around the cross section, etc.) will also influence both the component strengths and the shapes of the interactions. The estimation of the strength of a beam-column therefore requires a knowledge of the form of Fig. 8.2 appropriate to the particular set of parameters present. Succeeding sections of this chapter will address this problem in terms of both the design-oriented procedures that have evolved to meet this need and the evidence, both theoretical and experimental, on which these have been based.

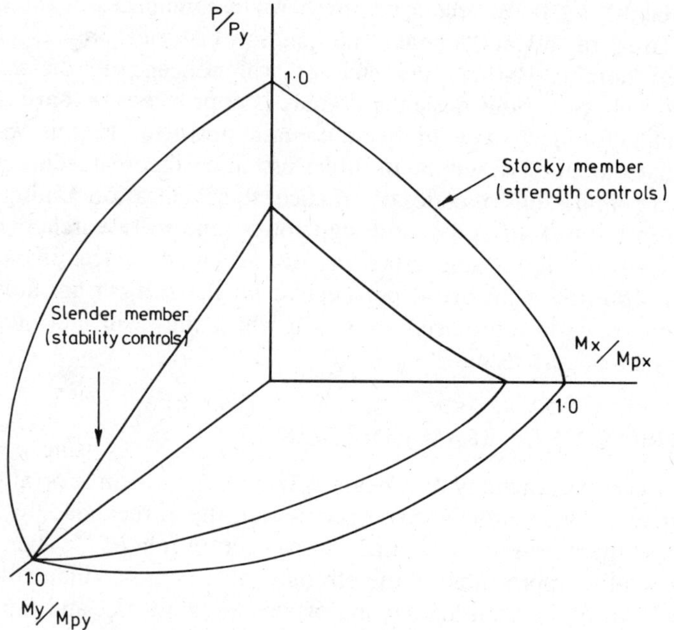

Fig. 8.2 Interaction surfaces for beam-columns.

Traditionally, design methods for beam-columns fall into one of two categories: (1) those that use charts or tables to provide safe combinations of the applied load components, and (2) interaction formulas of the type

$$f\left(\frac{P}{P_u}, \frac{M_x}{M_{ux}}, \frac{M_y}{M_{uy}}\right) \leq 1.0 \tag{8.1}$$

which provide a smooth transition between end points corresponding to strength under only one form of loading. More recently, the availability of inexpensive programmable calculators and microcomputers has meant that for special problems (e.g., beam-columns of unusual cross section for which the application of methods based on wide-flange shapes might be considered questionable), direct analysis is also a possibility.

Use of the interaction approach does, of course, imply that reliable estimates of the end points of the interaction, which depend on the strengths P_u, M_{ux}, and M_{uy} under only one component of the applied loading, are available. A word of caution is in order here, as changes in these end points (e.g., by the substitution of one column formula for another) can sometimes produce unexpectedly large changes in the predictions of an interaction equation. Although methods are available for obtaining accurate theoretical solutions to each of the problem types shown in Fig. 8.1, such methods always require the use of numerical procedures to follow the inelastic load-deflection behavior associated with the attainment of true maximum strength. They therefore cannot lead directly to design equations. Explicit forms of Eq. 8.1 for individual cases must therefore be developed either as modifications to formulas derived from elastic analysis or on a wholly empirical basis. The suitability of such formulas then requires verification against available theoretical and experimental data.

8.3 UNIAXIAL BENDING: IN-PLANE STRENGTH

A convenient form of Eq. 8.1, which is used as the starting point for several design formulas, is

$$\frac{P}{P_u} + \frac{M}{M_u} \leq 1.0 \tag{8.2}$$

where

P = thrust at failure
P_u = ultimate load for the centrally loaded column for buckling in the plane of the applied moment
M = maximum bending moment at failure
M_u = ultimate bending moment in the absence of axial load

If the design basis is to be the attainment of first yield in an initially stress-free member, then P_u and M_u should be chosen accordingly (Johnston, 1976). For ultimate load design P_u and M_u should be taken as the column strength as discussed in Chapter 3 and the in-plane bending capacity (i.e., M_p, the fully plastic moment capacity).

While the correct value for P is simply the applied axial load, the determination of the appropriate value for M in any particular case is more difficult since it will be affected by the deformation of the member. This in turn will be a function principally of the form of the applied loading producing the moments, the level of thrust, and the member slenderness. Figure 8.3 illustrates the difference between a stocky column for which the maximum moment may sensibly be taken as the maximum primary moment (i.e., that calculated neglecting the effects of the axial load) and a slender column. For the latter the action of the thrust acting through the deflections produced by the primary moment leads to so-called secondary moments, the effect of which is to increase the moments

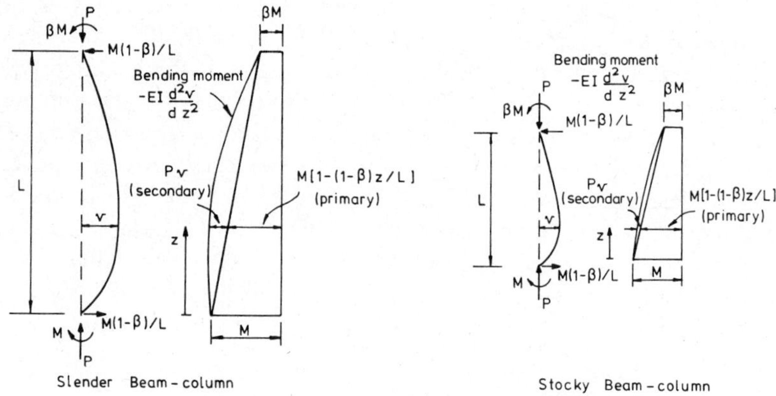

Fig. 8.3 Stocky and slender beam-columns.

as shown. The relative size of the primary and secondary moments will, of course, depend on the variables mentioned previously. In particular, for certain patterns of primary moment (e.g., uniform double curvature bending produced by end moments M_1 and $-M_1$), the point of maximum moment may be located at the end(s) of the member even though its slenderness is large.

The maximum moment at midlength in a beam-column subjected to compression P and equal and opposite end moments M_0 is given approximately by (Johnston, 1976)

$$M_{\max} = M_0\left(\frac{1}{1 - P/P_e}\right) \tag{8.3}$$

in which P_e is the elastic critical load for buckling in the plane of the applied moments. The term in parentheses in Eq. 8.3 may be regarded as an amplification factor by which the first-order moment M_0 is multiplied to obtain the second-order moment $M_{\max}$. It causes $M_{\max}$ to increase nonlinearly as shown in Fig. 8.4.

Although Eq. 8.3 is derived on the assumption of elastic behavior, its application in an ultimate load context is well established. Substituting it into Eq. 8.2 gives the design formula

$$\frac{P}{P_u} + \frac{M_0}{M_u(1 - P/P_e)} \leq 1.0 \tag{8.4}$$

which was first recommended by SSRC (Johnston, 1976) and which has been included in several national codes. A direct comparison between Eq. 8.4 and the early numerical solution of Galambos and Ketter (1961)

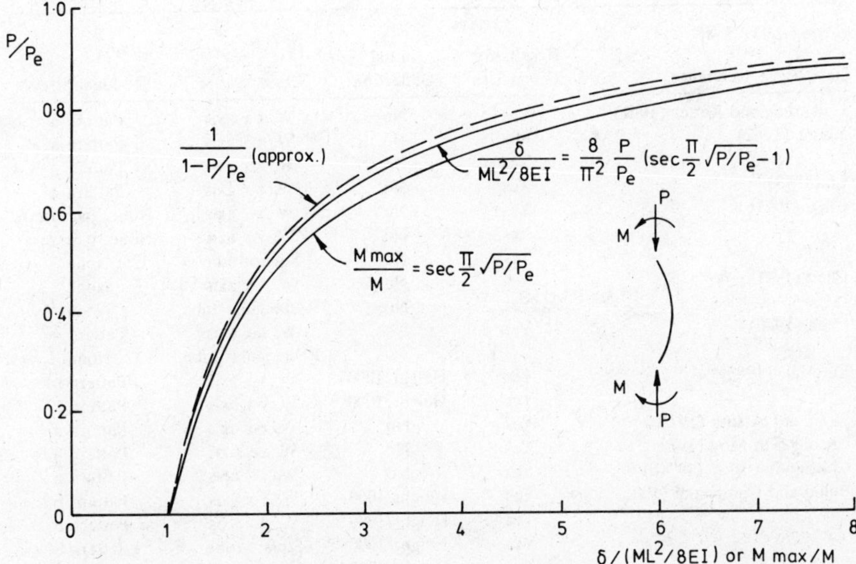

Fig. 8.4 Maximum deflection and moment in elastic beam-columns with equal end moments.

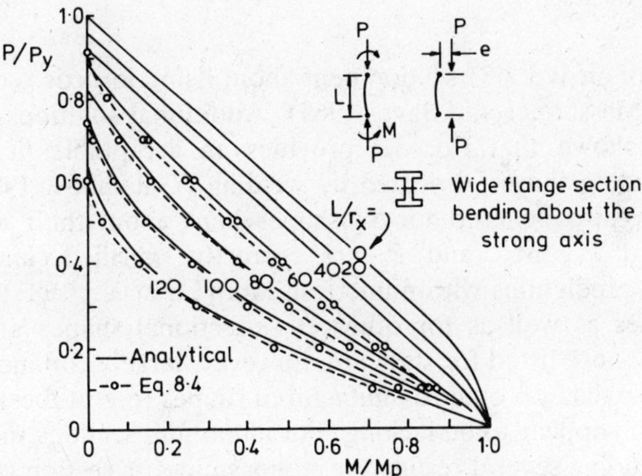

Fig. 8.5 Comparison of Eq. 8.4 with numerical results of Galambos and Ketter for major-axis bending of a W section (after Massonnet and Save, 1965).

Table 8.1A. Theoretical Solutions for the In-Plane Strength of Beam-Columns

Reference	Results for Direct Use	Initial Deflection	Cross Section	Residual Stress
Galambos and Ketter (1961)	Yes	No	W, x-x axis	Pattern a
Ketter (1962)	Yes	No	W, x-x axis	Pattern a
	Yes	No	W, x-x axis	Pattern a
Lu (1968)	Yes	No	W, x-x axis	Pattern a
Chen (1971b)	Yes	No	W, x-x axis	None, pattern a
	Yes	No	W, y-y axis	None, pattern a
	Yes	No	Square tube	None
Chen (1971a)	Yes	No	W, x-x axis	None
	Yes	No	Rectangular tube	None
Chen (1970)	Yes	No	W, x-x axis	Pattern a
			Rectangular tube	None
Lu et al. (1983)	Yes	Height/1000	W, x-x axis	Pattern a
	Yes	Height/1000	W, y-y axis	Pattern a
Chen and Atsuta (1972b)	Yes	No	W, x-x axis	Pattern a
Cheong-Siat-Moy (1979a)	Yes	No	W, x-x axis	Pattern a
Cheong-Siat Moy (1974b)	Yes	No	W, x-x axis	Pattern a
Ballio and Campanini (1981)	Yes	Height/1000	W, x-x axis	Pattern b
	Yes	Height/1000	W, y-y axis	Pattern b
	Yes	Height/1000	Square tube	Pattern b
	Yes	Height/1000	Circular tube	None
Ballio et al. (1973)	Yes	Height/1000	W, x-x axis	Pattern b
	Yes	Height/1000	W, y-y axis	Pattern b
	Yes	Height/1000	Square tube	Pattern a
	Yes	Height/1000	Square tube	Pattern b
Yu and Tall (1971)	Yes	No	W, x-x axis	Pattern a
			W, x-x axis	Pattern d
Young (1973)	Yes	Height/1000	W, x-x axis	Patterns b, c
			W, y-y axis	Patterns b, c
			Square tube	Pattern b

for the case of an W8 × 31 section bent about its major axis is reproduced as Fig. 8.5 (Massonnet and Save, 1965). Additional solutions (see Table 8.1A) have shown that Eq. 8.4 provides an acceptable fit for all W shapes, including those fabricated by welding (Galambos, 1964). Equation 8.4 may also be used for W shapes bent about their minor axis, provided that P_u, M_u, and P_e are correctly specified, and Fig. 8.6 compares its predictions with numerical data (Lu et al., 1983). Test data for both cases as well as for other cross-sectional shapes are listed in Table 8.1B. Those listed for steel are, however, largely confined to rolled W shapes; test data on other manufactured shapes (e.g., tubes) or for any welded section appear to be lacking. For aluminum sections the coverage is wider, including several results for monosymmetric section (Gilson and Cescotto, 1982; Klöppel and Barsch, 1973). It is also noticeable that only a small proportion of the theoretical studies of Table 8.1A include an

Load Cases	Comparison with Eq. 8.4
1 ($\beta = 1.0, 0$)	Partial
1 ($\beta = 1, 0.5, 0, -0.5, -1$)	Partial
4	Partial
4, 2, 5, 6	Yes
4, 1 ($\beta = 1$)	No
4, 1 ($\beta = 1$)	No
4, 1 ($\beta = 1$)	No
4, 1 ($\beta = 1$)	No
4, 1 ($\beta = 1$)	No
3, 1 ($\beta = 1$), 2	No
3, 1 ($\beta = 1$), 2	No
1 ($\beta = 1, 0.5, 0, -0.5, -1$)	No
1 ($\beta = 1, 0.5, 0, -0.5, -1$)	No
1 ($\beta = 1$)	No
2, 4, 6	No
2, 7, 13	No
1 ($\beta = 1, 0, -1$)	Yes
9, 2, 4, 8,	Yes
3, 10, 7,	Yes
12, 11	Yes
1 ($\beta = 1$)	Yes
1 ($\beta = 1$)	Yes
1 ($\beta = 1$)	Yes
1 ($\beta = 1$)	Yes
1 ($\beta = 1$)	No
1 ($\beta = 1$)	No
1 ($\beta = 1, 0, -1$)	No
1 ($\beta = 1, 0, -1$)	No
1 ($\beta = 1, 0, -1$)	No

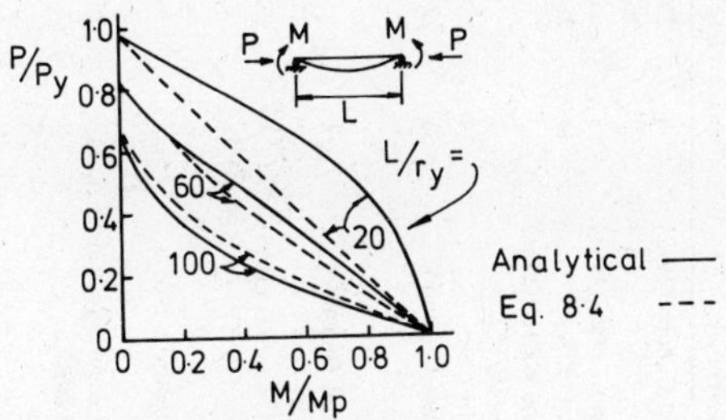

Fig. 8.6 Comparison of Eq. 8.4 with numerical results of Kanchanalai and Lu (1979) for minor-axis bending of a W section.

Table 8.1B. Experimental Data on the In-Plane Strength of Beam-Columns

Reference	Number of Tests	L/r Range	Cross Section	Residual Stress	Load Cases	Comparison with Eq. 8.4	Comments
Johnston and Cheney (1942)	30	23–122	W, x-x axis	Pattern a	1 ($\beta = 1$)	Galambos and Ketter (1961)	Major-axis failure
	30	26–126	W, y-y axis	Pattern a	1 ($\beta = 1$)	No	
Galambos and Ketter (1961)	5	11–90	W, x-x axis	Pattern a	1 ($\beta = 1$)	Yes	
Ketter et al. (1952, 1955) Galambos and Ketter (1961)	14	28–120	W, x-x axis	Pattern a	1 ($\beta = 1, 0$)	Yes	Some out-of-plane failures
Yu and Tall (1971)	2	40	W, x-x axis	Patterns a, d	1 ($\beta = 1$)	No	Braced about y-y axis
Mason et al. (1958)	24	36–117	Hat shape	Unknown pattern	1 ($\beta = 1$)	Yes	
van Kuren and Galambos (1964)	36	22–116	W, x-x axis	Pattern a	1 ($\beta = 1, 0.5, 0, -1$)	Yes	10 tests braced about y-y axis, several out-of-plane failures for the rest
van Kuren and Galambos (1964)	1	72	W, y-y axis	Pattern a	1 ($\beta = 1$)	Yes	
Dwyer and Galambos (1965)	3	38–80	Square tube	Unknown pattern	1 ($\beta = 1$)	Yes	
Bijlaard et al. (1955)	18	40-50	Square tube	Unknown pattern	1 ($\beta = 1$)	No	Elastic end restraints
			W, x-x axis	Pattern a	1 ($\beta = 1$)	No	
Lay and Galambos (1965)	7	30–60	W, x-x axis	Pattern a	1 ($\beta = 1$)	No	End restraint from beams, braced about y-y axis
Gilson and Cescotto (1982)	18	35–68	T-shape	Aluminum	1 ($\beta = 1$)	No	Comparisons with ECCS
Hill et al. (1956)	9		Circular tube	Aluminum	1 ($\beta = 1$)	Mazzolani and Frey (1983)	
Clark (1955)	28		Square tube Solid square	Aluminum	1 ($\beta = 1$)		
Klöppel and Barsch (1973)	48	30–70	W, x-x axis	Aluminum	1 ($\beta = 1$)	Mazzolani and Frey (1983)	
	48	30–82	Circular tube	Aluminum	1 ($\beta = 1$)	Mazzolani and Frey (1983)	
	27	30–80	T shape	Aluminum	1 ($\beta = 1$)	Mazzolani and Frey (1983)	

allowance for initial curvature and that existing data would appear to be confined to either W shapes or to square and circular tubes.

For beam-columns subjected to unequal end moments and/or transverse loading between points of support in the plane of bending, Eq. 8.4 may still be used provided that M_0 is replaced with an equivalent moment $M_{eq} = C_m M_0$, where C_m is a reduction factor and M_0 is taken as the maximum first-order moment. Since the C_m factor is also applicable to the other types of beam-column problems illustrated in Fig. 8.1, a full discussion of its basis will be delayed until Section 8.5. The more general form of the interaction equation is, therefore,

$$\frac{P}{P_u} + \frac{C_m M_0}{M_u(1 - P/P_e)} \leq 1.0 \tag{8.5}$$

Equation 8.5 refers to the conditions of failure resulting from instability due to excessive bending occurring within the member. However, when a beam-column is bent by moments producing plastic hinges at one or both ends, it is necessary to limit such terminal moments to within M_{pc}, the plastic-hinge moment modified to include the effect of axial compression. The determination of M_{pc}, by satisfying the requirements of equilibrium, is simple, if tedious, for rectilinear shapes. Expressions for M_{pc} for wide-flange shapes are available (ASCE, 1971a). These may be extended to other orthogonal shapes, in which case the superposition procedure developed by Chen and Atsuta for biaxially bent beam-columns can be applied (Chen and Atsuta, 1972a, 1974, 1977). Expressions for M_{pc}, applicable for most wide-flange sections, are plotted nondimensionally in Figs. 8.7 and 8.8. The accuracy of these curves has been verified by experiments. Simple approximate expressions to compute M_{pc} are also shown by dashed lines in Figs. 8.7 and 8.8. The limitation $M_0 \leq M_{pc}$ will lead to two additional interaction formulas as follows. For strong-axis bending,

$$\frac{P}{P_y} + 0.85\left(\frac{M_0}{M_p}\right) \leq 1.0 \qquad M_0 \leq M_p \tag{8.6}$$

and for weak-axis bending,

$$\left(\frac{P}{P_y}\right)^2 + 0.84\left(\frac{M_0}{M_p}\right) \leq 1.0 \qquad M_0 \leq M_p \tag{8.7}$$

An alternative to Eq. 8.7 suggested by Pillai (1974), which is easier to incorporate into the biaxial interaction formulas discussed in Section 8.6 due to its linear format, is compared with the "exact" results in Fig. 8.8.

$$\frac{P}{P_y} + 0.6\left(\frac{M_0}{M_p}\right) \leq 1.0 \qquad M_0 \leq M_p \tag{8.8}$$

Table 8.1C. Design-Oriented References on the In-Plane Strength of Beam-Columns

Reference	Direct Extension or Refinement of Eq. 8.4	Alternative to Eq. 8.4	Comments
Galambos and Ketter (1961)	Yes	—	Original basis for $\beta = 1, 0, -1$ cases
Ketter (1962)	Yes	—	Demonstrates dependence of C_m on axial load level
Lu (1968)	Yes	—	Interaction curves and C_m values for lateral load cases
Chen and Atsuta (1977)	Yes	Yes	Numerous interaction curves, elastic C_m factors
Ballio and Campanini (1981)	Yes	—	C_m factors for a variety of cases, includes dependence of C_m on axial load level
Kanchanalai and Lu (1979)	Yes	Yes	Improvements for minor-axis bending of I-sections, extension to members in unbraced frames
Massonnet and Save (1965)	—	—	Verification of Eq. 8.4 against theory and tests
Galambos (1964)	Yes	—	General discussion of Eq. 8.4, including direct comparison with tests, simple presentation of numerical approaches suitable for direct programming
Young (1973)	—	Yes	Alternative presentation of results as basis for a different type of design approach
Galambos (1981)	Yes	Yes	Discussion of Eq. 8.4 against more recent alternatives

Adams (1970)	Yes	—	Direct extension of Eq. 8.4 to 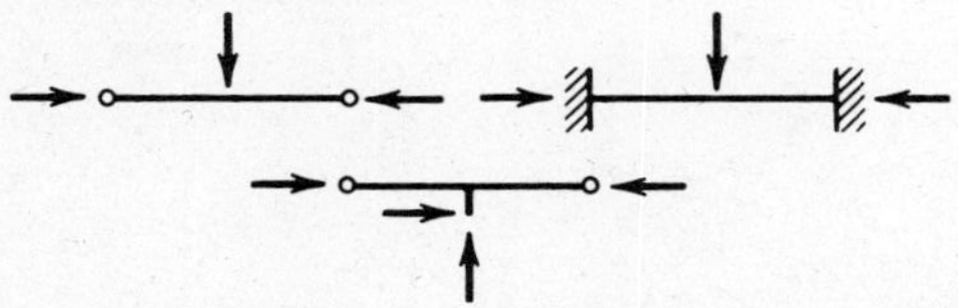
McLellan and Adams (1970)			
Adams (1974)	Yes	—	General discussion of Eq. 8.4 and its application to members in frames
Mazzolani and Frey (1983)	Yes	Yes	Comparison with test data for aluminum beam-columns
Ojalvo and Fukumoto (1962)	—	Yes	Graphical presentation of theoretical results for $\beta = +1.0$, including full moment-rotation behavior
Galambos and Prasad (1962)	Yes	Yes	Tabular presentation of theoretical results for full range of P, β, L/r
Kemp (1984)	No	Yes	Considers beam-columns in plastically designed frames, provides limits on P, L/r, and β which will ensure satisfactory performance (rotation capacity)
Chen and Cheong-Siat-Moy (1980)	Yes	Yes	Covers application of interaction formula to members in unbraced frames
Cheong-Siat-Moy and Downs (1980)	Yes	Yes	Provides improved formulas for major and minor axis bending of members in in unbraced frames
Djalaly (1975)	Yes	Yes	Suggests two modified amplification factors for major and minor axis bending of W sections, provides C factors for several cases
Roik (1976, 1977)	—	Yes	Interaction curves for 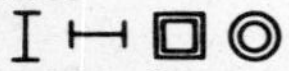 under various types of loading
Roik (1983)	—	Yes	Covers background to Eq. 8.30

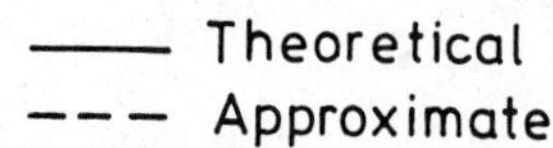

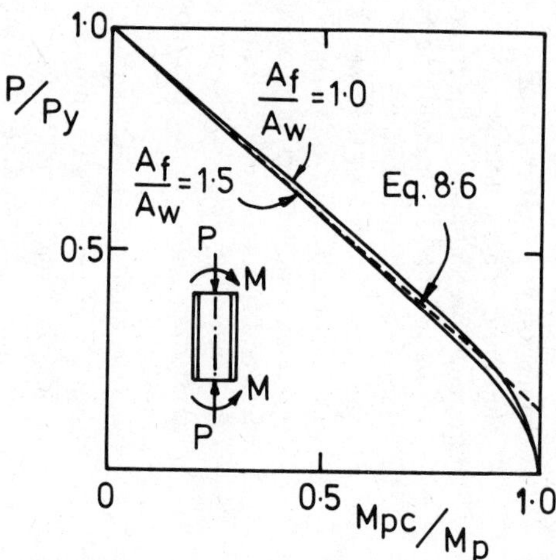

Fig. 8.7 Approximate interaction equation for a W section (strong-axis bending, short column).

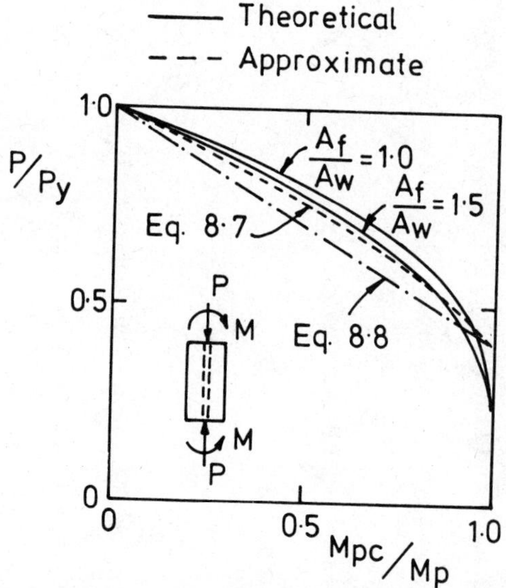

Fig. 8.8 Approximate interaction equation for a W section (weak-axis bending, short column).

Several of the references listed in Tables 8.1A and B contain material that could be of direct use to the designer confronted with a problem outside the scope of standard design treatments. The most useful of these, together with other design-oriented references, are summarized from this point of view in Table 8.1C.

Although Eqs. 8.6 through 8.8 permit the in-plane strength of a beam-column containing a plastic hinge to be determined, for such members to function satisfactorily in plastically designed structures they must also possess sufficient ductility. Quantitative assessments of this are usually expressed in terms of rotation capacity R, the ratio of plastic rotation to the hypothetical rotation in an elastic member at the moment capacity given by Eqs. 8.6 through 8.8, as appropriate. For I-sections in major-axis bending, assuming a minimum acceptable value for $R = 3$, Kemp (1984) has proposed that slenderness be limited to

$$\frac{L}{r_x} \leq \sqrt{\frac{\pi^2 E}{\sigma_y}}\,(0.6 - 0.4\beta)\,\frac{1 - P/P_y}{1.5P/P_y} \tag{8.9}$$

This expression is a development of that suggested by Lay (1974), which is used in the Australian code. A more general study of the interaction of axial force, end moment, slenderness, and rotation capacity is available in the work of van Manen (1982).

8.4 UNIAXIAL BENDING: LATERAL-TORSIONAL BUCKLING

When an I-section is bent about its major axis (i.e., in the plane of the web), there exists a tendency for it to fail by deflecting sideways and twisting, as explained in Chapter 5. The presence of an axial load when such a member is used as a beam-column will only serve to accentuate this tendency, since the preferred mode of failure under pure axial load would normally be by buckling about the minor axis. Beam-columns loaded in strong axis bending therefore exhibit an interaction between column buckling and beam buckling.

The elastic critical load for a member subject to compression P and equal and opposite end moments M_0 as shown in Fig. 8.1*b* is given by (Chen and Atsuta, 1977; Horne, 1956; Salvadori, 1956; Campus and Massonnet, 1956; Hill and Clark, 1951a, b)

$$\frac{M_0}{M_E} = \left[\left(1 - \frac{P}{P_{ey}}\right)\left(1 - \frac{P}{P_\phi}\right)\right]^{1/2} \tag{8.10}$$

where

M_E = elastic critical moment for lateral-torsional buckling as a beam (see Chapter 5)

P_{ey} = elastic critical load for minor-axis flexural buckling
P_ϕ = elastic critical load for pure torsional buckling (see Chapter 3).

Critical combinations of P and M_0 are shown plotted in Fig. 8.9.

If an approximate allowance is made for the effects of in-plane deflection (Chen and Atsuta, 1977), Eq. 8.10 becomes

$$\frac{M_0}{M_E} = \left[\left(1 - \frac{P}{P_{ex}}\right)\left(1 - \frac{P}{P_{ey}}\right)\left(1 - \frac{P}{P_\phi}\right)\right]^{1/2} \tag{8.11}$$

in which P_{ex} is the elastic critical load for major-axis flexural buckling. Noting that for W sections $P_{ey} < P_\phi$,

$$\left(\frac{1 - P}{P_\phi}\right) > \left(\frac{1 - P}{P_{ey}}\right)\left(\frac{1 - P}{P_{ex}}\right) \tag{8.12}$$

enables Eq. 8.11 to be simplified to

$$\frac{P}{P_{ey}} + \frac{M_0}{M_E(1 - P/P_{ex})} \leq 1 \tag{8.13}$$

which is of the same form as Eq. 8.4 for in-plane failure except that the quantities appearing in the denominator P_{ey} and M_E now relate to out-of-plane failure.

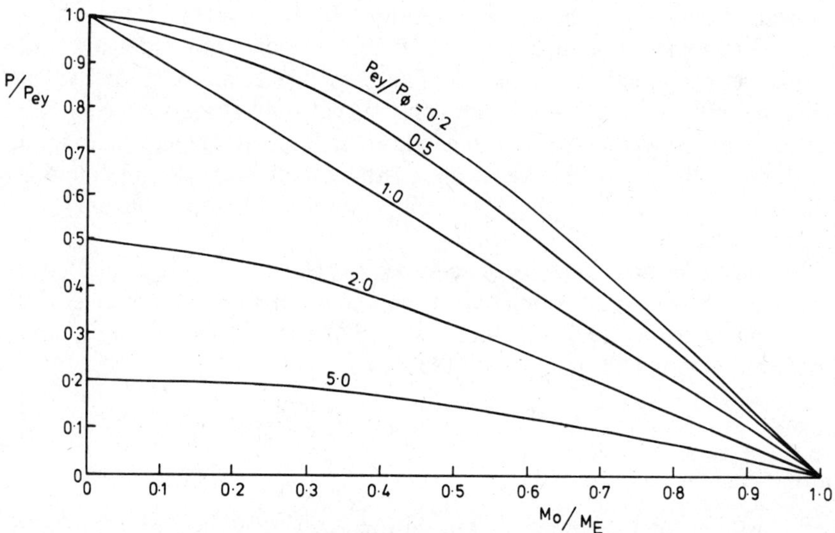

Fig. 8.9 Elastic critical-load combinations for beam-columns with equal end moments.

By analogy with Eq. 8.5 the design version of Eq. 8.13 may be written as

$$\frac{P}{P_{uy}} + \frac{C_m M_0}{M_u(1 - P/P_{ex})} \leq 1 \tag{8.14}$$

where

P = applied axial load
P_{uy} = axial load producing failure in the absence of bending moment, computed for weak-axis buckling
M_0 = maximum applied first-order moment
M_u = moment producing failure in the absence of axial load, allowing for lateral-torsional buckling
C_m = reduction factor as discussed in Section 8.3

In principle, the value of M_u in Eq. 8.14 should be that obtained from the design curve used to determine lateral-torsional beam buckling strength as explained in Chapter 5, thus ensuring a smooth transition to the correct end point. However, the AISC procedure recommends the following empirical formula for M_u, which is different from its beam formula:

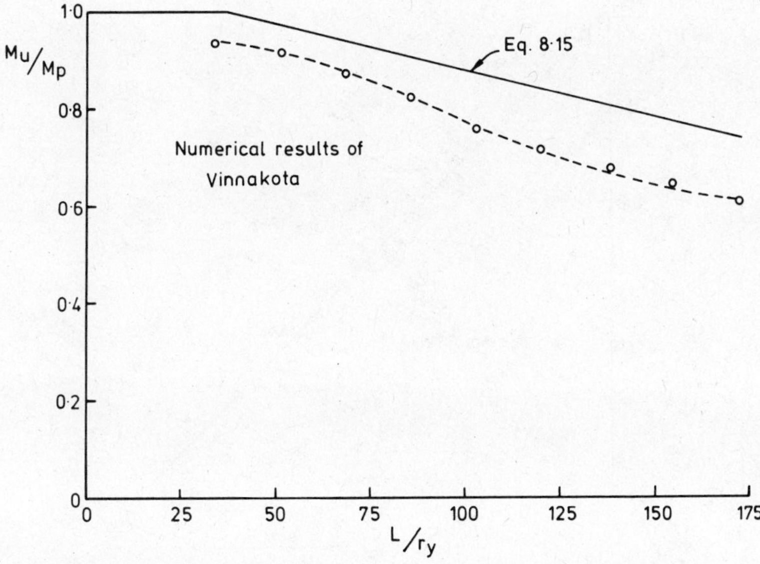

Fig. 8.10 Comparison of Eq. 8.15 with numerical results of Vinnakota for lateral-torsional beam strength.

$$M_u = \left[1.07 - \frac{(L/r_y)\sqrt{\sigma_y}}{3160}\right] M_P \leq M_P \tag{8.15}$$

where

L/r_y = weak-axis slenderness
σ_y = material yield stress, ksi

which plots significantly above the beam design curves discussed in Chapter 5, as shown in Fig. 8.10. A comparison of Eq. 8.14 using SSRC Curve 2 for P_u and Eq. 8.15 for M_u with the numerical results of Vinnakota (1977) is given in Fig. 8.11. It exhibits a tendency toward overprediction, as the degree of bending present increases, which does not happen for the similar comparison of Fig. 8.12, which replaces Eq. 8.15 with the values of M_u obtained by Vinnakota for zero axial load. The values of P_u for zero moment determined by Vinnakota are very close to

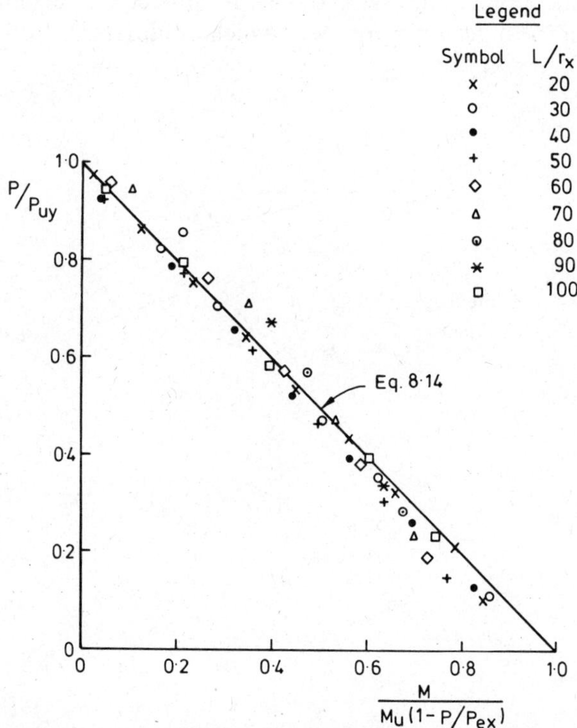

Fig. 8.11 Comparison of Eq. 8.14 with numerical data of Vinnakota for lateral-torsional strength of beam-columns, M_u from Eq. 8.15.

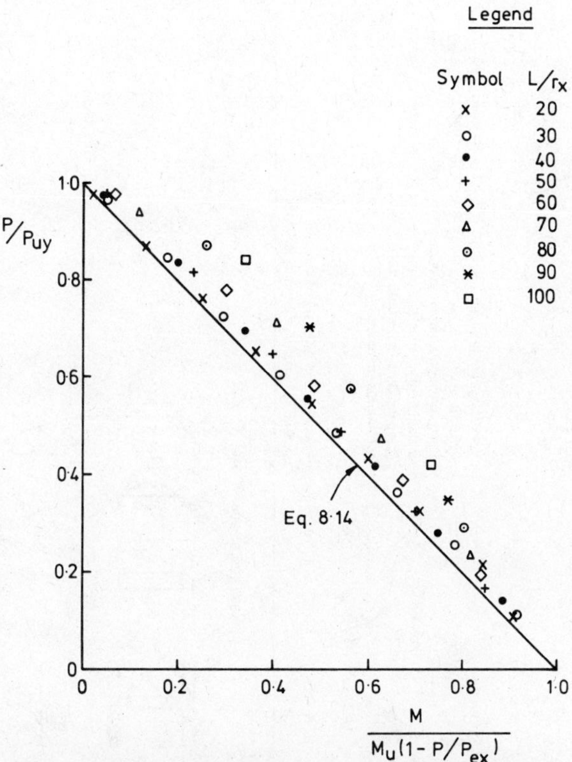

Fig. 8.12 Comparison of Eq. 8.14 with numerical data of Vinnakota for lateral-torsional strength of beam-columns, M_u values taken from Vinnakota (1977) (see Fig. 8.10).

those given by SSRC Curve 2 (Chen and Atsuta, 1977). Comparisons of Eq. 8.14 with the available test data (Campus and Massonnet, 1956) are generally satisfactory, although care is required in evaluating these, as different studies have used different expressions for P_u and M_u.

Vinnakota (1977) appears to be the only theoretical study of this problem in which a true maximum-strength approach has been used. However, a number of investigations have used the tangent-modulus approach, which neglects the effect of initial lateral deflection and/or initial bow, and their contributions are listed in Table 8.2A. It should be noted that most of the studies of biaxially loaded beam-columns discussed in Section 8.6 are capable of treating the laterally unbraced member subject to uniaxial bending as a special case.

Several series of tests have been performed on laterally unbraced

Table 8.2A. Theoretical Solutions for the Lateral-Torsional Buckling Strength of Wide-Flange Beam-Columns

Reference	Approach	Residual Stress	Load Cases	Comparison with Eq. 8.14	Comments	Results for Direct Use
Vinnakota (1977)	Maximum strength	Pattern a	1 ($\beta = 1$)	No		Yes
Fukumoto and Galambos (1966)	Tangent modulus	Pattern a	1 ($\beta = 1, 0$)	No		Yes
Galambos et al. (1965)	Tangent modulus	Pattern a	1 ($\beta = 1, 0$)	No		Yes
Miranda and Ojalvo (1965)	Tangent modulus	None	1 ($\beta = 1$)	No	Allows for prior in-plane deflection	Yes
Lim (1970)	Tangent modulus	None	1		Approx. extrapolation to maximum strength; includes minor-axis restraint, prior in-plane deflection, and continuous members	
Lindner (1978)	Maximum strength	Pattern b	1 ($\beta = 1, 0, -0.7$) 4, 2	No		Yes
Abdel-Sayed and Aglan (1973)	Tangent modulus	Pattern a	1 ($\beta = 1$)	No	Allows for prior in-plane deflection, sample results for aluminum section	

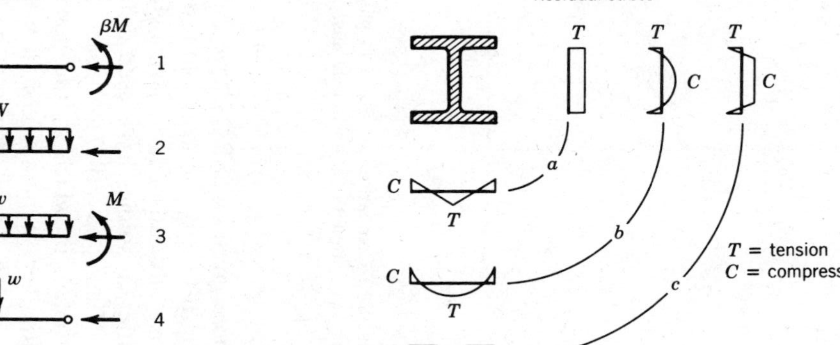

Table 8.2B. Experimental Data for the Lateral-Torsional Buckling Strength of Wide-Flange Beam Columns

Reference	Number of Tests	L/r Range	Load Cases	Comparison with Eq. 8.14	Comments
Von Kuren and Galambos (1964)	26	22–116	M … βM $\beta = 1, 0.5, 0, -1$	Yes	Some in-plane failures
Lindner (1982a)	70	58–169	$\beta = 1, 0.5, 0, -1$	No	High-strength steel
Campus and Massonnet (1956)	91	40–177	$\beta = 1, 0, -1$	Yes	See Nethercot (1983) for comparison
Djalaly (1971)	8	62–82	$\beta = 1, 0.5$	No	
Chagneau (1973)	8	60–80	$\beta = 1, 0.5$	No	
Hill (1951a, b)	59	28–183	$\beta = 1$	No	Aluminum
Gent and Sen (1977)	8	43–74	$\beta = 0.15$	No	Tests designed to investigate rotation capacity; some in-plane failures

Table 8.2C. Design-Oriented References for the Lateral-Torsional Buckling Strength of Wide-Flange Beam-Columns

Reference	Direct Extension of Refinement of Eq. 8.14	Alternative to Eq. 8.14	Comments
Galambos (1981)	Yes	Yes	Discussion of Eq. 8.14 against more recent alternatives
Campus and Massonnet (1956)	Yes	—	Original basis for Eq. 8.14, derivation of comparison with test data
Mazzolani and Frey (1983)	Yes	Yes	Comparison with test data for aluminum beam-columns
Galambos et al. (1965)	—	Yes	Considers importance of in-plane deflections, alternative design approach
Djalaly (1973)	Yes	Yes	Eq. 8.14 and several alternatives compared with test data
Cuk and Trahair (1981)	Yes	—	Suggests that C_m depends on axial load level; basis is elastic critical loads

beam-columns and these are reviewed in Table 8.2B. In addition to these investigations designed to study this problem, some of the test series reported in Table 8.1B contain instances of failure in a lateral-torsional mode, usually because inadequate bracing was provided.

When values of C_m of less than 1.0 are used in Eq. 8.14 it is, of course, also necessary to check cross-sectional strength using Eq. 8.6. References that expand on the use of Eq. 8.14 or present alternative design procedures are listed in Table 8.2C.

It should be remembered that lateral-torsional buckling is normally associated with torsionally weak sections bent in their stiffer principal plane. Therefore, certain classes of section (e.g., tubes) are not susceptible to this mode of failure. In such cases bending about both principal axis may be treated using the methods of Section 8.3.

8.5 EQUIVALENT UNIFORM MOMENT FACTOR

When a beam-column is loaded by unequal end moments M_0 and βM_0, where $-1.0 \leq \beta \leq 1.0$, it is usually overconservative to use the value M_0 directly in the design formulas of Eq. 8.4 or 8.14. This is particularly true as β approaches -1.0 and the member tends to be bent in double curvature, since the design formulas are based on the case of uniform primary moment (i.e., $\beta = +1.0$). However, studies for both the in-plane case (Austin, 1961; Chen and Atsuta, 1977) and the lateral-torsional buckling problem (Horne 1956; Salvadori, 1956; Campus and Massonnet, 1956) have shown that a simple and reasonably accurate correction results if M_0 is replaced by a reduced value $C_m M_0$, where

$$C_m = 0.6 + 0.4\beta \nless 0.4 \tag{8.16}$$

Equation 8.16 is a simplified version of the several different formulas that have been proposed by various authors; for design purposes it is particularly convenient to use the same expression for both problem types as well as to ignore the relatively small effect on C_m of member slenderness.

For beam-columns subjected to other forms of loading (e.g., transverse loads between supports) extensive numerical results are available (Chen and Atsuta, 1977; Lu, 1968; Ballio and Campanini, 1981) for the case where failure occurs by excessive bending. A simple approximation to these consists of replacing M_0 in Eq. 8.4 by the maximum value taken from a straight-line envelope of the actual (primary) moment diagram. More accurate values may, however, be obtained if the C_m factors given in Table 8.3 are used. Since these permit the use of moments below the maximum it becomes necessary to increase their values as the axial load

Table 8.3. C_m Factors for Use with Eq. 8.4 for In-Plane Strength of Beam-Columns

Load Case	Formula for C_m	Limits
M βM	$0.6 - 0.4\beta$	$P/P_u \geq 0.4$
	$1 - \dfrac{1-(0.6-0.4\beta)}{0.4}\dfrac{P}{P_u}$	$P/P_u < 0.4$
	$0.5 + 0.7M_m/M_0$	$P/P_e \geq 0.5$
	$1 - 2[1-(0.5+0.7M_m/M_0)]\dfrac{P}{P_e}$	$P/P_e < 0.5$
etc.	where M_0 = maximum moment in span, M_m = average moment	

decreases and the problem approaches a beam. Similar conclusions were reached by Djalaly (1975), who provides C_m factors for nine different load cases.

Similar refinements to Eq. 8.16 for the laterally unbraced case have been suggested by Cuk and Trahair (1981) based on studies of elastic buckling. However, it has yet to be demonstrated that such procedures lead to similar improvements in the prediction of ultimate strength.

8.6 BIAXIAL BENDING

Both of the cases discussed in Sections 8.3 and 8.4 are special cases of the more general beam-column problem illustrated in Fig. 8.1*c*. Contributions in this area prior to 1975 are reviewed fully in the third edition of the guide, while more recent general reviews are available (Massonnet 1976; Chen 1977a, 1981; Chen and Atsuta, 1977). For design purposes it is convenient to separate consideration of the problem into two phases: (1) short columns, and (2) intermediate and slender columns, corresponding to the outer and inner failure surfaces of Fig. 8.2, respectively. Phase (1) will therefore concern itself with the strength of the cross section under combined axial load and major- and minor-axis moments. This is relevant for stocky columns, particularly those subjected to nonuniform moments, for which failure is likely to be governed by the local strength of the most heavily stressed cross section, assuming the individual component plates to be such that local buckling does not occur. More slender columns are likely to fail in an overall fashion, which in the present case will involve the interaction of column buckling and beam biaxial bending.

8.6.1 Strength of Short Columns

Provided that the cross section meets the requirements for compactness necessary to ensure against local buckling, its local strength will be governed by the development of full plasticity. For the uniaxial case the location of the neutral axis may be determined from first principles (Chen and Atsuta, 1977); alternatively, Eqs. 8.6 and 8.7 for wide-flange sections or the formulas of Table 8.4 for other common shapes may be employed. No comparable simple expressions exist for the biaxial case. Moreover, determination of the location and angular disposition of the neutral axis, which satisfies the three conditions of equilibrium for applied load P and moments M_x and M_y, now involves so many possible cases as to be prohibitive for anything other than research purposes. Santathadaporn and Chen (1970) have derived interaction curves relating M_x to M_y for a series of values of P for wide-flange sections of varying width-to-depth

Table 8.4. Full Plastic Interaction for Uniaxial Bending

Cross Section		Interaction of P and M_0	
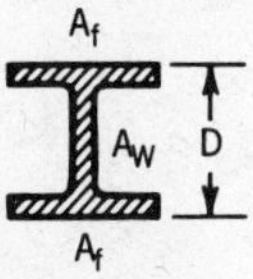 I-section (A_f, A_W, D)	$\sigma_y A_f D(1 - A_w/4A_f)$	$(P/P_y)^{1.2} + (M_0/M_p) = 1$	
 H-section (B)	$\sigma_y A_f B/2$	$(P/P_y)^2 + (M_0/M_p)^{1.2} = 1$	
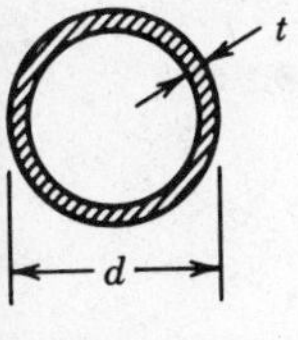 Circular tube (t, d)	$\frac{1}{6}\sigma_y d^3\left[1 - \left(1 - \frac{2t}{d}\right)^3\right]$	$(P/P_y)^{1.7} + (M_0/M_p) = 1$	
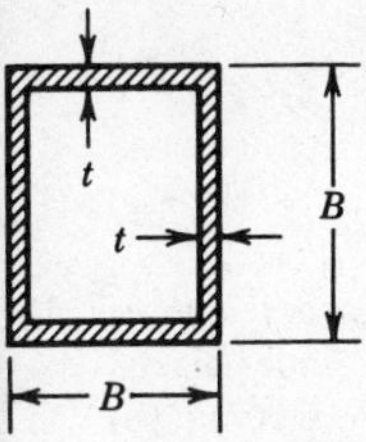 Rectangular tube (t, B, B)	$\sigma_y t D^2(B/D + 1/2)$	$D/B \geq 3$	$(P/P_y)^2 + M_0/M_p = 1$
		$1 \leq D/B < 3$	$(P/P_y)^{1.5} + M_0/M_p = 1$
		$D/B < 1$	$(P/P_y)^{1.2} + M_0/M_p = 1$

ratio and varying flange thickness-to-depth ratio. For design purposes the average set of curves given in Fig. 8.13 will normally prove adequate. If a single design equation is required, this can be provided only at the expense of accuracy, leading to significant underpredictions of the available strength in most cases. The most suitable linear expression is (Pillai, 1974)

$$\frac{P}{P_y} + 0.85\left(\frac{M_x}{M_{px}}\right) + 0.6\left(\frac{M_y}{M_{py}}\right) \leq 1 \tag{8.17}$$

in which the different coefficients for the two moment terms are a reflection of the different shapes of the uniaxial interactions as shown in Figs. 8.5 and 8.6. More accurate predictions may be obtained using (Chen and Atsuta, 1977; Tebedge, 1974)

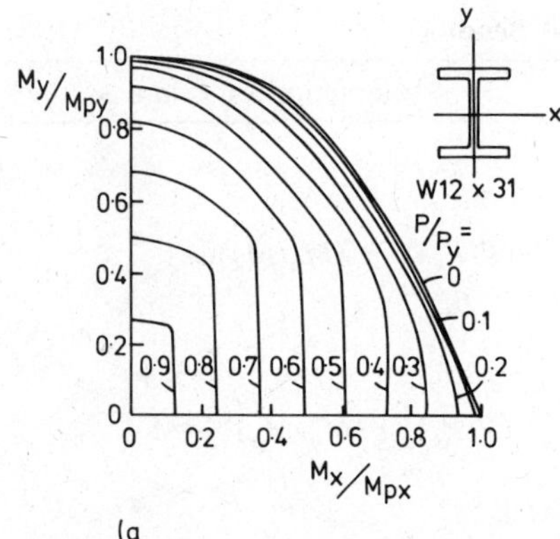

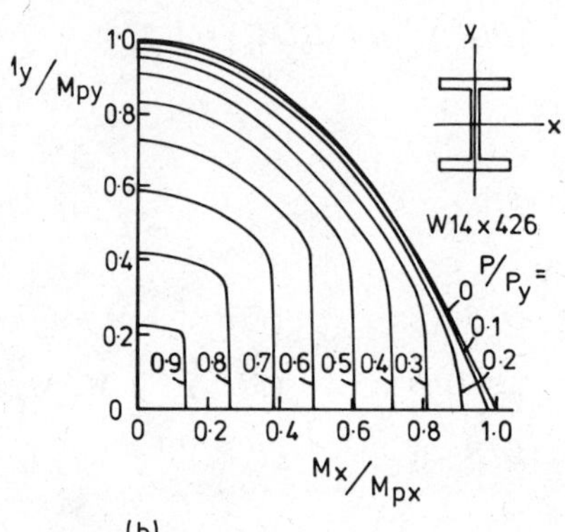

Fig. 8.13 Interaction curves for strength of short beam-columns under biaxial loading: (a) light W section; (b) heavy W section.

$$\left(\frac{M_x}{M_{px}}\right)^{\alpha} + \left(\frac{M_y}{M_{py}}\right)^{\alpha} \leq 1 \tag{8.18}$$

in which M_{pcx} and M_{pcy} are the moment capacities about the respective axes, reduced for the presence of axial load; they may be obtained from Eqs. 8.6 and 8.7. The value of the exponent α is given by

$$\alpha = 1.6 - \frac{P/P_y}{2 \ln (P/P_y)} \tag{8.19}$$

Figure 8.14, which compares Eqs. 8.17 and 8.18 with the average "exact" results of Santathadaporn (1970), shows how the interaction between the two moment terms becomes increasingly convex as the axial load increases. Equation 8.18 will therefore produce significant economies over Eq. 8.17 in situations where columns carrying very large axial loads must also withstand small moments.

Interaction curves for sections other than wide-flange shapes have been obtained by Chen and Atsuta (1974, 1977) using a superposition technique. Results for a number of different structural sections, including circular tubes and unsymmetrical shapes such as angles, are available in

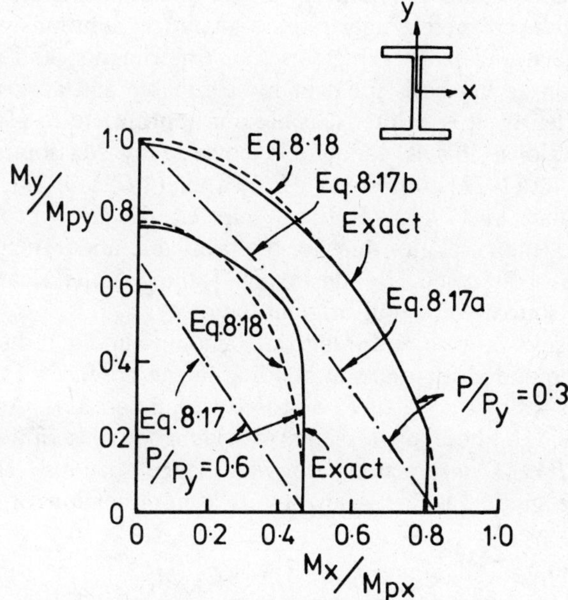

Fig. 8.14 Comparison of Eqs. 8.17 and 8.18 with numerical data of Chen and Atsuta for strength of biaxially loaded beam-columns.

Chen (1974, 1977). These show that the degree of symmetry present in the interaction surface parallels the degree of symmetry of the section. Chen and Atsuta (1977) also contains a procedure (suitable for programming) for developing these surfaces. For standard angles a full set of interaction curves is available (Bez, 1983).

All of the foregoing discussion has assumed that yielding occurs only as a result of the direct stresses produced by compression and bending. However, the presence of any shear stresses due to St.-Venant torsion will have the effect of reducing cross-sectional capacity. This may be allowed for in a particularly simple manner (Chen and Atsuta, 1972a, 1977; Morris and Fenves, 1969) if the material is assumed to follow the von Mises yield criterion, since all coordinates on the yield surface will be reduced by a factor.

$$\sqrt{1-t^2} \tag{8.20}$$

in which $t = \tau/\tau_y$ the ratio of applied shear stress to the shear shield stress of the material.

8.6.2 Strength of Intermediate and Slender Columns

Slender columns in biaxial bending constitute the most formidable example of the beam-column problem. Over the years, successive investigations have produced increasingly refined analytical solutions, have conducted expensive and painstaking series of experiments, and have used this information as the basis for ever more reliable and accurate design procedures. Clearly, it is neither feasible nor appropriate to review all of these contributions; this is adequately covered by Massonnet (1976), Chen and Atsuta (1977), Johnston (1976), Chen (1977a, 1981), and Chen and Santathadaporn (1968). Rather, selected design approaches are presented and their relationship to the available theoretical and experimental data discussed. By means of Table 8.5 the reader is also guided to the sources of useful original data.

Since a design procedure for biaxial bending should reduce to that already recommended for uniaxial bending in the absence of one of the moment terms, a logical starting point is Eqs. 8.5 and 8.14. As shown in Figs. 8.5 and 8.9, these expressions provide good descriptions of the $P - M_x$ and $P - M_y$ interactions for wide-flange sections. The second edition of the guide suggested an empirical combination of these two expressions to give

$$\frac{P}{P_u} + \frac{C_{mx}M_x}{M_{ux}(1 - P/P_{ex})} + \frac{C_{my}M_y}{M_{uy}(1 - P/P_{ey})} \leq 1 \tag{8.21}$$

In working stress format, this is the basis for the present AISC and

CSA design approaches. An improvement for the particular case of hollow circular and square box sections suggested by Pillai (1970) is

$$\frac{P}{P_u} + C\left[\frac{C_{mx}M_x}{M_{ux}(1 - P/P_{ex})} + \frac{C_{my}M_y}{M_{uy}(1 - P/P_{ey})}\right] \leq 1 \tag{8.22}$$

in which

$$C = \frac{(e_x^2 + e_y^2)^{1/2}}{e_x + e_y}$$

For such sections both M_{ux} and M_{uy} may, of course, be taken as the full plastic moment capacities.

An alternative way of writing Eq. 8.21 is to use algebraic transposition (Chen, 1977b) to obtain

$$\frac{C_{mx}M_x}{M_{ucx}} + \frac{C_{my}M_y}{M_{ucy}} \leq 1 \tag{8.23}$$

in which M_{ucx} and M_{ucy} are the uniaxial ultimate moment capacities according to Eqs. 8.14 and 8.5, respectively (i.e., the value of M for a given P). This implies that the interaction between the moment terms is linear, whereas both test data and theoretical solutions for biaxially loaded columns indicate a convex interaction, as illustrated in Fig. 8.15.

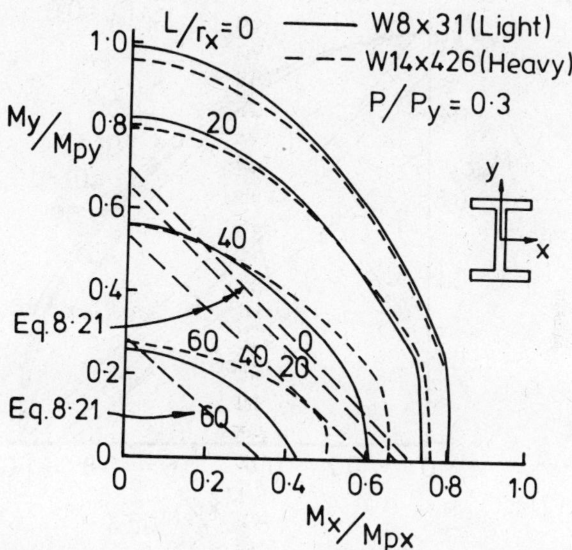

Fig. 8.15 Comparison of Eq. 8.21 with numerical data of Chen and Atsuta for stability of biaxially loaded beam-columns.

As a development of his work on short columns, Tebedge (1974) proposed that a nonlinear expression similar to Eq. 8.18 also be used for slender columns. Thus Eq. 8.23 becomes

$$\left(\frac{C_{mx}M_x}{M_{ucx}}\right)^{\eta} + \left(\frac{C_{my}M_y}{M_{ucy}}\right)^{\eta} \leq 1.0 \tag{8.24}$$

Based largely on numerical studies covering I-sections with $B/D = 0.3$ (Ross, 1976) as well as H sections with $B/D = 1.0$ (Tebedge, 1974) Chen recommends that η be determined from

$$\begin{aligned} \eta &= 0.4 + \frac{P}{P_y} + \frac{B}{D} \qquad &\text{for } B/D \geq 0.3 \\ \eta &= 1.0 \qquad &\text{for } B/D < 0.3 \end{aligned} \tag{8.25}$$

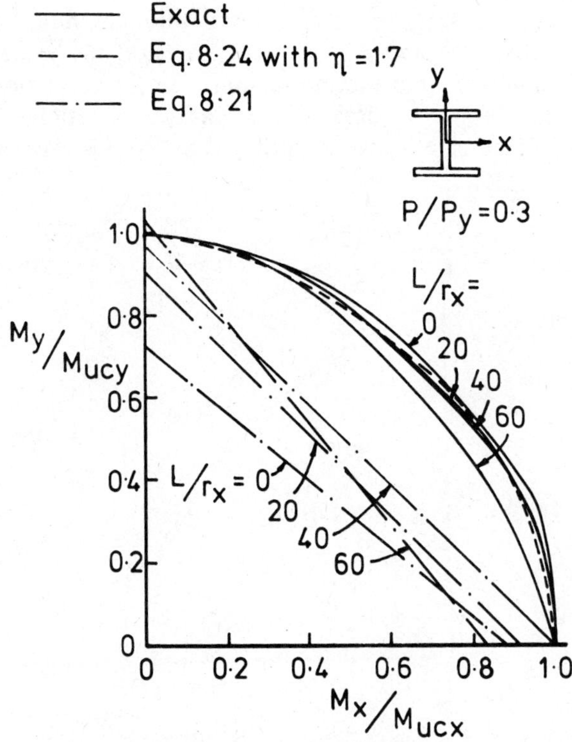

Fig. 8.16 Comparison of Eqs. 8.21 and 8.24 with numerical data of Chen and Atsuta for stability of biaxially loaded beam-columns.

Table 8.5A. Theoretical Solutions for the Biaxial Bending Strength of Beam-Columns[a]

Reference	Cross Section	Residual Stress	Slender Column	Load Cases	Results for Direct Use	Comments
Djalaly (1975)	W shape	Pattern b	Yes	M … M (end moments)	Yes	Tabulated results for one section
Virdi (1981)	Square and circular tubes	None	Yes	Any loading	No	Applicable to cases where torsional effects can be neglected
Lindner (1972, 1982b)	W shape	Pattern b	Yes	M … βM; $\beta = 1, 0$	Yes	Tabulated results for one section; method explained in Lindner (1972); comparison with Eqs. 8.18, 8.21, 8.24
Hancock (1977)	W shape	Pattern a	No		No	Cross section analysis only
Razzaq and McVinnie (1982)	Square tube	None	Yes	(axial load with end moment)	No	Column deflection curves
Vinnakota (1977)	W shape	Pattern a	Yes	Any loading	Yes	Detailed study of initial deflections
Rajasekaran (1977)	W shape	None	Yes	Any loading	No	General finite-element approach
Opperman (1983)	T, L, I, C shapes	None	Yes	Any loading	No	General finite-element approach
Ramm and Osterrieder (1983)	Any open section	None	Yes	Any loading	No	General finite-element approach

[a] This topic is discussed fully by Chen and Atsuta (1977); papers incorporated in the text have not been included in the list above.

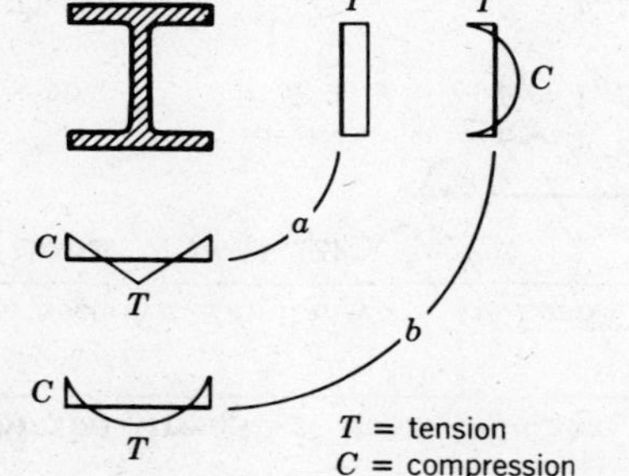

Table 8.5B. Experimental Data for the Biaxial Strength of Beam-Columns

Reference	Number of Tests	Cross Section	L/r Range	Load Cases	Comments
Birnstiel (1968)	12	W shapes	63–121	M … βM; $\beta_x = 1$, $\beta_y = 1$	Rolled, welded, and annealed sections
Chubkin (1959)	43	W shapes	50–150	$\beta_x = 1$, $\beta_y = 1$	Residual stress data not available
Klöppel and Winkelmann (1962)	69	W shapes	57–121	$\beta_x = 1$, $\beta_y = 1$	Variable residual stresses due to method of specimen manufacture
Anslijn (1983)	78	W shapes	40–96	Various β_x, β_y	
Matthey (1983)	15	W shapes		Various β_x, β_y	
Marshall and Ellis (1970)	29	Square tubes	27–135	Various β_x, β_y	Small-scale tests

Table 8.5C. Design-Oriented References for the Biaxial Bending Strength of Beam-Columns

Reference	Direct Extension or Refinement of Eqs.:				Alternative to Eqs.:				Comments
	8.17	8.18	8.21	8.24	8.17	8.18	8.21	8.24	
Chen and Atsuta (1977)	Yes	Yes	Yes	Yes	—	—	—	—	Full discussion in Chapter 13, including review of all proposals up to 1976
Pillai (1980, 1981)	—	Yes	—	Yes	—	—	—	—	Assessment of Eqs. 8.18 and 8.24 against test data using various component data
Lindner (1983)	—	Yes	Yes	Yes	Yes	Yes	Yes	Yes	Evaluation of various three-term formulas on basis of test data and theoretical results of Lindner (1982b)
Nethercot (1983)	—	Yes	—	Yes	Yes	Yes	Yes	Yes	Review of available interaction formulas with comparisons
Djalaly (1975)	—	Yes	—	Yes	Yes	Yes	Yes	Yes	
Chen and Liu (1983)	—	Yes	Yes	Yes	—	—	—	—	Proposed unification of beam-column and column design; discussion of C_m factor
Chen (1981)	—	Yes	Yes	Yes	Yes	Yes	Yes	Yes	Inclusion of external pressure effects
Maquoi and Rondal (1982)	Yes	Yes	Yes	Yes	Yes	Yes	Yes	Yes	Design-oriented reorganization of ECCS interaction formulas
[illegible]	[illegible]	[illegible]		Yes	Yes	Yes	Yes	Yes	Covers box sections and fabricated tubes,

where B is the flange width and D is the section depth. For square box sections (Chen, 1977b) η should be taken as

$$\eta = 1.3 + \frac{1000}{(L/r)^2}\frac{P}{P_y} \geq 1.4 \tag{8.26}$$

Use of Eq. 8.24 with η values greater than unity recognizes the convexity of the moment interaction as illustrated in Fig. 8.16.

A quantitative assessment of the accuracy of Eqs. 8.21 and 8.24, based on their ability to predict the strengths of the specimens in four series of tests (Birnstiel, 1968; Chubkin, 1959; Klöppel and Winkelmann, 1962; Anslijn, 1983) on biaxially loaded I-sections, has been made by Pillai (1980, 1981). In conducting this comparison, advantage was taken of the fact that the use of Eq. 8.24 is not tied to any particular method of determining M_{ucx} and M_{ucy}; the concept of combining uniaxial moment capacities can be used with any reasonable moment expression. For the 81 tests of Anslijn (1983) Pillai found that when used with Eqs. 3.15, 8.15, and 8.2 for P_u, M_{ucx}, and M_{ucy}, Eq. 8.24 was an almost perfect predictor in terms of the mean value for P_{test}/P_{calc} of 1.050. However, with a standard deviation of 0.101, this meant that 31% of the test results were overpredicted.

Theoretical procedures for assessing beam-column strength under biaxial loading are rather complex, so that even in the elastic range (Culver, 1966a, b) numerical solutions are required. Thus the results presented in any of the references listed in Table 8.5A will normally cover only a limited aspect of the problem. However, several series of large-scale tests have been conducted (see Table 8.5B), and these have proved invaluable in the evaluation of the various design approaches listed in Table 8.5C.

8.7 NEW DESIGN APPROACHES

The proposed AISC load and resistance factor design (LRFD, 1986) uses the following pair of interaction equations for the design of doubly or singly symmetric members:

$$\frac{P}{2P_u} + \frac{M_x^*}{M_{ux}} + \frac{M_y^*}{M_{uy}} \leq 1.0 \qquad \text{for } P/P_u < 0.2 \tag{8.27}$$

$$\frac{P}{P_u} + \frac{8}{9}\left(\frac{M_x^*}{M_{ux}} + \frac{M_y^*}{M_{uy}}\right) \leq 1.0 \qquad \text{for } P/P_u \geq 0.2 \tag{8.28}$$

where

$M^* = B_1 M_{NT} + B_2 M_{LT}$
M_{NT} = maximum end moments assuming no sidesway
M_{LT} = maximum end moment produced by forces causing lateral translation

$$B_1 = \frac{C_m}{1 - P/P_e} \geq 1.0$$

$$B_2 = \frac{1}{1 - \Delta_{0H} \sum P/L \sum H}$$

Δ_{0H} = lateral deflection (first-order analysis)
ΣP = story gravity load
ΣH = story shear

By keeping the value of B_1 above unity the need for a separate strength check in cases where $C_m < 1.0$ is avoided (Cheong-Siat-Moy and Downs, 1980). Determination of M_{ux} incorporates the allowance for lateral-torsional buckling given in Chapter 5.

In Europe the ECCS originally proposed (ECCS, 1976) a pair of formulas that utilized the concept of a parameter of imperfection e^* to link beam-column design to the philosophy of the European Column Curves. More recently, in association with the drafting of the European Code of Practice for structural steelwork (EEC, 1983) a number of alternative interaction formulas have been investigated (Lindner, 1982b; 1983; Nethercot, 1982, 1983). For in-plane failure without lateral-torsional buckling (case I) a rearrangement of the original e^* formula has been proposed as

$$\frac{P}{P_u} + \frac{C_m M}{(1 - P/P_u)\bar{\lambda}^2 M_u} \leq 1.0; \ \bar{\lambda} = \sqrt{P_y/P_e} \tag{8.29}$$

The values of C_m and M_u for use in Eq. 8.29 are similar but not identical to those used in Eq. 8.4. For example, M_u may be taken as $1.1M_p$ if $P/P_u > 0.1$ and C_m may be set as 1.1 for uniform single curvature bending provided that C_m always exceeds the value $(1 - P/P_e)$. A simpler alternative (Lindner, 1983; Roik, 1983) to Eq. 8.29 is also permitted as

$$\frac{P}{P_u} + \frac{C_m M}{M_u} \leq 1 - \Delta n$$

$$\left.\begin{aligned} \Delta n &= 0.25\left(\frac{P_u}{P_y}\right)^2 \bar{\lambda}^2 && \text{for } \frac{P}{P_y} > 0.1 \quad \text{or} \quad \frac{C_m M}{M_u} \geq 0.1 \\ \Delta n &= 0 && \text{for } \frac{P}{P_y} < 0.1 \quad \text{and} \quad \frac{C_m M}{M_u} < 0.1 \end{aligned}\right\} \tag{8.30}$$

For lateral-torsional buckling, a formula originally proposed by Lindner (1978) is used:

$$\frac{P}{P_u} + \left(\frac{M_x}{M_{ux}}\right)^{\beta_M} \leq 1.0 \tag{8.31}$$

in which $\beta_M = 1.8 - 0.7\beta$ for moment gradient. Equations 8.29 and 8.31 are combined in the style of Eq. 8.24 for the general case of biaxial bending, but η is specified as unity. In all cases P_u is obtained from the appropriate European Column Curve, use being made of the formula representation derived by Maquoi and Rondal (1982).

Equation 8.31 has also been proposed for use in West Germany (Vogel and Lindner, 1981) with the biaxial case being handled by a three-term expression; two basic formats in a number of different variants have recently been studied (Lindner, 1982a; Nethercot, 1983). In the United Kingdom (Nethercot, 1983) a different simplification of Eq. 8.29 is used together with Eq. 8.31 but taking β_M as unity. Biaxial bending is treated via Eq. 8.24 but with η generally taken as unity.

8.8 CYCLIC LOADING OF BEAM-COLUMNS

Determination of the response of beam-columns to wave or earthquake motions is an important factor in assessing the strength of offshore structures. Therefore, considerable effort has recently been directed at developing certain of the techniques for assessing static strength listed in Table 8.1A into efficient methods for the cyclic inelastic analysis of steel tubular beam-columns (Chen, 1981; Han and Chen, 1983a, b). Computer programs based on these techniques have been verified against the results of large-scale tests and are now being used to generate data on cyclic response (Han and Chen, 1983a, b). For a full account of this topic, the reader is referred to the monograph by Chen and Han (1983).

8.9 ECCENTRICALLY LOADED ANGLE STRUTS

Hot-rolled and cold-formed angles are used in many structural applications (e.g. diagonal braces in frames, members in space and plane trusses, main and bracing elements in communications and power transmission towers, etc.). Angle struts may be single elements (Fig. 8.17*a*) or double elements connected by fillers (Fig. 8.17*b*), gussets or stitch welds to form a tee (Fig. 8.17*c*) or a cruciform (Fig. 8.17*d*). For a full review of available research data, the reader is referred to the recent survey by Kennedy and Madugula (1982).

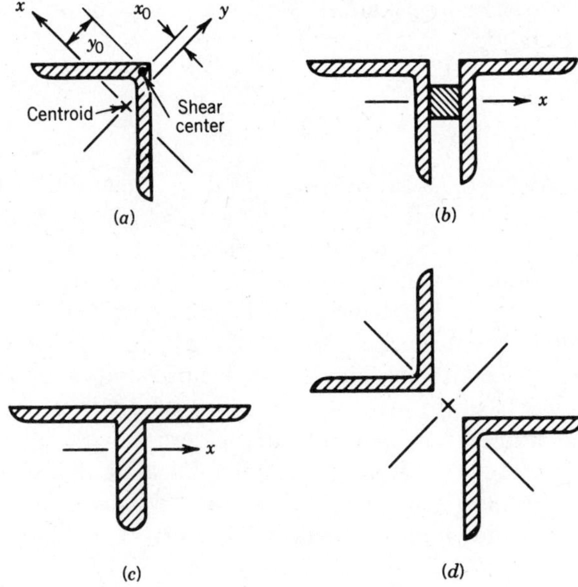

Fig. 8.17 Angle-columns.

8.9.1 Single-Angle Struts

In almost any application the end connection of single-angle compression members causes both eccentric loading and restraint. The elastic behavior is a special case of the stability of thin-walled members (see, e.g. Vlasov, 1961, Chap. 6; Timoshenko and Gere, 1961, Chap. 5; Bleich, 1952, Chap. 3). Since the shear center is located at the intersection of the two angle legs, there is practically no warping rigidity, and the warping constant C_w can be assumed to be zero.

The elastic bifurcation buckling load of an unequal-leg column which is loaded through the centroidal axis is the lowest root of the following cubic equation (Galambos, 1968).

$$(P_{cr} - P_x)(P_{cr} - P_y)(P_{cr} - P_\phi) - P_{cr}^2(P_{cr} - P_y)\frac{x_0^2}{\bar{r}_0^2} - P_{cr}^2(P_{cr} - P_x)\frac{y_0^2}{\bar{r}_0^2} = 0 \tag{8.32}$$

where

$$P_x = \frac{\pi^2 EI_x}{(K_x/L)^2} \qquad P_y = \frac{\pi^2 EI_y}{(K_y/L)^2} \qquad P_\phi = \frac{GJ}{\bar{r}_0^2} \tag{8.33}$$

$$\bar{r}_0^2 = x_0^2 + y_0^2 + \frac{I_x + I_y}{A} \tag{8.34}$$

The x, y subscripts refer to the principal-axis system (Fig. 8.17a) and x_0 and y_0 are the coordinates of the shear center; K_x and K_y are the effective length factors of flexural buckling. For equal leg angles ($x_0 = 0$ in Fig. 8.17a), Eq. 8.32 gives two critical loads, the lower load being the governing one. For standard hot-rolled angles, an equivalent slenderness, corresponding to the lowest root of Eq. 8.32 may be obtained from (Kitipornchai, 1983)

$$\left(\frac{L}{r}\right)_{eq} = \left[\left(\frac{L}{r_v}\right)^3 - 8(\alpha_1 - 0.5)\alpha_2^2\left(\frac{L}{r_v}\right) + 76\alpha_2^3\right]^{1/3}$$

for unequal-leg angles

$$\left(\frac{L}{r}\right)_{eq} = 0.05\left(\frac{L}{r_v}\right) + 4.8\alpha_2 \nless \left(\frac{L}{r_v}\right) \tag{8.35}$$

for equal-leg angles

where

r_v = radius of gyration about minor principal axis
α_1 = ratio of leg widths ($\alpha_1 \leq 1.0$)
α_2 = width-to-thickness ratio of the longer leg

Inelastic buckling strength can be approximated by using, for example, the method in the AISI Specification for Cold-Formed Steel Structural Members (AISI, 1986):

$$\text{For } P_{cr}/A = \sigma_e > 0.5\sigma_y \qquad P_{cr} = A\sigma_y(1 - \sigma_y/4\sigma_e) \tag{8.36}$$

Tests performed by Kennedy and Murty, (1972) on hot-rolled angles and tee shapes showed that this method provided a satisfactory prediction of strength. Similar confirmation has been provided by Marsh (1969), who found that his tests on slender, equal-leg aluminum angles with single- and double-bolt connections were adequately predicted by

$$P_{cr} = 0.9\pi^2 E/(L/r)_{eq}^2 \tag{8.37}$$

where

$$(L/r)_{eq} = \sqrt{\left(\frac{5b}{t}\right)^2 + \left(\frac{KL}{r_y}\right)^2}$$

Upper limits of $0.50P_y$ (single bolts) and $0.67P_y$ (double bolts) were placed on P_{cr}. Equations 8.32 and 8.35 apply also to tee and cruciform columns. Although not included as part of the original study for either the SSRC or ECCS multiple column curve concept, most codified adaptations of these have allocated angles to one of the lower curves (e.g., ECCS Curve C).

When the axial compressive load is applied with eccentricities e_x and e_y (Fig. 8.18) the problem is no longer a buckling problem because deformation will occur with any axial load. It then becomes necessary to determine the resulting stresses which may become amplified (or, perhaps, even reduced, as was found for single-angle beams by Thomas (1973), due to the deflection and twist of the section). The problem of eccentrically loaded and end-restrained single angle struts, for the special case of end restraint provided by tee stubs (Fig. 8.19) which approximate the double-angle chord of a truss, was examined for the elastic case by Trahair (1969) for the limit state $\sigma_{\max} = \sigma_y$, and for the inelastic case by Usami and Galambos (1971). Trahair compared his results with tests performed by Foehl (1948), and Usami and Galambos (1971) reported comparisons with a series of tests performed by himself, and with Foehl's tests. In both cases, good correlation was achieved between test and prediction. In a series of unpublished papers (Usami, 1970; Leigh, 1972), tests on the strength of single-angle eccentric webs in trusses were reported, and the applicability of the interaction equation

$$\frac{P}{P_y} + \frac{Pe_x}{M_{xY}(1 - P/P_x)} + \frac{Pe_y}{M_{yY}(1 - P/P_y)} = 1 \tag{8.38}$$

was examined against test results and theoretical predictions. It was found that Eq. 8.38 gave satisfactory if somewhat conservative prediction of the actual capacity, provided that the end eccentricities were reduced to account for end restraint. P_x and P_y in Eq. 8.38 are defined by Eqs. 8.33, e_x and e_y are the eccentricities shown in Fig. 8.19, and M_{xY} and M_{yY} are the moments required to produce compressive yield in the extreme fiber when $P = 0$.

Equation 8.38 has also been recommended for the design of single angle web members of trusses whose ends are connected to the chords by

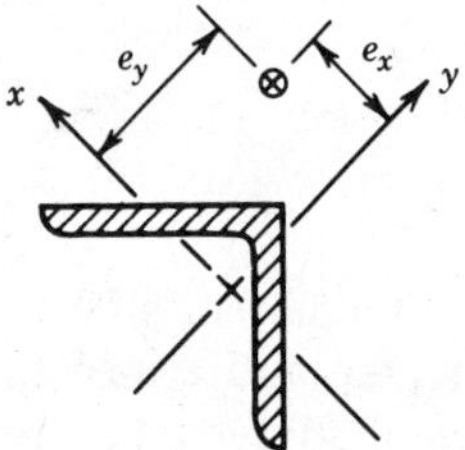

Fig. 8.18 Eccentrically loaded angle column.

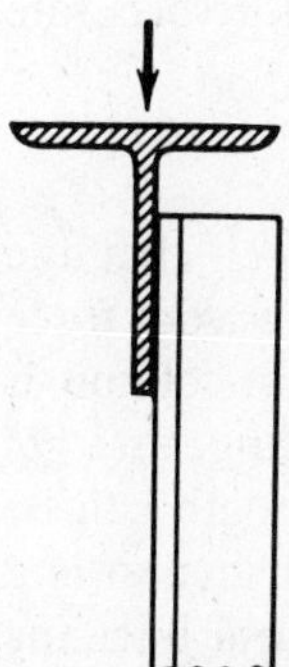

Fig. 8.19 Angle column loaded through tee stub.

welding or by using multiple bolted connections (Woolcock and Kitipoinchai, 1980, 1986). Once again, care is needed in deciding upon appropriate end moment values if these are accurately to reflect the effects of load eccentricity. A modification of Eq. 8.38 containing an additional torsional term has been suggested by Marsh (1972), who found that it gave good predictions of his tests on equal-leg angles under biaxially eccentric loading. While further clarification of the application of Eq. 8.38 would be useful, the available evidence does suggest that it constitutes a reasonable basis for the design of angles required to function as beam-columns.

The other alternative is quite different, and its origin is with the designers of electric transmission towers. This method applies to single-angle struts connected at their ends on one angle leg by bolts or by welding. The axial load is thus applied eccentrically, and the ends are restrained. The design solution is to modify the slenderness ratio for use in an appropriate column formula so as to account empirically for both the end eccentricity and the end restraints. Examples of this approach are given by both the ASCE (1967) Task Committee on Tower Design and the ECCS (1976) recommendations.

Support for the possible extension of this type of approach to a wider range of structures has recently been provided by Haaijer (1981) on the basis of detailed finite-element calculation for angles loaded (and restrained) through one leg.

In summary, present information on eccentrically loaded single-angle struts applicable for design office use indicates the utilization of the interaction equation (modified by the Q-factor reduction scheme of the AISI or the AISC Specification for local buckling, where this is appropriate) in general, and the use of an empirical effective-length factor in particular for triangulated towers. Theory, which has been validated by experiment, is available to be used for the development of more accurate design information.

8.9.2 Double-Angle Struts

Double-angle and tee-shaped struts are generally loaded such that the transverse load, the end moments, and the eccentricity of the axial force are in the plane of symmetry. The capacity of such a beam-column is determined by the limit states of (1) flexural-torsional buckling, and (2) in-plane strength. The elastic flexural-torsional buckling strength can be examined by the Theory of thin-walled open sections (Timoshenko and Gere, 1961), and the design method for both torsional flexural buckling and in-plane strength is treated satisfactorily and accurately in the AISI *Specification for the Design of Cold-Formed Steel Structural Members* (AISI, 1986).

In many applications (e.g., double angles with separators, double angles or tees with continuous lateral support) only in-plane strength needs to be considered in design, and the SSRC interaction equation is generally used in design specifications. Many angles are compact enough so that beam-column failure is due to in-plane inelastic instability. When bending is such that the flange is in compression, there is a considerable increase in the ultimate capacity over the capacity at first yield (the shape factor is approximately 1.8). Analytical and experimental studies on open-web steel joists with double-angle chords were reported by McDonald (1966), and Faravelli and Zanon (1980) published interaction curves for double-angle beam-columns with positive and negative bending.

The special problem of starred angles has been studied by Smith (1983), who has drawn attention to their low torsional buckling strength. Adequate design tools are available for double-angle and tee beam-columns, and analytical tools are also available to develop more refined design tools.

References

Abdel-Sayed, G., and Aglan, A. A. (1973), "Inelastic Lateral Torsional Buckling of Beam Columns," *Publ. IABSE*, No. 33-II, pp. 1–16.

Adams, P. F. (1970), "Segmental Design of Steel Beam-Columns," *Canadian Struct. Eng. Conf.*

Adams, P. F. (1974), *The Design of Steel Beam-Columns*, Canadian Steel Industries Construction Council.

American Iron and Steel Institute (1986), *Specification for the Design of Cold-Formed Steel Structural Members*, American Iron and Steel Institute, Washington, D.C.

Anslijn, R. (1983), "Tests on Steel I Beam-Columns in Mild Steel Subjected to Thrust and Biaxial Bending," *CRIF Rep. MT 157*, Brussels, August.

American Society of Civil Engineers (1971a), "Plastic Design in Steel—A Guide and Commentary," *ASCE Manual of Engineering Practice*, No. 41, 2nd Ed., ASCE, New York.

American Society of Civil Engineers (1971b), "Guide for Design of Steel Transmission Towers," *ASCE Manual of Engineering Practice*, No. 52, ASCE, New York.

Austin, W. F. (1961), "Strength and Design of Metal Beam-Columns," *ASCE J. Struct. Div.*, Vol. 87, No. ST4, pp. 1–34.

Ballio, G., and Campanini, C. (1981), "Equivalent Bending Moments for Beam-Columns," *J. Constr. Steel Res.*, Vol. 1, No. 3.

Ballio, G., Petrini, V., and Urbano, C. (1973), "The Effect of the Loading Process and Imperfections on the Load Bearing Capacity of Beam Columns," *Meccanica*, Vol. 8, No. 1.

Bez, R. (1983), "Diagrams d'interaction de cornières metalliques," *Publ. ICOM 111*, Swiss Federal Institute of Technology, Lausanne, March.

Bijlaard, P. P., Fisher, G. P., and Winter, G. (1955), "Eccentrically Loaded, End-Restrained Columns," *Trans. Am. Soc. Civ. Eng.*, Vol. 120, p. 1070.

Birnstiel, C. (1968), "Experiments on H-Columns under Biaxial Bending," *ASCE J. Struct. Div.*, Vol. 94, No. ST10, pp. 2429–2450.

Bleich, F. (1952), *Buckling Strength of Metal Structures*, McGraw-Hill, New York.

Campus, F., and Massonnet, C. (1956), "Recherches sur le flambement de colonnes en acier A37 a profil en double te sollicitées obliquement," *C.R. Rech. IRSIA*, April.

Chagneau, A. (1973), "la Barre dans la structure," *CTICM Rep.*, January.

Chen, W. F. (1970), "General Solution of Inelastic Beam-Column Problem," *ASCE J. Eng. Mech. Div.*, Vol. 96, No. EM4, pp. 421–442.

Chen, W. F. (1971a), "Approximate Solution of Beam Columns," *ASCE J. Struct. Div.*, Vol. 97, No. ST2, pp. 743–751.

Chen, W. F. (1971b), "Further Studies of an Inelastic Beam-Column Problem," *ASCE J. Struct. Div.*, Vol. 97, No. ST2, pp. 529–544.

Chen, W. F. (1977a), "Theory of Beam-Columns—The State-of-the-Art Review," *Proc. Int. Colloq. Stab. Struct. Under Static Dyn. Loads*, SSRC/ASCE Washington D.C., March.

Chen, W. F. (1977b), "Design of Box Columns under Biaxial Bending," *2nd. Int. Colloq. Stab. Steel Struct., Prelim. Rep.*, Liege.

Chen, W. F. (1981), "Recent Advances in Analysis and Design of Steel Beam-Columns in U.S.A.," *Proc. U.S.-Jpn. Sem. Inelastic Instab. Steel Struct. Struct. Elements*, Tokyo, May.

Chen, W. F. (1982), "Box and Cylindrical Columns under Biaxial Bending," in *Axially Compressed Structures—Stability and Strength* (ed. R. Narayanan), Applied Science Publishers, Barking, Essex, England, Chap. 3.

Chen, W. F., and Atsuta, T. (1972a), "Interaction Equations for Biaxially Loaded Sections," *ASCE J. Struct. Div.*, Vol. 98, No. ST5, pp. 1035–1052.

Chen, W. F., and Atsuta, T. (1972b), "Column-Curvature Curve Method for Analysis of Beam-Columns," *Struct. Eng.*, Vol. 50, No. 6.

Chen, W. F., and Atsuta, T. (1974), "Interation Curves for Steel Sections, under Axial Load and Biaxial Bending," *EIC Eng. J.*, Vol. 17, No. A-3.

Chen, W. F., and Atsuta, T. (1977), *Theory of Beam-Columns*, Vols. 1 and 2, McGraw-Hill, New York.

Chen, W. F., and Cheong-Siat-Moy, F. (1980), *Limit States Design of Steel Beam-Columns*, Solid Mechanics Archives, Vol. 5, Noordhoff International Publishing, Leyden, The Netherlands.

Chen, W. F., and Han, D. J. (1983), *Tubular Members in Offshore Structures*, Pitman, London.

Chen, W. F., and Lui, E. M. (1983), "Design of Beam-Columns in North America," *Proc. 3rd Int. Colloq. Stab. Met. Struct.*, Toronto.

Chen, W. F., and Santathadaporn, S. (1968), "Review of Column Behaviour under Biaxial Loading," *ASCE J. Struct. Div.*, Vol. 94, No. ST12, pp. 2999–3021.

Cheong-Siat-Moy, F. (1974a), "A Method of Analysis of Laterally Loaded Columns," *ASCE J. Struct. Div.*, Vol. 100, No. ST5, Proc. Pap. 10548, pp. 953–970.

Cheong-Siat-Moy, F. (1974b), "General Analysis of Laterally Loaded Beam-Columns," *ASCE J. Struct. Div.*, Vol. 100, No. ST6, Proc. Pap. 10625, pp. 1263–1278.

Cheong-Siat-Moy, F., and Downs, T. (1980), "New Interaction Equation for Steel Column Design," *ASCE J. Struct. Div.*, Vol. 106, No. ST5, pp. 1047–1062.

Chubkin, G. M. (1959), "Experimental Research on Stability of Thin Plate Steel Members with Biaxial Eccentricity," in *Analysis of Spatial Structures*, Vol. 5, Moscow, Paper 6.

Clark, J. W. (1955), "Eccentrically Loaded Aluminium Columns," *Trans. Am. Soc. Civ. Eng.*, Vol. 120, p. 1116.

Cuk, P., and Trahair, N. S. (1981), "Elastic Buckling of Beam-Columns with Unequal End Moments," *Civ. Eng. Trans. Inst. Eng. Aust.*, Vol. CE23, No. 2.

Culver, C. G. (1966a), "Exact Solution of the Biaxial Bending Equations," *ASCE J. Struct. Div.*, Vol. 92, No. ST2, pp. 63–84.

Culver, C. G. (1966b), "Initial Imperfections in Biaxial Bending," *ASCE J. Struct. Div.*, Vol. 92, No. ST3, pp. 119–138.

Djalaly, H. (1971), "la Barre dans la Structure, *CTIOM Rep.*, March.

Djalaly, H. (1973), "Calcul de la résistance ultime des barres au flambement par flexion-torsion," *Constr. Met.*, No. 4.

Djalaly, H. (1975), "Calcul de la résistance ultime des beams comprimees et flechies," *Constr. Met.*, No. 4.

Dwyer, T. J., and Galambos, T. V. (1965), "Plastic Behaviour of Tubular Beam-Columns," *ASCE J. Struct. Div.*, Vol. 91, No. ST4, pp. 125–152.

European Convention for Construtional Steelwork (1976), "Manual on the Stability of Steel Structures," *2nd Colloq. Stab. Steel Struct.*, *Introd. Rep.*, Liege.

European Economic Community (1983), *Eurocode 3—Design of Steel Structures*, EEC.

Faravelli, L., and Zanon, P. (1980), "Influence of the Model on the Behaviour of Trusses at Incipient Collapse," *Costr. Met.*, No. 3 (in English).

Foehl, F. P. (1948), "Direct Method of Designing Single Angle Struts in Welded Trusses," in *Design Book of Welding*, Lincoln Electric Co.

Fukumoto, Y., and Galambos, T. V. (1966), "Inelastic Lateral-Torsional Buckling of Beam-Columns," *ASCE J. Struct. Div.*, Vol. 92, No. ST2, pp. 41–62.

Galambos, T. V. (1964), "Combined Bending and Compression," in *Structural Steel Design*, (ed. L. Tall), Ronald Press, New York.

Galambos, T. V., and Ketter, R. L. (1961), "Columns under Combined Bending and Thrust," *Trans. Am. Soc. Civ. Eng.*, Vol. 126, Part I, pp. 1–25.

Galambos, T. V., and Prasad, J. (1962), "Ultimate Strength Tables for Beam-Columns," *Weld. Res. Counc. Bull.*, No. 78, June.

Galambos, T. V. (1968), *Structural Members and Frames*, Prentice-Hall, Englewood Cliffs, N.J.

Galambos, T. V. (1981), "Beam-Columns," *ASCE Conv.*, New York, May.

Galambos, T. V., Adams, P. F., and Fukumoto, Y. (1965), "Further Studies on the Lateral-Torsional Buckling of Steel Beam-Columns," *Fritz Eng. Lab. Rep. No. 205A.36*, Bethlehem, Pa.

Gent, A. R., and Sen, T. K. (1977), "The Plastic Deformation Capacity of H-Columns at High Axial Loads," *Prelim. Rep., 2nd Int. Colloq. Stab. Steel Struct.*, Liege.

Gilson, S., and Cescotto, S. (1982), "Experimental Research on the Buckling of Alu-Alloy Columns with Unsymmetrical Cross-Section," Laboratorie de Mécanique des Materiaux et Théorie des Structures, University de Liege, Liege, Belgium.

Haaijer, G. (1981), "Eccentric Load Test of Angle Column Simulated with MSC/Nastran Finite-Element Program, *Annu. Meet.*, SSRC, Chicago.

Han, D. J., and Chen, W. F. (1983a), "Behaviour of Portal and Strut Types of Beam-Columns," *Eng. Struct.*, Vol. 5, pp. 15–25.

Han, D. J., and Chen, W. F. (1983b), "Buckling and Cyclic Inelastic Analysis of Steel Tubular Beam-Columns," *Eng. Struct.*, Vol. 5, pp. 119–132.

Hancock, G. J. (1977), "Elastic-Plastic Analysis of Thin-Walled Cross-Sections," *6th Australian Conf. Mech. Struct. and Mater.*, Christchurch, N.Z., August.

Hill, H. N., and Clark, J. W. (1951a), "Lateral Buckling of Eccentrically Loaded I- and H-Section Columns," *Proc. 1st Natl. Cong. Appl. Mech.*, ASME.

Hill, H. N., and Clark, J. W. (1951b), "Lateral Buckling of Eccentrically Loaded I-Section Columns," *Trans. Am. Soc. Civ. Eng.*, Vol. 116, p. 1179.

Hill, H. N., Hartmann, E. C., and Clark, J. W. (1956), "Design of Aluminium Alloy Beam-Columns," *Trans. Am. Soc. Civ. Eng.*, Vol. 121.

Horne, M. R. (1956), "The Flexural-Torsional Buckling of Members of Symmetrical I-Section under Combined Thrust and Unequal Terminal Moments," *Q. J. Mech. Appl. Math.*, No. 4.

Johnston, B. G. (1976), *Guide to Stability Design Criteria for Metal Structures*, (SSRC) 3rd ed., Wiley, New York.

Johnston, B. G., and Cheney, L. (1942), "Steel Columns of Rolled Wide Flange Sections," *Prog. Rep. No. 2*, American Institute of Steel Construction, Chicago, November.

Kanchanalai, T., and Lu, L. W. (1979), "Analysis and Design of Framed Columns under Minor Axis Bending," *Eng. J. Am. Inst. Steel Constr.*, Second Quarter, pp. 29–41.

Kemp, A. R. (1984), "Slenderness Limits Normal to the Plane-of-Bending for Beam-Columns in Plastic Design," *J. Constr. Steel Res.*, Vol. 4, pp. 135–150.

Kennedy, J. B., and Madugula, M. K. S. (1982), "Buckling of Single and Compound Angles," *Axially Compounded Structures: Stability and Strength* (ed. R. Narayanan), Applied Science Publishers, Barking, Essex, England, Chap. 5.

Kennedy, J. B., and Murty, M. K. S. (1972), "Buckling of Steel Angle and Tee Struts," *ASCE J. Struct. Div.*, Vol. 98, No. ST11, pp. 2507–2522.

Ketter, R. L. (1962), "Further Studies of the Strength of Beam-Columns," *Trans. Am. Soc. Civ. Eng.*, Vol. 127, Part II, pp. 224–466.

Ketter, R. L., Beedle, L. S., and Johnson, B. G. (1952), "Column Strength under Combined Bending and Thrust," *Weld. J. Res. Suppl.*, Vol. 31, No. 12.

Ketter, R. L., Kaminsky, E. L., and Beedle, L. S. (1955), "Plastic Deformation of Wide-Flange Beam Columns," *Trans. Am. Soc. Civ. Eng.*, Vol. 120, p. 1028.

Kitipornchai, S. (1983), "Torsional-Flexural Buckling of Angles: A Parametric Study," *J. Constr. Steel Res.*, Vol. 3, No. 3.

Klöppel, K., and Barsch, W. (1973), "Versuche zum Kapitel Stabilitätsfalle der Neufassung von Din 4113," *Aluminium*, Vol. 10.

Klöppel, K., and Winkelmann, E. (1962), "Experimentale und theoretische Untersuchungen über die Traglast von zweiachsig assussermittig gedrückten Stahlstäben," *Stahlbau*, Vol. 31, p. 33.

Lay, M. G. (1974), *Source Book for the Australian Steel Strucures Code AS1250*, Australian Institute of Steel Construction.

Lay, M. G., and Galambos, T. V. (1965), "The Experimental Behaviour of Retrained Columns," *Weld. Res. Counc. Bull.*, No. 110, November.

Leigh J. M., and Galambos, T. V. (1972), "The Design of Compression Webs in Longspan Steel Joists," Research Report No. 21, Washington University Civil Engineering Department, August 1972.

Lim, L. C., and Lu, L. W. (1970), "The Strength and Behavior of Laterally Unsupported Columns," Lehigh University, Fritz Engineering Laboratories, Report No. 329.5, June 1970.

Lindner, J. (1972), "Theoretical Investigations of Columns Under Biaxial Loading," Proceedings International Colloquium on Column Strength, IABSE-CRC-ECCS, Paris, 1972.

Lindner, J., and Wiechart, G. (1978), "Zur Bemessung Gegen Biegedrillknickung," Beitrag in Festschrift Otto Jungbluth,"—60 Jahre. Technische Hochschule, Darmstadt, 1978.

Lindner, J., and Kurth, W. (1982a), "Zum Biegedrillknicken Von Stützen aus," Ste 690, Vol. 51, Der Stahlbau, 1982.

Lindner, J., and Gietzelt, R. (1982b), "Design of Biaxially Loaded Steel Beam-Columns," Technical University of Berlin, Report 2061E, November 1982.

Lindner, J., and Gietzert, R. (1983), "Discussion of Interaction Equations for Members in Compression and Bending," Proceedings Third International Colloquium on Stability of Metal Structures, Paris, November 1983.

Lu, L. W., and Kamalvand, H. (1968), "Ultimate strength of laterally loaded columns," *ASCE J. Struct. Div.*, Vol. 94, No. ST6, pp. 1505–1524.

Lu, L. W., Shen, Z. Y., and Hu, X. R. (1983), *Inelastic Instability Research at Lehigh University Instability and Plastic Collapse of Steel Structures* (ed. L. J. Morris), Granada, London.

Maquoi, R., and Rondal, J. (1982), "Sur la force portant des poutres colonnes," *Ann. Trav. Publics Belg.*, No. 5.

Marsh, C. (1969), "Single Angle Members in Tension and Compression," *ASCE J. Struct. Div.*, Vol. 95, No. ST5, p. 1043.

Marsh, C. (1972), "Lateral Stability of Single Angle Beam-Columns," *CSICC Project 712*, Sir George Williams University, March.

Marshall, P. J., and Ellis, J. S. (1970), "Ultimate Biaxial Capacity of Box Steel Columns," *ASCE J. Struct. Div.*, Vol. 96, No. ST9, pp. 1873–1888.

Mason, R. G., Fisher, G. P., and Winter, G. (1958), "Eccentrically Loaded, Hinged Steel Columns," *ASCE J. Eng. Mech. Div.*, Vol. 84, No. EM4, p. 1792.

Massonnet, C. (1976), "Forty Years of Research on Beam-Columns," *Solid Mech. Arch.*, Vol. 1, No. 2.

Massonnet, C., and Save, M. (1965), *Plastic Analysis & Design*, Vol. 1, *Beams and Frames*, Blaisdell Publishing Co., New York.

Matthey, P. A. (1982), "Flexion gauche non-symetrique de colonnes en double-te recherche experimentale et simulation,"*3rd Int. Colloq. Stab. Met. Struct., Prelim. Rep.*, Paris.

Mazzolani, F. M., and Frey, F. (1983), "ECCS Stability Code for Aluminium-Alloy Members: Present State and Work in Progress," *3rd Int. Colloq. Stab. Met. Struct., Prelim. Rep.*, Paris.

McDonald, W. S., Jr. (1966), "Inelastic Behaviour of the Compression Chord on open-Web Steel Joists, *Report No. 25*, Studies in Engineering Mechanics, University of Kansas, October.

McLellan, E. R., and Adams, P. F. (1970), "Design of Steel Crane Columns or Columns Subjected to Lateral Loads," *EIC Trans.*, Vol. 70, No. A-6.

Miranda, C., and Ojalvo, M. (1965), "Inelastic Lateral-Torsional Buckling of Beam-Columns," *ASCE J. Eng. Mech. Div.*, Vol. 91, No. EM6, pp. 21–38.

Morris, G. A., and Fenves, S. J. (1969), "Approximate Yield Surface Equations," *ASCE J. Eng. Mech. Div.*, Vol. 95, No. EM4, pp. 937–954.

Nethercot, D. A. (1982), "Evaluation of Interaction Formulae For Beam-Columns," Department of Civil and Structural Engineering, University of Sheffield, Reports to ad-hoc group of ECCS TWG 8.1, June 1982, Feb. 1983.

Nethercot, D. A. (1983), "Evaluation of Interaction Equations for Use in Design Specifications in Western Europe," *Proc. 3rd Int. Colloq. Stab. Met. Struct.*, Toronto.

Ojalvo, M., and Fukumoto, Y. (1962), "Nomographs for the Solution of Beam-Column Problems," *Weld. Res. Counc. Bull.*, No. 78, June.

Opperman, H. P. (1983), "Refined Beam-Column Analysis," *Proc. 3rd Int. Colloq. Stab. Met. Struct.*, Paris.

Pillai, U. S. (1970), "Review of Recent Research on the Behaviour of Beam-Columns under Biaxial Bending," *R. Mil. Coll. Can. Civ. Eng. Res. Rep.*, January.

Pillai, U. S. (1974), "Beam Columns of Hollow Sections," *J. Can. Soc. Civ. Eng.*, Vol. 1.

Pillai, U. S. (1980), "Comparison of Test Results with Design Equations for Biaxially Loaded Steel Beam-Columns," R. Mil. Coll. Can. *Dep. Civ. Eng. Res. Rep.*, No. 80-2, August.

Pillai, U. S. (1981), "An Assessment of CSA Standard Equations for Beam-Column Design," *Can. J. Civ. Eng.*, Vol. 8.

Rajasekaren, S. (1977), "Finite Element Method for Plastic Beam-Columns," in *Theory of Beam-Columns* (W. F. Chen and T. Atsuta), McGraw-Hill, New York, Chap. 12.

Ramm, E., and Osterrieder, P. (1983), "Ultimate Load Analysis of Three-Dimensional Beam Structures with Thin-Walled Cross Sections Using Finite Elements," *Proc. 3rd Int. Colloq. Stab. Met. Struct. Prelim. Rep.*, Paris.

Razzaq, Z., and McVinnie, W. W. (1982), "Rectangular Tubular Steel Columns Loaded Biaxially," *J. Strct. Mech.*, Vol. 10, No. 4, pp.

Roik, K., and Wagenknecht, R. (1976), "Traglastdiagramme Zur Bemessung Von Druckstaben Mit Doppellsymmetrischem Querschnitt Aus Baustahl," KIB/Berichte Heft 27, Vulkan, Verlag Essen 1976.

Roik, K., and Bergmann, R. (1977), "Steel Column Design," Preliminary Report Second International Colloquium on Stability of Steel Structures, Liege, 1977.

Roik, K., and Kindmann, R. (1983), "Design of Simply Supported Members by Means of European Buckling Curves for Uniaxial Bending with Compression," Preliminary Report Third International Colloquium on Stability of Metal Structures, Paris, 1983.

Ross, D. A., and Chen, W. F. (1976), "Design Criteria for Steel I-Columns Under Axial Load and Biaxial Bending," *Canadian Journal of Civil Engineering*, Vol. 3, No. 2, 1974.

Salvadori, M. G. (1956), "Lateral Buckling of Eccentrically Loaded I-Columns," *ASCE Transactions*, Vol. 121, 1956, p. 1163.

Santathadaporn, S., and Chen, W. F. (1970), "Interaction Curves for Sections Under Combined Biaxial Bending and Axial Force," Weld. Res. Counc. Bull., No. 148, February 1970.

Smith, E. A. (1983), "Buckling of 4 Equal-Leg Angle Cruciform Columns," *ASCE Journal of Structural Division*, Vol. 109, No. 2, February 1983, pp. 439–450.

Tebedge, N., and Chen, W. F. (1974), "Design Criteria for Steel H-Columns Under Biaxial Loading," *ASCE Journal of the Structural Division*, Vol. 100, No. ST3, March 1974, p. 579–598.

Thomas, B. F., Leigh, J. M., and Lay, M. G. (1973), "The Behaviour of Laterally Unsupported Angles," Civil Engineering Transactions Institution of Engineers, Australia, 1973.

Timoshenko, S. P., and Gere, J. M. (1961), *Theory of Elastic Stability*, McGraw-Hill, New York.

Trahair, N. S. (1969), "Restrained Elastic Beam-Columns," *ASCE J. Struct. Div.*, Vol. 95, No. ST12, pp. 2641–2664.

Usami, T. (1970), "Restrained Single-Angle Columns Under Biaxial Bending," *Res. Rep. No. 14*, Civil Engineering Department, Washington University, June.

Usami, T., and Galambos, T. V. (1971), "Eccentrically Loaded Single Angle Columns," IABSE Pub., Vol. 31-II, pp. 153–184.

Van Kuren, R. C., and Galambos, T. V. (1964), "Beam-Column Experiments," *ASCE J. Struct. Div.*, Vol. 90, No. ST2, pp. 223–256.

Van Manen, S. E. (1982), "Plastic Design of Braced Frames Allowing Plastic Hinges in the Columns," *Heron*, Vol. 27, No. 2, pp.

Vinnakota, S. (1977), "Finite Difference Method for Plastic Beam-Columns," in *Theory of Beam-Columns*, (W. F. Chen and T. Atsuta), McGraw-Hill, New York, Chap. 10.

Virdi, K. S. (1981), "Design of Circular and Rectangular Hollow Section Columns," *J. Constr. Steel Res.*, Vol. 1, No. 4.

Vlasov, V. Z. (1961), *Thin-Walled Elastic Beams, Israel Program for Scientific Translations*, Jerusalem.

Vogel, V., and Lindner, J. (1981), *Kommentar zur Din 18800*: Teil 2 (Gelbdraeh). *Stabilitätfälle in Stahlbau—Knicken von Stäben und Stabwerken*, Deutscher Ausschuss für Stahlbau, Stahlbau-Verlag, Cologne, West Germany, No. 11.

Woolcock, S. T., and Kitipornchai, S. (1980), "The Design of Single Angle Struts," *Steel Constr.*, Vol. 14, No. 4, p. 2.

Woolcock, S. T., and Kitipornchai, S. (1986), "Design of Single Angle Web Struts in Trusses," *ASCE J. Struct. Eng.*, Vol. 112, No. 6, pp. 1327–1345.

Young, B. W. (1973), "The In-Plane Failure of Steel Beam-Columns," *Struct. Eng.*, Vol. 51, No. 1.

Yu, C. K., and Tall, L. (1971), "A514 Steel Beam-Columns," *Publ. IABSE*, No. 31-II, pp. 185–213.

Chapter Nine

Tapered Structural Members

9.1 INTRODUCTION

Tapered beams, as utilized in continuous-frame construction, are those which have a continuous reduction in section from each end toward the center. In the limit the center section may become a hinge. Tapered columns may be hinged at the base with a section that varies continuously from top to bottom. In general, tapered-member framing results in more efficient utilization of structural material.

Analysis and design methods of tapered members have been presented by many authors, and a selection of such papers are listed at the end of this chapter. Lee (1959) and Boley (1963) have established that for small tapering angles (15° or less) the Bernoulli-Euler theory for beams yields satisfactory results. This has resulted in simplified analyses that have permitted the extension of design formulas for prismatic members to tapered members. Generally, there have been two basic types of studies: one dealt with the stability behavior of single tapered members, for example, column stability, lateral stability, and beam-column behavior, and the other was concerned with frame-analysis procedures, some of which are discussed in Section 9.3.

Although tapered steel members have been used in a large variety of structural applications for some time, and some analysis procedures were available prior to 1955, a joint task committee of the Structural Stability Research Council (then Column Research Council) and the Welding Research Council was formed in 1966 to study and formulate design information and recommendations relating to tapered members. Since then, numerous results, including those mentioned above, have been published. Along with the analytical studies, two test programs were carried out in recent years. The first experimental program was conducted at Columbia University under the direction of Butler (Butler and Anderson, 1963; Butler, 1966). In the Columbia test program, tapered I-shaped beams and channel sections tapered in both the web and flanges were tested as cantilever beam-columns. The primary interest was the elastic stability of these beams and their bracing requirements.

The second experimental program under the technical guidance of the SSRC-WRC joint task committee began in 1966 by Lee and Ketter at the State University of New York at Buffalo. The results of this series of tests were reported by Prawel et al. (1974). The primary interest in the second set of tests was the inelastic stability of tapered I-shaped beam-columns. Also in this experimental study residual stresses in welded tapered shapes were measured. The magnitude and distribution of the residual stresses are very similar to those in welded built-up prismatic members.

This chapter contains mainly the results of studies carried out at Buffalo since 1966. They provided the basis for the formulation of allowable stress formulas for tapered members. A rather detailed summary of all pertinent design information for tapered members is given in the book by Lee et al. (1981), which includes design examples of single-story rigid frames consisting of tapered members.

While the following portions of this chapter deal with a particular type of tapered member, namely, where only the depth of the web varies, other investigations dealt with the lateral stability of beams where both the web and the flanges were tapered. Kitipornchai and Trahair (1972) published a derivation of the differential equations and gave some numerical results of the torsion and lateral-torsional buckling of I-shaped beams with web and flange taper. The theoretical predictions compared well with the experimental buckling loads of 19 tests of aluminum beams. Nethercot (1973) presented extensive tabular solutions of the elastic lateral-torsional buckling strength of web and flange tapered I-shaped cantilever and simple beams under end moments, concentrated loads, and distributed loads. He used constant-section finite elements in his numerical solutions, and he developed simple algebraic multipliers by which the buckling load of an equivalent uniform-section beam can be multiplied to obtain the buckling strength of a tapered beam. Kitipornchai and Trahair (1975) also developed solutions for monosymmetric I-shaped flange-tapered beams. Four tests on aluminum beams confirmed the theoretical predictions.

9.2 FRAME ANALYSIS

Amirikian (1952) was the first to propose solution procedures for a special type of frame consisting of tapered members as shown in Fig. 9.1.

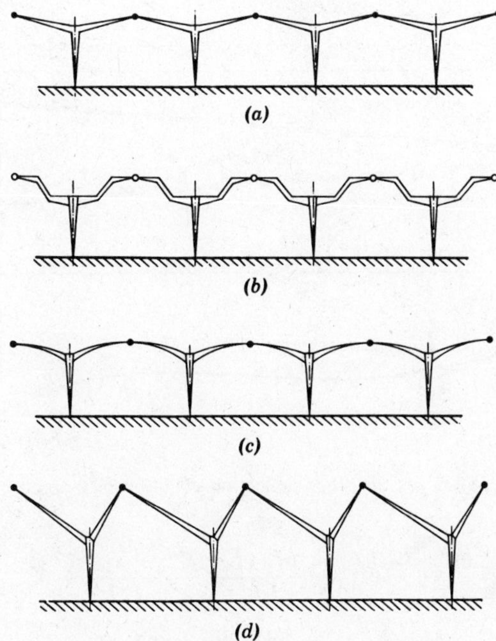

Fig. 9.1 Type of tapered member framing (Amirikian, 1952).

These structures contain hinges at each of the smaller ends of the various individual members and are fully continuous at the larger ends. As can be seen from Fig. 9.1 the framing assembly is subdivided into its component simple frames. Since each member of the simple frame has a rigid joint at one end and a hinged connection at the other end, the bending moment diagram of each member is similar to that of a cantilever beam. This permits computation of the deflection of a simple frame by summing the deflections of its component members. This method is well suited for solution by manual means of assemblies consisting of a few members or by computer of assemblies consisting of many members.

If hinged connections do not appear at all the small ends, other methods of analysis must be used. In particular, slope-deflection and finite-element methods can be used advantageously for computer analysis. Slope-deflection equations have been developed by Lee et al. (1972) for a linearly tapered beam-column (Fig. 9.2) where a continuous variation in the moment of inertia is considered (material concentrated in the flanges).

The slope-deflection equations for tapered beam-columns may be written as

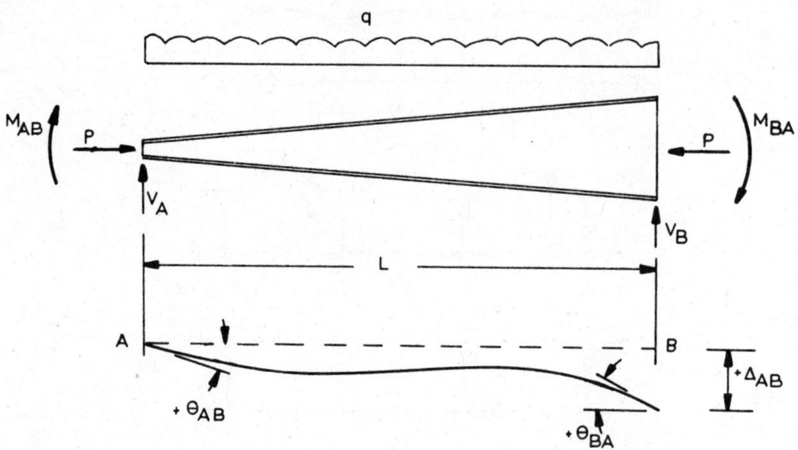

Fig. 9.2 Sign system for the slope-deflection equation of tapered members.

$$\theta_{AB} = \frac{M_{AB}L}{EI_0} C_{AA} - \frac{M_{BA}L}{EI_0} C_{AB} + \frac{\Delta_{AB}}{L} + \bar{\theta}_{AB} \tag{9.1}$$

$$\theta_{BA} = -\frac{M_{AB}L}{EI_0} C_{BA} + \frac{M_{BA}L}{EI_0} C_{BB} + \frac{\Delta_{AB}}{L} - \bar{\theta}_{BA} \tag{9.2}$$

where the sign convention in Fig. 9.2 applies and

$\theta_{AB}(\theta_{BA})$ = rotation of the tangent to the elastic curve at $A(B)$ with respect to the line $AB(BA)$

$\bar{\theta}_{AB}(\bar{\theta}_{BA})$ = rotation $\theta_{AB}(\theta_{BA})$ for a simply supported beam under applied load

$M_{AB}(M_{BA})$ = moment at $A(B)$ acting on member AB

Δ_{AB} = relative deflection of point A and B

L = length of member

I_0 = moment of inertia at the smaller end

For tapered beam-column analysis, the coefficients C_{AA}, C_{AB}, C_{BA}, and C_{BB} have different forms depending on the quantity a/γ^2, where

$$a = \pi^2 \frac{P}{P_{ex_0}}$$

$$P_{ex_0} = \frac{\pi^2 EI_0}{L^2}$$

$$\gamma = \frac{d_1 - d_0}{d_0}$$

The coefficients are given by the following if $a/\gamma^2 \leq \frac{1}{4}$:

$$C_{AA} = \frac{1}{a} - \frac{\gamma}{a}\left(\rho \frac{m+n}{m-n} - \frac{1}{2}\right) \tag{9.3}$$

$$C_{AB} = C_{BA} = \frac{\gamma}{a\sqrt{1+\gamma}}\frac{2\rho}{m-n} - \frac{1}{a} \tag{9.4}$$

$$C_{BB} = \frac{1}{a} - \frac{\gamma}{a(1+\gamma)}\left(\rho \frac{m+n}{m-n} + \frac{1}{2}\right) \tag{9.5}$$

where

$$\rho = \sqrt{\frac{1}{4} - \frac{a}{\gamma^2}} \tag{9.6a}$$

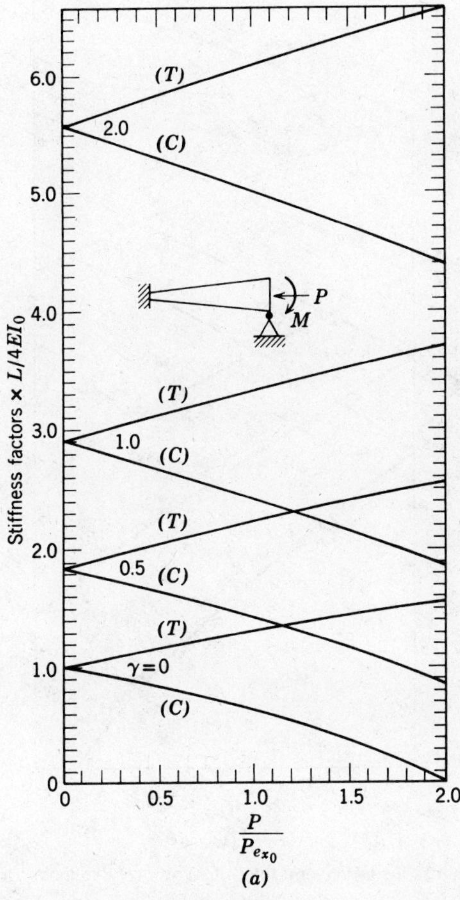

Fig. 9.3 (*a*) Stiffness factors for smaller end fixed, C, compressive axial force; T, tensile axial force.

$$m = (1 + \gamma)^{\rho} \tag{9.7a}$$

$$n = (1 + \gamma)^{-\rho} \tag{9.8a}$$

If $a/\gamma^2 \geqq \frac{1}{4}$,

$$C_{AA} = \frac{1}{a} - \frac{\gamma\beta}{a} \cot[\beta \ln(1 + \gamma)] + \frac{\gamma}{2a} \tag{9.6b}$$

$$C_{AB} = C_{BA} = \frac{\gamma}{a\sqrt{1 + \gamma}} \beta \csc[\beta \ln(1 + \gamma)] - \frac{1}{a} \tag{9.7b}$$

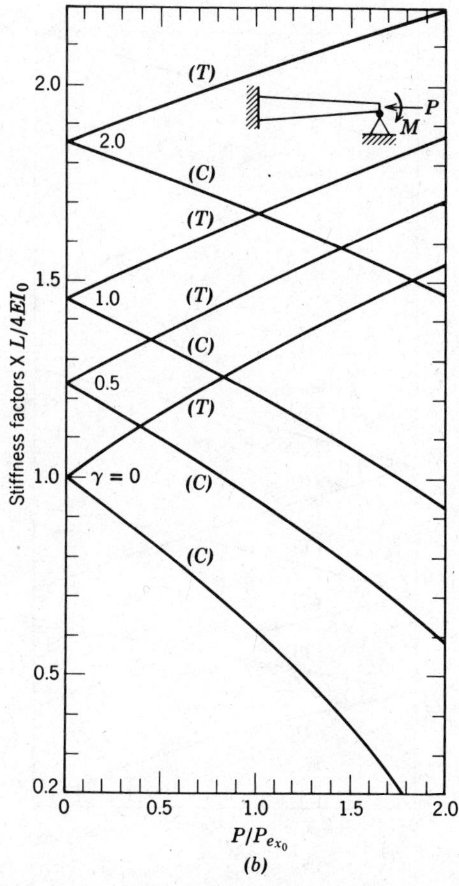

(b)

Fig. 9.3 (*b*) Stiffness factors for larger end fixed, *C*, compressive axial force; *T*, tensile axial force.

$$C_{BB} = \frac{1}{a} - \frac{\gamma\beta}{a(1+\gamma)} \cot[\beta \ln(1+\gamma)] - \frac{\gamma}{2a(1+\gamma)} \tag{9.8b}$$

where

$$\beta = \sqrt{\frac{a}{\gamma^2} - \frac{1}{4}}$$

Equations 9.6 through 9.8 reduce to the corresponding prismatic slope-deflection equations where $\gamma = 0$. If the member is prismatic and if there is no axial load, the standard coefficients are realized.

$$C_{AA} = \tfrac{1}{3} \tag{9.6c}$$

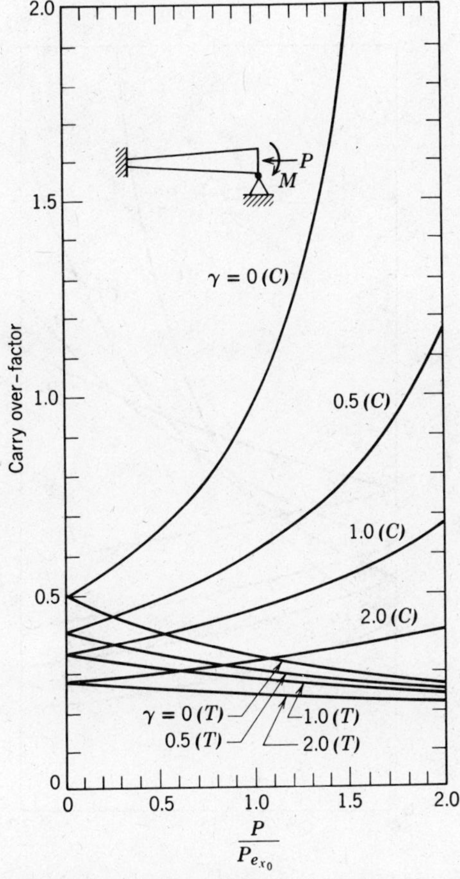

Fig. 9.4 (*a*) Carryover factors for smaller end fixed, *C*, compressive axial force; *T*, tensile axial force.

$$C_{AB} = C_{BA} = \tfrac{1}{6} \tag{9.7c}$$

$$C_{BB} = \tfrac{1}{3} \tag{9.8c}$$

Stiffness coefficients are plotted in Fig. 9.3 as functions of P/P_{ex_0} and γ for the cases: (1) the small end fixed, large end simple supported, and (2) the large end fixed, small end simply supported. The carryover factors are included in Fig. 9.4 for the same variables and cases.

Following the standard slope-deflection method, simultaneous equations can be generated and solved (an iterative method must be used if axial forces are considered) to yield the joint moments, rotations, and deflections. The slope-deflection method presents difficulties in program-

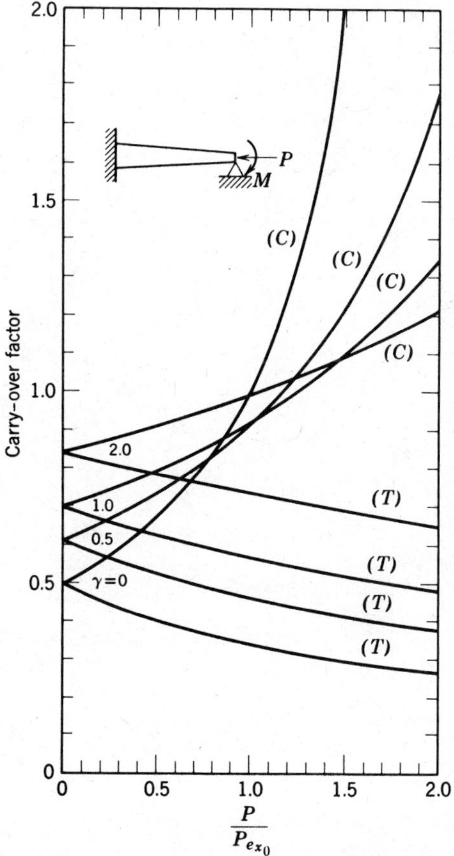

Fig. 9.4 (b) Carryover factors for larger end fixed, C, compressive axial force; T, tensile axial force.

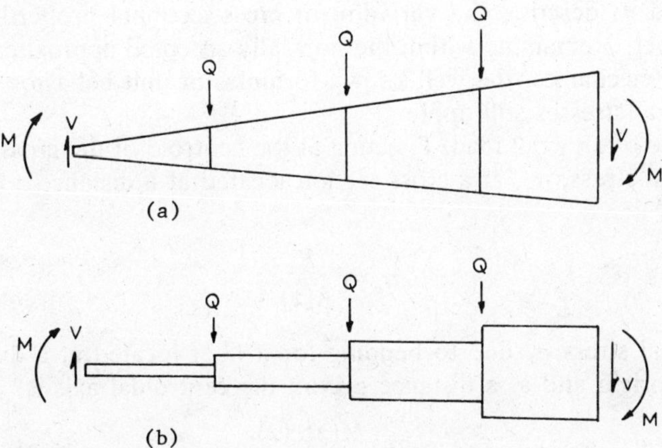

Fig. 9.5 Tapered beam discretizations: (*a*) tapered elements; (*b*) prismatic elements.

ming for highly redundant frames when sidesway is permitted. Additional shear equilibrium equations are necessary from geometry and free-body diagrams, which in general may be difficult for automated computation. In such instances the finite-element method of solution is the better approach. A summary of the methods of analysis for rigid frames consisting of tapered members is given in the book by Lee et al. (1981). Solutions obtained by the finite-element method can be approached in two ways: (1) elemental matrices can be developed for tapered elements (Fig. 9.5*a*), or (2) the tapered member can be considered as a series of prismatic elements (a step beam) (Fig. 9.5*b*).

Gallagher (1969) compared these two approaches by determining the axial buckling load of a tapered rectangular beam. He found that in the limit, as the number of elements is increased, both discretizations approached the same solution; however, by using a tapered element the solution was obtained with fewer elements. For simple cross sections, for example, a rectangle, and for simple behavior, for example, bending, tapered elements can easily be derived. For more complicated cross sections and behavior the elemental matrix may not be explicitly obtained; that is, the elemental matrix may have to be generated by the computer for practically each element. For these members a step-beam approach is more desirable.

9.3 STRESS ANALYSIS

The computation of stresses in tapered members is complicated by the fact that at least one new parameter i.e., the tapering ratio, must be introduced to describe the variation of cross-sectional properties along the member. Remaining within the normally accepted approximations of structural mechanics, the well-known formulas of unit behavior for axial and flexural stresses still apply.

Because of an axial load, P, acting at the centroid of the cross section, the normal stress, σ_a, at a cross section located at a distance z from the origin is

$$\sigma_a = \frac{P}{A(z)} \tag{9.9}$$

The normal stress σ_b due to bending for a fiber located at a distance z from the origin and at a distance c from the centroidal axis is

$$\sigma_b = \frac{M(z)c(z)}{I(z)} \tag{9.10}$$

where $I(z)$ is the moment of inertia at the cross section in question.

Due to a shearing force $V(z)$ the shear stress τ_v, located at a distance z from the origin and a point (x, y) on the cross section is

$$\tau_v = \frac{V(z)Q(z)}{I(z)t} \tag{9.11}$$

where $Q(z)$ is the first moment of area about the centroidal axis. It is also understood that $Q(z)$ will vary with x and y at a particular cross section.

If a transverse load is applied to a prismatic I-shaped beam at a point other than the shear center, the load can be replaced by a statically equivalent load system located at the shear center. This enables the flexural and torsional stresses to be computed separately and then superimposed to obtain the complete stress distribution. However, this procedure cannot be followed for an arbitrary cross-section tapered beam, since the shear center will be a function of the longitudinal coordinate. As an example, a channel section that has a linearly varying depth, Fig. 9.6*a*, has a shear center that converges towards the web in the plane of symmetry as one travels from the small end to the large end. If this channel is cantilevered at the large end and a concentrated force is applied parallel to the web, at the shear center of the small end, Fig. 9.6*b*, then the channel both bends and twists. The coupled flexural-torsional analysis for such beams is not available. However, if the force is applied perpendicular to the web at the shear center, Fig. 9.6*c*, then the channel will bend without twisting. Thus, depending on the cross section and loading, it is possible to uncouple the flexural-torsional behavior of tapered beams. Some examples are sections that have two axes of symmetry (I-shaped beams and rectangular beams), and sections that

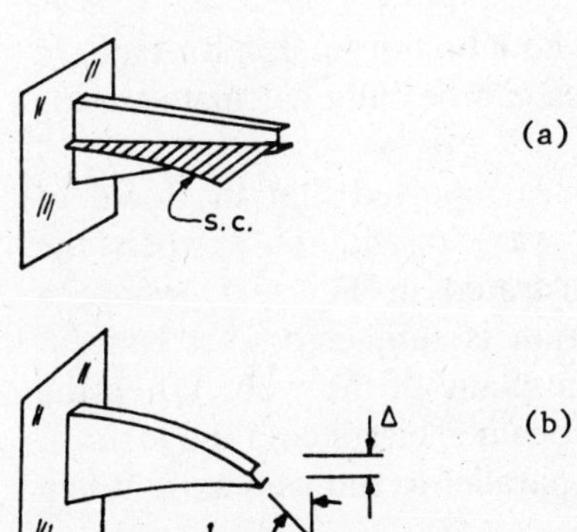

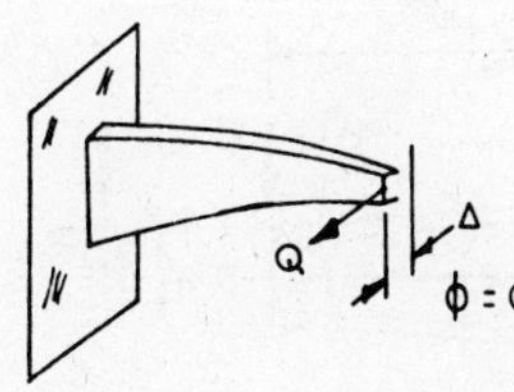

(c)

Fig. 9.6 Channel section with linearly tapered web: (*a*) locus of shear centers; (*b*) load applied at the smaller end shear center parallel to the web; (*c*) load applied at the smaller end shear center perpendicular to the web.

have one axis of symmetry and the loads are applied parallel to that axis (channels and T-beams).

Torsional stresses in tapered beams that satisfy the conditions above and have small tapering angles can be computed using the prismatic formulas for thin-walled beams. The St.-Venant's shearing stress at a location z is

$$\tau_{st} = Gt \frac{d\phi}{dz} \tag{9.12}$$

where G is the shear modulus, t the thickness, and ϕ the angle of twist.

The warping shearing stress is

$$\tau_w = -ES_w(z) \frac{d^3\phi}{dz^3} \tag{9.13}$$

where E is the elastic modulus and $S_w(z)$ is the warping statical moment.

The warping normal stress is

$$\sigma_w = EI_w(z) \frac{d^2\phi}{dz^2} \tag{9.14}$$

where $I_w(z)$ is the warping constant.

A more accurate torsional analysis has been reported by Lee and

Szabo (1967), who derived the differential equation of torsion for tapered I-girders. Elastic deformations and stresses can be fully calculated using their theory.

In a recent study Lee and Chang (1982) reported that the locus of shear centers for tapered channels can vary considerably when the member is partially yielded. This is illustrated in Fig. 9.7, where a 16-ft-long tapered cantilever channel section is subjected to a concentrated load at the free (shallow) end in the plane of the web. When the applied load V is smaller than 7500 lb, the beam is elastic and the locus of the shear centers is a straight line but not parallel to the web axis. When

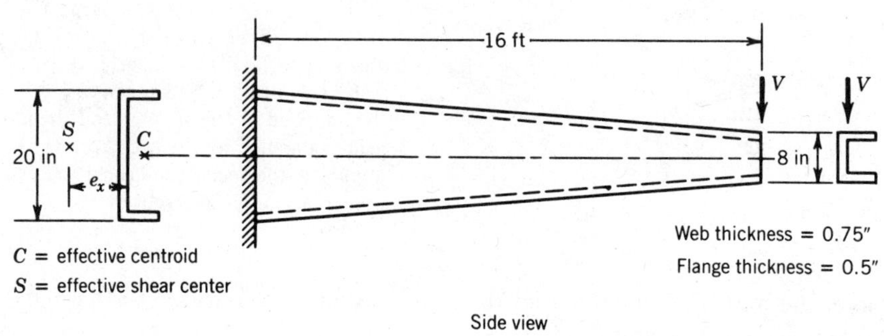

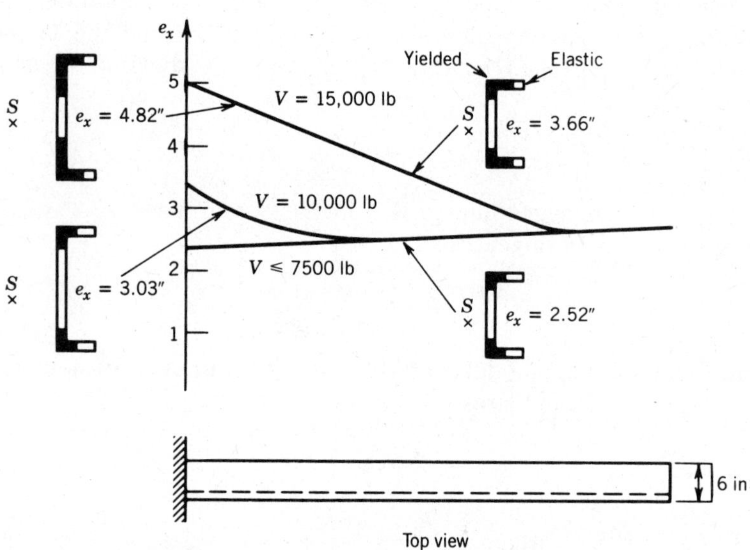

Fig. 9.7 Locations of effective shear centers.

the beam is partially yielded, the loci of shear centers in the yielded portion of the channel deviates drastically from their elastic solutions, as represented by the two curves marked for $V = 10{,}000$ lb and $V = 15{,}000$ lb, respectively. Also on these curves, the yielded zones of the cross sections at selected location are shown. The yielded zones are determined by a consideration of bending and warping normal stresses and the bending, warping and St.-Venant's shear stress simultaneously.

9.4 STABILITY OF TAPERED MEMBERS

For the linearly tapered member, the depth at any distance z from the smaller end can be expressed as

$$d_z = d_0\left(1 + \frac{z}{l}\gamma\right) \tag{9.15}$$

where d_0 represents the smallest depth at $z = 0$, and γ represents the tapering ratio. For a prismatic member, $\gamma = 0$; for a member whose depth at the large end is three times that of its smaller end, $\gamma = 2$. This geometry is defined in Fig. 9.8a and the loading condition is illustrated in Fig. 9.8b.

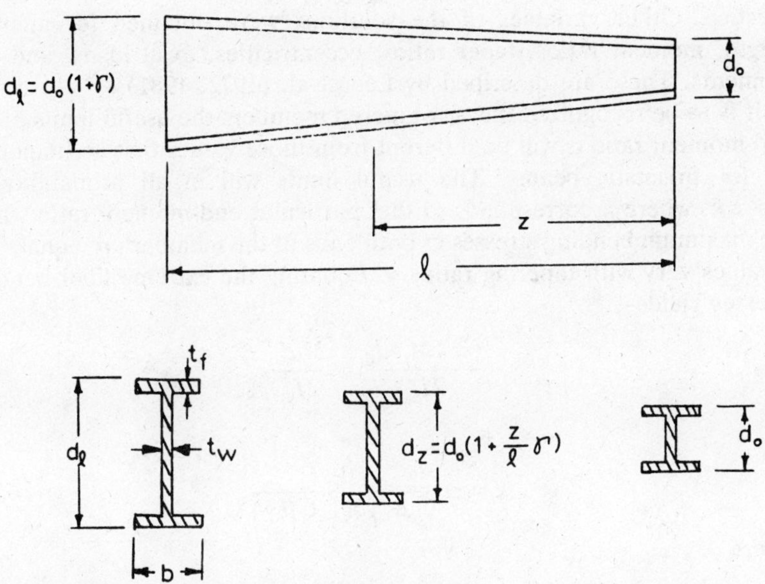

Fig. 9.8 (a) Geometry of linearly tapered I-beam (b, t_f, and t_w are all constant).

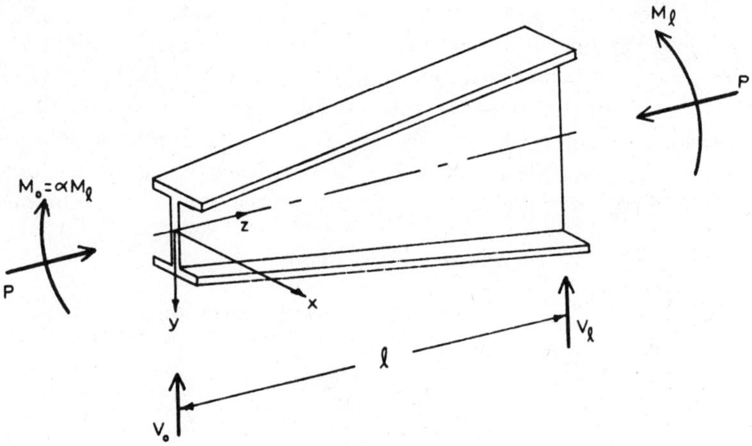

Fig. 9.8 (*b*) The presumed loading on a simply supported linearly tapered I-beam.

Since it was found to be virtually impossible to nondimensionalize the many parameters in tapered sections, solutions were obtained for members possessing five smaller end cross-sectional dimensions given in Fig. 9.9. These cover all the range of tapered members used in present-day practice. Other variables of the solutions were obtained for member length, moment ratios, taper ratios, eccentricities, axial loads, and end moments. These are described by Lee et al. (1972, 1981).

It is to be recognized that for tapered members the useful limits of the end-moment ratio α will be different from those values that are meaningful for prismatic beams. The useful limits will in all probability be $\alpha = \pm k$, where k corresponds to that particular end-moment ratio where the maximum bending stresses at both ends of the member are equal. The k values vary with tapering ratios γ. Equating the extreme fiber bending stresses yields

$$\frac{(\pm k M_l) d_0}{2 I_{x_0}} = \frac{M_l d_l}{2 I_{x_l}}$$

or

$$k = \pm \frac{1}{(1+\gamma)(1+\mu\gamma)} \tag{9.16}$$

where

$$\mu = \frac{t_w d_0^3}{12 I_{x_0}}$$

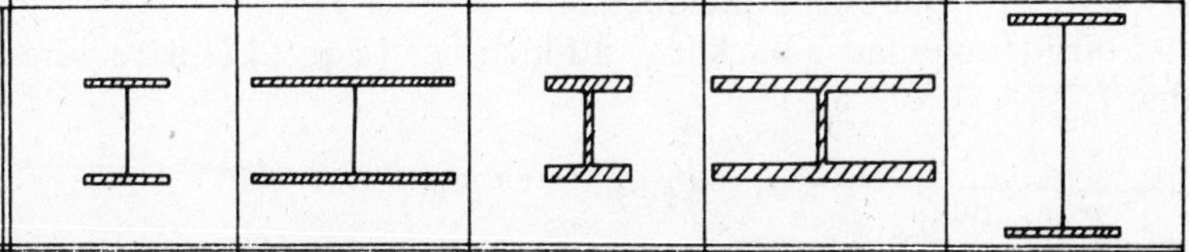

	I	II	III	IV	V
d_o	6.00 in.	6.00 in.	6.00 in.	6.00 in.	12.00 in.
b	4.00	12.00	4.00	12.00	6.00
t_f	0.25	0.25	0.75	0.75	0.25
t_w	0.10	0.10	0.25	0.25	0.10

Fig. 9.9 Dimensions of the smaller end of linear tapered I-beam presumed.

9.5 DEVELOPMENT OF AXIAL COMPRESSION FORMULAS

Several avenues of approach are available for the development of design approximations to the solution referred to in Section 9.4. First, using multivariable curve fitting techniques, polynomial expressions containing all the variables could be developed. A second approach could start from the basic assumption that adequate solutions are not available for prismatic members. Then modifying factors could be developed and introduced into the prismatic formulas. The factors can be calculated from

$$\frac{\text{strength of tapered member}}{\text{strength of prismatic member of the smaller cross section}} = f(\gamma, d_0, b, t, w, l) \qquad (9.17)$$

with the restriction that when $\gamma = 0$ (prismatic), $f = 1.0$.

Although both approaches require curve-fitting techniques, the second approach offers two appealing advantages to the designers. First, they will be using the same AISC (American Institute of Steel Construction, 1978) formulas for prismatic beams but with a factor added. Second, the factor will give them an intuitive feeling for the increase in strength over a prismatic section. This increase in strength may help in deciding the economy of the structure using tapered members.

There are three modes of general failure to be considered in the design of axially loaded tapered members of doubly symmetrical cross section: (1) strong-axis buckling, (2) weak-axis buckling, and (3) torsional buckling. The AISC has basically two formulas for flexural buckling: the elastic and inelastic buckling about the axis of largest effective slenderness ratio.

The effective length, $K_\gamma l_\gamma$ is discussed later; thus the formulas following are based on pin-ended members.

Considering elastic buckling, a function f (Eq. 9.17) is required, such that

$$P_{\text{taper}} = P_{\text{prismatic}} f(\gamma, d_0, b, t, w, l) \tag{9.18}$$

where

$$P_{\text{prismatic}} = \frac{\pi^2 E I_{x_0}}{l^2}$$

Since column-type buckling can occur about either the strong or weak axis of the cross section, the function f will be different for each case. Observing that the variation of the weak-axis radius-of-gyration along the length of a web-tapered member (flange width is constant) is small, no modification factor is necessary. Thus

$$(P_{ey})_\gamma = \frac{\pi^2 E I_{y_0}}{l^2} \qquad \text{(weak axis)} \tag{9.19}$$

If the member is braced against weak-axis buckling, then buckling will occur about the strong axis. For this case, let $f = 1/g^2$ so that

$$(P_{ex})_\gamma = \frac{\pi^2 E I_{x_0}}{(gl)^2} \tag{9.20}$$

Equation 9.20 implies that the buckling load for a tapered column of length l can be considered equivalent to that of a prismatic column having a cross section equal to that of the smaller end of the tapered column and a length equal to gl (Fig. 9.10). Note that for axially loaded tapered columns the critical stress will always occur at the smaller end and the stress will decrease toward the larger end.

Seeking a function of simple form with relatively small error, and assuming a curve fitted to the data,

$$g = \sqrt{\frac{(\pi^2 E I_{x_0}/l^2)}{(P_{ex})_\gamma}}$$

where the sections indicated in Fig. 9.9 were used to establish the range of variation. After examining the resulting fitted functions, the following was chosen (see Lee et al. 1972):

$$g = 1.000 - 0.375\gamma + 0.080\gamma^2(1.000 - 0.0775\gamma) \tag{9.21}$$

In general, these elastic formulas are applicable for slenderness ratios

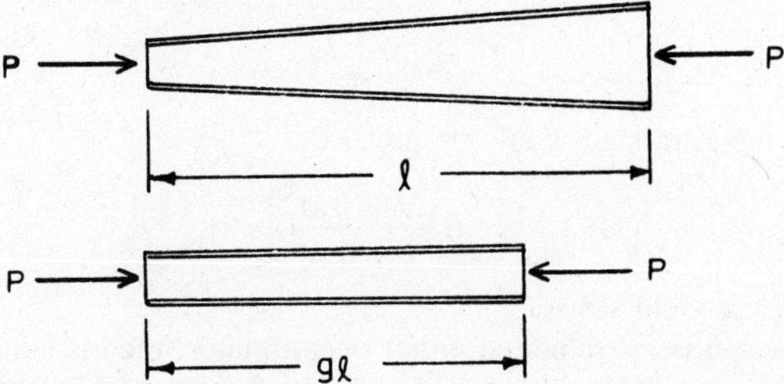

Fig. 9.10 Tapered column and prismatic column having the same smaller end cross section and the same critical load.

greater than a certain limiting ratio. For members with lesser slenderness values, buckling will occur in the inelastic range and the formula will overpredict the critical load. Since there are available at this date no inelastic solutions, one possible procedure for solving problems in this range is to adopt the current AISC philosophy, that is, adoption of a transition curve between the elastic linear situation and the fully yielded point ($l/r_0 = 0$). Using that procedure, if the slenderness ratio (l/r_0) is less than C_c, where

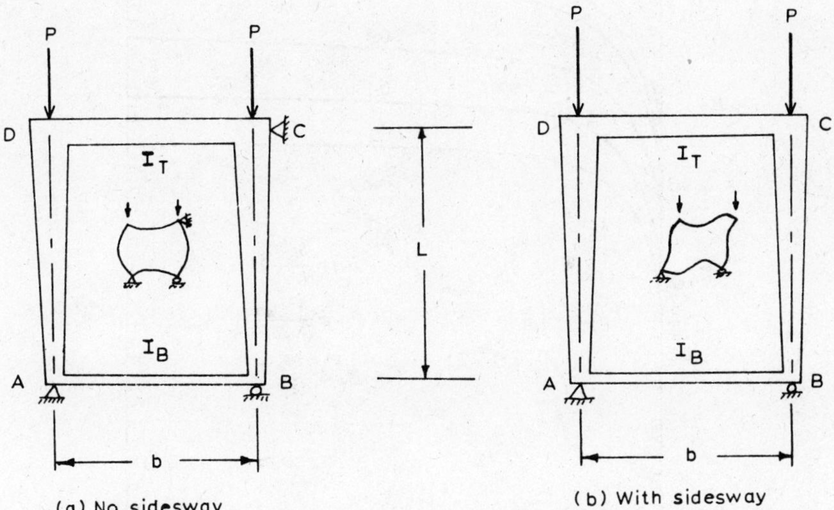

Fig. 9.11 Rectangular frame composed of linearly tapered I-shaped columns and prismatic I-beams used to determine the effective column length: (*a*) no sidesway; (*b*) with sidesway.

$$C_c = \sqrt{\frac{2\pi^2 E}{\sigma_y}}$$

the critical inelastic load would be given by

$$(P_x)_\gamma = \left[1.0 - \frac{(gl/r_0)^2}{2C_c}\right]\sigma_y A_0 \tag{9.22}$$

where σ_y is the yield stress.

If the column is restrained at either or both ends, the buckling load as developed above will be different. A restrained column of height l can be considered as a pin-ended column of length $K_\gamma l$, where K_γ is the effective

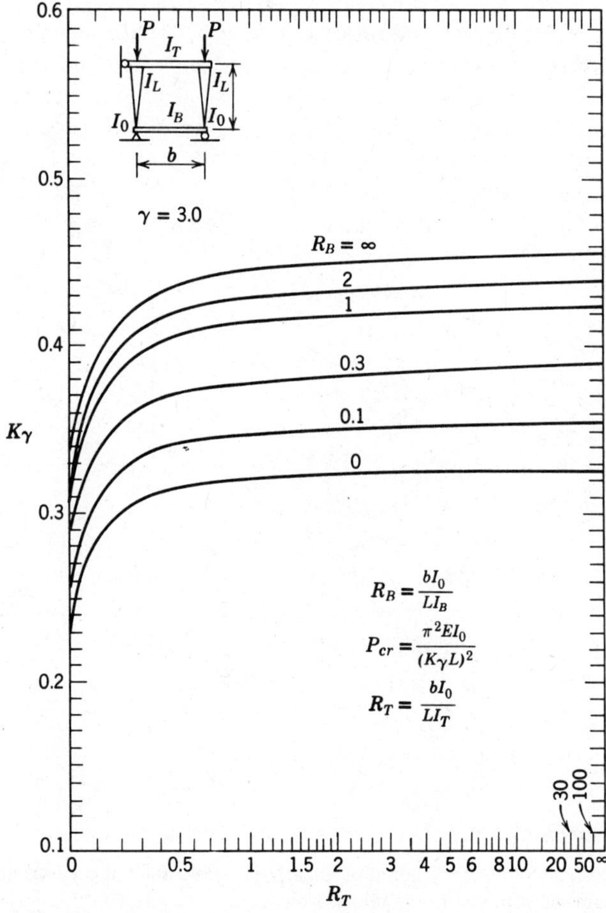

Fig. 9.12 (*a*) Effective-length factors for tapered columns: sidesway prevented ($\gamma = 3.0$).

length factor. In this consideration the function g has been absorbed into the effective length factor K_γ. Thus K_γ is interpreted as relating a restrained tapered column of height l to a pin-ended prismatic column of length $K_\gamma l$.

The effective length is determined by considering a rectangular rigid frame (Fig. 9.11) composed of prismatic beams and tapered columns. The top beam is assumed to have a moment of inertia I_t; the bottom beam, I_B. The loads are assumed to act at the centroid of the columns. The frame has two support conditions in the plane of the frame: (a) supported at A, B, and C (sidesway prevented) and (b) supported at A and B (sidesway permitted). Finally, it is to be understood that only

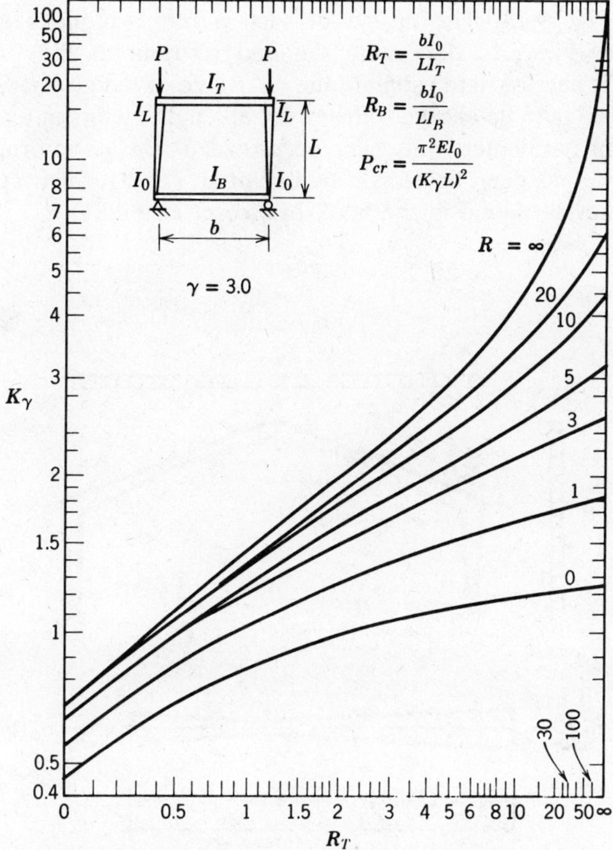

Fig. 9.12 (b) Effective-length factors for tapered columns: sidesway permitted ($\gamma = 3.0$).

buckling in the plane of the frame is considered (i.e., strong-axis buckling of the tapered columns).

The effective length $K_\gamma l$ was calculated from

$$K_\gamma = \sqrt{\frac{\pi^2 EI_{x_0}}{P_{cr} l^2}} \tag{9.23}$$

where P_{cr} is the critical load acting on the frame in Fig. 9.11. Thus Eq. 9.20 can be written to include end restraints,

$$(P_{ex})_\gamma = \frac{\pi^2 EI_{x_0}}{(K_\gamma l)^2} \tag{9.24}$$

In many single-story rigid frames the girders are designed to be multisegment members, for reasons of economy. A typical frame is shown in Fig. 9.13. The design of such a frame requires additional information. First, to determine the end restraint at the top of the column, it is necessary to estimate the I/L value of the girder. Second, if the girder AB is to be checked, the effective-length factors in terms of the end-restraint parameters must also be provided. Design information for both of these has been developed by Lee et al. (1979). Their application in design may be found in the book by Lee et al. (1981).

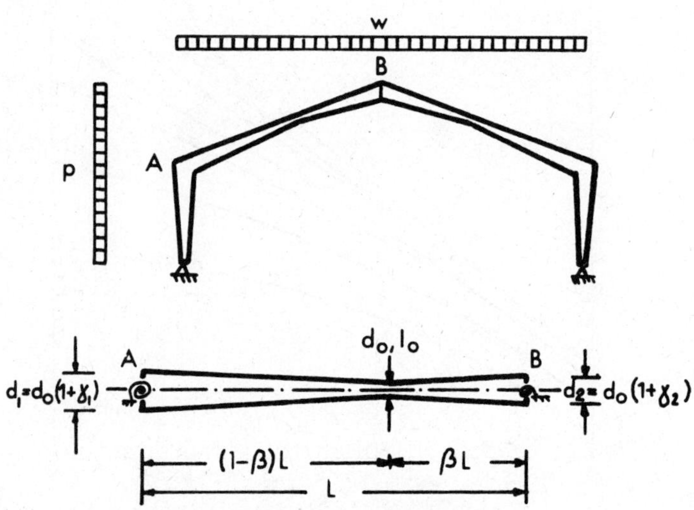

Fig. 9.13 Typical rigid frame with doubly tapered rafter.

9.6 DEVELOPMENT OF BENDING FORMULAS

Two modes of overall failure are of concern in the design of transversely loaded tapered beams or beam-columns: (1) lateral buckling, and (2) yielding. Formulas for determining the critical lateral-buckling moment are presented in this section, and the maximum bending strength at yield is considered in the following section.

To be able to take advantage of tapering, it is customary for a designer to choose a tapering ratio that results in the maximum stress being nearly constant along the length of the beam. In the case of a simply supported member, loaded with end moments only, a possible optimum design would require equal extreme fiber bending stresses at each end, that is, $(M/S_x)_0 = (M/S_x)_l$. For this to occur $M_0/M_l = S_{x_0}/S_{x_l}$, which is the limiting moment gradient discussed in Section 9.4 (Eq. 9.16). However, for sections with linearly tapered web depths the maximum stress under this limiting moment gradient ($\alpha = +k$) *does not* occur at the ends of the beam. It actually occurs somewhere within the length of the beam. It is to be noted, however, that while this is true for conventional tapered members, the stress at this point is seldom greater than 5% more than that at the ends.

Since there is no exact solution available for determining the lateral buckling strength of tapered members of the type considered, the approach used to formulate design recommendations is the same as that taken for the case of axial compression. An expression has been obtained for the critical moment in terms of the prismatic solutions with certain modification factors included. In the following discussion a modification factor is developed to handle members with the most severe condition. Later a moment gradient term is introduced to allow for the condition, $\alpha = 0$.

The critical lateral-torsional buckling stress for a prismatic beam bent into single curvature by equal end moments is given by

$$\sigma_{cr} = \frac{1}{S_x}\sqrt{\frac{\pi^2 EI_y GK_T}{l^2} + \frac{\pi^2 EI_y EI_w}{l^4}} \tag{9.25}$$

Instead of using Eq. 9.17 directly, the modifying factor is applied to the length as in Section 9.5. Thus the critical buckling stress for a tapered beam may be expressed by

$$(\sigma_{cr})_\gamma = \frac{1}{S_{x_0}}\sqrt{\frac{\pi^2 EI_y GK_{T_0}}{(hl)^2} + \frac{\pi^4 EI_y EI_{w_0}}{(hl)^4}} \tag{9.26}$$

where $h = h(\gamma, d_0, b, t, w, l)$.

Equation 9.26 seeks to find a prismatic beam of length, hl, with the cross section of the small end of a tapered beam [Fig. 9.14, which has the same small end moment (also same stress)].

Solving for h in Eq. 9.26 gives

$$h = \frac{\pi^2 EI_y GK_{T_0}}{l^2(\sigma_{cr})_\gamma}\left\{1.0 + \sqrt{1.0 + \left[\frac{(\sigma_{cr})_\gamma S_{x_0} d_0}{GK_{T_0}}\right]^2}\right\} \tag{9.27}$$

where $(\sigma_{cr})_\gamma$ was obtained in the analysis referred to in Section 9.4. This equation was used to calculate h for the given typical cross sections defined in Fig. 9.9. The resulting points plotted as a function of the tapering ratio and length were scattered. However, there was a strong dependence on the tapering ratio.

Since a modification factor to be applied to the prismatic beam formulas of the AISC is desired, it is helpful to review the development of these AISC formulas. At the critical stress level these have the following form: for

$$\left(\frac{l}{r_T}\right) \geqq \sqrt{\frac{510 \times 10^3 C_b}{\sigma_y}}$$

$$\sigma_{cr} = \frac{\pi^2 E C_b}{(l/r_T)^2} \qquad \text{(AISC 1.5-6b)}$$

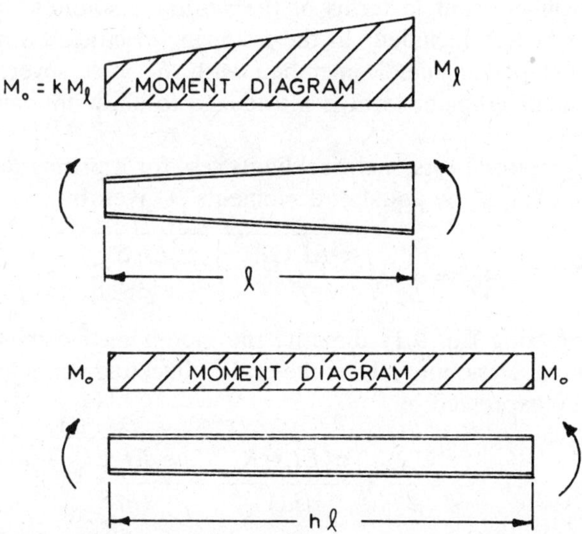

Fig. 9.14. Tapered beam and prismatic beam having the same small end critical moment.

or

$$\sigma_{cr} = \frac{0.65EC_b}{ld/A_f} \qquad \text{(AISC 1.5-7)}$$

These two equations were arrived at by considering in Eq. 9.25 the effects of St.-Venant's torsional resistance and warping resistance. For sections that are thin and deep (high warping resistance) the first formula will generally govern. For sections that are thick and shallow, the second formula will generally govern. Thus in the search for modification factors it appears desirable to allow two possibilities: one when the smaller end is thin and deep, and the other when the smaller end is thick and shallow.

Considering the five sections presumed in Section 9.4, it turns out that sections I through IV are basically thick and shallow for prismatic beams, that is, the second formula should govern. It should also be noted that sections I, III, and V are basically beams, whereas sections II and IV are basically columns. Thus sections I and III have been used to find the modification factor for use with the AISC Eq. 1.5-7, and section V has been used for the AISC Eq. 1.5-6b. Substituting the appropriate data into Eq. 9.27 and fitting curves through the points assuming a function of the tapering ratio and length, for thin, deep sections

$$h_w = 1.0 + 0.00385\gamma \sqrt{\frac{l}{r_{T_0}}} \tag{9.28}$$

for thick, shallow sections

$$h_s = 1.0 + 0.0230\gamma \sqrt{\frac{ld_0}{A_f}} \tag{9.29}$$

A conservative estimate of the buckling stress for tapered beams can be determined by selecting the larger of the two stresses computed by

$$(\sigma_{cr})_\gamma = \frac{\pi^2 E}{(h_w l / r_{T_0})^2} \tag{9.30}$$

or

$$(\sigma_{cr})_\gamma = \frac{0.65E}{h_s ld/A_f} \tag{9.31}$$

These presume that the yield stress has not been exceeded and $\alpha = +k$.

"Exact" solutions obtained by the analysis referred to in Section 9.4 are compared with Eqs. 9.30 and 9.31 by Lee et al. (1972). Although when $M_0 = kM_l$ the material is most effectively used, in many practical applications the moment at the smaller end is zero or near zero. To accommodate such cases, a moment gradient coefficient is introduced to Eqs. 9.30 and 9.31.

The moment coefficient can be defined as

$$C_{b_\gamma} = \frac{(\sigma_{cr})_\gamma|_{\alpha=0}}{(\sigma_{cr})_\gamma|_{\alpha=+k}}$$

After performing this calculation for sections III and V, the results were observed to depend on the cross section, length, and taper ratio. Keeping simplicity in mind, the AISC code was sought again; the moment gradient coefficient, C_b, equals 1.75 for a prismatic beam when $M_0/M_l = 0$. For a nonprismatic beam C_{b_γ}, calculated above, ranged from 1.54 to 1.84 for sections III and V. Moreover, for section V with a length of 36 ft and $\gamma = 0$, $C_{b_\gamma} = 1.75$. So that the curve fit would reduce to 1.75 (the AISC specifications) when $\gamma = 0$, data from section V was chosen at a length of 36 ft to fit C_{b_γ} as a function of the taper ratio only.

$$C_{b_\gamma} = \frac{1.75}{1.0 + 0.25\sqrt{\gamma}} \tag{9.32}$$

The elastic critical stress can now incorporate the zero moment gradient case: the larger of

$$(\sigma_{cr})_\gamma = \frac{\pi^2 E C_{b_\gamma}}{(h_w l / r_{T_0})^2} \tag{9.33}$$

or

$$(\sigma_{cr})_\gamma = \frac{0.65 E C_{b_\gamma}}{(h_s d_0 l / A_f)} \tag{9.34}$$

where

$$C_{b_\gamma} = \begin{cases} 1.0 & \alpha = +k \\ \dfrac{1.75}{1.0 + 0.25\sqrt{\gamma}} & \alpha = 0 \end{cases}$$

When comparing the exact stress to Eqs. 9.33 and 9.34 the order of magnitude is the same for the $\alpha = 0$ case as it is for the $\alpha = +k$ case.

Since inelastic buckling will occur, a transition curve in this range is also necessary. Thus Eq. 9.30 is applicable for (l/r_{T_0}) greater than some limiting (l/r_{T_0}). Using current thinking concerning residual stresses, the limiting value C_r occurs when the critical stress is equal to $\frac{1}{2}\sigma_y$. Imposing the condition that $\sigma_{cr} = \frac{1}{2}\sigma_y$ at $(l/r_{T_0}) = C_r$ gives

$$C_r = \sqrt{\frac{2\pi^2 E C_{b_\gamma}}{\sigma_y}} \tag{9.35}$$

Observing the similarity between Eqs. 9.30 and 9.20 (column buckling), the transition curve for the allowable inelastic lateral buckling stress can be assumed as, for $(l/r_{T_0}) < C_r$,

$$(\sigma_{cr})_\gamma = \left[1.0 - \frac{(h_w l/r_{T_0})^2}{2C_r^2}\right]^2 \sigma_y \tag{9.36}$$

The tapering ratio has a much smaller effect on the allowable stress for thin, deep beams than for thick, shallow ones as can be seen from Eqs. 9.28 and 9.29. It is important to keep in mind that there has been established a limiting moment gradient, which is a function γ (Eq. 9.16). Hence when using the curves, one must remember that the moment gradient changes for each tapering ratio. Thus although the critical stress is little affected, the large-end critical moment may be markedly different.

In continuous beams over several supports the critical span has added strength due to continuity with adjacent spans. The restraining factors are presented by Morrell and Lee (1974) for beams that were laterally braced at nearly equal intervals. The restraint factors would be used in place of C_{b_γ}.

For tapered girders whose tapering ratio is small, the end restraint factors may be estimated with reasonable accuracy from those that are available for continuous prismatic beams. One such design information is presented by Hsu and Lee (1981). In such case, Eqs. 9.33, 9.34, and 9.36 may be written as

$$(\sigma_{cr})_\gamma = \frac{\pi^2 E C_{b\gamma}}{(K_w h_w l/r_{T_0})^2} \tag{9.33a}$$

$$(\sigma_{cr})_\gamma = \frac{0.65 E C_b}{(K_s h_s d_0 l/A_f)} \tag{9.34a}$$

$$(\sigma_{cr})_\gamma = \left[1.0 - \frac{(K_w h_w l/r_{T_0})^2}{2C_\gamma^2}\right]^2 \sigma_y \tag{9.36a}$$

where K_w and K_s are the effective length factors for the warping and St.-Venant's torsional equations, respectively. Values for K_w and K_s are given in Fig. 9.15. Note that these are derived for prismatic beams with lateral-torsional end restraints, and therefore should only be used for tapered beams with small tapering ratio.

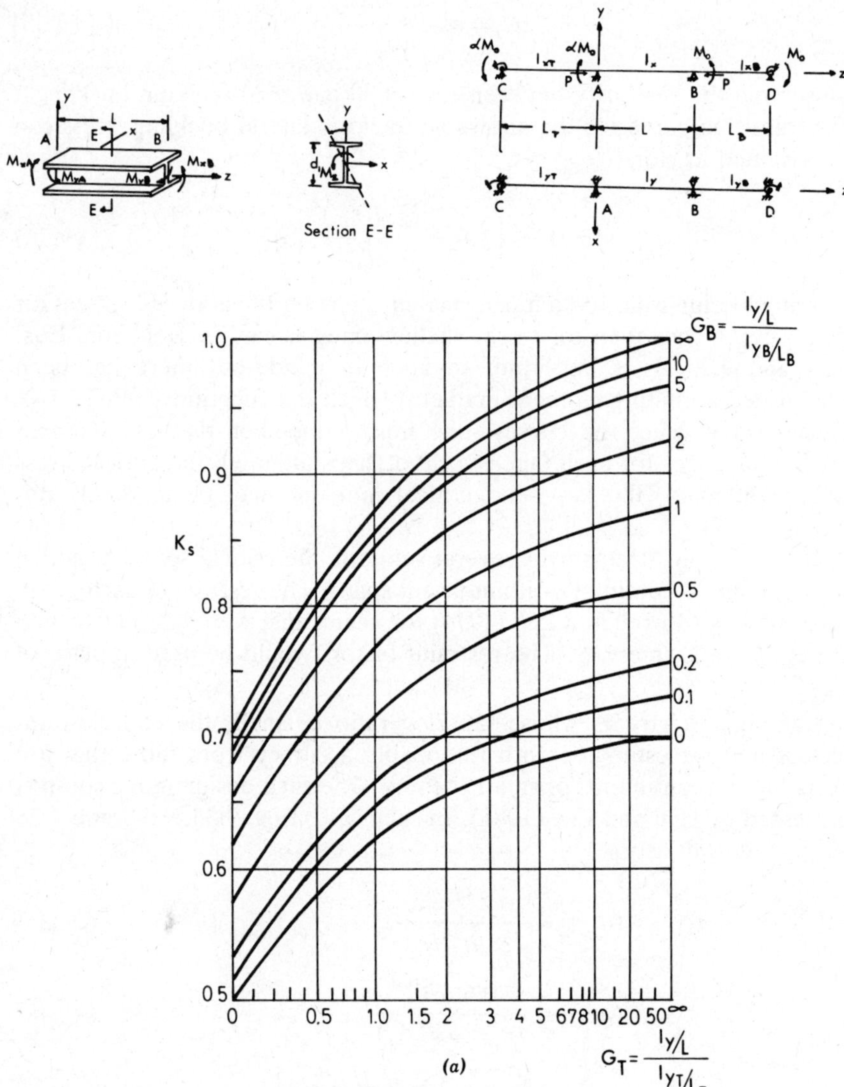

(a)

Fig. 9.15 (*a*) Effective length factors, K_s, for the "St.-Venant's" term of laterally unsupported beams.

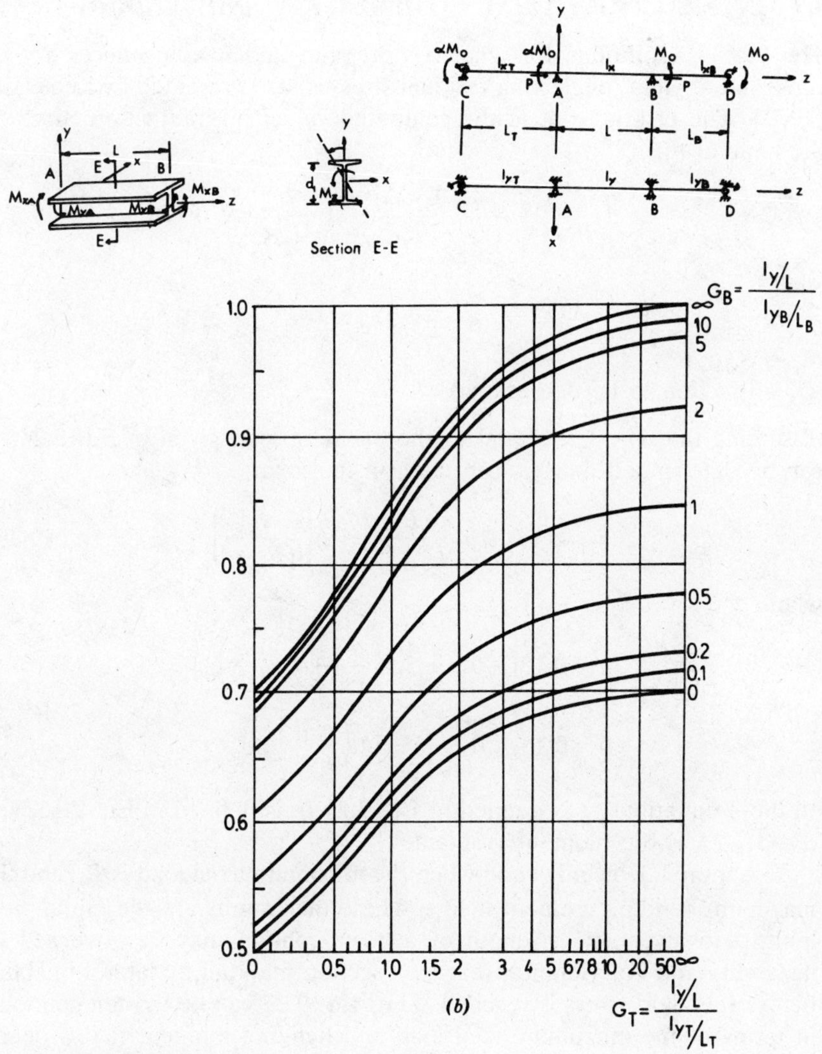

Fig. 9.15 (*b*) Effective length factors, K_w, for the "warping" term of laterally unsupported beams.

9.7 INTERACTION: AXIAL COMPRESSION AND BENDING

The "exact" solutions for axial compression and bending necessary to cause initial yield (neglecting residual stresses) are presented by Lee et al. (1972). The task is to fit to the computed values an interaction curve of the general form

$$\left(\frac{P}{P_{y0}}\right) + \frac{C_E}{1.0 - P/(P_{ex})_\gamma}\left(\frac{M}{M_{yl}}\right) = 1.0 \tag{9.37}$$

where

$P_{y_0} = A_0\sigma_y$
$M_{yl} = S_{xl}\sigma_y$
C_E = function to be determined

Assuming that $(P_{ex})_\gamma$ is equal to the buckling load given in Eq. 9.24, it can be determined that C_E can take on the form

$$C_E = 1.0 + a\left[\frac{P}{(P_{ex})_\gamma}\right] + b\left[\frac{P}{(P_{ex})_\gamma}\right]^2 \tag{9.38}$$

where

$$a = 0.10\left[-9.0 + 5.5\left(\frac{\alpha}{k}\right) + 4.5\left(\frac{\alpha}{k}\right)^2\right]$$

$$b = 0.15\left[4.0 + \frac{\alpha}{k} - 3.0\left(\frac{\alpha}{k}\right)^2\right]$$

In these equations, g is defined in Eq. 9.21, k is defined in Eq. 9.16, and $\alpha = M_0/M_l$ is the moment gradient.

Equation 9.37 is only valid when the maximum axial load is P_{y_0} and the maximum bending moment is M_{yl}. These maximums are the initial yield limits. However, the strength of a beam-column may be governed by instability and not yielding; that is, a section may fail by lateral buckling before the yield stress is reached. Thus Eq. 9.37 can be written generally in terms of the maximum axial load and bending moment that a beam-column can sustain individually without failure. Making thse adjustments gives

$$\left(\frac{P}{P_{u0}}\right) + \frac{C_E}{1.0 - P/(P_{ex})_\gamma}\left(\frac{M}{M_{ul}}\right) = 1.0 \tag{9.39}$$

where

P_{u_0} = axial load producing failure in the absence of bending moment, based on the small-end cross section

M_{ul} = bending moment producing failure in the absence of axial load, based on the large-end cross section

Equation 9.39 can be simplified for cases when P/P_u is small, less than 15%. For such cases the amplification factor is close to unity; thus

$$\left(\frac{P}{P_{u0}}\right) + \left(\frac{M}{M_{ul}}\right) = 1.0 \tag{9.40}$$

9.8 SUMMARY

Although this chapter presents some recent developments in tapered structural member design, it is hardly inclusive. There are a large number of questions that still need to be answered—for example, the local buckling requirements of web plate, the adaptability to hybrid design, and the correlations with haunched connection design. It is suggested that careful engineering judgments be exercised in conjunction with the application of the research results presented in this chapter. A more comprehensive description on the design of single-story rigid frames consisting of tapered members is given by Lee et al. (1981), published by the Metal Building Manufacturers Association. In this reference, additional topics such as the design of tapered members with unequal flanges is also included.

Further pertinent references are noted at the end of this chapter: Culver and Preg (1968), Falby and Lee (1976), Fogel and Ketter (1962), Gere and Carter (1962), Girijavallabhan (1969), Lee and Hsu (1981), Lee and Morrell (1974) and Lee and Morrell (1975).

References

American Institute of Steel Construction (1978), *Specification for the Design, Fabrication and Erection of Structural Steel for Buildings, with Commentary*, AISC, Chicago.

Amirikian, A. (1952), "Wedge-Beam Framing," *Trans. Am. Soc. Civ. Eng.*, Vol. 117, p. 596.

Appl, F. M., and Smith, J. O. (1968), "Buckling of Inelastic Tapered Pin-Ended Columns," *ASCE J. Eng. Mech. Div.*, Vol. 94, No. EM2, pp. 549–558.

Boley, B. A. (1963), "On the Accuracy of the Bernoulli-Euler Theory for Beams of Variable Section," *J. Appl. Mech.*, Vol. 30, pp. 373–378.

Butler, D. J. (1966), "Elastic Buckling Tests on Laterally and Torsionally Braced Tapered I-Beams," *Weld. J. Res. Suppl.*, Vol. 45, No. 1.

Butler, D. J., and Anderson, G. C. (1963), "The Elastic Buckling of Tapered Beam-Columns, "*Weld. J. Res. Suppl.*, Vol. 42, No. 1.

Culver, C. G., and Preg, S. M., Jr. (1968), "Elastic Stability of Tapered Beam-Columns," *ASCE J. Struct. Div.*, Vol. 94, No. ST2, pp. 455–470.

Falby, W. W., and Lee, G. C. (1976), "Tension-Field Design of Tapered Webs," *Eng. J. Am. Inst. Steel Constr.*, Vol. 13, No. 1, pp. 11–17.

Fogel, C. M., and Ketter, R. L. (1962), "Elastic Strength of Tapered Columns," *ASCE J. Struct. Div.*, Vol. 88, No. ST5, Proc. Pap. 3301, pp. 67–106.

Gallagher, R. H. (1969), "A Survey of Framework Finite Element Stability Analysis," *ASCE Natl. Meet.*, New Orleans, February.

Gere, J. M., and Carter, W. O. (1962), "Critical Buckling Loads for Tapered Columns," *ASCE J. Struct. Div.*, Vol. 88, No. ST1, Proc. Pap. 3045, pp. 1–11.

Girijavallabhan, C. V. (1969), "Buckling Loads of Nonuniform Columns," *ASCE J. Struct. Div.*, Vol. 95, No. ST11, pp. 2419–2431.

Hsu, T. L., and Lee, G. C. (1981), "Design of Beam Columns with Lateral-Torsional End Restraints," *Weld. Res. Counc. Bull.*, No. 272, November.

Kitipornchai, S., and Trahair, N. S. (1972), "Elastic Stability of Tapered I-Beams," *ASCE J. Struct. Div.*, Vol. 98, No. ST3, pp. 713–728.

Kitipornchai, S., and Trahair, N. S. (1975), "Elastic Stability of Tapered Monosymmetric Beams," *ASCE J. Struct. Div.*, Vol. 101, No. ST7, pp. 1333–1348.

Krefeld, W. J., Butler, D. J., and Anderson, G. B. (1959), "Welder Cantilever Wedge Beams," *Weld. J. Res. Suppl.*, Vol. 38, No. 3.

Lee, L. H. N. (1959), "On the Lateral Buckling of a Tapered Narrow Rectangular Beam," *J. Appl. Mech.*, Vol. 26, pp. 457–458.

Lee, G. C., and Chang, K. C. (1982), "On Shear Center Variations of Partially Yielded Channels," *Civ. Eng. Res. Rep.*, State University of New York, Buffalo.

Lee, G. C., and Hsu, T. L. (1981), "Tapered Columns with Unequal Flanges," *Weld. Res. Counc. Bull.*, No. 272, November.

Lee, G. C., and Morrell, M. L. (1974), "Stability of Space Frames Composed of Thin-Walled Structures," *Proc. ASCE Struct. Eng. Conf.*, Cincinnati, Ohio, April.

Lee, G. C., and Morrell, M. L. (1975), "Application of AISC Design Provisions for Tapered Members," *Eng. J. Am. Inst. Steel Constr.*, Vol. 12, No. 1, pp. 1–13.

Lee, G. C., and Szabo, B. A. (1967), "Torsional Response of Tapered I-Girders," *ASCE J. Struct. Div.*, Vol. 93, No. ST5, pp. 233–252.

Lee, G. C., Morrell, M. L., and Ketter, R. L. (1972), "Design of Tapered Members," *Weld. Res. Counc. Bull.*, No. 173, June.

Lee, G. C., Chen, Y. C., and Hsu, T. L. (1979), "Allowable Axial Stress of Restrained Multisegment Tapered Roof Girders," *Weld. Res. Counc. Bull.*, No. 248, May.

Lee, G. C., Ketter, R. L., and Hsu, T. L. (1981), "*The Design of Single Story Rigid Frames*," Metal Building Manufacturer' Association, Cleveland.

Morrell, M. L., and Lee, G. C. (1974), "Allowable Stress for Web-Tapered Beans with Lacteral Restraints," *Weld. Res. Counc. Bull.*, No. 192, February.

Nethercot, D. A. (1973), "Lateral Buckling of Tapered I-Beams," *IABSE Publ.*, Vol. 33-II, pp. 173–192.

Prawel, S. P., Jr., Morrell, M. L., and Lee, G. C. (1974), "Bending and Buckling Strength of Tapered Members," *Weld. Res. J. Suppl.*, p. 75S.

Tuma, J. J., and Munshi, R. K. (1971), *Advanced Structural Analysis*, Schaum's Outline of Theory and Problems, McGraw-Hill, New York.

Chapter Ten

Composite Columns

10.1 INTRODUCTION

Concrete can be used with strutural steel shapes, pipes, or tubes for compression members that are called composite columns. Steel pipe or tubing filled with concrete offers the most efficient use of the two basic materials. Steel at the perimeter of cross sections provides stiffness and triaxial confinement and the concrete core resists compression and prohibits local elastic buckling of the steel encasement. The toughness and ductility of concrete filled tube composite columns make them the preferred column type for earthquake resistance structures in Japan.

Concrete-encased structural shape composite columns currently are used in two applications. Perhaps the most popular application involves the perimeter columns in tube-type high-rise structures. The steel shape is designed to support several floors of the structural system, and concrete encasement is added around the steel shapes after construction of the building frame has proceeded several floors ahead of the concrete work. Steel shapes for such composite columns are relatively smaller than those used in the second application. The second application involves the use of heavy steel shapes inside a concrete column that would require more area of reinforcing bars than the maximum amount permited for reinforced concrete construction. There is no upper limit to the percentage of structural steel in a composite column cross section.

Structural shapes have been used in composite wall structures for both applications suggested above. Either the shapes form part of the vertical load system during construction or the shapes are used in lieu of excessive amounts of reinforcing bars to resist the loads applied to shear walls.

Figure 10.1 displays typical cross-section details for some composite columns and walls. North American practice has required auxiliary bar reinforcement in addition to steel shapes. The minimum amount of

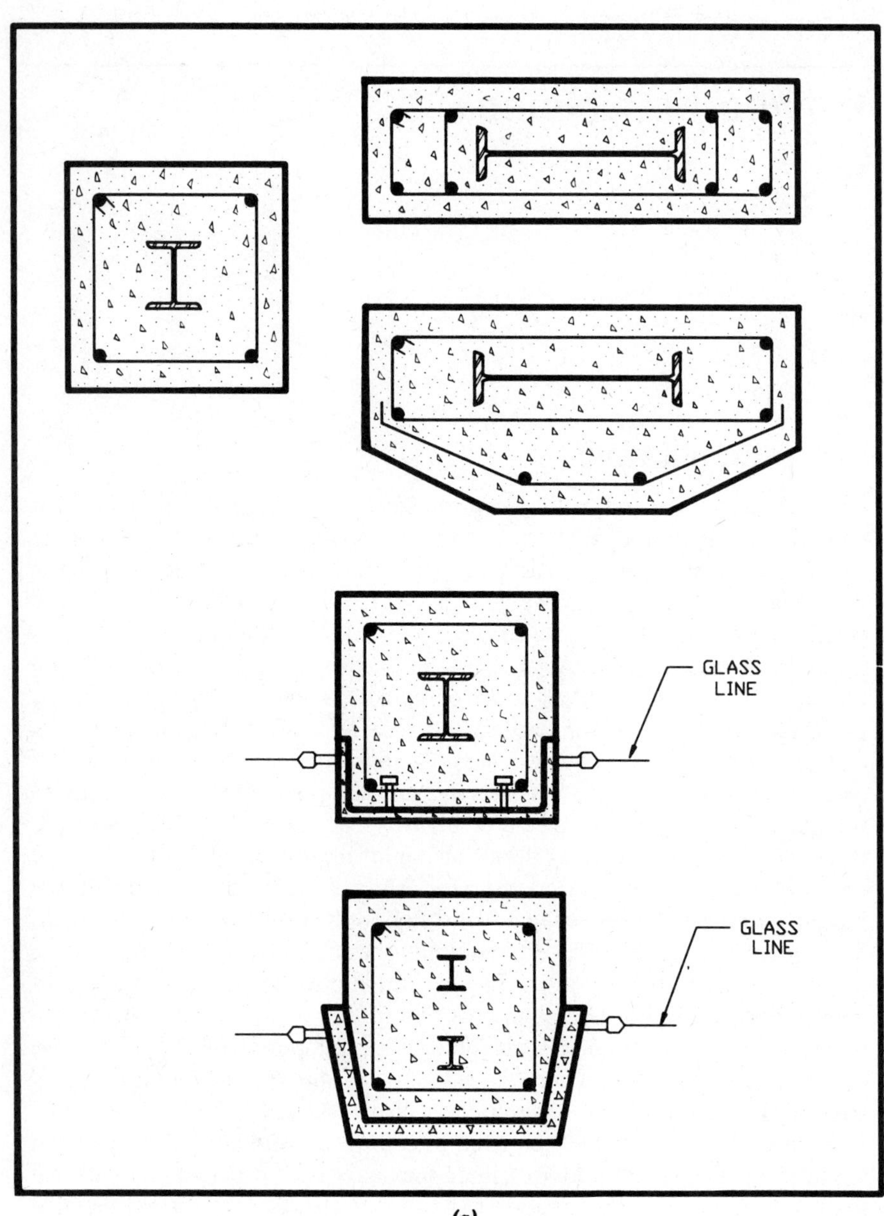

(a)

Fig. 10.1 Composite column types.

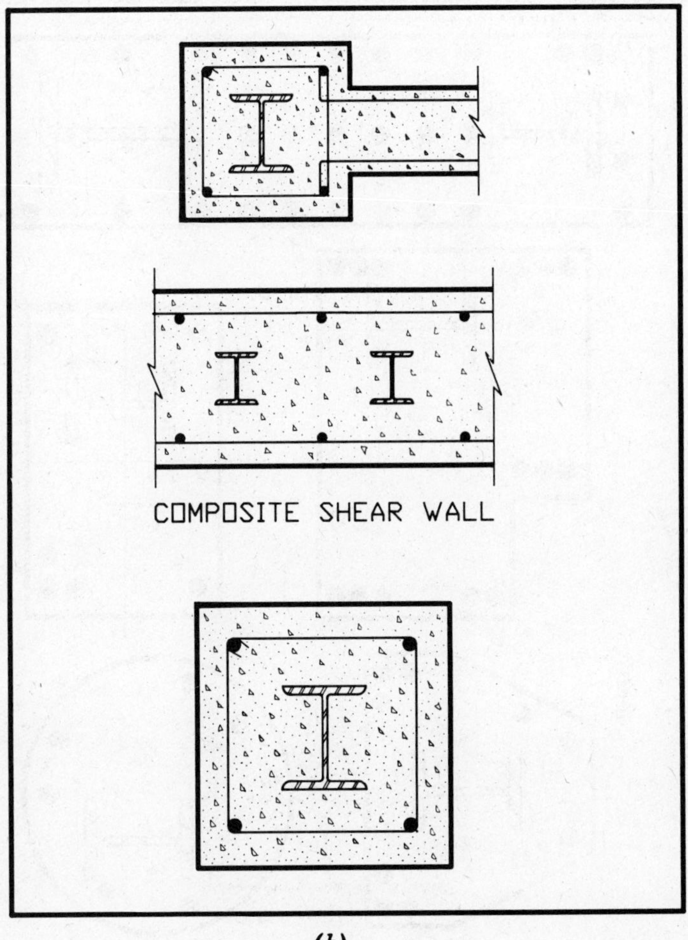

(b)

Fig. 10.1(*b*)

longitudinal and transverse reinforcement is roughly equal to the minimum specified for tied concrete columns.

A summary of the evolution in design methods for composite columns with brief discussions of the research that has lead to various formulas for strength is included in Chapter 19 of the third edition of this guide. The strength of composite column cross sections can be estimated with acceptable confidence. However, the flexural stiffness of composite columns that are loaded to the limit state of strength cannot be estimated with much precision, so some amount of caution is exercised in designing for the effects of slenderness.

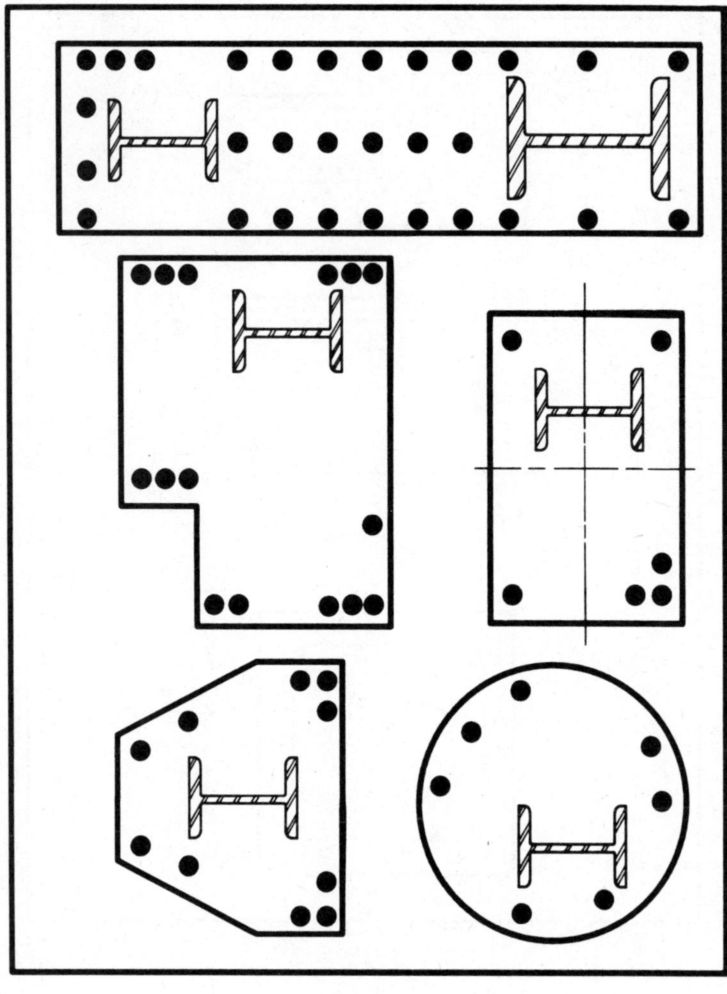

(c)

Fig. 10.1(c)

Composite columns involve both structural steel and reinforced concrete. Traditionally, design rules for steel have been promulgated by the steel industry, and rules for reinforced concrete by the cement and reinforcing bar industries. In North America a joint Structural Specifications Liaison Committee (SSLC) was formed in 1978 and chaired by the late George Winter. The committee was asked to recomment composite column design procedures that would be acceptable both to the steel and the concrete design groups.

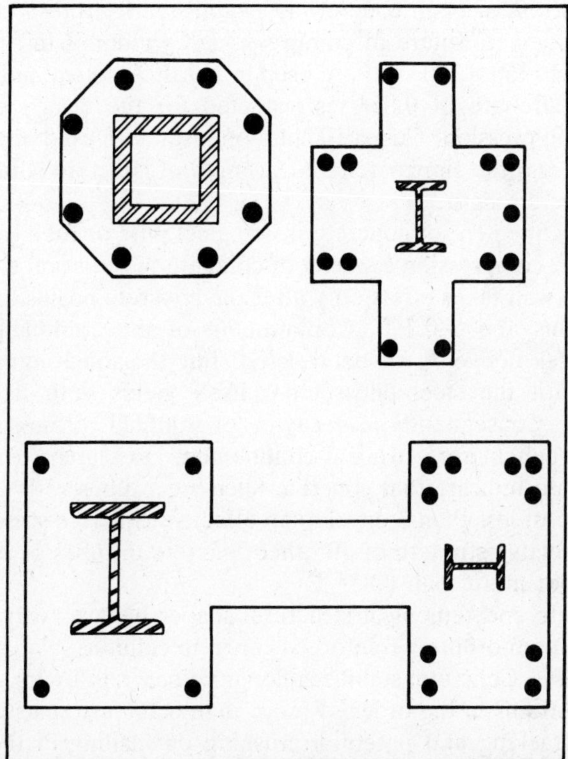

Fig. 10.1(*d*)

10.2 STRENGTH OF COMPOSITE CROSS SECTIONS

The strength of very short composite columns is the sum of the strengths of the materials that comprise the cross section (Basu, 1967; Emperger, 1907; Furlong, 1967; Klöppel and Goder, 1957; Knowles and Park, 1969; Mensch, 1917; Talbot and Lord, 1912). Structural steel is sufficiently ductile that its strength can be taken as the product of yield stress and cross-section area, with little attention to a limit strain. Concrete is less ductile, and its strength cannot be considered apart from strain limits. A lower-bound compression strain limit for concrete in flexure has been taken as 0.3% (ACI, 1980). In the absence of a strain gradient (from flexure) plain concrete cannot sustain significant strength at strains above 0.2%. Concrete under triaxial compression retains its strength at strains far in excess of 0.3%.

The limit stress of plain concrete is established from tests of standard specimens loaded to failure in compression. Cylinders 6 in. in diameter and 12 in. long (150×300 mm) are used in North America, and a reliable lower-bound strength of $0.85f'_c$ is accepted for the design strength of concrete in compression. Cubes 100 mm on a side are used in many parts of the world, and the similar reliable strength of concrete is taken as 0.7 times the cube strength.

Triaxial confinement of concrete inside steel pipe or tube can increase beyond f'_c, the compression capacity of concrete. The lateral confinement from the tube wall takes effect only after the concrete begins internally to crack at strains above 0.1%. Containment of the crumbling concrete permits stresses above f'_c to be resisted, but the longitudinal stiffness diminishes until the steel encasement itself yields both laterally and longitudinally. Consequently, a decay in longitudinal stiffness invalidates any practical benefit from triaxial confinement. Research (Tomii et al., 1973) results do indicate that concrete-filled steel tubes with a height-to-diameter ratio of less than 3 develop an effective concrete stress in excess of f'_c. The effective strength of all other concrete in tubes or pipe should be taken as not more than $0.85f'_c$.

The concrete encasement of structural shapes behaves very much like that of concrete in ordinary reinforced concrete columns (Varghese, 1961; Stevens, 1965). Concrete stabilizes longitudinal reinforcing bars until compression strains in the order of more than 0.2% are reached, beyond which microcracking and potential crushing or spalling of the concrete eliminate stability for the steel, which yields at strains of similar amount.

An analysis of cross-section strength can be made with the assumption that strains vary linearly across the section. For any assumed distribution of strain, the corresponding stresses on steel and concrete can be integrated in order to determine the axial force (thrust) and flexural force (moment) which created the assumed strains. Typical stress-strain characteristics for steel and for concrete are illustrated in Fig. 10.2. Analytical functions or numerical coordinates for stress-strain curves can be used for strength computations. The limit strength of column cross sections then can be displayed as a family of moment versus curvature graphs for various vales of thrust as illustrated in Fig. 10.3. Limit flexural strength is the peak value of a moment curvature graph. There are some simplifying assumptions that can be made to permit less tedious computations in order to determine limit moments for various values of thrust.

The principles for strength analysis of reinforced concrete can be applied to composite columns (Furlong, 1968; ACI, 1971). Limit strength is undervalued only slightly if it were assumed that cross section strength is exhausted whenever a strain of 0.3% exists on any concrete fiber. The

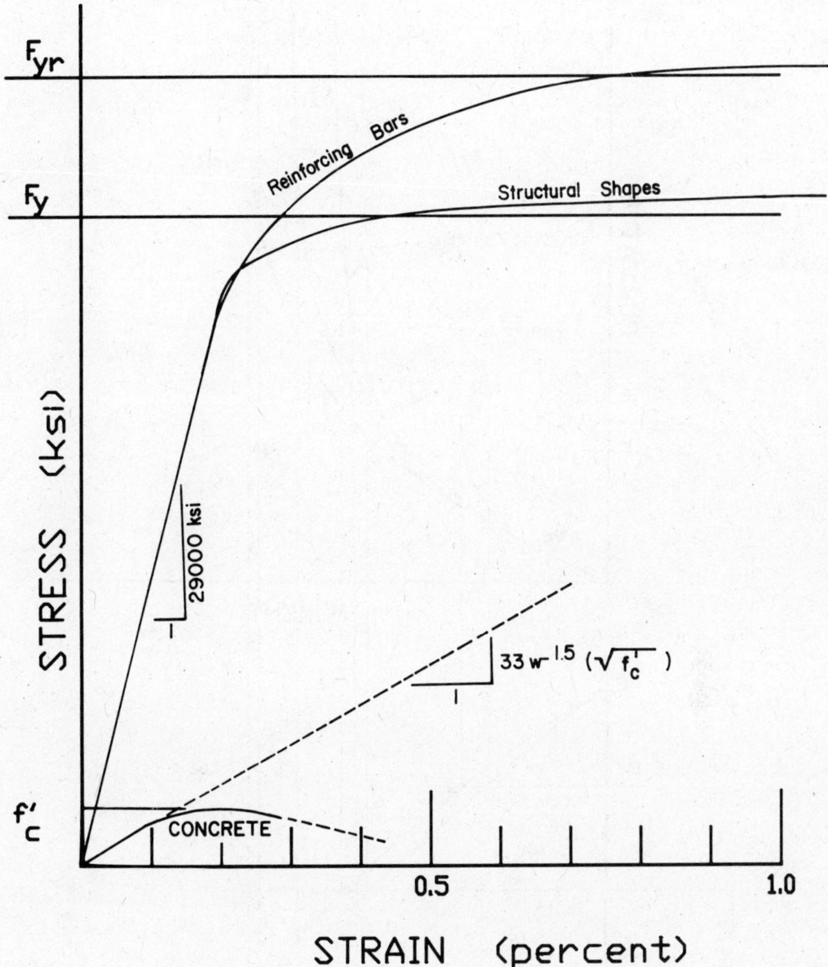

Fig. 10.2 Stress-strain characteristics for steel and concrete.

assignment of a neutral axis together with an edge strain of 0.3% on concrete defines a plane of failure strain. The stress-strain graph for concrete can be simplified by replacing the curve with a rectangular stress block that has a constant stress intensity of $0.85f'_c$ and a depth taken as a coefficient k_1 times the distance from the neutral axis to the fiber at 0.3% strain. A value $k_1 = 0.85$ should be decreased 0.05 for each 1000 psi by which f'_c exceeds 4000 psi. A lower-bound value for $k_1 = 0.65$ occurs when f'_c is greater than 8000 psi.

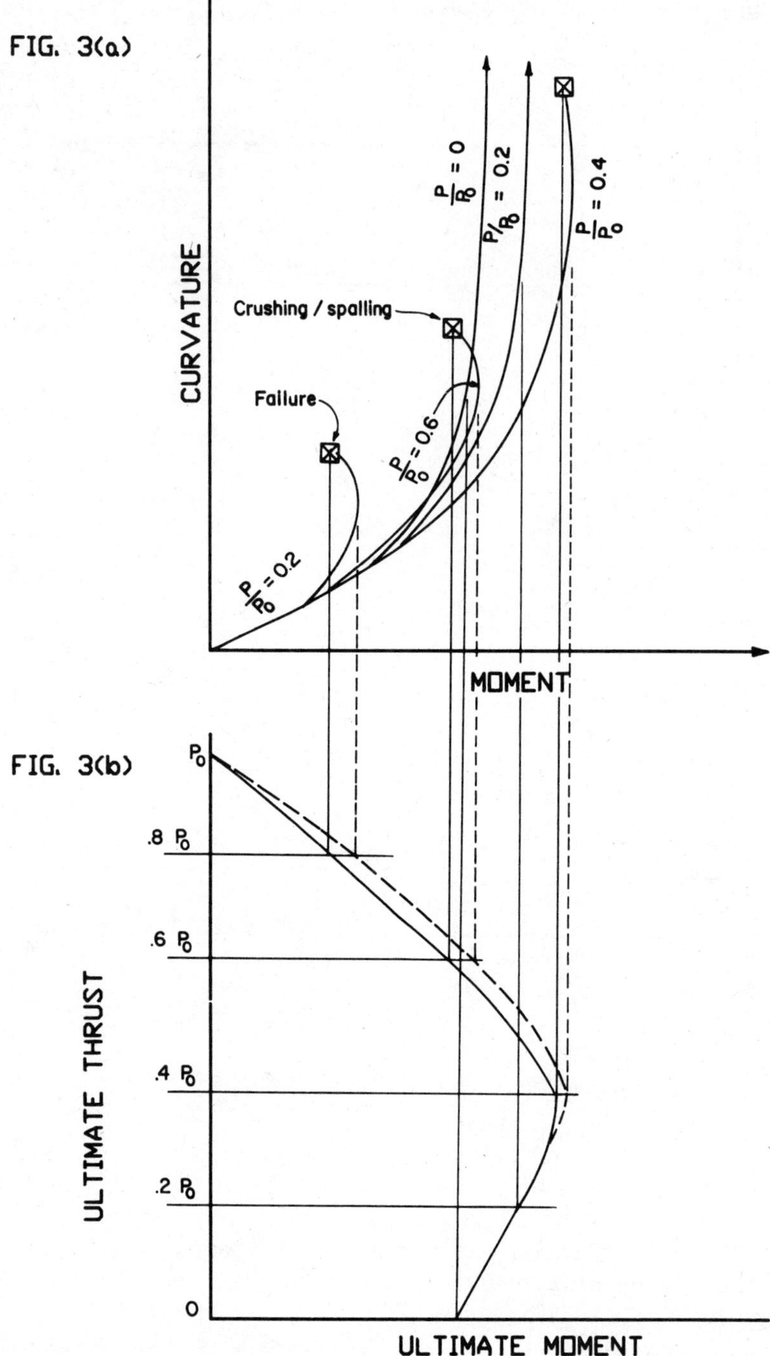

Fig. 10.3 Moment-curvature and moment-thrust strength graphs.

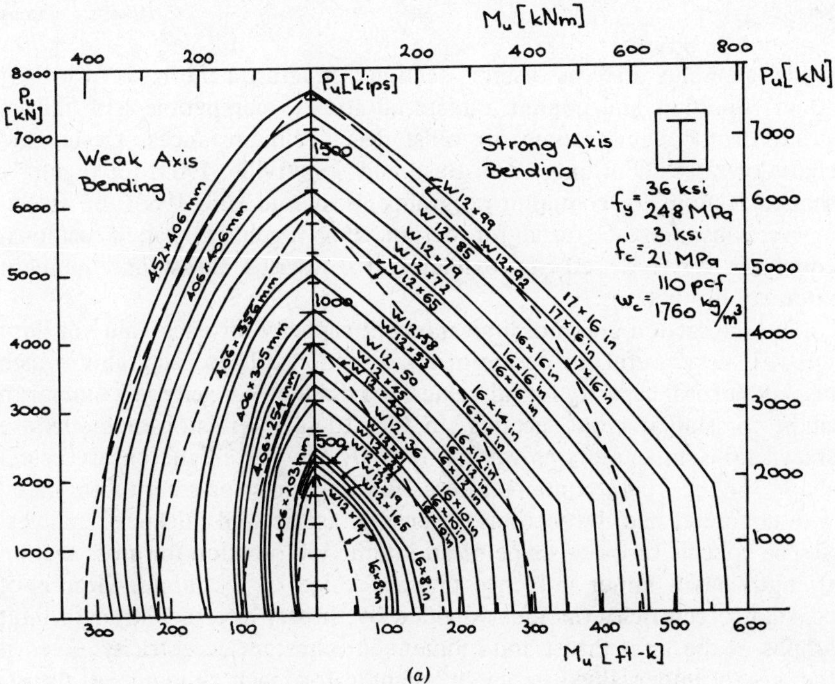

(a)

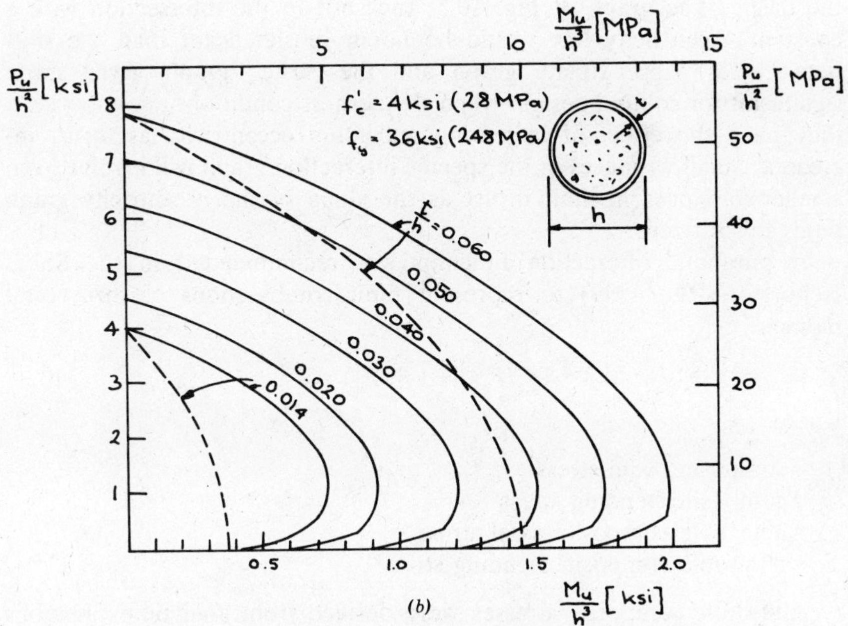

(b)

Fig. 10.4 Interaction plots for composite cross sections.

The complex analysis of cross-section strength on the basis of a limit strain condition and nonlinear material stresses compatible with failure strains is too cumbersome for most design circumstances. Design aid charts or tables (Furlong, 1974; Basu and Somerville, 1969) for regularly shaped sections, or computer programs (Basu and Hill, 1968) for irregular sections must be used for design. For regularly shaped sections, equations can be used approximately to describe thrust and moment interaction functions.

The interaction graph of limit moment capacity for each value of limit thrust is very useful for design of cross sections. These diagrams, which display flexural capacity for all values of axial load capacity, are similar in shape for similar cross sections. In Fig. 10.4*a* interaction diagrams are shown for similar sections varying only by the size of the encased steel shape. Figure 10.4*b* contains interaction diagrams for steel tubing filled with concrete, and the sections differ only as the wall thickness changes.

The specific or exact shape of an interaction function for cross-section strength has a minor influence on the design of a column. Graphs of constant eccentricity (moment divided by thrust) intersect strength limit graphs at the limit thrust and moment for constant eccentricity. Even if the graph undervalued moment slightly for each amount of thrust, shifting the graph to the left of the "true" strength curve, as suggested by the dashed-line graph of Fig. 10.5, the shift in the intersection with a constant eccentricity line would be no more significant than the shift between the approximate graph and the "true" graph. Even more significant for column design, when slenderness conditions are important, they have the effect of increasing maximum eccentricity as thrust increases. Small variances in the specific interaction graph will involve even smaller variances in limit thrust as the slope of the eccentricity graph tends toward zero.

A parabolic interaction function was recommended in the SSLC Report (SSRC, 1979) to represent limit combinations of thrust and moment.

$$(f_a/F_a)^2 + (f_b/F_b) = 1 \tag{10.1}$$

where

f_a = composite axial stress
f_b = composite bending stress
F_a = allowable composite axial stress
F_b = allowable composite bending stress

Allowable composite stresses were derived from analytic expressions for limit strength. The expressions for limit strength then were adopted

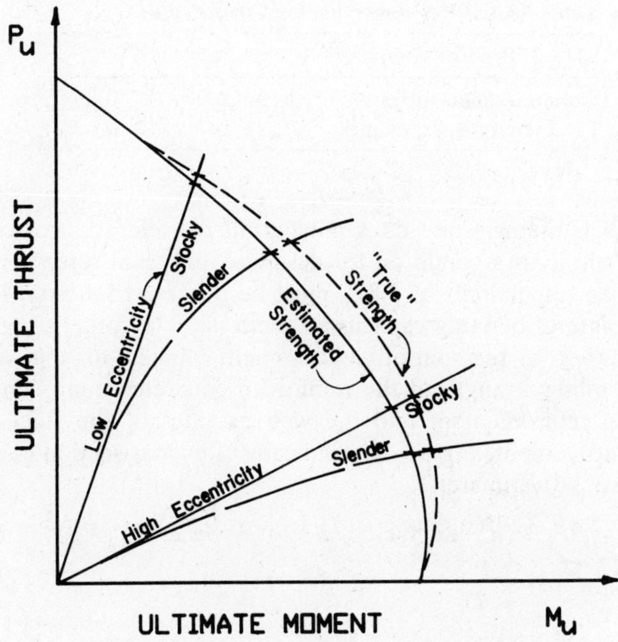

Fig. 10.5 Thrust-moment interaction limit strength.

for the AISC Load and Resistance Factor Design (LRFD) procedures (AISC, 1986). The LRFD document recommends a linear interaction relationship between P_u and M_u.

Limit concentric thrust P_0 can be expressed analytically:

$$P_0 = A_s F_y + c_1 A_r F_{yr} + c_2 f_c' A_c \tag{10.2}$$

where

A_s = area of structural steel
A_r = area of longitudinal reinforcement
c_1, c_2 = constants shown in Table 10.1
f_c' = strength of standard concrete cylinders
F_y = yield strength of structural steel
F_{yr} = yield strength of reinforcing bars

A composite axial stress limit F_{my} then can be derived from Eq. 10.2.

$$F_{my} = P_0/A_s = F_y + c_1 F_{yr} A_r/A_s + c_2 f_c' A_c/A_s \tag{10.2a}$$

Table 10.1. Resistance Factor Coefficients

Type of Column	c_1	c_2
Concrete-filled tubes	1.0	0.85
Concrete-encased shapes	0.7	0.6

The ACI Building Code does not permit consideration of concentric axial force alone as a condition for design. Limit axial force is taken not as P_0, but an upper limit of $0.8P_0$ must be observed for axial force.

An estimate of bending capacity M_0 with no axial force is suggested in the SSLC report as the sum of yield moment capacity for the steel shape plus the bending strength of the reinforced concrete beam, consisting of longitudinal reinforcement and the web or sides of the steel shape as tensile reinforcement. Figure 10.6 indicates the cross-section components used for the M_0 estimate.

$$M_0 = SF_y + (h - 2c_c)A_rF_{yr}/2 + A_wF_y(h/2 - A_wF_y/1.7f'_cb) \tag{10.3}$$

where

A_w = area of web of encased shape or sides of tube
b = width of rectangular cross section
c_c = distance from center of bar to edge of section
h = total thickness (depth) of section
S = section modulus of structural steel

Equation 10.3 is useful for regular and symmetrical cross sections. The equation can be used to produce an ultimate composite bending stress F_{mb} by dividing both sides by the steel shape section modulus S.

$$F_{mb} = M_0/S = F_y + (h - 2c_c)A_rF_{yr}/2S + A_wF_y(h/2 - A_wF_y/1.7f'_cb)/S \tag{10.3a}$$

The SSLC report employed an allowable stress

$F_{mb} = 0.60F_y$ for concrete encased composite columns
$F_{mb} = 0.75F_y$ for steel tubes filled with concrete

These values of F_{mb} can be used in Eq. 10.1 for F_b.

The analysis of bending strength for unsymmetric or irregular cross sections requires consideration of limit stresses in steel and concrete as failure occurs. The principles of strength analysis for reinforced concrete are applicable also to composite steel and concrete members. There are no test data for composite walls loaded to failure from in-plane bending,

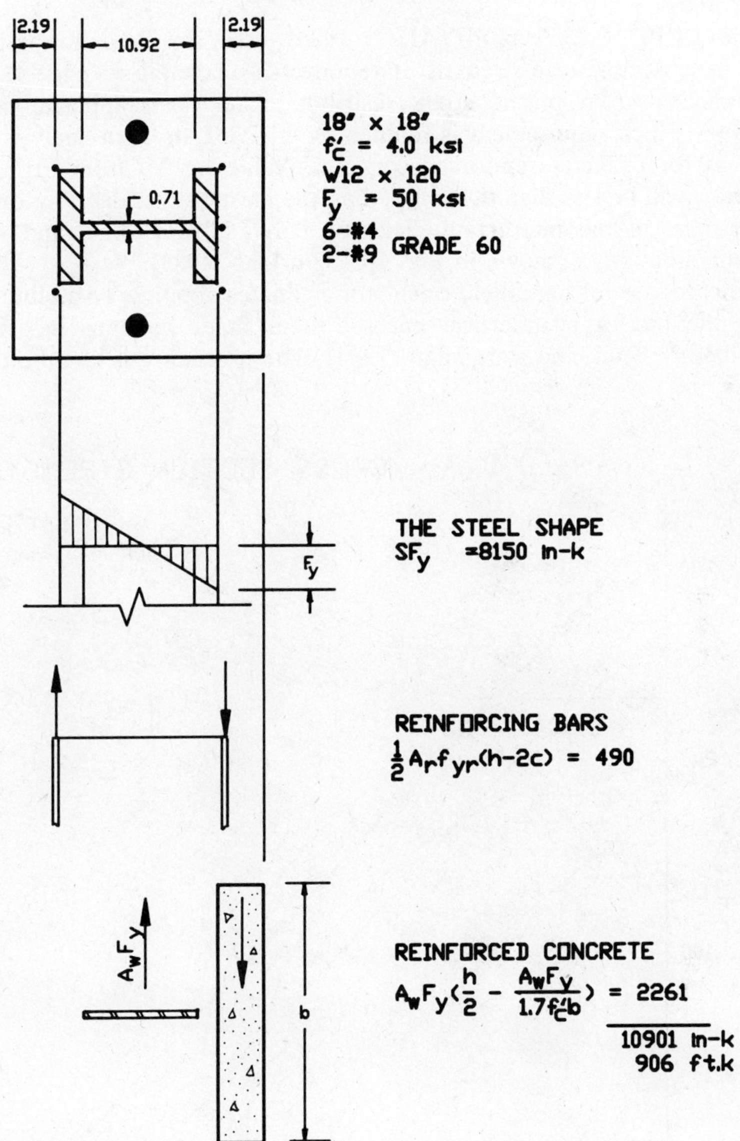

Fig. 10.6 SSLC procedure for M.

nor is there any laboratory evidence of design requirements that would ensure shear transfer between structural shapes and concrete in such applications.

The LRFD Specification (AISC, 1986) uses Eq. 10.2 for axially concentric compressive strength. It requires that flexural strength M_0 be determined "from plastic stress distribution on the composite cross section," which requirement is fulfilled by applying strength analysis the same as for ordinary reinforced concrete. Values of M_0 from Eq. 10.3 generally will be less than those based on the plastic stress distribution. A comparison of the interaction diagrams derived from SSLC and ACI recommendations is shown in Fig. 10.7 (Furlong, 1983).

A procedure for hand-held calculator estimates of points on an interaction function for symmetrical encased steel shape columns has been described by Roik and Bergmann (1984). The procedure is based on the

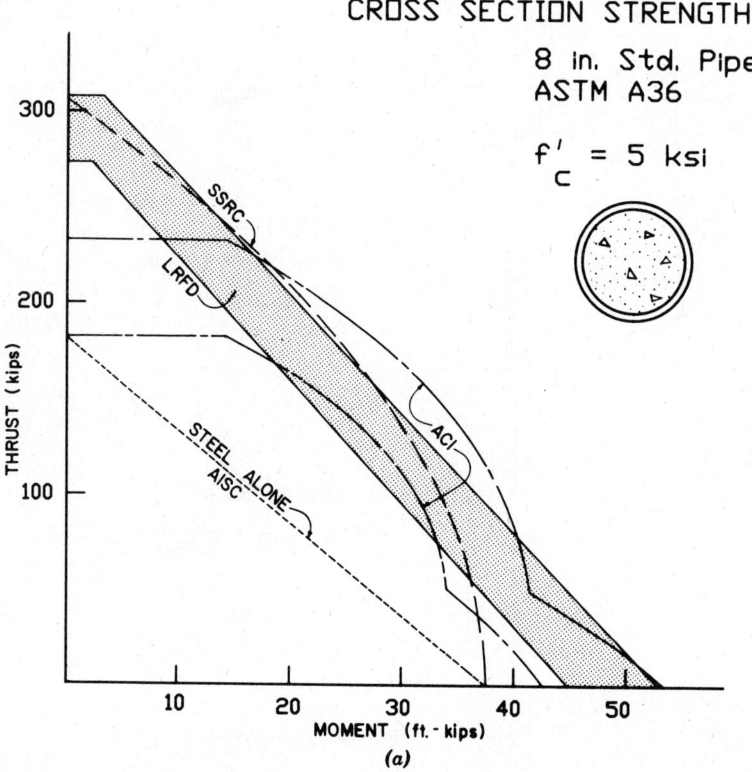

(a)

Fig. 10.7 Comparison of design methods.

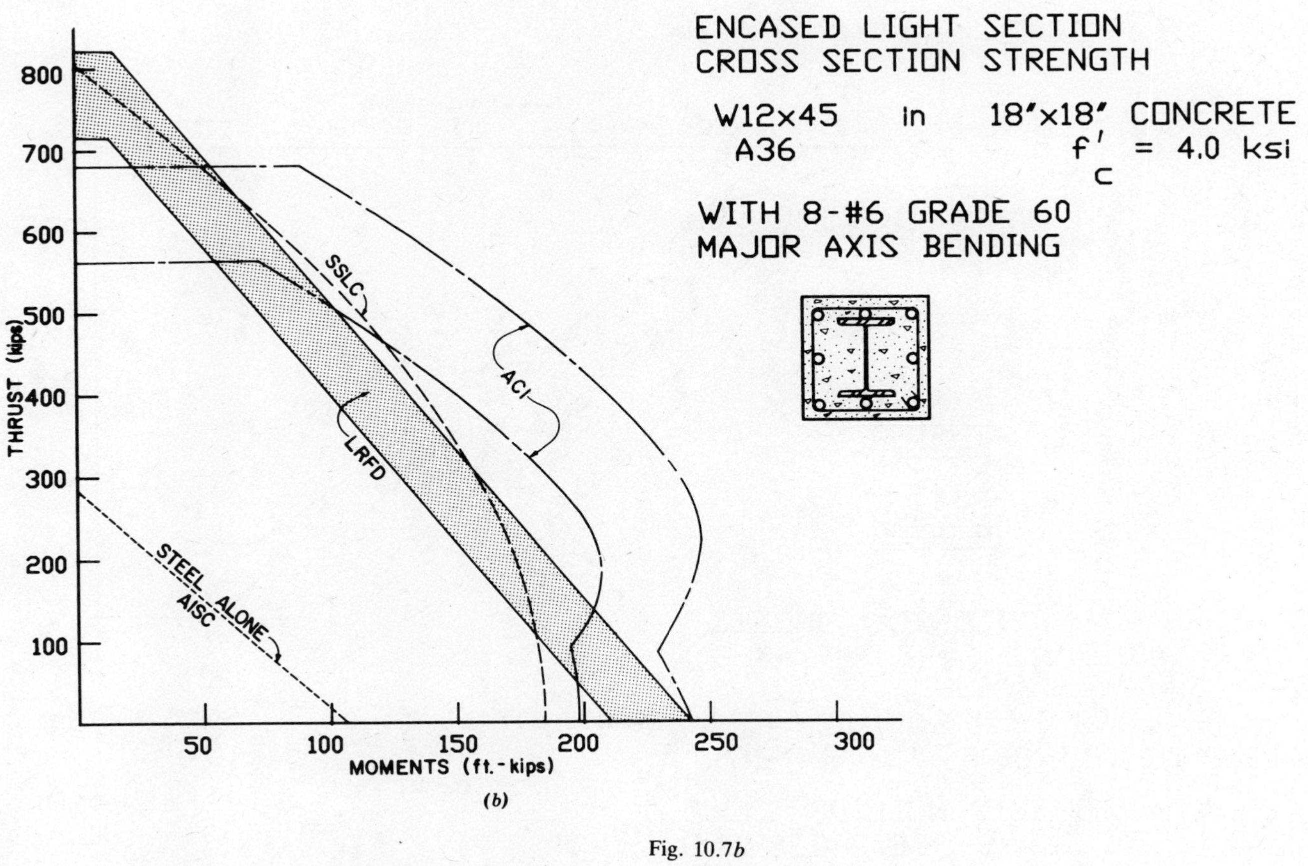
ENCASED LIGHT SECTION
CROSS SECTION STRENGTH
W12x45 in 18"x18" CONCRETE
A36 f'c = 4.0 ksi
WITH 8-#6 GRADE 60
MAJOR AXIS BENDING
THRUST (kips)
800
700
600
500
400
300
200
100
50
100
150
200
250
300
MOMENTS (ft.-kips)
SSLC
ACI
LRFD
STEEL ALONE
AISC
(b)

Fig. 10.7*b*

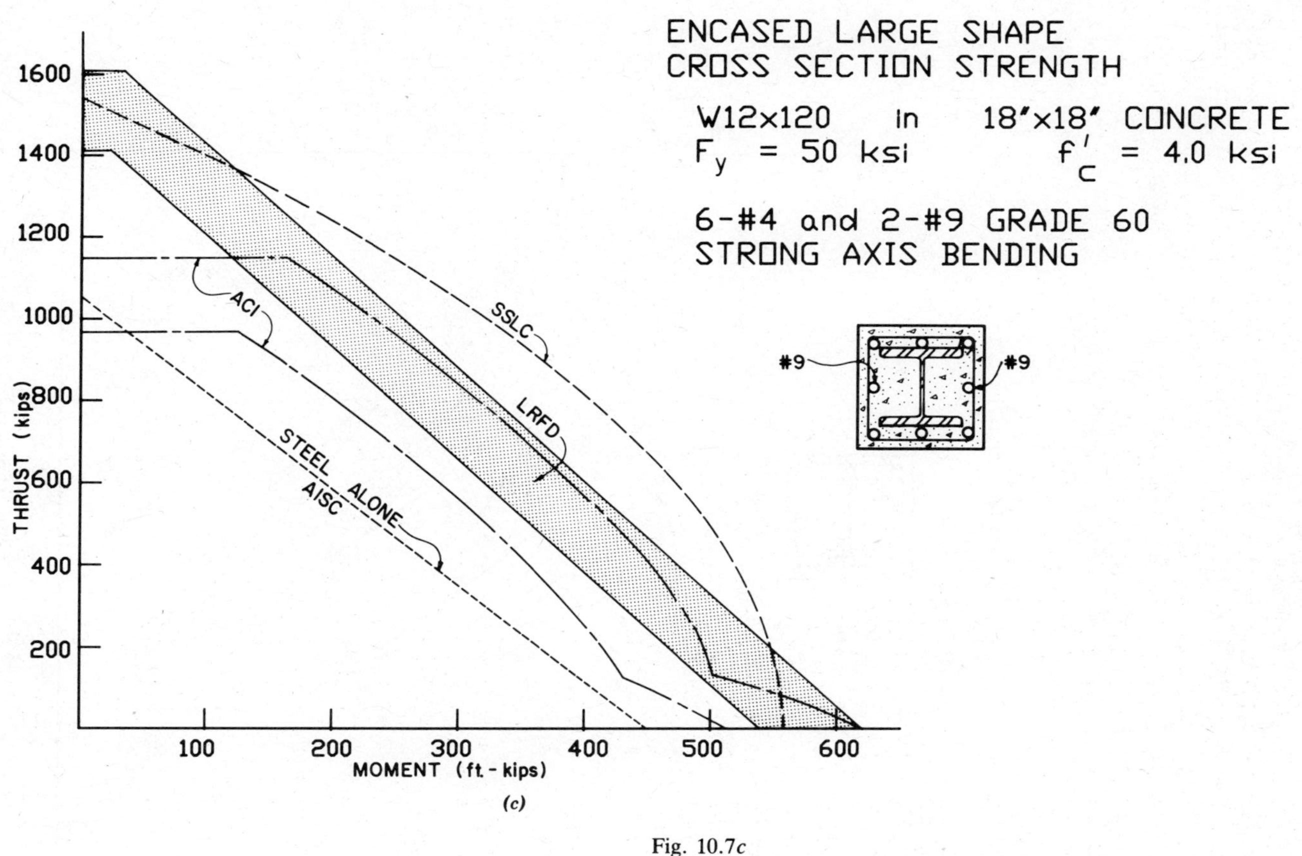
ENCASED LARGE SHAPE
CROSS SECTION STRENGTH
W12x120 in 18″x18″ CONCRETE
F_y = 50 ksi f′_c = 4.0 ksi
6-#4 and 2-#9 GRADE 60
STRONG AXIS BENDING
#9
#9
SSLC
LRFD
ACI
STEEL ALONE
AISC
THRUST (kips)
1600
1400
1200
1000
800
600
400
200
MOMENT (ft - kips)
100
200
300
400
500
600
(c)

Fig. 10.7*c*

assumption of a plastic stress distribution for both concrete and steel in order to simplify equilibrium relationships for hand calculation. The value of P_0 is obtained with Eq. 10.2 when $c_1 = 1$ and $c_2 = 0.85$. When the neutral axis is within the cross section, a "plastic" stress intensity, taken as 60% of the strength of standard cubes, is converted for this application to be 75% of the strength f_c' for standard cylinders.

The value of M_0 exists with a neutral axis located such that the total axial force on the section is zero. For symmetrical sections, the steel

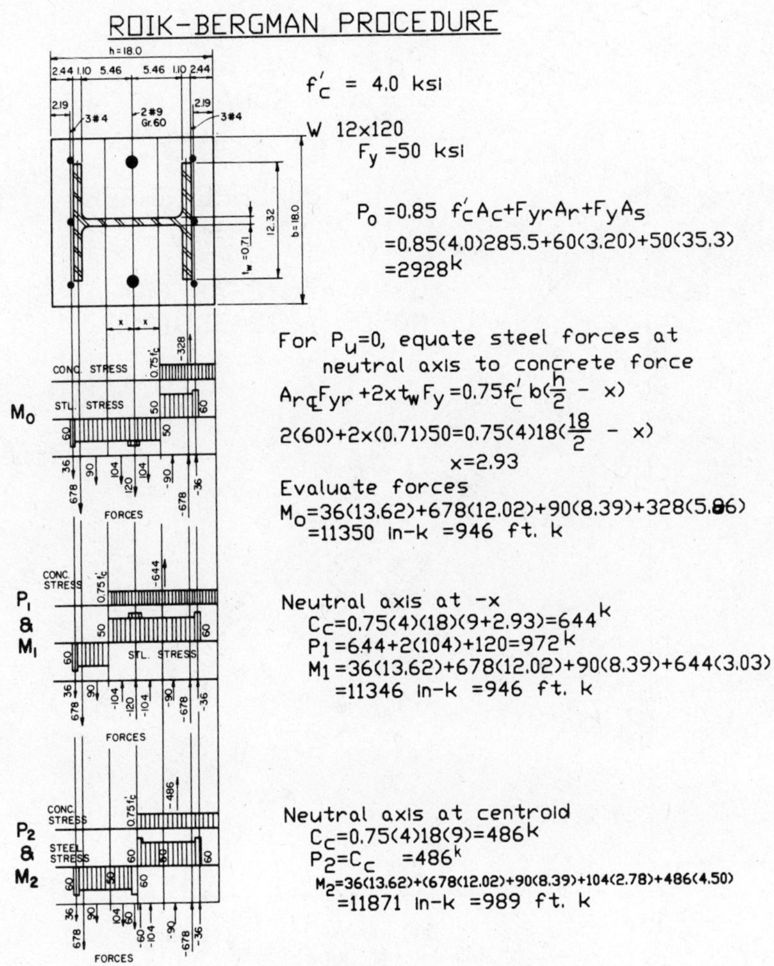

Fig. 10.8 Estimating points on interaction diagrams.

tensile and compressive components of force on opposite sides of the neutral axis will be equal and opposite. A portion of the steel at the neutral axis must equilibrate the concrete component of compression, and a neutral-axis location at a distance x from the section centroid can be computed. A plastic stress distribution for determining M_0 is illustrated in Fig. 10.8b. Two additional data points can be determined readily after M_0 is computed. The values of P_1 and M_1 are obtained as indicated in Fig. 10.8c when the neutral axis is located at the distance x toward the tension edge of the section. The values of P_2 and M_2 are then obtained as shown

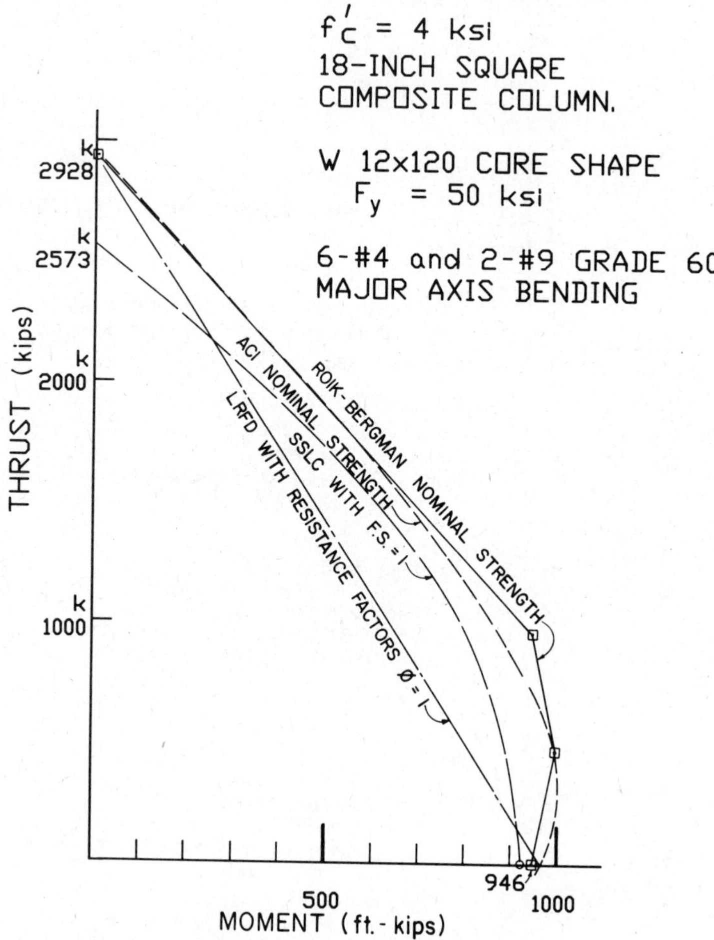

Fig. 10.9 Comparison of limit strength interaction functions.

in Fig. 10.8*d* with the neutral axis on the centroid of the cross section. The interaction diagram then consists of straight lines connecting P_0, P_1M_1, P_2M_2, and M_0.

Interaction diagrams determined for the example cross section are shown in Fig. 10.9. The Roik-Bergmann procedure is shown to produce a diagram very similar to that obtained from the much more complex ACI compatible strain procedure. The SSLC Eq. 10.1 with unity for the factor of safety and with M_0 determined with Eq. 10.3 was used also for a strength interaction graph shown in Fig. 10.9. The linear interaction relationship recommended in the LRFD specification is shown as a strength function with resistance factors taken as unity.

10.3 STRENGTH OF SLENDER COMPOSITE COLUMNS

The influence of slenderness in compression members is dependant on flexural stiffness and the effective length of the column. The flexural stiffness of structural steel components of a cross section are more readily estimated than are those for reinforced concrete components, and the composite interaction between the two is even less readily estimable. The extent and distribution of cracks due to the low tensile strength of concrete creates unpredictable variations in the effective stiffness EI through the height of composite columns before buckling failure occurs. Not only is concrete less homogeneous than steel, but the apparent value of E_c is reduced by sustained load. The flexural stiffness EI for a concrete-filled steel tube will be influenced much more by the tube wall than by the contained concrete. Conversely, the effective stiffness EI of a concrete-encased composite member will be influenced more by the concrete than by the structural steel.

The SSLC report (SSRC, 1979) recommended equations that produced for composite columns values of composite stiffness E_m that would be similar to those obtained from requirements of the ACI Building Code (ACI, 1980). The composite material stiffness E_m is expressed as

$$E_m = E_s + c_3 E_c A_c / A_s \tag{10.4}$$

where

E_c = modulus of elasticity for concrete
E_s = modulus of elasticity for steel
c_3 = 0.4 for filled steel tubes
= 0.2 for concrete encased steel sections

The relatively low values for the coefficient c are a lower-bound accommodation for the combined effects from flexural cracking and inelastic material response under limit compressive strains. The ACI Building Code further reduces the coefficient c by dividing it by a creep influence factor equal to 1 plus the ratio between the design dead load moment and the design total load moment. The influence of creep is ignored for concrete in composite columns according to both the SSLC report and the LRFD Specification, largely due to the fact that the relatively high percentage of steel in composite sections (4% minimum) makes small variations in concrete stiffness less significant than in less heavily reinforced cross sections.

It should be noted that Eq. 10.4 is in units of stress, and the equation does not recognize any contribution from longitudinal reinforcement in the section. The composite material stiffness E_m is then used in equations for buckling stress. Composite cross sections for which it is apparent that concrete and reinforcement have a predominant influence on flexural stiffness should be treated as reinforced concrete with an effective stiffness EI such that

$$EI = E_s(I_r + I_s) + c_3 E_c I_c/(1 + M_d/M_u) \tag{10.5}$$

where

I_r = moment of inertia for reinforcing bars
I_c = moment of inertia for concrete
M_d = design dead load moment
M_u = design total load moment

A slenderness ratio C_c at which Euler buckling can be assumed to occur can be determined with the composite material stiffness from Eq. 10.4, and an allowable axial stress F_a for composite columns was proposed by the SSLC report.

$$C_c = \pi\sqrt{2E_m/F_{my}} \tag{10.6}$$

For $L_u/r_m < C_c$,

$$F_a = F_{my}[1 - 0.5(L_u/r_m C_c)^2]/\text{F.S.} \tag{10.7a}$$

for $L_u/r_m > C_c$,

$$F_a = E_m(r_m/L_u)^2/\text{F.S.} \tag{10.7b}$$

where

L_u = effective length of column
r_m = composite radius of gyration
F.S. = factor of safety
$= 5/3 + 0.375L_u/r_m C_c - 0.125(L_u/r_m C_c)^3$

The radius of gyration r_m could be taken as the larger of the radius of gyration for the structural shape alone or the radius of gyration for the concrete alone. The ACI Building Code permits the calculation of a composite radius of gyration not greater than

$$r_m = (0.2E_cI_c + E_sI_s + E_sI_r)/(0.2E_cA_c + E_sA_s + E_sA_r) \qquad (10.8)$$

Strictly interpreted, the ACI Building Code uses a gross moment of inertia I and a gross area A for the cross section bounded by concrete. The Building Code does not use the concrete area A and concrete moment of inertia I as defined here. The difference is not significant for concrete members, as more than 96% of the cross sections are concrete. The difference could be significant for composite sections for which steel could occupy more than 20% of gross cross-section area. However, even in members with more than 20% of their cross section composed of steel, the stiffness contributed by concrete will be considerably less than that from steel. Possible variations in actual E values have more influence on the EI values than do minor changes in cross-section area assumed for calculations.

Equation 10.7 is expressed as an allowable service load compression stress. The LRFD Specification employs the same logic as that from the SSLC report, but the LRFD specification for limit strength uses a factor of safety F.S. equal to $[5/3 + 3L_u/8r_mC_c - (L_u/r_mC_c)^3/8]/1.76 > 1.00$. This modification to the allowable stress factor of safety has the same effect as that which occurs if the F.S. $= 1$ for ratios L_u/r_mC_c less than 0.25. For ratios L_u/r_mC_c greater than 0.25, the limit axial strength becomes 1.76 times the allowable stress value determined from Eqs. 10.7a and 10.7b.

The ACI Building Code does not provide direct guidance for axial force in terms of column length or slenderness. Indirectly, the ACI requirement of a minimum eccentricity combined with the magnification of moments with conservatively flexible concrete members does permit the determination of force versus slenderness graphs for concrete columns. The ACI Code employs a capacity reduction factor ϕ to reflect the reliablility of buckling load estimates that are based on a material stiffness E_c. The overall caution advocated by the Code reflecting slenderness appears to be exacerbated by the combined use of capacity reduction factors both for stiffness and for short-column strength. The graphs of Fig. 10.10 display service load axial stress as a function of column length. The graphs were determined in accordance with ACI, SSLC, and LRFD rules for designing composite columns. It is apparent that the ACI rules are significantly more cautious about slenderness than are the SSLC or LRFD recommendations.

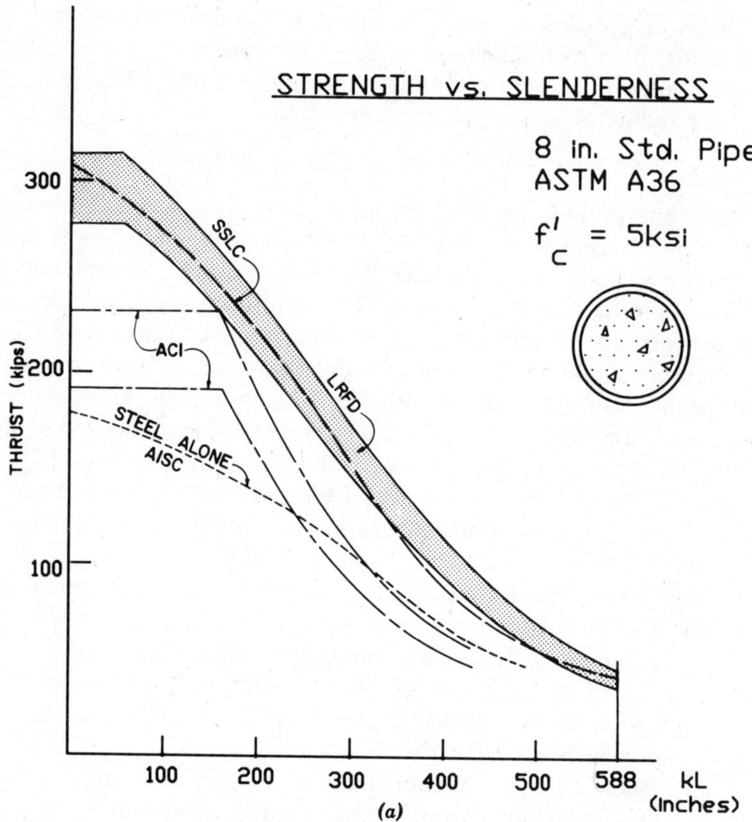

Fig. 10.10 Comparison of design methods for slenderness effects.

Under monotonic load tests until failure, the capacity of composite columns reflects a reliability much the same as that expected for structural steel or for reinforced concrete, including the effects of slenderness. Loading that creates load reversals with cracking of concrete has not been documented extensively except for studies of connections to floor systems. Studies have yet to be made to establish specifically the nature and methods for retention of shear transfer between strucural shapes and concrete along the length of composite columns.

Comparison of the various design methods are given in Fig. 10.10.

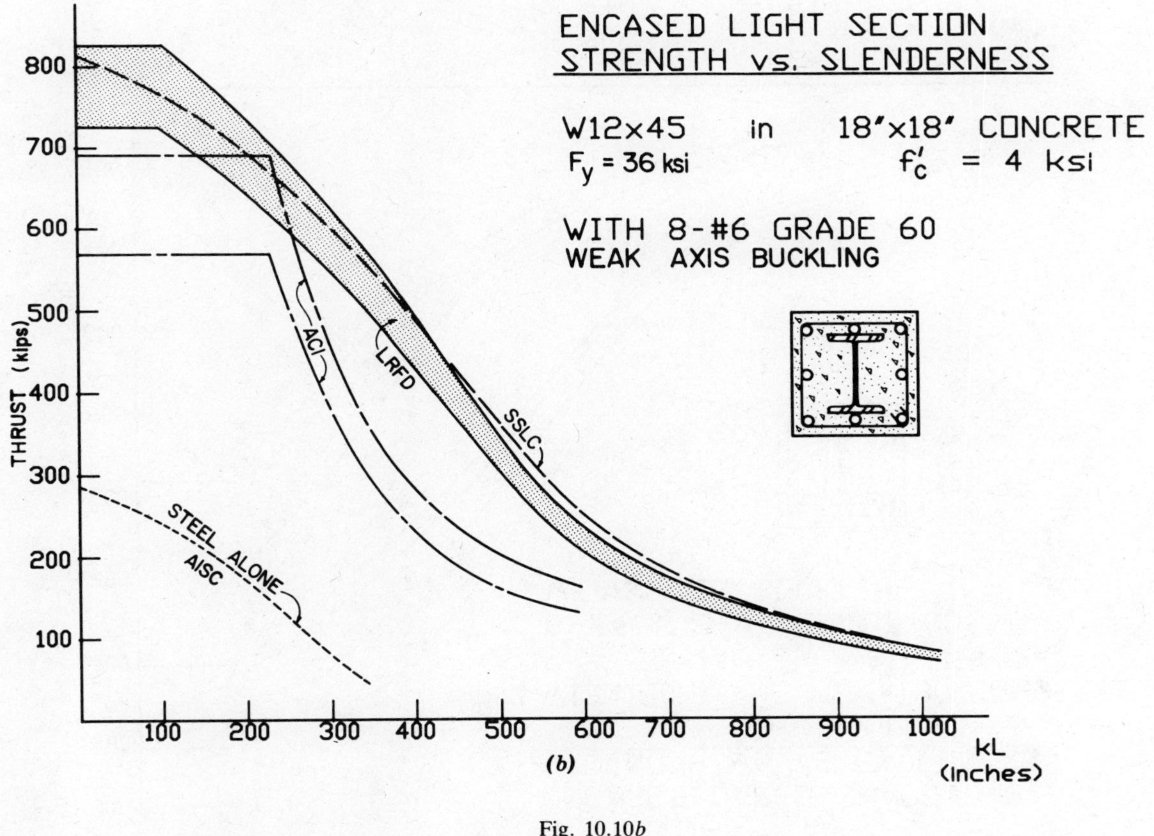
ENCASED LIGHT SECTION
STRENGTH vs. SLENDERNESS
W12×45 in 18″×18″ CONCRETE
F_y = 36 ksi
f′_c = 4 ksi
WITH 8-#6 GRADE 60
WEAK AXIS BUCKLING
SSLC
LRFD
ACI
STEEL ALONE
AISC
THRUST (kips)
800
700
600
500
400
300
200
100
100
200
300
400
500
600
700
800
900
1000
kL
(inches)
(b)

Fig. 10.10*b*

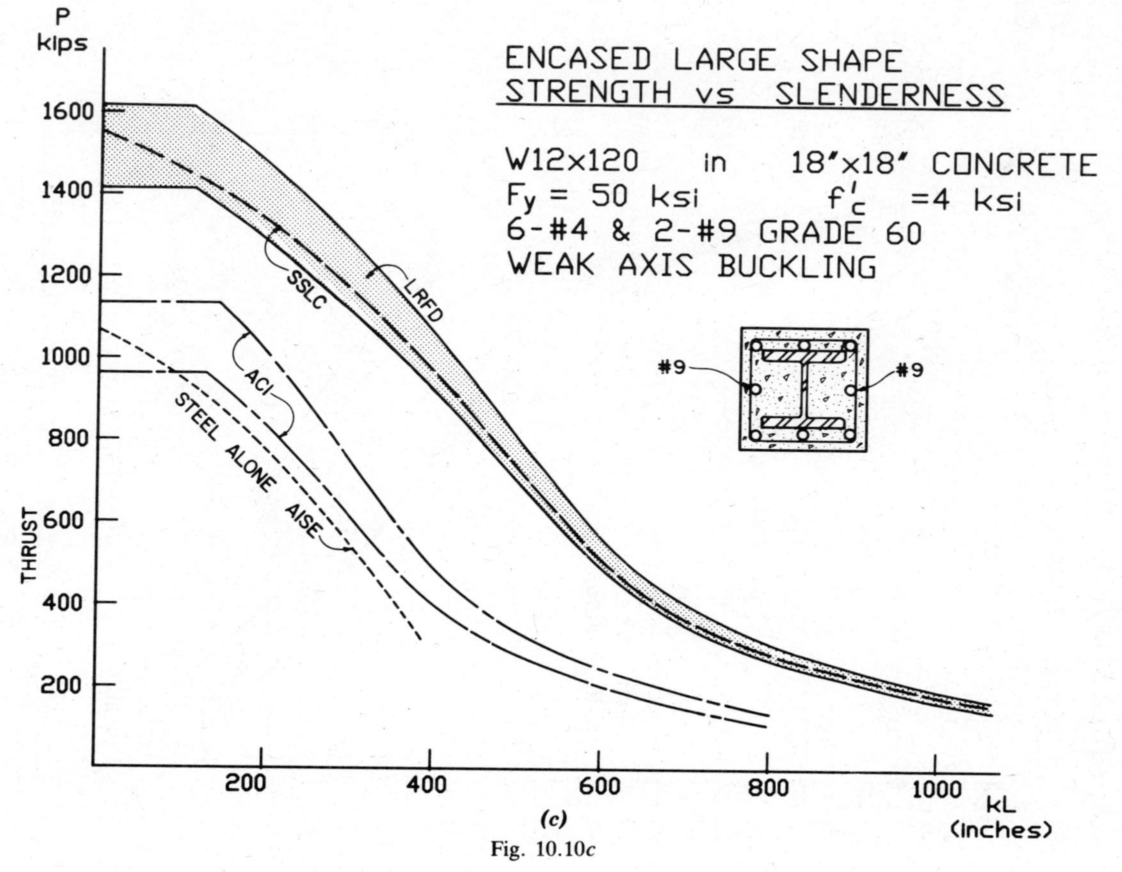

ENCASED LARGE SHAPE
STRENGTH vs SLENDERNESS
W12x120 in 18"x18" CONCRETE
Fy = 50 ksi f'c =4 ksi
6-#4 & 2-#9 GRADE 60
WEAK AXIS BUCKLING
#9
#9
LRFD
SSLC
ACI
STEEL ALONE
AISE
P
kips
THRUST
1600
1400
1200
1000
800
600
400
200
200
400
600
800
1000
kL
(inches)

(c)

Fig. 10.10c

10.4 SHEAR STRENGTH OF COMPOSITE COLUMNS

The shear strength of composite columns appears to be equal to the sum of shear capacity for the structural steel and the reinforced concrete components of the column, much the same as is the axial force capacity. (Kato, 1978). The shear strength of steel tubes filled with concrete should be taken the same as the shear strength of the tube with no additional strength from the plain concrete filling the tube. Concrete encasement that is reinforced with transverse and longitudinal bars around steel shapes does contribute to shear strength. The shear strength V_{us} of reinforcement in concrete can be taken as

$$V_{us} = F_{yr} A_v (h - c)/s \tag{10.9}$$

where

A_v = cross-section area (generally equal to two legs of a transverse tie) of transverse bars that intercept a diagonal shear crack.

s = longitudinal spacing of the transverse bars

The total shear capacity will be the shear strength of the structural steel encased shape plus the amount of V_{us}. Shear capacity for steel can be taken as the product of shear area and 60 percent of the yield strength of the steel shape.

10.5 CONNECTIONS TO COMPOSITE COLUMNS

Laboratory observations of composite columns involve loading arrangements carefully controlled to distribute applied forces appropriately into the steel and concrete components of the specimens. Strength estimates from such loading are acceptably accurate. Test loads applied only to the structural steel can cause the steel to yield locally, and test loads applied only to the concrete can cause local crushing (Furlong, 1967; Wakabayashi, 1984). In either case, the cross-section capacity is less than the sum of material strengths that provide a basis for member strength estimates.

If floor loads can be applied without local distress to the concrete or the steel part of a composite member, it is reasonable to assume that the entire cross section can be made to resist the floor loads if shear transfer mechanisms between steel and concrete exist along the length of the column. Effective transfer of shear between concrete and steel is required not only for the distribution of applied loads, but also to ensure the strength condition for which all components of cross sections strain at the same curvature before failure begins.

Shear is transferred between concrete and steel through several "mechanisms." Some adhesion and shear strength exists along the steel-concrete contact surfaces. The contact surface shear strength is improved by lateral binding reinforcement that helps maintain pressure of concrete against steel. If service loading or overload breaks the adhesion, slip can occur, and the contact surface shear transfer is significantly less effective thereafter. Shear transfer occurs also through any bearing surfaces on steel, such as those of deformation patterns on reinforcing bars. Physical "lugs" or deformation patterns maintain effective shear transfer even after overload events cause concrete to crack. Welded studs provide bearing surfaces for effective shear transfer, and they can help maintain effectiveness of the steel-concrete surface as well, even after severe overload and concrete cracking.

Tests and corresponding analysis of combined shear transfer mechanisms have yet to be made in order to reveal quantitative estimates of effectiveness through the entire range of pre-failure strain conditions. The strength of welded studs is well documented for shear transfer to composite concrete floor slabs. The same stud strength (limited by the bearing strength of the concrete against the stud) should apply equally well for concrete encasement of steel shapes. Shear transfer through the "development" or anchorage length of deformed bar reinforcement likewise is well documented for reinforced concrete, and the same effectiveness can be expected in composite compression members. The bond strength of weldable deformed bars attached to structural shapes should be assessed with due consideration that the steel-to-steel contact surfaces are not available for bond and shear transfer to concrete.

Structural steel floor members can be attached safely to the structural steel component of composite columns in the same way that such connections would be made with all-steel columns (Roeder, 1984). Subsequently, through shear transfer along the column, the concrete can be made to help resist the applied loads. Actually, the flanges of the steel floor beams may be in direct bearing against enough concrete to develop the concrete contribution of column strength. One specification (Roik and Bergmann, 1984) requires that web plate shear connections for steel tube columns filled with concrete be continuous through the concrete inside the tube. The reason given for the requirement is stated as an improvement to fire-resistance strength as the connection plate within the concrete core is protected against a rise in temperature.

Concrete floor systems connected to composite columns can develop the floor reaction by bearing against column concrete and by bearing against steel surfaces (brackets) attached to the steel core of the composite column. Concrete flat plate systems probably should contain shear head devices that can be attached directed to the steel core.

Isolated studies and reports of composite column connections indicate that strength is developed readily if weldments and steel and concrete bearing surfaces are proportioned adequately in accordance with principles of steel design or reinforced concrete design. The sharp difference in stiffness between steel brackets and reinforced concrete elements is always revealed by the location of initial cracks where the stiffness changes.

Reversed loading tests of butt welded steel connections encased in reinforced concrete at beam-to-column joints indicate excellent hysteresis loops and energy dissipation in force-deformation graphs (Wakabayashi and Minami, 1978). The toughness and damping influence of composite columns has been well documented from studies in Japan (JSCE, 1984). Japanese standards employ considerably more transverse binding reinforcement than that required by North American nonseismic standards.

Further documentation of steel-concrete connection studies will be necessary before analytic models can be recommended generally for design. In the absence of such studies, connections can be designed, probably conservatively, on the basis of steel-to-steel or concrete-to-concrete components added together for strength.

Rererences

American Concrete Institute (1971), *Commentary on Building Code Requirements for Reinforced Concrete*, ACI 318-71, ACI, Detroit, p. 40.

American Concrete Institute (1980), *Building Code Requirements for Reinforced Concrete*, ACI 318-83, ACI, Detroit.

American Institute of Steel Construction (1986), *Load and Resistance Factor Design Specification for Structural Steel Buildings*, AISC, Chicago.

Basu, A. K. (1967), "Computation of Failure Loads of Composite Columns," *Proc. Inst. Civ. Eng.*, Vol. 336, pp. 557–578.

Basu, A. K., and Hill, W. F. (1968), "More Exact Computation of Failure Loads of Composite Columns," *Proc. Inst. Civ. Eng.*, Vol. 37.

Basu, A. K., and Sommerville, W. (1969), "Derivation of Formulae for the Design of Rectangular Composite Columns," *Proc. Inst. Civ. Eng.*, Suppl. Vol., Pap. 7206s, pp. 233–280.

Emperger, F. V. (1907), "Welche statische Bedeutung hat die Einbetonierung einer Eisensäule?" *Beton Eisen*, Berlin, pp. 172–174.

Furlong, R. W. (1967), "Strength of Steel Encased Concrete Beam Columns," *ASCE J. Struct. Div.*, Vol. 93, No. ST5, pp. 113–124.

Furlong, R. W. (1968), "Design of Steel Encased Concrete Beam Columns," *ASCE J. Struct. Div.*, Vol. 94, No. ST1, pp. 267–281.

Furlong, R. W. (1974), "Concrete Encased Steel Columns—Design Tables," *ASCE J. Struct. Div.*, Vol. 100, No. ST9, pp. 1865–1883.

Furlong, R. W. (1983), "Comparison of AISC, SSLC, and ACI Specifications for Composite Columns," *ASCE J. Struct. Div.*, Vol. 109, No. ST9, pp. 1784–1803.

Japan. Society of Civil Engineers (1984), *Steel-Concrete Composite Structures in Japan—State of the Art 1984*, Subcommittee on Steel-Concrete Composite Construction, JSCE, July.

Kato, B. (1978), "The Ultimate Strength of Encased Columns under Combined Loading," *1st. US-Jpn. Sem. Compos. Struct.*, Tokyo, January.

Klöppel, K., and Goder, W. (1957), "Traglastversuche mit ausbetonierton Stahl-rohren und Aufstellung einer Bemessungsformel," *Stahlbau*, Vol. 26.

Knowles, R. B., and Park, R. (1969), "Strength of Concrete Filled Steel Tubular Columns," *ASCE J. Struct. Div.*, Vol. 95, No. ST12, pp. 2565–2587.

Mensch, J. L. (1917), "Tests on Columns with Cast Iron Core," *ACI Proc.*, Vol. 13, pp. 22 et seq.

Roeder, C. W. (1984), "Bond Stress of Embedded Steel Shapes in concrete," *2nd US-Jpn. Sem. Comp. Struct.*, Seattle, July.

Roik, K., and Bergmann, R. (1984), "Composite Columns—Design Examples for Construction," *2nd US-Jpn. Sem. Compos. Struct.*, Seattle, July.

Stevens, R. F. (1965), "The Strength of Encased Stanchions," *Res. Pap. No. 38*, Natl. Build. Stud., H.M. Stationery Office, London.

Structural Stability Research Council, (Task Group 20) (1979), "A Specification for the Design of Steel-Concrete Composite Columns," *Eng. J. Am. Inst. Steel Constr.*, Vol. 16, No. 4, pp. 101–115.

Talbot, A. N., and Lord, A. R. (1912), "Tests of Columns: An Investigation of the Value of Concrete as Reinforcement for Structural Steel Columns," *Eng. Exp. Stat. Bull.*, No. 56, University of Illinois, Urbana.

Tomii, M., Matsui, C., and Sakino, K. (1973), "Concrete Filled Steel Tube Structures," *Tokyo Reg. Conf., IABSE-ASCE Tall Build. Conf.*, August.

Varghese, P. C. (1961), "Ultimate Strength of Encased Steel Members Subject to Combined Bending and Axial Loads," *J. Inst. Civ. Eng.* India, Vol. 49, No. 6, Part 1, pp. 225–237.

Wakabayashi, M. (1984), "Recent Research Activities on Composite Steel Reinforced Concrete Building Structures and Design Practices in Japan," *2nd US-Jpn. Sem. Compos. Struct.*, Seattle, July.

Wakabayashi, M., and Minami, K. (1978), "Experimental Study of the Hysteretic Characteristics of Beam to Column Connections in Composite Structures," *1st US-Jpn. Sem. Compos. Struct.*, Tokyo, January.

Chapter Eleven

Columns, with Lacing, Battens, or Perforated Cover Plates

11.1 INTRODUCTION

The importance of proper design of shear resisting elements in built-up columns was tragically demonstrated in 1907 by the failure of the first Quebec Bridge during construction. Bridge-design practice in this country today reflects the lessons learned from that failure and the extensive research that followed. A descriptive review of column failures prior to 1940 was presented by Wyly (1940), who concluded that about three-fourths of the early recorded failures of laced structural columns resulted from local rather than general column failure.

The three types of built-up columns treated herein are illustrated in Fig. 11.1. Laced or "latticed" columns are no longer common in modern bridge construction, having been largely replaced (when a "built-up" column is required) by the use of perforated cover plates (Fig. 10.1*c*). However, latticed columns are used in guyed radio and TV towers, in derrick booms, and for projected use in space exploration vehicles.

Battened columns are the least resistant to shear of any of the tree types and because of this may experience appreciable reduction in axial column strength. Battened columns are not generally used in the United States for bridge or building construction. Box columns with perforated cover plates require no special consideration of shear effects if proportioned by specification rules.

11.2 SHEAR IN BUILT-UP COLUMNS

An evaluation of shear in built-up columns is needed for two reasons: (1) as a possible cause of reduction in buckling load, and (2) as a basis for the design of lacing bars, battens, and their connections.

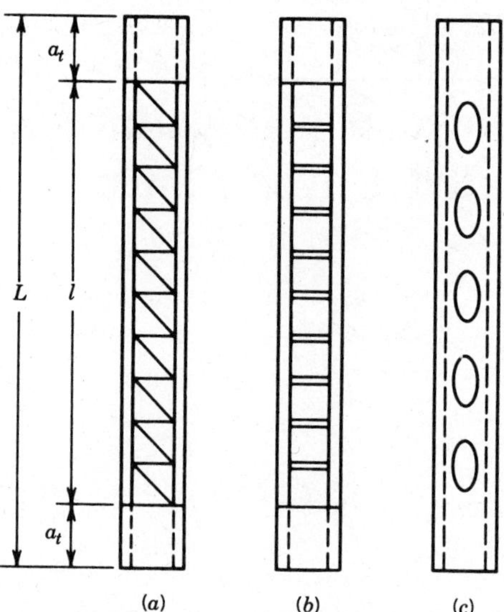

Fig. 11.1 Built-up columns.

Engesser (1891) was the first to consider the effect of shear in modifying the Euler analysis for critical load of an axially loaded column. Bleich (1952) reviews the extension of Engesser's work to the critical load analysis of built-up columns (see also Timoshenko and Gere, 1961). Additional references are listed in the third edition of the SSRC guide (Johnston, 1976). Ziegler (1982) evaluates recent studies of Engesser's initial work.

Shear in a column is caused by a combination of three factors:

1. Lateral load, resulting from wind, earthquake gravity, or other cause.
2. Slope, with respect to the line of thrust, due both to unintentional initial curvature and added curvature developed during the buckling process.
3. End eccentricity of load, introduced either by the end connections or by fabrication imperfections.

The shear from factor 1 should always be calculated and combined with the estimated allowance for shear caused by factors 2 and 3. Cause 2 is increasingly important in slender columns, and cause 3 in short columns.

There is a wide divergence worldwide as to design requirements for shear in built-up columns, as evidenced in "Stability of Metal Structures—A World View" (AISC, 1981). In Western Europe ECCS (1978) recommends evaluating shear on the basis of the end slope due to a specified initial out-of-straightness, magnified by the effect of axial load, and added, of course, to shear due to known applied loads.

In the United States, AASHTO and AREA Bridge Specifications provide an empirical formula for shear to be added to that due to the weight of the member or the external forces:

$$V = \frac{P}{100}\left[\frac{100}{(l/r)+10} + \frac{(l/r)F_y}{3{,}300{,}000}\right] \quad \text{or} \quad \frac{P}{100}\left[\frac{100}{(l/r)+10} + \frac{(l/r)F_y}{22{,}754}\right] \tag{11.1}$$

In the expresision above:

V = normal shearing force, lb (N)
P = allowable compressive axial load on members, lb (N)
l = length of member, in. (m)
r = radius of gyration of section about the axis perpendicular to plane of lacing or perforated plate, in. (m)
F_y = specified minimum yield point of type of steel being used, psi (MPa)

The AISC Specifications for buildings (1978) call for the addition of a shear of 2% of the total axial stress to be added to shear due to estimated known causes. In Canada, highway bridge specification (Canadian Standards Association (1980)) require a shear force of 2.5% of the axial force to be added to that due to lateral loads.

11.3 EFFECT OF SHEAR DISTORTION ON CRITICAL LOAD

A nondimensional parameter μ will be introduced to characterize the shear flexibility of battened or laced structural members (Lin et al., 1970). The parameter μ takes account of the added distortion due to axial force or bending in the web elements. It is assumed that the end stay plates do not contribute to the shear flexibility. The effect of shear on the elastic critical load is depicted in Fig. 11.2 for three basic end conditions:

1. Both ends hinged (subscript h).
2. One end fixed, one hinged (subscript f, h).
3. Both ends fixed (subscript f).

One enters Fig. 11.2 with a chosen a_t/L and a calculated shear parameter value of μ. The load ratio P_{cr}/P_e can then be read for the appropriate end conditions, after which the equivalent length factor may be calculated by Eq. 11.2.

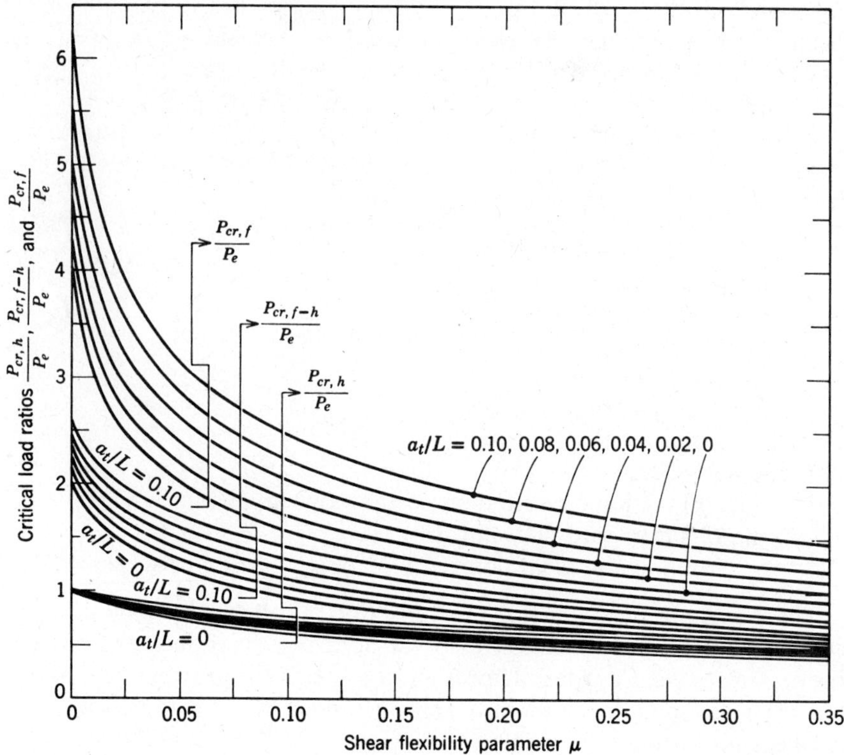

Fig. 11.2 Critical loads of columns with various end conditions, constant shear flexibility μ, and rigid stay plates.

$$K = \sqrt{\frac{P_e}{P_{cr}}} \tag{11.2}$$

In Eq. 11.2 and in Fig. 11.2, P_{cr} is the elastic critical load of the given open-web or open-flange column, and $P_e = \pi^2 EI/L^2$, the Euler load for a solid-wall column with both ends hinged. Allowable stresses may then be obtained for a given specification as a function of KL/r by formula or table.

11.4 LACED STRUCTURAL MEMBERS

For a typical laced member (Fig. 11.1a), consisting of two main longitudinal elements, two planes of diagonal lacing and transverse struts, Lin et al. (1970) provide the following formula for the shear flexibility factor μ:

$$\mu = \frac{\xi_b}{1+\xi_a}\left(\frac{b}{l}\right)^2 \frac{A_c}{A_d}\left\{\frac{b}{\xi_a a}\left[1+\left(\frac{\xi_a a}{b}\right)^2\right]^{3/2} + \frac{b}{\xi_a a}\frac{A_d}{A_c}\right\} \tag{11.3}$$

The notation used in Eq. 11.3 is described by Figs. 11.1*a* and 11.3. The third edition of the SSRC Guide (Johnston, 1976) provides an illustrative example of the application of Eq. 11.3 to a typical design. Laced columns are rarely used today in bridge and building construction, nor do U.S. and Canadian specifications call for any reduction of allowable load because of shear, hence the illustrative example will be omitted herein. The general procedure is similar to that for battened columns, as will be illustrated in Section 11.5.

In view of the usual small effect of shear in laced columns, Bleich (1952, p. 174) has suggested that a conservative estimate of the influence of 60° or 45° lacing, as generally specified in bridge-design practice, can be made by modifying the effective-length factor K (determined by end-restraint conditions) to a new factor K', as follows:

For $KL/r > 40$:

$$K' = K\sqrt{1 + \frac{300}{(KL/r)^2}}$$

For $KL/r \leq 40$: (11.4)

$$K' = 1.1K$$

The change in K is significant only for small KL/r values, in which case there is little change in the allowable stress for the lower strength structural carbon steels. The reduction in the AISC allowable stress by Eq. 11.4 agrees well with te analysis of the illustrative example presented in the third edition of the SSRC Guide.

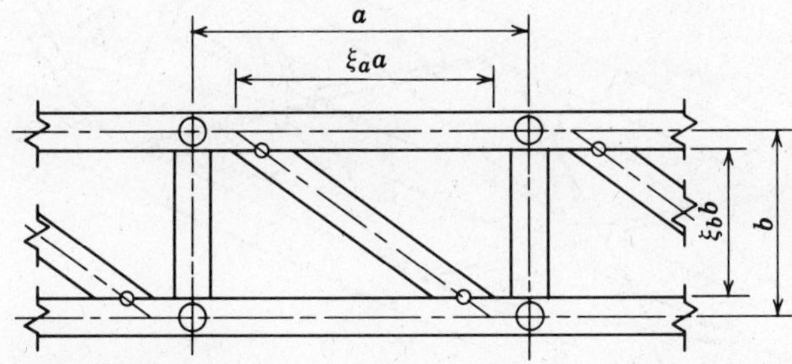

Fig. 11.3 Typical panel of a laced structural member. Circles indicate where hinges were assumed in the analysis.

A number of countries, including Eastern Europe and Japan, require a reduction in allowable load because of shear using formulas similar to Eq. 11.4 that empirically increases the effective KL/r.

Bridge design practice in the United States (AASHTO 1979; AREA 1979) requires that the slenderness ratio of the portion of the flange between lacing bar connections have a slenderness ratio of no more than 40 or more than two-thirds that of the entire member. Canadian bridge specifications (CSA, 1978) change the foregoing limits to 60 and three-fourths, respectively.

The lacing bars and their connections must be designed to act either in tension or compression, and the rules for general column design apply to them as well. In exceptional cases, such as very large members, double diagonals can be designed as tension members and the truss system completed by compression struts. The importance of adequate and tight-lacing bar connections has been demonstrated by test (Hartmann et al. 1938).

In recent years renewed interest in laced members has been activated primarily by needs in the space industry to develop members of minimal weight to sustain very small loads. A type under current study is illustrated in Fig. 11.4, in which three longitudinal tubular chords (longerons) are held in position by struts and are braced for shear by diagonal cross-bracing. Designs under consideration may have very large overall slenderness ratios with the individual chords between struts of comparable slenderness.

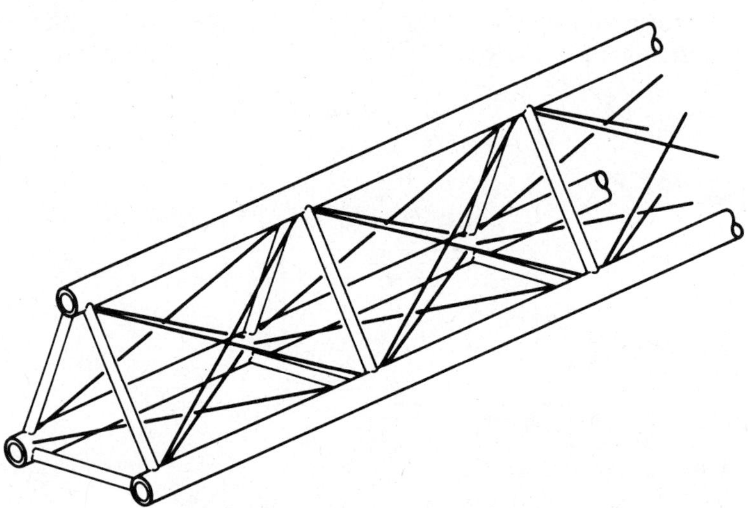

Fig. 11.4 Triangular latticed column.

Miller and Hedgepeth (1979) studied the deleterious combined effects of initial out-of-straightness in the overall column coupled with out-of-straightness of the individual chords between struts. Elastic behavior was assumed and the effect of shear was neglected. Overall initial curvature, amplified by load-induced bending moments, causes local increase of stress in the chord or chords on the concave side of the bent column. The local moment in the chord between struts causes added local deflection and resultant chord shortening increases the curvature of the entire member. Assuming sinusoidal initial curvature between struts and parabolic overall initial curvature, Miller and Hedgepath provide charts that predict the ratio of maximum buckling load to the critical Euler bifurcation load of a perfectly straight member. A wide range of the pertinent variables is covered by the charts and the analysis is based on the assumption of the same initial out-of-straightness of all chords locally between struts. Subsequently, their report also considers the modified effect assuming random variation of imperfections.

In an independent study of the same type of member, Crawford and Benton (1980), using similar assumptions as to initial curvature, also provide a chart to determine critical load-reduction factors. In a specific example design of the diagonal spars of an 800-m^2 solar sailing spacecraft, they estimated initial ratios of center deflection to length to be 1/2500 for individual chords between struts and 1/1000 for the overall member. The slenderness ratios were 214 for the chords between struts and 279 for the overall member.

The Miller and Hedgepeth charts indicate a maximum buckling load of about 59% of the critical bifurcation load for a straight member. With both overall and local slenderness ratios of 279, the reduction drops to less than 50% and is confirmed by the Crawford and Benton chart, which is only valid for equal overall and local slenderness ratios.

The foregoing studies indicate that for built-up columns of extremely large slenderness ratios the interaction of local and overall initial curvature has a deleterious effect on buckling strength far greater than the effect of shear on shorter columns having proportions common in bridge and building construction.

Latticed crane booms represent another important application of built-up laced columns. Vroonland (1971) has provided analyses of nonuniform latticed booms under combined lateral and axial load and including effects of deflection and intermediate lateral support. Brolin et al. (1972) report on tests to destruction of four booms varying in length from 60 to 200 ft with predicted failure loads in close agreement with Vroonland's analyses. Failure occurs when the maximum compressive force in an individual chord reaches the critical failure load for the unsupported length between brace points.

11.5 COLUMNS WITH BATTENS

Figure 11.5 shows the basic elements of the battened structural member, which consist of two main longitudinal shafts connected by batten elements in one or two (or more) planes by means of rigid-joint connections. The batten elements serve as the web of the member, transmitting shear force by virtue of their own shear and moment resistance, in combination with local bending of the longitudinal elements.

The battened column has greater shear flexibility than either the laced column or the column with perforated cover plates; hence the effect of shear distortion must be taken into account in calculating the effective length of the member. Battened columns generally are not permitted by current U.S. design specifications for bridges and buildings. However, small television and radio towers are frequently made of battened columns, and some specifications permit such columns for use as secondary members. The lower "crane shaft" portions of columns in some steel mill buildings, supporting crane runway girders, are superficially similar to

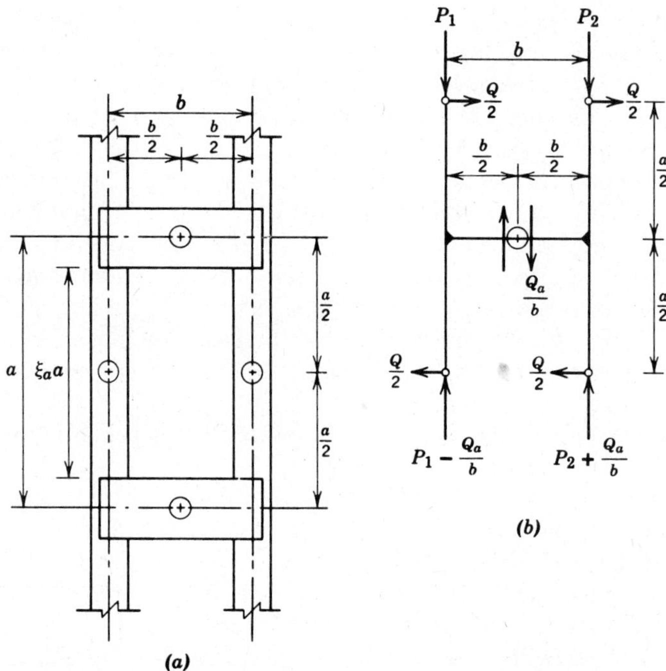

Fig. 11.5 Typical panel of battened structural member: (*a*) geometrical configuration; (*b*) force equilibrium. Circles indicate assumed points of inflection.

battened columns. In many instances, however, these do not have rigidly framed connections beftween the battens and the longitudinal elements and are more aptly described as "spaced columns" and as such are treated in Section 11.6. The end stay plates have an important role in enhancing the strength of such a weakly battened column.

The geometric arrangement of the battened structural member is based on the typical unit of length a and width b between centroids of the longitudinals. The properties of the battened structural members are characterized by the shear shape factors and moments of inertia of the batten and longitudinal elements.

Battened structural members are highly redundant. In the development of the shear flexibility effect (Lin et al., 1970) it was assumed that points of inflection in the batten elements of a symmetric member are at the batten midpoints, and that for the longitudinal elements points of inflection occur midway between battens. These assumptions result in the battened structural member becoming statically determinate. The analysis is conservative, since it neglects the overall continuity of the longitudinals.

The following expression is obtained for the shear flexibility parameter μ of a battened structural member:

$$\mu = \left[\frac{1}{(l/r_c)^2} + \left(\frac{b}{2l}\right)^2\right]\left[\frac{A_c}{A_b}\left(\frac{ab}{6r_b^2} + 5.2\,\frac{a}{b}\,\eta_b\right) + 2.6\xi_a\eta_c + \frac{\xi_a^3}{12}\left(\frac{a}{r_c}\right)^2\right] \tag{11.5}$$

The nomenclature for Eq. 11.5 is shown in Fig. 11.5 and (additionally) as follows:

A_c, A_b = area of a single longitudinal and all batten elements within a length a, respectively

r_c, r_b = radius of gyration of longitudinal and batten elements, respectively

η_c, η_b = shear shape factors for the longitudinal and batten elements, respectively

l = length of column between end-tie plates

Equation 11.5 for μ includes terms to account for the amplification of deflection in the column segments between battens as well as the increased column deflections due to the semirigid behavior of the connections between battens and longitudinals. However, the local amplification effects may be neglected if the slenderness ratio of the local segments of the longitudinal meets the following requirement:

$$\frac{\xi_a a}{r_c} \leq \frac{80}{\sqrt{f_a}} \tag{11.6}$$

(f_a is the average column stress in ksi in the fully loaded column.) It may be noted that Eq. 11.6 is considerably more conservative than the batten spacing requirements in some specifications.

Generally, it would be impractical to make the necessary experimental determination of partial rigidity of semirigid connections. It is recommended that batten connections be designed for full continuity; otherwise, the column should be treated simply as a spaced column as discussed in Section 11.6.

Example 11.1 Battened Column

As shown in the Fig. 11.6 batten flange stiffeners have been included to eliminate local web distortion and to ensure that the battens are fully effective. If such stiffeners were not provided, the column could be treated as a "spaced column" as covered by Section 11.6. From the AISC Manual,

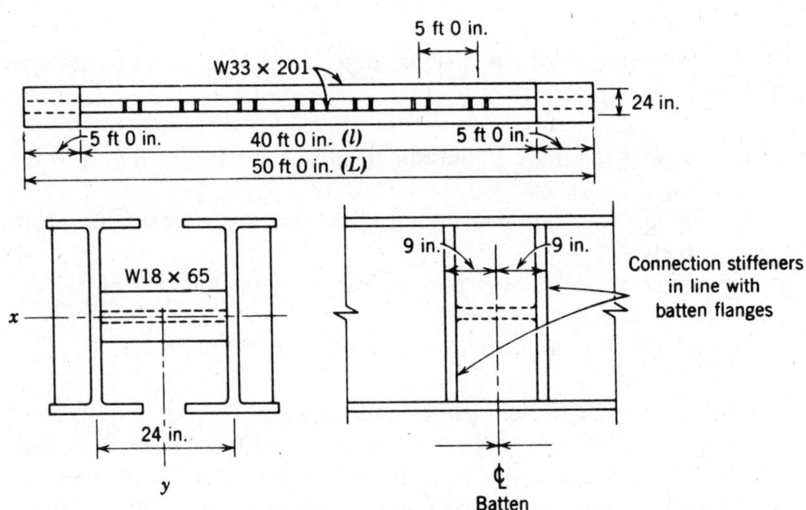

Fig. 11.6 Example details.

$$W33 \times 201 \text{ (longitudinal)}$$

$$A_c = 59.1 \text{ in.}^2$$

$$r_c = r_y = 3.56 \text{ in.}$$

$$r_x = 14.0 \text{ in.}$$

$$I_y = 749 \text{ in.}^4$$

$$W18 \times 65 \text{ (batten)}$$

$$A_b = 19.1 \text{ in.}^2$$

$$r_b = r_x = 7.49 \text{ in.}$$

$$a = 60 \text{ in.}$$

$$b = 24 \text{ in.}$$

$$\xi_a = \frac{60 - 18}{60} = \frac{42}{60} = 0.70$$

$$l = 480 \text{ in.}$$

$$L = 600 \text{ in.}$$

$$\eta_c = 1.6$$

$$a_t/L = 0.10$$

$$\eta_b = 2.6$$

For the combined cross section,

$$I_y = \frac{59.1 \times 24^2}{2} + 2 \times 749 = 18{,}519 \text{ in.}$$

$$r = \sqrt{\frac{18{,}519}{2 \times 59.1}} = 12.5 \text{ in.} < r_x$$

Substituting Eq. 11.5 yields

$$\mu = \left[\frac{1}{(480/3.56^2} + \left(\frac{24}{2 \times 480} \right)^2 \right]$$

$$\times \left[\frac{59.1}{19.1} \left(\frac{60 \times 24}{6 \times 7.49^2} + 5.2 \times \frac{60}{24} \times 2.6 \right) + 2.6 \times 0.7 \times 1.6 \right.$$

$$\left. - \frac{0.70^3}{12} \left(\frac{60}{3.56} \right)^2 \right]$$

$$= 0.088$$

Table 11.1

End Conditions	P_{cr}/P_c (Fig. 11.2)	K	KL/r	F_a (AISC) $F_y = 36$ ksi	$F_y = 50$ ksi
Hinged-hinged	0.80	1.12	53.8	18.01	23.75
Hinged-fixed	1.45	0.83	39.9	19.20	25.84
Fixed-fixed	2.52	0.63	30.3	19.92	27.1

Referring to Fig. 11.2, the critical load ratio is read: then by Eq. 11.2, the effective-length factor is evaluated. Summarizing the results for three sets of end conditions are presented in Table 11.1.

If no reduction were made, the value of KL/r for the hinged-hinged case would be $600/12.5 = 48$ with allowable stress of 18.53 and 24.66 ksi for the 36 and 50 ksi yield strengths. The corresponding stress reductions are 2.8 and 3.7%.

Bleich (1952) gives the following approximate formula for the effective length of a battened column with both ends hinged:

$$\frac{KL}{r} = \sqrt{\left(\frac{L}{r}\right)^2 + \frac{\pi^2}{12}\left(\frac{a}{r_c}\right)^2} \tag{11.7}$$

where L/r is the slenderness ratio of the column as a whole and a/r_c is the slenderness ratio of one chord center-to-center of battens. Bleich estimates that the buckling strength of a steel column having an L/r of 110 is reduced by about 10% when $a/r_c = 40$, and by greater amounts for larger values of a/r_c.

For Example 11.1, $a/r_c = 60/3.56 = 16.9$ and Eq. 11.7 gives

$$\frac{KL}{r} = \sqrt{\left(\frac{600}{12.5}\right)^2 + \frac{\pi^2}{12}(16.9)^2} = 50.4$$

which is lower than the 53.8 determined by Eqs. 11.5 and 11.2.

The unconservative assumption has been made that the addition of two stay plates gives the end regions full rigidity with respect to shear. At the same time the effective KL/r is slightly decreased because of the increase in bending resistance due to the stay plates in the end regions. The two effects are offsetting and may be neglected.

The design of the individual chords and the batten plates and their connections should take account of the local bending resulting from specified shear forces, shown acting longitudinally and transversely on the free-body diagram of the column shown in Fig. 11.5*b*.

The batten plates and their connections to the chords should be designed for the combination of shear Qs/nb and moment at the connection of

$$M_b = \frac{Qa}{2n} \tag{11.8}$$

where

Q = shear required by specification plus shear due to any transverse loading
a = distance center-to-center of battens
n = number of parallel planes of battens (Fig. 11.5*b*)

The maximum combined bending and axial compression stress in each individual chord should not exceed the maximum permissible stress for a zero-length column. For this calculation, the chord bending moment should be taken as

$$M_c = \frac{Q\xi_a a}{4} \tag{11.9}$$

These combined stresses are not secondary stresses and therefore should not be neglected.

11.6 STAY PLATES AND SPACED COLUMNS

End stay plates in battened columns may contribute significantly to the buckling strength. Their importance is revealed by the study of a "spaced column," defined herein as the limiting case of a battened column in which the battens are attached to the longitudinal column elements by hinged connections. The battens then act simply as spacers, with no shear transmitted between the longitudinal elements. Without end-stay plates, the buckling strength of such a spaced column is no greater than the sum of the critical loads of the individual longitudinal components of the built-up member. The strengthening effect of the end-stay plates arises from two sources: (1) a shortening of the length within which the column components can bend about their own axes, and (2) the forcing of the longitudinal components to buckle in a modified second-mode shape and thus have an elastic-buckling coefficent that may approximate four times that of the first mode. The buckling load of a spaced column with end-tie plates is a lower bound to the buckling load of the battened column (also with tie plates) but with low or uncertain moment resistance in the connections between battens and the longitudinal components. Such columns are sometimes used in mill-building construction. In addition to their contribution to column strength, end-tie plates perform their usual role of distributing the applied forces or moment to the component elements of either laced or battened columns. They also provide a means of transmitting load to another member or to a footing.

With regard to the distance along the column between spacer elements, the same rule as for battens (Eq. 11.6) should provide a conservative basis for design.

Example 11.1 involved a battened column. If the stiffener plates to provide rigidity of the batten connections were omitted, the behavior would approach the conditions assumed for the spaced column because the attachment of the battens to the column webs would not transmit moment effectively.

Using the notation adopted for battened columns, the moment of inertia of the two longitudinals about the *y-y* axis would be, as in Example 11.1.

$$I = \frac{A_c b^2}{2} + 2I_c$$

where I_c is the moment of inertia of one of the individual longitudinal elements.

The ratio I/I_c in mill building columns is usually at least 40 and could be greater than 100. This ratio is used as a parameter for the determination of the buckling load of a spaced column with end-tie plates.

Four modes of buckling for spaced columns without sidesway are treated in Johnston (1971), as illustrated in Fig. 11.7. Spaced columns with sideways permitted also are treated briefly.

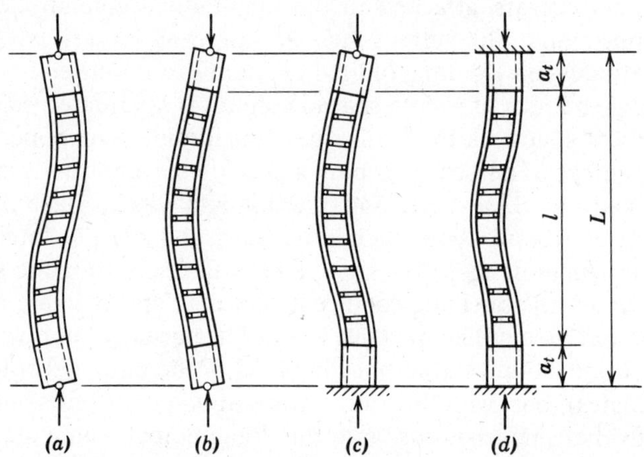

Fig. 11.7 Spaced column buckling modes: (*a*) mode A, hinged-hinged; (*b*) mode E, hinged-hinged; (*c*) mode C, hinged-fixed; (*d*) mode D, fixed-fixed.

For the hinged-end condition the spaced column with end-stay plates buckles either in S-curvature (mode A) (Fig. 11.7*a*) or in mode B curvature (Fig. 11.7*b*), depending on the values of I/I_c and a_t/L. It will be noted that in S-curvature there is no differential change of length of the two longitudinal column components between the end-tie plates; thus they buckle under identical loads $P/2$, and the critical load is independent of the ratio I/I_C. When the column buckles in mode B curvature (Fig. 11.7*b*), the component on the concave side (at the center) shortens more than on the convex; thus there arises overall moment resistance due to the difference in the direct component forces. This is added to the separate moments induced in the components as a result of their own curvature. The critical loads for S-curvature buckling could be less than those for mode B curvature when the ratio I/I_C is relatively small and a_t/L is large.

In practice, the base of a column will usually be fixed to a footing, and S-curvature buckling cannot take place. Buckling will be in mode C as illustrated in Fig. 11.3*c*, but as I/I_C gets large it will tend toward the shape with both ends fixed, as illustrated in Fig. 11.6*d*. In fixed-end buckling (Fig. 11.7*d*), as in mode A, the moment resistance is simply the sum of the moments in the component parts, with no contribution due to differential direct forces as in Fig. 11.7*b* or *c*. The fixed-end case is the simplest to evaluate, since the critical load is simply twice the critical load of a longitudinal component, length l, with both ends fixed, that is, the Euler load with an equivalent length of $0.5l$, multiplied by 2.

Added moment resistance in a spaced column due to differential changes in component length occurs only when the end rotations are different in magnitude and/or sense, as in Fig. 11.7*b* and *c*. Within length l between the tie plates there is no shear transfer between longitudinals; hence the differential direct forces and the resisting moment that they contribute must remain constant within l.

Although the cited reference gives critical load information for a variety of end conditions, including the four shown in Fig. 11.7, the fixed-base and hinged-top case, Fig. 11.7*c* is possibly of greatest practical application. In terms of the overall length $L = l + 2a_t$ the equation for elastic critical load is written:

$$P_{cr} = \frac{CEI_c}{L^2} \tag{11.10}$$

in which C is termed the elastic-buckling coefficient. For the determination of approximate critical loads in the inelastic range or for evaluation of permissible stresses by column-design formulas, it is convenient to determine the effective-length factor K, for use in the equation:

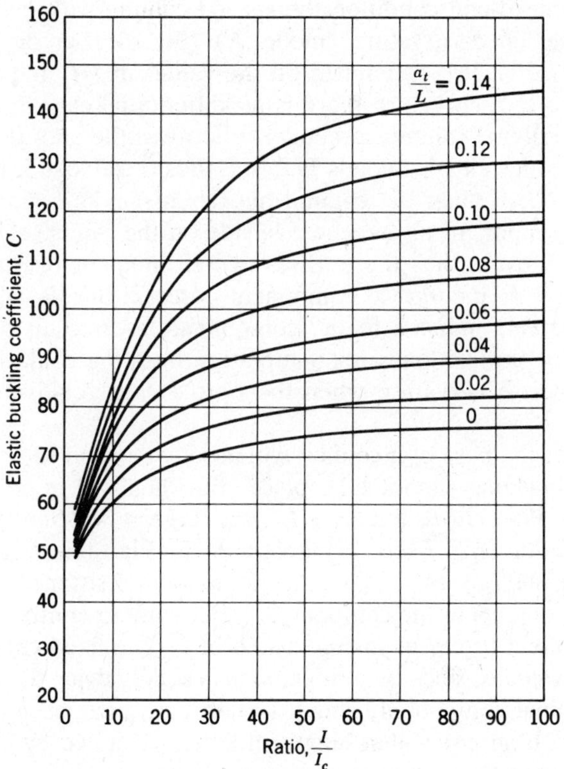

Fig. 11.8 Elastic buckling coefficients, one end hinged, one end fixed (mode C of Fig. 11.7).

$$P_{cr} = \frac{2\pi^2 EI_c}{(KL)^2} \tag{11.11}$$

where

$$K = \pi \sqrt{\frac{2}{C}} \tag{11.12}$$

Elastic-buckling coefficients C for the hinged-fixed case are plotted in Fig. 11.7, the use of which will now be illustrated.

Example 11.2

Identical with Example 11.1 but omitting batten-flange stiffeners and considering the design as a spaced column,

$$\frac{a_t}{L} = \frac{5}{50} = 0.10 \qquad \frac{I}{I_c} = \frac{18{,}519}{749} = 24.7$$

From Fig. 11.7, $C = 100$. By Eq. 11.12,

$$K = \pi\sqrt{\frac{2}{100}} = 0.445$$

$$\frac{KL}{r_c} = \frac{0.445 \times 600}{3.56} = 75.0$$

(Note that K is now referenced to L/r_C, not L/r.) By AISC Specification (1978, Appendix A, Tables 3-36 and 3-50)

	F_a (ksi)	
F_y	Spaced Column	Battened (Example 11.1)
36	15.90	19.20
50	19.99	25.84

In a limited number of tests of spaced columns (Freeman, 1973) the maximum failure loads fell short of the predicted, due in part to open holes in the bolted specimens and to deformation of the end stay plates. Pending further tests, it is recommended that only half of the length of the end stay plates be considered effective and that 90% of the theoretical failure loads be used as a basis for design. In Example 11.2 this recommendation would reduce the allowable stresses from 15.90 and 19.99 to 13.59 and 16.70 for F_y of 36 and 50 ksi, respectively.

11.7 COLUMNS WITH PERFORATED PLATES

White and Thürlimann (1956) provide (in addition to the results of their own research) a digest of investigations at the National Bureau of Standards (Stang and Greenspan, 1948) and give recommendations for the design of columns with perforated cover plates. The following design suggestions for such columns are derived from the White-Thürlimann study and from AASHTO Specifications (1983).

When perforated cover plates are used, the following provisions govern their design:

1. The ratio of length, in direction of stress, to width of perforation, should not exceed 2.
2. The clear distance between perforations in the direction of stress, should not be less than the distance between points of support [that is $(c - a) \geq d$ in fig. 11.9*a*].

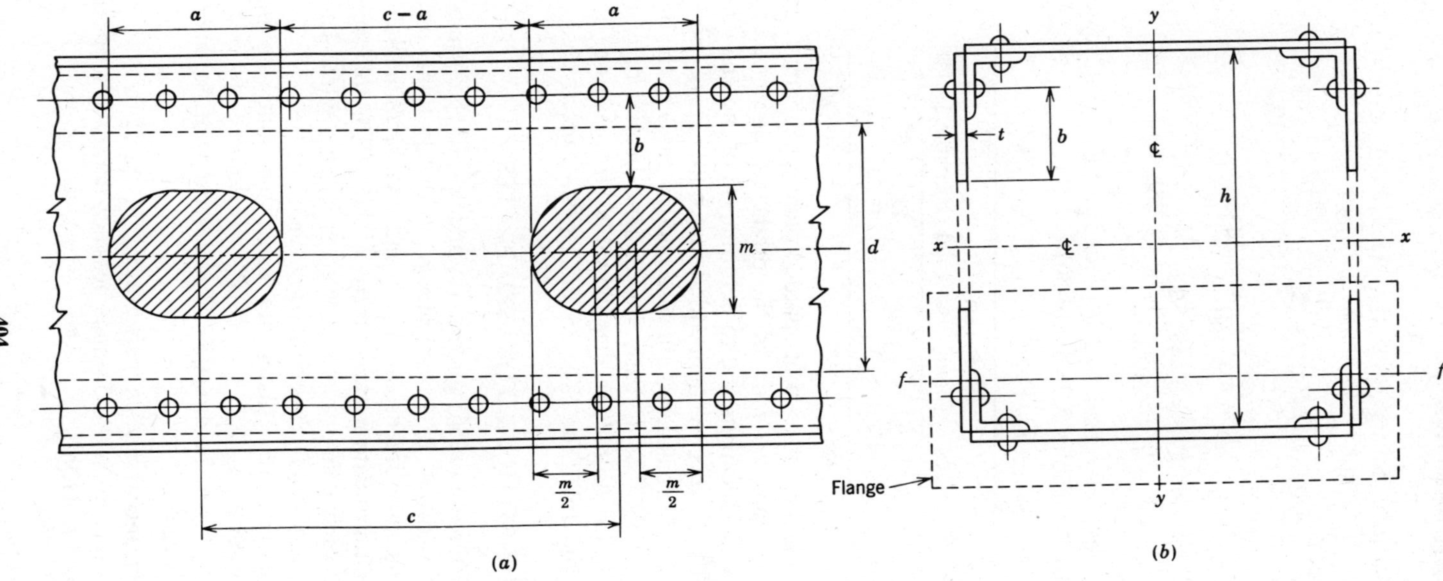

Fig. 11.9 Column with perforated web plates.

3. The clear distance between the end perforation and the end of the cover plate should not be less than 1.25 times the distance between points of support.
4. The point of support should be the inner line of fasteners or fillet welds connecting the perforated plate to the flanges. For plates butt welded to the flange edge of rolled segments the point of support may be taken as the weld whenever the ratio of the outstanding flange width to flange thickness of the rolled segment is less than 7. Otherwise, the point of support should be the root of the flange of the rolled segment.
5. The periphery of the perforation at all points should have a minimum radius of $1\frac{1}{2}$ in. (38.2 mm).
6. The transverse distance from the edge of a perforation to the nearest line of longitudinal fasteners, divided by the plate thickness, that is, the b/t ratio of the plate adjacent to a perforation (see Fig. 11.9), should conform to minimum specification requirements for plates in main compression members.

11.8 DESIGN OF LATTICE TOWERS AND MASTS

The classical treatment of lattice struts has been directed primarily to overall behavior of pin-ended assemblages and the effect on these of strain in the lattice web system. The effect of lateral loads which occur in transmission towers and guyed communication structures is to produce considerable bending moments. In addition, guyed towers are subject to end restraint and nonrigid supports, requiring careful assessment of the effective length.

In design of these structures, the individual chord elements may be checked for axial load between brace points, arising from overall axial load and applied moment on the structure. The structure as a whole may be evaluated for combined axial load and moment, using an interaction equation. A shortcoming of this approach, in the case of steel structures, is that the column curve used in the overall check will have been derived, in all probability, for a hot-rolled shape of wide-flange or solid-bar form, reflecting residual stresses and lack of straightness appropriate to that shape only, not the overall structure. Furthermore, while the amplification factor will have been applied to the overall moment, the effect of this on the individual leg element may not have been considered.

The concept suggested by Marsh and verified by full-scale tests is that a lattice structure will fail only when an individual chord element fails. Provided that the maximum force to be resisted by a chord is determined correctly, it is only necessary to design this element, which design will be for axial load on the chord between brace points of the lattice. The concept above is used in the design formulae given in the ASCE *Guide to the Design of Aluminum Transmission Towers* (1972). To the chord mean axial load is added the *amplified* load due to overall bending and overall initial out-of-straightness. The amplification factor $(1 - P/P_e$ is for the tower as a whole; the Euler load should take account of both equivalent length and web shear effects, if significant.

References

American Association of State Highway and Transportation Officials (1983), *Standard Specifications for Highway Bridges*, 13th ed., AASHTO, Washington, D.C.

American Institute of Steel Construction (1978), *Specification for the Design, Fabrication and Erection of Structural Steel for Buildings*, AISC, Chicago.

American Institute of Steel Construction (1981), "Stability of Metal Structures—A World View," 1981, *Eng. J. Am. Inst. Steel Constr.*, Vol. 18, No. 4, p. 154.

American Railway Engineering Association (1979), "Specifications for Steel Railway Bridges," *AREA Manual for Railway Engineering*, AREA, Chicago, Chap. 15.

American Society of Civil Engineers (1972), *Guide for the Design of Aluminum Transmission Towers*, Task Committee on Lightweight Alloys of the Committee on Metals of the Structural Division, *Proc. Am. Soc. Civ. Eng.*, Vol. 98, No. ST12.

Bleich, F. (1952), *Buckling Strength of Metal Structures*, McGraw-Hill, New York.

Brolin, C. A., Durscher, H. E., and Serentha, G. (1972), "Destructive Testing of Crane Booms," *SAE Trans.*, Vol. 81, Pap. 720784.

Canadian Standards Association (1980), *Specification for Design of Highway Bridges*, CSA Standard CAN3-S6-M78, CSA, Rexdale, Ontario.

Crawford, R. F., and Benton, M. D. (1980), "Strength of Initially Wavy Lattice Columns," *AIAA J.*, Vol. 18, No. 5, p. 581.

Engesser, F. (1891), "Die Knickfestigkeit gerader Stabe," *Zentralbl. Bauverwaltung*, Vol. 11.

European Convention for Constructional Steelwork (1978), *European Recommendations for Steel Structures*, ECCS, Brussels.

Freeman, B. G. (1973), "Tie Plate Effects in Weakly Battened Columns," Ph.D. dissertation, University of Arizona, Civ. Eng. Dept.

Hartmann, E. C., Moore, R. L., and Holt, M. (1938), "Model Tests of Latticed Structural Frames," *Alcoa Res. Labs. Tech. Pap. No. 2.*

Johnston, B. G. (1971), "Spaced Steel Columns," *ASCE J. Struct. Div.*, Vol. 97, No. ST5, p. 1465.

Johnston, B. G., ed. (1976), *Guide to Stability Design Criteria for Metal Structures* (SSRC) 3rd ed., Wiley, New York.

Lin, F. J., Glauser, E. C., and Johnston, B. G. (1970), "Behavior of Laced and Battened Structural Members," *ASCE J. Struct. Div.*, Vol. 96, No. ST7, p. 1377.

Miller, R. K., and Hedgepeth, J. M. (1979), "The Buckling of Latticed Columns with Stochastic Imperfections," *Int. J. Solids Struct.*, Vol. 15, pp. 71–84.

Stang, A. H., and Greenspar, M. (1948), "Perforated Cover Plates for Steel Columns: Summary of Compressive Properties," *U.S. Nat. Bur. Stand. J. Res.*, Vol. 40, No. 5, RP 1880, p. 347.

Timoshenko, S. P., and Gere, J. M. (1961), *Theory of Elastic Stability*, 2nd ed., McGraw-Hill, New York.

Vroonland, E. J. (1971), "Analysis of Pendant-Supported Latticed Crane Booms," *SAE Trans.*, Vol. 80, Pap. 710697.

White, M. W., and Thürlimann, B. (1956), "Study of Columns with Perforated Cover Plates," *AREA Bull.*, No. 531.

Wyly, L. T. (1940), "Brief Review of Steel Column Tests," *J. West. Soc. Eng.*, Vol. 45, No. 3, p. 99.

Ziegler, H. (1982), "Arguments for and against Engesser's Buckling Formulas," *Ing. Arch.*, Vol. 52, pp. 105–113.

Chapter Twelve

Mill Building Columns

12.1 INTRODUCTION

Mill buildings are industrial structures within which machinery or products are moved about by overhead traveling cranes. The cranes travel on runway girders that are supported by column brackets or by stepped columns.

Mill building columns are designed by a variety of semiempirical procedures which, on the basis of experience, have produced generally satisfactory results. The design procedure presented herein is based on a modification of AISC Building Specification procedures. It is assumed that the user of this chapter also has available the AISE Technical Report No. 13, "Guide for the Design and Construction of Mill Buildings" (AISE, 1979) which classifies mill building structures, suggests design-load combinations, defines terms, and provides design information such as the equivalent length factors of stepped columns. Some common types of mill building columns are illustrated in Fig. 12.1. Figure 12.1*a* shows a column of uniform section for the entire length with the crane runway girders supported by column brackets. Figure 12.1*b* shows a stepped column, with the lower shaft a single heavy wide-flange section. The upper shaft, supporting the roof structure is a lighter wide-flange shape.

Figure 12.1*c* represents, in the lower shaft, either a "battened" column or a "spaced" column, as differentiated in Chapter 11. The battened column is one in which the two longitudinal column elements are connected by battens with moment-resisting end connections. These are to be spaced close enough and designed with sufficient strength and rigidity as to make the two longitudinals act very nearly as segments of a single section and thus achieve integral beam-column action. But, in some mill building columns, the battens are little more than spacers, and the longitudinal elements are simply forced to deflect in the same curve.

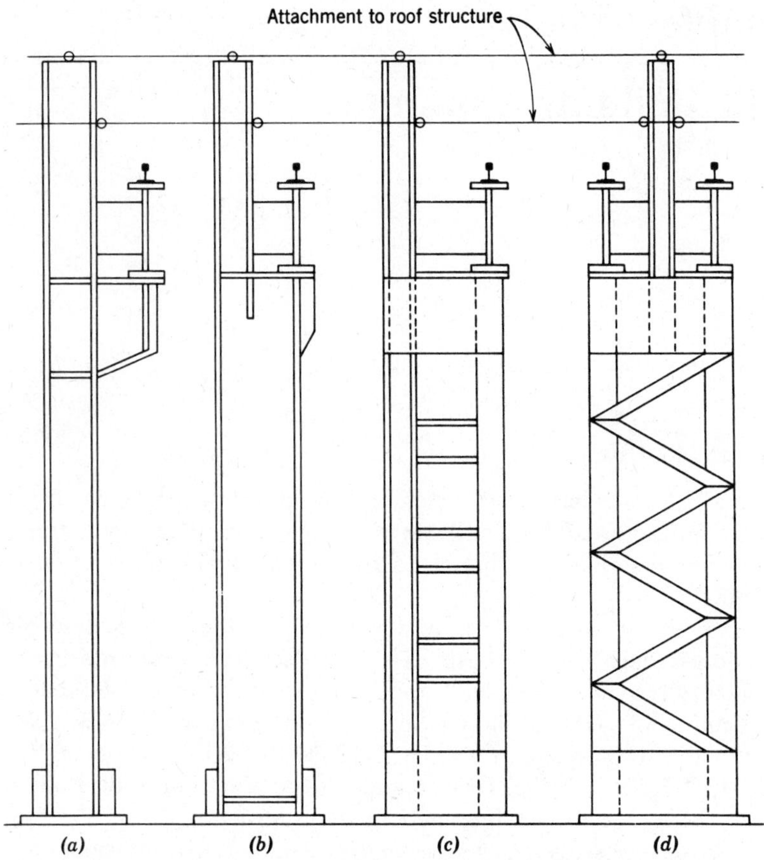

Fig. 12.1 Types of mill building columns.

Their effective lengths as separate columns are, however, appreciably reduced by end-tie plates (see Chapter 11).

Figure 12.1*d* shows a laced column, often used between two crane aisles. With adequate lacing bars the two lower shaft segments act substantially with integral action (see Chapter 11). Lacing is particularly appropriate when the longitudinal elements are spaced some distance apart, in which case eccentricity of load may be large. However, any of the types illustrated in Fig. 12.1, may be used for either interior or exterior columns.

The determination of internal moments, shears, and direct forces is a preliminary to the design of a stepped column. For manual analysis, Murray and Graham (1957) have developed a widely used method in which it is assumed, for crane loadings, that the column is hinged at the top, midway between points of attachment to the roof structure. The neglect of top rotational restraint increases the calculated moment at the footing, which is assumed fixed, and thus compensates roughly for sidesway that may take place at the top. Huang (1968a) has developed graphs that facilitate analysis by the Murray procedure. Alternatively, the complete bent may be analyzed by the slope-deflection or moment-distribution procedures, for which coefficients for stepped members are given by Maugh (1963). A more complete analysis must take account of the fact that individual bents are not free to sway independently, but are partially restrained by their neighbors through cross-bracing in the plane of the roof. Usually, for wind loads, the columns are either assumed rotationally fixed at the top, or else considered to be supported by two hinges as shown in Fig. 12.1, with the hinges assumed to have equal sideways displacements.

Given an analysis of bending moments resulting from a particular load condition, the design check may then be made by a modification of the AISC three-part interaction formula for combined stress due to compression and biaxial bending. It is assumed herein that "integral" column action is achieved in the lower shaft by means of a continuous web plate or by properly designed battens or lacing. Stepped "spaced" columns, on the other hand, do not achieve integral action, but nevertheless can provide an economical and satisfactory solution for mill building application, especially if adequate end stay plates are introduced to reduce the effective lengths of the components (see Chapter 11).

12.2 EFFECTIVE LENGTH OF STEPPED COLUMNS

The determination of the effective length of a stepped column is a necessary preliminary to the calculation of the terms F_a and F'_e which are, respectively, the allowable column stress under axial load and the Euler critical stress divided by 23/12, the AISC factor-of-safety for axially loaded slender columns.

Tables for K, needed to determine the effective length KL, are provided by the AISE Guide (1979) in terms of three parameters which are a, the ratio of the length of the reduced section to the total length of the column; B, the ratio of the maximum moment of inertia of the combined column cross section to that of the reduced section about their respective centroidal axes perpendicular to the plane of the lacing or battens; and P_1/P_2, the ratio of the axial force in the upper shaft (roof and wall load) to the axial force that is added in the lower shaft (crane girder reactions plus allowance for lower shaft and wall dead load). The three parameters as well as other notation are defined in Fig. 12.2. The tables cover a range of a from 0.10 to 0.50, B from 1.0 to 100.0, and P_1/P_2 from 0 to 0.25, with the column base assumed either fixed or hinged and the column top assumed hinged. Huang (1968b) also provides values of K, presented in the form of graphical charts, over a somewhat different range of parameters, for members with fixed base and hinged top, covering values of P_1/P_2 up to unity.

For buckling about the strong axis (x-x of Fig. 12.2), the stepped column is usually laterally unsupported over its entire length, and it is for this condition that the afore referenced tables and charts are applicable. For buckling about the weak y-y axis, lateral support is usually provided by the crane runway girder at its seat, location B in Fig. 12.2; therefore, the upper and lower shaft segments need to be checked separately for the determination of F_a (if it is governed by y-y axis bending and for the determination of F'_{cy}).

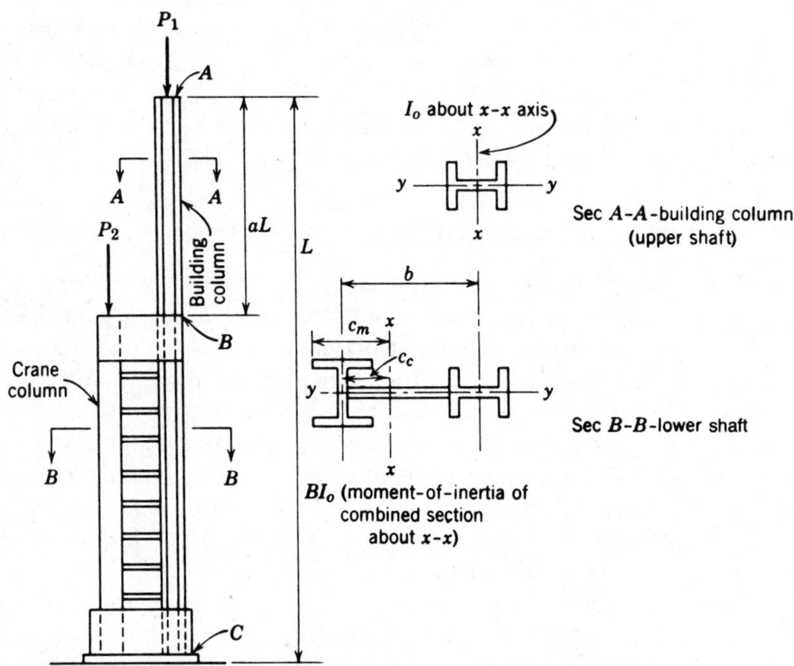

Fig. 12.2 Notation for stepped columns (AISE, 1979).

12.3 DESIGN PROCEDURE FOR STEPPED COLUMNS

The AISC Specification (1978, Sec. 1.6.1) provides that members subjected to both axial compression and bending stresses be proportioned to satisfy the following requirements:

$$\frac{f_a}{F_a} + \frac{C_{mx} f_{bx}}{[1 - (f_a/F'_{ex})]F_{bx}} + \frac{C_{my} f_{by}}{[1 - (f'_a/F'_{ey})]F_{by}} \leq 1.0 \qquad (12.1a)$$

$$\frac{f_a}{0.60F_y} + \frac{f_{bx}}{F_{bx}} + \frac{f_{by}}{F_{by}} \leq 1.0 \qquad (12.1b)$$

Equations 12.1a and 12.1b are nearly identical to Eqs. 1.6a and 1.6b of the AISC Specification, except for the introduction of f'_a as differentiated from f_a. In addition, the other terms are in some cases evaluated in a manner adopted especially to the stepped-column problem.

As applied to mill building columns, referring again to Fig. 12.2 the terms in Eq. 12.1 are discussed and defined as follows:*

f_a In the lower shaft $f_a = (P_1 + P_2)/A$, where A is the area of the entire lower shaft. In the upper shaft, $f_a = P_1/A$, with A the area of the upper (building) shaft.

f'_a In checking the lower shaft for bending about the *y-y* axis, it is conservatively assumed that the crane support segment resists all of the bending introduced by eccentricity of crane girder reactions. Thus the amplification of f_{by} as a result of deflection is dependent on the average axial stress (f'_a) in the crane segment alone. The stress f'_a is determined by adding (or subtracting) the *average* stress due to moment about the *x-x* axis, calculated at the centroid of the crane segment, to (or from) the average stress f_a of the entire lower shaft.

F_a The allowable axial stress under axial load. It may be for buckling of the entire stepped column about the *x-x* axis, based on the equivalent length KL/r_x, or it may be determined by buckling about the *y-y* axis for whatever column length is unsupported, in either the upper or lower shaft. It is to be taken as the minimum of the two values in each of the two sets pertinent to the upper and lower shafts, respectively. Interior wall girts usually are not assumed to provide longitudinal (lateral) support to the columns in mill buildings because building alterations may result in their removal.† If support in the *x* direction is provided only at

* Essentially abstracted from the AISE Guide (1979) but with some amplification and clarification.

† Recommendation of the Subcommittee on Mill Building Design of the Association of Iron and Steel Engineers.

locations A, B, and C, the equivalent length KL for buckling about the y-y axis should be taken as the full unsupported length AB in checking the upper shaft. In checking the lower shaft for the y-y axis, the equivalent length KL should be taken as 0.8 of length BC if the base is assumed to be fully fixed, or as length BC if base fixity against rotation cannot be assured.

C_{mx} For bending about the x-x axis, use a value of 0.85 when all bents are under simultaneous wind load and sidesway is assumed to take place. For specified crane load combinations, when only one bent is under consideration, use a value of 0.95 for C_{mx}.

C_{my} Since the crane segment of the lower shaft is assumed to resist all of the bending about the y-y axis, this term applies only to the lower shaft (f_{by} is assumed zero in the upper shaft). Assuming fixity at the base but no interaction with the building column, half of the moment introduced at B as a result of unequal reactions from adjacent girders will be carried down to the base, in which case $C_m = 0.4$. If base fixity cannot be assumed, take $C_{my} = 0.6$ (hinged condition at base), or, in intermediate situations, interpolate between 0.4 and 0.6.

f_{bx} Maximum stress due to bending about the x-x axis, assuming integral action of crane and building column segments in the lower shaft, and the building column alone in the upper shaft.

f_{by} Maximum stress due to bending about the y-y axis in the crane column segment of the lower shaft. Usually zero in upper shaft.

F_{bx} For compression on the crane column side of the lower shaft. F_{bx} is the permissible extreme fiber stress due to bending about the x-x axis, reduced if necessary below $0.6F_y$ because of lack of lateral support. The reduced allowable stress may be based on the permissible axial stress in the crane column segment for buckling about the y-y axis as shown in Fig. 12.2. (The y-y axis in this sketch would correspond to the x-x axis of the individual wide flange segment in the AISC Manual.) The permissible column stress, so determined, should be multiplied by the ratio c_m/c_c as defined by section B-B in Fig. 12.2. In no case is the allowable stress to be greater than $0.6F_y$.

F_{by} Since this component of bending is about the weak axis of the combined crane and building columns, no reduction in permissible stress need be made for lateral buckling. Also, since the bending resistance is assumed to be provided solely by the crane segment of the lower shaft, the allowable stress for a compact section may be used if the provisions of Sect. 1.5.1.4 of the AISC Specification are met.

F'_{ex} Since this stress is used as a basis for the determination of the amplification of column deflection in the plane of bending it should be based on the equivalent length of the complete stepped column, as in the case of F_a, for bending about the x-x axis.

F'_{ey} If the base may be assumed as fixed let $K = 0.8$ for the crane column segment alone; otherwise, assume that $K = 1.0$. The length in the determination of KL would be that of the crane column segment BC.

Illustrative Design Example 12.1

To clarify the foregoing modifications of the AISC (1978) terms as applied to stepped mill building columns, and to demonstrate their usage, a typical design check will be made. For this example the structural dimensions and components of A36 steel, illustrated in Fig. 12.2 are taken as follows:

$$aL = 14.5 \text{ ft}$$
$$L = 36.0 \text{ ft}$$
$$b \text{ (see Fig. 12.2)} = 3 \text{ ft } 10 \text{ in.}$$

Crane column segment is a W14 × 109*

$$I_{max} = 1240 \text{ in.}^4$$
$$I_{min} = 447 \text{ in.}^4$$
$$S_{max} = 173 \text{ in.}^3$$
$$A = 32.0 \text{ in.}^2$$
$$r_{max} = 6.22 \text{ in.}$$

Building column segment is a W12 × 45:

$$I_{max} = 350 \text{ in.}^4$$
$$I_{min} = 50 \text{ in.}^4$$
$$S_{max} = 58.1 \text{ in.}^3$$
$$A = 13.2 \text{ in.}^2$$
$$r_{max} = 5.15 \text{ in.}$$
$$r_{min} = 1.94 \text{ in.}$$

The routine calculations required for the calculation of the properties of the combined lower shaft section will be omitted.

Combined lower shaft—section B-B of Fig. 12.2.

$$I_{x(max)} = 20{,}571 \text{ in.}^4$$
$$I_{y(min)} = 1290$$
$$A = 45.2 \text{ in.}^2$$
$$r_{max} = 21.33 \text{ in.}$$
$$r_{min} = 5.34 \text{ in.}$$
$$c_m = 20.73 \text{ in.}$$
$$c_c = 13.43$$

* Subscripts max and min are used in place of x-x and y-y, respectively, to avoid confusion with x-x and y-y as defined in Fig. 12.2 for the combined section.

It is assumed that an analysis has been made for a particular load combination that results in the following:

Column loads and bending moments:

$$P_1 = 20 \text{ kips}$$
$$P_2 = 385 \text{ kips}$$

Numerical values for entering the effective length tables of the AISE Guide as follows:

$$B = 20571/350 = 58.8$$
$$a = 14.5/36 = 0.40$$
$$P_1/P_2 = 20/385 = 0.05$$

Equations 12.1 will first be evaluated for the lower shaft, *BC* (see Fig. 12.3), later for the upper shaft, *AB*. In the lower shaft:

$$f_a = \frac{P_1 + P_2}{A} = \frac{20 + 335}{43.2} = 8.96 \text{ ksi}$$

The value F_a may now be determined by the overall column action about *x-x* of the full length *AC*, or about *y-y* of the lower shaft segment *BC* which has a length of 21 ft 6 in. It is assumed conservatively that the column is hinged at *A*. From Table 16 of the AISE Guide, for the calculated values of *B*, *a*, and P_1/P_2, the effective length factor *K* is found to be 1.14. Thus

$$KL = (1.14)(36) = 41.0$$

$$\frac{KL}{r} = \frac{(41.0)(12)}{21.33} = 23.1$$

$F_a = 20.40$ (AISC (1978), Appendix A, Table 3-36, p. 5-74)

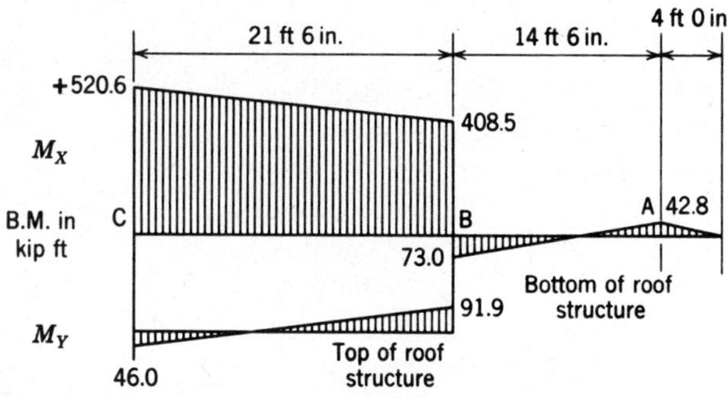

Fig. 12.3

But F_a may be controlled by the y-y axis, in which case, assuming a fixed base at C and a hinge at B, assume that $K = 0.8$:

$$\frac{KL}{r} = \frac{(0.8)(21.5)(12)}{5.34} = 38.7$$

The larger value (38.7) of KL/r determines the smaller value of F_a, which, from Table 3.36 of AISC (1978), again, is found to be 19.30 ksi. The first term in Eq. 12.1a can now be calculated:

$$\frac{f_a}{F_a} = \frac{8.96}{19.30} = 0.464$$

The second term in Eq. 12.1a will now be evaluated. Reference should be made to the prior explanations of C_{mx}, f_{bx}, F_{bx}, and F'_{ex}. Assume $C_{mx} = 0.95$. The initial step in the calculation of F_{bx} parallels the determination of F_a for buckling about the y-y axis, except that the r of the individual crane segment is used instead of r for the entire lower shaft.

$$\frac{KL}{r} = \frac{(0.80)(21.5)(12)}{6.22} = 33.2$$

Referring again to AISC (1978), Table 3.36, p. 5-74, $F_a = 19.7$ for $KL/r = 33.2$. Thus for this allowable column stress, F_a,

$$F_{bx} = F_a\left(\frac{C_m}{C_c}\right) = \frac{(19.7)(20.73)}{13.43} = 30.42 \text{ ksi}$$

Since 30.42 is greater than 22, the allowable stress in bending for built-up members, F_{bx} is taken as 22 ksi.

Referring to Table 9, p. 5-79 of AISC (1978), for the previously determined KL/r of 23.1,

$$F'_{ex} = 279.0 \text{ ksi}$$

The maximum compressive stress due to bending moment about x-x is

$$f_{bx} = \frac{(520.6)(12)(20.73)}{20{,}571} = 6.30 \text{ ksi}$$

The second term in Eq. 12.1a can now be evaluated:

$$\frac{C_{mx} f_{bx}}{[1 - (f_a/F'_{ex})]F_{bx}} = \frac{(0.95)(6.30)}{[1 - (8.96/279.0)]22} = 0.281$$

Finally, the third term in Eq. 12.1a will be evaluated.

$$C_{my} = 0.4 \text{ (base assumed fixed)}$$

$$f_{by} = \frac{(91.9)(12)}{173} = 6.37 \text{ ksi}$$

In the calculation of f'_a, the *average* bending moment in the crane segment of the lower shaft is required:

$$M_{x(\text{avg})} = \frac{520.6 + 408.5}{2} = 464.6 \text{ kip-ft}$$

$$f'_a = 8.96 + \frac{(464.6)(12)(13.43)}{20571} = 12.60 \text{ ksi}$$

$F'_{ey} = 135.5$ ksi (from Table 9, p. 5-79 of AISC (1978) for $KL/r = 33.2$ as previously calculated)

The W14 × 111 crane column segment of the lower shaft is a "compact" shape for A36 steel. Hence for the crane column segment acting alone in bending, and assuming lateral support as provided by battens attached to the building column segment,

$$F_{by} = 24 \text{ ksi}$$

The third term in Eq. 12.1a can now be evaluated:

$$\frac{C_{my} f_{by}}{[1 - (f'_a/F'_{ey})]F_{by}} = \frac{(0.4)(6.37)}{[1 - (12.60/135.5)]24} = 0.117$$

All three terms in Eq. 12.1a have now been evaluated and the lower shaft column check can be made:

$$0.464 + 0.281 + 0.117 = 0.862 < 1 \qquad \text{O.K.}$$

Now check stresses at B and C by Eq. 12.1b. At B, as previously determined:

$$f_a = 8.96 \text{ ksi}$$
$$f_{by} = 6.37 \text{ ksi}$$
$$F_a = F_{bx} = 22 \text{ ksi}$$
$$F_{by} = 24 \text{ ksi}$$

The bending moment M_{xx} at B is 408.5 kip-ft:

$$f_{bx} = \frac{(408.5)(12)(20.73)}{20{,}571} = 4.94 \text{ ksi}$$

Substituting in Eq. 12.1b:

$$\frac{8.96}{22} + \frac{4.94}{22} + \frac{6.37}{24} = 0.897 < 1 \qquad \text{O.K.}$$

At C, add 6 kips to allow for dead weight of column between B and C.

$$f_a = \frac{411}{45.2} = 9.09 \text{ ksi}$$

$$f_{by} = 3.19 \text{ ksi (half of } f_{by} \text{ at top)}$$
$$f_{bx} = 6.30 \text{ ksi (calculated previously)}$$

Substituting in Eq. 12.1b, yields

$$\frac{9.09}{22} + \frac{6.30}{22} + \frac{3.19}{24} = 0.832 < 1 \qquad \text{O.K.}$$

The upper shaft, AB, will now be checked. Since $f_{by} = 0$ in this part of the column, Eq. 12.1 has only two terms to the left of the equal sign. It is assumed that knee braces make the unsupported length AB equal to 14.5 ft, and it is assumed that $K = 1$ in the weak direction; hence

$$\frac{KL}{r} = \frac{(14.5)(12)}{1.94} = 89.7$$

$$F_a = 14.24 \text{ ksi (AISC (1978), Table 3-36, p. 5-74)}$$

$$f_a = \frac{20}{13.2} = 1.52 \text{ ksi}$$

Calculate F_{bx}. (Refer to AISC (1978), Sec. 1.5.1.4.5)

$$\frac{M_1}{M_2} = \frac{42.8}{73.0} = 0.59$$

$$C_b = 1^*$$

$53C_b = 53$ (AISC (1978), Table 7, p. 5-78). For the W12 × 45, the AISC Manual lists $r_T = 2.15$

$$\frac{L}{r_T} = \frac{(14.5)(12)}{2.15} = 80.9 > 53$$

Hence

$$F_{bx} = 24.0 - \frac{36^2 . 80.9^2}{1{,}530{,}000} = 18.5 \text{ ksi}$$

$$f_{bx} = \frac{(73.0)(12)}{58} = 15.08 \text{ ksi}$$

$$C_{mx} = 0.4$$

Check by Eq. 12.1a.

* The AISC Specifications of Sec. 1.5.1.4.5, p. 5-22, provides that C_b shall be taken as unity in computing the value of F_{bx} and F_{by} to be used in Eq. 1.6-1a.

$$\frac{1.52}{14.24}+\frac{(0.4)(15.08)}{[1-(1.52/279)]18.5}=0.43<1 \qquad \text{O.K.}$$

Check by Eq. 12.1b, at location B.

$$\frac{1.52}{22.0}+\frac{15.08}{24.0}=0.70<1 \qquad \text{O.K.}$$

All critical locations in the columns have now been checked and found satisfactory for the particular illustrative load condition of the example.

12.4 DESIGN TRENDS AND RESEARCH NEEDS

This chapter has presented a conventional allowable stress design procedure for stepped mill building columns that is based on AISE modifications of AISC Specifications. It is applicable to members for which the lower shaft can be considered to behave as an integral unit, with complete or nearly complete interaction of the longitudinal column elements provided by adequate web plate, lacing bars, or battens.

When heavy longitudinals are required for the separate components of the lower shaft of a mill building column, it may be good economy *not* to provide integral action in battened columns. The lower shaft may then be assumed to consist of two separated elements with the battens acting as spacers. In such a spaced column the use of adequate end-tie plates will increase the critical load.

The design of the column with constant cross section having lateral loads and moments at a bracket, as in Fig. 12.1*a*, has been considered by Adams (1970). The upper and lower segments are treated separately as individual beam-columns.

Further research is needed as to the transfer of crane impact loads into columns, the structural significance of such loads, and the degree to which repeated load and the possibility of fatigue failure should be factors in mill-building column design.

It has been assumed herein that girts do not provide lateral support to the building column because of the possibility that they may be removed. In situations where premanency of girts is assured, the work of Horne and Ajmani (1971) on the lateral bracing of girts attached to one flange may be of interest.

References

Adams, P. F. (1970), "Segmental Design of Steel Beam Columns," *Proc. Can. Struct. Eng. Conf.*, Toronto.

American Institute of Steel Construction (1978) "Specification for the Design, Fabrication and Erection of Steel Buildings" AISC, Chicago.

Association of Iron and Steel Engineers (1979), *Guide for the Design and Construction of Mill Buildings*," AISE, Tech. Rep. No. 13, Pittsburgh.

Horne, M. R., and Ajmani, J. L. (1971), "Design of Columns Restrained by Side-Rails and the Post-buckling Behavior of Laterally Restrained Columns," *Struct. Eng.*, Vol. 49, No. 8.

Huang, H. C. (1968a), "The Design of Mill Building Columns," *Iron Steel Eng.*, Vol. 45, No. 3, p. 97.

Huang, H. C. (1968b), "Determination of Slenderness Ratios for Design of Heavy Mill Building Stepped Columns," *Iron Steel Eng.*, Vol. 45, No. 11, p. 123.

Maugh, L. C. (1963), *Statically Indeterminate Structures*, 2nd ed., Wiley, New York.

Murray, J. J., and Graham, T. C. (1957), "The Design of Mill Buildings," *Iron Steel Eng.*, Vol. 34, No. 2, p. 159.

Chapter Thirteen

Thin-Walled Metal Construction

13.1 INTRODUCTION

Thin-walled metal members are used as framing for light construction and as secondary members in heavy construction. Thin-walled metal panels and decks are widely used as floors, roofs, and walls.

The following specifications cover the design of this type of construction in the United States: (1) *Specification for the Design of Cold-Formed Steel Structural Members*, published by the American Iron and Steel Institute (AISI, 1980, 1986); (2) *Specification for the Design of Cold-Formed Stainless Steel Structural Members* (AISI, 1974); (3) *Specification for Aluminum Structures*, published by the Aluminum Association (AA, 1982), and (4) *Specification for the Design, Testing and Utilization of Industrial Steel Storage Racks*, published by the Rack Manufacturers Institute (RMI, 1979). In Canada thin-walled design is covered by the following standards of the Canadian Standards Association (CSA): *Cold-Formed Steel Structural Members*, CAN3-S136-M84 (CSA, 1984) and *Strength Design in Aluminum*, CAN3-S157-M83, (CSA, 1983).

New developments in research, design, and construction are presented in the proceedings of international specialty conferences, which meet periodically to discuss advances in this type of structural system (Yu and Senne, 1984; Rhodes and Walker, 1979).

Thin-walled steel members are usually cold-formed to shape from hot or cold-rolled sheet, strip, or plate. Aluminum sections are generally extruded or cold-formed. Cold-forming to shape may increase the yield and ultimate strength of the member because of strain hardening and strain aging (Karren and Winter, 1967; Uribe and Winter, 1970). This results in a higher yield strength for the material in the more severely cold-worked zones, such as corners, and in gradual yielding behavior of a cross section under load, even though the virgin material may exhibit sharp yielding characteristics. The strengthening effect of cold forming can be determined by test for any member, and the AISI Specification (1980, 1986) and CSA Standard (1984) contain provisions for analytically determining the strength increase due to cold forming of cold-formed carbon and low-alloy steel members. Whether testing or analysis is used, the design approach is to modify ordinary procedures to permit utilization of this increase in strength, recognizing the end service intended.

This chapter covers the behavior of flexural and compression members that is unique to thin-walled construction. Because of the thinness of the member elements, local buckling is an important consideration, and emphasis is given to aspects of buckling that affect member behavior.

13.2 FLEXURAL MEMBERS

13.2.1 General

Thin-walled sections such as tubular members (Fig. 4.1*b*), I-sections (Fig. 4.1*h*), channels, Z-sections, hat sections (Fig. 4.1*i*), T-sections (Fig. 4.1*m*), and panels (Fig. 4.1*o*, *p*, *q*) are often used as flexural members in thin-walled construction. The depth of such members generally ranges from 1 to 12 in., and the thickness of material ranges from about 0.012 to about 0.5 in. or thicker.

In the design of thin-walled flexural members, consideration should be given to the following (Yu, 1985):

1. Moment-resisting capacity and stiffness of the member.
2. Web design.
3. Bracing requirements.

13.2.2 Moment Capacity

In thin-walled metal construction, the moment-resisting capacity of a flexural member is governed by one or a combination of the following factors:

1. Yielding of material.
2. Local buckling of compression flange or web.
3. Lateral buckling.

When the section geometry and loading result in bending in the plane of loading, and when local buckling or lateral buckling does not occur, flexural yielding is the limiting factor and traditional design concepts apply. However, the width-to-thickness ratios are frequently great enough to result in local buckling before the ultimate load is reached, and attention must be given to a cross section in which the normal stress distribution in a plane perpendicular to the longitudinal axis of the member varies because of local buckling. One may either consider the actual stress distribution and resort to the fundamental integrals governing flexural behavior, or replace the actual distribution with an assumed distribution, which will give the same results.

Where local buckling of individual elements is involved, two different approaches are commonly used to facilitate design—one based on effective-width, the other on average or reduced allowable stress—as described in Chapter 4. The Aluminum Association Specification (AA, 1982) generally uses the average stress method for strength determination and the effective width for computing deflections. The 1980 AISI Specification uses both methods, depending on whether the compression flange is "stiffened" or "unstiffened." A stiffened compression element is defined as one that is stiffened by a web, edge stiffener, or intermediate stiffener at both edges parallel to the direction of stress. An unstiffened compression element is one that is stiffened at only one edge parallel to the direction of stress. The CSA (1984) uses the effective width approach for stiffened and unstiffened compression elements. The 1986 AISI Specification extends the procedure to include partially stiffened elements.

The treatment of the stiffened and unstiffened beam elements in the 1980 AISI Specification is described in detail in Chapter 9 of the third edition of this guide and will not be repeated here. The AISI Advisory Group on the Specification has approved a new specification in 1986. In this document stiffened, partially stiffened, and unstiffened elements are designed according to a unified effective width method, as follows.

Equation 4.15 is used as the basis for determining the effective width of locally buckled elements. This equation is expressed by the following nondimensional form:

$$b_e = \frac{b(1 - 0.22/\lambda)}{\lambda} \tag{13.1}$$

where

b_e = effective width
b = flat width of plate
$\lambda = (\sigma_e/\sigma_{cr})^{1/2}$
σ_e = for ultimate-strength calculations, the maximum stress on the plate element computed for the moment causing first yield in the effective cross section; for deflection calculations, the maximum stress due to service loads
σ_{cr} = plate buckling stress defined by Eq. 4.1, i.e.,

$$\sigma_{cr} = \frac{\pi^2 Ek}{12(1-\nu^2)}\left(\frac{t}{b}\right)^2 \tag{13.3}$$

E = modulus of elasticity (E = 29,500 ksi for cold-formed steel)
ν = Poisson's ratio ($\nu = 0.3$)
t = plate thickness
k = plate-buckling coefficient ($k = 4$ for stiffened elements; $k = 0.43$ for unstiffened elements; for partially stiffened elements an intermediate value for k is calculated)

When $b = b_e$, $\lambda = 0.673$, which is the limiting slenderness parameter below which the effective width is equal to the actual flat width. The effective-width criteria in the 1986 AISI Specification are thus expressed by the following equations:

$$b = b \qquad \text{for } \lambda \le 0.673 \tag{13.4a}$$

$$b = \rho b \qquad \text{for } \lambda > 0.673 \tag{13.4b}$$

where

$$\rho = \frac{1 - 0.22/\lambda}{\lambda} \tag{13.5}$$

and

$$\lambda = \frac{1.052}{\sqrt{k}}\,\frac{b}{t}\sqrt{\frac{\sigma_e}{E}} \tag{13.6}$$

This equation derives from Eqs. 13.2 and 13.3 with $\nu = 0.3$. Equation 13.5 is identical to the following formula in the 1980 AISI Specification:

$$\frac{b_e}{t} = \frac{253}{\sqrt{\sigma}}\left(1 - \frac{55.3}{b\sqrt{\sigma}/t}\right) \tag{13.7}$$

where σ is the stress due to service loads, and E = 29,500 ksi, $\nu = 0.3$, $k = 4$, and the factor of safety is $\frac{5}{3}$. Equation 13.7 is the effective-width formula for a stiffened element, and is the same in both the 1980 and the 1986 versions of the AISI Specifications. An average reduced allowable stress is used for unstiffened elements in the 1980 and earlier rules, while an effective-width criterion (with $k = 0.43$) is used in the 1986 criteria. The extension of the effective-width approach is based on research by DeWolf et al. (1974), Kalyanaraman et al. (1977), Kalyanaraman and Peköz (1978), Peköz et al. (1981a, b), and Milligan and Peköz (1983).

Research has been conducted on the inelastic reserve strength of cold-formed steel beams whose compression flanges are stiffened along both longitudinal edges. Results indicate that this inelastic reserve strength due to partial plastification of the cross section can be significant for many practical shapes (Reck et al., 1975). Design provisions in the AISI Specification permit use of this reserve in design. Additional inelastic reserve capacity due to the redistribution of moments in statically indeterminate beams was studied by Yener and Peköz (1980).

The Aluminum Association Specifications (1982) do not take advantage of any postbuckling strength in angle struts and do take postbuckling strength into account in defining allowable stresses for unstiffened flanges. The Aluminum Association treats stiffened flanges similarly to unstiffened flanges; that is, moment capacity is determined using the full section properties and a reduced stress that is based on the postbuckling strength.

The post-local-buckling behavior of continuous beams is discussed by Wang and Yeh (1974). Sloping edge stiffeners of beams have been studied by Peköz and He (1981) and LaBoube (1983) and are the subject of current further research.

13.2.3 Lateral Buckling

In addition to yielding and local buckling, as discussed previously, the moment-resisting capacity of a thin-walled flexural member may be limited by lateral buckling of the beam between lateral supports. Theoretical and experimental treatments for lateral buckling of hot-rolled shapes and built-up members were discussed in Chapter 5. For thin-walled metal construction, the critical stress for lateral buckling of an I-beam having unequal flanges can be determined by the following formula (Winter, 1943, 1970):

$$\sigma_e = \frac{\pi^2 Ed}{2S_{xc}L^2}\left(I_{yc} - I_{yt} + I_y\sqrt{1 + \frac{4GJL^2}{\pi^2 EI_y d^2}}\right) \tag{13.8}$$

where S_{xc} is the compressive section modulus of the entire section about the major axis; I_{yc} and I_{yt} are moments of inertia of the compression and tension portion, respectively, of a section about its centroidal axis parallel to the web; E is the modulus of elasticity; G is the shear modulus; J is the torsional constant of the section; d is the depth of the section; and L is the unbraced length.

For thin-walled steel sections, the first term under the radical in Eq. 13.8 usually exceeds the second term by a considerable margin (Winter, 1947). If the second term is omitted and considering that $I_y = I_{yc} + I_{yt}$, the following equation (Eq. 13.9) can be obtained for determination of critical stress for lateral buckling in the elastic range:

$$\sigma_c = \pi^2 EC_b\left(\frac{dI_{yc}}{L^2 S_{xc}}\right) \tag{13.9}$$

where C_b is a bending coefficient that can conservatively be taken as unity or calculated from $C_b = 1.75 + 1.05(M_1/M_2) + 0.3(M_1/M_2)^2$ but not more than 2.3. Here $M_1 < M_2$ and the ratio of M_1/M_2 is positive when M_1 and M_2 have the same sign (reverse curvature bending) and negative when they are of opposite signs (single curvature bending).

The lateral buckling criteria for I-, Z-, and channel section beams in the 1980 and 1986 AISI Specification are essentially identical. In the elastic range they are based on Eq. 13.9. The 1980 rules are described in the third edition of this guide. The 1986 criteria are the following:

$$M_u = M_{cr}(S_c/S_f) \tag{13.10}$$

where

M_u = ultimate lateral-torsional buckling moment
S_f = elastic section modulus of the full reduced section for the extreme compression fiber
S_c = elastic section modulus of the effective section calculated at a stress M_{cr}/S_f in the extreme compression fiber

The critical moment is determined as follows:

$$M_{cr} = M_y \qquad \text{for } M_e \geq 2.78M_y \tag{13.11}$$

$$M_{cr} = \left(\frac{10}{9}\right)\left(1 - \frac{10M_y}{36M_e}\right)M_y \qquad \text{for } 2.78M_y > M_e > 0.56M_y \tag{13.12}$$

$$M_{cr} = M_e \qquad \text{for } M \leq 0.56M_y \tag{13.13}$$

where

M_y = moment causing yielding at the extreme compression fiber of the full section
M_e = elastic critical moment (Eq. 13.9 times S_f)

The solid curve in Fig. 13.1 shows the variation of the critical moment M with the unbraced length. The curve consists of three regions: yielding (Eq. 13.11), inelastic buckling (Eq. 13.12), and elastic buckling (Eq. 13.13). CSA (1984) uses a similar approach to lateral buckling except that the second term under the radical in Eq. 13.8 is retained. Previous study has indicated that equations developed for I-beams can also be used for channels with reasonable accuracy (Hill, 1954). For channels and other singly symmetric shapes, the 1986 AISI Specification has deleted the yield plateau.

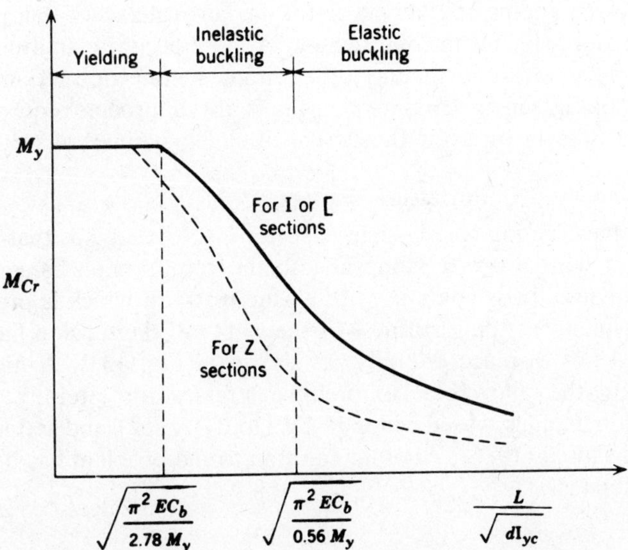

Fig. 13.1 Allowable compressive stress for lateral buckling of beams (Winter 1970).

For a given $L/\sqrt{dI_{yc}}$ ratio, a Z-section will buckle laterally at a lower stress than will an I-beam or a channel section. A conservative design approach has been used in the AISI Specification (1986), in which the critical moments for Z-sections in the elastic range are one-half of those permitted for I-beams or channels with the same $L/\sqrt{dI_{yc}}$ ratio. The lateral buckling curve for Z-shaped beams is shown as the dashed line in Fig. 13.1.

Laterally unbraced steel box sections with length-to-width ratios up to $2500/\sigma_y$ and with web plates no less than $h/6$ apart can be used as beams without any stress reduction for lateral buckling because the St.-Venant torsional stiffness of box sections is high, and the I_y/I_x ratio is higher than for I-sections (Winter 1970).

For laterally unbraced hat sections bent about the X-axis, no stress reduction is necessary if $I_y > I_x$, because there is no tendency to buckle. When $I_y < I_x$ a conservative estimate of the critical elastic stress may be determined by regarding the compression portion of the section as an independent strut, which gives

$$\sigma_c = \frac{\pi^2 E}{(L/r_y)^2} \tag{13.14}$$

where r_y is the radius of gyration about the vertical axis of that portion of the hat section which is in compression. A more accurate analysis for such hat-shaped sections and for any other singly symmetric section is to use the equations given in Chapter 5. This is the approach required in the 1986 AISI Specification for the design of singly symmetric section.

13.2.4 Unbraced Compression Flanges

When a hat section or U-shaped beam is loaded so that the two outstanding flanges are in compression, the compression flanges tend to buckle individually as columns with elastic restraint which is provided by the out-of-plane bending stiffness of the web and the tension flange. The latter does not displace laterally, as shown in Fig. 13.2. A method for determining the allowable compressive stresses for laterally unbraced compression flanges was developed by Douty (1962) and is included in the AISI Manual (1986). Another solution to the problem has been given by Haussler (1964).

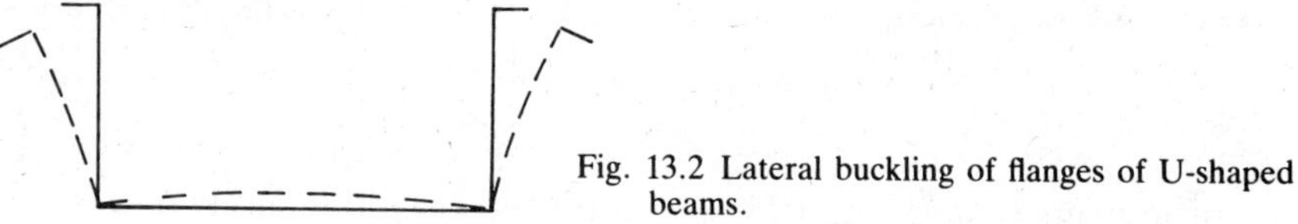

Fig. 13.2 Lateral buckling of flanges of U-shaped beams.

13.2.5 Design of Webs

The design of webs of thin-walled beams is somewhat different than for hot-rolled members, because the width-to-thickness ratio of webs of thin-walled members is usually large and the use of bearing stiffeners is in most cases impractical. Nevertheless, new design provisions for webs with transverse stiffeners are included in the AISI Specification (AISI, 1986; Nguyen and Yu, 1978).

In the design of webs, consideration should be given to (1) shear strength, (2) local buckling due to bending stress in the web, (3) effect of combined bending and shear, (4) web crippling, and (5) combined bending and web crippling. For webs having small width-to-thickness ratios, the shear strength of the web will be governed by the yield stress in shear determined by the Hencky–von Mises yield criterion; that is,

$$\tau_y = \sigma_y/\sqrt{3} \tag{13.15}$$

As discussed in Chapter 4, for webs having relatively large width-to-thickness ratios, the strength of the web may be governed by shear buckling. The theoretical critical shear-buckling force in the elastic range can be determined using the values of k_s given in Eqs. 4.24a and 4.24b. For carbon steel ($E = 29{,}500$ ksi and $\nu = 0.3$).

$$V_{cr} = \frac{\pi^2 E k_s t^3}{12(1-\nu^2)h} = \frac{0.904 k_s E t^3}{h} \tag{13.16}$$

where h is the flat width of the web and t is the web thickness. The shear force V_a permitted by the AISI Specification (1986) for flat webs with h/t ratios exceeding $1.38\sqrt{Ek_s/\sigma_y}$ is determined by a factor of safety of 1.71 applied to the elastic critical shear-buckling stress; that is,

$$V_a = 0.53 k_s E t^2/h \tag{13.17}$$

For the intermediate range of h/t ratios ($0.95\sqrt{EK_s/\sigma_y} < h/t < 1.38\sqrt{Ek_s/\sigma_y}$) where shear buckling occurs above the proportional limit in shear, the allowable shear force is determined by the following formula:

$$V_\partial = 0.38 t^2 \sqrt{E k_s \sigma_y} \tag{13.18}$$

Figure 13.3 shows the allowable shear stress for different ranges of h/t of unreinforced webs for a steel with $\sigma_y = 33$ ksi. For this case $k_s = 5.34$.

Webs of beams can buckle not only in shear but also because of the compressive stress caused by bending. For beams with the neutral axis at

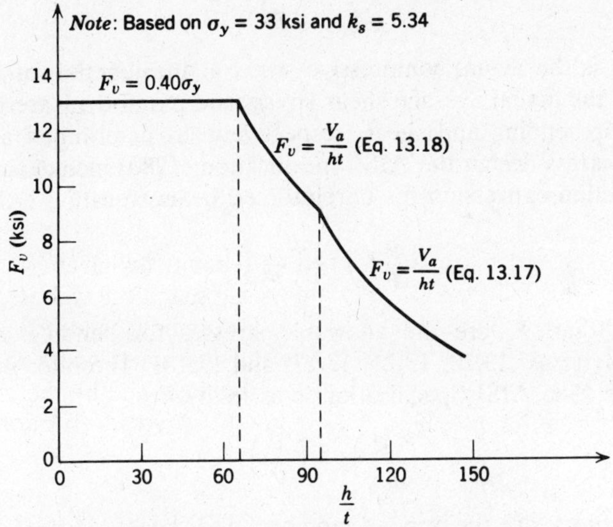

Fig. 13.3 Allowable shear stress for webs of carbon steel beams (AISI, 1986).

middepth, the web-buckling stress can be determined by Eq. 4.1 with $k = 23.9$. For $E = 29{,}500$,

$$\sigma^*_{cb} = \frac{640{,}000}{(h/t)^2} \tag{13.19}$$

Considering the available postbuckling strength, the AISI (1980) allowable stress for bending in webs, F_{bw}, is obtained by using a factor of safety of $\frac{5}{3}$ applied to the experimentally determined maximum stress (LaBoube and Yu, 1982). For beams having stiffened compression flanges,

$$F_{bw} = [1.21 - 0.00034(h/t)\sqrt{\sigma_y}](0.60\sigma_y) \le 0.60\sigma_y \tag{13.20}$$

and for beams having unstiffened compression elements,

$$F_{bw} = [1.26 - 0.00051(h/t)\sqrt{\sigma_y}](0.60\sigma_y) \le 0.60\sigma_y \tag{13.21}$$

In the 1986 Specification an effective width approach is used to determine the allowable moment M_a of the member with a buckled web.

When high bending stresses and high shear stresses act simultaneously, as in cantilever beams and at supports of continuous beams, the webs of beams will buckle at a lower stress than if only one stress were present. For a combination of bending and shear, Eq. 4.30 can be used to predict buckling:

$$\left(\frac{\sigma_{cb}}{\sigma^*_{cb}}\right)^2 + \left(\frac{\tau_c}{\tau^*_c}\right)^2 \le 1 \tag{13.22}$$

where σ_{cb} is the actual compressive stress at the junction of flange and web, τ_c is the actual average shear stress, and σ^*_{cb} and τ^*_c are the critical stresses for bending and shear, respectively, as determined above. For allowable stress design the AISI Specification (1980) includes the following interaction expression for unreinforced beam webs:

$$\left(\frac{\sigma_{cb}}{F_{bw}}\right)^2 + \left(\frac{\tau_c}{F_v}\right)^2 \le 1 \tag{13.23}$$

where F_{bw} and F_v are the allowable stresses for bending and shear, respectively (Eqs. 13.20, 13.21, 13.17, and 13.18). The interaction equation in the 1986 AISI Specification is as follows:

$$\left(\frac{M}{M_a}\right)^2 + \left(\frac{V}{V_a}\right)^2 \le 1 \tag{13.24}$$

where M and V are the applied moments and shears, and M_a and V_a are the allowable forces.

Webs with transverse stiffeners also are treated in this Specification. The economy of thin flat webs is reduced by the fabrication costs associated with attaching stiffeners necessary to prevent premature local buckling. The use of "self-stiffened' corrugated webs could reduce fabrication costs and provide increased material efficiency. Design guidelines for both local and overall shear buckling are given by Shanley (1968) and Mertz and Matthiesen (1970). A study of the optimum attachment for a corrugated web to the beam flange was conducted by Sherman and Fisher (1971) and Hlavacek (1972).

The specifications of the Aluminum Association (1982) also base allowable shear stresses in webs on the shear buckling stress, applying a factor of safety of 1.65; a factor of safety of 1.65 is also applied to the post buckling or crippling strength of a web in bending. Postbuckling strength of webs contributes appreciably to the bending strength of some formed aluminum sections used for roofing and siding (Jombock and Clark, 1968). The Aluminum Association also uses the interaction equation 4.29 for handling combined shear and bending.

The use of web stiffeners is frequently impractical in cold-formed steel construction, and without such stiffeners thin webs of beams may cripple due to the high local stresses caused by concentrated loads or reactions. Figure 13.4 shows the types of deformation that occur due to crippling of unrestrained single webs and restrained double webs.

Theoretical analysis of web crippling is rather complicated. Experimental investigation (Winter and Pian, 1946) has indicated that the crippling strength of cold-formed steel beams depends on N/t, h/t, R/t, and σ_y, where t is the web thickness, N the actual length of bearing but is not greater than h, h the clear distance between flanges, σ_y the yield point of steel, and R the inside bend radius. The AISI provisions to avoid web crippling are based on this experimental investigation and further study of this problem is reported by Rockey et al. (1972) and Hetrakul and Yu (1978). CSA (1984) provisions for allowable reaction are similar to the

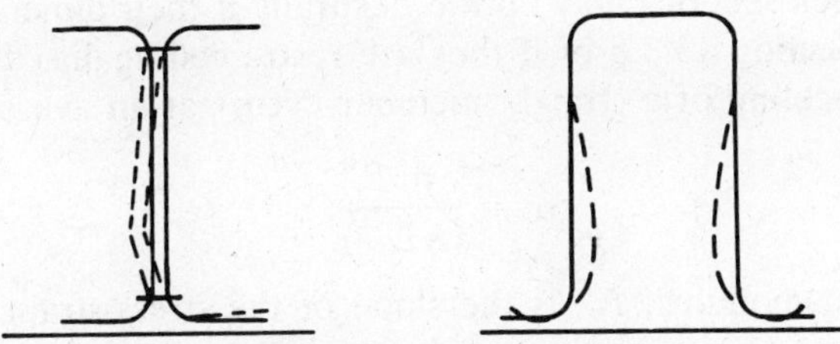

Fig. 13.4 Web crippling of beams.

AISI (1980) for shapes having single webs. For multiweb deck sections a different set of expressions are used based on the work of Wing (1981), including the data from Yu (1980). Wing extended the earlier work on web crippling to include the effect of varying web slope and large bend radius-to-thickness ratios on deck-type profiles.

13.3 COMPRESSION MEMBERS

13.3.1 General

Thin-walled open sections are susceptible to flexural-torsional buckling and, in some unusual cases, torsional instability, whereas this mode of behavior is rather exceptional in hot-rolled steel construction. Due to the thinness of the component plate elements, the effect of local buckling may be of critical importance. Another problem peculiar to thin-walled construction is the sensitivity to imperfections in both the member and the connections. Imperfections in connections may lead to local crippling and hence reduce the overall load-carrying capacity. Fortunately, as discussed in Chapter 4, such members may have considerable post-local-buckling strength. With regard to overall flexural buckling, the behavior of cold-formed carbon and low-alloy steel members is similar to that of hot-rolled sections. This behavior, which is caused by the presence of residual cooling stresses in hot-rolled sections, is brought about by the effects of cold forming, which results in increased yield strength at affected zones such as corners. This results in an apparent gradual yielding behavior, as discussed in section 13.1. However, a complication less easily dealt with may be the presence of through-the-wall residual stresses around the periphery of the section (McDermott, 1981).

13.3.2 Flexural Buckling

Slender compression members that are not susceptible to torsional, or flexural-torsional buckling will lose their stability by flexural buckling. Doubly symmetric sections and closed sections, axially loaded, do not have any tendency to twist if they are of dimensions commonly used in structures. Open sections may buckle flexurally if their dimensions are as discussed in Section 13.3.5 of if they are restrained against twist.

Flexural buckling of a straight member occurs at an average stress of

$$\sigma_c = \frac{\pi^2 E_t}{(KL/r)^2} \tag{13.25}$$

The tangent modulus, E_t, is the slope of the stress-strain diagram at the buckling stress σ_c. The exact shape of the stress-strain diagram—hence the variation of the tangent modulus with the stress level—depends on the process of forming the section as well as the properties of the virgin material. Thus it is impossible to predict the details of the shape of the stress-strain diagram for cold-formed members. For columns of small or moderate slenderness ratios (i.e., when the buckling stress is at a level where the stress-strain relationship is nonlinear) a simple but sufficiently accurate approach has been followed, similar to the CRC Basic Column Curve treatment of the influence of residual cooling stresses in hot-rolled column behavior. Early tests (Karren and Winter, 1967) have shown that for the range of slenderness ratios in question, the flexural buckling stress for carbon and low-alloy steel members can be approximated by

$$\sigma_c = \sigma_y - \frac{\sigma_y^2}{4\pi^2 E}\left(\frac{KL}{r}\right)^2 \tag{13.26}$$

If the column is slender enough to buckle at a stress level where the stress-strain relation is linear, the buckling stress can be obtained from

$$\sigma_c = \frac{\pi^2 E}{(KL/r)^2} \tag{13.27}$$

The limiting value of the slenderness ratio $(KL/r)_{\text{lim}}$, determining which type of behavior is to be expected can be found by equating the buckling stresses according to Eqs. 13.26 and 13.27. Thus

$$\left(\frac{KL}{r}\right)_{\text{lim}} = \pi\sqrt{\frac{2E}{\sigma_y}} \tag{13.28}$$

Cold-formed thin-walled columns are sensitive to imperfections and connection details for all slenderness ratios; thus a factor of safety of 1.92 is required by the AISI Specification (1980, 1986).

Recent studies (Peköz, 1980; Dat and Peköz, 1980) have shown that for certain sections Eq. 13.26 may overestimate the strength. It was found that even though the cold-forming increases the yield stress, the proportional limit is lowered and hence the buckling load is lowered. Consequently, for some ranges of slenderness ratios, the RMI specification (1979) requires factors of safety that are up to 15% more than 1.92. Dat and Peköz (1980) give new column curves based on analytical and experimental studies involving cold-formed steel columns. Guiaux (1974) and Ingvarsson (1975, 1977) present the results of research on cold-formed tubular sections. The Aluminum Association Specifications (1982), which cover both thick- and thin-walled members, use a constant factor of safety of 1.95 in determining the allowable column stress for building structures.

13.3.3 Effect of Local Buckling on Column Strength

Results of early studies on interaction between local and overall buckling were presented by Bijlaard and Fisher (1952a,b). More recently, this problem has received a significant amount of attention. Graves-Smith (1969) treats the post-local-buckling behavior within the scope of the large deflection plate theory and accounts for the inelasticity effects both locally and overall. The overall buckling load is computed using the tangent modulus approach based on the stiffnesses of the locally buckled plate elements. Although the method is basically general the author treats only the case of a square tubular columns. Hancock (1981) and Sridharan and Benito (1984) have investigated the interaction problem using the finite strip method. Konig and Thomasson (1980) and Thomasson (1978) treat the post-local-buckling behavior with an effective-width approach and the column strength is determined on the basis of an initial column imperfection. The ultimate load is defined as either the load that causes yielding or the maximum load that can be sustained. Mulligan and Peköz (1983) studied singly symmetric columns.

The foregoing approaches necessitate computer calculations. A computationally simpler approach is presented by Kalyanaraman et al. (1977) and DeWolf et al. (1974, 1976). In these studies an effective-width approach is used to find stiffnesses that depend on the value of the axial load. The stiffnesses thus obtained are used with a modified tangent modulus approach to obtain the overall buckling load. A more recent experimental study is presented by Braham et al. (1980).

A simple but conservative treatment is provided by the use of a form factor as recommended by the AISI Specifications (1980). Sections consisting of stiffened elements, sections consisting of unstiffened elements, and combinations of these two types of sections exhibit different types of behavior. This method is described in detail in the third edition of this guide. The AISI (1980) form factor method has the advantage of simplicity and appears to have performed well historically. Recent tests have indicated, however, that its performance in predicting maximum strength is uneven, ranging from the excessively conservative to the unconservative.

The singly symmetric column which is not fully effective is a unique and difficult problem. Not only are the effective section properties reduced by local buckling, but the effective centroid shifts along the axis of symmetry. Thus an initially concentrically loaded column becomes a beam column. Testing such a column which is truly concentrically loaded throughout is loading history appears difficult if not impossible. Furthermore, the centrally loaded singly symmetric column appears to exist in practice only if it is fully effective and is loaded at its ends uniformly around the periphery. In practice, it may be difficult to be assured that such conditions will exist. Consequently, many columns that have no obvious moment applied to their ends may be, in actual fact, beam columns.

Springfield and Trestain developed a design method for the CSA Standard CAN3-S136-M84, which differed radically from the essentially similar methods of the 1974 CSA and 1980 AISI. Those methods tended to overpredict stiffened column strength for slender columns in which local buckling supposedly was not a factor but clearly was having an effect. For design office use, the iterative approach utilizing effective section properties was unsuitable. The key features of the new method were the use of gross section properties to determine slenderness and buckling stress, and the determination of effective area at this buckling stress rather than at F_y as used previously. The method predicted the De Wolf et al. (1974) test results with remarkable accuracy. The column curve developed by Lind for the 1974 CSA Standard was retained. This is geometrically similar to the AISC column curve, in which a Johnson parabola extends from F_y, tangential to an Euler curve lowered to 0.833 (equivalent to 1.6/1.92) of the standard value. This reflects the unlikely event of real columns ever reaching the Euler load of a perfect column. Other important aspects were the adoption of the effective-width approach for all sections, stiffened and unstiffened, and, based on Trestain's (1982) evaluation of available test data, the use of a lower resistance factor for single symmetrical sections ($\emptyset = 0.75$ versus 0.90). In recognition of the dramatic reduction in stiffness of unstiffened channels as the effective flange width is reduced, the effective stress is limited to the unstiffened flange local buckling stress.

The 1984 CSA method was incorporated into the unified approach proposed by Peköz (1986) for the 1986 AISI Specification. Further evaluation by Peköz using more recent test data has lead to further refinement: in particular, the use of the effective centroid rather than the gross centroid as the origin for determination of eccentricity of load.

The 1986 AISI Specification uses an effective cross-sectional area, as follows:

$$P_u = A_e F_u \tag{13.29}$$

where

A_e = effective area at stress F_u
F_u = the ultimate stress determined by the following equations:

$$F_u = F_y(1 - F_y/4\sigma_e) \qquad \text{for } F_u > F_y/2 \tag{13.30}$$

$$F_u = \sigma_e \qquad \text{for } F_u \leq F_y/2 \tag{13.31}$$

where σ_e is the elastic flexural buckling stress (Eq. 13.26) or the elastic flexural-torsional buckling stress. Equation 13.30 is identical to Eq. 13.25 for the case of flexural buckling. The form factor approach is used in the RMI specification (1979) for design of columns with perforations. The form factor Q is determined by stub column tests.

The treatment of the effect of local buckling of stiffened elements on column strength in the Aluminum Association Specifications (1982) is based on the following empirical interaction formula, which has been found to agree well with the results of tests on aluminum members (Sharp, 1970):

$$\sigma_u = \sigma_e \left(\frac{\sigma_c}{\sigma_e} \right)^{2/3} \tag{13.32}$$

where σ_u is the average stress at failure, σ_c the local buckling stress, and σ_e the Euler buckling stress.

13.3.4 Wall Studs

Thin-walled sections are frequently used as wall studs. Wall studs carry axial forces as well as support the wall sheathing. For such a column, the buckling strength in the plane parallel to the sheathing may be increased greatly by the sheathing it supports. This is quite beneficial, since, in general, this direction is the weaker direction for the wall stud. Full advantage of the bracing action can be taken provided that the following criteria are satisfied:

1. The strength and rigidity of the wall material should be adequate to prevent excessive deflection or buckling of the wall studs.
2. The spacing between fasteners joining each stud to the sheathing should be such that the stud will not buckle between the attachments.
3. The fasteners should be capable of developing the required forces and rigidity. A further description of bracing requirements is given in Section 13.5.

13.3.5 Flexural-Torsional Buckling

Concentrically loaded columns can buckle by (1) flexure about one of the principal axes, (2) twisting about the shear center (torsional buckling), or (3) a combination of both flexure and twisting, called flexural-torsional buckling. Torsional buckling is a possible failure mode for point symmetric sections. Flexural-torsional buckling must be checked for open sections that are singly symmetric and for sections that have no symmetry. Open sections that are doubly symmetric or point symmetric are not subject to flexural-torsional buckling because their shear center and centroid coincide. Closed sections also are immune to flexural-torsional buckling.

One can explain the nature of flexural-torsional buckling with the aid of Fig. 13.5. At buckling, the axial load can be visualized to have a lateral component (qdz) as a consequence of the column deflection. The torsional moment of this lateral component about the shear center of the open section shown in the figure causes twisting of the column. The degree of interaction between the torsional and flexural deformations determines the amount of reduction of the buckling load in comparison to the flexural buckling load. Therefore, as the distance between the shear center and the point of application of the axial load increases, the twisting tendency increases and therefore the flexural-torsional buckling load decreases. Flexural-torsional buckling can be a critical mode of failure for thin-walled open sections because of their low torsional rigidity. The theory of elastic flexural-torsional instability is well developed (Goodier, 1942; Timoshenko and Gere, 1961; Vlasov, 1959; Galambos, 1968). Flexural-torsional buckling of singly symmetric thin-walled open sections

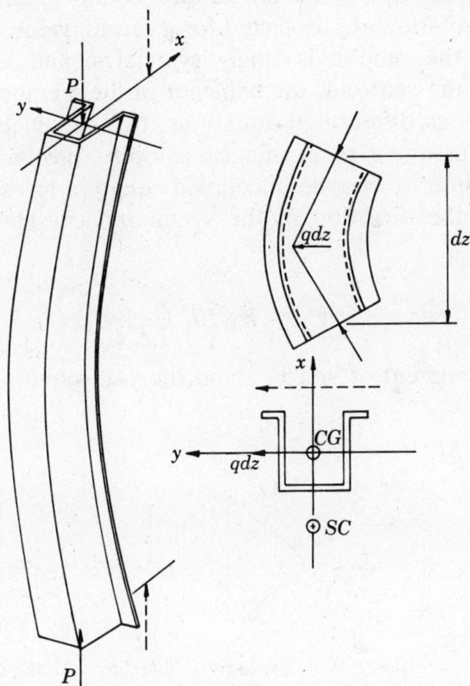

Fig. 13.5 Forces causing flexural-torsional buckling.

under concentric and eccentric loading also has been studied in detail, and design aids have been devised (Chajes and Winter, 1965; Klöppel and Schardt, 1958; Pfluger, 1961; Chilver, 1967; Peköz, 1969; Peköz and Winter, 1969). For the inelastic domain, approximate approaches have been developed and adequately substantiated by tests (Chajes et al., 1966). The AISI Specification (1980 and 1986) for carbon steel contains flexural-torsional buckling provisions based on the work of Chajes and Winter (1965), Peköz (1969) and Peköz and Winter (1969).

Equations for Elastic Flexural-Torsional Buckling. Differential equations of equilibrium for the general case of biaxial eccentricities have been solved by Vlasov (1959), Culver (1966), Dabrowski (1961), Thürlimann (1953), Prawel and Lee (1964), and Peköz and Winter (1969) using different procedures of solution. If the section is singly symmetric, such as the sections shown in Fig. 13.6, and is acted on by an axial load not in the plane of symmetry; or if the section is not symmetric, the solution of the differential equations indicates that as the axial load increases the member continuously twists and deflects biaxially. The principal axes, twist angle ϕ and deflections u and v are shown in Fig. 13.7. Analogous to small deflection flexural beam-column theory, infinite deflections and rotation are predicted for a certain value of the axial load.

However, if the section is singly symmetric and the axial load is applied through the centroid, the behavior of the member is described by three homogeneous differential equations, two of which are coupled. If the member is assumed to be hinged at both ends, namely, $u'' = v'' = \phi'' = 0$, the solution of the one uncoupled equation gives the critical load for buckling in the direction of the symmetry axis (taken here as the x-axis):

$$P_{ye} = K_{11} E I_y \frac{\pi^2}{L^2} \tag{13.33}$$

where I_y is the moment of inertia about the y-axis and L is the length of

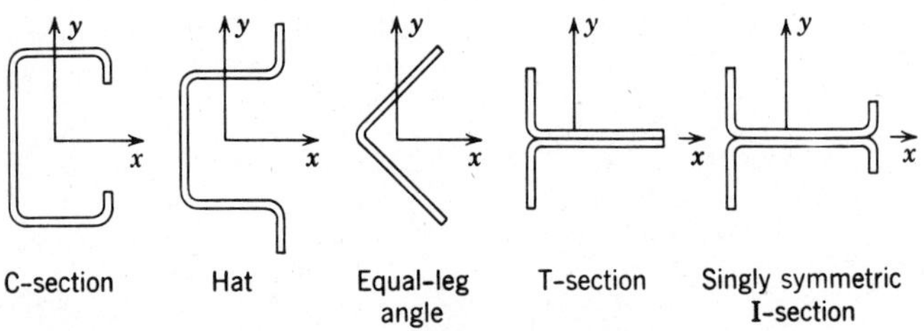

Fig. 13.6 Some singly symmetric sections and coordinate axis orientation.

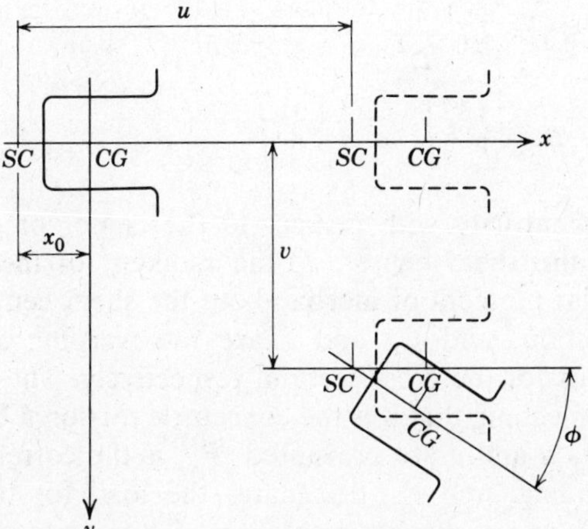

Fig. 13.7 Principal axes and deflection components.

the column. K_{11} and other K values determined by the Galerkin method for various boundary conditions are given by Peköz (1969). The discussion here will be limited to hinged ends.

The two coupled equations describing deformations v and ϕ result in a single buckling load P_{TF} for the flexural-torsional mode. The same buckling mode also occurs in the more general case of the load acting eccentrically in the plane of symmetry. Then the member continuously deflects as a beam-column in the plane of symmetry (x-direction), but is subject to flexural-torsional buckling out of this plane under load P_{TF} given in this case by Eq. 13.34. (The solution for a concentric load is obtained by setting $e_x = 0$ in this equation.)

$$P_{TF} = \frac{(P'_{\phi e} + \alpha P_{xe}) \pm \sqrt{(P'_{\phi e} + \alpha P_{xe})^2 - 4\gamma P_{xe} P'_{\phi e}}}{2\gamma} \tag{13.34}$$

where

$$\alpha = 1 + e_x \beta_2 \frac{A}{I_0} \tag{13.35}$$

$$\gamma = \frac{A}{I_0}(x_0 - e_x)^2 + e_x \beta_2 \frac{A}{I_0} + 1 \tag{13.36}$$

$$P'_{\phi e} = P_{\phi e} \alpha \tag{13.37}$$

$$P_{\phi e} = \frac{A}{I_0}\left(EC_w \frac{\pi^2}{L^2} + GJ\right) \tag{13.38}$$

$$P_{xe} = EI_x \frac{\pi^2}{L^2} \tag{13.39}$$

$$\beta_2 = \frac{1}{I_y}\left(\int_A x^3\, dA + \int_A xy^2\, dA\right) - 2x_0 \tag{13.40}$$

and e_x is the eccentricity with respect to the center of gravity, x_0 the x-coordinate of the shear center, I_x the moment of inertia about the x-axis, I_0 the polar moment of inertia about the shear center, A the area of the cross section, and C_w and J are the warping and St.-Venant torsional constants for the cross section, respectively. The parameter $P_{\phi e}$ has the physical meaning that it is the concentric torsional buckling load if the displacements u and v are prevented, $P'_{\phi e}$ is the corresponding value for eccentric loading, and P_{xe} designates the load for buckling in the direction of the y-axis if displacements ϕ and u are prevented.

For discussion purposes infinite elasticity is assumed in this section. With this assumption P_{ye} is the limiting eccentric axial force for member capacity in the plane of symmetry, and the following types of behavior can be defined depending on relative magnitudes of P_{ye} and P_{TF}. If P_{ye} is less than P_{TF}, the member will behave purely as a flexural beam-column without twisting, and deflection in the plane of symmetry will increase gradually. At an axial load equal to P_{ye}, infinite deflection without twisting is indicated by the solution. For concentric loading, if P_{ye} is less than P_{TF} the member will buckle flexurally about the y-axis. On the other hand, for concentric loading or eccentric axial loading in the plane of symmetry, if P_{ye} is greater than P_{TF}, buckling will occur by the lateral deflection (v) and twisting (ϕ) at an axial load equal to P_{TF}, the flexural-torsional buckling load.

When there is no axial load, the Galerkin method solution (Peköz, 1969) of the general differential equations of equilibrium gives the following expression for the critical moments, M_{CR}:

$$M_{CR} = -\frac{P_{xe}\beta_2}{2}\left(1 \pm \sqrt{1 + \frac{4I_0R}{\beta_2^2 A}}\right) \tag{13.41}$$

where $R = P'_{\phi e}/P_{xe}$. This equation yields one positive and one negative value of M_{CR} which will be denoted M_{CR+} and M_{CR-}, respectively.

Design Simplifications. The preceding section has been restricted to the elastic flexural-torsional stability of single symmetric thin-walled sections. In this section the general behavior of such sections is discussed on the basis of an assumed elastic-plastic stress-strain diagram.

In the precritical stages, the member does not twist; therefore, the maximum fiber stresses at subcritical loads can be found simply from the secant formula or any other appropriate beam-column formula. It is assumed that for the thin-walled sections in question, the attainment of the yield stresses represents the limit of load-carrying capacity; that is, the plastic reserve capacity, if any, is negligible. This point has been verified experimentally (Peköz, 1969; Peköz and Winter, 1969). Therefore, elastic flexural torsional buckling is a possible mode of failure only if the axial load P_{yd} that causes incipient yielding (e.g., as predicted by the secant formula) is larger than P_{TF}.

Extensive numerical studies were carried out on a variety of singly symmetric open sections and are reported by Peköz (1969). Figure 13.8 is a typical sample of the plots given in that reference that illustrates the complex behavior of such compression members.

For positive eccentricities, numerical studies indicate that both yielding and instability need to be considered. The following expression is shown in the reference to give very satisfactory results:

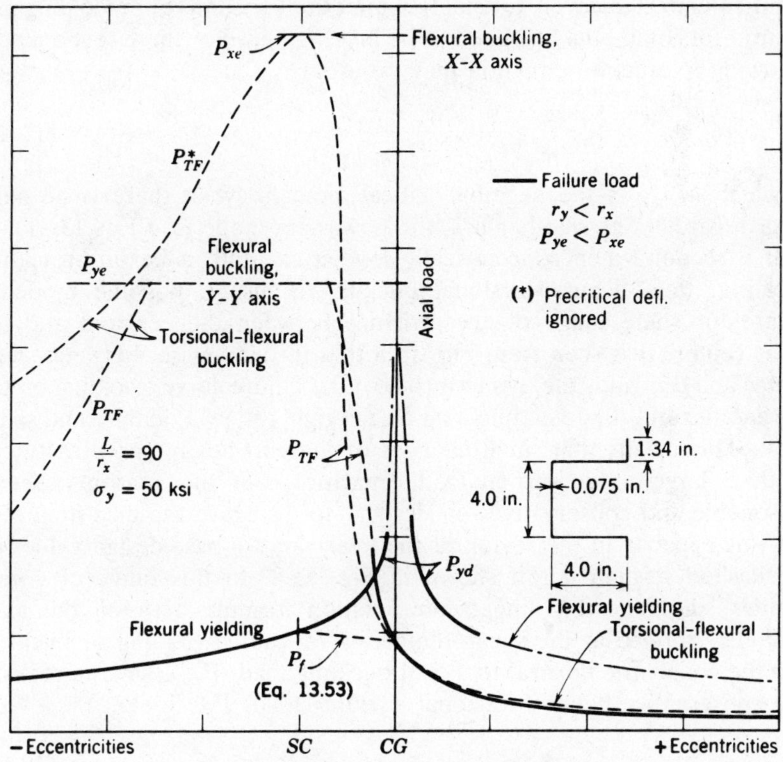

Fig. 13.8 Failure modes and loads (Peköz, 1969).

$$\frac{1}{P_{TFO}} + \frac{e_x}{M_{CR+}} = \frac{1}{P_{TF}} \tag{13.42}$$

in which P_{TF} is the flexural-torsional buckling load for an eccentricity e_x; P_{TFO} the flexural-torsional buckling load for concentric loading, regardless of whether it is the governing mode; and M_{CR+} the positive critical moment when there is no axial load, regardless of whether it is the governing mode (see Eq. 13.41). With the aid of charts given by Peköz (1969) for computing P_{TFO} and M_{CR+}, this equation is much more convenient to use than Eq. 13.35.

For negative eccentricities greater than x_0—that is, if the point of application of the axial load is on the side of the shear center opposite from the center of gravity—numerical studies on hat, channel, lipped channel, angle, and lipped angle sections of typical dimensions and yield stresses below 50 ksi indicate that flexural-torsional buckling is not a governing mode of failure. For such eccentricities, these members fail by yielding after deflecting in the direction of the symmetry axis as a beam-column. However, for singly symmetric I-sections both yielding and flexural-torsional buckling need to be investigated. For these sections, the following interaction equation may be used:

$$\frac{1}{P_{xe}} + \frac{e_x - x_0}{M_{CR-}} = \frac{1}{P_{TF}} \tag{13.43}$$

in which M_{CR-} is the negative critical moment when there is no axial load, regardless of whether it is the governing mode (see Eq. 13.41).

If a section when concentrically loaded can fail in flexural-torsional buckling, then flexural-torsional buckling is also a possible mode of failure for some range of eccentricities between the centroid and the shear center. It is seen from Fig. 13.8 that in this region, between shear center and centroid, the two branches of the failure curve (yielding on the left and flexural-torsional buckling on the right) show a definite and sharp peak. This means that small changes or inaccuracies in eccentricity can produce large reductions in load capacities. For this reason it seems reasonable and conservative—in design—to disregard the uncertain high carrying capacity in the region of the peak, and to base design values on the dashed straight cutoff shown in Fig. 13.8. In this range of eccentricities, the following linear interpolation formula between the axial load, P_3, applied at the shear center, which causes yielding or buckling and the concentric flexural-torsional buckling load, P_{TFO}, gives a realistic and conservative flexural-torsional buckling load, P_F.

$$P_F = P_{TFO} + \frac{e_x}{x_0}(P_S - P_{TFO}) \tag{13.44}$$

For singly symmetric I-sections, P_s is the smaller of the yield load P_{yd} or the buckling load P_{xe}, whereas for the other open sections, only yielding need be considered, as explained previously.

Additional Design Considerations. In addition to the points discussed in the preceding section, the following need to be considered in the design of thin-walled members to resist flexural torsional buckling.

First is the inelastic stability behavior for members of relatively low slenderness ratios. Chajes et al. (1966) studied this problem and reported that an expression similar to the CRC column curve is satisfactory for concentric flexural-torsional buckling. The AISI Specifications reflect this approach for both concentric and eccentric loading.

Second is the frequent case of unequal eccentricities at opposite ends of the member. Peköz (1969) presents the results of an extensive study of this subject and makes the conclusion that application of a modification factor, C_{TF}, to the second term of Eq. 13.42 is quite accurate. The value C_{TF} is the same as C_m discussed in Chapter 8, except that it does not have 0.4 as its lower limit.

Third, the influence of precritical beam-column deflections on the flexural-torsional buckling load is an important consideration. Again, on the basis of an analytical and experimental treatment of the subject, Peköz (1969) recommends the use of an amplification factor $1/(1 - P/P_{ye})$ for the moments.

Fourth, is the wandering centroid problem where centrally loaded, singly symmetric columns become beam-columns upon local buckling and the shifting of the neutral axis. In an extensive statistical study, Peköz (1986) established good correlation with test results if a concentrically loaded column is defined as a member loaded through its effective centroid. The effective centroid is calculated at the reduced column stress F_u from Eq. 13.30 or 13.31. This approach is used in the 1986 AISI Specification. The CSA (1984) Standard defines a concentrically loaded column as a member loaded through its fully effective centroid. The capacity of singly symmetric sections is reduced by the ratio 0.75/0.9 to account for the wandering centroid effect.

Fifth, is the behavior of biaxially loaded beam-columns. The 1980 AISI Specification did not permit the calculation of singly symmetric beam-columns bending about the symmetry axis. The designer had to resort to tests. Based primarily on the work of Peköz (1986), Loh (1985), and Mulligan and Peköz (1983), the 1986 AISI Specification determines the capacity of biaxially loaded open sections using an interaction equation with eccentricities measured from the effective centroid. The method is similar to procedures adopted previously by the RMI (1969) Standard and followed in the CSA (1984) Standard. Lipped channel sections used in the endwalls of metal buildings and the columns in industrial storage racks are a few examples of members subjected to such loads.

13.4 DIAPHRAGM ACTION OF THIN-WALLED PANELS

Thin-walled metal panels are often used as wall cladding, roof decking, and floor decking, where their primary structural function is to carry loads acting normal to their surface. Properly designed and interconnected metal roof, wall, and floor systems are also capable of resisting shear forces in their own planes, referred to as diaphragm action. Considerable progress has been made in recent years in predicting the shear strength and stiffness of a diaphragm composed of such panels. Publications by Bryan and Davies (1981), the Steel Deck Institute (1981), the European Convention for Constructional Steelwork (1977), Atrek and Nilson (1980), Easley (1977), and Ha et al. (1979) are among the recent additions to the literature on the subject. The strength and stiffness of a diaphragm also can be determined from the load-deflection relationship obtained from a shear diaphragm test as described by AISI (1967). Procedures have been established for making use of this shear resistance in designing buildings to resist forces caused by wind, seismic action, and other lateral loads (AISI, 1967; Bryan and Davies, 1981). The shear strength and rigidity of thin-walled panels can be utilized in folded plate structures (Nilson, 1960), hyperbolic paraboloids (Gergely et al., 1971), and other shell roof structures (Bresler, 1968). Procedures have been developed that recognize the ability of diaphragms assembled from such panels to transfer load from a heavily loaded frame to less heavily loaded adjacent frames in a single-story structure, thus reducing the required maximum frame size (Bryan and Davies, 1981; Luttrell, 1967). Research has also assessed the influence of thin-walled metal cladding on the behavior of multistory buildings (Miller, 1972; El-Dakhakhni and Daniels, 1973). In addition, theory and test results both have shown that the shear strength and rigidity of properly connected diaphragms can be very effective as bracing for individual beams and columns.

13.5 BRACING REQUIREMENTS

13.5.1 Tyes of Bracing

Structural bracing may be divided into two general types, according to its function: (1) bracing provided to resist secondary loads on structures, such as wind bracing, and (2) bracing provided to increase the strength of individual structural members by preventing them from deforming in their weakest direction (Winter, 1958). In the latter instance, there are again two different cases: (1) bracing applied to prevent buckling and

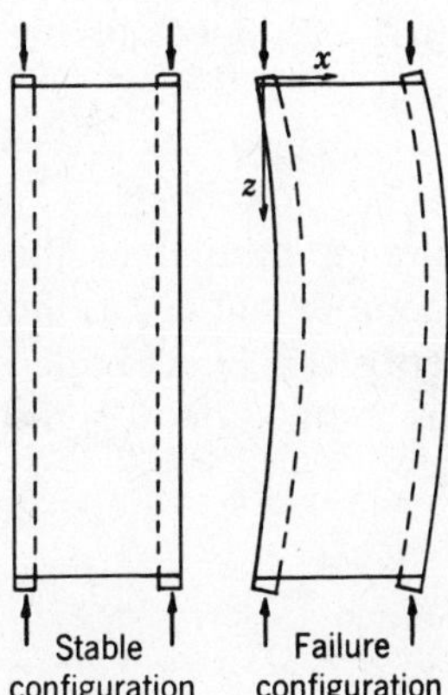

Fig. 13.9 Members with diaphragm bracing.

thereby increase the unstable strength of the member, and (2) bracing applied to counteract stable but detrimental types of deformation. As examples of the latter, channel- and Z-shaped beams loaded in the plane of the web twist or deflect laterally, with consequent loss of strength unless they are properly braced.

For bracing against buckling to be effective in an actual situation, it must possess not only the requisite strength but also a definite minimum rigidity. However, the required strength cannot be computed uniquely, except on the basis of assumed imperfections of shape and/or loading of the member to be braced (Winter, 1958). Bracing may be "continuous," such as that provided by wall panels, roof decking, or floor systems, or it may be noncontinuous or "discrete," such as cross-bracing. For discrete bracing the spacing of the braced points also is important. Finally, bracing may also be distinguished according to its behavior: (1) that which provides restraint through resistance to axial deformation, as does cross-bracing, and (2) that which provides restraint through resistance to shear deformation, as do diaphragms (Fig. 13.9).

13.5.2 Diaphragm-Braced Columns

In the elastic range the predicted weak-axis buckling load of an ideal axially loaded I-section column or wall stud with directly attached symmetrial diaphragm bracing (Larson, 1960; Pincus and Fisher, 1966) is determined as

$$P = P_{yy} + Q \tag{13.45}$$

where P_{yy} is the weak-axis buckling load of the unbraced column and Q is the shear rigidity of the diaphragm contributing to the support of the column. The shear rigidity can be expressed as

$$Q = A_d G_{\text{eff}} \tag{13.46}$$

where A_d is the cross-sectional area of the diaphragm normal to the column axis and contributing to the support of the member and G_{eff} is the effective shear modulus of the diaphragm. The theory can be extended into the inelastic range using the tangent-modulus expression for the weak-axis buckling load (Errera et al., 1967). Thus

$$P = \frac{\pi^2 E_t^* I_y}{L^2} + Q \tag{13.47}$$

in which E_t^* is the tangent modulus associated with the increased buckling load, P. It is emphasized that Eqs. 13.45 and 13.47 predict the increased weak-axis buckling loads of ideal members; only the stiffness of the bracing is considered. For real members, initial imperfections and the strength of the bracing must also be taken into consideration.

The theory and experimental verification have been extended to the case of diaphragm bracing connected to girts, which, in turn, are connected to columns (Apparao et al., 1969; Massicotte et al., 1985), and to diaphragm-braced columns of singly-symmetric and point-symmetric sections, such as channels and zees used as wall studs (Simaan, 1973).

Errera and Apparao (1976b) have presented a procedure for the design of I-shaped columns with diaphragm bracing provided by cold-formed steel panels. The procedure covers axially loaded I-section columns with shear diaphragms on both flanges or on one flange only. It also covers the case of axially loaded I-section columns braced by girts, which in turn are braced by diaphragms, as usually encountered in most metal building construction.

The AISI Specification bases the bracing requirements for load bearing wall studs on the diaphragm behavior of the sheathing. Closed-form expressions are given if identical sheathing is attached to both flanges of the stud. For other cases, reference is made to the theoretical investigations of these types of stability problems as documented by Simaan and Peköz (1976). A recent study by Zhang and Peköz (1982) explored the case of wall studs subjected to axial and lateral loads.

13.5.3 Diaphragm-Braced Beams

The same type of diaphragm action is useful also in counteracting lateral-torsional buckling of beams. For ideal I-section beams braced directly by diaphragms on the compression flange, the critical lateral-buckling moment can be estimated as (Errera et al., 1967)

$$M_{cr} = M_o + 2Qe \tag{13.48}$$

where M_0 is the lateral-buckling moment of the unbraced beam, e the distance between the center of gravity of the beam cross section and the plane of the diaphragm, and Q the shear rigidity of the diaphragm contributing to the support of the member, as defined previously. Again, it is emphasized that Eq. 13.48 predicts the buckling load of an ideal member. For real members the initial imperfections and the strength of the bracing must be taken into consideration. Design procedures for real beams with diaphragm bracing have been presented by Nethercot and Trahair (1975) and Errera and Apparao (1976a).

Because of their asymmetry channels and Z-sections used as beams and loaded in the plane of the web will continuously twist or deflect laterally with consequent loss of strength unless they are properly braced. When both flanges of such beams are connected to deck or sheathing material in such a manner as to restrain lateral deflection effectively, no further bracing is required. For other cases the AISI Specification defines the maximum spacing and the required strength of extensional braces to prevent such deformations. The spacing is such that rotation between bracing points will be small enough to be unobjectionable, and the additional stresses associated with the rotation will not reduce the carrying capacity of the member. The required strength of the bracing is based on statics, using an approximate method of analysis which was checked against test results (Winter et al., 1949; Zetlin and Winter, 1955). Here again, more recent research (Celebi et al., 1971) has led to procedures for analyzing the strength and deflections of diaphragm-braced channels and Z-sections, and current efforts are aimed at developing suitable design procedures for both gravity and uplift loading (Peköz and Soroushian, 1982).

13.6 STAINLESS STEEL STRUCTURAL MEMBERS

Cold-formed stainless steel sections have been widely used architecturally in buildings because of their superior corrosion resistance, ease of maintenance, and pleasing appearance. Typical applications include column covers, curtain wall panels, mullions, door and window framing, roofing and siding, stairs, elevators and escalators, flagpoles, signs, and many others. Since 1968, their use for structural load-carrying purposes has been increased due to the availability of an AISI design specification (AISI, 1974).

The first edition of the *Specification for the Design of Cold-Formed Stainless Steel Structural Members* was issued by American Iron and Steel Institute in 1968 on the basis of the extensive research conducted by

Johnson and Winter (1966) at Cornell University. This Specification was revised in 1974 to reflect the results of additional research (Wang and Errera, 1971) and the improved knowledge of material properties and structural applications. The current Specifications (AISI, 1974) contains design information on annealed and cold-rolled grades of sheet and strip stainless steels, types 201, 202, 301, 304, and 316. The main reason for having a different specification for stainless steel structural members is because stainless steel has the following differing characteristics as compared with carbon steel:

1. Anisotropy.
2. Nonlinear stress-strain relationship.
3. Low proportional limit.
4. Pronounced response to cold work.

Figure 13.10 shows the stress-strain curves of annealed, half-hard, and full-hard stainless steels.

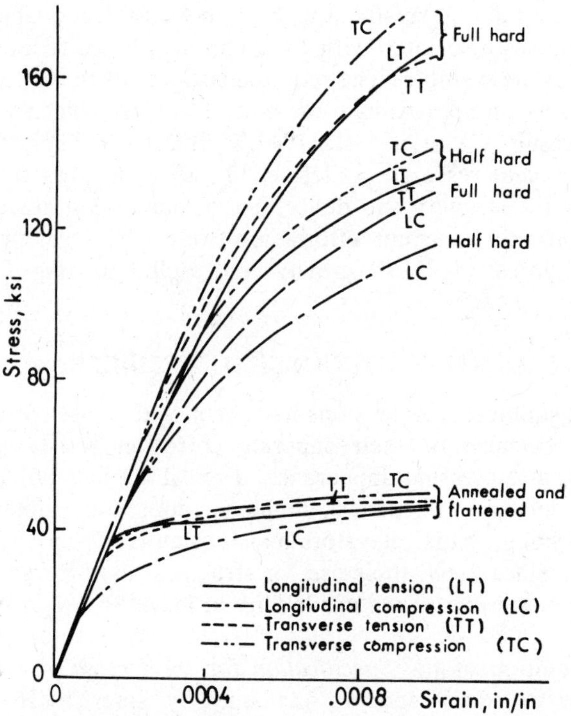

Fig. 13.10 Stress-strain curves of annealed half-hard, and full-hard stainless steels (Wang, 1969; Johnson and Kelsen, 1969).

Because of the differences in mechanical properties and structural uses between stainless steel and carbon steel, the AISI specification for stainless steel design contains modified design provisions for (1) local buckling of flat elements, (2) w/t limitations, (3) deflection calculations, (4) service stress limitations, (5) lateral buckling of beams, (6) column buckling, and (7) connections. In general, the factors of safety used for the design of stainless steels are somewhat larger than those used for carbon steel. AISI (1974) contains the Specification, commentary, illustrative examples, design tables, and design charts.

13.7 ALUMINUM MEMBERS

13.7.1 Effective Widths

Postbuckling strengths of thin aluminum plate elements are generally based on the von Kármán concept that the width, b_e, for which the elastic buckling stress (Eq. 4.1)

$$\sigma_{cr} = \frac{\pi^2 Ek}{12(1-\nu^2)(b/t)^2} = \frac{\pi^2 E}{(mb/t)^2} \tag{13.49}$$

is equal to the yield stress gives a limiting capacity which remains constant for all other widths. Thus, from Eq. 13.49,

$$b_e = \frac{\pi t}{m}\sqrt{\frac{E}{\sigma_y}} \tag{13.50}$$

For a simply supported plate the buckling coefficient $k = 4$, so $m = 1.63$ and $b_e = 1.93t\sqrt{E/\sigma_y}$.

This is not entirely consistent with the treatment for buckling of thin walls, which for the elastic-plastic range uses the equivalent slenderness ratio, mb/t, in an expression of the type

$$\sigma_c = (B_p - D_p mb/t) \tag{13.51}$$

This expression represents the true limiting stress (Jombock and Clark, 1968), and the maximum load carried by a plate element in the elastic-plastic range is

$$P = \sigma_c bt = (B_p - D_p mb/t)bt \tag{13.52}$$

The effective width, defined as that width which when multiplied by the yield stress and the thickness, gives the failing load for the element, then becomes

$$b_e = (b/\sigma_y)(B_p - D_p mb/t) \tag{13.53}$$

As b increases the highest load that can be carried occurs when

$$b/t = B_p/2D_pm \tag{13.54}$$

giving

$$P = B_p^2t^2/4D_pm \tag{13.55}$$

This load capacity remains reasonably constant for all higher values of b/t. Jombock and Clark (1968) provide values for B_p and D_p which can be represented as

$$B_p = \sigma_y\alpha^2 \tag{13.56}$$

$$D_p = \sigma_y k\alpha/g \tag{13.57}$$

where

$g = (E/\sigma_y)^{1/2}$
$\alpha = (1 + 2/g^{2/3})^{1/2}$ for fully heat-treated alloys
$\alpha = (1 + 3/g^{2/3})^{1/2}$ for other alloys
$k = 0.1$ for fully heat-treated alloys
$k = 0.12$ for other alloys

The effective width is then

$$b_e = b\alpha(\alpha - kmb/t) \tag{13.58}$$

with a maximum value of

$$b_e = \alpha^3gt/4mk \tag{13.59}$$

Jombock and Clark (1968), on the basis of limiting strain, provided a theoretical foundation for the model above, which has been adopted in North American codes (AA, 1982; CSA, 1984). This model gives a continuous transition from compact to thin-walled elements without discontinuities or changes in the form of the design expression.

Postbuckling behavior of elements supported on the long, unloaded edges, to which the model above applies, differs from that of outstanding, flange-type elements, in that, for the latter, initial elastic buckling precipitates the collapse of single unsymmetrical elements, while there may be some reserve capacity in symmetrical sections which can be represented by the treatment for edge-supported elements using an appropriate value of the coefficient m, usually 5.

The limiting stress on unstiffened elements of an unsymmetrical section such as a channel or Z, is obtained using the expression

$$mb/t < C \qquad \sigma = B - D(mb/t) \tag{13.60}$$

$$mb/t > C \qquad \sigma = \pi^2E/(mb/t)^2 \tag{13.61}$$

where the value of m lies between 3 and 5, depending on the degree of edge restraint, and C is the slenderness parameter separating elastic and inelastic buckling.

In symmetrical sections, such as I- and double-channel shapes the effective width will be given by Eqs. 13.58 and 13.59 using $m = 5$, unless local buckling can precipitate overall flexural or lateral-torsional buckling, which will occur when the critical stresses for the different buckling modes are close in value.

13.7.2 Effective Section at Service Loads

A margin of over 1.5 between service loads and the ultimate load is usual; thus the extent of any local buckling in service will be small and confined to zones of maximum moment. For this reason, the influence of local buckling on deflections under service loads has been neglected in some codes. If a calculation is to be made, the difficulty of computing deflections with a varying and initially unknown effective moment of inertia is usually resolved by assuming that the effective section at the point of highest moment applies throughout. The effective width b_e of the elements comprising the effective section are obtained using the actual width b and the ratio of the critical stress σ_c to the applied stress σ computed on the basis of a fully effective section, thus:

$$b_e = b(\sigma_c/\sigma)^{1/2} \tag{13.62}$$

13.7.3 Torsional Buckling

Angles. A single angle may fail by flexure or torsional buckling; only by a special proportion of heavy root bulbs and very thin legs can local buckling be made to occur. To optimize shapes, bulbs and root fillets are added to increase the torsional rigidity such that the equivalent slenderness ratio for torsional buckling is around 60. Because of the interaction of torsion with flexure about the stronger axis, it is not effective to design sections of equal inertia about the two principal axes, and the optimum is an equal-leg right-angle section for both plain and bulb shapes. This is also true of double angles, designed to balance torsional and flexural buckling, in which case equal-leg angles are again very close to the optimum.

Eccentrically Loaded Columns. Unsymmetrical open sections loaded axially fail in combined torsion and flexure. Should they be loaded through the shear center, the modes are uncoupled and torsional buckling can be eliminated. Use has been made of this in T sections for diagonals which, when bolted through the flanges, are loaded through the shear center. This allows much thinner sections to be used with a considerable increase in efficiency in spite of the moment due to the eccentricity. The optimum form is a lipped shape to control local buckling of the flanges.

Single angles loaded through one leg fail by lateral-torsional buckling in the manner of a beam-column (Marsh, 1969) and the design procedure adopted by CSA (1983) and ASCE (1972) treat this interaction by using an effective slenderness ratio:

$$(KL/r)_{\text{eff}} = [(5b/t)^2 + (KL/r_v)^2]^{1/2} \tag{13.63}$$

where

$5b/t$ = slenderness ratio for torsional buckling of angles
KL/r_v = slenderness ratio for flexural buckling

Postbuckling Strength. In general, the critical stress for a column failing by torsional buckling represents the maximum capacity of the member. This is always true of pin-ended unsymmetrical sections, as the application of the load through the centroid requires a uniform stress in the section for equilibrium. Should the column be loaded by fixed platens, the axis of load application can shift as the member twists, causing an increase in stress toward the shear center (Smith, 1955). Symmetrical sections, such as a cruciform, even when pin-ended, can accept a higher stress at the center as the member twists, giving a higher critical load than that obtained for a uniform stress.

References

Aluminum Association (1982), *Specifications for Aluminum Structures*, AA, Washington, D.C.

American Institute of Steel Construction (1978), *Specification for the Design, Fabrication and Erection of Structural Steel for Buildings*, AISC, Chicago.

American Iron and Steel Institute (1967), *Design of Light Gage Steel Diaphragms*, AISI, Washington, D.C.

American Iron and Steel Institute (1974), *Stainless Steel Cold-Formed Structural Design Manual*, AISI, Washington, D.C.

American Iron and Steel Institute (1980, 1986), *Specification for the Design of Cold-Formed Steel Structural Members*, AISI, Washington, D.C.

American Iron and Steel Institute (1986), *Cold-Formed Steel Design Manual*, AISI, Washington, D.C.

Apparao, T. V. S. R., Errera, S. J., and Fisher, G. P. (1969), "Columns Braced by Girts and a Diaphragm," *Proc. Am. Soc. Civ. Eng.*, Vol. 95, No. ST5, pp. 965–990.

American Society of Civil Engineers, Task Committee on Light-Weigh Alloys (1972), "Guide for the Design of Aluminum Transmission Towers," *ASCE J. Struct. Div.*, Vol. 98, No. ST12, pp. 2785–2804.

Atrek, E., and Nilson, A. H. (1980), "Non-linear Analysis of Cold-Formed Steel Shear Diaphragms," *ASCE J. Struct. Div.*, Vol. 106, No. ST3, pp. 693–710.

Bijlaard, P. P., and Fisher, G. P. (1952a), "Interaction of Column and Local Buckling in Compression Members," *NACA Tech. Note No. 2640*, March.

Bijlaard, P. P., and Fisher, G. P. (1952b), "Column Strength of H-Sections and Square Tubes in Post-Buckling Range of Component Plates," *NACA Tech. Note No. 2994*, August.

Braham, M., Rondal, J., and Massonnet, C. E. (1980), "Large Size Buckling Tests on Steel Columns with Thin-Walled Rectangular Hollow Sections," in *Thin-Walled Structures* (ed. J. R. Rhodes and A. C. Walker), Granada, London.

Bresler, B., Lin, T. Y., and Scalzi, J. B. (1968), *Design of Steel Structures*, Wiley, New York.

Bryan, E. R., and Davies, J. M. (1981), *Steel Diaphragm Roof Decks*, Wiley, New York.

Canadian Standards Association (1983), *Strength Design in Aluminum*, CAN3-S157-M83, CSA, Rexdale, Ontario.

Canadian Standards Association (1984), *Cold-Formed Steel Structural Members*, CAN3-S136-M84, CSA, Rexdale, Ontario.

Celebi, N., Peköz, T., and Winter, G. (1971), "Behavior of Channel and Z-Section Beams Braced by Diaphragms," *Proc. 1st Spec. Conf. Cold-Formed Steel Struct.*, University of Missouri at Rolla, August.

Chajes, A., and Winter, G. (1965), "Torsional-Flexural Buckling of Thin-Walled Members," *ASCE J. Struct. Div.*, Vol. 91, No. ST4, pp. 103–124.

Chajes, A., Fang, P. J., and Winter, G. (1966), "Torsional-Flexural Buckling, Elastic and Inelastic, of Cold Formed Thin-Walled Columns," *Cornell Eng. Res. Bull.*, No. 66-1.

Chilver, A. H. (1967), *Thin-Walled Structures*, Wiley, New York.

Culver, C. G. (1966), "Exact Solution of the Biaxial Bending Equations," *ASCE J. Struct. Div.*, Vol. 92, No. ST2, pp. 63–84.

Dabrowski, R. (1961), "Dünnwändige Stäbe unter zweiachsig aussermittigem Druck," *Stahlbau*, Vol. 30, p. 360.

Dat, D. T., and Peköz, T. (1980), "Strength of Cold-Formed Steel Columns," *Report No. 80-4*, Cornell University, Dept. of Struct. Eng., Ithaca, N.Y. February.

DeWolf, J., Peköz, T., and Winter, G. (1974), "Local and Overall Buckling of Cold-Formed Steel Members," *ASCE J. Struct. Div.*, Vol. 100, No. ST10, pp. 2017–2036.

DeWolf, J., Peköz, T., and Winter, G. (1976), Closure to "Local and Overall Buckling of Cold-Formed Steel Members," *ASCE J. Struct. Div.*, Vol. 102, No. ST2, pp. 451–454.

Douty, R. T. (1962), "A Design Approach to the Strength of Laterally Unbraced Compression Flanges," *Cornell Univ. Bull.*, No. 37, April.

Easley, J. T. (1977), "Strength and Stiffness of Corrugated Metal Shear Diaphragms," *ASCE J. Struct. Div.*, Vol. 103, No. ST1, pp. 169–180.

El-Dakhakhni, W. M., and Daniels, J. H. (1973), "Frame-Floor-Wall System Interaction in Buildings," *Rep. No. 376.2*, Lehigh University, Fritz Eng. Lab., Bethlehem, Pa.

Errera, S. J., and Apparao, T. V. S. R. (1976a), "Design of I-Shaped Beams with Diaphragm Bracing," *ASCE J. Struct. Div.*, Vol. 102, No. ST4, pp. 769–783.

Errera, S. J., and Apparao, T. V. S. R. (1976b), "Design of I-Shaped Columns with Diaphragm Bracing," *ASCE J. Struct. Div.*, Vol. 102, No. ST9, pp. 1685–1702.

Errera, S. J., Pincus, G., and Fisher, G. P. (1967), "Columns and Beams Braced by Diaphragms," *ASCE J. Struct. Div.*, Vol. 93, No. ST1, pp. 295–318.

European Convention for Constructional Steelwork (1977), "European Recommendations for the Stressed Skin Design of Steel Structures," *ECCS-XVII-77-1E*, March. (English version published by Constrado, Croydon, England.)

Galambos, T. V. (1968), *Structural Members and Frames*, Prentice-Hall, Englewood Cliffs, N.J.

Gergely, P., Banavalkar, P. V., and Parker, J. E. (1971), "The Analysis and Behavior of Thin-Steel Hyperbolic Paraboloid Shells," *Cornell Univ. Dept. Struct. Eng. Rep.*, No. 338, September.

Goodier, J. N. (1942), "Flexural-Torsional Buckling of Bars of Open Section," *Cornell Univ. Eng. Exp. Sta. Bull.*, No. 28, January.

Graves-Smith, T. R. (1969), "The Ultimate Strength of Locally Buckled Columns of Arbitrary Length," in *Thin Wall Steel Structures* (ed. K. C. Rockey and H. V. Hill), Crosby Lockwood, London, pp. 35–60.

Guiaux, P. (1974), "Buckling Test on Cold Formed Hollow Sections, Square and Circular," *Rep. No. 74/18/E(2C)*, Structural Test Laboratories of the Civil Engineering Faculty of the University of Liege, Belgium.

Ha, K. H., El-Hakim, N., and Fazio, P. P. (1979), "Simplified Design of Corrugated Shear Diaphragms," *ASCE J. Struct. Div.*, Vol. 105, No. ST7, pp. 1365–1378.

Hancock, G. J. (1981), "Nonlinear Analysis of Thin Sections in Compression," *ASCE J. Struct. Div.*, Vol. 107, No. ST3, pp. 455–472.

Haussler, R. W. (1964), "Strength of Elastically Stabilized Beams," *ASCE J. Struct. Div.*, Vol. 90, No. ST3, pp. 219–264.

Hetrakul, N., and Yu, W. W. (1978), "Structural Behavior of Beam Webs subjected to Web Crippling and a Combination of Web Crippling and Bending," *Rep. No. 78-4*, University of Missouri at Rolla, June.

Hill, H. N. (1954), "Lateral Buckling of Channels and Z-Beams," *Trans. Am. Soc. Civ. Eng.*, Vol. 119, p. 829.

Hlavacek, V. (1972), "The Effect of Support Conditions on the Stiffness of Corrugated Sheets Subjected to Shear," *Acta Technica CSAV*, Prague, November 2.

Ingvarsson, L. (1975), "Cold-Forming Residual Stress, Effect on Buckling," *Proc. 3rd Int. Spec. Conf. Cold-Formed Steel Struct.*, Vol. I, Department of Civil Engineering, University of Missouri at Rolla, pp. 85–120, November, 1975.

Ingvarsson, L. (1977), "Cold-Forming Residual Stresses and Box Columns Built up by Two Cold-Formed Channels Sections Welded Together," *Bull. No. 121*, Department of Statics and Structural Engineering, Royal Institute of Technology, Stockholm.

Johnson, A. L., and Kelsen, G. A. (1969), "Stainless Steel in Structural Applications," Stainless Steel for Architecture, *ASTM STP*, 454.

Johnson, A. L., and Winter, G. (1966), "Behavior of Stainless Steel Columns and Beams," *ASCE J. Struct. Div.*, Vol. 92, No. ST5, pp. 97–118.

Jombock, J. R., and Clark, J. W. (1968), "Bending Strength of Aluminum Formed Sheet Members," *ASCE J. Struct. Div.*, Vol. 94, No. ST2, pp. 511–528.

Kalyanaraman, V. and Peköz, T. (1978), "Analytical Study of Unstiffened Elements," *ASCE J. Struct. Div.*, Vol. 104, No. ST9, pp. 1507–1524.

Kalyanaraman, V., Peköz, T., and Winter, G. (1977), "Unstiffened Compression Elements," *ASCE J. Struct. Div.*, Vol. 105, No. ST9, Proc. Pap. 13197, pp. 1833–1848.

Karren, K. W., and Winter, G. (1967), "Effects of Cold-Forming on Light Gage Steel Members," *ASCE J. Struct. Div.*, Vol. 93, No. ST2, pp. 433–470.

Klöppel, K., and Schardt, R. (1958), Beitrag zur praktischen Ermittlung der Vergleichsschlankheit von mittig gedruckten Stäben mit einfachsymmetrischem offenem dünnwändigem Querschnitt, *Stahlbau*, Vol. 27, p. 35.

Konig, J., and Thomasson, P. O. (1980), "Thin-Walled C-Shaped Panels in Axial Compression or in Pure Bending," in *Thin-Walled Structures* (ed. J. R. Rhodes and A. C. Walker, Granada, London.

LaBoube, R. A. (1983), "Z-Purlin Edge Stiffeners," *Res. Rep.*, Butler Manufacturing Company, Kansas City, Mo, March.

LaBoube, R. A., and Yu, W. W. (1982), "Bending Strength of Webs of Cold-Formed Steel Beams," *ASCE J. Struct. Div.*, Vol. 108, No. ST7, pp. 1589–1604.

Larson, M. A. (1960), discussion of "Lateral Bracing of Columns and Beams" by George Winter, *Trans. Am. Soc. Civ. Eng.*, Vol. 125, pp. 830–838.

Loh, T. S. (1985), "Combined Axial Load and Bending in Cold-Formed Steel Members," Cornell Univ. *Dept. Struct. Eng. Rep.*, No. 85-3, February.

Luttrell, L. C. (1967), "Strength and Behavior of Light Gage Steel Shear Diaphragm," *Cornell Univ. Eng. Res. Bull.*, No. 67-1.

Marsh, C. (1969), "Single Angles in Tension and Compression," *ASCE J. Struct. Div.*, Tech. Note, Vol. 95, No. ST5, pp. 1043–1049.

Massicotte, B., Beaulieu, D., and Picard, A. (1985), "Bracing of Columns by Girt-Diaphragm Systems in Light Industrial Buildings," *Proc. SSRC Ann. Tech. Sess.*

McDermott, R. J. (1981), "Column Strength of Cold Formed Tubular Sections," M.S. thesis, Lehigh University, Bethlehem, Pa.

Mertz, K. L., and Matthiesen, R. B. (1970), "Design of Corrugated Shear-Transmitting Members," *AISI Project 149*, School of Engineering and Applied Science, University of California, Los Angeles, December.

Miller, C. J. (1972), "Analysis of Multistory Frames with Light Gage Steel Panel Infills," Dissertation (also *CU Rep. No. 349*), Cornell University, Ithaca, N.Y., August.

Mulligan, G. P., and Peköz, T. (1983), "Influence of Local Buckling on the Behavior of Singly Symmetric Cold-Formed Steel Columns," *Cornell Univ. Dep. Struct. Eng. Rep.*, No. 83-2, March.

Nethercot, D. A., and Trahair, N. S. (1975), "Design of Diaphragm-Braced I-Beams," *ASCE J. Struct. Div.*, Vol. 101, No. ST10, pp. 2045–2062.

Nguyen, R. P., and Yu, W. W. (1978), "Webs for Cold-Formed Steel Flexural Members—Structural Behavior of Transversely Reinforced Beam-Webs," *Final Rep.*, University of Missouri at Rolla, July.

Nilson, A. H. (1960), "Shear Diaphragms of Light Gage Steel," *ASCE J. Struct. Div.*, Vol. 86, No. ST11, pp. 111–140.

Peköz, T. B. (1969) (with a contribution by N. Celebi), "Torsional-Flexural Buckling of Thin-Walled Sections under Eccentric Load," *Cornell Eng. Res. Bull.*, No. 69.1, September.

Peköz, T. (1980), "Design of Cold-Formed Steel Storage Racks," in *Thin-Walled Structures* (ed. J. R. Rhodes and A. C. Walker), Granada, London.

Peköz, T. (1986), "Developent of a Unified Approach to the Design of Cold-Formed Steel Members," *Res. Rep.*, Committee of Sheet Steel Producers, American Iron and Steel Institute, Washington, D.C., May.

Peköz, T., and He, B. K. (1981), "Experiments on Z-Purlins to Explore Effectiveness of Sloping Edge Stiffeners," *Progress Rep.*, Cornell University, Dep. Struct. Eng., Ithaca, NY., January.

Peköz, T., and Soroushian, P., (1982), "Behavior of C- and Z-Purlins under Wind Uplift," *Proc. 6th Int. Conf. Cold-Formed Steel Struct.*, St. Louis, November.

Peköz, T. B., and Winter, G. (1969), "Torsional-Flexural Buckling of Thin-Walled Sections under Eccentric Load," *ASCE J. Struct. Div.*, Vol. 95, ST5, pp. 941–964.

Peköz, T., Desmond, T. P., and Winter, G. (1981a), "Intermediate Stiffeners for Thin-Walled Members," *ASCE J. Struct. Div.*, Vol. 107, No. ST4, pp. 627–648.

Peköz, T., Desmond, T. P., and Winter, G. (1981b), "Edge Stiffeners for Thin-Walled Members," *ASCE J. Struct. Div.*, Vol. 107, No. ST2, pp. 329–353.

Pfluger, A. (1961), "Thin-Walled Compression Members," *Mitt. Inst. Statik Techn. Hochsch. Hannover*, Part 1 (1959); Part 2 (1959); Part 3 (1961).

Pincus, G., and Fisher, G. P. (1966), "Behavior of Diaphragm-Braced Columns and Beams," *ASCE J. Struct. Div.* Vol. 92, No. ST2, pp. 323–350.

Prawel, S. P., Jr., and Lee, G. C. (1964), "Biaxial Flexure of Columns by Analog Computers," *ASCE J. Eng. Mech. Div.*, Vol. 90, No. EM1, Proc. Pap. 3805, pp. 83–112.

Rack Manufacturers Institute (1979), *Specification for the Design, Testing and Utilization of Industrial Steel Storage Racks*, RMI,

Reck, H., Peköz, T., and Winter, G. (1975), "Inelastic Strength of Cold-Formed Steel Beams," *ASCE J. Struct. Div.*, Vol. 101, No. ST11, pp. 2193–2204.

Rhodes, J., and Walker, A. C., eds. (1979), "Thin-Walled Structures," *Proc. Int. Conf. Univ. Strathclyde*, Glasgow, Granada, London.

Rockey, K. C., Elgaaly, M. A., and Bagchi, D. K. (1972), "Failure of Thin-Walled Members under Patch Loading," *ASCE J. Struct. Div.*, Vol. 98, No. ST12, pp. 2739–2752.

Shanley, F. R. (1968), "Investigation of Shear-Transmitting Members: Design of a Test Beam," *Report No. 68-20*, Department of Civil Engineering, University of California, Los Angeles, April.

Sharp, M. L. (1970), Strength of Beams or Columns with Buckled Elements," *ASCE J. Struct. Div.*, Vol. 96, No. ST5, pp. 1011–1015.

Sherman, D. R., and Fisher, J. M. (1971), "Beam with Corrugated Webs," *Proc. 1st Int. Spec. Conf. Cold-Formed Steel Struct.*, University of Missouri at Rolla, August.

Simaan, A. (1973), "Buckling of Diaphragm-Braced Columns of Unsymmetrical Sections and Application to Wall Studs Design," Dissertation, Cornell University, Ithaca, N.Y.

Simaan, A., and Peköz, T. (1976), "Diaphragm-Braced Members and Design of Wall Studs," *ASCE J. Struct. Div.*, Vol. 102, No. ST1, pp. 77–92.

Smith, R. E. (1955), "Column Tests on Some Proposed Aluminum Standard Structural Sections," *Rep. No. 25*, Aluminum Development Association, London.

Sridharan, S., and Benito, R. (1984), "Columns: Static and Dynamic Interactive Buckling," *ASCE J. Eng. Mech. Div.*, Vol. 110, No. 1, pp. 49–65.

Steel Deck Institute (1981), *Diaphragm Design Manual*, SDI, St. Louis.

Thomasson, P. O. (1978), "Thin-Walled C-Shaped Panels in Axial Compression," *Res. Document D1: 1978*, Swedish Council for Building, Stockholm.

Thürlimann, B. (1953), "Deformation of and Stresses in Initially Twisted and Eccentrically Loaded Columns of Thin-Walled open Cross Section," *Rep. No. E696-3*, Brown University, Providence, R.I., June.

Timoshenko, S. P., and Gere, J. M. (1961), *Theory of Elastic Stability*, 2nd ed., McGraw-Hill, New York.

Trestain, T. W. J. (1982), "A Review of Cold Formed Steel Column Design," *Rep. No. 81109-1*, CSCC Project DSS 817, Canadian Sheet Steel Building Institute, Toronto, Ontario, December.

Uribe, J., and Winter, G. (1970), "Cold-Forming effects in Thin-Walled Steel Members," *Cornell Eng. Res. Bull.*, No. 70-1.

Vlasov, V. Z. (1959), *Thin-Walled Elastic Beams*, 2nd ed. (translation from Russian). (Available from the Office of Technical Services, U.S. Department of Commerce.)

Wang, S. T. (1969), "Cold-Rolled Austenitic Stainless Steel: Material Properties and Structural Performance," *Cornell Univ. Dept. Struct. Eng. Rep.*, No. 334, July.

Wang, S. T., and Errera, S. J. (1971), "Behavior of Cold Rolled Stainless Steel Members," *Proc. 1st. Spec. Conf. Cold-Formed Steel Struct.*, University of Missouri at Rolla, August.

Wang, S. T., and Yeh, S. S. (1974), "Post-Local-Buckling Behavior of Continuous Beams," *ASCE J. Struct. Div.*, Vol. 100, No. ST6, pp. 1169–1188.

Wing, B. A. (1981), "Web Crippling and the Interaction of Bending and Web Crippling of Unreinforced Multi-Web Cold Formed Steel Sections," Masters thesis, University of Waterloo, Waterloo, Ontario.

Winter, G. (1943), "Lateral Stability of Unsymmetrical I-Beams and Trusses in Bending," *Trans. Am. Soc. Civ. Eng.*, Vol. 108, pp. 247–260.

Winter, G. (1947), discussion of "Strength of Beams as Determined by Lateral Buckling," by Karl deVries, *Trans. Am. Soc. Civ. Eng.*, Vol. 112, pp. 1272–1276.

Winter, G. (1958), "Lateral Bracing of Columns and Beams," *ASCE J. Struct. Div.*, Vol. 84, No. ST2, p. 1561.

Winter, G. (1970), *Commentary on the 1968 Edition of Light Gage Cold-Formed Steel Design Manual*, American Iron and Steel Institute, Washington, D.C.

Winter, G., and Pian, R. H. J. (1946), "Crushing Strength of Thin Steel Webs," *Cornell Univ. Eng. Exp. Sta. Bull.*, No. 35, Part 1.

Winter, G., Lansing, W., and McCalley, R. B., Jr. (1949), "Performance of Laterally Loaded Channel Beams," *Colston Pap.*, Vol. 11. (Reprinted in "Four Papers on the Performance of Thin Walled Steel Structures," *Cornell Univ. Eng. Exp. Sta. Rep.*, No. 33.)

Yener, M., and Peköz, T. (1980), "Inelastic Load Carrying Capacity of Cold-Formed Steel Beams," *Proc. 5th Int. Conf. Cold-Formed Steel Struct.*, St. Louis, November.

Yu, W. W. (1980), "Web Crippling and Combined Web Crippling and Bending of Steel Decks," *1st Prog. Rep.* (May 1980); *2nd Prog. Rep.* (August 1980), University of Missouri at Rolla.

Yu, W. W. (1985), *Cold-Formed Steel Design*, Wiley, New York.

Yu, W. W., and Senne, J. H. eds. (1984), "Recent Research and Development in Cold-Formed Steel" *Proc. 7th Int. Conf. Cold-Formed Steel Struct.*, University of Missouri at Rolla.

Zetlin, L., and Winter, G. (1955), "Unsymmetrical Bending of Beams with and without Lateral Bracing," *Proc. Am. Soc. Civ. Eng.*, Vol. 81, pp. 774-1–774-20.

Zhang, Y., and Peköz, T. (1982), "An Exploratory Study on the Behavior of Cold-Formed Steel Wall Studs," *Cornell Univ. Dep. Struct. Eng. Rep.*, No. 82-14, September.

Chapter Fourteen

Circular Tubes and Shells

14.1 INTRODUCTION

A variety of structures consist of or include thin-walled cylinders that are subject to buckling. The round cylinder provides the most efficient shape available for centrally loaded columns with no lateral support between ends. Such columns are used in three-dimensional loading applications such as transmission towers, reticulated shells, and offshore platforms. Stiffened and unstiffened cylindrical shells (cylinders with large diameter-to-thickness ratios) are used as grain storage bins, liquid storage tanks, pressure vessels, and caissons for underwater construction.

Tubes and shells may be subject to axial compression, bending, twisting, or external or internal pressure, any one of which can cause failure. Depending of the dimensions of the cylinder, either local or overall buckling failures can occur. If the diameter of the cylinder is relatively large, longitudinal and/or ring stiffeners are often used to provide additional strength. Notation for the geometric parameters defining both an unstiffened and stiffened cylinder is given in Fig. 14.1.

Cylinders with relatively small diameter-to-thickness (D/t) ratios are usually referred to as tubes or pipes, and cylinders with large (D/t) ratios most often are called shells. Typically, shells are stiffened. When a single descriptive term is required, "cylinder" or "tubular member" is utilized.

14.1.1 Production Practice

The behavior of a tubular member is influenced to some extent by whether it is manufactured in a pipe or tubing mill or fabricated from plate. The distinction is important primarily because of the differences in geometric imperfections and residual stress levels that usually result from

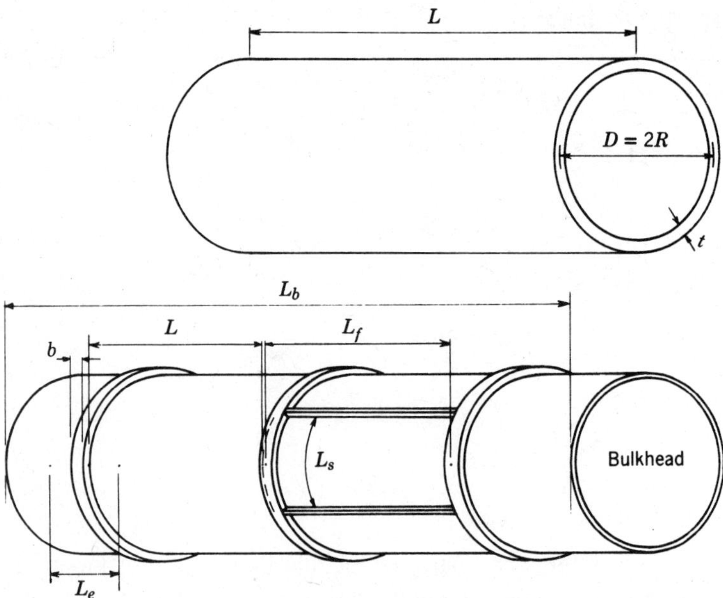

Fig. 14.1. Dimensional notation.

the two production methods (Schilling, 1965). In general, fabricated cylinders may be expected to have considerably larger magnitudes of imperfections (in diameter, ovality, and lack of straightness) than the mill products.

Manufactured cylinders are made as seamless pipe, or with continuous seam welds of various types or in the case of nonferrous metals, by extrusion. Each of these methods includes a finishing treatment to obtain the proper size and circular shape (Graham, 1965; U.S. Steel, 1964).

Fabricated tubes and shells are produced by welding or mechanically joining plates of cold- or hot-formed materials such as carbon steel, high-strength low-alloy steel, constructional alloy steel, or structural aluminum alloys. Fabricated structural members are frequently made by butt welding a series of short cans with the longitudinal welds on adjacent cans separated by rotating the cans.

14.1.2 Stress-Strain Curves and Residual Stresses

The basic stress-strain curve of a tubular section can be either (1) linear up to a yield-point stress with subsequent plastic straining at essentially constant stress, or (2) linear up to a proportional limit, less than yield strength, with subsequent gradual nonlinear transition to a yielding plateau or nonlinear softening prior to failure (Schilling, 1965). These two general categories of stress-strain behavior are illustrated by the solid curves in Fig. 14.2. The presence of residual stresses will change the effective stress-strain relationships to the dashed curves in Fig. 14.2. Generally, hot-finished cylinders will have sharp-yielding stress-strain curves, whereas cold-finished cylinders exhibit gradual-yielding behavior. Cold work in any cold-finished operation causes a change in stress-strain behavior from the basic material properties. Cylinders made of materials such as certain strainless steels and aluminum alloys will also have gradual-yielding stress-strain curves regardless of the production practice employed.

Residual stresses most commonly arise from the cooling effects after hot finishing, from the welding practices employed, or by the prevention

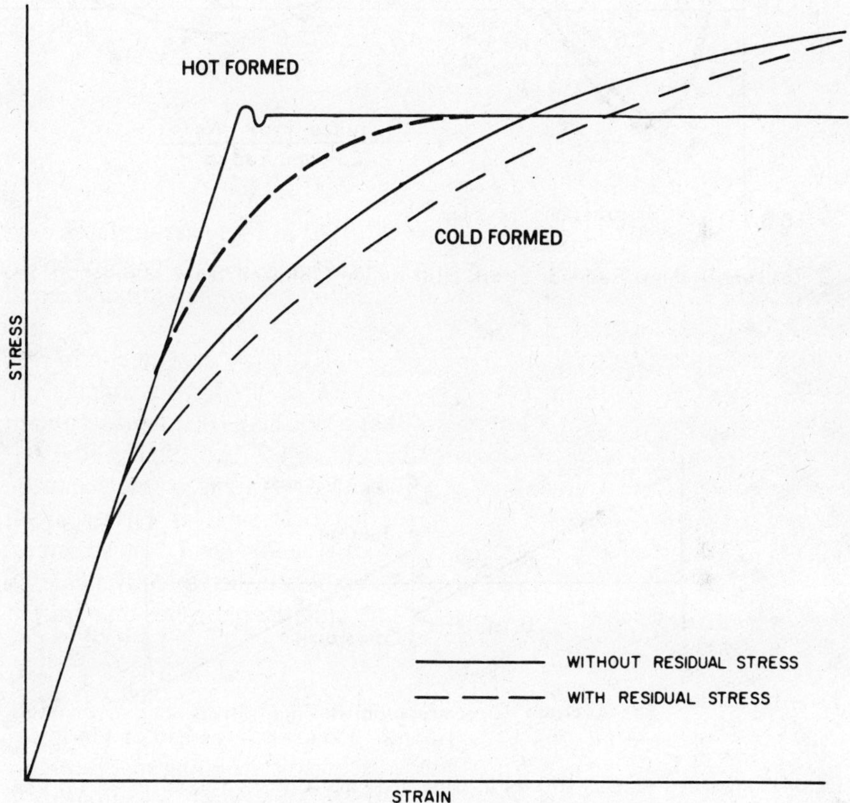

Fig. 14.2 Two general types of stress-strain curves.

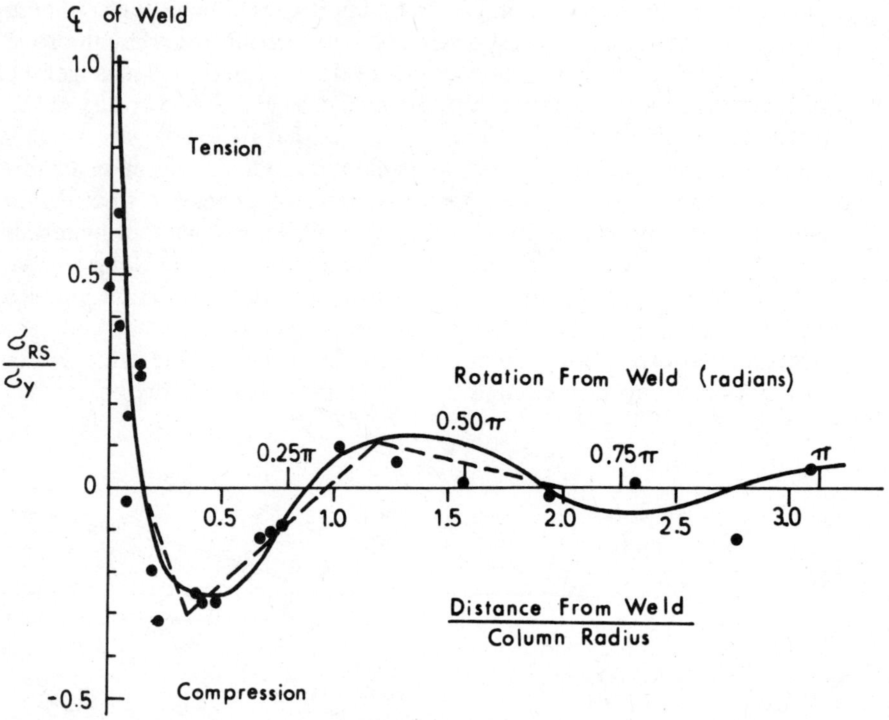

(a) Longitudinal Residual Stress Distribution Obtained from Method of Sectioning

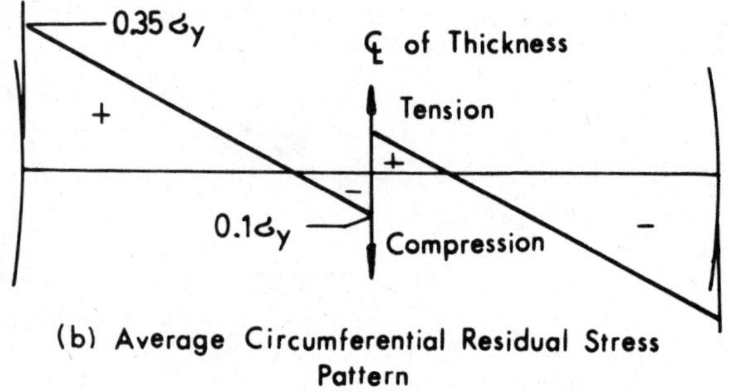

(b) Average Circumferential Residual Stress Pattern

Fig. 14.3 Measured residual stresses in fabricated pipe (Chen and Ross, 1978). (*a*) Longitudinal residual stress distribution obtained from method of sectioning; (*b*) Average circumferential residual stress pattern.

of springback introduced during forming operations. Longitudinal residual stresses in manufactured cylinders may result from nonuniform plastic flow through the thickness of the cylinder wall. Because of the foregoing, the exact shape of the stress-strain curve, the proportional limit, and the yield strength of a tubular member are rather unpredictable.

Residual stresses can be measured (Denton and Alexander, 1963a,b; Sherman, 1969) and in certain cases, the effects of cold work can be determined (Kuper and Macadam, 1969). Measurements on members fabricated for a column testing (Chen and Ross, 1978) program gave the longitudinal and through-thickness circumferential residual stress patterns shown in Fig. 14.3. Manufactured cold-formed pipes have similar patterns unless the finishing operation has a stress-relieving effect. Hot-formed and hot-finished cylinders generally have very low residual stresses.

14.2 DESCRIPTION OF BUCKLING BEHAVIOR

14.2.1 Factors Affecting Buckling

Buckling load predictions for cylindrical members are more complicated than for wide-flange sections due to the larger number of variables. The variety of geometric proportions is quite large since the cylindrical shape is used as a structural member with large length-to-diameter (L/D) and small D/t ratios or as a shell structure where L/D is frequently less than 1 and D/t may be 100 times larger than for some unstiffened members. In many applications, circumferential and/or longitudinal stiffeners are used to increase the buckling strength. Their size, spacing, and position on the inside or outside of the cylinder will affect the performance of the cylinder.

Cylinders are made by several different procedures, which result in different magnitudes and patterns of both longitudinal and circumferential residual stresses. As in any structural element, the residual stresses can play an important role in determining the buckling strength. The method of producing the cylinder and the acceptable tolerances also influence the degree of initial imperfection that is present. Tests and theory have shown that the elastic local buckling strength of thin-walled cylinders is very sensitive to geometric imperfections and in some cases, to the boundary conditions. Tests of thin cylinders under axial loading resulted in failures at buckling strengths considerably less than the theoretical elastic capacity. These discrepancies have generally been attributed to specimen imperfections and poorly modeled boundary conditions; however, in many cases the observed buckling capacities cannot be totally correlated with measured imperfection levels.

Quantitative predictions of buckling strengths must include all the loadings associated with metal members: axial compression, flexural bending and shear, and torsion. In addition, the enclosed nature of the cylinder causes pressure to be an important type of loading. Only axial loadings cause an overall general buckling of a cylindrical member since circular beams are not subject to lateral-torsional buckling. Tubular columns and beam-columns are included in Chapters 3 and 8. This chapter therefore concerns only the local buckling modes, except as they may interact with general buckling. The local buckling mode is a type of shell buckling in unstiffened cylinders. For cylinders stiffened with circumferential and/or longitudinal stiffeners, the local buckling mode may be shell or panel buckling between heavy stiffeners or may include buckling of light stiffeners in a wave configuration.

14.2.2 Buckling Equations

The stability of cylinders has been studied analytically and experimentally for a number of years. As a result of these studies and the wide variation in types of cylinders, a large number of empirical and semiempirical expressions have been proposed for predicting the buckling strength. Many of these were presented in the third edition of the guide as part of a historical discription of progress. This edition presents only those that are best supported by recent research. These equations represent the classical approach to the stability of cylinders.

Since the third edition, Miller (Miller, 1983; Miller et al., 1983) has proposed the use of a unified equation format. In this format, the linear elastic bifurcation buckling stress is reduced by a series of "knockdown" (reduction) factors to obtain a realistic critical stress:

$$F_i = \eta K_i \alpha_{ij} \sigma_{iej} \tag{14.1}$$

In this equation, σ_{iej} is the theoretical elastic buckling stress for a particular loading, i, with j being either x or c representing axial and circumferential stresses, respectively. The terms η, K, and α are the three "knockdown" factors that are used.

- η is a "plasticity factor" that reflects the residual stress levels and shape of the stress-strain curve. This factor is unity if buckling is purely elastic.
- K is a "slenderness factor" that accounts for the length and theoretical boundary conditions.
- α is a "capacity reduction factor" used to adjust for deviations between theory and tests. It accounts for the effects of imperfections in the boundary conditions and shell dimensions.

The "knockdown" factors are derived empirically and are normally a lower bound to test data. This approach requires an extensive data base to account for all the parameters that exist in cylinders. However, the effects of various parameters are included in specific terms. In addition to clarifying the influence of the parameters, this approach also facilitates modifications as additional research becomes available. Miller (1984) presents a large number of design equations for the factors applicable to various loading conditions in unstiffened and stiffened cylinders. Sections 14.3 through 14.7 represent this classical approach.

14.3 UNSTIFFENED OR HEAVY-RING-STIFFENED CYLINDERS

14.3.1 Axial Compression

The instability modes for an axially compressed cylinder are overall column buckling and local wall buckling, either of which can be elastic or inelastic. The type of buckling to which a particular cylinder is susceptible is dependent both on the ratio of length to radius of gyration, L/r, and the ratio of cylinder diameter to wall thickness, D/t. Generally, column buckling is controlled by the L/r ratio, while shell buckling is dependent on the D/t ratio. For example, a cylinder with a large L/r and sufficiently small D/t will buckle as an elastic column, whereas a cylinder with a moderate to large D/t can buckle in either an inelastic or an elastic shell buckling mode. In many cases it is difficult to predict which of the four types of buckling a partricular cylinder will exhibit.

Cylinders with heavy ring stiffeners that do not distort with the local buckle exhibit behavior between the rings which is similar to that of unstiffened cylinders. Closely spaced rings can enhance the local buckling strength. They will influence column buckling only in slenderness ranges where an interaction between column and local buckling exists. The general elastic stability theory for cylinders includes the spectrum from pure column to pure local buckling. The specific topic of realistic column buckling that reflects the stress-strain characteristics, residual stresses, and out-of-straightness of cylindrical columns is covered in Chapter 3. Therefore, this section is devoted to the discussion of local buckling of unstiffened cylinders and of cylindrical shells between theoretically rigid ring stiffeners. Guidelines for sizing fully effective ring stiffeners are provided in Section 14.4, and Section 14.7 countains a discussion of the interaction between column and local buckling.

Elastic Shell Buckling. The buckling of axially compressed cylindrical members was first approximately analyzed by Lorenz in 1908, and then in succeeding years more accurately by Flügge (1932), Southwell (1914), and Timoshenko (1910). Test results indicated, however, that actual cylinders buckled at loads well below those predicted by these early theoretical solutions. All of these solutions were based on small-deformation theory. In 1934, Donnell realized that a linear analysis was inadequate and suggested the need for a method of analysis that would account for large deformations. The first correct large-deflection solution was obtained by von Kármán and Tsien in 1941. Since then, numerous large-deflection investigations of axially compressed cylinders have been carried out. Of paramount importance among these was the analysis of Donnell and Wan (1950), who showed that initial imperfections are responsible for much of the discrepancy between linear theory and experimental results. The entire development of the theory of axially compressed cylinderical members is reviewed by Hoff (1966), who has himself made several important contributions to the subject. A comprehensive study of all aspects of the buckling of cylinderical shells has been prepared by Gerard and Becker (1957). It includes theories, test results, and design recommendations.

The surface of a short cylinder buckles like an infinitely wide plate. The critical stress depends on L/D, D/t, and on the boundary conditions of the edges. As L/D decreases, the critical stress approaches that for a plate strip of unit width discussed in Chapter 4. Longer cylinders buckle into a series of diamond shaped bulges and the critical stress depends only on D/t. Still longer cylinders buckle as Euler columns where L/r is the parameter.

The theoretical elastic buckling stresses are summarized below with

$$Z = 2\left(\frac{L}{D}\right)^2\left(\frac{D}{t}\right)\sqrt{1-\mu^2} \tag{14.2}$$

used as a parameter to delineate the regions of behavior. For plate-like buckling,

$$Z < 2.85 \qquad \sigma_{xc} = k_c \frac{\pi^2 E}{12(1-\mu^2)(L/t)^2} \tag{14.3a}$$

$$k_c = \frac{12Z^2}{\pi^4} \qquad \text{for simply supported edges} \tag{14.3b}$$

$$k_c = 4 + \frac{3Z^2}{\pi^4} \qquad \text{for fully clamped edges} \tag{14.3c}$$

For diamond-shaped bulges,

$$2.85 \leq Z < \frac{1.2(D/t)^2}{C} \qquad \sigma_{xc} = \frac{2CE}{D/t} \tag{14.4a}$$

where

$$C = \frac{1}{\sqrt{3(1-\mu^2)}} \tag{14.4b}$$

For the column buckling,

$$Z \geq \frac{1.2(D/t)^2}{C} \qquad \sigma_{xc} = \frac{\pi^2 E}{(L/r)^2} \tag{14.5}$$

Boundary conditions at the ends have little influence on Eq. 14.4 except when the edges are simply supported but can move freely in the tangential direction (Almroth, 1966; Batdorf et al., 1947a), in which case the buckling stress is half as large as the classical one.

Imperfections. The classical buckling stress given by Eq. 14.4 is a theoretical value assuming a geometrically perfect cylinder. The results of numerous compression tests (Donnell, 1956; Lundquist, 1933; Stein, 1968) show that actual cylinders may buckle elastically and fail at stresses as low as 30% of the critical stress given by Eq. 14.4. This discrepancy is due to the unstable postbuckling strength of such shells, which makes them extremely sensitive to small initial imperfections (such as deviations from the perfect geometrical shape) or due to residual stresses. The load that an axially loaded shell can support drops sharply subsequent to the onset of buckling, and the maximum load attained by the imperfect specimen is significantly below the critical load given by classical theory. Consequently, different values of C are recommended for applications where the normal degree of imperfections differ.

For manufactured or fabricated structural members, the value of C in Eq. 14.4a should be 0.165. This is approximately one-fourth of the theoretical value of Eq. 14.4b and was recommended by Plantema in 1946. Plantema's value was based on tests of manufactured members and recent tests of fabricated members with D/t in the range 350 to 450 (Stephens et al., 1982, 1983) correlate with this value. The test members that form the basis for the recommended values had an out-of-roundness limit of 1%, $(D_{max} - D_{min})/D_{nominal}$. Usually, structural members do not have proportions for which the critical local buckling stress would be increased by boundary conditions as in Eq. 14.3.

Considerable research has been conducted to determine realistic values of C to be used in place of Eq. 14.4b for cylinders with large D/t ratios and aerospace-quality tolerances. Donnell and Wan (1950) developed theoretical curves for imperfect cylinders that are shown in Fig. 14.4. The parameter U in their curves is a measure of the initial imperfection of the cylinder.

An empirical curve (also in Fig. 14.4) developed by Batdorf et al. (1947a) for cylinders with D/t greater than 1000 merges with the Donnell-

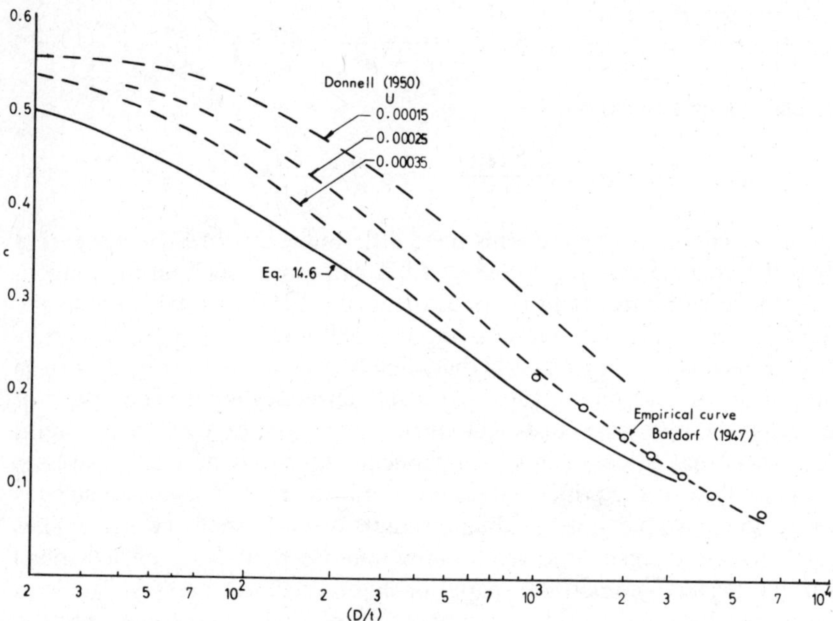

Fig. 14.4 Elastic buckling coefficient, C, for axially compressed cylinders.

Wan curve for $U = 0.00025$. The NASA publication (Weingarten, 1968) suggests

$$C = 0.6[1.0 - 0.9(1.0 - e^{-\theta})] \tag{14.6a}$$

where

$$\theta = 0.0442\sqrt{\frac{D}{t}} \tag{14.6b}$$

for D/t less than 3000. This relation is also plotted in Fig. 14.4 and is somewhat more conservative than the empirical curve by Batdorf and associates. Figure 14.4 also shows that the value of 0.165 recommended for C in structural members is conservative for D/t values less than 1000.

The NASA publication (Weingarten et al., 1965) also suggests that the effect of typical imperfections in short cylinders can be included by using $1.67CZ$ in place of Z in the K coefficients of Eqs. 14.3b and 14.3c. An earlier NACA Guide (Gerard, 1957) presented different expressions for the short cylinder coefficients, but these differ by only a few percent from the values obtained using the modified Z.

A design procedure is proposed by Clark and Rolf (1964) for aluminum tubes. Based on the empirical relation,

$$C = \frac{1}{\sqrt{3(1-\mu^2)}(1+0.02\sqrt{D/t})^2} \tag{14.7a}$$

and when $\mu = 0.33$ for aluminum

$$C = \frac{0.612}{(1+0.02\sqrt{D/t})^2} \tag{14.7b}$$

They suggest the following formula for moderately long cylinders:

$$\sigma_{xc} = \frac{\pi^2 E}{8(D/t)(1+0.02\sqrt{D/t})^2} \tag{14.8}$$

For a cylinder whose length is less than its mean radius, they suggest using either Eq. 14.8 or the buckling equation for a flat-plate column, whichever gives the higher stress.

In the Aluminum Association Specification (1982), Eq. 14.8 is presented in terms of an "equivalent slenderness ratio," which is substituted into the Euler column formula to obtain Eq. 14.9. This equivalent slenderness ratio is

$$(kL/r)_{\text{equiv}} = 2.828\sqrt{\frac{D}{t}}\left(1+0.02\sqrt{\frac{D}{t}}\right) \tag{14.9}$$

By substituting this value in the column formula (Eq. 14.5), the ultimate buckling strength can be determined for alloys for which ready-to-use formulas are not available.

Inelastic Shell Buckling. The inelastic buckling stress of cylindrical shells and tubes is usually obtained in one of two ways. Either the elastic formula is used with an effective modulus in place of the elastic modulus, or empirical relations are developed for specific classes of materials. The former approach is applicable only when the material stress-strain curve varies smoothly. This method has a long history of use and discussion (Clark and Rolf, 1964; Gerard, 1956; Harris et al., 1957; Weingarten, 1968). When the cylinder geometry and material properties are such that the computed buckling stress is in the plastic range, it is suggested that E in the elastic buckling equations be replaced by an effective modulus

$$E_{\text{eff}} = \sqrt{E_s E_t} \tag{14.10}$$

where E_s and E_t are the secant and tangent moduli, respectively.

As is usually the case when the effective modulus approach is used, a direct solution for the critical stress is not possible, and a graphical trial-and-error approach is helpful.

The effective-modulus approach is applicable to homogeneous materials such as aluminum alloys and stainless steels, while the inelastic buckling of cylinders made from structural steels is more conveniently handled with an empirical formula. Even for aluminum alloys, however, Clark and Rolf (1964) point out that a shortcoming of the effective-modulus approach is that each different alloy requires its own design curve. They therefore propose that the following single equation be used for all aluminum alloys:

$$\sigma_{xc} = B_t - D_t \sqrt{\frac{D}{2t}} \tag{14.11a}$$

where

$$B_t = \sigma_2 \left[1 + 4.6 \left(\frac{1000\sigma_2}{E} \right)^{0.2} \left(\frac{\sigma_2}{\sigma_1} - 1 \right) \right] \tag{14.11b}$$

$$D_t = \frac{B_t}{0.9} \left(\frac{B_t}{E} \right)^{1/3} \sqrt{\left(\frac{\sigma_2}{\sigma_1} \right) - 1} \tag{14.11c}$$

The quantities σ_1 and σ_2 are the values of the compressive yield strength at 0.1 and 0.2% offset (kips per square inch), respectively. To avoid curves that do not intersect the elastic curve, the ratio of σ_2/σ_1 is taken to be 1.06 in Eqs. 14.11b and 14.11c for those cases where the actual ratio exceeds this value. Equation 14.11 is applicable to any material whose stress-strain curve can be represented by the Ramberg-Osgood-Hill three-parameter relation

$$\varepsilon = \frac{\sigma}{E} + 0.002 \left(\frac{\sigma}{\sigma_2} \right)^n \tag{14.12a}$$

where

$$n = \frac{0.301}{\log_{10}(\sigma_2/\sigma_1)} \tag{14.12b}$$

Equation 14.11a is shown by Clark and Rolf (1964) to agree well with both test results and the inelastic design criteria developed by Gerard (1956).

When dealing with cylinders made from carbon or low-alloy steel, the empirical approach is used exclusively. Several major experimental programs have been conducted since 1933 and test progams involving axially loaded cylinders are currently under way, with additional ones planned. Figure 14.5 shows the experimental test data from Birkemoe et al. (1983), Marzullo and Ostapenko (1978), Ostapenko and Guzelman (1976, 1978), Ostapenko and Grimm (1980), Plantema (1946) and Wilson (1937), and Wilson and Newmark (1933) for thin-wall cylinders. All data

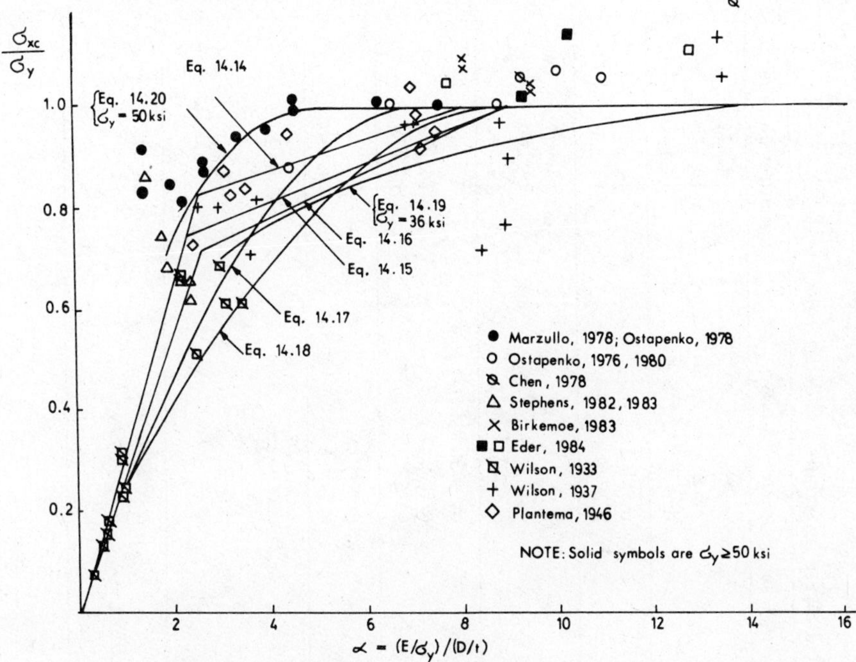

Fig. 14.5 Inelastic buckling equations and data for axially loaded cylinders.

from Wilson (1937) and Wilson and Newmark (1933) are shown for cylinders having thicknesses greater than or equal to $\frac{1}{4}$ in., while only the data for thicknesses less than $\frac{1}{4}$ in. that failed at a stress below the proportional limit are shown.

The considerable scatter in the data is probably due to differences in methods of making the cylinder that result in variation in imperfections, residual stresses, and material characteristics. There may also have been differences in test methods, data interpretation, and particularly in how the yield strength was determined.

As a result of the scatter in the test data, a number of different equations applicable to intermediate-length cylinders have been proposed for predicting the critical stress as a function of the wall slenderness. The equations vary in complexity and values of the critical stress depending on the philosophy of the proposer and the data base used. The slenderness parameters are usually either D/t or a nondimensional local-buckling parameter α, which for a circular cylinder can be expressed as

$$\alpha = \frac{E/\sigma_y}{D/t} \tag{14.13}$$

Table 14.1. Various Equations Proposed for the Critical Stress in Axially Loaded Steel Cylinders

References	Equation for σ_{xc}/σ_y	Limits	Equation Number
Plantema (1946)	1.0	$\alpha \geq 8$	14.14a
	$0.75 + 0.031\alpha$	$2.5 \leq \alpha < 8$	14.14b
	0.33α	$\alpha < 2.5$	14.14c
AISI Spec. (1986)	1.0	$\alpha \geq 9.1$	14.15a
	$0.665 + 0.0368\alpha$	$2.27 \leq \alpha < 9.1$	14.15b
SSRC Guide, 3rd ed. (1976)	1.0	$\alpha \geq 9.1$	14.16a
	$0.61 + 0.043\alpha$	$2.57 \leq \alpha < 9.1$	14.16b
	0.28α	$\alpha < 2.57$	14.16c
AWWA (1967)		$\sigma_y = 30$ ksi, $t \geq 1/4''$	
Timoshenko (1910)	1.0	$\alpha \geq 7.25$	14.17a
Tokugawa (1929)	$0.27\alpha - 0.019\alpha^2$	$\alpha < 7.25$	14.17b
Marshall (1971)	1.0	$\alpha \geq 9.1$	14.18a
	$0.22\alpha - 0.0121\alpha^2$	$\alpha < 9.1$	14.18b
API (1982)	1.0	$\alpha \geqslant \dfrac{E/\sigma_y}{60}$	14.19a
	$1.64 - 0.23/(\sigma_y \alpha/E)^{0.25}$ not to exceed 0.3α	$\alpha < \dfrac{E/\sigma_y}{60}$	14.19b
Ostapenko and Grimm (1980)	1.0	$\alpha \geq 0.07/\gamma$	14.20a
	$38(\gamma\alpha) - 400(\gamma\alpha)^2 + 2020(\gamma\alpha)^3$ where $\gamma = (\sigma_y/E)^{2/3}$	$\alpha < 0.07/\gamma$	14.20b

Table 14.1 contains a listing of the various proposed equations. They are compared to one another and to the test data in Fig. 14.5. These equations have been converted from the original sources to express the ultimate strength in terms of α as required.

Equations 14.14 through 14.16 have a similar form with a linear dependence of α. Plantema proposed Eq. 14.14 in 1946 based on tests of manufactured tubes, and the AISI adopted the more conservative Eq. 14.15 after consdering a larger data base that included manufactured cylinders. Equation 14.16 is an even more conservative relation fitted to the pre 1976 data. The American Water Works Association adopted Eq. 14.17 with a safety factor of 2 on the basis of the fabricated cylinder tests conducted in the 1930s. Equation 14.18 also uses the quadratic form but is closer to a lower bound on the test data. The API equation 14.19 and the most recent equation 14.20, which was derived as a best fit to a series of tests on fabricated cylinders with several strength levels, contain the yield strength as part of the slenderness parameter in addition to α.

The tests on fabricated cylinders conducted since 1976 fall considerably higher in Fig. 14.5 than the 1930s tests. This may reflect improved fabricating technology and indicates that lower-bound equations for the total data base may be overly conservative. At the same time, the scatter in the data probably does not warrant a complex best-fit expression for predicting the strength. Therefore, Eq. 14.15 is recommended as a reasonable estimation of a lower bound for the inelastic axial buckling stress of currently produced fabricated or manufactured steel tubes and pipe.

14.3.2 Cylindrical Shells Subjected to Bending

The buckling behavior of cylinders in bending differs from that of axially compressed cylinders in that bent cylinders have a stress gradient which is not present in axially compressed cylinders and the cross section tends to ovalize (Brazier, 1927; Ades, 1957; Gellin, 1980). Donnell (1934) found that the elastic buckling stress in bending is somewhat higher than the critical stress for axial compression. Flügge (1932) and Timoshenko and Gere (1961) reached the same conclusion. Other investigators (Wilson and Olson, 1941; Weingarten and Seide, 1961; Yao, 1962) have indicated that there is not much difference between the critical stress in bending and in axial compression. Until this disagreement is resolved, it is recommended that Eq. 14.4 with $C = 0.165$ as in axial compression be used for determining the critical bending stress of cylinders that buckle elastically.

Inelastic buckling in flexure includes not only the nonlinear behavior below the material yield stress as in axial compression, but also the region between the yield moment and where the maximum stress is at yield while the strain level increases. This is a significant region because of the relatively high shape factors in cylinders, that is,

$$\frac{Z}{S} = \frac{4}{\pi}\left(1 + \frac{t}{D}\right) \tag{14.21}$$

In the above, Z is the plastic-section modulus and S is the elastic-section modulus. Typical values for tubes listed in the AISC Manual range from 1.30 upward. About 96% of the plastic moment can be developed at only twice the yield strain.

Data for the inelastic flexural capacity of steel tubes with a uniform bending moment are plotted in Fig. 14.6. Since local buckling after the yield moment has been reached is a function of strain rather than stress, the plot is in terms of normalized moment capacity rather than critical stress. The earliest tests on hot-formed pipe reported by Schilling (1965) indicated that the plastic moment could be achieved for $\alpha > 8.33$, which is

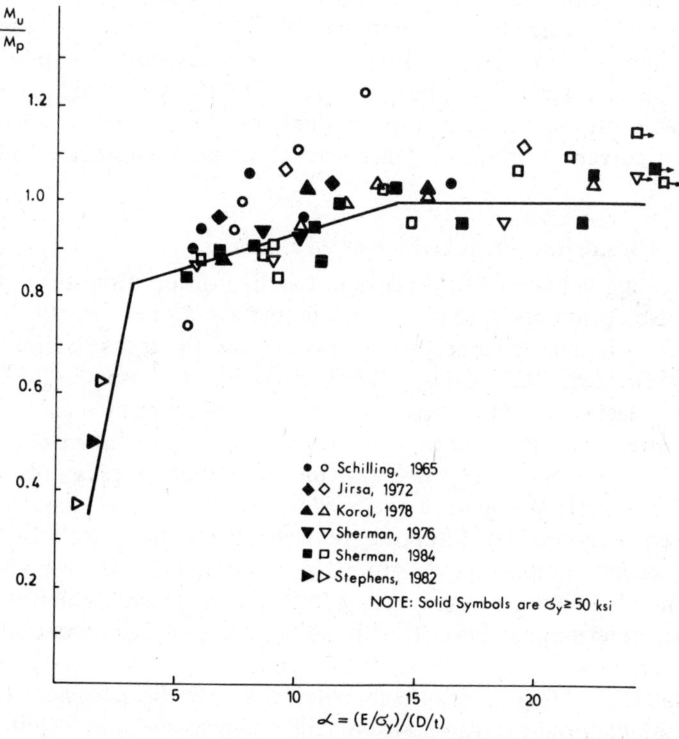

Fig. 14.6. Moment-capacity data.

approximately the same limit as that used in Eq. 14.14 for achieving the full yield strength of axially loaded cylinders. Later tests that also included cold-formed (Jirsa et al., 1972; Korol, 1978; Sherman, 1976) and fabricated pipe (Sherman and Tanarde, 1984) did not always reach the plastic moment when α exceeded 8.33. This is logical because the inelastic strain required for a plastic moment is much greater than that required for the yield capacity under axial load. In addition, pipes ovalize somewhat in bending, thereby producing additional geometric imperfection and biaxial stress conditions.

The moment capacities in Fig. 14.6 can be reasonably represented by a "best-fit" linear expression in terms of α similar to the form of Eq. 14.15 for critical axial stresses.

$$M_u/M_p = 1.0 \qquad \text{for } \alpha \geq 14 \tag{14.22a}$$

$$M_u/M_p = 0.775 + 0.016\alpha \qquad \text{for } \alpha < 14 \tag{14.22b}$$

Equation 14.22b is valid until the moment capacity is determined by elastic buckling,

$$M_u = \sigma_{xc} S \tag{14.22c}$$

where σ_{xc} is from Eq. 14.4. A lower bound to the data is obtained if the values in Eqs. 14.22a and 14.22b are multiplied by 0.9.

Clark and Rolf (1964) developed the following criterion for aluminum and materials with a similar stress-strain curve.

$$\sigma_{tb} = B_{tb} - 0.707 D_{tb} \sqrt{\frac{D}{t}} \tag{14.23}$$

where σ_{tb} is the bending stress in the tube (maximum value of bending moment divided by section modulus) and B_{tb} and D_{tb} are coefficients given by

$$B_{tb} = 1.5\left[1 + 4.6\left(\frac{1000\sigma_y}{E}\right)^{0.2}\left(\frac{\sigma_2}{\sigma_1} - 1\right)\right]\sigma_y \tag{14.24a}$$

$$D_{tb} = \frac{B_{tb}}{2.7}\left(\frac{B_{tb}}{E}\right)^{1/3} \tag{14.24b}$$

In the formulas above, σ_y (kips per square inch) is either the tensile or compressive yield strength in the axial direction, whichever is lower, and σ_1 and σ_2 (kips per square inch) are the values of the compressive yield strength at 0.1 and 0.2% offset, respectively. The ratio σ_2/σ_1 should be taken to be 1.06 for those cases where the actual ratio exceeds this value. When $(D/2t) > (B_{tb} - B_t)^2/(D_{tb} - D_t)^2$, the formulas for axial compression should be used instead of Eq. 14.23. The coefficients B_t and D_t are defined in Eqs. 14.11b and 14.11c.

14.3.3 Cylindrical Shells Subjected to Torsion

The elastic shear buckling stress, τ_c, due to torsion in cylinders of any length can be expressed as (Batdorf et al., 1947b)

$$\tau_c = K_s \frac{\pi^2 E}{12(1-\mu^2)(L/t)^2} \tag{14.25a}$$

As in the case of axial loading, the value of K_s depends on the cylinder proportions and, for short cylinders, on the boundary conditions. The value of K_s can be expressed in terms of the parameter Z defined in Eq. 14.2.

For short cylinders ($Z < 50$), end conditions are of major importance. For a simply supported short cylinder,

$$K_s = 5.35 + 0.213Z \tag{14.25b}$$

and for a short tube having full end fixity,

$$K_s = 8.98 + 0.10Z \tag{14.25c}$$

where Z is defined by Eq. 14.2. With intermediate-length cylinders, for $100 \le Z \le 19.2(1-\mu^2)(D/t)^2$,

$$K_s = 0.85Z^{0.75} \tag{14.25d}$$

for all end conditions. For very long cylinders, as recommended by Timoshenko and Gere (1961),

$$K_s = \frac{0.406Z}{(1-\mu^2)^{0.25}(D/t)^{0.5}} \tag{14.25e}$$

Schilling (1965) presents test data for some alloy steels whose stress-strain curve approches that of a sharp-yielding material. From these data he concludes that the critical shear stress for these steels can be approximated by the shear yield strength τ_y when the parameter $(\tau_y/E)(D/t)^{1.25}(L/D)^{0.5}$ is less than 1.076 and by the elastic shear buckling stress when the same parameter exceeds 1.076.

Studies of postbuckling behavior of intermediate-length cylinders loaded in torsion (Nash, 1957) show that the maximum load that an initially imperfect cylinder can resist is less than the classical shear buckling load. However, the drop in load subsequent to buckling is very small compared to that which takes place with axial compression. Hence the failure stress of an actual imperfect cylinder is only slightly lower than the critical stress predicted by linear theory. Schilling (1965) recommends that the theoretically obtained torsional buckling stress of intermediate-length cylinders loaded in torsion be reduced by 15% to account for initial imperfections.

For materials such as aluminum alloys and stainless steels, which have gradually yielding stress-strain curves, the buckling stress in the inelastic range can be obtained by replacing E, in the elastic formulas, by E_s, the secant modulus at $\sigma = 2\tau_c$ (Gerard, 1957).

In the Aluminum Association Specification (1982), an equivalent slenderness ratio is used to obtain the shear buckling stress for tubes subjected to torsion. In the elastic range the buckling stress is given by

$$\tau_c = \frac{\pi^2 E}{\lambda^2} \tag{14.26a}$$

in which λ, the equivalent slenderness ratio, is approximated by

$$\lambda = 3.73\left(\frac{D}{t}\right)^{0.75} W^{0.5} \tag{14.26b}$$

The coefficient W is equal to unity for long, unstiffened tubes, and

$$W = 0.561\,\frac{\sqrt{L/D}}{(D/t)^{0.25}} \tag{14.26c}$$

for tubes where the clear length L between circumferential stiffeners is such that the value of W from Eq. 14.26c is less than unity.

For the inelastic buckling of aluminum tubes, Clark and Rolf (1964) propose the following relation:

$$\tau_c = B_s - D_s\lambda \tag{14.27a}$$

in which λ is given by Eq. 14.26b, and B_s and D_s are coefficients defined as follows:

$$B_s = \tau_2\left[1 + 5.8\left(\frac{1000\tau_2}{E}\right)^{1/3}\left(\frac{\tau_2}{\tau_1} - 1\right)\right] \tag{14.27b}$$

$$D_s = \frac{B_s}{2}\sqrt{\frac{B_s}{E}\left(\frac{\tau_2}{\tau_1} - 1\right)} \tag{14.27c}$$

in which τ_1 and τ_2 are the shear yield strengths at 0.1 and 0.2% offsets, respectively expressed in kips per square inch. To avoid inelastic buckling curves that do not intersect the elastic curve, the ratio τ_2/τ_1 is taken to be 1.06 for those cases where the actual ratio exceeds this value. It is permissible to substitute σ_2/σ_1 for τ_2/τ_1 where σ_1 and σ_2 are determined from the compressive stress-strain curve. Good agreement is shown to exist between test results for five different aluminum alloys and Eq. 14.27.

14.3.4 Cylindrical Shells Subjected to Transverse Shear

Little information is available on the subject of local instability of shells or tubes subjected to transverse shear. It seems logical, however, that because of the presence of a stress gradient, tubes loaded in transverse shear will buckle at a higher stress than similar tubes loaded in torsion. Schilling (1965) suggests that for manufactured tubes the critical shear stress in transverse shear be taken as 1.25 times the critical stress in torsion when buckling is elastic. In the inelastic range, he advises using the same critical stress for transverse shear as is used for torsion.

14.3.5 Cylindrical Shells Subjected to Uniform External Pressure

The strength of a shell under external pressure depends on its L/D and D/t ratios and upon the physical properties of the material. It also depends on the amount of deviation of the shell from a true circular form. Failure of a shell can occur by yielding or by buckling at stresses which may be considerably below the yield point. The effective length of the shell can be reduced by the addition of circumferential stiffeners.

A thinness factor has been presented (Windenburg and Trilling, 1934) that is indicative of the mode of failure to be expected.

$$K = \left(\frac{D}{t}\right)^{0.75} \sqrt{\left(\frac{L}{D}\right)\left(\frac{\sigma_y}{E}\right)} \tag{14.28}$$

where L is the unsupported length of shell between stiffeners or between the ends of the cylinder (Fig. 14.1) and D is the diameter to midthickness ratio of the cylindrical shell. For approximation purposes, elastic instability is likely to occur in the range $K > 1.2$, plastic-shell instability in the range 0.8 to 1.2, and shell yielding if $K < 0.8$.

If the shell stiffeners are placed a large enough distance apart, the shell region between stiffeners will behave under pressure as though no stiffeners were present. The shortest length of cylinder for which the strengthening effect of the stiffeners can be ignored is defined as the "critical length."

A distinction can also be made in regard to the support conditions at the ends of the cylinder. If the pressure produces longitudinal stresses in addition to circumferential stresses, the cylinder is hydrostatically loaded. However, if end conditions do not produce longitudinal pressure stresses, the cylinder is defined as being loaded by lateral pressure only.

Elastic Buckling. Solutions for the critical elastic pressure of cylinders with finite length were first developed in the early 1900s (Southwell, 1915; von Mises, 1931, 1933). In the intervening years, modifications have been made to account for realistic boundary conditions (Von Sanden, 1949) and to provide simpler equations that closely approximate the exact solution in certain ranges of cylinder proportions. The most exact theory is that of Reynolds (1962), which includes the influence of elastic stiffening rings at the boundaries on both the prebuckling and buckling deformations. This theory agrees well with the test results presented by Hom and Couch (1961) and by Reynolds (1962) and should be used for comparison of theory and experiment. However, the simpler but more conservative equations of von Mises are recommended for general use.

The von Mises equation for lateral pressure (Windenburg and Trilling, 1934, Eq. 2) is given by

$$p_c = 2E\frac{t}{D}\left\{\frac{(t/D)^2}{3(1-\mu^2)}\left[n^2 - 1 + \frac{\lambda^2(2n^2 - 1 - \mu)}{n^2 - \lambda^2}\right] + \frac{\lambda^4}{(n^2-1)(n^2+\lambda^2)^2}\right\} \tag{14.29a}$$

where n is the number of circumferential lobes formed at collapse and

$$\lambda = \frac{\pi D}{2L} \tag{14.29b}$$

The von Mises equation for the more common hydrostatic pressure case can be approximated by

$$p_c = \frac{2E(t/D)}{n^2 + (\lambda^2/2) - 1}\left\{\frac{(t/D)^2}{3(1-\mu^2)}[(n^2+\lambda^2)^2 - 2n^2 + 1] + \frac{\lambda^4}{(n^2+\lambda^2)^2}\right\} \tag{14.30}$$

Equation 14.30 is a simplified version of Eq. 6 of the Windenburg and Trilling (1934) paper and agrees with it closely for all shell geometries. The correct value of n in Eq. 14.29 or 14.30 is that which makes p_c a minimum.

In certain ranges of L/D, the number of lobes in the buckling mode may be known or one of the terms in the expression for critical pressure becomes negligible. This leads to reasonable simplifications of the more complex equations so that iterative solutions for n can be avoided. Table 14.2 contains a summary of these equations when Poisson's ratio is equal to 0.3. Simplified equations are provided only for the case of hydrostatic external pressure. The range of validity for each of the equations is defined in terms of L/D, or more conveniently by a parameter

$$\theta = [12(1-\mu^2)]^{0.25}\frac{L}{D}\sqrt{\frac{D}{t}} \tag{14.37a}$$

or

$$\theta = 1.818\frac{L}{D}\sqrt{\frac{D}{t}} \qquad \text{for } \mu = 0.3 \tag{14.37b}$$

In a few cases, the equations are still relatively complex, but approximate values of the critical pressure can be obtained by using a simpler expression from an adjacent range, as is evident in Fig. 14.7.

Table 14.2. Summary of Equations for the Elastic Buckling of a Perfect Cyinder with Poisson's Ratio (ν) = 0.3[a]

Pressure				Reference		Critical Pressure, p_c	
Lateral External				Windenburg and Trilling (1934)		$\frac{2E}{D/t}\left\{\frac{0.366}{(D/t)^2}\left[n^2-1+\frac{\lambda^2(2n^2-1.3)}{n^2-\lambda^2}\right]+\frac{\lambda^4}{(n^2-1)(n^2+\lambda^2)^2}\right\}$	Eq, 14.29a
Hydrostatic External							
θ		L/D					
From	To	From	To			Equation	Number
0	0.8	0	$\frac{0.44}{\sqrt{D/t}}$		Very short	$\frac{3.615E}{(D/t)^3(L/D)^2}$	14.31
0.8	1.4	$\frac{0.44}{\sqrt{D/t}}$	$\frac{0.77}{\sqrt{D/t}}$	SSRC Guide 3rd ed. (1976)		$\frac{3.615E}{(D/t)^2}\left[0.448(L/D)^2(D/t)+\frac{1}{(L/D)^2(D/t)}\right]$	14.32
1.4	2.0	$\frac{0.77}{\sqrt{D/t}}$	$\frac{1.1}{\sqrt{D/t}}$	Windenburg and Trilling (1934)	Short	$\frac{2.0E}{(D/t)(n^2+0.5\lambda^2-1)}\left\{\frac{0.367}{(D/t)^2}[(n^2+\lambda^2)^2-2n+1]+\frac{\lambda^4}{(n^2+\lambda^2)^2}\right\}$	14.30
2.0	10.0	$\frac{1.1}{\sqrt{D/t}}$	$\frac{5.5}{\sqrt{D/t}}$	Windenburg and Trilling (1934)		$\frac{2.6E}{(D/t)^{2.5}[(L/D)-0.45(t/D)^{0.5}]}$	14.33
10	D/t	$\frac{5.5}{\sqrt{D/t}}$	$0.55\sqrt{D/t}$		Intermediate	$\frac{2.6E}{(L/D)(D/t)^{2.5}}$	14.34
D/t	$4D/t$	$0.55\sqrt{D/t}$	$2.1\sqrt{D/t}$	SSRC Guide 3rd ed. (1976)		$\frac{2.0E}{(D/t)(3+0.5\lambda^2)}\left\{\frac{0.367}{(D/t)^2}[(4+\lambda^2)^2-7]+\frac{\lambda^4}{(4+\lambda^2)^2}\right\}$	14.35
$4D/t$	—	$2.1\sqrt{D/t}$	—	Bryant (1954)	Long	$\frac{2.2E}{(D/t)^3}$	14.36

[a] L is the distance between a major bulkhead and a ring stiffener or successive ring stiffeners, n is the integer at which p_c is a minimum, $\lambda = 0.5\pi/(L/D)$, $\theta = 1.818(L/D)(\sqrt{D/t})$.

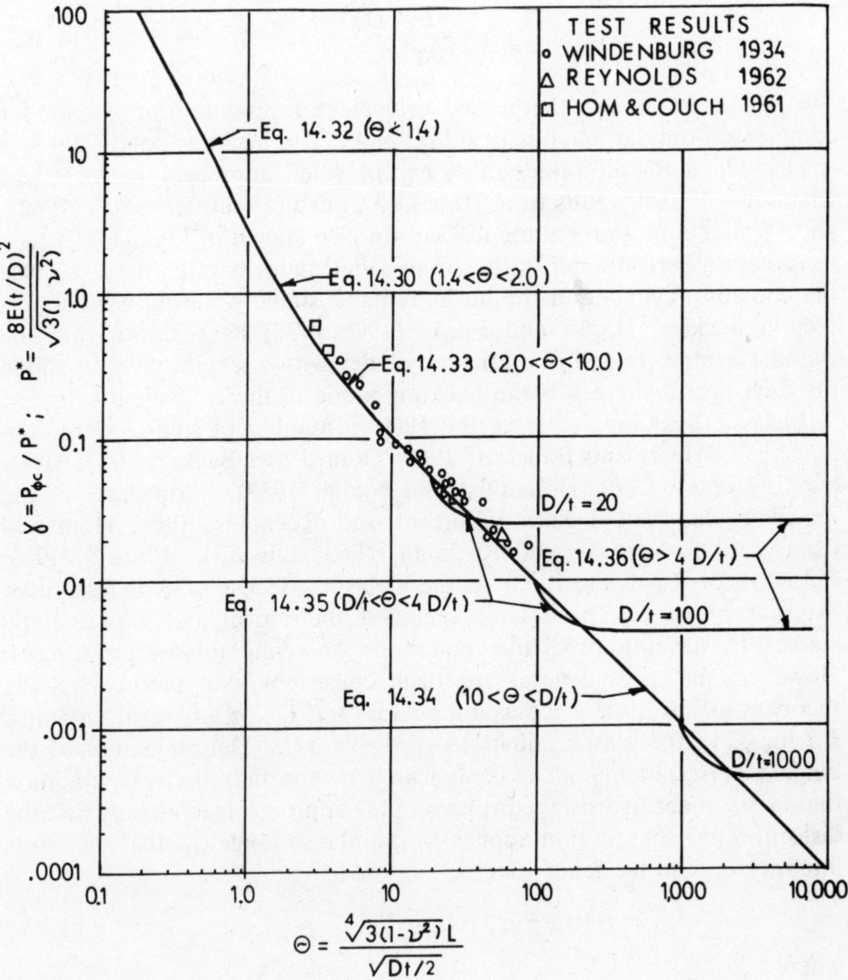

Fig. 14.7. Elastic buckling coefficients for circular cylinders under hydrostatic pressure ($\nu = 0.3$).

Figure 14.7 is a plot of the buckling pressure coefficient γ as determined from Eqs. 14.30 through 14.36 for values of the shell geometry parameter θ where

$$\gamma = \frac{p_{\phi c}}{p^*} \tag{14.38a}$$

and

$$p^* = \frac{8E(t/D)^2}{3(1-\mu^2)} \tag{14.38b}$$

The value p^* is the theoretical critical end pressure for a cylinder compressed only at its ends (see Eq. 14.4). The buckling coefficient γ is single valued for all values of θ, except when approaching D/t values where $n = 2$. Test results from Hom and Couch (1961), Reynolds (1962), and Windenburg and Trilling (1934) are also shown in Fig. 14.7.

A hemispherical head, rather than a flat plate, is often used to close the ends of a cylinder. If the heads remains stable while the cylinderical section buckles, Harari and Baron (1970, 1971) have shown that the cylinder can be treated as a longer cylinder with a length of $D/2n$ added for each head, where n is the buckling mode of the equivalent cylinder.

Inelastic Buckling. During the 1960s a number of studies were conducted (DeHeart and Badekas, 1960; Gerard and Becker, 1957; Heise and Esztergar, 1970; Holmquist and Nadai, 1939), particularly at the David Taylor Model Basin (Boichot and Reynolds, 1964; Hom and Couch, 1961; Krenzke and Kierman, 1963; Lunchick, 1961a,b, 1963; Pulos, 1963; Reynolds, 1960) on the inelastic collapse of cylinders under external pressure. These have included theoretical and experimental studies to develop predictive equations or semigraphical procedures. However, the most general approach consistent with plastic buckling theory is to substitute a reduced modulus, $\sqrt{E_t E_s}$, or a tangent modulus, E_t, for E in the elastic equations of Table 14.2. Determination of the tangent or secant moduli is complicated by the biaxial stress condition that exists under hydrostatic pressure. One approach is to assume that the distortion energy criterion applies to the plastic range, so that the stress intensity σ_i, can be defined as

$$\sigma_i = (\sigma_\phi^2 + \sigma_x^2 - \sigma_\phi \sigma_x)^{0.5} \tag{14.39a}$$

where

$$\sigma_\phi = -\frac{pD}{2t} \quad \text{and} \quad \sigma_x = -\frac{pD}{4t} \tag{14.39b}$$

From a representative stress-strain curve of the material used in the cylinder, the values of the desired moduli that correspond to σ_i are determined.

As in the case of columns, the use of the tangent modulus provides the lower limit load when it is used in the elastic buckling expression (Heise and Eszterger, 1970). Another limit is the plastic collapse pressure when the circumferential stress is at yield:

$$p_p = \frac{2\sigma_y}{D/t} \tag{14.40}$$

One practical way of determining the critical pressure is to use the charts in the ASME Code (1980). The charts include elastic buckling, elastic-plastic buckling, and plastic collapse for several different materials. They have been developed from the theoretical elastic equations, with the tangent modulus for the particular material and a uniform factor of safety of 3. Figure 14.8*a* is a chart that includes the length and D/t terms in the critical stress equations. Knowing L/D and D/t for the cylinder, the factor A is determined, and this corresponds to σ_y/E. The material curve in Fig. 14.8*b* is for carbon steel or low-alloy steels with yield strength greater than 30 ksi. Entering the chart with factor A, factor B is determined. In the ASME Code, the allowable pressure is computed from B. However, multiplying by 3 to remove the factor of safety, the critical pressure is

$$p_c = \frac{4B}{D/t} \tag{14.41}$$

or the critical circumferential stress is

$$\sigma_{\phi c} = 2B \tag{14.42}$$

The reduced modulus is given by $E_R = 2B/A$.

The horizontal portion of the material line in Fig. 14.8*b* represents the plastic collapse at a circumferential yield stress of 35 ksi. The other extreme of elastic buckling of long cylinders is also represented. The vertical portions of the D/t lines in Fig. 14.8*a* correspond to long cylinders, and the value of A for this case is given by

$$A = \frac{1.1}{(D/t)^2} \tag{14.43}$$

If this is substituted in the equation for the elastic portion of the material line in Fig. 14.8*b*, Eq. 14.36 in Table 14.2 is obtained.

The difficulty with the ASME charts is that a class of materials is included in one chart regardless of the yield strength. For example, Fig. 14.8*b* is for carbon and low-alloy steels with specified yield strengths of 30 ksi and over. To obtain a better estimate for the inelastic buckling pressure of cylinders with a different yield strength, new curves would have to be estimted between the elastic limit and yield.

Effects of Imperfections. The effect of initial imperfections has been studied by several investigators. Most analyses of the effects of initial imperfections are based on the assumption that the intial out-of-roundness is similar in form to one of the assumed buckling modes and the critical pressure is assumed to be the pressure at which the extreme fibers of the shell begin to yield.

Timoshenklo and Gere (1961, Eq. e, p. 296) have developed a formula

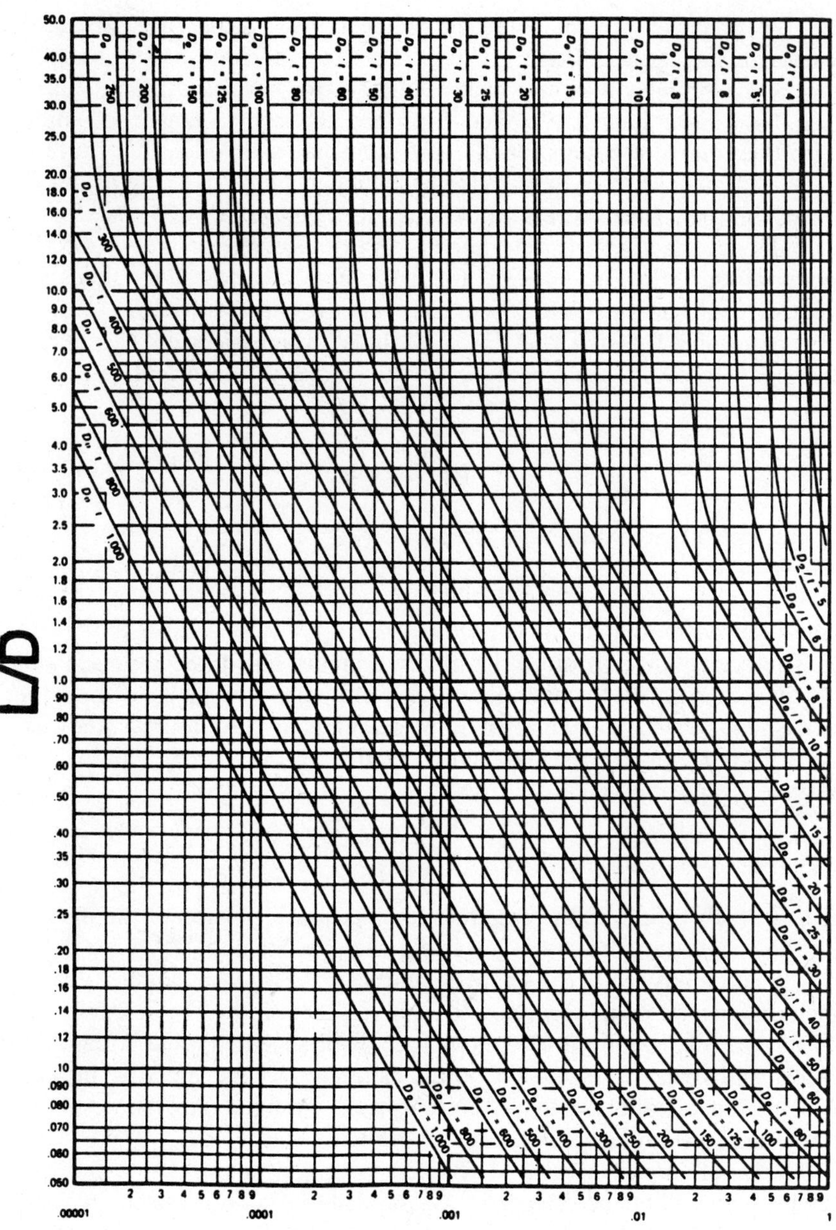

FACTOR $A = \sigma_\theta / E_{eff}$

(a)

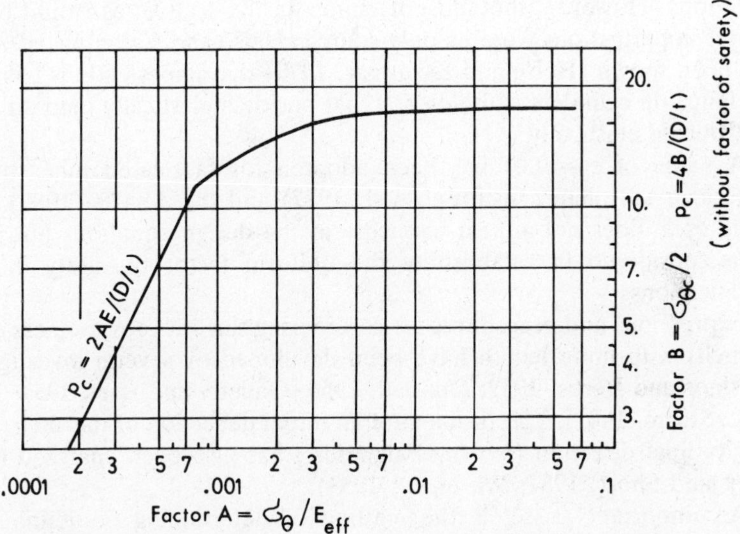

(*b*)

Fig. 14.8 ASME charts for buckling under hydrostatic pressure: (*a*) geometry terms from elastic equations; (*b*) material curve for carbon and low alloy steels with $\sigma_y \geq$ 30 ksi.

for determining the elastic critical pressure for cylinders of infinite length having a definable eccentricity. The hydrostatic pressure p_y at which yielding begins can be determined from the following equation:

$$p_y^2 - \left[\frac{2\sigma_y t}{D} + \left(1 + \frac{1.5De_0}{t}\right)p_c\right]p_y + \frac{2\sigma_y t}{D}\, p_c = 0 \qquad (14.44)$$

where

e_0 = out-of-roundness = $(D_{max} - D_{min})/D = 4e/D$ (see Fig. 14.2), where e is the radial eccentricity

p_c = critical pressure determined by Eqs. 14.30 through 14.36 (with E or E_t as appropriate)

This equation is applicable only when the buckling mode is such that $n = 2$. For situations where the critical n is greater than 2, Kendrick's (1953b) equations should be used in place of Eq. 14.44.

The use of the equations with out-of-roundness produces more conservative approximations of test data than the theoretical elastic or inealstic equations. However, they do not eliminate the scatter, as would be the case if imperfections were its only source (Heise and Esztergar, 1970). It has been shown (Heise and Esztergar, 1970) that imperfections have less influence on cylinders with low D/t that buckle inelastically than on those that buckle elastically.

A value of $e_0 = 0.01$ has been adopted for fabricated tubes by the American Petroleum Institute (API, 1977) and the ASME (1980). This value of e_0 does not appear explicitly in the design equations but it has been considered in establishing the uniform factor of safety in both specifications.

Expressions and procedures for considering the out-of-roundness effect in shells with finite length have been developed by several investigators (Bodner and Berks, 1952; Donnell, 1956; Galletly and Bart, 1957; Holt, 1952; Sturm, 1941). The problem of an initial deflection of the shell in the longitudinal direction (out-of-straightness) has also been analyzed (Lunchick and Short, 1957; Wu et al., 1953).

An important aspect is the method of determining or defining the initial imperfection. Several methods have been proposed (Galletly and Bart, 1957), but the most satisfactory from a theoretical point of view is to use the maximum radial deviation from a perfect circle, e, measured over an arc length, A, corresponding to one-half lobe length.

$$A = \frac{\pi D}{2n} \tag{14.45}$$

where n is the lobar number in Eq. 14.30. Based on this imperfection measurement, the corrective method of Galletly and Bart (1957) is the most effective, although it does not account for all test data scatter (Heise and Esztergar, 1970).

For design purposes, imperfections are usually considered by specifying permissible out-of-roundness. An empirical expression (Windenburg, 1960) for the maximum e/t value so that the collapse pressure will not be less than 80% of that corresponding to a perfect shell is

$$\frac{e}{t} = \frac{0.018}{n}\frac{D}{t} + 0.015n \tag{14.46}$$

This procedure has been incorporated into the ASME Code (1980), where the maximum permissible deviation and the length over which it is to be measured are presented in graphical form.

14.4 GENERAL INSTABILITY OF RING-STIFFENED CYLINDERS

Under static conditions, a ring-stiffened shell may fail in one or more of the three possible instability modes shown in Fig. 14.9. These are:

(a)

(b) (c)

Fig. 14.9 Buckling modes for ring-stiffened cylinders: (a) local buckling (axisymmetric); (b) local buckling (asymmetric); (c) general instability (overall collapse).

1. Axisymmetric collapse of the shell between adjacent ring stiffeners. This mode is a combination of yielding and axisymmetric buckling and is characterized by accordion-shaped pleats around the circumference of the shell.
2. Asymmetric or lobar buckling of the shell between adjacent ring stiffeners. This mode is characterized by the forming of two or more dimples or lobes around the circumference.
3. General instability or overall collapse of the entire shell and stiffeners (not column buckling). This mode is characterized by large dished-in portions of the stiffened shell, wherein the shell and stiffeners deflect together. The occurrence of this mode is strongly influenced by the shape, moment of inertia, and circularity of the stiffening rings and the ratio of the overall length to the radius of the cylinder.

The first two instability modes were the topic of the preceding section of this chapter and the third is the topic of this section.

Ring stiffeners have little influence on column buckling strengths of cylinders and the column curves for fabricated cylinders in Chapter 3 are applicable. Therefore, the purpose of ring stiffeners is primarily to enhance the local buckling strength of a cylinder. To accomplish this, their spacing L must be small enough to influence the strength predictions in the equations of Section 14.3 (i.e., L must be a term in the equation).

If the stiffeners have sufficient stiffness and strength, the failure will be in one of the first two instability modes and the collapse load a function of L. Lighter stiffeners, on the other hand, will lead to the third instability mode and its collapse load will be between that of an unstiffened long cylinder and one whose stiffeners are spaced L.

14.4.1 Nonpressure Loadings

Very little information is available on the general instability of ring-stiffened cylinders under nonpressure loads. This is probably due to the infrequent occurrence of conditions that would produce this mode. Under axial load, the critical length implied by the limit for Eq. 14.3 is $L/D < 1.22/\sqrt{D/t}$. This is an extremely small spacing relative to the critical length for pressure loading and would seldom be encountered in practice. Ring stiffeners are a very inefficient method of reinforcing a cylinder for axial loads. Ring stiffeners can be more effective in flexure because they reduce ovalization prior to local buckling. However, the spacing would have to be very small, as in the case of axial load, and would be considered only for regions of steep moment gradients. In the event that closely spaced ring stiffeners are present, their stiffness should be considered as uniformly distributed and the general buckling strength under axial load or flexure estimated using orthotropic shell theory.

The maximum (critical) spacing beyond which properly sized rings are no longer effective indicated in Eq. 14.25 for torsional loading is $L/D = 3.03\sqrt{D/t}$, which is considerably greater than that for the axial load requirements of Eq. 14.5. Therefore, it would appear that stiffeners could be effective in increasing the local buckling strength of torsion and similarly cylinders subject to transverse shear. However, little published information is readily available to determine the minimum size for a stiffener to be fully effective or to determine the general buckling strength if lighter stiffeners are used for either case. As a practical matter, it would seem that ring stiffeners are seldom used, or needed, to increase the buckling capacity of cylinders in torsion or shear.

14.4.2 Uniform External Pressure

Ring stiffeners are most frequently used to increase the local buckling strength of cylinders subject to external pressure. According to Table 14.2, the critical spacing is $L/D = 2.1\sqrt{D/t}$ for the hydrostatic loading case. A review of the state of the art in the design of ring-stiffened cylindrical shells under hydrostatic pressure was presented by Pulos in 1963. This report includes those formulations that have found extensive use. A later survey was made by Basdekas (1966) of the analytical and empirical procedures for the determination of dynamic as well as static strengths of cylindrical, spherical, and spheroidal shells. Meck (1965) also presents, in a concise form, the elastic and inelastic solutions for shell buckling and general instability.

Elastic General Instability. The elastic instability of ring-stiffened cylinders has been considered in the orthotropic shell theory (Timoshenko and Gere, 1961, p. 499; Ball, 1962; Becker, 1958; Bodner, 1957) and by a "split-rigidity" method (Bryant, 1954; Tokugawa, 1929) in which the formula for critical pressure consists of a shell term and a stiffener term. The most acceptable solution is that of Kendrick (1953a), whose theoretical predictions have been confirmed experimentally (Reynolds, 1958). However, these equations cannot be readily solved without the aid of a computer or graphical procedures (Ball, 1962; Reynolds, 1957; Reynolds and Blumenberg, 1959).

Although somewhat less exact than Kendrick's solution, the most widely used design equation is that of Bryant (1954) modified for the position of the stiffener on the inside or outside of the shell (Krenzke and Kierman, 1963).

$$p_c = \frac{2E}{D/t} \frac{\lambda^4}{(n^2 + (\lambda^2/2) - 1)(n^2 + \lambda^2)^2} + \frac{EI_e(n^2 - 1)}{L_f R_o R_c^2} \qquad (14.47)$$

where

$\lambda = \pi D/2L_b$
R_o = outside radius of shell
R_c = radius to centroidal axis of the combined stiffeners and shell of effective width L_e (see Fig. 14.1)
n = number of circumferential lobes existing at collapse
L_f = center-to-center spacing of stiffening rings
L_b = length of shell between bulkheads (or stiffening elements with sufficient stiffness to act as bulkheads) (see Fig. 14.1)
I_e = effective moment of inertia about the centroid of a section comprising one stiffener plus an effective width of shell, L_e

The effective width of a shell acting as part of the stiffener may be determined by

$$L_e = F_1 L + b \tag{14.48a}$$

where L is the unsupported length of the shell between stiffeners and

$$F_1 = \frac{2}{\theta} \frac{\cosh\theta - \cos\theta}{\sinh\theta - \sin\theta} \tag{14.48b}$$

where $\theta = 1.818\ L/D\sqrt{D/t}$ when $\mu = 0.3$ (Eq. 14.37) and b = contact width between stiffener and shell. The value L_e can be approximated by $1.1\sqrt{Dt} + b$ when $\theta > 2$, and $L + b$ when $\theta < 2$.

The first term in Eq. 14.47 is the shell term and is identical to the last term of Eq. 14.30. It is important only for shells with large D/t and low L/D ratios. The second term of Eq. 14.47 is the Levy formula for determining the critical uniform radial load on a circular ring [see discussion of Timoshenko and Gere (1961, pp. 287–302)] and in many cases is sufficiently accurate for design purposes.

The correct value of n is that which gives a minimum value of p in Eq. 14.47. The number of waves in a buckle pattern is determined by the restraint of adjoining stiffeners, heads, or diaphragms and the distance between them (spacing). When there are no effective restraints, $n = 2$. However, where heads, diaphragms, and/or large stiffeners are used to restrain the ends or are spaced along the length of the cylinder between which intermediate smaller rings are attached, n becomes greater but will be less than 10 for most shells of interest. The ASME Code (1980) does not recognize the restraint at the ends of a vessel and assumes that $n = 2$. Use of this approximation in design can lead to very conservative ring sizes.

Except for large values of θ where n equals 2 and the second term of Eq. 14.47 is dominant, the desired stiffener size for a particular pressure must be determined by trial procedures. However, the largest effective stiffener size achieved is when the critical general buckling pressure of Eq. 14.47 is equal to that from the appropriate equation in Table 14.2 for buckling of the shell between stiffeners.

The formula for general instability developed by Bryant (1954) is dependent on the distance between large rings (any diaphragm or bulkhead is equivalent to a large ring in preventing buckling). In practice, many designs involve small, uniform rings evenly spaced, and at greater intervals of spacing there may be incorporated intermediate heavy rings, again of uniform but different cross section. Although theoretical solutions for the critical pressure with this stiffening arrangement have been formulated (Kendrick, 1953a,b; Reynolds, 1957), an empirical equation (Blumenberg, 1965) that agrees quite well with test results is recommended:

$$p_c = \frac{(I_E - I_c)(p_F - p_B)}{I_{FE} - I_e} + p_B \tag{14.49}$$

where

$$I_{FE} = \frac{p_F L_f R_o R_d^2}{E(n^2 - 1)}$$

and n is the number of buckling lobes as determined by using p_B.

I_E = moment of inertia of large stiffener plus effective width of shell L_E
I_{FE} = value of I_E that makes the large stiffener fully effective, that is, equivalent to a bulkhead
$L_e = L_e$ of Eq. 14.48
$L_E = F_1 L(A_s/A_S) + b$
L_f = center-to-center spacing of large stiffeners
p_F = critical pressure determined by Eq. 14.47, where $L_b = L_f$
p_B = critical pressure determined by Eq. 14.47 assuming the large stiffeners are the same size as the small stiffeners
R_d = radius to centroidal axis of the large stiffener plus effective width of shell L
A_s = area of small stiffener plus total area of shell between small stiffeners
A_S = area of large stiffener plus total area of shell between small stiffeners
R_o, b, and I_c as in Eq. 14.47 and F_1 from Eq. 14.48b

Inelastic General Instability. Reynolds (1971) states that all the inelastic results can be summed up by one simple formula that came from the work of Bijlaard (1949). If the elastic buckling pressure of a structure can be expressed in the form

$$p_c = E(C_a + C_b) \tag{14.50}$$

in which C_a is a term that represents the membrane stiffness and C_b the bending stiffness of a stiffened cylinder, the inelastic buckling pressure can be written as

$$p_c = \sqrt{E_s E_t} C_a + E_t C_b \tag{14.51}$$

where E_s and E_t are the secant and tangent moduli, respectively, in the inelastic region. Therefore, $\sqrt{E_s E_t}$ would be used in place of E in the first term of Eq. 14.47, whereas E_t would replace E in the second term of Eq. 14.47.

Determination of the moduli to be used in Eq. 14.51 requires a knowledge of the stress field at the stiffener and in the shell midway between stiffeners. Since first yielding at the stiffeners is localized, it does not appreciably alter the elastic distribution of stresses between the stiffener region and the shell between stiffeners (Lunchick, 1959). Therefore, using $\mu = 0.3$ and assuming that the stiffener width is negligible compared to the spacing between stiffeners, the circumferential stresses are given by (Pulos and Salermo, 1961)

$$\sigma_\phi = -\frac{0.85 p R_o L_e}{A_f + L_e t} \quad \text{at the stiffener} \tag{14.52a}$$

$$\sigma_\phi = -\frac{pD}{2t}\left[1 - \frac{0.85 A_f}{A_f + L_e t} F(\theta)\right] \text{in the shell} \tag{14.52b}$$

where

$$F(\theta) = \begin{cases} 1 & \text{for } \theta \le 1 \\ 1.33 - 0.33\theta & \text{for } 1 < \theta < 4 \\ 0 & \text{for } \theta \ge 4 \end{cases}$$

and θ is defined in Eq. 14.37.

The longitudinal stresses at both locations are given by

$$\sigma_x = -\frac{pD}{4t} \tag{14.52c}$$

As in the case of inelastic buckling of unstiffened cylinders, the stress intensity of Eq. 14.39 based on the distortion energy yield criterion is calculated at the two locations and the moduli are determined from a representative stress-strain curve of the material. The ASME charts similar to Fig. 14.8*b* may also be used to determine a reduced modulus $E_r = 2B/A$.

More precise equations and charts for determining the circumferential stress when the stiffener width is not negligible are available from

Krenzke and Short (1959) and Pulos and Salerno (1961) to be used in place of Eqs. 14.52a and 14.52b. Also, a more exact formulation of a collapse pressure based on a three-hinge mechanism of failure of the shell between stiffeners is presented by Lunchick (1959).

Effect of Imperfections. In addition to the effects of shell imperfections discussed under unstiffened cylinders, the initial out-of-roundness of the stiffener rings is equally important in determining the reduction in the general instability strength of a stiffened shell under external pressure. Analytical studies of the effects of stiffener out-of-roundness were first considered by Kendrick (1953a,b). Hom (1962), assuming a more realistic out-of-roundness function, developed a method for determining the maximum bending stress introduced in the stiffener flange by stiffener eccentricity. Hom also developed an approximate formula for this bending stress.

$$\sigma_b = \pm \frac{16}{\pi}\left[\frac{Ee_f e}{D^2}(n^2-1)\frac{p}{p_c - p}\right] \tag{14.53}$$

where

n = number of circumferential buckling lobes (corresponding to n from Eqs. 14.47 and 14.49

e_f = distance from midthickness of the shell to the extreme fiber of stiffener, positive for internal stiffener and negative for external stiffener

p = applied pressure

p_c = critical pressure for perfect ring-stiffened cylinder (Eq. 14.47 or 14.49)

e = radial eccentricity from a true circle

Equation 14.53 is applicable only in the elastic range.

Another type of stress is introduced if the stiffener is initially tilted or twisted. In this case the radial forces do not pass through the centroid of the cross section of the stiffener, and a twisting moment results. When the radial stress in the web and the circumferential stress in the flange become excessive, yielding can result and cause crippling of the stiffeners, thereby precipitating collapse in the general instability mode. For outside stiffeners these radial forces induce a twist that tends to reduce the initial tilt, whereas for internal stiffeners this radial force tends to increase the tilt.

The total stress in the flange is the sum of the hoop stress σ_ϕ, the bending stress σ_b in the plane of the ring, and any tilt-induced stress. A method for calculating the latter stress is given by Wenk and Kennard (1956).

Torsional buckling of the stiffeners has also been considered by Wah (1967) and Farmer (1966). They make the overly conservative assumptions that the shell offers no resistance to twisting and that the torsional failure of the stiffener is independent of the in-plane stiffener buckling mode. A comparison of their predicted buckling pressures with test results (Blumenberg, 1965) indicates that their formula is ultraconservative. Experience indicates that if the stiffeners are symmetrical sections and are proportioned to satisfy the compact section requirements of the AISC specification, torsional buckling will not occur.

Inelastic Action. Because of the complexity of the effects that inelastic action has on the correct prediction of critical buckling pressures of imperfect cylinders, no definitive guidance is available to offer the designer at this time.

14.5. STRINGER- OR RING-AND-STRINGER-STIFFENED CYLINDER

Stringer (longitudinal) stiffeners are primarily used to increase the axial or bending load capacities of cylinders. They can be used either alone or in combination with ring stiffeners. Due to the greater number of parameters involved and the multiple potential mechanisms of failure, it is virtually impossible to achieve a universal set of design formulas substantiated by tests. Therefore, this section of the chapter is limited to a general discussion of the methods that can be used to evaluate the critical axial stress and critical pressure.

14.5.1 Nonpressure Loadings

Stiffeners may be placed on the outside of the shell (positive eccentricity), on the inside of the shell (negative eccentricity), or a combination of both may be employed, such as stringers on the outside and rings on the inside. However, Hutchinson and Amazigo (1967) point out that based on large-deflection theory, shells with external stiffening are more sensitive to imperfections. Quite to the contrary, Singer (1967) claims that stiffened shells are relatively insensitive to imperfections. The conflicting results are typical and illustrate why only a general discussion of the action of stiffened cylinders under axial loading is included.

To check the design of a stiffened cylindrical shell subjected to axial compression, the following forms of failure should be considered:

1. Overall column buckling or yielding.
2. Local buckling encompassing several stiffeners.
3. Local buckling between stiffeners (panel buckling).
4. Buckling of individual stiffeners.
5. Local yielding of the shell or the stiffeners.

The possibility of overall column buckling can be studied by including the stringers in the calculation of the radius of gyration of the column cross section. Circumferential stiffeners have no direct effect on the overall column-buckling mode.

In many cases, close stringer spacing allows their being treated as if they were uniformly distributed on the shell's circumference. Under this assumption, the stiffened shell can be modeled as an equivalent orthotropic shell. This approach is often used to investigate the possibility of local buckling encompassing several stiffeners. If closely spaced, ring stiffeners can similarly be considered as being uniformly distributed along the axial direction; however, their spacing is normally such that they should be considred discretely spaced.

Spacing of longitudinal stiffeners (stringers) must be close enough to prevent panel buckling between the stiffeners. If both the stringers and the rings are sufficiently rigid, the cylinder can be treated as a series of curved panels each of which is supported along four edges. The buckling behavior of these curved panels is very similar to that of an entire cylinder. If the panel is short and its curvature is small, the panel buckles essentially as if it were a flat plate (as treated in Chapter 4) and the critical stress is given by Eq. 14.54,

$$\sigma_{xc} = \frac{k\pi^2 E}{12(1-\mu^2)(L_s/t)^2} \tag{14.54a}$$

where for $L/L_s < 1$,

$$K = \left(\frac{L_s}{L} + \frac{L}{L_s}\right)^2 \tag{14.54b}$$

and for $L/L_s \geq 1$,

$$K = 4.0 \tag{14.54c}$$

In the above, L is the axial distance between circumferential stiffeners and L_s is the circumferential distance between longitudinal stringers (Fig. 14.1).

Long panels of sufficient curvature behave in the same manner as do moderately long cylinders. To obtain a rough estimate of the elastic buckling strength of moderately curved panels, Timoshenko and Gere (1961) suggest that Eq. 14.4 be used. When the axial and circumferential dimensions of the panel are about equal and the central angle subtended by the panel is less than one-half radian, C may be taken equal to 0.6.

Buckling of individual stiffeners can be investigated by treating the stiffener and an effective width of the shell as a column. For example, the critical load of a typical stringer and the effective shell skin is obtained by assuming the stringer to behave like a column on an elastic foundation.

The circumferential rings act as the foundation, and depending on their spacing and area, the foundation is considered to be continuous or made up of elastic or rigid point supports.

Local yielding of the shell or stiffeners is perhaps more of a stress-analysis problem than a stability problem, but it must be given consideration. The designer must keep one point in mind when adding stiffeners to overcome a shell-stability problem. It is possible that the added stiffener is so rigid compared to the shell (that it is supposed to reinforce) that the stiffener becomes the main load-carrying component of the assemblage. For a further discussion of stiffened-plate behavior, refer to Chapter 4.

14.5.2 Uniform External Pressure

Stringer stiffeners are not generally used to increase the critical external pressure of a cylinder. Ring stiffeners are preferred for pressure loading, but in cases when the cylinder must withstand several loading conditions, stringer stiffeners may also be present. If stringers are spaced more closely than the buckling wave length of the shell, they will increase the critical pressure. Test programs are currently under way to study the influence of size and spacing of stringer stiffeners on the critical pressure, but little definitive information is presently available.

14.6 EFFECTS ON COLUMN BUCKLING

The basic column curves for tubular member subjected to axial compression are discussed in Chapter 3. The designer should, however, be aware of situations when these require some modification or interpretation to ensure that the proper critical axial stress is predicted.

14.6.1 Interaction Between Column and Local Buckling

For many practical applications, when the D/t ratio of a member is such that local buckling is probable, the KL/r ratio of the member may be such that column buckling is also an important consideration. One way to establish the allowable load for such a member is to use an allowable load based on the smaller of the loads derived from local buckling and overall column behavior.

A different and recommended approach (Marshall, 1971) is also used when designing light-gage cold-formed steel structural members for determining allowable loads for sections whose geometry and material properties are such that both local buckling and column buckling are important considerations. This approach is best explained with an example. Referring to Fig. 14.10, assume that a column is to be designed according to curve A; however, assume that theoretically, local-buckling limits the useful column loading to 30 ksi. In this approach, the local-buckling

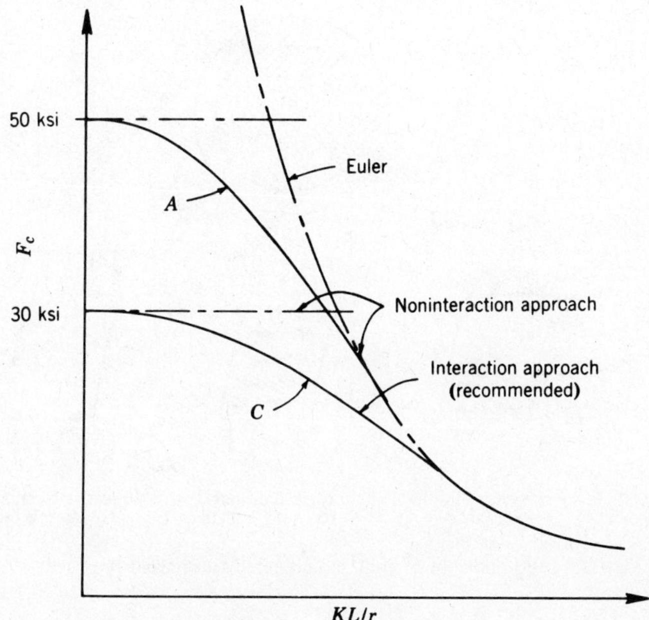

Fig. 14.10 Interaction of local buckling and column buckling.

strength is substituted for σ_y in the appropriate column formula to yield the curve C. A similar approach is taken by AISC in Appendix C of its Specification. For a more complete discussion of both possible methods of analysis, refer to Chapter 4 for a simple treatment with respect to flat plates.

14.6.2 Effect of Fluid Pressure on Overall Column Buckling

The influence of pressure on the overall stability of tubular members is normally not a structural consideration; however, in the oil industry, tubular members transmitting fluids from deep oil and gas wells have experienced failures of this type (Lubinski, 1951; Lubinski et al, 1962). In this type of problem, the cylinder is not closed ended and the pressure does not create longitudinal stresses.

The fact that pressures can influence column stability was recognized by Prescott (1946) and was discussed more recently in publications with theoretical derivations (Flügge, 1973; Seide, 1960) and has been experimently confirmed (Palmer and Baldry, 1974).

A free-body diagram of an element of an open-ended cylindrical member subjected to internal pressure is shown in Fig. 14.11. It can be

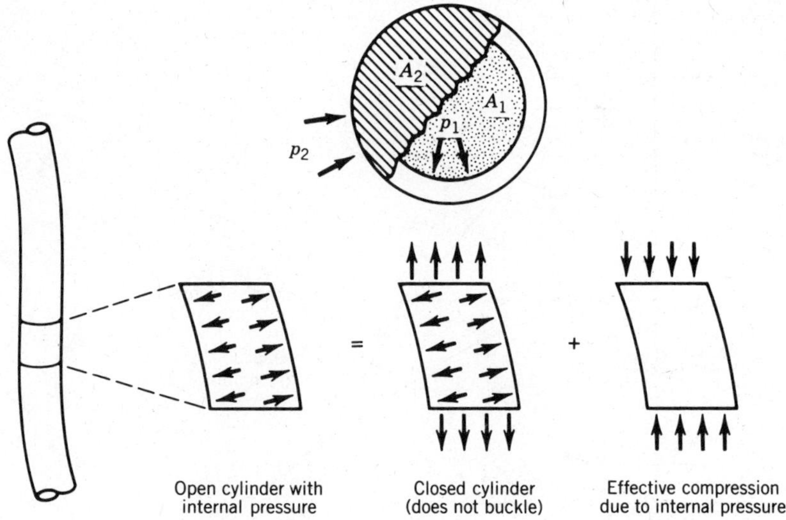

Fig. 14.11 Equilibrium of column subjected to internal pressure.

seen that an effective axial compressive force on the element results from the internal pressure. This force can influence the stability of the column by adding to or subtracting from any preexisting axial force.

Figure 14.12 shows an example where internal pressurization alone can cause a cylindrical tube to buckle. In this illustration, two frictionless pistons at the tube ends contain the fluid within the tube. The pressure force on the pistons is restrained by the cable connecting the two pistons. With no external pressure or applied load, the effective buckling force is p_1A_1. In this illustration the end conditions are essentially pinned and the elastic buckling pressure will be given by the corresponding Euler formula:

$$p_1 = \frac{\pi^2 EI}{A_1 L^2} \tag{14.55}$$

The buckled shape will be sinusoidal until, with continued fluid injection, the inner wall contacts the cable. At that point the buckling will stop in the plane shown and additional pressurization will cause a deformation out of plane that leads to a helical shape for the buckled tube.

In contrast, a closed-ended cylinder subjected to internal pressure will not suffer general instability as the compressive force will be balanced by the tensile force from the end closures. The resulting effective force is zero.

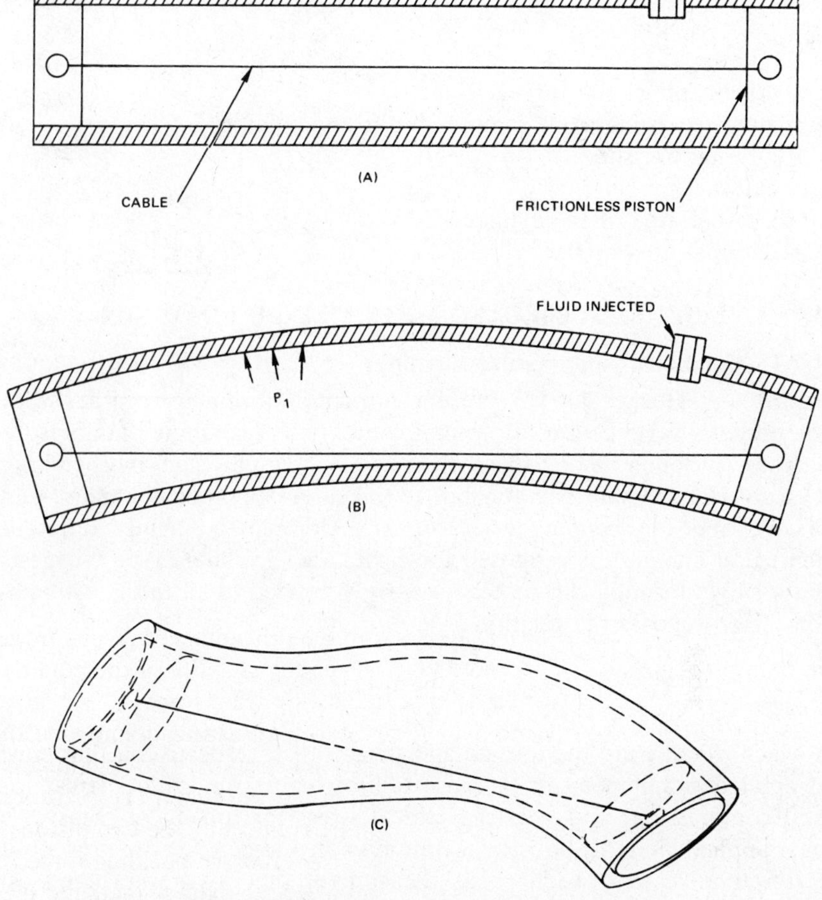

Fig. 14.12 Buckling under internal pressure.

In a membrane shell where the walls of the shell are incapable of maintaining compression, internal pressurization will allow the pressurized membrane to support a column load by virtue of a prestressing effect in the membrane. A good example is a cylindrical rubber balloon, which has column strength only if inflated.

The axial tension force p_2A_2 and other effects resulting from external pressure acting alone can be deducted from the free-body diagram of Fig. 14.11. Hence in the presence of external and internal pressures, the overall stability of the column will depend on an effective compressive force given by

$$P_e = P + p_1 A_1 - p_2 A_2 \tag{14.56}$$

where

P_e = effective-buckling force
P = any preexisting axial compression in the column
p_1 = internal pressure
p_2 = external pressure
A_1 = internal area of the cylinder
A_2 = external area of the cylinder

14.7 CYLINDERS SUBJECTED TO COMBINED LOADINGS

14.7.1 Combined Nonpressure Loadings

Gerard and Becker (1957) present intraction formulas together with experimental vertification for several conditions of combined nonpresure loadings. Included among these are: (1) axial compression and bending; (2) axial compression and torsion; (3) axial compression, bending, and torsion; and (4) bending and torsion. Although a slightly different interaction equation is proposed for each case, Schilling (1965) suggests that a single formula can be conservatively applied to all these combinations. He proposes the relation

$$\frac{\sigma}{\sigma_{xc}} + \left(\frac{\tau}{\tau_c}\right)^2 = 1 \tag{14.57}$$

in which σ and τ are the normal and shear stress, respectively, that must be applied simultaneously to cause failure; σ_{xc} is the critical stress for axial compression; and τ_c is the critical stress for torsion or transverse shear applied alone. If both axial compression and bending are present, it is conservative to consider σ as the sum of the two normal stresses. Schilling implies that this procedure is sufficiently accurate for use in both the elastic and inelastic stress ranges.

14.7.2 Axial Stress in Combination with Internal or External Pressure

Since radial stresses may be neglected in structural cylinders of usual proportions (D/t exceeding about 12), this topic is concerned with the interaction of hoop stresses caused by pressure and axial stresses resulting from a combination of axial load, bending, and hydrostatic pressure. In the general case, shear stresses may also be present. The overall problem is illustrated in Fig. 14.13, where the general yield failure of thick-walled cylinders based on the maximum strain energy theory is shown. Ellipses have been plotted for Poisson's ratio of 0.3 and 0.5.

The applicability of this criterion to the general yielding situation has

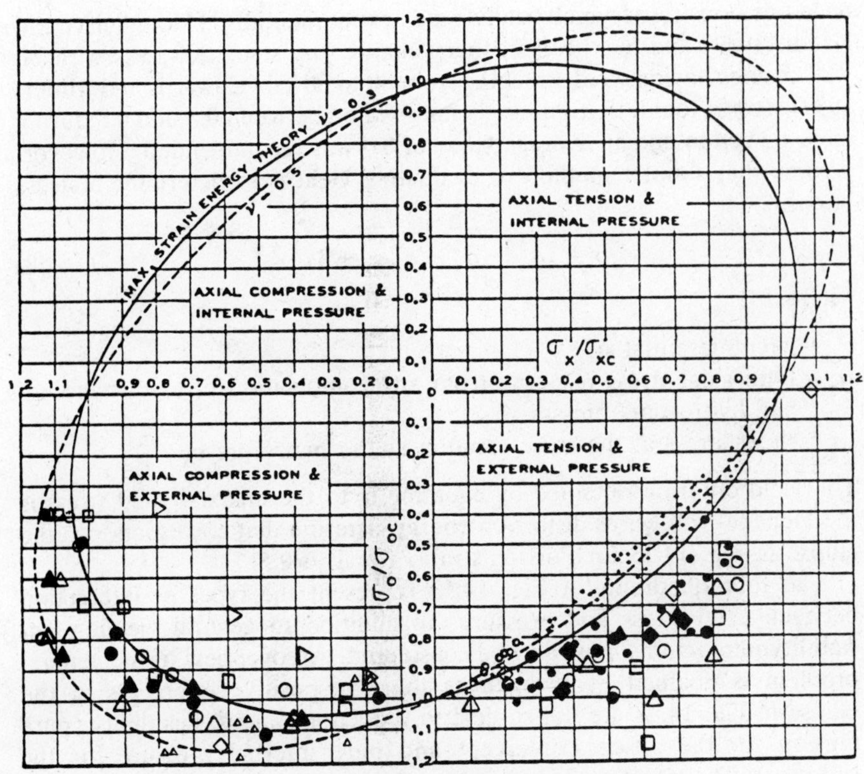

NOTE: σ_{xc} and $\sigma_{\phi c}$ are yield stresses or collapse stresses from uniaxial stress tests in the Group with corresponding D/t and σ_y

Armco, 1966 (Data from Edwards, 1930 and Holmquest, 1938)

+ stretch failures

○ D/t = 16.5

● D/t = 21.7

3rd Ed. SSRC "Guide"

△

Miller, 1982a,b: Kizilug, 1985; Vojta, 1983
(includes unstiffened and ring stiffened cylinders)

◇ ◆ D/t = 33

○ ● D/t = 47

△ ▲ D/t = 62   solid symbols are $\sigma_y \geq 50$ ksi

□ D/t = 96

▷ D/t = 1000

Fig. 14.13 Normalized ellipse of biaxial yield stresses and data for combined pressure and axial load.

been demonstrated analytically and experimentally (Holmquist and Nadai, 1939) and has been further validated by some 200 tests with oil well casings having yield strengths from 30 to 80 ksi (Edwards and Miller, 1939). It was found that Poisson's ratio varied between 0.3 and 0.5 during inelastic action and an average value of 0.4 was recommended. However, a somewhat simpler method of evaluating yield is to determine a stress intensity

$$\sigma_i = (\sigma_\phi^2 + \sigma_x^2 - \sigma_\phi \sigma_x + 3\tau^2)^{0.5} \tag{14.58}$$

where

σ_ϕ = circumferential stress
σ_x = total axial stress resulting from any combination of load, bending, and hydrostatic pressure
τ = shear stress at the same location as the maximum σ_x

The yield criterion obtained by equating σ_i to the uniaxial yield stress is the Hencky–von Mises distortion energy criterion that corresponds to the ellipse in Fig. 14.13 for Poisson's ratio equal to 0.5.

The four quadrants of Fig. 14.13 represent the types of interaction between axial stress and pressure. In addition to general yielding, instability interactions must also be considered. An overview of the general problem is obained by recognizing that the positive intercepts of the ellipse in Fig. 14.13 are not subject to reductions due to instability (pure tension for the abscissa intercept and pure internal pressure for the ordinate). The negative intercepts, however, will not be achieved in thin cylinders due to buckling under axial compression or external pressure. The instability interaction in the four regions is discussed in the foilowing.

Tension in Combination with Internal Pressure. Since there is no instability mode in this region, only the yield criterion need be considered. The general criterion of Eq. 14.58 may be used, or it would be reasonable and only slightly conservative to neglect the interaction and consider the effects separately.

Tension in Combination with External Pressure. This combination of loading can lead to a phenomenon referred to as a propagating buckle. An indentation develops at a point around the circumference and spreads along the length of the cylinder.

Since the tension component does not lead directly to a stability failure, the results of the interaction are usually expressed in terms of a reduced critical external pressure stress, $\sigma'_{\phi c}$. Two approaches are used to obtain $\sigma'_{\phi c}$, one being a standard interaction equation and the other a reduced yield strength in the critical pressure (or stress) equations.

The interaction approach assumes an elliptic relation similar to the form for general yielding. However, the normalization of the hoop stress is based on the critical stress for pressure only rather than on yield. This results in

$$\left(\frac{\sigma_x}{\sigma_y}\right)^2 + \left(\frac{\sigma'_{\phi c}}{\sigma_{\phi c}}\right)^2 + \frac{\sigma_x}{\sigma_y}\frac{\sigma'_{\phi c}}{\sigma_{\phi c}} = 1 \qquad (14.59)$$

where

σ_x = imposed axial stress
$\sigma'_{\phi c}$ = reduced critical hoop stress = $p_c'D/2t$
σ_y = yield stress
$\sigma_{\phi c}$ = critical hoop stress when no axial loading is imposed

This normalization has been used to plot the data in Fig. 14.13, where it can be observed that it is conservative for the collapse failures.

The API (1982) procedure uses an effective hoop yield stress σ_{yr}, obtained by solving

$$\left(\frac{\sigma_{yr}}{\sigma_y}\right)^2 + \frac{\sigma_{yr}}{\sigma_y}\frac{\sigma_x}{\sigma_y} + \left(\frac{\sigma_x}{\sigma_y}\right)^2 = 1 \qquad (14.60)$$

The critical pressure can be obtained as the value of p_y in Eq. 14.44 using σ_{yr} in place of σ_y. It is also possible to use in empirical design curves that include inelastic and out-of-roundness effects in place of Eq. 14.44.

A variation on the reduced-yield approach is to base the inelastic properties on the stress intensity:

$$\sigma_i = [(\sigma'_{\phi c})^2 + \sigma_x^2 - \sigma_x\sigma'_{\phi c}]^{0.5} \qquad (14.61)$$

Again the collapse pressure is based on Eq. 14.44 and the p_c term is based on the equations of Table 14.2 using a reduced modulus of elasticity.

The approaches used with Eqs. 14.60 and 14.61 are not as conservative as the interaction equation (Eq. 14.59). It is also important to note that all of the methods just discussed are for ultimate conditions. Actual design is more complicated in that different factors of safety are desirable for tension yielding and pressure collapse. Iterative solutions are required to obtain optimal designs from given tension and hydrostatic design loads.

Compression in Combination with External Pressure. For very stocky cylinders for which failure occurs by general yielding, Fig. 14.13 indicates that it would be conservative and reasonable to ignore the interaction between axial compression and external pressure. However, for cylinders the larger D/t ratios, for which both external pressure and axial loading may result in a collapse failure mode, the two stresses acting on a preexisting imperfection often have an additive effect. The construction shown in Fig. 14.14 has been suggested (Johns et al., 1975, 1976). This construction is conservative when compared to available bending and

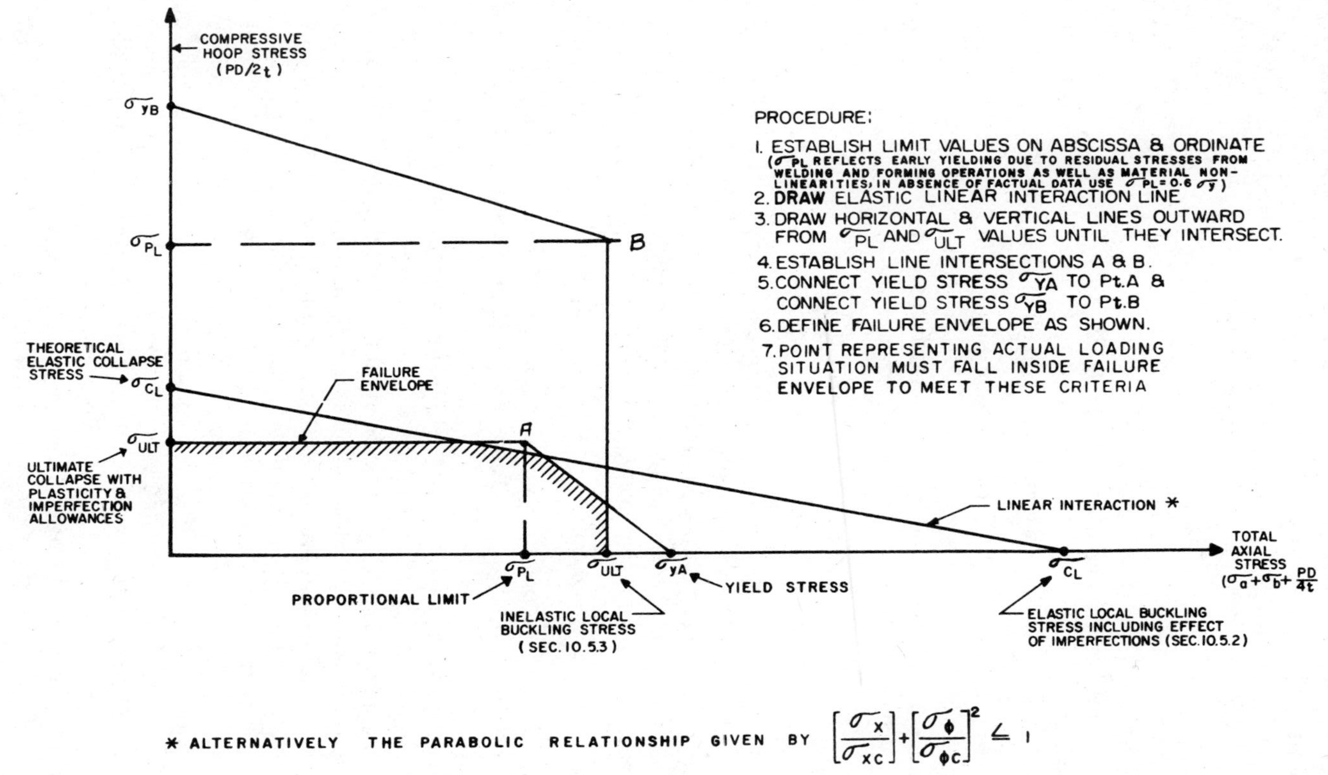

Fig. 14.14 Construction of composite failure envelope for interaction of combined compressive loadings.

collapse test data (Johns et al., 1975, 1976) and with axial compression and external pressure tests (Miller and Vojta, 1984).

Compression in Combination with Internal Pressure. The exact design of cylinders subjected to a combination of axial compression and internal pressure is a very complicated procedure. The following offers an approximate procedure that should be sufficiently accurate for most design purposes.

Many tests have been conducted (Baker et al., 1968; Fung and Sechler, 1957; Gerard, 1957; Lo et al., 1951; Mungan, 1974; Seide, 1960) which show that internal pressure increases the axial-buckling stress as long as the failure stress remains elastic. A graphical method has been

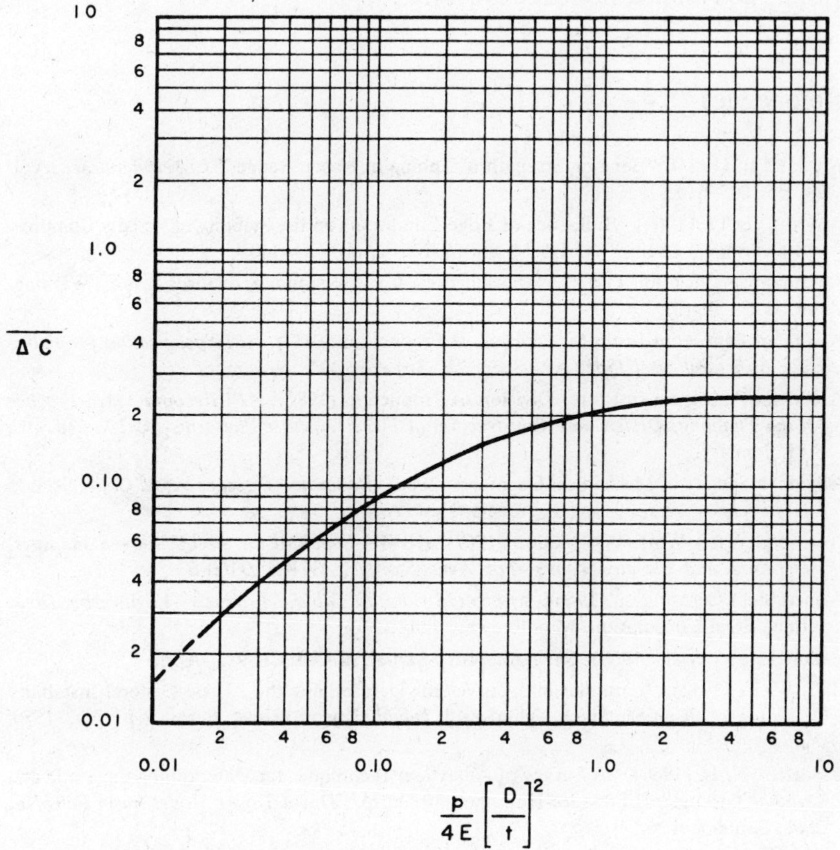

Fig. 14.15 Increase in axial compressive elastic buckling stress due to internal pressure (Baker et al., 1968).

developed by Baker for determining this increase. The following equation, when used in conjunction with Fig. 14.15 (taken from equation, when used in conjunction with Fig. 14.15 (taken from Baker, 1951), can be used for cylinders when the effective stress σ_i (from Eq. 14.61) is less than the proportional limit.

$$\sigma_x < \sigma_{xc} + 2\,\overline{\Delta C}\,\frac{Et}{D} \tag{14.62}$$

with σ_{xc} determined from Section 14.3.1 and $\overline{\Delta C}$ from Fig. 14.15. When σ_i is greater than the proportional limit, the second term is no longer valid and therefore

$$\sigma_x \leq \sigma_{xc} \tag{14.63}$$

with σ_x determined from Section 14.3.1.

References

Ades, C. F. (1957), "Bending Strength of Tubing in Plastic Range," *J. Aeronaut. Sci.*, Vol. 24, pp. 605–610.

Almroth, B. O. (1966), "Influence of Edge Conditions on the Stability of Axially Compressed Cylindrical Shells," *AIAA J.*, Vol. 4, No. 1, pp. 134–140.

Aluminum Association (1982), "Specifications for Aluminum Structures," AA, Washington, D.C., April.

American Petroleum Institute, Division of Production (1977), *API Specification for Fabricated Structural Steel Pipe*, API Spec. 2B, 3rd ed., API, November.

American Petroleum Institute, Division of Production (1982), *API Recommended Practice for the Planning, Design, and Construction of Fixed Offshore Platforms*, RP2A, 13th ed., API, January.

American Society of Mechanical Engineers (1980), *Boiler and Pressure Vessel Code*, Section VIII, *Pressure Vessels*, Div. 1, ASME, New York.

American Water Works Association (1967), *AWWA Standard for Steel Tanks—Standpipes, Reservoirs, and Elevated Tanks—For Water Storage*, *AWWA D*100-67.

Armco Steel Corporation (1966), *Armco Oil Country Tubular Products—Engineering Data*, Armco Steel Corporation, Middletown, Ohio,.

Baker, E. H., et al. (1968), *Shell Analysis Manual*, NASA CR-912, April.

Ball, W. E. (1962), "Formulas and Curves for Determining the Elastic General-Instability Pressures of Ring-Stiffened Cylinders," *David Taylor Model Basin Rep. No. 1570*, January.

Basdekas, N. L. (1966), "A Survey of Analytical Techniques for Determining the Static and Dynamic Strength of Pressure-Hull Shell Structure," *David Taylor Model Basin Rep. No. 2208*, September.

Batdorf, S. B. (1947), "A Simplified Method of Elastic Stability Analysis for Thin Cylindrical Shells," *NACA Rep. No. 874*.

Batdorf, S. B., Schildcrout, M., and Stein, M. (1947a), "Critical Stress of Thin-Walled Cylinders in Axial Compression," *NACA Tech. Note No. 1343.*

Batdorf, S. B., Schildcrout, M., and Stein, M. (1947b), "Critical Stress of Thin-Walled Cylinders in Torsion," *NACA Tech. Note No. 1344*, July.

Becker, H. (1958), "General Instability of Stiffened Cylinders," *NACA Tech. Note No. 4237*, July.

Bijlaard, P. P. (1949), "Theory and Tests on the Plastic Stability of Plates and Shells," *J. Aeronaut. Sci.*, Vol. 16, No. 9.

Birkemoe, P. C., Prion, H. G. L., and Sato, J. A. (1983), "Compression Behavior of Unstiffened Fabricated Steel Tubes," *ASCE Ann. Convention Struct. Cong.*, October.

Blumenberg, W. F. (1965), "The Effect of Intermediate Heavy Frames on the Elastic General-Instability Strength of Ring-Stiffened Cylinders under External Hydrostatic Pressure," *David Taylor Model Basin Rep. No. 1844*, February.

Bodner, S. R. (1957), "General Instability of Ring-Stiffened Circular Cylindrical Shell under Hydrostatic Pressure," *J. Appl. Mech.*, Vol. 24, No. 2, pp. 269–277.

Bodner, S. R., and Berks, W. (1952), "The Effect of Imperfections on the Stesses in a Circular Cylindrical Shell under Hydrostatic Pressure," *PIRAL Rep. No. 210*, Polytechnic Institute of Brooklyn, December.

Boichot, L., and Reynolds, T. E. (1964), "Inelastic Buckling Tests of Ring-Stiffened Cylinders under Hydrostatic Pressure," *David Taylor Model Basin, Rep. No. 1992*, May.

Brazier, L. G. (1927), "On the Flexure of Thin Cylindrical Shells and Other Thin Sections," *Proc. R. Soc. London*, Vol. A116, pp. 104–114.

Bryant, A. R. (1954), "Hydrostatic Pressure Buckling of a Ring-Stiffened Tube," *Rep. No. R-306.* Naval Construction Research Establishment.

Chen, W. F., and Ross, D. A. (1978), "Tests of Fabricated Tubular Columns," *Rep. No. 393.8*, Lehigh University, Fritz Eng. Lab., Bethlehem, Pa., September.

Clark, J. W., and Rolf, R. L. (1964). "Design of Aluminum Tubular Members," *ASCE J. Struct. Div.*, Vol. 90, No. ST6, p. 259.

DeHeart, R. C., and Basdekas, N. L. (1960), "Yield Collapse of Stiffened Circular Cylindrical Shells," *Report under NOMR Contract NR 2650(00), Project NR 064-435*, Southwest Research Institute, September.

Denton, A. A., and Alexander, J. M. (1963a), "On the Determination of Residual Stresses in Tubes," *J. Mech. Eng. Sci.*, Vol. 5, No. 1, pp. 75–88.

Denton, A. A., and Alexander, J. M. (1963b), "New Method for Measurement of Local Residual Stresses in Tubes," *J. Mech. Eng. Sci.*, Vol. 5, No. 1, pp. 89–90.

Donnell, L. H. (1934), "A New Theory for the Buckling of Thin Cylinders under Axial Compression and Bending," *Trans. ASME*, Vol. 56.

Donnell, L. H. (1956), "Effect of Imperfections on Buckling of Thin Cylinders under External Pressure," *J. Appl. Mech.*, Vol. 23, No. 4, p. 569.

Donnell, L. H., and Wan, C. C. (1950), "Effect of Imperfections on Buckling of Thin Cylinders and Columns under Axial Compression," *J. Appl. Mech.*, Vol. 17, No. 1.

Eder, M. F., Grove, R. B., Peters, S. W., and Miller, C. D. (1984), "Collapse Tests of Fabricated Cylinders under Combined Axial Compression and External Pressure," *Final Rep., American Petroleum Institute Project 83-46*, CBI Industries, Inc., Research Laboratory, Plainfield, Ill., February.

Edwards, S. H., and Miller, C. P. (1939), "Discussion on the Effect of Combined

Longitudinal Loading and External Pressure on the Strength of Oil-Well Casing," *Drilling and Production Practice*, American Petroleum Institute, pp. 483–502.

Farmer, L. E. (1966), "Method for Determining Buckling Loads of Circular Stiffening Rings Including Consideration of Torsional Displacement," *AEDC-TR-66-188*, Engineering Support Facility, September.

Flügge, W. (1932), "Die Stabilität der Kreiszylinderschale," *Ing. Arch.*, Vol. 3.

Flügge, W. (1973), *Stresses in Shells*, 2ed., Springer-Verlag, New York, pp. 459–463.

Fung, Y. C., and Sechler, E. E. (1957), "Buckling of Thin Walled Circular Cylinders under Axial Compression and Internal Pressure," *J. Aeronaut. Sci.*, Vol. 24, pp. 351–356.

Galletly, G. D., and Bart, R. (1957), "Effect of Boundary Conditions and Initial Out-of-Roundness on the Strength of Thin-Walled Cylinders Subjected to External Hydrostatic Pressure," *David Taylor Model Basin Rep. No. 1066*, November.

Gellin, S. (1980), "The Plastic Buckling of Long Cylindrical Shells under Pure Bending," *Int. J. Solids Struct.*, Vol. 16, pp. 397–407.

Gerard, G. (1956), "Compressive and Torsional Buckling of Thin-Walled Cylinders in Yield Region," *NACA Tech. Note No. 3726.*

Gerard, G. (1957), "Plastic Stability Theory of Thin Shells," *J. Aeronaut. Sci.*, Vol. 24, No. 4, p. 269.

Gerard, G., and Becker, H. (1957), "Handbook of Structural Stability: Part 3. Buckling of Curved Plates and Shells," *NACA Tech. Note No. 3783*, August.

Graham, R. R. (1965), "Manufacture and Use of Structural Tubing," *J. Met.*, Vol. 17.

Harari, A., and Baron, M. L. (1970/1971), "Buckling of Vessels Composed of Combinations of Cylindrical and Spherical Shells," *J. Appl. Mech.*, July 1970, pp. 393–398; July 1971, pp. 571–572.

Harris, L. A., et al. (1957), "The Stability of Thin-Walled Unstiffened Circular Cylinders under Axial Compression Including the Effects of Internal Pressure," *J. Aeronaut. Sci.*, Vol. 24, No. 8, pp. 587–596.

Heise, O., and Esztergar, E. P. (1970), "Elestoplastic Collapse of Tubes under External Pressure," *J. Eng. Ind.*, Vol. 92, pp. 735–742.

Hoff, N. J. (1966), "The Perplexing Behavior of Thin Cylindrical Shells in Axial Compression," *Isr. J. Technol.*, Vol. 4, No. 1.

Holmquist, J. L., and Nadai, A. (1939), "A Theoretical and Experimental Approach to the Problem of Collapse of Deep-Well Casing," *Drilling and Production Practice*, American Petroleum Institute, pp. 392–420.

Holt, M. (1952), "A Procedure for Determining the Allowable Out-of-Roundness for Vessels under External Pressure," *ASME Trans.*, Vol. 74, p. 1225.

Hom, K. (1962), "Elastic Stess in Ring-Frames of Imperfectly Circular Cylindrical Shells under External Pressure Loading," *David Taylor Model Basin Rep. No. 1505*, May.

Hom, K., and Couch, W. P. (1961), "Hydrostatic Tests of Inelastic and Elastic Stability of Ring-Stiffened Cylindrical Shells Machined from Strain-Hardening Material,"*David Taylor Model Basin Rep. No. 1501*, December.

Hutchinson, J. W., and Amazigo, J. C. (1967), "Imperfection-Sensitivity of Eccentrically Stiffened Cylindrical Shells," *AIAA J.*, Vol. 5, No. 3, pp. 392–401.

Jirsa, J. O., et al. (1972), "Ovaling of Pipelines under Pure Bending," Pap. 1569, *4th Annu. Offshore Technol. Conf.*, Houston, May.

Johns, T. G., Mesloh, R. E., Winegardner, R., and Sorenson, J. E. (1975), "Inelastic Buckling of Cylinders under Combined Loads," Pap. 2209, *7th Annu. Offshore Technol. Conf.*, Houston, May.

Johns, T. G., Sorenson, J. E., Mesloh, R. E., and Attenbury, T. J. (1976), "Buckling Strength of Offshore Platforms Program: Phase I," Battelle Institute.

Kendrick, S. (1953a), "The Deformation under External Pressure of Circular Cylindrical Shells with Evenly Spaced Equal Strength Nearly Circular Ring Frames," *Rep. No. R-259*, Naval Construction Research Establishment, October.

Kendrick, S. (1953b), "The Buckling under External Pressure of Circular Cylindrical Shells with Evenly Spaced Equal Strength Circular Ring Frames; Part III," *Rep. No. R-259*, Naval Construction Research Establishment, October.

Kiziltug, A. Y., Grove, R. B., Peters, S. W., and Miller, C. D. (1985), "Collapse Tests of Short Tubular Columns Subjected to Combined Loads," *Final Rep.*, *Contract No. UI-851604*, CBI Industries, December.

Korol, R. M. (1978), "Inelastic Buckling of Circular Tubes in Bending," *ASCE J. Eng. Mech. Div.*, Vol. 104, No. EM4, Pap. 13944, pp. 939–952.

Krenzke, M. A., and Kiernan, T. J. (1963), "Structural Development of a Titanium Oceanographic Vehicle for Operating Depths of 15,000 to 20,000 feet," *David Taylor Model Basin Rep. No. 1677*, September.

Krenzke, M. A., and Short, R. D. (1959), "Graphical Method for Determining Maximum Stesses in Ring-Stiffened Cylinder under External Hydrostatic Pressure," *David Taylor Model Basin Rep. No. 1348*, October.

Kuper, E. J., and Macadam, J. N. (1969), "Quantitative Effects of Strain Hardening on the Mechanical Properties of Cold-Formed Electric Welded Steel Tubing," *Proc. 10th Mech. Working Steel Process. Conf.*, AIME.

Lo, H., Crate, H., and Schwartz, E. B. (1951), "Buckling of Thin Walled Cylinders under Axial Compression and Internal Pressure," *NASA Rep. No. 1027* (formerly 2021).

Lorenz, R. (1908), "Achsensymmetrische Verzerrungen im dünnwandigen Hohlzylinder," *Z. Ver. Dsch. Ing.*, Vol. 52, No. 43.

Lubinski, A. (1951), "Buckling of Rotary Drilling Strings: Part III," World Oil, May, pp. 122-132.

Lubinski, A., Althouse, W. S., and Logan, J. L. (1962), "Helical Buckling of Tubing Sealed in Packers," *J. Pet. Technol.*, June, pp. 655–670.

Lunchick, M. E. (1959), "Yield Failure of Stiffened Cylinders under Hydrostatic Pressure," *David Taylor Model Basin Rep. No. 1291*, January.

Lunchick, M. E. (1961a), "Plastic Axisymmetric Buckling of Ring-Stiffened Cylindrical Shells Fabricated from Strain-Hardening Materials and Subjected to External Hydrostatic Pressure," *David Taylor Model Basic Rep. No. 1393*, January.

Lunchick, M. E. (1961b), "Graphical Methods for Determining the Plastic Shell-Buckling Pressures of Ring-Stiffened Cylinders Subjected to External Hydrostatic Pressure," *David Taylor Model Basin Rep. No. 1437*, March.

Lunchick, M. E. (1963), "Plastic General-Instability Pressure of Ring-Stiffened Cylindrical Shells," *David Taylor Model Basin Rep. No. 1587*, September.

Lunchick, M. E., and Short, R. D. (1957), "Behavior of Cylinders with Initial Shell Deflection," *David Taylor Model Basin Rep. No. 1130*, July.

Lundquist, E. E. (1933), "Strength Tests of Thin-Walled Duraluminum Cylinders in Compression," *NACA Rep. No. 473.*

Marshall, P. W. (1971), "Design Criteria for Structural Steel Pipe," *Column Res. Counc. Proc.*

Marzullo, M. A., and Ostapenko, A. (1978), "Tests on Two High-Strength Steel Large-Diameter Tubular Columns," Pap. 3086, *Proc. 10th Offshore Technol. Conf.*, Houston, May.

Meck, H. R. (1965), "A Survey of Methods of Stability Analysis of Ring-Stiffened Cylinders under Hydrostatic Pressure," *J. Eng. Ind.*, August, p. 385.

Miller, C. D. (1982), "Summary of Buckling Tests on Fabricated Steel Cylindrical Shells in USA," in *Buckling of Shells in Offshore Structures* (ed. Harding, Dowling, and Angelidis), Granada, London, pp. 429–472.

Miller, C. D. (1983), "Research Related to Buckling of Nuclear Containment," *7th Int. Conf. Struct. Mech. Reactor Technol.*, Chicago, August.

Miller, C. D. (1984), "API Bulletin on Stability Design of Shells" (draft), *Am. Pet. Inst. Bull.*, Vol. 2u, September.

Miller, C. D., and Vojta, J. F. (1984), "Strength of Stiffened Cylinders Subjected to Combinations of Axial Compression and External Pressure," *Proc. SSRC Annu. Tech. Sess.*

Miller, C. D., Kinra, R. K., and Marlow, R. S. (1982), "Tension and Collapse Tests of Fabricated Steel Cylinders," Pap. 4218, *Proc. 14th Offshore Technol. Conference*, Houston, May.

Miller, C. D., Grove, R. B., and Vojta, J. F. (1983), "Design of Stiffened Cylinders for Offshore Structures," *AWS Weld. Offshore Struct. Conf.*, New Orleans, December.

Mungan, I. (1974), "Buckling Stress States of Cylindrical Shells," *ASCE Struct. Div.*, Vol. 100, No. ST11, pp. 2289–2306.

Nash, W. A. (1957), "Buckling of Initially Imperfect Shells Subject to Torsion," *J. Appl. Mech.*, Vol. 24, No. 1.

Ostapenko, A., and Grimm, D. F. (1980), "Local Buckling of Cylindrical Tubular Columns Made of A-36 Steel," *Fritz Eng. Lab. Rep. No. 450.7*, Lehigh University, Bethlehem, Pa., February.

Ostapenko, A., and Gunzelman, S. X. (1976), "Local Buckling of Tubular Steel Columns, *Proc. Methods Struct. Analysis*," Vol. 2, American Society of Civil Engineers, New York, p. 549.

Ostapenko, A., and Gunzelman, S. X. (1978), "Local Buckling Tests on Three Steel Large-Diameter Tubular Columns," *Proc. 4th Int. Conf. Cold-Formed Steel Struct.*, St. Louis, p. 409, June.

Palmer, A. C., and Baldry, J. A. S. (1974), "Lateral Buckling of Axially Constrained Pipelines," *J. Pet. Technol.*, November, pp. 1283–1284.

Plantema, F. J. (1946), "Collapsing Stress of Circular Cylinders and Round Tubes," *Rep. No. S.280*, Nat. Luchtfaart Laboratorium, Amsterdam.

Prescott, J. (1946), *Applied Elasticity*, 1st American ed., Dover, New York, pp. 552–554.

Pulos, J. G. (1963), "Structural Analysis and Design Considerations for Cylindrical Pressure Hulls," *David Taylor Model Basin Report No. 1639*, April.

Pulos, J. G., and Salerno, V. L. (1961), "Axisymmetric Elastic Deformations and Stresses in a Ring-Stiffened, Perfectly Circular Cylindrical Shell under External Hydrostatic

Pressure," Structural Mechanics Laboratory, *David Taylor Model Basin Rep. No. 1497*, September.

Reynolds, T. E. (1957), "A Graphcal Method for Determining the General Instability Strength of Stiffened Cylindrical Shells," *David Taylor Model Basin Rep. No. 1106*, September.

Reynolds, T. E. (1960), "Inelastic Local Buckling of Cylindrical Shells under External Hydrostatic Pressure," *David Taylor Model Basin Rep. No. 1392*, August.

Reynolds, T. E. (1962), "Elastic Local Buckling of Ring-Supported Cylindrical Shells under Hydrostatic Pressure," *David Taylor Model Basin Rep. No. 1614*, September.

Reynolds, T. E. (1971), "Stability of Cylindrical Shells," *Graduate Course on Analysis and Design of Cylindrical Shells under Pressure*, The Catholic University of America, Washington, D.C., June.

Reynolds, T. E., and Blumenberg, W. F. (1959), "General Instability of Ring-Stiffened Cylindrical Shells Subjected to External Hydrostatic Pressure," *David Taylor Model Basin Rep. No. 1324*, June.

Schilling, C. G. (1965), "Buckling Strength of Circular Tubes," *ASCE J. Struct. Div.*, Vol. 91, No. ST5, p. 325.

Seide, P. (1960), "The Effect of Pressure on the Bending Characteristics of an Actuator System," *J. Appl. Mech.*, September, p. 429.

Sherman, D. R. (1969), "Residual Stress Measurement in Tubular Members," *ASCE J. Struct. Div.*, Vol. 95, No. ST4, Pap. 6502, pp. 635–637.

Sherman, D. R. (1976), "Tests of Circular Steel Tubes in Bending," *ASCE J. Struct. Div.*, Vol. 102, No. ST11, Pap. 12568.

Sherman D. R., and Tanavde, A. S. (1978), "Comparative Study of Flexural Capacity of Pipes," *Civ. Eng. Dept. Rep.*, University of Wisconsin at Milwaukee. March.

Singer, J. (1967), "The Influence of Stiffener Geometry and Spacing on the Buckling of Axially Compressed Cylindrical and Conical Shells," *Prelim. Preprint Pap., 2nd IUTAM Symp. Theory Thin Shells*, Copenhagen, September.

Southwell, R. V. (1915), "On the Collapse of Tubes by External Pressure," *Philos. Mag.*, Part 1, Vol. 25, May 1913, pp. 687–698; Part 2, Vol. 26, September 1913, pp. 502–511; Part 3, Vol. 29, January 1915, pp. 66–67.

Southwell, R. V. (1914), "On the General Theory of Elastic Stability," *Philos. Trans. R. Soc. London*, Ser. A, No. 213.

Stein, M. (1968), "Some Recent Advances in the Investigation of Shell Buckling," *AIAA J.*, Vol. 6, No. 12.

Stephens, M. J., Kulak, G., and Montgomery, C. J. (1982), "Local Buckling of Thin-Walled Tubular Steel Members," *Struct. Eng. Dept. Rep. No. 103*, University of Alberta, Edmonton, February.

Stephens, M. J., Kulak, G., and Montgomery, C. J. (1983), "Local Buckling of Thin-Walled Tubular Steel Members," *Proc. 3rd Int. Colloq. Stab. Met. Struct.* (George Mem. Sess.), Toronto, May.

Sturm, R. G. (1914), "A Study of the Collapsing Pressure of Thin-Walled Cylinders," *Univ. Ill., Eng. Exp. Sta. Bull.*, No. 329.

Timoshenko, S. (1910), "Einige Stabilitäts-Probleme der Elastizitäts-Theorie," *Z. Math. Phys.*, Vol. 58, No. 4.

Timoshenko, S., and Gere, J. M. (1961), *Theory of Elastic Stability*, McGraw-Hill, New York.

Tokugawa, T. (1929), "Model Experiments on the Elastic Stability of Closed and Cross-Stiffened Circular Cylinders under Uniform External Pressure," *Proc. World Eng. Cong.*, Tokyo, Vol. 29, pp. 249–279.

U.S. Steel Corporation (1964), *The Manufacture of Steel Tubular Products*, U.S. Steel Corporation, Pittsburgh, Pa.

Vojta, J. F., and Miller, C. D. (1983), "Buckling Tests on Ring and Stringer Stiffened Cylindrical Models Subject to Combined Loads," *Final Rep., Contract No. 11896*, Vol. 1, Main Report, 3 Volume Appendix, CBI Industries, Inc., April.

Von Kármán, T., and Tsien, H. S. (1941), "The Buckling of Thin Cylindrical Shells under Axial Compression," *J. Aeronaut. Sci.*, Vol. 8, No. 8.

Von Mises, R. (1931), "The Critical External Pressure of Cylindrical Tubes" (Der kritische Aussendruck zylindrischer Röhre), *VDI-Z.*, Vol. 58, No. 19, 1914, pp. 750–755; David Taylor Model Basin Trans. 5, August 1931.

Von Mises, R. (1933), "The Critical External Pressure of Cylindrical Tubes under Uniform Radial and Axial Load" (Der kritische Aussendruck für allseits belastete zylindrische Röhre), Stodolas Festschrift, Zurich 1929, pp. 418–430; David Taylor Model Basin Trans. 5, August 1933.

Von Sanden, K., and Tolke, F. (1949), "On Stability Problems in Thin Cylindrical Shells" (Über Stabilitätsprobleme dünner, kreiszylindrischer Schalen), *Ing. Arch.*, Vol. 3, 1932, pp. 24-66; David Taylor Model Basin Transl. 33, December 1949.

Wah, T. (1967), "Buckling of Thin Circular Rings under Uniform Pressure," *Int. J. Solids Struct.*, Vol. 3, Pergamon Press, Oxford, pp. 967–974.

Weingarten, V. I., and Seide, P. (1961), "On the Buckling of Circular Cylindrical Shells under Pure Bending," *Trans. ASME, J. Appl. Mech.* Vol. 28, No. 1.

Weingarten, V. I., et al. (1965), "Elastic Stability of Thin-Walled Cylindrical and Conical Shells under Combined Internal Pressure and Axial Compression," *AIAA J.*, Vol. 3, No. 6, pp. 118–125.

Weingarten, V. I. et al. (NASA) (1968), "Buckling of Thin-Walled Circular Cylinders," *NASA SP-8007*, August 1968.

Wenk, E., and Kennard, E. H. (1956), "The Weakening Effect of Initial Tilt and Lateral Buckling of Ring Stiffeners on Cylindrical Pressure Vessels," *David Taylor Model Basin Rep. No. 1073*, December.

Wilson, W. M. (1937), "Tests of Steel Columns," *Uni. Ill. Eng. Exp. Sta. Bull.*, No. 292.

Wilson, W. M., and Newmark, N. M. (1933), "The Strength of Thin Cylindrical Shells as Columns," *Univ. Ill. Eng. Exp. Sta. Bull.*, No. 255.

Wilson, W. M., and Olson, F. D. (1941), "Tests on Cylindrical Shells," *Univ. Ill. Eng. Exp. Sta. Bull.*, No. 331, September.

Windenburg, D. F. (1960), "Vessels under External Pressure," *Pressure Vessel and Piping Design*, American Society of Mechanical Engineers, New York, pp. 625–632.

Windenburg, D. F., and Trilling, C. (1934), "Collapse by Instability of Thin Cylindrical Shells under External Pressure," *Trans. ASME*, Vol. 56, No. 11, p. 819.

Wu, T. S., Goodman, L. E., and Newmark, N. M. (1953), "Effect of Small Initial Irregularities on the Stresses in Cylindrical Shells," *Univ. Ill. Struct. Res. Ser.*, No. 50, April.

Yao, J. C. (1962), "Large Deflection Analysis of Buckling of a Cylinder under Bending," *Trans. ASME J. Appl. Mech.*, Vol. 29, No. 4.

Chapter Fifteen

Members with Elastic Lateral Restraints

15.1 INTRODUCTION

Compression members are sometimes restrained laterally between their ends by intermittent elastic lateral supports. Typical examples include: (1) the unbraced compression flange of a girder whose tension flange is laterally restrained by a bridge-floor or building-roof system; and (2) the top chord of a "pony truss" for which vertical clearance requirements prohibit direct lateral bracing. The methods of design presented in this chapter can also be used for guyed towers.

In the case of the girder, the web prevents the top flange from buckling in the vertical plane and intermittent elastic lateral restraints may be provided in the plane normal to the web by means of vertical stiffeners in combination with the contiguous framing elements in the floor or roof system adjacent to the tension flange. In a truss supporting a floor or roof system in the plane of the tension chord members, the panel points provide vertical support and elastic lateral restraint for the compression chords as a whole.

The pony truss, while no longer in extensive use in new construction, has served as the prototype through the years for the development of theory and design procedures that currently may find applications in other similar situations. The behavior during buckling of a member with intermittent elastic lateral supports lies between two limiting extremes. If the elastic restraints are very stiff, nodal points can be induced at each restraint location—if the restraints are very flexible, buckling can be in the shape of a single half-wave over the full member length. The actual buckled shape consists of a number of half-waves not greater than the total number of spaces between supports.

The design of a compression member with intermittent elastic lateral supports may be based on the computed critical load; or because of initial crookedness and because of moments introduced by bending of the floor beams, it may be based on combined-stress calculations that include the effect of deflection. The latter approach is a rational one but has not as yet been simplified sufficiently to make it a practical design procedure. On the other hand the critical-load analysis gives only an upper bound to the actual strength of the member. Current design rules are based on a semiempirical procedure in which adequate stiffness of the compression-chord lateral supports is obtained by designing them for fictitious horizontal loads introduced at the top-chord panel points normal to the plane of the truss. The critical load of the chord with elastic lateral supports at the panel points is then determined, and the design load is found by dividing this critical load by a suitable factor of safety.

Using the pony truss as the prototype, a procedure of analysis is developed to determine the critical load of a member with discrete elastic lateral restraints. To a lesser extent the combined-stress procedure is reviewed in this chapter. The design of pony truss transverse frames (floor beams, truss verticals, and connecting knee braces) has a direct bearing on both procedures. Proper design of transverse frames is essential to the safety of the pony-truss bridge.

Toward the end of the nineteenth century the failure of several pony-truss bridges focused attention on the top-chord buckling problem. Engesser, (1884, 1893), was the first to present a simple, rational, and approximate formula for the required stiffness, C_{req}, of elastic supports equally spaced between the ends of a hinged-end column of constant section. As equivalent uniform elastic support was assumed in the Engesser analysis.

Early developments are reviewed by Bleich (1952). See, also, the list of references appended to Chapter 14 of the third edition of the SSRC Guide (1976).

Hu (1952), using the energy method, has studied the problem of elastically supported chords. He considered nonuniform axial forces, various chord cross sections, and variable stiffness spring supports for both simple and continuous pony-truss bridges.

Holt (1951, 1952, 1956, 1957), in work sponsored by Column Research Council, has presented a method of analysis for determining the critical load of a pony-truss top chord which is essentially "exact" in that it includes most of the secondary effects which influence the behavior of the pony truss. In a similar manner, Lee and Clough (1958) and more recently Elgaaly and Khalifa (1970) studied the stability of pony-truss bridges.

The effect of floor-system deflections on the top-chord stresses was studied in another CRC-sponsored project by Barnoff and Mooney (1957). Tests on models of pony-truss bridges have been conducted by Holt (1957). Oliveto (1980) describes a computer program and considers both elastic and inelastic buckling of columns partially restrained at intervals. Medland and Segedin (1979) evaluate brace forces for an initially crooked member.

15.2 BUCKLING OF THE COMPRESSION CHORD

The buckling problem of the compression chord of a pony truss can be reduced to that of a column braced at intervals by elastic springs whose spring constants correspond to the stiffness of the truss transverse frames. The top-chord axial compression and the top-chord stiffness vary from panel to panel, and the stiffness of the transverse frames varies from panel point to panel point, thus complicating the theoretical problem In addition, there are secondary factors such as the following:

1. The stiffening effect of the truss diagonals.
2. The torsional stiffness of the chord and web members.
3. The initial crookedness of the chord and the eccentricity of the axial load.
4. For non-parallel-chord trusses, the effect of chord curvature.

Engesser's solution (1885, 1893) is based on the following simplifying assumptions:

1. The top chord, including the end posts, is straight and of uniform cross section.
2. Its ends are taken as pin-connected and rigidly supported.
3. The equally spaced elastic supports have the same stiffness and can be replaced by a continuous elastic medium.
4. The axial compressive force is constant through the chord length.

Engesser's analysis can be applied with reasonable accuracy to the case where the lateral support is supplied by equally spaced springs, provided that the half-wavelength of the buckled shape of the continuously supported bar is at least 1.8 times the spring spacing; and this will be true if the bar is stable as a two-hinged column carrying the same axial load and having a length no less than 1.3 times the spring spacing. Since the flexibility of the end supports is neglected, the Engesser solution can best be used as a preliminary design tool with a more accurate subsequent evaluation by Table 15.1 as hereinafter discussed.

Table 15.1. $1/K$ for Various Values of Cl/p_c and n

	n						
$1/K$	4	6	8	10	12	14	16
1.000	3.686	3.616	3.660	3.714	3.754	3.785	3.809
0.980		3.284	2.944	2.806	2.787	2.771	2.774
0.960		3.000	2.665	2.542	2.456	2.454	2.479
0.950			2.595				
0.940		2.754		2.303	2.252	2.254	2.282
0.920		2.643		2.146	2.094	2.101	2.121
0.900	3.352	2.593	2.263	2.045	1.951	1.968	1.981
0.850		2.460	2.013	1.794	1.709	1.681	1.694
0.800	2.961	2.313	1.889	1.629	1.480	1.456	1.465
0.750		2.147	1.750	1.501	1.344	1.273	1.262
0.700	2.448	1.955	1.595	1.359	1.200	1.111	1.088
0.650		1.739	1.442	1.236	1.087	0.988	0.940
0.600	2.035	1.639	1.338	1.133	0.985	0.878	0.808
0.550		1.517	1.211	1.007	0.860	0.768	0.708
0.500	1.750	1.362	1.047	0.847	0.750	0.668	0.600
0.450		1.158	0.829	0.714	0.624	0.537	0.500
0.400	1.232	0.886	0.627	0.555	0.454	0.428	0.383
0.350		0.530	0.434	0.352	0.323	0.292	0.280
0.300	0.121	0.187	0.249	0.170	0.203	0.183	0.187
0.293	0						
0.259		0					
0.250			0.135	0.107	0.103	0.121	0.112
0.200			0.045	0.068	0.055	0.053	0.070
0.180			0				
0.150				0.017	0.031	0.029	0.025
0.139				0			
0.114					0		
0.100						0.003	0.010
0.097						0	
0.085							0

Engesser's solution for the required stiffness of a pony-truss transverse frame is

$$C_{\text{req}} = \frac{P_c^2 l}{4EI} \tag{15.1}$$

where C_{req} is the elastic transverse frame stiffness at a panel point that is required to ensure that the overall chord having panel lengths l and flexural rigidity EI will attain buckling load P_c. If the proportional limit of the column material is exceeded, E should be replaced by the tangent modulus E_t.

Equation 3.2 for critical stress can be written as follows for a column of length l:

$$E_t I = \frac{P_c (Kl)^2}{\pi^2} \tag{15.2}$$

Taking l in this equation as the panel length of the pony-truss compression chord, we can substitute Eq. 15.2 into Eq. 15.1, obtaining the required spring constant as

$$C_{\text{req}} = \frac{\pi^2 P_c}{4K^2 l} \tag{15.3}$$

This equation has been shown* to be adequate when the half-wavelength of the buckled chord is no less than $1.8l$; and this limiting value corresponds to a K factor of 1.3. It is not applicable to short bridges with a small number of panels.

It should be noted that P_c is both of the following:

1. The buckling load of the entire compression chord laterally supported by the transverse frames (and assumed to be pin-ended).
2. The buckling load of the portion of the compression chord *between* the transverse frames (with end restraints consistent with the factor K.

According to the Engesser theory, the maximum compression-chord buckling load and the corresponding required spring constant of each support can be determined as follows for a member of given cross section having area A:

1. Determine the critical load P for the member between spring supports, using the expression

$$P_c = A\sigma_c \tag{15.4}$$

   Obtain σ_c from an appropriate column-strength curve, taking the equivalent column slenderness ratio as Kl/r, with $K = 1.3$ and r estimated on the basis of probable shape and size of member.
2. Determine the spring constant C_{req} such that the buckling load of the chord member as a whole is equal to P_c:

$$C_{\text{req}} = 1.46 \frac{P_c}{l} \tag{15.5}$$

It may be noted that Eq. 15.5 follows from Eq. 15.3, taking

$$\frac{\pi^2}{4K^2} = \frac{\pi^2}{4(1.3)^2} = 1.46$$

* See Hu (1952, p. 275).

The Engesser simplifying assumption of taking the chord ends as pin-connected may result in significantly unsafe errors in C_{req} in the case of short pony trusses. Holt (1952, 1956) provides an alternative design procedure which does not require this simplifying assumption. Holt's solution for the buckling load of the compression chord of a pony truss is based on the following assumptions (see Fig. 15.1):

1. The transverse frames at all panel points have identical stiffness.
2. The radii-of-gyration of all top-chord members and end posts are identical.
3. The top-chord members are all designed for the same allowable unit stress, hence their areas, and (from step 2) their moments of inertia are proportional to the compression forces.
4. The connections between the top chord and the end posts are assumed pinned.
5. The end posts act as cantilever springs supporting the ends of the top chords.
6. The bridge carries a uniformly distributed load.

The results of Holt's studies are presented in Table 15.1, which gives the reciprocal of the effective-length factor K as a function of n (the numbers of panels) and of Cl/P_c (where C is the furnished stiffness at the top of the least-stiff transverse frame).

Table 15.1, where applicable, provides a rapid design aid in checking the stability of a pony-truss compression chord. The procedure is as follows:

1. Design the floor beams and web members for their specified loads.
2. Calculate the spring constant C furnished at the upper end of the cross frame having the least transverse stiffness.

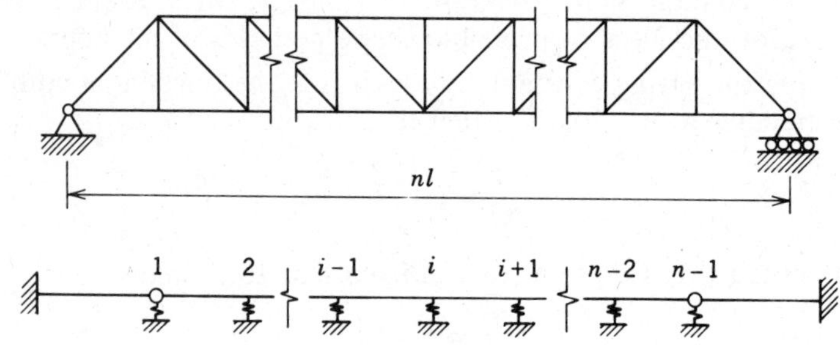

Fig. 15.1 Pony truss and analogous top chord.

3. Calculate the value of parameter Cl/P_c, where P_c is the maximum design chord stress multiplied by the desired factor of safety.
4. Enter the table with n and Cl/P_c, and find the corresponding value of $1/K$ for a compression-chord panel, interpolating as necessary.
5. Determine the value of Kl/r for the compression-chord panel (note that this value of Kl/r is to be applied to all panels).
6. Determine the allowable top-chord compressive unit stress corresponding to this value of Kl/r, using the appropriate column curve or table.

Values of $1/K$ less than 0.5 (i.e., $K > 2$) are only of academic interest, since usual bridge proportions and transverse-frame stiffnesses lead to values of $1/K$ reasonably near 1.0, and this results in economical use of material.

Hu (1952) developed the curves shown in Fig. 15.2. These curves give the stiffness of the compression-chord transverse supports that is required to make each panel of chord buckle as one half wave. Hu assumed that both the flexural rigidity (EI) and the axial force (P_c) of the compression chord vary as a symmetrical second-degree parabola along the chord length.

Hu's results for a chord of constant section [$(EI)_{\text{end}}/(EI_{\text{middle}} = 1.0$ in Fig. 15.2] can be compared with Holt's work for $1/K = 1$ (first line of Table 15.1) for the cases $n = 4$, 6, and 8, which were considered by both investigators. Hu's results give stiffness requirements approximately 7%

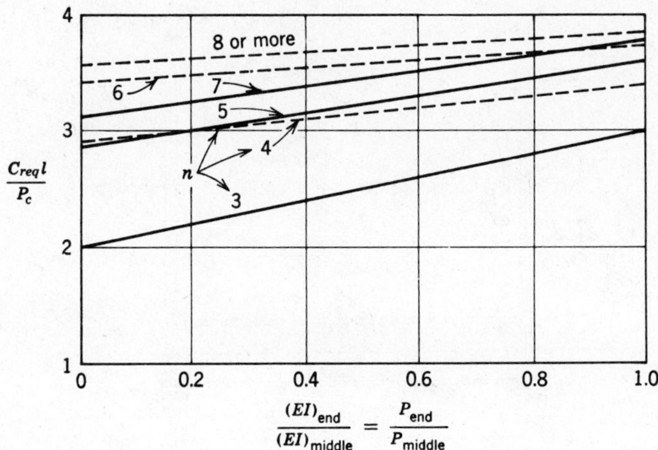

Fig. 15.2 Effect of variation in compression chord on transverse stiffness requirements.

less than those of Holt for $n = 4$, and 5% greater for $n = 6$ or 8. Thus the results are in reasonable agreement, even though the procedures are somewhat different.

Hu (1952) also studied the effect of the variation of C_{req} caused by parabolic variation of the length of the pony-truss verticals, and the effect of parabolic variation of C_{req}. In both cases the value of C_{req} will be less than that for the case where C_{req} has the same value at each transverse frame.

Because of the uncertainties involved in the analysis of pony-truss top chords, it is reasonable to require a factor of safety for overall top-chord buckling somewhat greater than that used for designing hinged-end columns.

The transverse-frame spring constant C that is actually furnished can be determined for the frame loaded as shown in Fig. 15.3 by means of the following equation:

$$C = \frac{E}{h^2[(h/3I_c) + (b/2I_b)]} \tag{15.6a}$$

The first term within the denominator brackets represents the contribution of the truss vertical, and the second term represents the contribution of the floor beam. Thus the contributions of the top-chord torsional strength and the web-diagonal bending strength to the frame stiffness are neglected in this equation. It is evident that if the floor beam is very stiff in comparison with the truss vertical, the frame stiffness is approximately

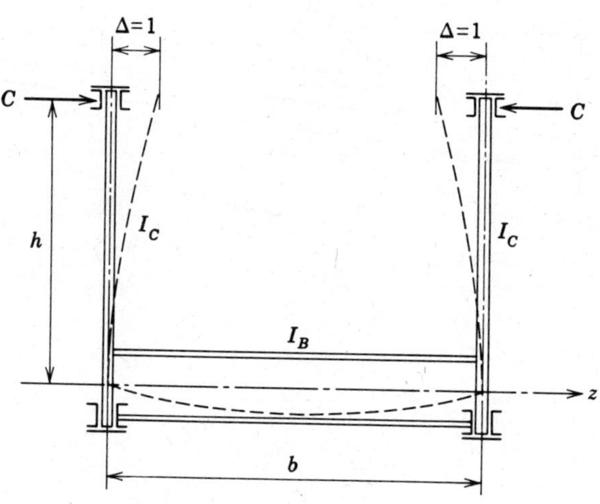

Fig. 15.3

$$C = \frac{3EI_c}{h^3} \tag{15.6b}$$

When the two chords tend to move in the same direction, the stiffness C will be greater than that given by Eq. 15.6a; therefore, C as found from Eq. 15.6a is always the lower bound.

If the diagonals of the truss system are effectively fixed at their base, their contribution to C in Eq. 15.6a may be included by introducing the additional term $L_d^3/3I_d$ into the denominator. The values L_d and I_d represent the length and moment of inertia, respectively, of the diagonal members. In this case C becomes

$$C = \frac{E}{h^2\{h/(3I_c + 3I_d(h/L_d)^3) + (b/2I_b)\}} \tag{15.6c}$$

15.3 EFFECT OF SECONDARY FACTORS ON BUCKLING LOAD

The consideration of secondary factors involves procedures that require a large amount of computation. Most of these procedures use the usual methods of indeterminate structural analysis to set up a system of simultaneous, linear, homogeneous equations. The stability criterion is that the determinant of the coefficients of this system of equations must vanish.

Holt (1952) considered the following secondary factors in his analysis:

1. Torsional stiffness of the chord and the web members.
2. Lateral support given to the chord by the diagonals.
3. Effect of web-member axial stresses on the restraint provided by them.
4. Effect of non-parallel-chord trusses.
5. Error introduced by considering the chord and end posts to be a single straight member.

Holt's analyses show that the error in the critical load introduced by neglecting all of these factors is quite small, and that satisfactory results in calculating the compression-chord buckling load can be obtained by assuming that the chord is a straight elastically braced column whose length is the total length of the chord and end posts. These conclusions are in agreement with those reached by Schibler (1946), who finds that the torsional stiffness of the top chord and support furnished by the web diagonals increase the chord buckling strength only slightly.

15.4 TOP-CHORD STRESSES DUE TO BENDING OF FLOORBEAMS AND TO INITIAL CHORD ECCENTRICITIES

The compression chord of a pony truss is displaced laterally at some panel points as a result of live load on the bridge, and because of initial crookedness and unintentional eccentricities of the chord. Such lateral deflections will, of course, reduce the maximum load capacity of the chord (and of the bridge), just as end eccentricity and initial curvature will reduce the compression strength of any column.

Design procedures that take account of such imperfections are not presently available. It is difficult to take them into account, because of both the complexity of the necessary calculations and the lack of knowledge with regard to probable initial imperfections. The calculations involve the top-chord stiffness in both bending and torsion.

Holt (1957) has developed an empirical procedure for estimating bending moments in the top chord and end posts that is in agreement with his test results. He recommends that the end post be designed as a simple cantilever beam to carry the axial load combined with a transverse load of 0.5% of the axial load, applied at the upper end. Tentatively, a value of 1% should probably be used.

5.5 DESIGN PROCEDURES

In the design of half-through truss spans, AASHTO Specifications (1983) require that:

> The top chord shall be considered as a column with elastic lateral supports at the panel points. The critical buckling force of the column, so determined, shall exceed the maximum force from dead load, live load and impact in any panel of the top chord by not less than 50 percent.

Thus a load factor of 1.5 is considered adequate, and this is less than that required by the same specification in the determination of allowable compressive stresses in hinged-end columns. This is presumably justified on the basis that all pony-truss top-chord compression members cannot be stressed simultaneously up the same proportion of critical buckling load. However, it seems evident that in cases where maximum compressive stress may occur simultaneously in the entire length of the compression chord, the safety factor should be higher than that used for general column design, rather than lower.

German Buckling Specifications base the design of pony-truss compression chords on the Engesser solution for the buckling load, with the recommendation that K be kept the same for all panels and between limits of 1.2 and 3.0. The formula for required transverse-frame stiffness is given as:

$$C_{\mathrm{req}} = \frac{2.50}{K_m^2} \frac{P_c}{l} \tag{15.7}$$

For $K_m = 1.3$. Eq. 15.7 gives very nearly the same result as Eq. 15.5, which is to be expected since both equations are based on Engesser's solution.

For design of the web verticals, the AASHTO Specification (1983) reads:

> The vertical truss members and the floorbeams and their connections in half-through truss spans shall be proportioned to resist a lateral force of not less than 300 pounds per linear foot (4350 N/m) applied at the top-chord panel points of each truss.

The problem of the lateral stability of a pony-truss compression chord will now be illustrated by means of a design example, using, in part, AASHTO Specifications (1983).

Design Example 15.1 (see Fig. 15.4)

Consider a pony truss that has 12 panels of 13 ft 4 in. for a span of 160 ft. The transverse frame is shown in the sketched cross section. The W27 × 84 floorbeams are required by bridge deck loads. The top chord is a 10 in. by 10 in. box section, with wall thickness to be determined by design requirements for a maximum compressive force of 360 kips (dead load, live load, and impact). The verticals are W10 rolled sections.

The approximate properties of the compression-chord cross sections are as follows:

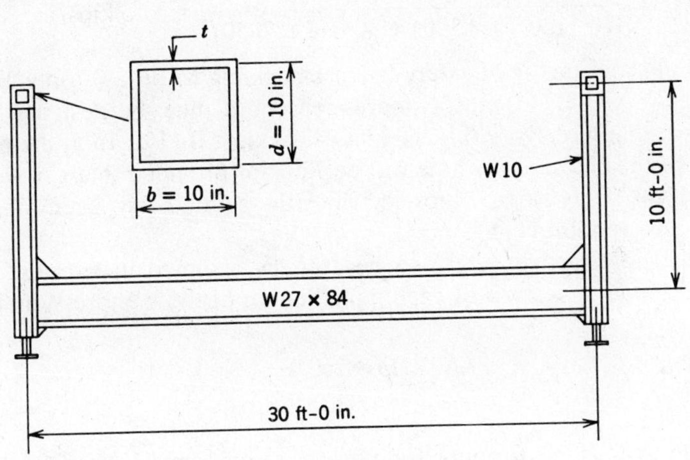

Fig. 15.4

Area: $A = 4td = 40\,t\ \text{in.}^2$

Moment of inertia: $I = \dfrac{Ad^2}{6} = 670t\ \text{in.}^4$

Radius of gyration: $r = \dfrac{d}{\sqrt{6}} = 4.08\ \text{in.}$

By AASHTO requirements, the lateral force to be resisted at the upper panel points is

$$0.3\ \text{kips/ft} \times 13.33\ \text{ft} = 4\ \text{kips}$$

The maximum moment in the transverse frame will be at the joint of the web vertical and the floor beam and is

$$M = 4\ \text{kips} \times 120\ \text{in.} = 480\ \text{kip-in.}$$

Assume that the maximum tension in the web vertical due to the bridge load is 24 kips in the region where the compression chord is most highly stressed. Taking r as approximately $0.42d$, and the allowable stress as 20 ksi due to combined bending and direct stress, the required cross-sectional area of the vertical is

$$A = \frac{1}{f_a}\left(P + \frac{Mc}{r^2}\right) = \frac{24 + (480 \times 5)/(4.2)^2}{20} = 80\ \text{in.}^2$$

Try a W10 × 33, for which $A = 9.71\ \text{in.}^2$ and $I = 170\ \text{in.}^4$. The I of the floor beam is 2850 in.^{4}, and from Eq. 15.6a the transverse-frame spring constant is

$$C = \frac{29{,}000}{120^2[120/(3 \times 170) + 360/(2 \times 2850)]} = 6.75\ \text{kips/in.}$$

Assuming the factor of safety against buckling to be 2.0 (only 1.50 is required by AASHTO), the compression chord must be designed for a buckling strength of $P = 2.0 \times 360\ \text{kips} = 720\ \text{kips}$. By Eq. 15.5, the buckling load of the chord as a whole will be 720 kips provided that C is no less than $(1.46 \times 720/160 = 6.57$ kips/in. Since the actual C has been found to be 6.75, this requirement is met.

Taking the effective length to be $1.3l$, as assumed in Eq. 15.5, the effective slenderness ratio of each panel length of the compression chord is

$$\frac{Kl}{r} = \frac{1.3 \times 160}{4.08} = 51$$

Using the AASHTO formula for pin-ended columns of A36 steel, the allowable compression is

$$F_a = 16{,}930 - 0.53(51)^2 = 15{,}601 \text{ psi} = 15.6 \text{ ksi}$$

The required chord area for a 360-kip load is, therefore,

$$A = \frac{360}{15.6} = 23.1 \text{ in.}^2$$

Since $A = 40t$, the required wall thickness of the compression chord is readily found as

$$t = \frac{23.1}{40} = 0.577 \text{ in.} \qquad (\text{use } \tfrac{5}{8}\text{-in. wall thickness})$$

As a more accurate alternative and preferred procedure to the use of Eq. 15.5, Kl/r can be determined from Table 15.1. Using the actual supplied C and the value of $P_c = 720$ kips yields

$$\frac{Cl}{P_c} = \frac{6.75 \times 160}{720} = 1.50$$

Entering Table 15.1 in the column for $n = 12$ panels and $Cl/P_c = 1.50$, $1/K$ falls between 0.804 and 0.850. The interpolated value is $1/K = 0.81$, or $K = 1.23$. The slenderness ratio of the chord between panel points is then

$$\frac{Kl}{r} = \frac{1.23 \times 160}{4.08} = 48.2$$

which is very close to the value of 51 found previously. The allowable compressive stress and the required wall thickness are not appreciably different from the previous values; however, it should be kept in mind that for bridges having less than 10 panels, C_{req} as found by Holt's procedure (Table 15.1) could be appreciably greater than the value found by the Engesser method.

15.6 PLATE GIRDER WITH ELASTICALLY BRACED COMPRESSION FLANGE

Although most of the research and references presented herein concern the pony-truss bridge, the design recommendations that have been given are also applicable to the design of plate girders with elastically braced compression flanges. Such girders will customarily have full-depth vertical stiffeners serving the dual function of web stiffener and top-flange transverse support. The lack of girder diagonal members is of no concern since Eq. 15.6a does not include the contribution of truss diagonals. In applying the design procedure to girders, one-third of the compression area of the web should be added to the area of the compression flange and introduced as the equivalent area of the top chord in the formulas as developed for the pony truss. The same section may be used in calculating I_c. If girders are used in an arrangement differing from the section shown in Fig. 15.3, an independent calculation of C must be made appropriate to the framing arrangement that is used.

15.7 GUYED TOWERS

In determing the effective-length factor K for use in the design of a guyed tower, Springfield* has used Table 15.1 together with an adaptation of the pony-truss analysis. He reports on the application to a "mast" with guys at nine levels, as follows:

> Stiffnesses (of restraints) are not constants, but vary according to the magnitude of lateral displacement. The procedure we have adopted is as follows:
>
> 1. Compute the spring constants C at each guy level. This is done by dividing the horizontal guy reaction by the displacement under maximum wind load. The average spring stiffness is chosen rather than the instantaneous stiffness because the buckling mode could reduce the displacements at guy points, thus reducing the spring stiffness.
> 2. Calculate the parameter Cl/P_c for each column segment between guys. The smaller value of C is chosen from the top or bottom of the segment. l is the length of the column segment and P_c is the maximum column segment design load multiplied by the appropriate factor-of-safety.
> 3. The value of K for each segment is determined in turn from Table 15.1 (a total of nine K values). In its application, we have assumed a continuous compression strut with nine spans. Each span has identical spring supports of magnitude equal to the C value for the column segment being considered. The spacing of these supports is uniform, this distance being the length of column segment under consideration.
>
> Holt assumes that the strut is fixed at each end and pinned at the first interior supports. Our approximation would be fairly good for the interior spans, but could be in error for the discontinuous ends—especially at the top end where both joint displacement and rotation are possible.
>
> The range of CL/P_c for our mast is 1.3 to 2.3 and the corresponding range of K is 1.7 to 1.1. We have also applied Engesser's analysis for comparison which shows close agreement of C required for K values between 1.7 and 1.3. For K values smaller than 1.3 a rapid divergence appears as expected.

It is of course recognized that the complete design of guyed towers is a very complex problem, involving effects that may include preloading of guys, vertical movement of gusts, vortex shedding, and other vibration

* June 9, 1971, letter to the editor, from J. Springfield, of C. D. Carruthers & Wallace Consultants Limited, Toronto, Canada.

and dynamic problems. The foregoing is only a brief summary of the column-buckling aspects of the design problem as approached in an approximate and simplified manner.

References

American Association of State Highway and Transportation Officials (1983), *Standard Specifications for Highway Bridges*, 13th ed., AASHTO, Washington, D.C.

Barnoff, R. M., and Mooney, W. G. (1957), "The Effect of Floor System Participation on Top Chord Stresses in Single Span Pony Truss Bridges," Stability of Bridge Chords without Lateral Bracing, *Column Res. Counc. Rep.*, No. 5.

Bleich, F. (1952), *Buckling Strength of Metal Structures*, McGraw-Hill, New York.

Elgaaly, M. A., and Khalifa, M. K., Jr. (1970), "Stability of Pony-Truss Bridges," on file, Institution of Structural Engineers, United Kingdom, June.

Engesser, F. (1885), "Die Sicherung offener Brücken gegen Ausknicken," *Zentralbl. Bauverwaltung*, 1884, p. 415; 1885, p. 93.

Engesser, F. (1893), *Die Zusatzkräfte und Nebenspannungen eiserner Fachwerkbrücken*, Vol. II, Berlin.

Holt, E. C. (1951), "Buckling of a Continuous Beam-Column on Elastic Supports," Stability of Bridge Chords without Lateral Bracing, *Column Res. Counc. Rep. No. 1.*

Holt, E. C. (1952), "Buckling of a Pony Truss Bridge," Stability of Bridge Chords without Lateral Bracing, *Column Res. Counc. Rep. No. 2.*

Holt, E. C. (1956), "The Analysis and Design of Single Span Pony Truss Bridges," Stability of Bridge Chords without Lateral Bracing, *Column Res. Counc. Rep. No. 3.*

Holt, E. C. (1957), "Tests on Pony Truss Models and Recommendations for Design," Stability of Bridge Chords without Lateral Bracing, *Column Res. Counc. Rep. No. 4.*

Hu, L. S. (1952), "The Instability of Top Chords of Pony Trusses," Dissertation, University of Michigan, Ann Arbor, 1952.

Lee, S. L., and Clough, R. W. (1958), "Stability of Pony Truss Bridges," *Pub. Int. Assoc. Bridge Struct. Eng.*, Vol. 18, p. 91.

Medland, I. C., and Segedin, C.M. (1979), "Brace Forces in Interbraced Column Structures," *ASCE J. Struct. Div.*, Vol. 105, No. 7, pp. 1543–1556.

Oliveto, G. (1980), "Inelastic Buckling of Columns Partially Restrained against Sway and Rotation," *Eng. Struct.*, Vol. 2, pp. 97–102.

Schibler, W. (1946), "Das Tragvermogen der Druckgurte offener Fachwerkbrücken mit parallelen Gurten," *Inst. Baustatik E.T.H., Zurich, Mitt.*, No. 19.

Chapter Sixteen

Frame Stability

16.1 INTRODUCTION

In designing columns as parts of frames, it is convenient to isolate the columns from the frame and treat their design as separate problems with each column subjected to end moments and loads as well as to the forces applied along their length. Because of inelastic action within the frame, the manner in which a column is loaded affects the displacements and ultimate load. That is, the displacements for a given value of load depend on the order of application of the axial force and end moments (i.e. the structural response is load-path dependent). However, studies of biaxially loaded columns have shown that drastic variations in load path have negligible effect on the values of the displacements at a particular load and little effect on the ultimate load (Harstead et al. 1971). Thus the load path effect can be neglected and the framed column can be considered as an isolated hinged-end column subjected to end moments and loads provided that these forces have been determined so that they are consistent with the joint rotations and sway angle of the column as determined from frame analysis.

With this concept of isolating a column for design purposes, there are two approaches for designing the frame for stability. The first approach attempts to modify the design of the individual column to account for action of the rest of the frame. This approach has led to column formulas that include terms to adjust them for frame instability. A second approach is based on SSRC Technical Memorandum No. 5 (Appendix B), which states:

> Although the maximum strength of frames and the maximum strength of the component members are interdependent (but not necessarily coexistent), it is recognized that, in many structures it is not practical to take this interdepend-

ence into account rigorously. At the same time, it is known that difficulties are encountered in complex frameworks when attempting to compensate automatically in column design for the instability of the entire frame, (for example, by adjustment of column effective length). Therefore, SSRC recommends that, in design practice, the two aspects, stability of individual members and elements of the structure and stability of the structure as a whole, be considered independently.

Separation of column stability from frame stability results in simpler column formulas but leaves a designer the additional problem of evaluating the stability of the entire frame.

This chapter attempts to cover both of these approaches to column design. Since it is essential that frame behavior be understood, the chapter first deals with this subject and then leads into a discussion of analytic models for frame behavior followed by an outline of most of the factors that can influence the correctness of these models.

After this introductory background material, various procedures for designing frames will be considered. Some of these procedures are applicable to the framed-column approach, while others relate to the isolated-column approach. Also, some are analysis approaches, while others are design approaches. Finally, the chapter will treat stability of frames subjected to dynamic loads, followed by a number of special topics.

16.2 BEHAVIOR OF FRAMES

The primary mode of frame failure is instability (Beedle et al. 1969). Hence an understanding of how frames respond to external loads is essential to the subject of frame stability.

Frames are geometrically imperfect, that is, story levels are not perfectly plumb. This imperfection, although within specified tolerance levels, results in spatial displacement from the beginning of loading. It is convenient to treat the movement of any point on a frame as composed of two parts, the framework joint displacements (global displacements) plus the member displacements (local displacements). Some cases of frame behavior will now be reviewed. For simplicity, consider only the joint displacements of planar frames (frames in which out-of-plane joint movements are prevented by lateral bracing) that are rectangular and are permitted to sway in their plane.

Figure 16.1*a* depicts an imperfect, symmetrical, rigid-jointed portal frame subjected to symmetrical joint loading. The columns are erected out-of-plumb with the mislocation of the right joint denoted by Δ_0. As shown in Fig. 16.1*b* the behavior of this imperfect frame (curve *abc*) is

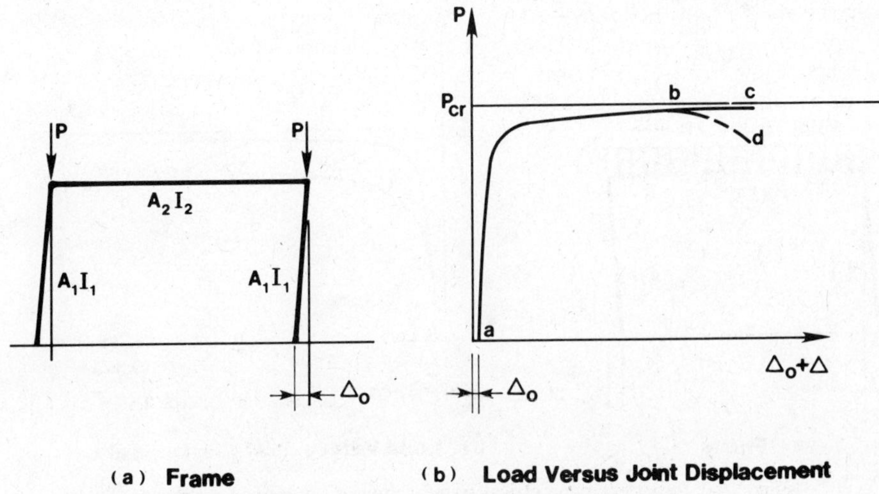

Fig. 16.1 Imperfect symmetrical joint-loaded frame.

similar to that of an imperfect centrally loaded column. The smaller the imperfection, the closer the instability load (point *b*) to the critical load P_{cr} corresponding to the sideway buckling mode, computed on the basis of inelastic action.

If the loading is distributed along the girder as in Fig. 16.2*a*, both the girder and the columns are subjected to primary bending. The effect of this is to introduce axial force into the girder and possibly flexural yielding in local regions of the frame. As a consequence the stiffness of the frame is reduced and the instability load is reduced to curve $ab'c'$, in Fig. 16.2*b*.

An unsymmetrical portal frame subjected to horizontal and vertical joint loads, and distributed vertical loading, is shown in Fig. 16.3*a*. The response of this frame is shown as curve $ab''c''$ in Fig. 16.3*b*. The frame translates from the beginning of loading due to the following:

1. The horizontal load.
2. The overturning caused by the gravity loads acting on the displaced frame (the P-Δ effect). Some analysts consider this a horizontal destabilizing force.
3. The sidesway due to dissymmetry of column flexural stiffnesses.

The ultimate load at the point of instability (b'' in Fig. 16.1*b*) is less than the critical load (computed on the basis of only the vertical joint loading) for the same frame with only column top vertical loads. The

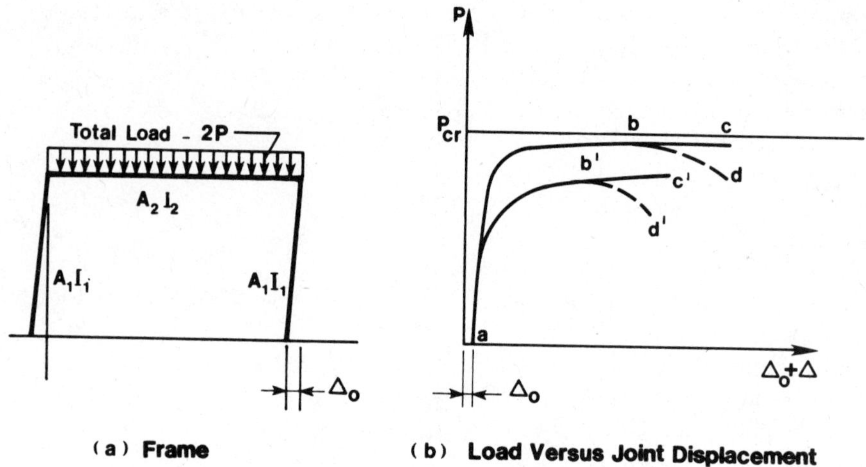

Fig. 16.2 Imperfect symmetrical frame subject to primary bonding.

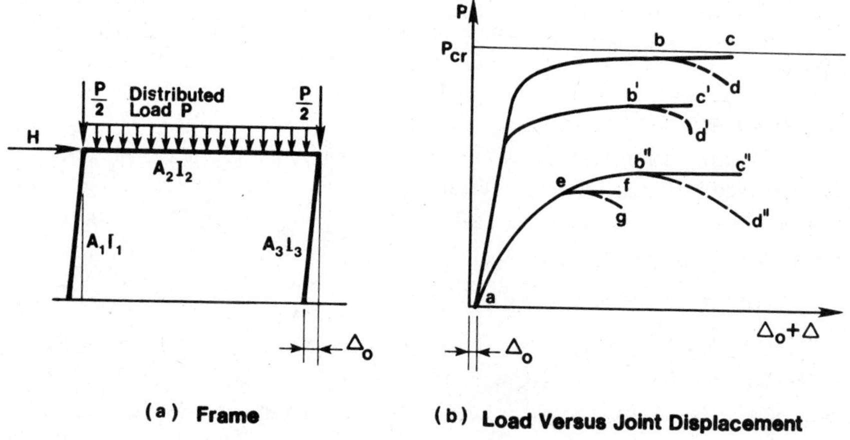

Fig. 16.3 Imperfect unsymmetrical frame subject to vertical and horizontal loading.

instability point is reached when the combination of second-order member forces and inelastic action reduce the sway stiffness of the structure to zero.

The response of all three previous frame examples was predicated on the assumption of no local buckling, no member lateral-torsional buckling, and no beam mechanism formation in the girder. As such, the curves shown are upper bounds on load-carrying capacity. If the dimensions of a column are such that lateral-torsional buckling occurs then the response of the frame in Fig. 16.3*a* would be as shown by curve *aef* in Fig. 16.3*b* for gravity-type loads. The out-of-plane translation and twist of a column can substantially change the frame response.

The behavior of multistory rectangular sway-permitted frames is similar to that just described for the portal frames. The factors causing instability are the same; the only difference is in their relative importance.

The behavior of frames just described has been verified to a limited extent by experiments on laboratory models. Because of physical restrictions there are very few such tests of full-size prototype structures. As a result, some of the models used in tests have been quite small. For example, in one set of experiments the girder spans were 15 in and the story heights 7 in. Other frame experiments have been conducted on larger models (ASCE, 1971), but few could really be termed "full size." Nevertheless, some factors which were observed in these experimental programs are worth noting.

Most rectangular frame experiments were designed to provide data that could be used to verify analytical predictions of frame behavior, or at least of ultimate load capacity. The most widely used analytical assumptions are:

1. Either completely rigid joints or frictionless hinged joints.
2. An elastic-plastic material, with cross-section inelasticity modeled as concentrated plastic hinges.
3. No out-of-plane displacements, neither global nor local.
4. Absence of initial residual stresses.
5. No participation of the frame with the floor system.
6. Proportional loading.
7. Perfect geometry.

In order to serve the purpose of verification, the test frames were designed to correspond to the analytical model as closely as possible. The preferred alternative, analytically modeling the arbitrary frame, is extraordinarily difficult. Hence the columns of the frames were selected to preclude lateral-torsional buckling, semirigid behavior of girder-to-column connections was avoided, and loads were applied proportionally.

In reality, most building frames do not have perfectly rigid joints, are geometrically imperfect, are subjected to nonproportional loading, and have some columns that may buckle laterally. Furthermore, some frames may have local areas of weakness such that the ultimate proportional load may be limited: for example, proportional load-carrying capacity may be governed by beam mechanisms or individual column buckling in its braced mode. These factors all tend to reduce the load capacity of the frames below the computed predictions. On the other hand, some neglected features, such as the stiffness contributed by the floor system and cladding attached to the frame, and strain hardening of the material, tend to increase the instability load of the frame.

16.3 ANALYTICAL MODELS

Consider the unbraced frame in Fig. 16.4 and some possible analytical models for predicting failure. Curve (*a*) represents a first-order elastic analysis. It is a straight line. This linearity accrues from the facts that all material behavior is taken as elastic, and that equilibrium is based on the original undeformed frame geometry. Curve (*b*) represents a second-order elastic analysis. In this model the deformations of the structure are considered in the formulation of the equilibrium equations. Curve (*c*) represents a first-order elastic-plastic analysis. This analysis is based on a stress-strain curve that has a constant slope up to the yield point and a horizontal line at the yield point. Curve (*d*) represents a second-order inelastic analysis. This analysis can include many second-order effects which will be discussed later. If there is a local buckling failure or a torsional buckling failure, curve (*d*) would be aborted at a lower load as shown by curve (*e*). If the structure is analyzed as a mechanism using rigid plastic analysis, the failure load is shown by curve (*f*).

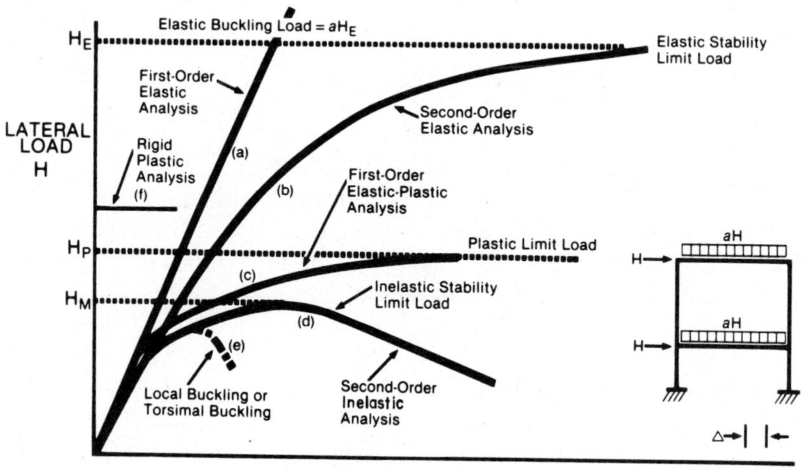

Fig. 16.4 Analytical models.

To illustrate that curve (*d*) (assuming that local buckling and torsional buckling are prevented) is the best failure model, consider the frame shown in Fig. 16.5. The inset to Fig 16.5 shows a single-story single-bay frame subjected to constant vertical loads and to a horizontal load, *H*. The horizontal displacement of the beam is denoted by Δ. The upper broken line depicts the behavior predicted by a first-order inelastic analysis, while the lower dashed curve represents that predicted by a similar second-order analysis. The test results are represented by the full lines joining the solid circles. Although both analyses account for the inelastic action of the frame, the second-order analysis considers the additional moments and shears produced by the *P*-Δ effect, whereas the first-order analysis does not. The second-order analysis also makes an allowance for the reduction in the flexural stiffness of the individual columns by axial load effects.

As load is increased on a frame, members are stressed elastically until parts of the frame are strained into the inelastic region. The ultimate capacity is reached when the combination of progressive yielding, axial force, and joint displacements reduce the stiffness of the structure so that the frame becomes unstable.

The significant stages in the response of the frame are denoted by *A*, *B*, and *C*. Stage *A* on the actual response curve is the stage corresponding to the (nominal) formation of the first plastic hinge. In this test the

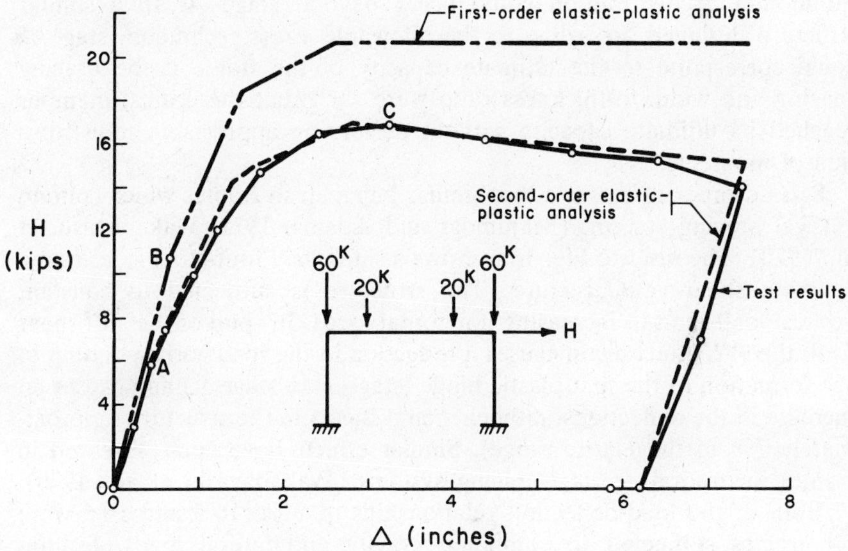

Fig. 16.5 Frame free to sway (Arnold et al., 1968).

transverse bracing and width-to-thickness ratios were selected to prevent premature out-of-plane or local deformations. In a frame designed according to the allowable stress provisions, stage A would correspond to the attainment of initial yielding. The corresponding stage, as predicted by first-order analysis, is denoted as B. Thus the inclusion of second-order effects has significantly reduced the load corresponding to the attainment of initial yielding. These effects have also produced increases in the deflections and corresponding moments and shears throughout the frame. In other words, second-order analysis leads to amplification of all structural effects.

The ultimate strength of the frame is shown as stage C. This stage corresponds to the load level at which the rate of increase of induced shear is greater than the rate of increase of the resisting shear in the story. An increase in the induced story shear involves two components; the increase in the applied lateral load, H, and the increased effective shears developed by the moments due to the vertical loads on the structure acting through the laterally deformed shape (the P-Δ effect). At stage C both the applied lateral load and the vertical loads are constant, but an increase in deflection will increase the P-Δ effect so that any increase in the resisting shear is "used up" by this effect: to sustain any further deformation, the lateral force H would have to be reduced.

Stage C is assumed to correspond to failure of the structure. The structure shown in the inset to Figure 16.5 was designed plastically so that full-moment redistribution could occur beyond stage A. In a similar structure designed according to the allowable stress technique, stage A could correspond to the ultimate capacity of the frame if the bracing spacing and width-to-thickness ratio were such that the critical member reached its ultimate capacity without permitting appreciable redistribution of moments.

Second-order effects produce similar behavior in frames which contain vertical bracing systems (Majumdar and Adams, 1971; Wakabayashi et al., 1970). The inset to Fig. 16.6 shows a single-bay four-story steel-frame concrete shear-wall structure. The structure is subjected to constant vertical loads and to increasing horizontal loads. In spite of the stiff shear wall, the P-Δ effect again causes a reduction in the load corresponding to the formation of the first plastic hinge (stage B to stage A) and causes an increase in the deflections, moments, and shears in the structure (approximately 7% in the elastic range). Similar effects have been observed in frames incorporating steel-bracing systems (Wakabayashi et al., 1970).

Both of the load-deflection relationships of Figs. 16.5 and 16.6 were for frames subjected to combined gravity and lateral loads. Frames subjected to vertical loads only exhibit somewhat different relationships which approach the bifurcation solution of the ideal axially loaded frame.

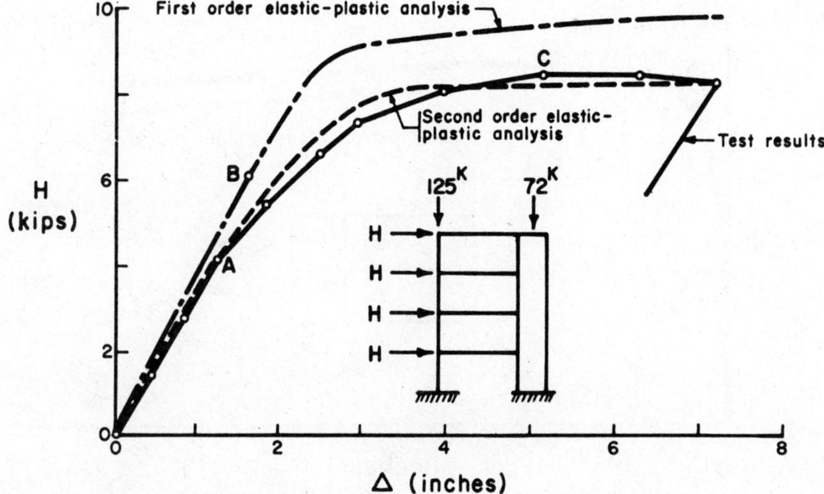

Fig. 16.6 Braced Frame (Majumdar and Adams, 1971).

The inset to Fig. 16.7 shows a three-story single-bay frame subjected only to vertical loads. Since the frame is symmetrical, a first-order analysis will not predict lateral deflections and the ultimate strength will be reached when a beam mechanism forms at $P = 27$ kips. The actual frame is not truly symmetrical (as shown by the load-deflection relationship) because of unavoidable eccentricities from fabrication and erection. For the second-order analysis shown by the dashed line in Fig. 16.7, it was assumed that the frame is subjected to small lateral loads of $\frac{1}{4}$ of 1% of P as well as to the vertical loads to simulate these eccentricities.

During the test the first plastic hinge formed at stage A, shown in Fig. 16.7, and inelastic action progressed throughout the frame, until the ultimate load-carrying capacity was reached at stage C. The action of the frame is similar in most respects to that of a frame under combined loading.

In all three examples shown in Figs. 16.5 through 16.7, failure is represented by point C. It is called the stability limit load since it represents the point at which the frame becomes unstable. Also, in all cases, second-order inelastic analysis, as represented by curve (d) in Fig. 16.4, was the best analytical model representation of frame response. The lateral stiffness of the frame is the important parameter regardless of whether the frames are braced or not.

Referring again back to Fig. 16.4 it should be noted that most contemporary frame designs are based on linear elastic analysis, curve (a). The use of the amplification factor in member design is a means of

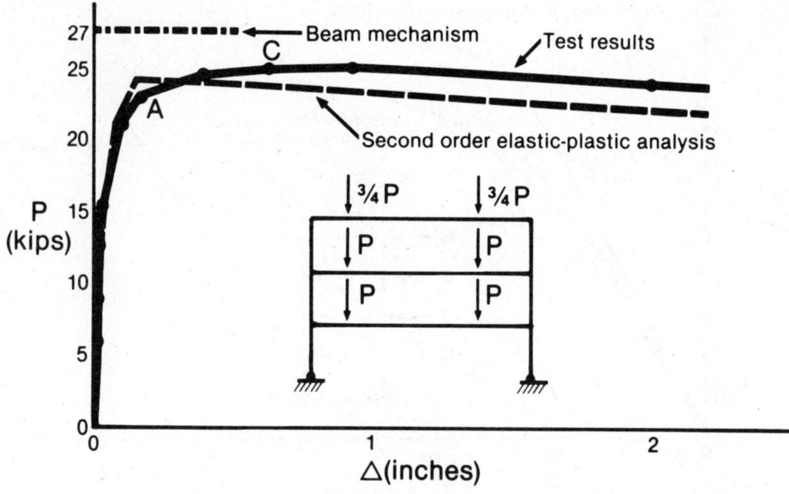

Fig. 16.7 Gravity-loaded frame (McNamee, 1967).

approximating curve (*b*). The use of P_u, C_m, and M_u and the interaction equations given in Chapter 8 are a means of approximating curve (*c*). Finally, the combined attempts to reach curves (*b*) and (*c*) by this method lead to an approximation for curve (*d*), the target for design, but all within the context of a column as an isolated member.

16.4 FACTORS INFLUENCING FRAME STABILITY

For the frames shown in Figs. 16.5 through 16.7, the second-order inelastic analysis included only three second-order effects. These were the moments and shears produced by the P-Δ effect, the reduction in the stiffness of the individual columns by axial loads, and the changes in the lengths of beams and columns resulting from axial strain. It was further assumed that inelastic action was concentrated at specific points in the members. These assumptions were adequate for these models as shown by comparison of the models to the test results.

Much of the literature on the subject of frame stability has emphasized the need to evaluate the effect of horizontal joint displacements, commonly called the P-Δ effect, and the influence of axial force on stability. Undue emphasis on either one or both of these factors could result in a restricted approach to the frame stability problem. Actually, there are other factors that also influence the stability limit of structural frameworks. The various factors may be summarized as follows:

1. Geometrical effects. These consist of:
   a. The influence of axial force on member bending stiffness.
   b. The effect of relative horizontal joint displacements, commonly called the P-Δ effect.
   c. Changes in member chord length resulting from axial strain and bowing, commonly called curvature shortening.
   d. Initial crookedness of members and out-of-plumb erection of the frame.
   e. Finite sizes of joints and panel zone deformations.
2. Material effects. These consist of:
   a. Nonlinear stress-strain relationship.
   b. Residual stresses present in members prior to loading as a result of manufacturing and fabricating processes.
   c. Spread of inelastistic zone in members as member forces increase.
   d. Variations in member strength due to variations from the theoretical cross-sectional dimensions listed in catalogs.
   e. Shearing deformations.
   f. Local buckling or other local distortions.
   g. Out-of-plane movement of frames.
   h. Connection flexibility.
   i.* Strain hardening.
   j.* Contributions of the slab to strength and stiffness.
3. Loading effects. These consist of:
   a.* Nonproportional loading.
   b. Variable repeated loading.
   c. Dynamic effects

All the listed factors influence the stiffness and strength of the frame, and although the first two of the geometric effects are of primary importance, there are situations where one or several of the others could significantly reduce the frame stiffness and thereby reduce the factor of safety with respect to instability. It is of course assumed in the list above that if individual members are tapered or have haunches or other variations in properties along their length, these variations are included in evaluating member stiffness properties.

Designers should be aware of these factors. Many can be designed around such as local buckling and out-of-plane movement. There are analytical procedures available for considering most of these factors if

*Effects 2i, 2j and 3a are beneficial in that they tend to counteract the destabilizing influences of the other effects.

desired (Birnstiel and Iffland, 1980). In any case, the first two factors listed under geometric effects are the prime factors. These are the ones that control the behavior of usual frames.

16.5 STABILITY DESIGN PROCEDURES

Structural steel building frames of usual proportions are likely to fail by instability before a plastic collapse mechanism is formed under increasing loads (Beedle et al, 1969). This mode of failure has been demonstrated both experimentally and numerically. Therefore, the factor of safety with respect to frame instability should be evaluated in every rational frame design procedure. The lateral stiffness of a frame under a particular loading is a measure of the factor of safety with respect to instability at the loading.

Five basic approaches for designing frames to ensure against instability will be discussed herein. These are:

1. Critical load concept (effective length).
2. Second-order, inelastic analysis.
3. P-Δ methods.
4. Merchant-Rankine formula.
5. Moment amplification method.

Methods 2 and 4 are analysis methods wherein frames already designed are checked for the instability limit load. Methods 1, 3, and 5 are design methods wherein the design of the frame is modified to account for instability effects.

16.5.1 Critical Load Concept

Historically, the problem of instability has been approached from the consideration of ideally loaded elastic columns and frames. The solutions for ideal elastic frames are consequently numerous and simple, and cover a relatively extensive variation of frame types and loadings. Although an ideal elastic solution may not even approximate the ultimate load response of a real frame, it can provide a definitive measure of instability.

If an ideal column member is loaded in a manner that does not produce direct bending, the performance under increasing load is that pictured in Fig. 16.8. That is, the only displacements that occur at low loads are those in the direction of the applied loads, until the load reaches its critical value P_{cr}. This load is uniquely determined by an eigenvalue analysis. If P_{cr} lies in the elastic range, then at that load lateral displacements of the column begin and will increase without bound (according to the linearized theories of elastic stability). This is shown by the horizontal

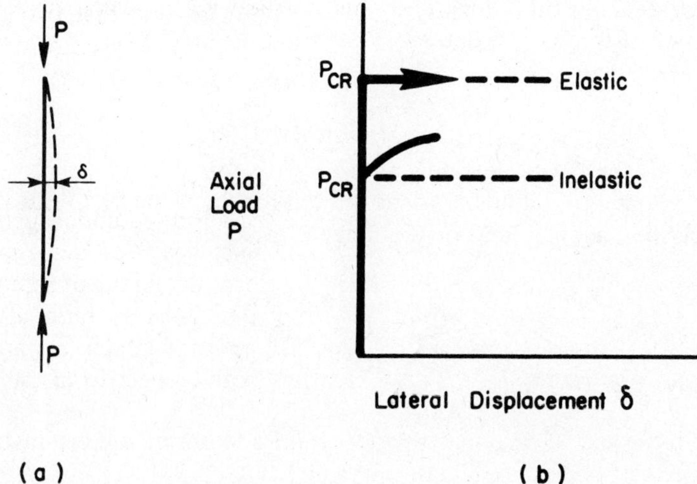

Fig. 16.8 Load-displacement curves for axially loaded perfect column.

line designated "elastic"; and the load at which these displacements are possible is known as the buckling load or, more accurately, as the bifurcation load. If P_{cr} lies in the inelastic range of the member or frame, a similar bifurcation load P_{cr} exists, but lateral displacements are possible only under slightly increasing load, as is also shown in Fig. 16.8, due to strain reversal associated with finite curvature in the postbuckling configuration.

The solution for a column with hinged ends was first published by Euler (1744) and it was expressed as

$$P_{cr} = P_e = \frac{\pi^2 EI}{L^2} \tag{16.1}$$

where

P_e = elastic buckling load
E = modulus of elasticity
I = cross-sectional moment of inertia
L = length of the column

Equation 16.1 can be modified for elastic columns having other than hinged ends by substituting the value of KL for L, where K is defined as the effective length factor. The product KL is the effective length. For a column with hinged ends $K = 1$.

Equation 16.1 can be written in terms of stress by replacing I with Ar^2, where A is the cross-sectional area of the column and r is the radius of gyration. Making this substitution and introducing the concept of effective length, Eq. 16.1 becomes

$$\sigma_e = \frac{P_e}{A} = \frac{\pi^2 E}{(KL/r)^2} \tag{16.2}$$

where σ_e is the elastic buckling stress.

If the tangent modulus E_t is substituted for the value of E in Eq. 16.1, the inelastic buckling load or the tangent modulus load, P_t is determined as

$$P_{cr} = P_t = \frac{\pi^2 E_t I}{L^2} \tag{16.3}$$

Similarly, Eq. 16.2 becomes

$$\sigma_c = \frac{P_t}{A} = \frac{\pi^2 E_t}{(KL/r)^2} \tag{16.4}$$

where σ_c is the critical stress. Procedures have been proposed for obtaining the effective length factors for inelastic columns (Yura, 1971; Disque, 1973).

The critical load for a column is the load at which bifurcation occurs as determined by the theoretical stability analysis. The term "bifurcation" refers to the load-deflection behavior of a perfectly straight compression element at critical load as illustrated in Fig. 16.8. Bifurcation can occur in the inelastic range only if the pattern of postyield properties and/or residual stresses is symmetrically disposed so that no bending moment is developed at subcritical loads. At the critical load, a member can be in equilibrium in either a straight or slightly deflected configuration, and a bifurcation results at a branch point in the plot of axial load versus lateral displacement from which two alternative load-displacement plots are mathematically valid.

The critical load for an ideal axially loaded frame can also be determined by an eigenvalue analysis, resulting in load-displacement curves as shown in Fig. 16.9 similar to those for axially loaded ideal columns. Since the axially loaded frame is ideal, no primary bending of the members can exist. This requires that the differential axial shortening during buckling of all the columns in Fig. 16.9*a* be ignored.

The critical-load solution for frames also results in the concept of effective length. In a beam-column interaction formula for a framed column the effective length factor must be included in the formula if it is to evaluate properly the column strength in the limit as any prebuckling moments induced in the column approach zero.

The problem of determination of the effective lengths of members in a

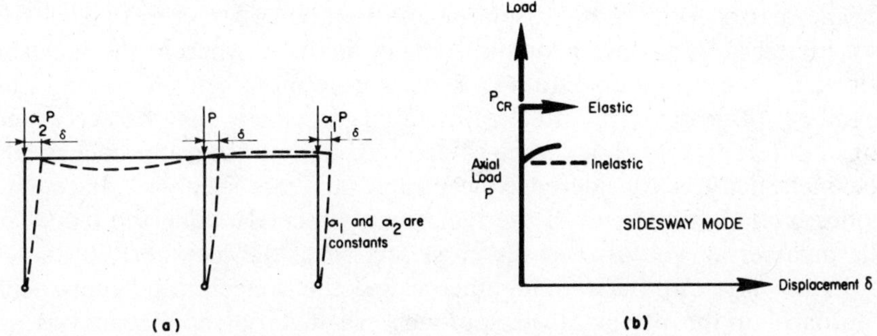

Fig. 16.9 Load-displacement curve for ideal joint-loaded frame.

frame was treated in detail by Johnston (1976) and will not be repeated herein. The subject has also been treated extensively by CTBUH (1979). Reference is made to these documents for further details.

The basis for using the critical load concept for stability design can be seen graphically by comparing the response curves of the gravity loaded frame in Fig. 16.7 to the response curve of the ideal joint-loaded frame of Fig. 16.9. The similarity indicates that for gravity-loaded-only frames, the critical load approach should provide a reasonable solution.

16.5.2 Second-Order Inelastic Analysis

The proportions and material properties of building frames are such that they are usually stressed into the inelastic range prior to attaining the stability limit load. The inelastic action may be due to the gravity loads on the girders, or to horizontal forces acting on the structure, or to a combination of both. If it is assumed that the members are geometrically perfect and that out-of-plane displacements and twist of the members is precluded, the frames may be represented by a set of equations:

$$EI \frac{d^4 v}{dz^4} + P \frac{d^2 v}{dz^2} = q(z) \tag{16.5}$$

where

$q \neq 0$ for some of the members
v = transverse or lateral displacement of a member in the plane of the frame
z = direction of axis along length of the member
EI = flexural rigidity
q = transverse load distributed along member length
P = axial compression of the member

The matrix form of the displacement method of structural analysis is considered most practical for the stability analysis, wherein the inelastic zones of the members are taken into account when evaluating the member stiffnesses. The effect of material inelasticity may be accounted for in different ways, but can be categorized by two basic approaches: (1) the inelasticity is considered concentrated at cross sections adjacent to nodes, or (2) the spread of the inelasticity proceeds along the length of the member as well as over the cross section of the member.

In the first approach, sometimes called the concentrated inelasticity approach or the plastic-hinge approach, all material nonlinearity is assumed concentrated at the ends of the elements that comprise the model. If the real member is not loaded along its length, the element length is equal to the member length. For transversely loaded members, nodes are assigned at load points and elements are defined between nodes. The element is considered elastic between the nodes. An end moment-rotation function is selected that adequately describes the inelastic cross-section response at a node.

The second approach considers the member to be divided into segments along its length and the cross section is subdivided into elemental areas in order to "keep track" of the inelastic strains. Because the strain for each elemental area must be monitored, the distributed inelasticity approach requires large computer memory capacity.

Many analyses of plane frames have been presented using variations of the concentrated inelasticity approach. Because the problem is highly nonlinear in the inelastic range, it is customary to load the model of the frame incrementally satisfying equilibrium after each increment of the load.

The distributed inelasticity approach is really a research tool, because of the high cost of such solutions. The first frame analysis of this type appears to be that by Wright (1966). A more recent analysis of the distributed inelasticity type was presented by Swanger and Emkin (1979), wherein it was termed the fiber element model.

In the foregoing it was assumed that no out-of-plane motion of any member of the framework was possible. In reality the columns of frames are usually unrestrained normal to the column webs between floors. Such columns could buckle in the lateral-torsional mode under certain circumstances, with the result that the critical load of the frame would be less than the same frame with that type of buckling prevented.

Inelastic three-dimensional bare frameworks, that is, frameworks in which the cladding and the floor system were neglected, have been analyzed by both the concentrated and distributed inelasticity approaches. Jonatowski and Birnstiel (1969) utilized the concentrated inelasticity approach with incremental loading to determine the stability limit load of a variety of three-dimensional structures. It appears that the only distributed inelasticity solution for such frames is that by Swanger and Emkin (1979).

Table 16.1 gives a broad classification of some of the existing works. It is possible that some important contributions have been overlooked. A brief review of some of the works follows.

Majid and Anderson (1968) have developed a second-order inelastic analysis based on a matrix displacement method and a concentrated inelasticity approach to analyze irregular frames loaded by proportional concentrated loads. The joint displacements (three per joint) are the basic unknowns, and hinge rotations are extra unknowns in the presence of plastic hinges, so the matrix size varies as plastic hinges form. The program can deal with very large frames.

Parikh (1966) developed a computer method using a concentrated inelasticity approach for second-order inelastic analysis of regular frames under proportional loading. He included the effect of axial shortening of columns. The slope deflection equations are generalized to consider the behavior of members after the formation of one or more plastic hinges in the members. The structure is entirely reanalyzed at each load level and

Table 16.1. Classification of Some Inelastic Stability Analyses

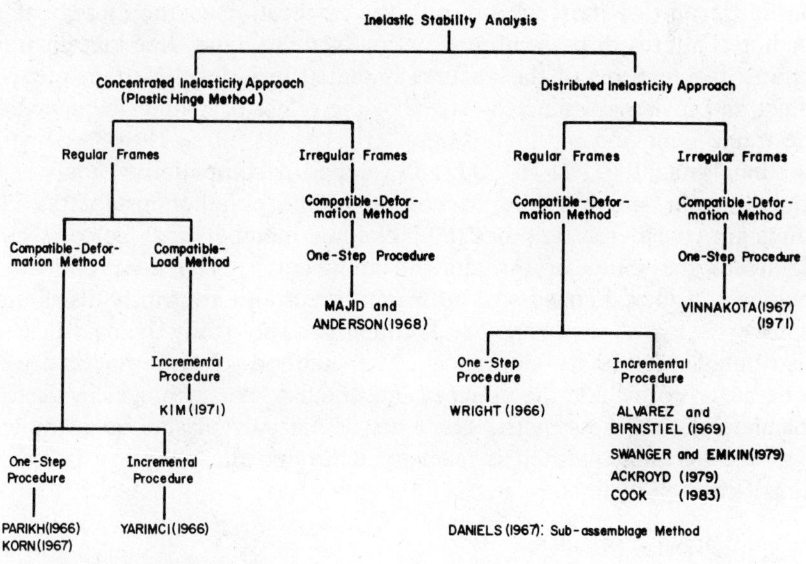

plastic hinges are added, as needed, until the procedure diverges. The program is effective in the use of computer time and storage space and can handle structures up to 30 stories and five bays in size. Korn (1967) made a behavioral study of single-bay and two-bay multistory regular rectangular frames. He considered the effect of axial deformation and curvature shortening due to bending.

Wright (1966) presented a rigorous compatibility analysis of regular multistory frames under nonproportional loads. The distributed inelasticity approach is utilized for the study. Each member of the frame is divided into discrete intervals and the effective stiffness of each segment is computed using the best available estimate of moment and axial force in each cycle of iteration. Using these effective stiffnesses, the stiffness coefficients of partially plastic members are obtained by the finite-difference method. The stiffness matrix of the structure is solved by iteration. To avoid the divergence of the solution near the stability limit load and to obtain the unloading part of the load-deformation relationship, he introduced special numerical stabilization techniques. Axial shortening of members was neglected.

Alvarez and Birnstiel (1969) presented a compatibility analysis to study the behavior of plane unbraced rigid frames under proportional and nonproportional loading, up to the point of instability. The unknown incremental displacement vector was determined from the incremental load vector and the stiffness matrix based on the average of the remaining elastic portion of the frame during the application of the increment of loading. Failures of both bifurcation and stability type were investigated. One of the features of the analysis is that it includes the strain reversal effect and makes possible the study of the effect of loading sequence on the frame behavior.

Vinnakota (1967, 1971, 1972) developed a compatibility analysis to study regular and irregular frames up to the point of instability. The joints are considered rigid but the individual members may be connected rigidly to the joints or through pin connections. The nonproportional loading considered consists of both joint loads and arbitrarily distributed transverse loads on beams and columns. The study is based on a distributed inelasticity approach. The method of transfer matrices, generalized to include the effect of intermediate plastic hinges, is used to calculate the stiffness matrix elements of partially plasticized members. The displacement method is used to determine the unknown deformations (three per joint).

16.5.3 *P*-Delta Methods

Consider the frame in Fig. 16.10, which depicts a frame subjected to both horizontal and vertical loads. For illustration, these loads are applied

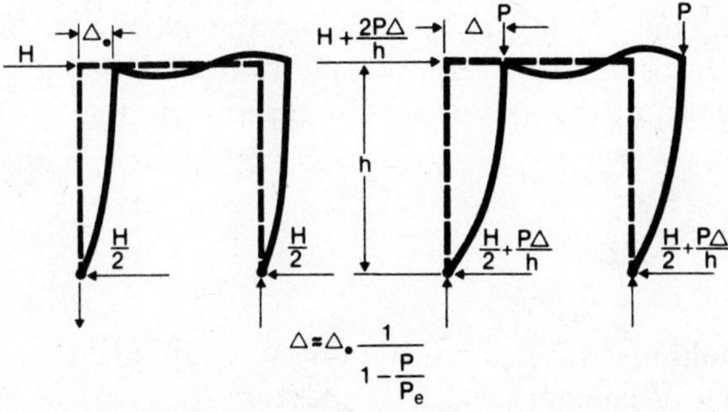

Fig. 16.10 Effect of joint displacements on bending moments and forces.

separately. The horizontal force induces an initial horizontal deflection. The vertical load is carried through this initial displacement with the resulting P-Δ moment creating additional horizontal deflection, which in turn increases the P-Δ moment. The horizontal deflection increases progressively to instability of the structure unless the structure has sufficient internal sway stiffness to stabilize the movement at some amplified initial deflection. If this stiffness exists, the structure is stable. From this description of the failure process it is seen that formulation of the equilibrium equations on the deformed structure is a primary consideration in assessing a structure's stability. A derivation of the P-Δ method follows.

The method is based on the assumption that the P-Δ effect is the primary consideration in stability. If, in any ultimate strength determination of a frame, working loads are determined by dividing ultimate loads by a load factor F, it is consistent to determine the fictitious lateral force due to the P-Δ effect by dividing the lateral force computed at ultimate load by the same load factor. This is shown in the following equation:

$$V' = \frac{FP\Delta_{\text{ult}}}{h}\,\frac{1}{F} = \frac{P\Delta_{\text{ult}}}{h} \tag{16.6a}$$

Assume that

$$\Delta_{\text{ult}} = \bar{F}\Delta \tag{16.6b}$$

and then

$$V' = \bar{F}\,\frac{P\Delta}{h} \tag{16.6c}$$

In Eq. 16.6c the derivation of the working load story shear V' on a column is shown, remembering that the moment, equal to the axial force P, times the displacement Δ, divided by the column height h, gives the additional column shear V. To eliminate the ultimate deflection term, introduce $\bar{F}$, a deflection factor of safety, as shown in Fig. 16.11. The actual derivation of the P-Δ equations is shown in Fig. 16.12 and in the following equations:

$$V'_i = \frac{\bar{F} \sum P_i}{h_i} (\Delta_{i+1} - \Delta_i) \tag{16.7}$$

where

V'_i = additional shear in story i due to sway forces
P_i = sum of column axial loads in story i
$\bar{F}$ = ratio of factored load lateral deflection to design load lateral deflection
h_i = height of story i
Δ_{i+1}, Δ_i = displacements at levels $i+1$ and i, respectively

Sway forces from Fig. 16.12 are then

$$H'_i = V'_{i-1} - V'_i \tag{16.8}$$

The steps in the P-Δ procedure, as given in the third edition of this guide, are as follows:

1. Determine sum of axial loads in all the columns at each story.
2. Determine incremental story shears due to P-Δ effects using Δ_i from first-order analysis in Eq. 16.7.

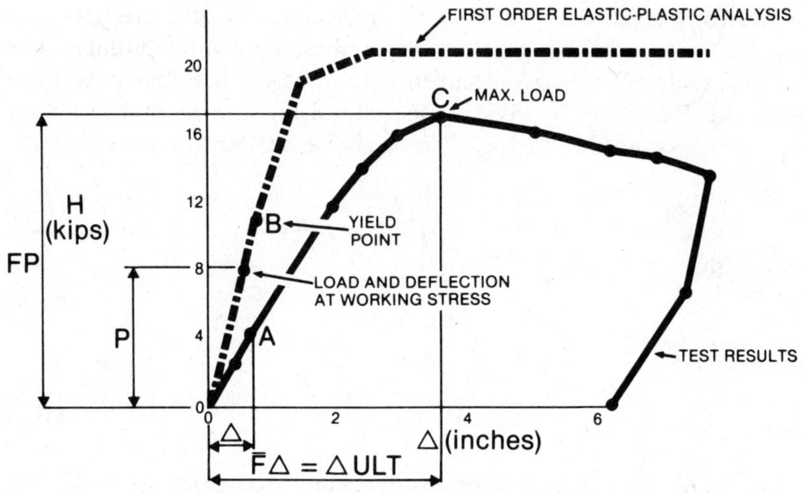

Fig. 16.11 Load and deflection factors of safety.

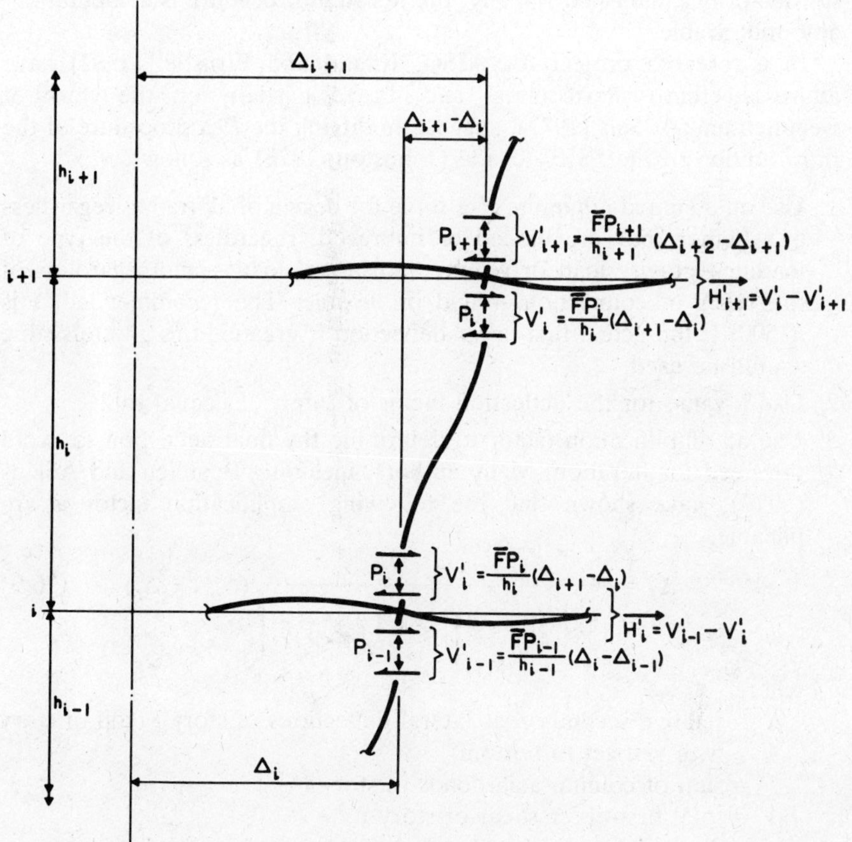

Fig. 16.12 Sway forces.

3. Determine sway forces due to P-Δ.
4. Perform first-order analysis with horizontal loads plus sway forces.
5. Check new drift with that assumed in step 2 and repeat steps 2 through 4 if necessary.
6. Design columns as pin-ended.

There are several shortcomings to the P-Δ procedure as just presented. The method does not include any means of assessing the effect of axial load on member stiffness (Kanchanalai, 1977). This is important for gravity-loaded frames, as shown by the previous comparison of Fig. 16.9 to Fig. 16.7. The value of the deflection factor of safety, $\bar{F}$, has not been satisfactorily established. Finally, the iteration procedure is computationally undesirable.

In a research project for AISC, Iffland and Birnstiel (1982) have addressed these shortcomings and, based partially on the work of Beaulieu and Adams (1977), suggest modifying the P-Δ procedure of the third edition of the SSRC Guide (Johnston, 1976) as follows:

1. Use an assumed minimum P-Δ force for design of all frames regardless of whether they are braced or unbraced, regardless of the type of loading—gravity load or gravity and lateral loads—and regardless of the types of connections—rigid or flexible. The recommended Δ is $h/500$. If the actual first-order deflection is greater, this greater value should be used.
2. Use a value for the deflection factor of safety, $\bar{F}$, equal to 2.
3. Use an amplification factor to determine the final deflection to avoid the need for iteration. Many authors, including Beaulieu and Adams (1977), have shown that the following amplification factor is applicable:

$$\Delta_i'' = \frac{1}{1 - \left[\sum P_i(\Delta_{i+i} - \Delta_i) \Big/ \sum V_i h_i\right]} (\Delta_{i+1} - \Delta_i) \qquad (16.9)$$

where

Δ_i'' = relative second-order lateral deflections of story i (top of story with respect to bottom)
ΣP_i = sum of column axial loads in story i
ΣV_i = total first-order shear of story i

Equation 16.7 can then be modified to

$$V_i' = \frac{\bar{F} \sum P_i \Delta_i''}{h_i} \qquad (16.10)$$

and iteration can be avoided.

The suggested procedure of applying P-Δ forces to frames loaded with only gravity loads or to braced frames and elimination of the effective length factor in the bending moment term of the interaction formulas can be justified theoretically. It should be remembered that the elastic critical-load solution is just an upper-bound solution to the real problem. The fact that the 1978 AISC Specification (AISC, 1978) is based on this solution (i.e., the effective-length concept is used to evaluate F_e' in the moment amplification term) is a simplification of how real columns in a building frame behave. Application of P-Δ forces on gravity loaded or braced frames converts the analyses into bending problems. Thus the use of the moment from the P-Δ procedure in the beam-column design formula eliminates the need of the use of $1/(1 - fa/F'_e)$ in the AISC design formula.

The value of Δ to use in the P-Δ procedure for gravity loaded or braced frames is almost immaterial from the viewpoint of changing the problem from the simplified bifurcation solution to a bending solution. The value of $h/500$ has been selected since this value will account for erection tolerances—a stability consideration quite often ignored. Beaulieu and Adams (1977) have made an extensive study of out-of-plumbness and erection tolerances and their effect on frame stability. In the P-Δ procedure they recommend the following value of Δ for frames without lateral loads:

$$\Delta = \frac{0.006\,h}{n^{0.455}} \tag{16.11}$$

where h is the story height and n is the number of columns in the building. Their formula is based on a statistical study of out-of-plumbness for a number of buildings. It can be seen that as soon as n gets as large as 12, $h/500$ is a conservative value. Since this value is also the AISC erection tolerance for columns, it has been recommended as the minimum Δ to use in the P-Δ procedure.

The recommended value of $\bar{F}$ equal to 2 is somewhat arbitrary. In Canadian practice, where the P-Δ procedure is commonly used, $\bar{F}$ is taken equal to F, the load factor of safety, which is 1.67. A value of 2 has been selected because of the relationship illustrated in Fig. 16.11. It is believed that the larger value of 2 will account for some of the many other nonlinear considerations not usually included in inelastic second-order analyses and discussed in further detail by Birnstiel and Iffland (1980).

The P-Δ method is generally the best suited for design use since most designers first carry out a linear elastic analysis of a proposed structure or system and then design each member one at a time. Interaction beam-column formulas are also ideally suited to the use of this method. However, as noted in Section 16.5.1, frames must be evaluated in the limit as any prebuckling moments in the column approach zero. This requires that for frames with high axial loads but low bending moments in the columns, the effective column length should be used in the first (axial load) term, f_a/F_a, of any interaction equations used for design of beam-columns if either of the design methods 1, 3, or 5 is used. When the magnitude of this first term is less than some critical value, classical (bifurcation) buckling of the frame does not govern and the effective length of the column need not be used. This critical value can be conservatively estimated as $f_a/F_a = 0.85$, where F_a is evaluated using an effective length. This value is based on a review of the work by Kanchanalai (1977), which presented a comparison between exact solutions of beam-columns, effective length solutions and P-Δ solutions. For values below this figure, the effective length need not be used.

In the P-Δ method, designers should be careful to include all the vertical load in their P-Δ computations. Loads on unbraced neighboring frames within the building must be included in determining $P\Delta$ effects on the sway resistant frames or other bracing elements.

16.5.4 Merchant-Rankine Formula

Consider Fig. 16.13, which is a comparison of experimental failure loads

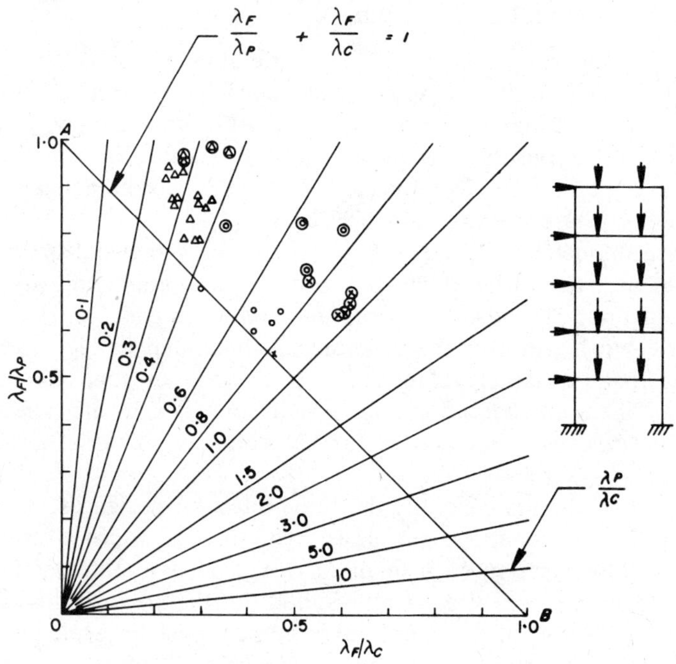

Fig. 16.13 Comparison of Rankine formula with experimental failure loads (Horne, 1963).

on steel frames with the Merchant-Rankine formula, taken from Horne (1963). The Merchant-Rankine formula is

$$\frac{\lambda_F}{\lambda_p} + \frac{\lambda_F}{\lambda_c} = 1 \tag{16.12a}$$

where

λ_F = failure load multiplier
λ_p = collapse multiplier obtained by rigid plastic theory
λ_c = multiplier for elastic critical load

Figure 16.13 indicates that the Merchant-Rankine formula (Eq. 16.12a) is a safe approximate method of obtaining the failure load if the rigid plastic load and the elastic-critical loads are known. The latter of these two values can be relatively easily calculated.

Next consider Fig. 16.14, taken from ECCS (1976), which gives a comparison of the Merchant-Rankine formula (curve 1) with results from elastic-plastic second-order calculations. The form of the Merchant-Rankine formula used is

$$\lambda_F = \frac{\lambda_p}{1 + \lambda_p/\lambda_c} \tag{16.12b}$$

which is identical to Eq. 16.12a. Figure 16.14 shows that Eq. 16.12b is

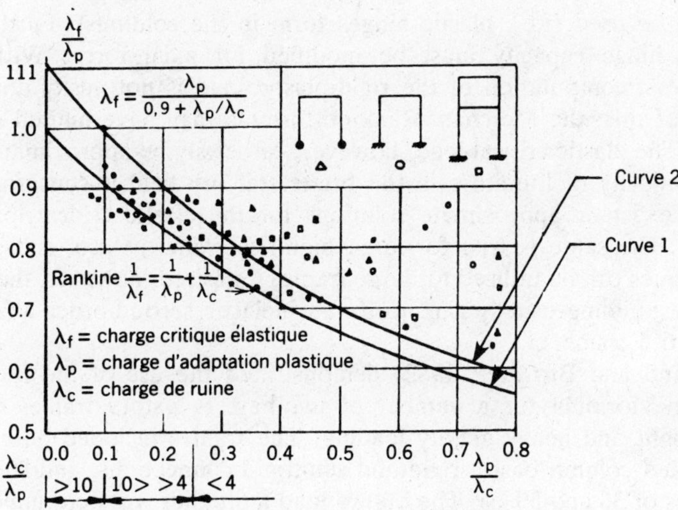

Fig. 16.14 Comparison of merchant-Rankine formula with second-order calculation (ECCS, 1976).

rather conservative. Further, it was established by comparison with the theoretical behavior of bare frames following an ideal rigid-plastic material behavior. Real frames benefit from the strengthening effect of cladding and of the strain hardening of steel. For these reasons the following modified version of Eq. 16.12b, due to Wood (1974),

$$\lambda_F = \frac{\lambda_p}{0.9 + \lambda_p/\lambda_c} \tag{16.13}$$

has been tentatively adopted in the ECCS recommendations for the design of unbraced frames. Equation 16.13 is shown as curve 2 in Fig. 16.14. The simplified design method adopted is as follows:

1. If λ_c/λ_p is greater than 10, then λ_F is limited to λ_p and the frame may be designed according to first-order theory.
2. If $4 < \lambda_c/\lambda_p < 10$, then λ_F is given by Eq. 16.13, provided that:
   a. Plastic hinges are allowed only in beams and not in the columns.
   b. The frames are braced in the perpendicular direction, unless it may be shown that the frames remain elastic under wind loading in that direction.
3. If λ_c/λ_p is less than 4, an elastic-plastic second-order method must be used (eventually computer based).

The rigid-plastic (mechanism) load is easily calculated for most small frames provided that hinges do not form in the columns. However, for stability computations, only collapse mechanisms that involve sidesway should be used (i.e., plastic hinges form in the columns). Further, the plastic hinge capacity must be modified for axial force. With these restraints, computation of the rigid-plastic load is not easily done. Because of this, the Merchant-Rankine formula may have limited applications. The elastic critical load, however, can easily be approximated. The vast majority of literature in the frame stability field is concerned with either exact or approximate solutions for the elastic critical load of a frame. The same is true for large high-rise frames. Also, subassembly techniques can be utilized for large frames if desired; however, the cost of a large building usually can justify an inelastic second-order analysis of the entire frame.

Iffland and Birnstiel (1982) demonstrated the use of the Merchant-Rankine formula for a number of two-bay, two-story frames carrying both light and heavy gravity loading. The frames included both hinged and fixed column bases, rigid and semirigid connections, and steel yield stresses of 36 and 50 ksi. The failure load multiplier was determined from both the Merchant-Rankine formula and from a second-order inelastic analysis. This analysis utilized the method developed by Jonatowski and

Birnstiel (1969). Based on this study it was concluded that the Merchant-Rankine formula gave conservative estimates of the stability limit load. Collapse mechanisms involving sideway were used for computing the rigid-plastic load.

16.5.5 Moment Amplification Method

The moment amplification method (Kanchanalai and Lu, 1979; Lu, 1983; LeMessurier, 1977) is a design method utilizing the isolated-column approach. The method incorporates the two primary second-order effects that influence frame stability—the influence of axial force on member bending stiffness and the effect of horizontal joint displacements. It is an indirect method wherein the structure is analyzed by first-order theory, and the instability effects are accounted for in the member selection phase of the design process by amplifying the design moments.

The following form of this method for beam-column design was adopted for the AISC Load and Resistance Factor Design Specification (1986). The ultimate-strength interaction equations are

$$\frac{P}{P_{cr}} + \frac{8M^*}{9M_u} \leq 1 \qquad \text{for} \quad P/P_{cr} \geq 0.2 \tag{16.14a}$$

$$\frac{P}{2P_{cr}} + \frac{M^*}{M_u} \leq 1 \qquad \text{for} \quad P/P_{cr} \leq 0.2 \tag{16.14b}$$

These equations are special forms for the case where bending is only about a major axis of the member and where lateral-torsional and minor-axis buckling are prevented by suitable lateral bracing. [The general formulas (AISC, 1986) consider biaxial bending and lateral-torsional buckling and minor-axis buckling also.]

The terms in Eqs. 16.14 are defined as follows:

P: axial force.

P_{cr}: critical buckling load of member, computed for an effective length $Kh > h$ in the case of unbraced frames.

M_u: the flexural capacity of the member in the absence of axial force (e.g., for a laterally braced compact section $M_u = M_p$, the plastic moment).

$$M^* = B_1 M_{nt} + B_2 M_{lt} \tag{16.15}$$

M^* is the amplified moment, and the terms in Eq. 16.15 are defined as:

M_{nt}: Moment due to a first-order elastic structural analysis assuming no story translation

$$B_1 = \frac{C_m}{1 - P/P_{e1}} \geq 1 \tag{16.16}$$

where C_m was defined p.:eviously, and

$$P_{e1} = \frac{\pi^2 EI_x}{h^2} \tag{16.17}$$

M_{lt}: Moment due to a first-order elastic analysis for the forces causing story translation

$$B_2 = \frac{1}{1 - \left\{\sum P \Big/ \sum H\text{-}\right)(\Delta/h)} \tag{16.18}$$

where
ΣP = sum of gravity loads acting on the story
ΣH = story shear
h = story height
Δ = first-order story sway deflection

Alternatively B_2M_{lt} may be taken as the second-order moment at the column ends as determined by a second-order elastic analysis or the revised P-Δ analysis given in Section 16.5.3. In lieu of Eq. 16.18 it is also permitted to use $B_2 = 1(1 - \Sigma P/\Sigma P_e)$, when P_e is computed for a sway buckling effective length.

16.6 STABILITY OF FRAMES SUBJECTED TO DYNAMIC LOADS

The behavior of dynamic stabiity can be demonstrated by reference to the elastic column shown in Fig. 16.15. Let the column be subjected to a periodic longitudinal force, $N(t) = (\alpha + \beta \cos \theta t)N_0$, in which N_0 is the static buckling load of the member, and α and β are the fractional numbers that ensure that the applied axial loads are always less than the

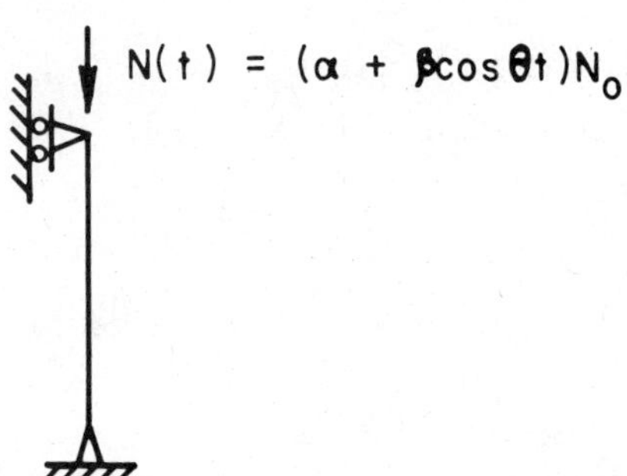

Fig. 16.15 Column member.

buckling load. It can then be shown that for a certain relationship between the longitudinal forcing frequency, θ, and the natural frequency of transverse vibration, ω, the column becomes dynamically unstable and transverse vibrations occur. The amplitudes of the transverse vibrations increase rapidly to large values, and resonance develops. The resonance, which is called parametric resonance, is different from the resonance resulting from ordinary forced vibrations. The general theory of dynamic stability and it application were extensively presented by Bolotin (1956) (see also Chapter 19). Related studies on the cyclic and dynamic stability of bars and structures are given by Arnold et al. (1966a,b), Krajcinovic and Herrmann (1968), Brown et al. (1968), Herrmann (1967), Wakabayashi (1970), and Barr and McWhannell (1971). The purpose of the following discussion is to demonstrate the behavior of dynamic instability of frameworks and the P-Δ effect on dynamic response.

Assume that the typical structure shown in Fig. 16.16 is subjected to a time-dependent axial force, $N(t)$, and a lateral force, $F(t)$, or foundation movement, $G(t)$, and has a superimposed mass, m, and a concentrated mass, M_i, in addition to its own weight. Consider now that this structure is subjected to the externally applied forces, $N(t)$ and $F(t)$, only. The force $F(t)$ can be any type of forcing function, but the longitudinal force is confined to $N(t) = (\alpha + \beta \cos \theta t)N_0$ in which N_0, α, and β have been defined and are in terms of the structural system. The motion equations of the system may be expressed in a matrix form as

$$[M]\{\ddot{x}\} + ([K] - \alpha N_0[S_s] - \beta N_0[S_t])\{x\} = \{F(t)\} \qquad (16.19)$$

where

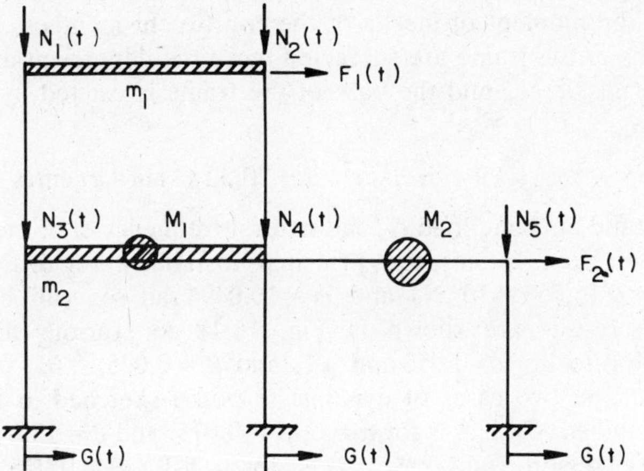

Fig. 16.16 General loading condition (Cheng and Tseng, 1973).

$[M]$ = mass matrix
$[K]$ = structural stiffness matrix
$[S_s]$ = geometric matrix occasioned by the P-Δ effect of static loads
$[S_t]$ = geometric matrix occasioned by the P-Δ effect of the dynamic axial loads having periodic functions, cos θt
$\{\ddot{x}\}, \{x\}$ = acceleration and displacement of the structural nodal coordinates, respectively

When the structure is subjected to earthquakes, Eq. (16.19) may be rewritten as

$$[M]\{\ddot{u}\} + ([K] - \alpha N_0[S_s] - \tau[M][S_g])\{u\} = -[M]\ddot{x}_g \qquad (16.20)$$

in which $[S_g]$ is the geometric matrix occasioned by the P-Δ effect of vertical earthquake accelerations, τ the percentage of acceleration of gravity, $\ddot{x}_g$ the horizontal earthquake acceleration, and $\{\ddot{u}\}$ and $\{u\}$ are the acceleration and displacement of the structural nodal coordinates relative to the base.

The dynamic instability may be defined (Bolotin, 1956; Cheng and Tseng, 1973) by a region in relation to transverse natural frequency, longitudinal forcing frequency, and the magnitude of axial dynamic force. The principal instability region is important in practice and may be obtained from the following determinant:

$$\left|[K] - \left(\alpha \pm \frac{1}{2}\,\beta\right)N_0[S_s] - \frac{\theta^2}{4}\,[M]\right| = 0 \qquad (16.21)$$

The instability region may be shown for the two-story steel framework shown in Fig. 16.17 for which the masses lumped at the floor and the length and the moment of inertia of the constituent members are given. The columns of the frame are subjected to a time-dependent axial force, $N_t = (\alpha + \beta \cos \theta t)N_0$, and the base of the frame is excited by a ground acceleration,

$$\ddot{x}_g = -8\pi^2 \sin 4\pi t \text{ in./sec}^2 \quad (-20.32\pi^2 \sin 4\pi t \text{ cm/s}^2)$$

After the static buckling load N_0 and natural frequency ω of the structual system have been found, the principal instability regions for $N_0 =$ 3766.27 kips ($16{,}753 \times 10^3$ N) and $\omega = 10.0494$ rad/sec can be investigated. The results are shown in Fig. 16.18 for various axial loads corresponding to $\alpha = 0$, 0.05, and 0.1, and $\beta = 0.025$, 0.05, 0.075, 0.1, and 0.1025. The two cases of dynamic response sketched in Fig. 16.18 have been studied. Case A is for $\alpha = 0$, $\beta = 0.075$, and $\theta = 15.0$ rad/sec in the stability region, and case B is for $\alpha = 0$, $\beta = 0.075$ and $\theta =$ 20.1 rad/sec in the instability region. The incremental numerical method

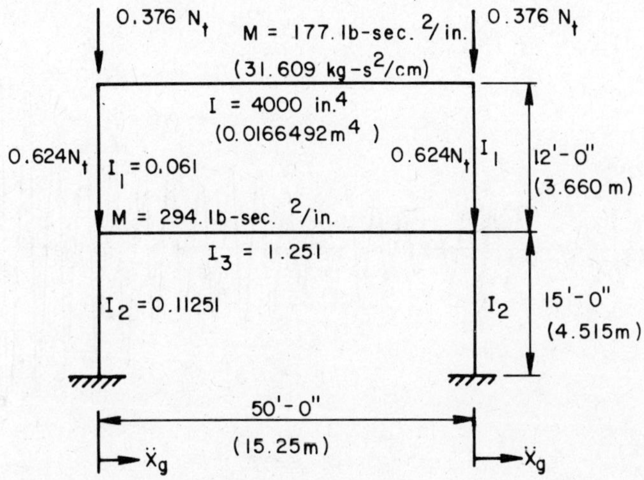

Fig. 16.17 Given loading for instability analysis (Cheng and Tseng, 1973).

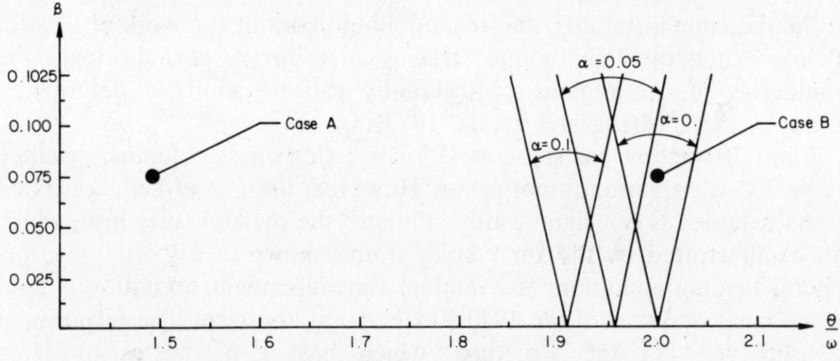

Fig. 16.18 Dynamic instability region of Fig. 16.17 (Cheng and Tseng, 1973).

with a time interval of $\Delta t = 0.025$ sec has been employed for studying the relative displacement at the top floor (Cheng and Oster, 1976a). The results associated with cases A and B are shown in Fig. 16.19. The response behavior illustrates that dynamic instability can occur for a certain relationship of θ and ω and that the magnitude of an axial load can influence the domain of an instability region but is not a primary factor for the instability developement. The discussion above is based on an elastic structure subject to undamped vibrations resulting from a pulsating periodic axial load. For the effect of damping and nonlinearlity

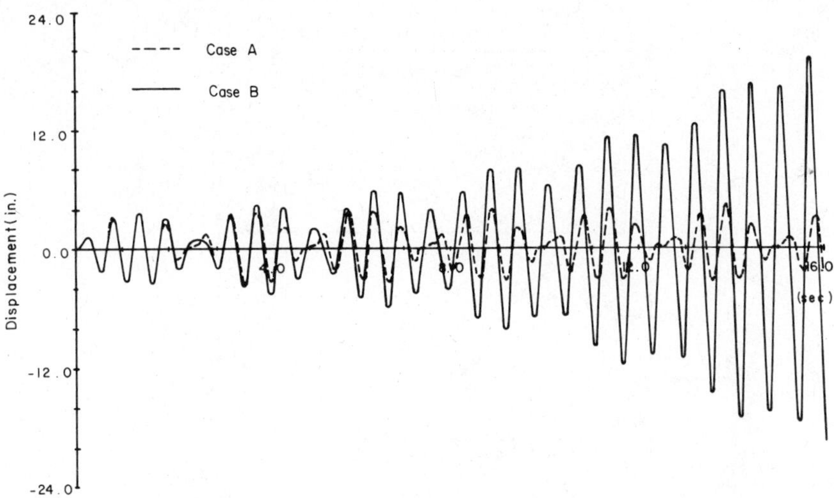

Fig. 16.19 Displacement at top floor of Fig. 16.17 (Cheng and Tseng, 1973).

on the dynamic instability, the reader should consult the work of Bolotin (1956). When the longitudinal force is an arbitrary periodic load, the boundaries of the regions of instability can be similarly determined (Bolotin, 1956; Cheng and Tseng, 1973).

When a structure has static axial loads acting on its columns, dynamic instability will apparently not occur. However, the *P*-Δ effect occasioned by the axial loads can significantly influence the dynamic response, which can be illustrated by the three-story frame shown in Fig. 16.20. For a general loading condition, the random timedependent excitation is used on the basis of 30 sec of the 1940 El Centro earthquake. The response at the top floor of the structure, which has floor masses of M_1 = 14.715 kips/sec^2/ft (219.0 kg-s^2/cm), M_2 = 13.749 kip-sec^2/ft (204.6 kg-s^2/cm), and M_3 = 6.883 kip-sec^2/ft (102.4 kg-s^2/cm), is shown in Fig. 16.21 with and without the *P*-Δ effect included. These floor masses result in a first-mode natural period of T_n = 2.426 sec when one does not consider the *P*-Δ effect and in a natural period of T_s = 2.731 sec when one does consider the *P*-Δ effect. As shown in Fig. 16.21, the maximum floor displacement is realized when the *P*-Δ effect is included.

The maximum displacements of the top floor of the structure are obtained for different natural frequencies, which are evaluated by multiplying each floor mass by equivalent factors. Figure 16.22*a* shows the variation in the maximum responses with and without the *P*-Δ effect. The natural period, T_n, used in the analysis does not include the *P*-Δ effect of

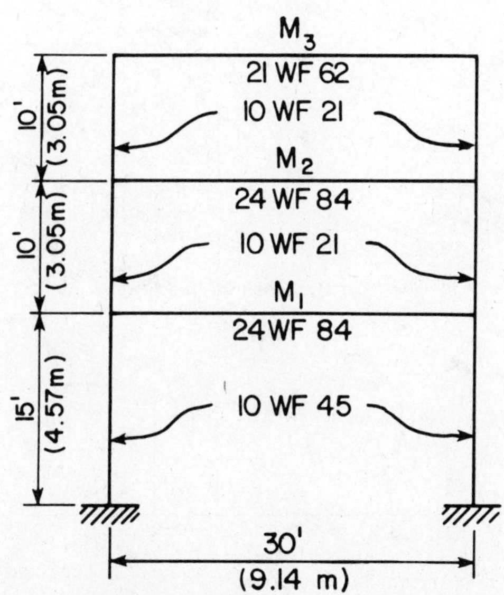

Fig. 16.20 Three-story frame for P-Δ effect analysis.

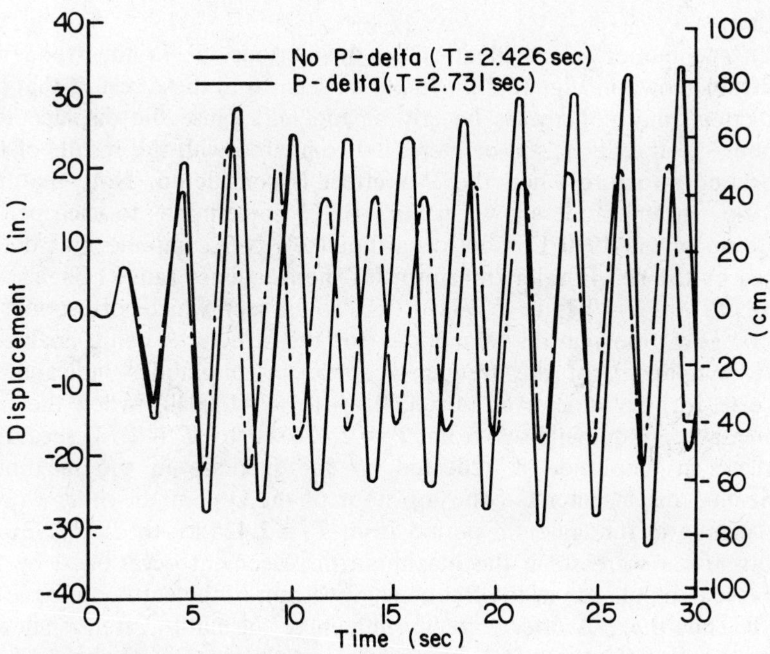

Fig. 16.21 Top-floor response of Fig. 16.20 for the N-S component of 1940 El Centro (Cheng and Tseng, 1973).

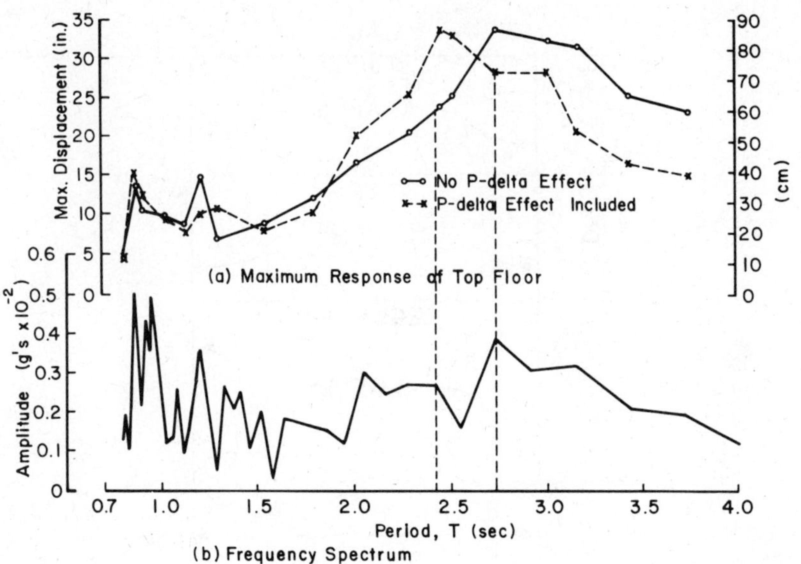

Fig. 16.22 Comparison of maximum response with frequency spectrum of earthquake (only for $T = 0.7–4$ of N-S, 1940 El Centro)(Cheng and Tseng, 1973).

which the numerals are identical to the periods, T, in the frequency spectrum shown in Fig. 16.22*b*. It can be seen from these results that the P-Δ effect may not always be critical and may cause the displacement response to the decreased or increased compared with the results of the associated structure when the P-Δ effect is considered. Note that the variation in amplitude shown in Fig. 16.22*b* corresponds to each period made up of the 1940 El Centro earthquake's N-S component as determined by the fast Fourier transform for the frequency range (Cheng and Oster, 1976b) $p = 0.25$ to 1.25 Hz ($T = 4$ to 0.8 sec), which corresponds to the range of the natural frequencies for the structures being analyzed.

A comparison of the frequency spectrum amplitudes indicates an increase in amplitude from $A = 0.0026g$ to $A = 0.0040g$ when the fundamental period increases from $T_n = 2.426$ sec to $T_s = 2.731$ sec. This increase in amplitude is reflected by the increase in the maximum horizontal displacement of the top floor of the system. In other words, the change in fundamental period from $T_n = 2.426$ sec to $T_s = 2.731$ sec results in an increase in the maximum displacement occasioned by the increase in amplitude of the frequency spectrum of the earthquake record used. Thus the P-Δ effect directly influences the natural frequency and consequently affects the dynamic forces and the response behavior.

The discussion above relates to plane frames. Three-dimensional frames have also been studied (Cheng and Kitipitayangkul, 1979): the influence of the P-Δ effect of the frame response should also be considered.

16.7 SPECIAL TOPICS

16.7. Ultimate Capacity of Flexibly Connected Steel Frames

For ease of computation, designers quite often assume that beam-to-column connections are either fully rigid or are frictionless hinges. Actually, all beam-to-column connections have some degree of flexibility. The degree of flexibility of these connections affects the structural stiffness of the frame and hence its stability limit load. One way to approximate roughly this reduction in stiffness of the frame is arbitrarily to reduce the beam stiffnesses used in computing the total stiffness of the frame. The actual reduction of stiffness is a function of the moment-rotation characteristics of the connections. Only limited experimental information on connection moment-rotation characteristics is available (Goverdhan, 1983).

As an alternative to the availability of moment-rotation test data, sensitivity studies have been conducted to provide a better evaluation of the stability limit load of flexibly connected unbraced steel frames. Much of this work has been done by Ackroyd (1979), Gerstle et al. (1979), Ackroyd and Gerstle (1983a,b), and Chen and Lui (1985). A necessary requirement for these studies is development of inelastic second-order analytical models that include the characteristics of the flexible beam-to-column connections. The following statements are adapted from Ackroyd and Gerstle (1983b). The study also included limited experimental verification.

A broad range of realistic building frames were studied to determine the effects of nonlinear beam-to-column connections and inelastic member instability. Simple subassemblages from the frames were analyzed by computer, and the results were represented in terms of two nondimensional parameters, ψ and p, as shown in Fig. 16.23. The following conclusions apply to the subassemblages studied.

1. Excess connection stiffness *can*, in exceptional cases, cause a reduction in the ultimate capacity of flexibly connected steel frames due to premature column yielding from gravity moments.
2. The parameter ψ appears to be a reliable index for determining the susceptibility of a frame to increases or decreases in factors of safety against collapse due to changes in connection stiffness, where ψ is given by

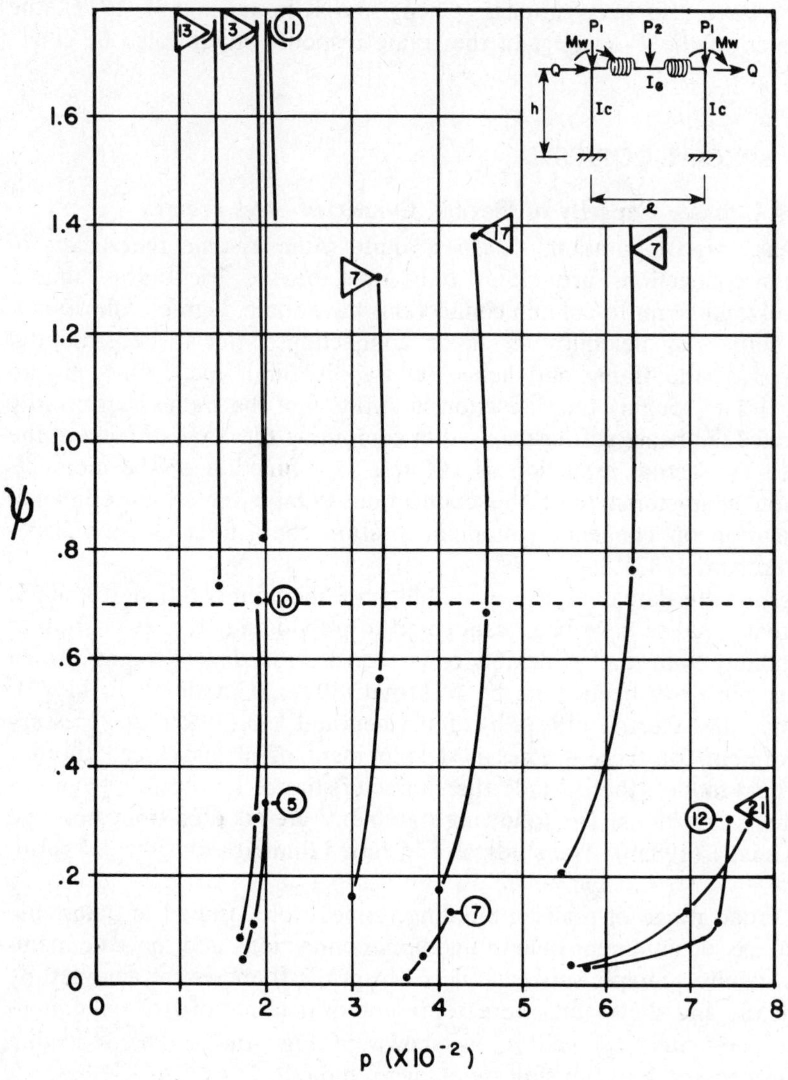

Fig. 16.23 Flexibly connected frame parametric study (Ackroyd and Gerstle, 1983a,b).

$$\psi = \frac{I_G}{I_C} \frac{l}{h} \frac{P_2 l}{M_W + Qh} \frac{k_0}{EI_G/l} \left(\frac{r}{h}\right)^2 \tag{16.22}$$

where

ψ = nondimensional structure/loading parameter
E = Young's modulus of elasticity
I_C = cross-sectional moment of inertia of column
I_G = cross-sectional moment of inertia of girder
M_W = column top wind moment on subassemblage
P_1 = ultimate vertical load acting at top of each column
P_2 = ultimate girder midspan load on subassemblage
Q = column top wind shear on subassemblage
h = story height
p = nondimensional frame capacity $= (P_1 + 0.5P_2)/P_{CR}$, where $P_{CR} = \pi^2 EI_C/h^2$
r = radius of gyration
k_0 = initial tangent stiffness of flexible connection
l = bay span

Note from Eq. 16.22 that ψ depends only on the initial, elastic stiffnesses of all structural components and on the relative magnitudes of the applied loads. Thus the ψ parameter provides an a priori evaluation of frame sensitivity.

1. For values of ψ less than 0.15, substantial increases in frame capacity can be attained by providing additional connection stiffness. Increases on the order of 20 to 30% can be realized.
2. For values of ψ greater than 0.15, the increases in frame capacity due to additional connection stiffness become progressively smaller until ψ attains values between 0.6 and 0.7, when little increase in frame capacity can be realized by providing extra connection stiffness.
3. For values of ψ greater than 0.7, providing overstiff connections may result in decreased frame capacity. Capacity reductions on the order of 1 to 8% have been observed.

Ongoing research is testing the validity of this subassemblage approach as well as evaluating the stability effects in entire flexibility-connected unbraced frames.

16.7.2 Threshold Stiffness

Structural frames, in accordance with the prevailing specifications, are designed on the basis of an a priori classification as braced or unbraced frames. The presumption is that if bracing physically exists, the frame is braced—that is, it will not buckle in a sidesway mode. This assumption could be in error. The problem of determining when the frame may be considered braced or unbraced has been treated by Biswas (1983) in terms of threshold stiffness.

Threshold stiffness, in the case of elastic stability of a structural frame, may be defined as a set of limiting values of lateral bracing stiffness that would allow the frame to sustain a set of critical loads corresponding to a nonsway bifurcation buckling mode rather than a much lower value of the set of critical loads corresponding to a sway bifurcation buckling mode, which would result in the absence of any bracing external lateral stiffness. Recognizing that the so-called elastic critical loads are only reference values and are not of real concern for ordinary building frames, a more practical design threshold stiffness may be defined as a set of limiting values of lateral bracing stiffness that would allow a frame to sustain the design load of a "braced" frame. The design load on the frame is calculated on the basis of design force of the compression members with a computed allowable stress corresponding to a factor of safety of 1.0. In comparison with load and resistance factor design (LRFD), the design load would compare with factored load.

Using these concepts, Biswas (1983) has studied the threshold stiffness of two-story frames. He shows that the threshold stiffness of a frame is the sum of the threshold stiffnesses of each story. These parts can be optimized to obtain a unique value for the frame. A significant conclusion of the study was that the usual 2% rule for design of bracing (design bracing for 2% of total axial load in columns) was conservative for the frames studied.

16.7.3 *P*-Δ Analysis Using a Fictitious Column with Negative Stiffness

As an alternative to an iterative *P*-Δ analysis, a direct *P*-Δ analysis can be carried out if advantage is taken of the observation that the geometric stiffness of a multistory building structure has the same mathematical form as the stiffness matrix of a shear beam: both are tridiagonal. Therefore, the *P*-Δ effect can be modeled as a fictitious column in the building having, however, negative shear stiffness properties (Rutenberg, 1981, 1982) and acting in parallel with the actual structure (Fig. 16.24). Considering Fig. 16.12, it is seen that the story shears are proportional to the factored axial column forces $\bar{F}\,\Sigma\,P_i$. Therefore, the shear rigidity, GA_i^{Δ}, of the fictitious column can be obtained from

$$\bar{F} \sum P_i/h_i = -\,GA_i^{\Delta}/h_i \rightarrow GA_i^{\Delta} = \bar{F} \sum P_i \tag{16.23}$$

The negative shear rigidity of the fictitious column is thus numerically equal to the factored total axial load in the story. Usually, it is more

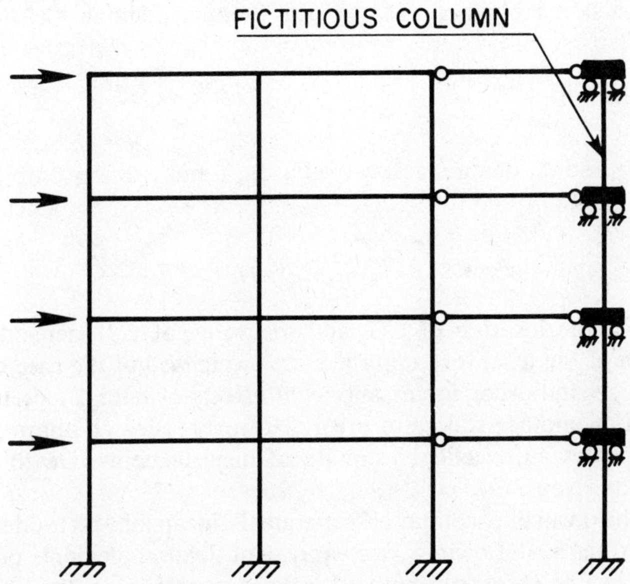

Fig. 16.24 Fictitious column (Rutenberg, 1982).

convenient to model the fictitious column as a rotation-fixed translation-free flexural column with flexural rigidity EI_i^{Δ}, so that

$$\bar{F} \sum P_i / h_i^2 = -12EI_i^{\Delta} / h_i^3 \rightarrow EI_i^{\Delta} = -\bar{F}h_i^2 \sum P_i / 12 \qquad (16.24)$$

Once a fictitious column with these properties is incorporated into the analytical model of the structure, as shown in Fig. 16.24, any first-order computer program will automatically generate all the terms of the second-order stiffness matrix.

It will be observed that in the foregoing formulas it was assumed that the deflected shape of the column within a story is a straight line. Since columns are bent due to end moments, additional shear forces are generated due to the eccentricity of the axial force about the deflected centroidal axis. It has been shown (e.g., LeMessurier, 1977), that an amplification factor of 1.2 usually gives a conservative estimate of this effect.

In the three-dimensional case, the planar coordinates of the fictitious column and its torsional properties must also be determined for every story (Rutenberg, 1982). It is evident that the fictitious column is to be located at the centroid of the column axial loads in the story (CG_i). The negative torsional rigidity GJ_i^Δ of this column, about CG_i, can be determined from

$$\bar{F}\sum P_i(d_{ix}^2 + d_{iy}^2)/h_i = \bar{F}r_{Gi}^2 \sum P_i/h_i = -GJ_i^\Delta h_i \rightarrow GJ_i^\Delta = -\bar{F}r_{Gi}^2 \sum P_i \tag{16.25}$$

in which d_{ix} and d_{iy} denote, respectively, the x and y distance of the story columns from CG_i, and

$$r_{Gi}^2 = \sum P_i(d_{ix}^2 + d_{iy}^2) \Big/ \sum P_i \tag{16.26}$$

Note that the location of CG_i and the value of GJ_i^Δ depend on the distribution of the axial forces in the story, which is not the case of GA_i^Δ. Neglecting second-order forces and axial effects of lateral forces on this distribution is another source of error. However, such an approximation usually provides an excellent estimate of displacements due to second-order effects (Nair, 1975).

When the available computer program is incapable of modeling the torsional properties of beams, two equivalent flexural elements per story, as defined in Eq. 16.24, can be used instead, each having one-half of the negative story stiffness, that is,

$$EI_i^\Delta/2 = -\bar{F}h_i^2 \sum P_i/24 \tag{16.27}$$

To satisfy Eq. 16.25, each of the two columns must be located a distance r_{Gi} on opposite sides of GC_i.

Because P-Δ effects are automatically accounted for by the introduction of the fictitious column, this approach is also readily applicable to the second-order analysis of frames subjected to dynamic loads.

It will be observed that in applying P-Δ methods to perpendicular frames sharing common columns, it is tacitly assumed that plastic hinges form in girders rather than in columns, so that the effect of biaxial yield condition can be disregarded. This appears to be a reasonable assumption, particularly in earthquake resistant design. It is also assumed that the effect of twist on yielding of columns and girders may be ignored.

16.8 CONCLUDING REMARKS

The reader of this chapter may well wonder about the profusion of analysis methods and design criteria presented in the preceding discussion. The fact is that the subject of frame stability is a very complex one which has received extensive research treatment for the past century. Thus it was possible only to present part of this work; many developments had to be omitted because of space limitations. Furthermore, it must be realized that in frame behavior there is a confluence of effects arising from problems of member strength (plastic moment capacity, plate local buckling, and lateral-torsional and minor-axis buckling between points of bracing) and overall stability. Eventually, the designer must deal with all of these effects. Following any of the methods presented here will result in safe designs. As to which of the methods is the one true, the very best, the optimal approach, there is yet no general agreement among the leaders of the structural engineering profession. Research, debating, analysis, and experimentation continues, and the reader is asked for patience until a resolution of all questions is achieved.

References

Ackroyd, M. H. (1979), "Nonlinear Inelastic Stability of Flexibly-Connected Plane Steel Frames," Thesis, University of Colorado, Boulder, Colo.

Ackroyd, M. H., and Gerstle, K. H. (1983a), "Elastic Stability of Flexibly Connected Frames," *ASCE Struct. Div.*, Vol. 109, No. ST1, Proc. Pap. 17602, pp. 241–245.

Ackroyd, M. H. and Gerstle, K. H. (1983b), "Strength of Flexibly-Connected Steel Frames," *Eng. Struct.*, Vol. 5, pp. 31-37.

Alvarez, R. J., and Birnstiel, C. (1969), "Inelastic Analysis of Multistory Frames," *ASCE J. Struct. Div.*, *ASCE*, Vol. 98. No. ST11, Proc. Pap. 6922, pp. 2477–2503; with Closure, Vol. 100, No. ST4, 1971.

American Institute of Steel Construction (1978), *Specifications for the Design, Fabrication and Erection of Structural Steel for Buildings*, effective November 1, 1978, AISC, Chicago.

American Institute of Steel Construction (1986), Specifications for Structural Steel Buildings, Load and Resistance Factor Design, AISC, Chicago.

American Society of Civil Engineers (1971), *Plastic Design in Steel—a Guide and Commentary*, ASCE Manual No. 41, ASCE, New York.

Arnold P., Adams, P. F., and Lu, L. W. (1966a), "Experimental and Analytical Behavior of a Hybrid Frame," *Fritz Eng. Lab. Rep. No.* 297.18, Lehigh University, Bethlehem, Pa.

Arnold, P., Adams, P. F., and Lu L. W. (1966b), "The Effect of Instability on the Cyclic Behavior of a Frame," *RILEM Int. Symp. Effect Repeat. Load. Mater. Struct. Elements*, Mexico City, September.

Arnold, P., Adams, P. F., and Lu, L. W. (1968), "Strength and Behavior of an Inelastic Hybrid Frame," *ASCE J. Struct. Div.* Vol. 94, No. ST1, Proc. Pap. 5759, pp. 243–266.

Barr, A. D. S., and McWhannell, D. C. (1971), "Parametric Instability of Structures under Support Motion," *J. Sound Vib.*, Vol. 14, No. 4.

Beaulieu, D., and Adams, P. F. (1977), "The Destabilizing Forces Caused by Gravity Loads Acting on Initially Out-of-Plumb Members in Structures," *Struct. Eng. Rep. No.* 59, Department of Civil Engineering, University of Alberta, Edmonton, February.

Beedle, L. S., Lu, L. W., and Lim, L. C. (1969), "Recent Developments in Plastic Design Practice," *ASCE J. Struct. Div.*, Vol. 95, No. ST9, Proc. Pap. 6781, pp. 1911–1937.

Birnstiel, C., and Iffland, J. S. B. (1980), "Factors Influencing Frame Stability," *J. Struct. Div.*, Vol. 106, No. ST2, Proc. Pap. 15196, pp. 491–504.

Biswas, M. (1983), "Threshold Bracing Stiffness for Two Story Frames, *Proc. 3rd SSRC,*" *Int. Colloq. Stab. Met. Struct.*, Toronto, May, pp. 389–409.

Bolotin, V. V. (1956), *The Dynamic Stability of Elastic Systems*, Russian ed., Gostekhizdat, Moscow; German ed., Deutscher Verlag der Wissenschaften, East Berlin, 1961; English ed., Holden-Day, San Francisco, 1964.

Brown, J. E., Hutt, J. M., and Salama, A. E. (1968), "Finite Element Solution to Dynamic Stability of Bars," *AIAA, J.* Vol. 6, No. 7.

Chen, W. F., and Lui, E. M. (1985), "Columns with End Restraint and Bending in Load and Resistance Factor Design," *Eng. J. Am. Inst. Steel Constr.*, Third Quarter, pp. 105–132.

Cheng, F. Y., and Kitipitayangkul, P. (1979), "Investigation of the Effect of 3-D Parametric Earthquake Motions on Stability of Elastic and Inelastic Building Systems," *Tech. Rep.*, National Science Foundation. (Available from U.S. Department of Commerce, National Technical Information Service, Springfield, VA 22151. NTIS Access No. 80-17693).

Cheng, F. Y., and Oster, K. B. (1976a), "Ultimate Instability of Earthquake Structures," *ASCE J. Struct. Div.* Vol. 102, No. ST5, Proc. Pap. 12117, pp. 961–972.

Cheng, F. Y., and Oster, K. B. (1976b), "Dynamic Instability and Ultimate capacity of Inelastic Systems Parametrically Excited by Earthquakes: Part II," *Tech. Rep.*, National Science Foundation, August. (Available from U.S. Department of Commerce, National Technical Information Service, Springfield, VA 22151. NTIS Access No 261097/AS.)

Cheng, F. Y., and Tseng, W. H. (1973), "Dynamic Instability and Ultimate Capacity of Inelastic Systems Parametrically Excited by Earthquakes: Part I," *Tech. Rep.* National Science Foundation, August. (Available from U.S. Department of Commerce, National Technical Information Service, Springfield, VA 22151. NTIS Access No. PB261096/AS.)

Cook, N. E., Jr. (1983), "Strength and Stiffness of Type 2 Steel Frames," Thesis, University of Colorado, Boulder, Colo.

Council on Tall Buildings and Urban Habitat (1979), *Monograph on Planning and design of Tall Buildings*, Vol. SB, *Structural Design of Tall Steel Buildings*, CTBUH, ASCE, pp. 239–342.

Daniels, J. H. (1967), "Combined Load Analysis of Unbraced Frames," Thesis, Lehigh University, Bethlehem, Pa.

Euler, L. (1744), "Method of Investigating Curved Lines Which Have the Property of Maxima and Minima" (Methodus inveniendi lineas curvas maximi minimive proprietate gaudentes, Additamentum primum: De curvis elasticis), Lausanne and Geneva, Switzerland.

European Convention for Constructional Steelwork (1976), *Manual on the Stability of Steel Structures*, Chapter 8, "Stability of Frames," ECCS Committee on Stability. Published in *2nd Int. Colloq. Stab., Introd. Rep.* Tokyo, September 9, 1976, Liege, April 13–15, 1977, Washington, May 17–19, 1977, pp. 229–252.

Gerstle, K. H., Kahl, T. L., and Vanaskie, W. F. (1979), "Preliminary Analysis of Building Frames with Flexible Connections," *ASCE Conv. Expos.*, *Preprint* 3572, Boston, April 2–6, pp. 1–28.

Goverdhan, A. (1983), "Experimental Moment-Rotation Curves and Evaluation of Prediction Equations for Semirigid Connections," M.S. thesis, Vanderbilt University, Nashville, Tenn., December.

Harstead, G. A., Bernstiel, C., and Leu, K. C. (1971), closure to "Inelastic Behavior of H-Columns under Biaxial Bending," *ASCE, J. Struct. Div.*, Vol. 97, No. ST7, Proc. Pap. 8225, pp. 2011–2016.

Herrmann, G., ed. (1967), *Dynamic Stability of Structures*, Pergamon Press, Elmsford, N.Y.

Horne, M. R. (1963), "Elastic-Plastic Failure Loads of Plane Frames, *Proc. R. Soc. London*, Vol. A274, No. 1358, pp. 343–364.

Iffland, J. S. B., and Birnstiel, C. (1982), "Stability Design Procedures for Building Frameworks," *Res. Rep., AISC Project No.* 21.62, American Institute of Steel Construction, September 7.

Johnston, B. G., ed. (1976), *Guide to Stability Design Criteria for Metal Structures*, 3rd ed., (SSRC), Wiley, New York, pp. 410–454.

Jonatowski, J. J., and Birnstiel, C. (1969), "Elasto-Plastic Analysis of Space Frameworks," School of Engineering and Science, New York University, New York.

Kanchanalai, T. (1977), "The Design and Behavior of Beam-Columns in Unbraced Steel Frames," *CESRL Rep. No.* 77-2, (AISC Project No. 189, Column Design in Unbraced Frames, Rep. No. 2), Department of Civil Engineering, University of Texas at Austin, October.

Kanchanalai, T., and Lu, L. W. (1979), "Analysis and Design of Framed Columns under Minor Axis Bending," *Eng. J. Am. Inst. Steel Constr.*, Vol. 16, No. 2, pp. 29–41.

Kim, S. W. (1971), "Elastic-Plastic Analysis of Unbraced Frames," Thesis, to Lehigh University, Bethlehem, Pa.

Korn, A. (1967), "The Elastic-Plastic Behavior of Multistory, Unbraced Planar Frames," Thesis Washington University, St. Louis.

Krajcinovic, D. P., and Herrmann, G. (1968), "Stability of Straight Bars Subjected to Repeated Impulsive Loads," *AIAA J.*, Vol. 6, No. 10.

LeMessurier, W. J. (1977), "A Practical Method of Second Order Analysis; Part 2. Rigid Frames," *Eng. J. Am. Inst. Steel Constr.*, Vol. 4, No. 2, pp. 49–67.

Lu, L. W. (1983), "Frame Stability Research and Design Procedures," *Proc. 3rd SSRC, Int. Colloq. Stab. Met. Struct.*, Toronto, May, pp. 369–375.

Majid, K. I., and Anderson, D. (1968), "The Computer Analysis of Large Multistory Framed Structures," *Struct. Eng.*, Vol. 46, No. 11.

Majumdar, S. N. G., and Adams, P. F. (1971), "Tests on Steel-Frame, Shear-Wall Structures," *ASCE J. Struct. Div.*, Vol. 97, No. ST4, Proc. Pap. 8031, pp. 1097–1111.

McNamee, B. M. (1967), "The General Behavior and Strength of Unbraced Multistory Frames under Gravity Loading," *Fritz Eng. Lab. Rep. No.* 276.18, Lehigh University, Bethlehem, Pa., June.

Nair, R. S. (1975), "Overall Elastic Stability of Multistory Buildings," *ASCE J. Struct. Div.*, Vol. 101, No. ST12, Proc. Pap. 11762, pp. 2487–2503.

Parikh, B. P. (1966), "Elastic-Plastic Analysis and Design of Unbraced Multistory Steel Frames," Thesis, Leigh University, Bethlehem, Pa.

Rutenberg, A. (1981), "A Direct *P*-Delta Analysis Using Standard Plane Frame Computer Programs," *Comput. Struct.* Vol. 14, Nos. 1–2, pp. 97–102.

Rutenberg, A. (1982), "Simplified *P*-Delta Analysis for Asymmetric Structures," *ASCE J. Struct. Div* Vol. 108, No. ST9 Proc. Pap. 17327, pp. 1995–2013.

Swanger, M. H., and Emkin, L. Z. (1979), "A Fiber Element Model for Nonlinear Frame Analysis," *Proc. 7th Cong. Electron. Comput.*, ASCE, St. Louis, August 6–8, pp. 510–536.

Vinnakota, S. (1967), "Inelastic Stability Analysis of Rigid Jointed Frames" (Flambage des cadres dans le domaine elasto-plastique), Thesis, Federal Institute of Technology, Lausanne, Switzerland.

Vinnakota, S. (1971), "Elastic-Plastic Stability of Frames" (Stabilité elasto-plastique des cadres), *Publ. No.* 121, Ecole Polytechnique Federale de Lausanne, Switzerland, October, and *Int. Civ. Eng. Isr.*, Vol. 3, 1974, pp. 37–48.

Vinnakota, S. (1972), "Elastic-Plastic Instability of Multistory Frames," *ASCE-IABSE Int. Conf. Tall Build., Preprints: Discussion and Summary*, pp. 568–574, August.

Wakabayashi, M. (1970), "The Behavior of Steel Frames with Diagonal Bracings under Repeated Loading," *Proc. US-Jpn. Sem. Earthquake Eng. With Emphasis Safety Sch. Build.*, Sendai, Japan, p. 328.

Wakabayashi, M., Matsui, C., Minami, K., and Mitani, I. (1970), "Inelastic Behavior of Full Scale Steel Frames," *Disaster Prev. Res. Inst., Kyoto Univ. Bull.* No. 13-A.

Wood, R. H. (1974), "Effective Lengths of Columns in Multistory Buildings: Part 3. Features Which Increase the Stiffness of Tall Frames against Sway Collapse and Recommendations for Designers," *Struct. Eng.*, Vol. 52, No. 9, pp. 341–346.

Wright, E. W. (1966), "Analysis of Multistory Steel Rigid Frames Subject to Sidesway," Thesis, University of Illinois at Urbana.

Yarimci, E., (1966), "Incremental Inelastic Analysis of Framed Structures and Some Experimental Verifications," Thesis, Lehigh University, Bethlehem, Pa.

Yura, J. A. (1971), "The Effective Length of Columns in Unbraced Frames," *Eng. J. Am. Inst. Steel Constr.*, Vol. 8, No. 2, pp. 37–42.

Chapter Seventeen

Arches

17.1 IN-PLANE STABILITY OF ARCHES, INTRODUCTION

The analysis and design of arches may be broadly classified according to their response to load and their in-plane mode of failure. These characteristics depend on the type of arch, such as:

- Slender arches, generally of solid-web rolled or built-up sections, subject primarily to axial force in the arch rib (analogous to axially loaded columns).
- Slender arches subject to significant bending and deformation, due primarily to asymmetric loading (analogous to beam-columns).
- Stocky arches, frequently of truss form, used because of high bending loads, in which failure will be due primarily to chord or flange failure arising from axial load and excessive bending (similar to any truss subject to axial load and moment).
- Arches composed of arch ribs and deck stiffening girders.

These categories are treated separately in this chapter. Finally, attention is given to out-of-plane buckling.

When the loads acting on an arch are increased proportionally, it loses its stability as a certain critical value of the load is attained. In the case of elastic structures under conservative loads the critical load always corresponds to either a bifurcation point or a stability limit point. Various possible load-deformation plots for arches, called equilibrium paths, are shown in Fig. 17.1. Each point on a path represents an equilibrium configuration of the structure.

Figure 17.1*a* shows an equilibrium path for an unsymmetric problem in which the arch boundary conditions and/or the manner of loading may be

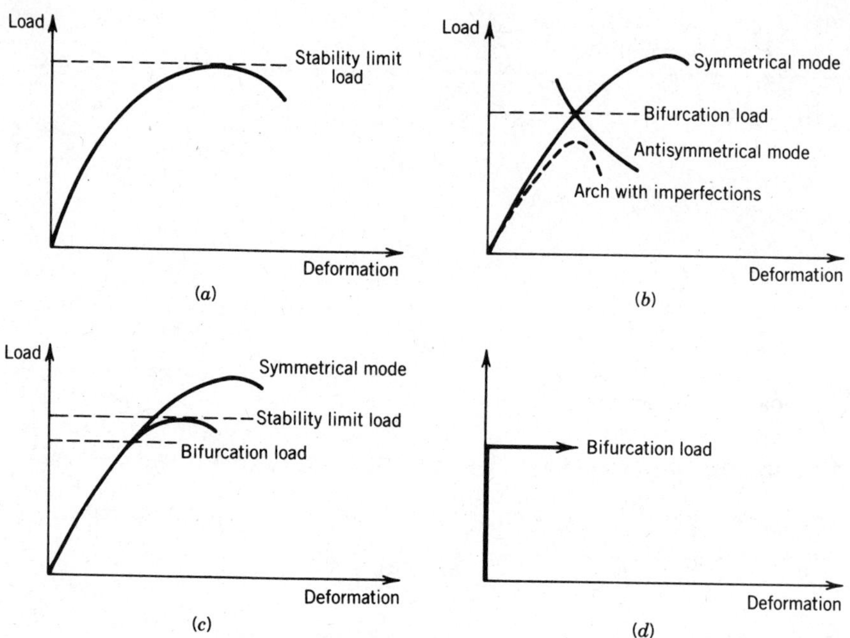

Fig. 17.1 Some possible load-deformation paths for arches. (*a*) Nonlinear stability: Unsymmetrical or unsymmetrically loaded arch. (*b*) Nonlinear stability: symmetrical and symmetrically loaded arch. (*c*) Nonlinear stability: symmetrical and symmetrically loaded arch. (*d*) Linear stability: negligible prebuckling deformation.

unsymmetrical. The point on the equilibrium path at which the load is a relative maximum is called a limit point.

Figure 17.1*b* illustrates possible load-deformation relationships for a symmetrical and symmetrically loaded arch. Here the primary or fundamental equilibrium path is intersected by a secondary path. The primary branch represents a symmetric deflected mode and the secondary branch an antisymmetrical mode. Points at which equilibrium paths intersect one another are called bifurcation points. If an antisymmetrical mode does not first become dominant, the arch eventually will become unstable when the load-deflection curve reaches a limit point. Then the arch buckles in a symmetrical mode. However, Fig. 17.1*b* shows the antisymmetrical mode dominating. Note that the load drops beyond the bifurcation point. This is the usual behavior when the arch buckles by sidesway, but it is also possible for the load to increase slightly after bifurcation in which case the maximum load is attained at a limit point after large movements in the plane of the arch (Fig. 17.1*c*). In case of an unstable, rapidly descending postbuckling path the arch is sensitive to geometric imperfections or load eccentricity. The associated drop in critical load is shown in Fig. 17.1*b* (dashed curve), and the buckling load at the bifurcation point is replaced by the snap load at the limit point on the dashed curve. The degeneration of a bifurcation point into a limit point due to the presence of imperfections suggests that bifurcation is an exception rather than the rule.

Arch structures are most efficient if they carry their load in such a way that the funicular curve coincides with the centroidal axis of the rib, so that there will be axial compression but no bending of the arch rib. Such a rib experiences only very small displacements before buckling. Circular arches subjected to uniform normal pressure, commonly called hydrostatic loading, parabolic arches subjected to uniform load on a horizontal projection and catenary arches with load uniformly distributed along the arch axis are some examples of arches under pure axial compression. In the ideal situation such structures undergo bifurcation buckling. These bifurcation problems are far more amenable to analysis because they permit a linearization of the equilibrium equations for the prebuckling, or fundamental state. The linearization is based on the assumption of very small displacements and the equilibrium path is horizontal (Fig. 17.1*d*), as in small-deflection column theory. For practical design loadings, however, the funicular curves normally do not coincide with the centroidal axis and in such cases the arch experiences substantial bending moments and displacements before buckling. Studies that take into account these prebuckling deformations are nonlinear stability problems, as illustrated by Fig. 17.1*a–c*.

17.2 IN-PLANE LINEAR STABILITY

Early papers on arch stability were devoted to linear stability problems. This work by Gaber, Stüssi, Kollbrunner, Hilman, Dischinger, and Dinnik is summarized by Austin (1971) and Timoshenko and Gere (1961).

Consider the buckling of arches of constant cross section (herein termed uniform arches) in which the arch is the funicular curve for the loading. The critical values of distributed load and horizontal reaction of fixed two-hinged and three-hinged symmetric arches are summarized in Table 17.1 for the following cases: (1) parabolic arches subjected to vertical load uniformly distributed on a horizontal projection, (2) catenary arches under uniform vertical load along the arch axis, and (3) circular arches subjected to uniform normal pressure, commonly called hydrostatic loading. In each of these cases, for geometrically perfect

Table 17.1. Critical Load Parameter and Critical Horizontal Reaction Parameter for Uniform Elastic Arches in Pure Compression[a]

$\frac{h}{L}$	Three-Hinged Arch		Two-Hinged Arch		Fixed Arch	
	qL^3/EI	HL^2/EI	qL^3/EI	HL^2/EI	qL^3/EI	HL^2/EI
Parabolic arches subjected to vertical load uniformly distributed on a horizontal projection						
0.10	22.5	28.1	29.1	36.3	60.9	76.2
0.15			39.5	32.9	85.1	70.9
0.20	39.6	24.8	46.1	28.8	103.1	64.5
0.25			49.2	24.6	114.6	57.3
0.30	49.5	20.6	49.5	20.6	120.1	50.0
0.35			47.8	17.1	120.6	43.1
0.40	45.0	14.1	45.0	14.1	117.5	36.7
0.50	38.2	9.6	38.2	9.6	105.3	26.3
Catenary arches subjected to vertical load uniformly distributed along the arch axis						
0.10			28.7	36.3	60.1	76.2
0.15			38.3	32.8	82.7	70.9
0.20			43.5	28.5	98.0	64.3
0.25			44.8	24.1	105.9	56.9
0.30			43.2	19.8	107.4	49.4
0.35			39.7	16.1	104.0	42.1
0.40			35.3	12.9	97.2	35.4
0.50			26.5	8.2	79.3	24.5
Circular arches subjected to normal load uniformly distributed along the arch axis						
0.10	22.2	26.7	28.4	34.1	58.9	70.7
0.20	33.5	17.6	39.3	20.6	90.4	47.5
0.30	34.9	9.3	40.9	10.9	93.4	24.9
0.40	30.2	3.4	32.8	3.7	80.7	9.1
0.50	24.0	0	24.0	0	64.0	0

[a] h, rise; L, span; q, critical intensity of distributed load; H, critical horizontal reaction at supports; E, Young's modulus of elasticity; I, moment of inertia of the cross section.

arches, the loading causes pure axial compression (no bending) at every cross section of the arch. The arches are free to buckle in their plane without restraint. Axial compressive strain has been neglected in the analyses reported since it has a very small effect for slender arches. Buckling is a bifurcation phenomenon from an undeflected position, as shown in Fig. 17.1*d*. The critical values in Table 17.1 are given for a range of rise-to-span ratios from 0.10 to 0.50.

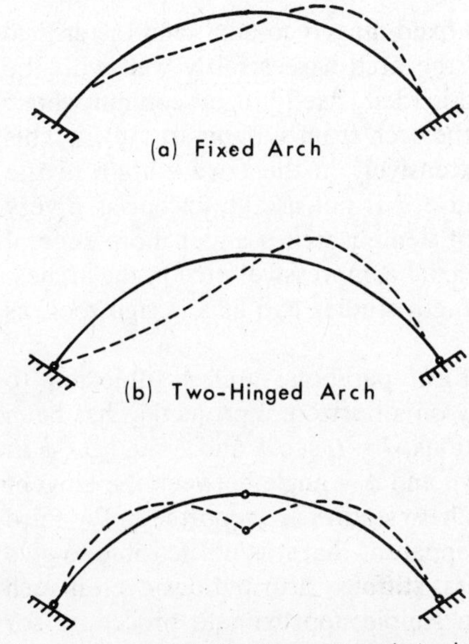

Fig. 17.2 Fundamental buckling-mode shapes.

In Table 17.1 the critical values for two-hinged and fixed parabolic and catenary arches were taken from Austin and Ross (1976). The remaining values were taken from Timoshenko and Gere (1961). Kollbrunner (1936, 1942) showed that the elastic critical values obtained experimentally are in close agreement with the theoretical values reported in Table 17.1.

The fundamental modes that correspond to the critical values given in Table 17.1 are shown in Fig. 17.2. The fixed and two-hinged arches always buckle into an antisymmetrical mode in which the arch sways sideways, with the crown moving horizontally and becoming a point of contraflexure (Fig. 17.2*a* and *b*). In the case of a three-hinged arch, for low rise-to-span ratios the buckling is symmetric, as shown in Fig. 17.2*c*; the crown moves downward. For high rise-to-span ratios the three-hinged arch buckles in an antisymmetrical mode like that of the two-hinged arch and at the same critical load.

In the buckling behavior of uniform arches of low and moderate span-to-depth ratios under pure axial compression which buckle by sidesway there is a similarity with the buckling behavior of straight columns. For example, a fixed arch that buckles into two waves with a point of contraflexure at the crown (Fig. 17.2*a*) has a mode shape from support to crown similar to that of a fixed-hinged column, and the critical axial thrust at the quarter-point of the arch agrees fairly well with the critical compressive force in the end-loaded, fixed-hinged column whose length is equal to the arc length of the arch from support to crown. This interpretation is developed fairly extensively in the third edition of the guide but is not repeated here because it is not useful, except as a very broad concept, for elastic buckling of slender arches under more general loadings which do not produce pure axial compressive force in the arches. It is used, however, in ultimate strength studies and as a design tool, as will be described.

The elastic buckling of nonuniform parabolic arches subjected to vertical loading distributed uniformly on a horizontal projection has been studied for moment-of-inertia variations, $I = I_c \sec \phi$ and $I = I_c \sec^3 \phi$ in which I_c = moment of inertia at crown and ϕ = angle between the tangent to the arch axis and the horizontal. These studies are reported in the third edition of the guide. However, it is apparent that it is not feasible to give tables for all possible practical flexural stiffness distributions even if such values were available. Therefore, a simple approximate procedure for extrapolating uniform arch-buckling information to nonuniform arches for antisymmetrical buckling is useful. In this procedure the critical load on a nonuniform arch is considered to be equal to that for a similar uniform arch with an "equivalent" moment of inertia computed in the following manner. Imagine that one-half of the nonuniform arch (from support to crown) is straightened out to form a simply supported beam and loaded with a concentrated load at midspan. Compute the deflection at midspan. The equivalent uniform moment of inertia of this nonuniform arch then is equal to the moment of inertia of a simply supported prismatic beam of the same span and loading that deflects at the center the same amount. The same equivalent value of moment of inertia is recommended for fixed, two-hinged and three-hinged arches. The result of such a computation is discussed in the third edition of the guide. Similar procedures developed by Aas-Jakobsen are described by Forrester in a discussion of Austin's paper (1971).

Stiffened arches are deck structures that consist of an arch connected to an overhead horizontal girder by closely spaced columns. As discussed in the third edition of the guide, for two- and three-hinged, low-rise parabolic arches subjected to uniform load on a horizontal projection the critical load and critical horizontal reaction can be closely approximated by the use of the values in Table 17.1, using for the moment of inertia, I, an equivalent I equal to the sum of the arch and girder moments of inertia. This simple concept recognizes that the arch and girder buckle together.

17.3 IN-PLANE NONLINEAR ELASTIC STABILITY

Section 17.2 of this chapter has treated arches subjected to special loadings that produced only axial compression. Since the arches were considered to be inextensible, there were no prebuckling deflections for these cases. However, actual design loadings on arches usually produce both axial compression and bending moment on a general cross section of the arch rib. These internal forces cause a change in shape of the arch before buckling occurs. This makes the problem nonlinear even though the material is deforming elastically.

17.3.1 Symmetrical Loadings on Symmetrical Arches

The general deformational behavior of symmetrical arches symmetrically loaded is shown in Fig. 17.1*b* and has ben described above. Buckling data are given in Table 17.2 for two-hinged and fixed parabolic and circular arches subjected to uniform vertical load uniformly distributed along the arch axis. These data are from Austin and Ross (1976). Similar data are availale in this paper also for a single concentrated load at midspan. In these studies axial strains have been neglected since their influence is very small for slender arches.

The antisymmetrical buckling critical loads for circular arches are shown to be less than the symmetrical buckling critical loads in Table 17.2. The same is true for parabolic arches, although the symmetrical buckling critical loads are not available. Antisymmetrical buckling usually governs but not always. Fixed circular arches with a single concentrated load at the crown buckle symmetrically. This is also true for fixed parabolic arches with concentrated load at the crown, except for deep arches with a rise-to-span ratio greater than about 0.40. Note in Table 17.2 that the ratio h_c/h_i is close to unity for the arches that buckle by sidesway, which indicates that the profile at the instant of buckling was nearly the same as the unloaded profile for these cases.

The critical conditions for antisymmetrical buckling often are expressed in terms of horizontal reactions; this is especially meaningful for flatter arches. In Fig. 17.3 graphs are shown of the critical horizontal reaction coefficient versus the rise-to-span ratio for the antisymmectrical buckling solutions given in Tables 17.1 and 17.2. Note in Fig. 17.3 that the critical horizontal reactions are quite insensitive to the arch shape and to a lesser extent to the loading; they vary primarily with the rise-to-span ratio. This is particularly true for the fixed arches; the curves for three different shapes all with uniform rib loading lie on a single line and the parabolic arch with uniform deck loading case is very close. The rise-to-span ratios vary from 0.10 to 0.30, which includes the most practical

Table 17.2. Elastic Buckling Coefficients for Uniform Arches with Vertical Load Uniformly Distributed Along Arch Axis[a]

		Two-Hinged Arch			Fixed Arch		
θ (deg)	h_i/L	qL^3/EI	HL^2/EI	h_c/h_i	qL^3/EI	HL^2/EI	h_c/h_i
		Parabolic arches—antisymmetrical modes					
	0.10	28.6	36.3	1.002	60.4	76.1	1.003
	0.15	38.2	32.9	1.004	83.4	70.8	1.006
	0.20	43.4	28.8	1.006	99.3	64.3	1.010
	0.25	44.9	24.5	1.008	107.7	57.0	1.013
	0.30	43.5	20.5	1.009	109.6	49.5	1.015
	0.35	40.4	16.96	1.009	106.4	42.5	1.017
	0.40	36.5	13.98	1.009	100.0	36.2	1.017
	0.50	28.5	9.51	1.008	83.0	25.9	1.017
		Circular arches—antisymmetrical modes					
50	0.1109	31.2	35.6	0.994	64.8	75.2	0.992
70	0.1577	39.5	32.0	0.989	83.4	70.0	0.983
90	0.2071	44.0	27.4	0.981	95.5	63.3	0.969
106.26	0.2500	44.5	23.2	0.974	99.9	57.0	0.966
120	0.2887	42.8	19.34	0.968	99.8	51.2	0.938
140	0.3501	37.1	13.75	0.959	93.8	42.1	0.910
160	0.4196	28.9	8.72	0.953	81.9	32.6	0.880
180	0.5000	20.0	4.78	0.950	66.0	23.4	0.854
		Circular arches—symmetrical modes					
50	0.1109	63.0	74.6	0.88	90.9	110.6	0.94
70	0.1577	77.6	66.8	0.81	110.0	98.5	0.91
90	0.2071	85.6	57.9	0.75	119.4	85.7	0.88
106.26	0.2500	87.7	50.5	0.69	120.3	75.2	0.85
120	0.2887	86.7	44.3	0.64	117.2	66.7	0.82
140	0.3501	81.7	35.1	0.57	107.5	54.2	0.78
160	0.4196	73.8	27.1	0.45	93.5	43.1	0.73
180	0.5000	64.4	19.07	0.34	77.2	32.5	0.67

[a] h_i, initial rise of arch; h_c, height of arch at crown at instant of buckling; θ, angle of opening of the circular arch.

range for bridges. The curves separate more for larger rise-to-span ratios. These results suggest that the curves of Fig. 17.3 can be used to make close estimates of the critical horizontal reaction for other symmetrical loadings. Note that the horizontal reaction in Fig. 17.3 is the exact value obtained by nonlinear analysis for some of the cases. However, the horizontal reactions given by a first-order solution for the same load agree closely with these values.

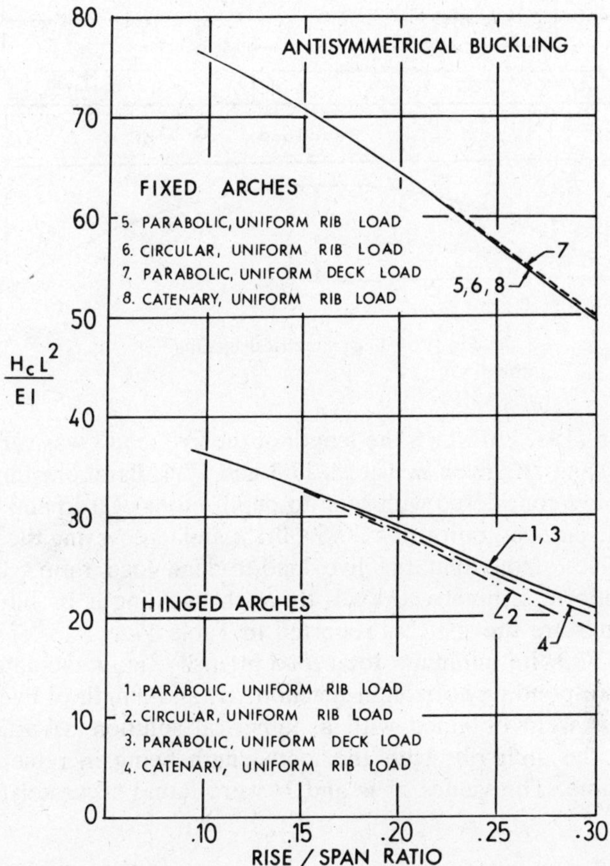

Fig. 17.3 Variation of critical horizontal reaction with rise-to-span ratio.

17.3.2 Unsymmetrical Loading

The general deformational behavior of unsymmetrically loaded arches is shown in Fig. 17.1*a* and has been described above. Several studies have been made of the most important practical loading, which is comprised of a uniformly distributed dead load (on a horizontal projection), q, plus a uniformly distributed live load, p, extending a variable distance from one abutment, as shown in Fig. 17.4. These studies were for parabolic arches.

The first studies by Deutsch (1940), Kuranishi and Lu (1972), and Harries (1970) used half-span live load. Kuranishi and Lu noted that for elastic buckling the total dead plus live load intensity, $w = p + q$, at buckling seemed roughly equal to the buckling value for uniform load over the entire span. Recently, studies have been made by Chang (1973)

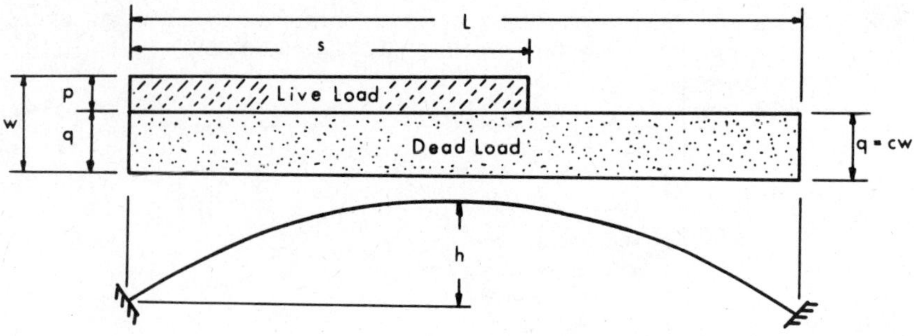

Fig. 17.4 Unsymmetrical loading.

and Harrison (1982) in which the length of the live load s was varied. The results by Chang are given in Tables 17.3 and 17.4. Parabolic hinged and fixed arches are considered with span-to-depth ratios of 0.15 and 0.25 and live load-to-dead load ratios of ∞, 0.50, 0.15, and 0, covering the range of practical values. Note that the live load-to-dead load ratio = 0 corresponds to uniform symmetrical load, the arch buckling is by bifurcation, and the values are the same as reported in Table 17.1.

In Table 17.3 the minimum total load intensity, w, is tabulated along with the corresponding horizontal reaction, H, and length of live load, s. These results were obtained with a numerical solution using 24 equal divisions in the arch rib, with the load length being incremented one panel at a time. The values of w and H were found accurately, but the

Table 17.3. Minimum Elastic Buckling Coefficients for Uniform Hinged and Fixed-Ended Parabolic Arches Under Distributed Dead and Live Loads

		Hinged Ends			Fixed Ends		
$\frac{h}{L}$	$\frac{p}{q}$	s/L	wL^3/EI	HL^2/EI	s/L	wL^3/EI	HL^2/EI
0.15	∞	0.72	34.0	26.6	0.72	71.7	58.1
	0.50	0.67	35.5	29.4	0.66	75.9	62.2
	0.15	0.63	36.8	30.8	0.63	78.4	66.1
	0	—	39.5	32.9	—	85.1	70.9
0.25	∞	0.77	44.7	22.5	0.73	99.3	49.5
	0.50	0.68	46.0	23.3	0.66	104.0	51.9
	0.15	0.64	47.0	23.9	0.64	107.5	54.4
	0	—	49.2	24.6	—	114.6	57.3

Table 17.4. Elastic Buckling Coefficients for Uniform Hinged and Fixed-Ended Parabolic Arches Under Distributed Dead and Half-Span Live Load ($s = 0.50L$)

		Hinged Ends		Fixed Ends	
$\frac{h}{L}$	$\frac{p}{q}$	wL^3/EI	HL^2/EI	wL^3/EI	HL^2/EI
0.15	∞	42.4	22.4	96.2	46.7
	0.50	37.1	28.4	80.2	60.4
	0.15	37.3	30.6	80.0	65.2
	0	39.5	32.9	85.1	70.9
0.25	∞	56.4	19.3	131.4	41.0
	0.50	48.3	22.8	109.8	50.4
	0.15	47.7	23.8	109.1	53.7
	0	49.2	24.6	114.6	57.3

loaded length, s, is only approximate. Table 17.4 gives the corresponding values for half-span live loading ($s = 0.50$). The following conclusions can be drawn for practical ranges. The minimum total buckling load is obtained when the live load acts over from 63 to 68% of the span. The buckling value of the total load w is minimized for high live-to-dead load ratios but does not drop below about 89% of the full-span buckling value for the practical range. This is remarkably constant. Half-span loading is a good approximation for the partial loading case, as the corresponding elastic buckling load is not greater than about 6% more than the absolute minimum, in the practical range.

It should be noted that high live loads acting over partial spans produce very large moments and deformations. The elastic limit load is reached only after very large displacements which completely distort the arch, as discussed by Harrison (1982)and by Yabuki and Vinnakota (1984). When inelastic behavior is considered the limit loads for these cases are very much less than the elastic buckling values reported above, as discussed in the next section.

17.4 IN-PLANE ULTIMATE LOAD

17.4.1 Limit Analysis of Stocky Arches

The limit analysis of arches was presented by Onat and Prager in 1953 and by Galli and Franciosi in 1955. Following their investigations several researchers reported on the collapse load of arches (Stevens, 1957; Coronforth and Childs, 1967). However, in these studies the collapse loads were obtained by the upper and lower bound methods assuming rigid-plastic behavior. These methods, based on the localized plastic hinge concept, cannot consider the longitudinal spread of yielding zones nor the effect of deflections on moments. Hence the studies are valid only for structures with very stocky cross sections subjected to predominant flexural moments. In 1957, Stevens experimentally checked the validity of this approach on models of circular, elliptic, and parabolic arches with two-hinged and fixed support conditions. He found that a reasonably accurate estimate of the collapse load can be obtained by limit analysis for arches under a concentrated load. However, the experimental collapse load for an arch under half-span loading is considerably less than the calculated value.

17.4.2 Slender Arches in Pure Compression

Slender two-hinged and fixed arches with loadings that produce pure axial compression buckle by sidesway with a node at the crown (Fig. 17.2*a* and *b*). The behavior is very much like that of a column and it is common to express the buckling strength of such arches in terms of the axial thrust at the quarter point of the arch, *P*, as described earlier. The elastic critical value can be expressed as

$$P_c = \pi^2 EI(kS)^2 \tag{17.1}$$

where

S = length of curved centroidal axis of arch rib from support to crown
E = Young's modulus
I = moment of inertia of the arch rib

The "effective-length factor" k depends on the support fixity condition, the rise-to-span ratio, and the shape of the arch profile. The values of k for arches are tabulated in the third edition of this guide and have values close to the well-known corresponding values for a column with a hinge at one end (crown) and a fixity condition at the other end which corresponds to the arch support condition.

Komatsu and Shinke (1977) have made inelastic ultimate strength studies for a two-hinged parabolic arch subjected to uniform load. Residual stress and initial crookedness were considered, and rise-to-span ratios = 0.10, 0.15, and 0.20 were used with a wide range of other parameters. It was found that the ultimate value of the thrust at the quarter point of the arch was accurately predicted with the usual column curves adjusting for yield points, effective lengths, and so on. Thus one could use for arches the standard relationships employed to relate the inelastic ultimate strength of columns to the elastic critical values.

17.4.3 Slender Arches under Symmetrical Load

In this case the loading produces on a general cross section combined axial compressive force and bending moment. Then the arch acts much like a beam-column. It has been shown by Shinke et al. (1975) that the most demanding practical loading for low bridge arches of uniform cross section is the unsymmetrical loading discussed in the preceding section and shown in Fig. 17.4 with $s = 0.50$. The general deformational behavior is shown in Fig. 17.1*a*, in which the buckling failure is a limit-point phenomenon. Arches of practical proportions develop extensive inelastic action before failure. However, the limit load is reached before the displacements become so large that consideration of nonlinear geometry is necessary. To accurately predict the load-carrying capacity of steel arch structures it is necessary to consider the effects of the spread of yielding zones in the cross section and along the longitudinal axial direction of the rib, the effects of initial residual stresses, and, of course, the amplification of the moments caused by the displacements. The following studies considered uniform arches subjected to uniform dead load over the entire span and half-span live load (as in Fig. 17.4, except as noted).

In 1970 Harries reported some analytical results on the ultimate strength of two-hinged, parabolic steel arches in which the effects of prebuckling deformations and the spread of yielding zones were considered. Elastic-plastic deformations were calculated. Harries considered rectangular sections and circular tubes.

Kuranishi and Lu in 1972 reported extensive parametric studies on the load-carrying capacity of two-hinged parabolic steel arches having either rectangular or idealized sandwich cross sections (approximating a wide-flange or box cross section). Residual stresses and strain hardening were taken into account. They found, as did Harries, that when the effect of yielding is considered, the strength of an arch can be drastically reduced under unsymmetrical loads. The buckling load is less when the live load acts over only one-half the span instead of over the entire span, and the larger the live load-to-dead load ratio the smaller the buckling load becomes.

In a restudy of the ultimate solutions of the previous paper, Kuranishi (1973) examined the ratio of the ultimate load to the elastic limit load found by a second-order analysis for the sandwich cross section which approximates I and box cross sections. He shows that the inelastic ultimate load lies between the load at initial yielding and about 93% of this value for a practical range of live load-to-dead load ratios. He concludes that one could design efficiently by using second-order elastic analyses and keeping the maximum stress below 90% of the yield stress. The same could not be said for a solid rectangular section where the ultimate load always exceeded the load at first yielding.

Shinke et al. (1975) studied analytically the behavior of two-hinged and fixed parabolic arches of solid rectangular, pipe, and box cross sections. In this first study residual stresses and strain hardening were not considered. They compare the ultimate strength of several cases of uniform live load symmetrically distributed about the centerline with arches having an unsymmetrically distributed load in the pattern of Fig. 17.4, and find that for the same dead- and live-load intensities unsymmetrically distributed load always governs. They also investigated the effect of length of live load, s, and found that the buckling load is least when the length of live load, s, is roughly equal to one-half the span. With this loading, then, a parametric study was made of the ultimate strength of a box section. In a second study the same authors (Shinke et al., 1977) consider the effect of residual stresses and initial crookedness of the arch rib. Residual stresses are shown to be important, while initial crookedness is not important for unsymmetrical loading. An extensive series of experiments on two-hinged and fixed parabolic arches of solid rectangular cross section are also reported. The analytical results and experiments agree well.

More recently, Kuranishi and Yabuki (1979) and Yabuki (1981) have published numerical studies on the load-carrying capacity of two-hinged parabolic steel arches with thin-walled box cross sections. These works have been summarized by Yabuki and Vinnakota (1984).

The strength of arches with stiffening girders has been studied by Shinke et al. (1980), Kuranishi et al. (1980), and Yabuki (1981). Both propose formulas to estimate the ultimate load.

Studies of the effects of the unsymmetrical distributed load of Fig. 17.4 supplemented by a single concentrated load at the quarter point of the span as per Japanese highway specifications have been made by Yabuki and are summarized by Yabuki and Vinnakota (1984).

The limit load depends on the following factors: rise-to-span ratio, live load-to-dead load ratio, slenderness ratio, yield stress, type of cross section (box section, wide-flange, rectangle, etc.), and residual stress magnitudes and patterns. There are many variables, so that it does not seem possible to present a comprehensive and definite coverage in the space available. Therefore, only a brief review of the major points will be given. This has been extracted from the paper by Yabuki and Vinnakota (1984).

The effects of some of these parameters on the ultimate load-carrying capacity of two-hinged, parabolic, uniform arches subject to uniform dead load plus half-span live load are shown in Figs. 17.5 through 17.7. In these figures the ultimate or maximum load is expressed in terms of the ordinate w_{max}/w_y, in which w_y is the magnitude of a uniformly distributed

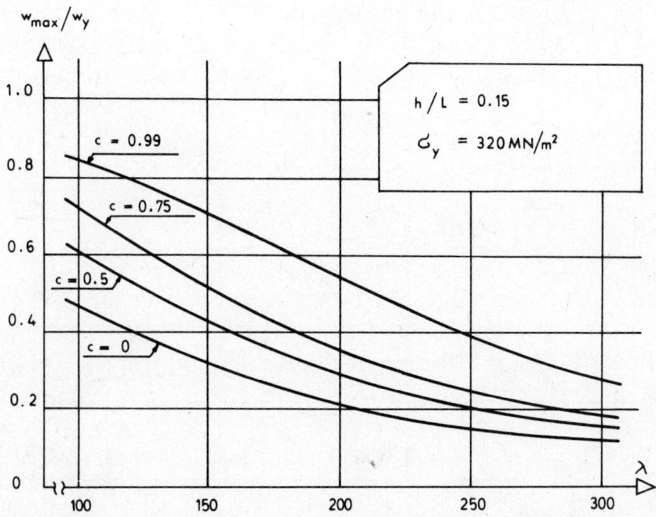

Fig. 17.5 Variation of load-carrying capacity as a function of slenderness ratio.

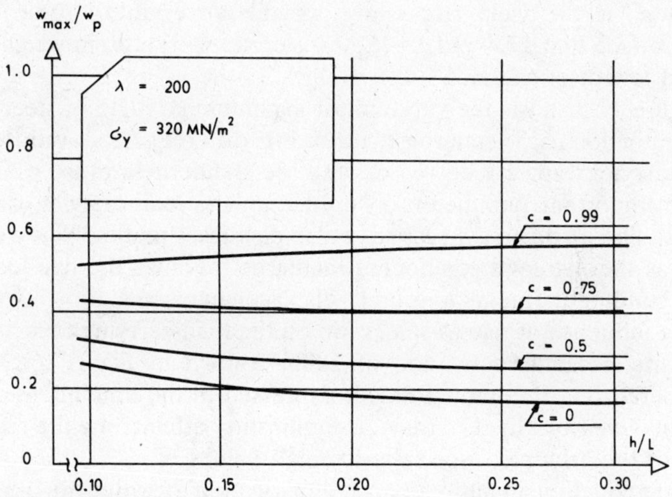

Fig. 17.6 Variation of load-carrying capacity as a function of rise-to-span ratio.

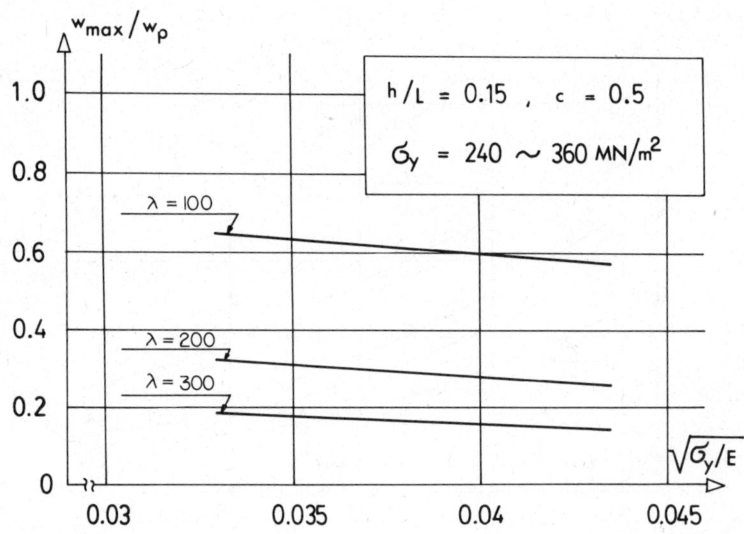

Fig. 17.7 Influence of yield-stress level of material on the load-carrying capacity of arches.

load ($p = 0$) that would cause the arch to yield at the springings under the direct axial thrust produced. Hence

$$w_y = \frac{2A\sigma_y}{L\sqrt{\frac{1}{16}(L/h)^2 + 1}} \tag{17.2}$$

where σ_y is the yield stress level of the material. In these diagrams $c = q/w$ (see Fig. 17.4 and $\lambda = L_s/r$ = slenderness ratio, in which L_s is the curved length of the arch axis).

Figure 17.5 shows the variation of maximum strength of steel arch ribs as a function of slenderness ratio for different load ratios, c. The load-carrying capacity decreases with the slenderness ratio, which is one of the important parameters in predicting the load-carrying capacity of arches. The load-carrying capacity also decreases markedly as c decreases (i.e., as the live-load component increases) because the live load causes large bending moments and large displacements.

The influence of rise-to-span ratio on the load-carrying capacity of arch ribs with a slenderness ratio of 200 is shown in Fig. 17.6. It can be considered that the ultimate load, expressed in the nondimensional form with w_y given by Eq. 17.2, is not significantly affected by the rise-to-span ratio of the arch rib, especially for high values of c.

The reduction in ultimate strength of arch ribs due to the effect of residual stresses can be as much as 20%. The reduction becomes especially noticeable when the arch is subjected to uniformly distributed load covering the entire span. The maximum variation of the ultimate strength using various distribution patterns for the residual stresses is within 10%. The reduction in ultimate strength becomes more significant as the level of compressive residual stress increases, and especially so when the compressive residual stress is greater than $0.4\sigma_y$.

The initial deformations of arch ribs result in a reduction of their ultimate strength. This reduction is important when the arch rib is subjected to uniformly distributed load covering the entire span. However, in the case of unsymmetrical load, which is the critical loading condition for arch rib structures, the deformations due to the loading become large so that the effects of the initial deformations are not significant. The nondimensional ultimate load intensity, w_{max}/w_y, decreases in proportion to the square root of the yield stress level of the material as shown in Fig. 17.7.

The maximum compressive strains occurring in the flange plates of box sections of arch ribs under the ultimate load were examined. Based on these computed strains, the following remarks are made. The maximum strain occurring at a cross section of the arch subjected to unsymmetrical load is higher than its value when the arch rib is subjected to uniformly distributed load covering the entire span. The difference between the two becomes significant when the slenderness ratio becomes large. If the arch rib has to keep its load-carrying capacity up to the ultimate state without premature local buckling of the flange plates, they are required to endure strains four times as large as the yield strain ε_y. However, the flange plates of practical thin-walled box ribs might buckle by strains under $4\varepsilon_y$. Assuming that the critical strain of the cross section is $2\varepsilon_y$, tentatively, the ultimate strength reduction due to this limitation on critical strain becomes large as the loading pattern becomes unsymmetrical and the slenderness ratio becomes small.

The ultimate strength of stiffened two-hinged arch structures is analogous to that of two-hinged arches, if the arch and the stiffening girder behave as an entirely integrated structure. Let λ_T be the slenderness ratio of the stiffened arch structure defined by

$$\lambda_T = \frac{L_s}{\sqrt{(I + I_g)/A}} \tag{17.3}$$

where I_g is the moment of inertia of stiffening girder. The ultimate strength of a stiffened arch structure with a slenderness ratio $\lambda = \lambda_T$ is always somewhat greater than that of a two-hinged arch whose slenderness ratio $\lambda = \lambda_T$. However, judging from the analytical results the two may be considered to be equal to each other for all practical purposes. It is generally not required that attention be given to the "local failure" of arch rib members (buckling between the columns connecting the arch and the stiffening girder) for the unsymmetrical loading case, if a check is made for their strength for constant uniformly distributed load. The "local failure strength" of arch rib members can be determined by the so-called basic column strength curve, when they have straight members between columns. However, for curved members, it is advisable to reduce this strength by 15%.

17.5 DESIGN OF ARCHES FOR IN-PLANE STABILITY

Arches of considerable span are normally not of uniform cross section but are composed of segments with different cross-sectional properties. Each of the segments must be designed for the loads which produce the maximum stresses in that segment. Many arch structures are complex structural systems such as deck bridge arches in which the arch rib is connected to the roadway girder by columns rigidly attached. Studies of the behavior of uniform free standing arches may be of limited usefulness in these cases. A general approach is to factor the loads, use a second-order elastic analysis for the entire system and keep the maximum combined stresses for each segment below some reference stress. Kuranishi (1973) recommends this procedure with the maximum stress less than 90% of the yield stress, as stated previously.

Design procedures for arches based on the ultimate inelastic strength studies reviewed in Section 17.4 have been proposed by Kuranishi (1973), Komatsu and Shinke (1977), and Kuranishi and Yabuki (1984b).

Kuranishi proposed an interaction-type formula similar to beam-column formulas for two-hinged parabolic arches under unsymmetrical loading. Komatsu and Shinke presented practical formulas for the planar ultimate load intensity of two-hinged and fixed parabolic steel arches as a function of the normal thrust calculated at a quarter point of the arch rib by first-order elastic analysis. They also recommended that the ultimate load capacity of arch ribs with varying and/or hybrid cross section can be evaluated accurately by using mean values of the cross-sectional area and/or yield stress level of the material, calculated by averaging along the entire axial length of the rib. Kuranishi and Yabuki also presented accurate practical formulas for the in-plane ultimate strength of parabolic two-hinged steel arch ribs and steel arch bridge structures with a stiffening girder. These formulas are expressed in term of bending moment and axial thrust (or stresses provided by these cross-sectional forces) at a quarter point of the rib, which are all calculated by the first-order elastic analysis.

17.6 OUT-OF-PLANE STABILITY OF ARCHES, INTRODUCTION

When applied forces acting in the plane of a curved member reach a certain critical level, a combination of twisting and lateral bending causes the member to deform out of its original plane. The critical load is influenced by the nature and distribution of the loads, the shape of the axis of the member, the variation of the flexural and torsional stiffness of the cross sections along the axis of the member, and restraint available at the supports and elsewhere. For a steel arch, residual stresses resulting from uneven cooling after rolling and yielding of the material are also important factors. The multiplicity of parameters defeats attempts at formulating simply and widely applicable rules for the determination of buckling loads.

Subjects of research have progressed with time from elastic linear buckling of a single arch to ultimate strength of total systems—that is, from simple and idealized theoretical models to more practical models to be encountered in actual structures. Some of the available solutions that may be useful as a guide to the design of practical structures are presented in formulas and tables. Results from elastic linear buckling theory are shown in Sections 17.7 and 17.9, and in Sections 17.10 and 17.11 more recent results from ultimate strength analyses are presented.

Figure 17.8 shows notations and a coordinate system used for the descriptions in the following sections. Unless stated otherwise, the results presented apply to symmetric arches of constant cross section with doubly symmetric axes. The effect of warping torsional restraint was not included in the results presented except in Fig. 17.9. Warping torsional resistance is a negligible factor for closed-profile cross sections, while it is of

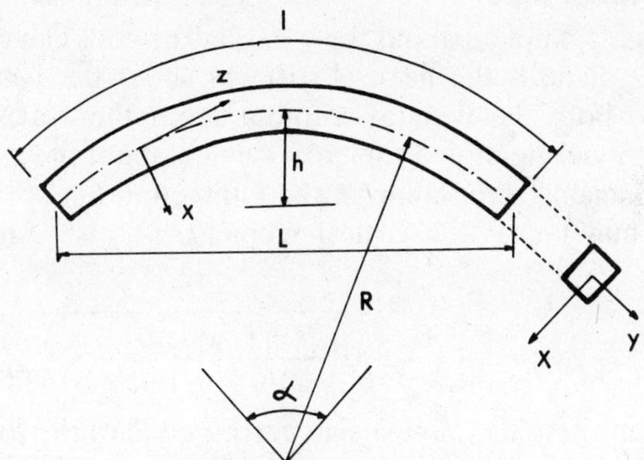

Fig. 17.8 Notation and coordinate system.

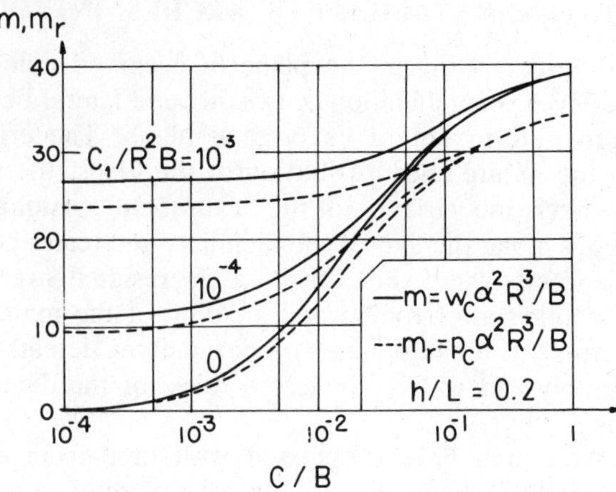

Fig. 17.9 Effects of the warping rigidity C_1.

importance for open profile cross sections. Results obtained by ignoring warping torsional restraint are conservative for open profile sections.

Isolated arches supported so that their ends are free to rotate about the x-axis have very little lateral stability and should be avoided. Although results for this end condition were not given here in most cases, the practical significance of this fact should be emphasized, since in actual arches, completely rigid supports are difficult, if not impossible, to realize.

17.7 OUT-OF-PLANE BUCKLING OF CIRCULAR ARCHES

17.7.1 Uniform Moment

Equal couples, M, applied about the y-axis at the ends can cause buckling out of the x-z plane if the flexural stiffness about the y-axis is large in comparison to both the flexural stiffness about the x-axis, B, and the St.-Venant torsional rigidity of the cross section about the z-axis, C. If the ends are fixed against translation in the y direction but are free to rotate about the x- and y-axes the critical moment, M_c, is (Timoshenko and Gere, 1961)

$$M_c = \frac{B + C}{2R} \pm \sqrt{\left(\frac{B - C}{2R}\right)^2 + \frac{BC}{R^2}\frac{\pi^2}{\alpha^2}} \tag{17.4}$$

In this equation the plus (minus) sign corresponds to the lowest buckling moment which compresses the fibers on the concave (convex) side of the bar. The effects of warping torsional restrain and other boundary conditions are investigated by Kollár and Iványi (1966).

17.7.2 End Forces Directed along the Chord

Two collinear forces applied to the ends of circular members can cause lateral buckling (Klöppel and Protte, 1961; Ojalvo et al., 1969). Forces directed away from each other (pull loads) will generally cause antisymmetric buckling about the center of the member, whereas forces directed toward each other (push loads) cause a symmetric buckling. The buckling load for ends hinged about the x-axis may be expressed in the form (Klöppel and Protte, 1961)

$$F_c = m\,\frac{B}{(R\alpha/2)^2} \tag{17.5}$$

When the central angle α is less than π and the warping torsional stiffness is negligible, the value m (minus sign for pull loads) was given by the following approximate equations:

$$\left.\begin{aligned} &\text{for pull loads:} && m = a_0 + a_1/\alpha + a_2/\alpha^2 \\ &\text{for thrust loads:} && m = 2.47 + a_3\alpha + a_4\alpha^2 \end{aligned}\right\} \tag{17.6}$$

Some of the coefficients a_0 through a_4 are given in Table 17.5. The effect of the warping torsional stiffness was also studied by Klöppel and Prote (1961).

Table 17.5. Coefficients for Use in Eq. 17.6

C/B	a_0	a_1	a_2	a_3	a_4
0.01	0.052	−0.217	−0.36	−0.24	0.09
0.5	0.815	−6.313	−7.411	−0.034	0.069

17.7.3 Uniformly Distributed Radial Forces

If the radial forces, p, are direted inward along the centroidal axis and the ends of the bar are free to rotate with respect to their principal axes x and y but unable to rotate with respect to the axis z, the buckling load p_c is (Timoshenko and Gere, 1961)

$$p_c = \frac{B}{R^3}\,\frac{(\pi^2 - \alpha^2)^2}{\alpha^2[\pi^2 + \alpha^2(B/C)]} \tag{17.7}$$

An approximate equation for the case of built-in ends can be obtained by replacing the terms π^2 with $4\pi^2$ in Eq. 17.7, but this would not be accurate unless α is smaller than $\pi/2$ and C/B is greater than or equal to 0.5.

More comprehensive studies, including the effect of warping torsional restraint, are reported by Fukasawa (1983) and Kollár and Iványi (1966). Some of the theoretical results from Fukasawa (1983) are discussed in the next section.

17.7.4 Uniformly Distributed Vertical Forces

When the arch is fixed against rotation about the x-axis at its ends, the uniform load w_c at buckling may be expressed in the form (Demuts, 1969)

$$w_c = m \frac{B}{L^3} \tag{17.8}$$

Values of m are shown in Table 17.6.

The usual closed-profile cross section used in arches has a C/B ratio in the range 0.5 to 1.5 and for the open-profile cross section C/B is in the range 0.01 to 0.001. If the arch cross section has a thin-walled open profile, the warping torsional rigidity, C_1, also has a primary effect on the buckling loads, as shown in Fig. 17.9 (Fukasawa, 1963; Namita, 1968). In this figure the buckling coefficient $m = w_c\alpha^2R^3/B$ for uniform vertical loads, w, is shown by solid lines and the buckling coefficient $m_r = p_c\alpha^2R^3/B$ for uniform radial loads, p, is shown by broken lines. It is also found that the values m and m_r are related to each other by an approximate relation $m_r \simeq m[2(\sin\alpha/2)/\alpha]$ when the height span ratio, h/L, is not large (for $h/L \leqq 0.2$).

Table 17.6. Approximate Values of *m* for use in Eq. 17.8

	Circular Arch		Parabolic Arch					
$m = \frac{w_cL^3}{B}$	Vertical Loads		Vertical Loads		Hanger Loads (Symmetric)		Column Loads (Antisymmetric)	
$C/B =$ H/L	0.01	0.5	0.01	0.5	0.01	0.5	0.01	0.5
0.1	18	28	16	28	39	70	12	18
0.2	17	39	15	39	35	110	12	29
0.3	13	38	13	37	28	116	10	32
0.4	9	30	11	30	24	104	8	31
0.5	5	20	9	24	20	87	6	28

17.8 OUT-OF-PLANE BUCKLING OF PARABOLIC ARCHES

17.8.1 Uniformly Distributed Vertical Loads

The uniformly distributed buckling load, w_c, again may be expressed in the form of Eq. 17.8. Representative values of m are given in Table 17.6 (Tokarz and Sandhu, 1972). The ratio $C/B = 0.01$ is typical of thin-walled open-profile cross sections for which warping torsional restraint is of importance. The effect of the warping torsional rigidity may be similar to what can be observed in a circular arch (see Fig. 17.9). Additional information is available in the papers by Stüssi (1943), Tokarz (1971), and Kee (1961).

17.8.2 Parabolic Arches with Tilting Loads

It has been recognized by Stüssi (1943) and others that the buckling load of an arch is increased if the loads are applied to the arch by a system of vertical hangers connected to a laterally stiff roadway at the elevation of the chord as in the case of the through-type arch bridges. If the loads are applied by means of columns connected to a laterally stiff roadway above, the lateral deformations of the buckled arch will be antisymmetric about the crown when the arch is connected to the roadway at this point or if the columns are very short near the crown. Both hanger-loaded and column-loaded parabolic arches were studied theoretically by Östlund (1954), Godden (1954), and Shukla and Ojalvo (1971). The buckling load w_c may be expressed by Eq. 17.8 using appropriate values of m from Table 17.6. It is assumed there that the ends of the arch are fixed against rotation about the x-axis. Östlund (1954) and Almeida (1970) reported that certain braced arches subjected to hanger loads are also controlled by antisymmetric buckling.

17.9 BRACED ARCHES AND REQUIREMENTS FOR BRACING SYSTEMS

Most arches used in practice are braced against lateral movement either continuously or at regularly spaced intervals. An arch rib connected continuously to a curved roof is an example of continuous bracing. Unless such arch ribs are unusually deep, lateral buckling will not be a problem. Twin arch ribs with a lateral bracing system often occur in arch bridges.

The elastic lateral buckling of twin arch ribs braced with transverse bars normal to the plane of the ribs has been studied by several investigators (Tokarz, 1971; Östlund, 1954; Almeida, 1970; Kuranishi, 1961a; and Sakimoto and Namita, 1971). The following properties of the bracing system are important in the suppression of lateral buckling: (1) the location and spacing of the transverse bars; (2) the distance between the arch ribs, b; (3) the flexural stiffness of the bars, D_x, about the x-axis; and (4) the flexural stiffness of the bars, D_z, about the z-axis.

It was found that increasing the flexural stiffness D_x was a more effective way of suppressing lateral buckling than increasing the flexural stiffness D_z. The flexural stiffness D_z is less important except when the arch ribs have an open-profile cross section and the stiffness D_x can not be provided. When either of the flexural stiffness D_x or D_z is increased independently, the buckling load of twin arches increases and tends to attain an asymptotic value. While the magnitude of the asymptotic value depends on the properties of the bracing system listed above, the asymptotic value of the buckling load for a closely braced arch system can be obtained when $D_x \geq 10b\ B/R$ or $D_z \geq b\ B/R$, respectively (Östlund, 1954; Sakimoto and Namita, 1971).

When pairs of arch ribs are braced closely with such stiff transverse bars, the critical load per one arch rib of a set of braced arches can be increassed up to 250% or more of that for the identical isolated single arch so far as elastic buckling is concerned. For actual bridge arches, however, this increase of the buckling strength will be limited by yielding of the material (see Fig. 17.13).

Wästlund (1960) and others have suggested a simple approximate method for the determination of the lateral buckling load of braced arches by utilizing a planar system which is obtained by straightening out the arch in a horizontal plane. The compressive force required to buckle the longitudinal ribs of the assumed planar system would be computed as for a column with battens, and these compressive forces would be the approximations to the rib forces required to buckle the actual arch structure. The approximate method can be accurate for the arches of a torsionally stiff (closed) cross section, but it is unconservative for the arches of an open-profile cross section (Östlund, 1954).

Although twin arches braced with transverse bars might be a favorable structure from an aesthetic point of view, transverse bars cannot be a more effective bracing system than a bracing system of diagonal members for suppressing lateral buckling (Sakimoto, 1979a; Sakimoto and Komatsu, 1982). If the lateral bracing consists of either diagonals, diagonals in combination with transverse bars, or a K system, out-of-plane buckling of the arch tends not to be a problem except for through-type arch bridges, which cannot have lateral bracing over the entire arch length. Through-type steel arches with double diagonal bracing system are discussed in the following section.

The stiffness of diagonal bracing members required for the necessary lateral stability of an arch system was studied by Sakimoto and Komatsu (1977a,b) and Kuranishi and Yabuki (1981) in relation to the ultimate strength of braced steel arches subjected to the combination of vertical and horizontal uniform loads.

17.10 ULTIMATE STRENGTH OF STEEL ARCHES SUBJECTED TO UNIFORMLY DISTRIBUTED VERTICAL LOADS

Since the arch is basically a compressive member, residual stresses resulting from welding and initial out-of-plane deflections have a significant influence on the strength of steel arches limited by lateral instability. The computer results obtained for typical theoretical models of square box cross sections show that (1) the residual stress may cause a reduction of strength of at most 20% for mild-steel arches and 10% for high-strength-steel arches, and (2) the initial out-of-plane deflections may reduce, at maximum, 15% of the strength of a perfectly plane arch (Sakimoto and Komatsu, 1977b; Komatsu and Sakimoto, 1977).

Some of the results are shown in Figs. 17.10 and 17.11. In these figures

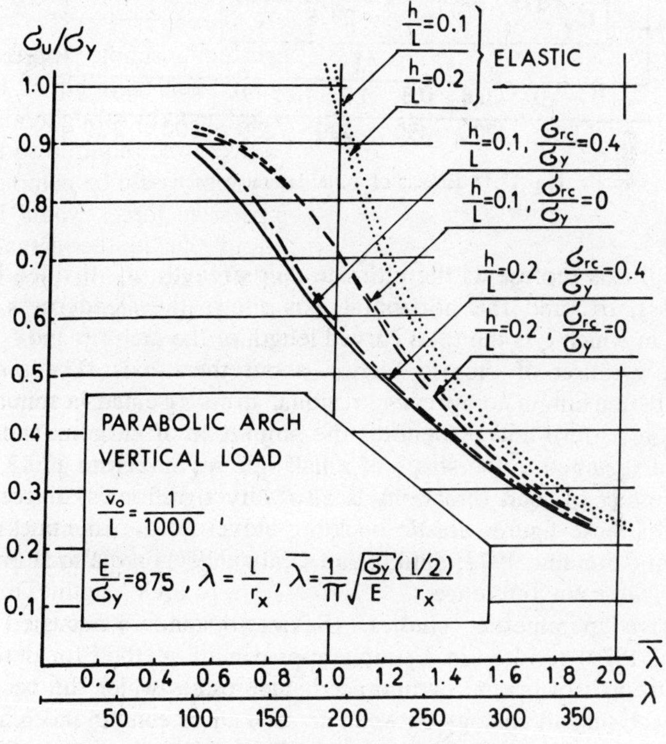

Fig. 17.10 Effects of residual stresses.

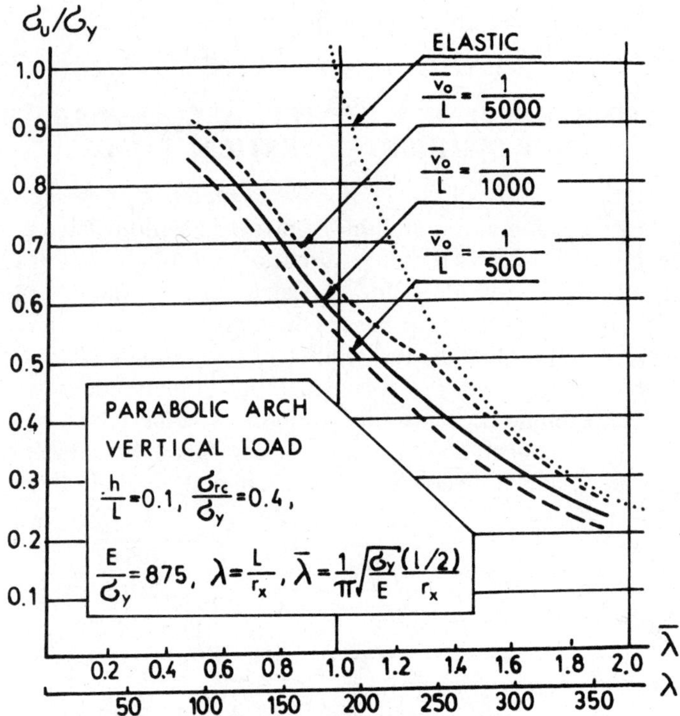

Fig. 17.11 Effects of initial lateral deflections.

the vertical axis represents the ultimate unit strength, σ_u, divided by the yield stress, σ_y, and the horizontal axis shows the slenderness ratio, $\lambda = l_s/r_x$, in which l_s is the total curved length of the arch rib and r_x is the radius of gyration of the rib section about the x-axis. The term σ_{rc} denotes the maximum compressive residual stress assumed in trapezoidal distribution pattern and $\bar{v}_0$ denotes the amplitude of initial out-of-plane deflections assumed in the shape of a half-sine wave. Figure 17.12 shows that the effect of load directions is also very significant for steel arch bridges. In these figures elastic buckling curves of the identical arches (Tokarz and Sandhu, 1972; Shukla and Ojalvo, 1971) are also shown and noted as *elastic* for reference.

Extensive parametric studies (Sakimoto and Komatsu, 1982; Sakimoto, 1979) resulted in a simple approximate method for determining the strength of braced or unbraced steel arches which fail by lateral instability. Using an analogy between an arch and a column an equivalent slenderness function λ_a for the determination of the ultimate strength of

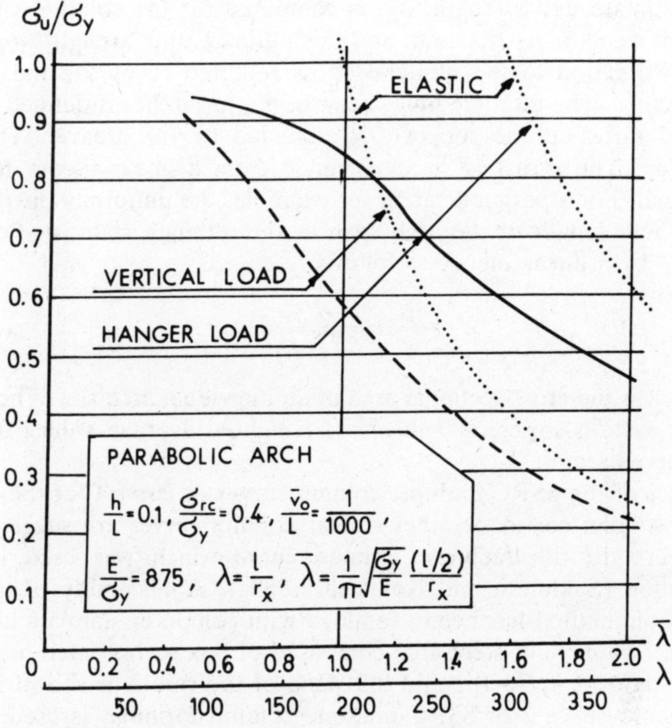

Fig. 17.12 Effects of load directions.

through-type steel arches of box cross sections is defined as follows (Sakimoto, and Komatsu, 1983a):

$$\lambda_a = \frac{1}{\pi} \sqrt{\frac{\sigma_y}{E}} \frac{l}{r_x} K_e K_l K_\beta \tag{17.9}$$

where σ_y is the lowest yield stress among those of different steel grades used in the arch rib and K_e, K_l, and K_β are effective-length factors.

The coefficent K_e relates to the rotational fixity of an arch rib at its ends with respect to the centroidal x-axis. For the clamped condition K_e is 0.5 and for the hinged condition it is 1.0. The coefficient K_l relates to the direction of the loads and it is 0.65 for the tilting hanger case and 1.0 for nontilting hangers (i.e., for vertical loads). The coefficient K_β is related to the lateral restraint supplied by the bracing system and given by $K_\beta = 1 - \beta + (2r_x\beta / K_e b)$. The term β denotes a ratio of the length of braced portion to the total length of the arch rib. Since β is equal to zero for arches without bracing, K_β equals 1.0 for an isolated single arch.

The ultimate unit strength, σ_u, is computed for the column with the equivalent slenderness function, and this ultimate unit strength would be the approximation to the ultimate stress required to buckle the actual arch structure. The ultimate unit strength, σ_u, for arches is defined as the tangential thrust at the support, N_u, divided by the area A. That is, $N_u = A \cdot \sigma_u$. The thrust N_u is determined from a linear theory for the loaded arch. For a parabolic arch, for example, the uniformly distributed load per unit length of the arch span at the ultimate state, w_u, can be computed by a linear theory as follows

$$w_u = \frac{2A\sigma_u}{L\sqrt{\frac{1}{16}(L/h)^2 + 1}} \tag{17.10}$$

in which A is the cross-sectional area of an individual arch rib. Where the rib cross section varies, A and r_x are weighted average values for the entire curved length l.

Curve 2 of the SSRC multiple column curves or curve C of the ECCS multiple column curves or other similar column curves are suggested as counterparts to the Japanese column curve which was used in this investigation (Sakimoto and Komatsu, 1983a). Applicability of the approximation method has been examined with computer simulations for a parabolic or a circular steel arch composed of box section arch members with the ratio of $h/L = 0.1$ and 0.2. One of the results is shown in Fig. 17.13, where curve 2 of SSRC multiple column formulas is used in the approximation method.

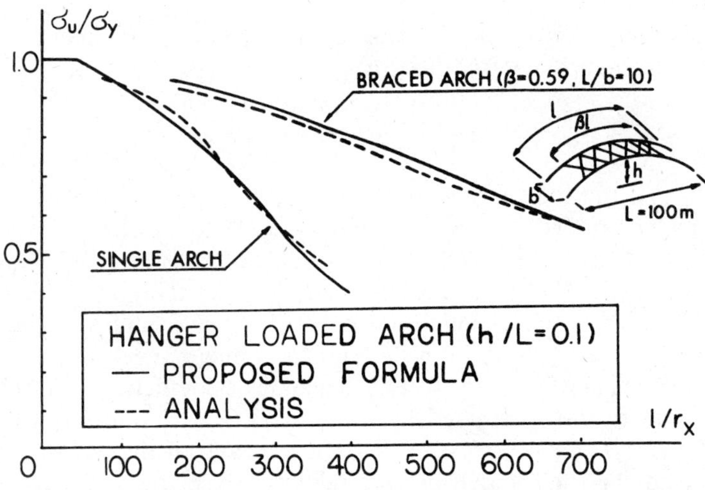

Fig. 17.13 Accuracy of the approximation method.

The effect of constraint to the lateral movement of an arch, provided by a bridge deck, was studied by Sakimoto (Sakimoto and Komatsu, 1983b; Sakimoto and Yamao, 1983). He reports that the ultimate strength of through-type arch bridges with hangers clamped at their ends to the arch rib and the bridge deck increases considerably, because the lateral bending stiffness of the hangers constrains the lateral movement of the arch. Even for deck-type arch bridges, constraint against lateral movement can increase the lateral buckling strength. The stabilizing effect of this constraint can afford more than mere compensation for the destabilizing effect produced by inclined columns. It should be noted that this result is contrary to the result of elastic bifurcation theory shown in Table 17.6. The main reason for this contradiction is thought to be caused by the fact that in the elastic bifurcation theory, the effect of constraint is not considered, while the destabilizing effect produced by inclined columns is considered.

When overall lateral buckling is suppressed, only the buckling of the portions of the ribs between points of bracing need be considered. The slenderness limit for which the local instability of the arch segment may occur prior to overall lateral instability can be determined approximately from Eq. 17.9 as follows (Sakimoto and Komatsu, 1983a)

$$\sqrt{\sigma_y/E}(l_p/r_x)_{\max} \geq \sqrt{\sigma_y/E}(l/r_x)K_eK_lK_\beta \tag{17.11}$$

Here $(l_p/r_x)_{\max}$ denotes the slenderness ratio for the most slender of the arch segment having an unbraced length, l_p, between points of bracing, and the term σ_y in the left side of Eq. 17.11 is the yield stress of the arch segment where $(l_p/r_x)_{\max}$ occurs.

17.11 ULTIMATE STRENGTH OF STEEL ARCH BRIDGES SUBJECTED TO VERTICAL AND LATERAL UNIFORM LOADS

The spatial elastic-plastic behavior and the ultimate load capacity of the through-type braced arch bridges of box cross sections was studied for a combination of the vertical and horizontal uniform loads by Sakimoto (Sakimoto and Komatsu, 1977a, 1979; Sakimoto et al., 1979). Computer analyses for various theoretical models resulted in a simple approximation method for the determination of the ultimate lateral strength of bridge arches braced partly over the central portion. By utilizing an analogy between a laterally loaded arch and a beam-column an interaction formula is proposed as follows:

$$\frac{N}{N_u} + \frac{M}{M_y[1-(N/N_e)]} \leq 1.0 \tag{17.12}$$

Although this formula has a form similar to the interaction formula for a beam column, the meanings of the quantities should be understood according to the following definitions:

N = tangential end thrust computed by a linear theory for uniformly distributed design loads

N_u = tangential end thrust at inelastic lateral buckling of the arch under uniformly distributed vertical loads, which is determined by $N_u = A\sigma_u$ from the result of the former section

M = lateral end moment of the individual arch rib subjected to uniformly distributed horizontal loads, not including contribution of axial load interacting with deflections; this can be approximated by the value computed for the planar system which is obtained by straightening out the arch in a horizontal plane; a simple approximation formula for the determination of this moment is also given by Sakimoto and Komatsu (1979)

M_y = yield moment of the arch rib at the ends with respect to the x-axis

N_e = Euler buckling load computed for a hinged column of which the length and the cross section are identical to those of the arch rib at its unbraced end part

The validity of the interaction formula has been verified by extensive computer simulations for various braced bridge arches of practical proportions. If the twin ribs are braced closely over the entire length of the arch, the second term of the left side of Eq. 17.12 becomes negligibly small. This means such a braced arch tends not to be a problem against usual lateral loads.

Deflections and stresses due to forces normal to the plane of the arch are increased by a contribution of axial load interacting with lateral deflections. Stüssi (1943) and others (Östlund, 1954; Wästlund, 1960) suggest that increased deflections and stresses may be estimated by multiplying the deflections and stresses due to transverse force acting alone by the amplification factor $1/(1 - w/w_c)$, in which w is the intensity of the uniform load in the plane of the arch and w_c is the intensity of the in-plane load that would cause lateral buckling. While this amplification factor is at best only roughly approximate (Layranguez, 1959; Donald and Godden, 1961a, b), it may be convenient for use in a preliminary design.

The effect of lateral horizontal force on the in-plane strength of arch bridges was studied by Kuranishi and Yabuki (Kuranishi, 1961b; Yabuki and Kuranishi, 1973; Kuranishi and Yabuki, 1977). It was found that the in-plane strength of arch bridges braced closely over the entire length of the arch is not significantly affected by the lateral loads usually encountered in actual structures. Therefore, through-type steel arch bridges can be designed with a lateral bracing system (between arch ribs) of sufficient out-of-plane stiffness that the arch rib design will be determined primarily by the in-plane loads. For such a case lateral loads can be taken into account for practical purposes as a set of additional in-plane vertical loads. Thus the arch rib design can be made on the basis of a quasi-planar model subjected to the principal vertical loads with a small modification of these loads resulting from wind loads. The lateral bracing stiffness required to ensure that two-hinged arch bridges act basically as in-plane structures has been defined by Kuranishi, Yabuki and Vinnakota (Kuranishi and Yabuki, 1981, 1984b; Yabuki et al., 1983).

References

Almeida, P. N. (1970), "Lateral Buckling of Twin Arch Ribs with Transverse Bars," Dissertation, Ohio State University, Columbus, Ohio.

Austin, W. J. (1971), "In-Plane Bending and Buckling of Arches," *ASCE J. Struct. Div.*, Vol. 97, No. ST5, pp. 1575–1592.

Austin, W. J., and Ross, T. J. (1976), "Elastic Buckling of Arches under Symmetrical Loading," *ASCE J. Struct. Div.*, Vol. 102, No. ST5, pp. 1085–1095.

Chang, C. K. (1973), "Effect of Loaded Length on the Buckling Strength of Slender Arches," Thesis, Rice University, Houston.

Coronforth, R. C., and Childs, S. B. (1967), "Computer Analysis of Two Hinged Circular Arches," *ASCE J. Struct. Div.*, Vol. 93, No. ST2, pp. 319–338.

Demuts, E. (1969), "Lateral Buckling of Circular Arches Subjected to Uniform Gravity Type Loading," Dissertation, Ohio State University, Columbus, Ohio.

Deutsch, E. (1940), "Das Knicken von Bogenträgern bei unsymmetrischer Belastung," *Bauingenieur*, December, pp. 353–360.

Donald, P. T. A., and Godden, W. G. (1961a), "A Numerical Solution to the Curved Beam Problem," *Struct. Eng.*, Vol. 41, pp. 179–186.

Donald, P. T. A., and Godden, W. G. (1961b), "The Transverse Behavior of Laterally Unsupported Parabolic Arches," *Struct. Eng.*, Vol. 41, pp. 187–191.

Fukasawa, Y. (1963), "Buckling of Circular Arches by Lateral Flexure and Torsion under Axial Thrust," *Trans. Jpn. Soc. Civ. Eng.*, No. 96, pp. 29–47 (in Japanese).

Galli, A., and Franciosi, G. (1955), "Limit Analysis of Thin Arch Bridges with Stiffening Girders," *G. Genio Civ.*, Vol. 93, No. 11.

Godden, W. G. (1954), "The Lateral Buckling of Tied Arches," *Proc. Inst. Civ. Eng.*, Vol. 3, Part III, pp. 496–514.

Harries, H. (1970), "Traglast stahlerner Zweigelenkbogen mit ausgebreiteten Fliesszonen," *Stahlbau*, Vol. 6, pp. 170–177; Vol. 8, pp. 248–257.

Harrison, H. B. (1982), "In-Plane Stability of Parabolic Arches," *ASCE J. Struct. Div.*, Vol. 108, No. ST1, pp. 195–205.

Kee, C. F. (1961), "Lateral Inelastic Buckling of Tied Arches," *ASCE J. Struct. Div.*, Vol. 87, No. ST1, pp. 23–39.

Klöppel, K., and Protte, W. (1961), "A Contribution to the Buckling Problem of Circular Curved Bars," *Stahlbau*, No. 30, pp. 1–15 (in German).

Kollár, L. U., and Iványi, G. (1966), "Buckling of Shell Arches by the Energy Method," *Bautechnik Arch.*, pp. 1–23 (in German).

Kollbrunner, C. F. (1936), "Versuche über die Knicksicherheit und die Grundschwingungzahl vollwandiger Bogen," *Bautechnik*, March, pp. 186–188.

Kollbrunner, C. F. (1942), "Versuche über die Knicksicherheit und die Grundschwingungzahl vollwandiger Dreigelenkbogen: *Schweiz. Bauzg.*, Vol. 120, No. 10, pp. 113–115.

Komatsu, S., and Sakimoto, T. (1977), "Ultimate Load Carrying Capacity of Steel Arches," *ASCE J. Struct. Div.*, Vol. 103, No. ST12, pp. 2323–2336.

Komatsu, S., and Shinke, T. (1977), "Practical Formulas for In-Plane Load Carrying Capacity of Arches," *Proc. Jpn. Soc. Civ. Eng.*, No. 267, pp. 39–51 (in Japanese).

Kuranishi, S. (1961a), "The Torsional Buckling Strength of Solid Rib Arch Bridge," *Trans. Jpn. Soc. Civ. Eng.*, No. 75, pp. 59–67 (in Japanese).

Kuranishi, S. (1961b), "Analysis of Arch Bridge under Certain Lateral Forces," *Trans. Jpn. Soc. Civ. Eng.*, No. 73, pp. 1–6 (in Japanese).

Kuranishi, S. (1973), "Allowable Stress for Two-Hinged Steel Arch," *Proc. Jpn. Soc. Civ. Eng.*, No. 213, pp. 71–75.

Kuranishi, S., and Lu, L. W. (1972), "Load Carrying Capacity of Two-Hinged Steel Arches," *Proc. Jpn. Soc. Civ. Eng.*, No. 204, pp. 129–140 (in English).

Kuranishi, S., and Yabuki, T. (1977), "In-Plane Strength of Arch Bridges Subjected to Vertical and Lateral Loads," *2th Int. Colloq. Stab. Steel Struct., Prelim. Rep.*, Liege, April, pp. 551–556.

Kuranishi, S., and Yabuki, T. (1979), "Some Numerical Estimations of Ultimate In-Plane Strength of Two-Hinged Steel Arches," *Proc. Jpn. Soc. Civ. Eng.*, No. 287, pp. 155–158.

Kuranishi, S., and Yabuki, T. (1981), "Required Out-of-Plane Rigidities of Steel Arch Bridges with Two Main Ribs Subjected to Vertical and Lateral Loads," *Tech. Rep. Tohoku Univ.*, Sendai, Japan, Vol. 46, No. 1.

Kuranishi, S., and Yabuki, T. (1984a), "Ultimate Strength Design Criteria for Two-Hinged Steel Arch Structures," *Proc. Jpn. Soc. Civ. Eng.*, No. 3005/I-2 (Div. Struct. Earthquake Eng.).

Kuranishi, S., and Yabuki, T. (1984b), "Lateral Load Effect on Arch Bridge Design," *ASCE J. Struct. Eng.* Vol. 110, No. 9, pp. 2263–2274.

Kuranishi, S., Sato, T., and Otsuki, M. (1980), "Load Carrying Capacity of Two-Hinged Steel Arch Bridges with Stiffening Deck," *Proc. Jpn. Soc. Civ. Eng.*, No. 300, pp. 121–130.

Layrangues, P. (1959), "Elastic Deformations and Forces in a Fixed End Symmetrical Circular Arch Subjected to a Constant Normal Force and a Uniformly Distributed Lateral Force," *Ann. Ponts Chaussees*, Vol. 129, pp. 323–343 (in French).

Namita, Y. (1968), "Second Order Theory of Curved Bars and Its Use in the Buckling Problem of Arches," *Trans. Jpn. Soc. Civ. Eng.*, No. 155, pp. 32–41 (in German).

Ojalvo, M., Demuts, E., and Tokarz, F. J. (1969), "Out-of-Plane Buckling of Curved Members," *ASCE J. Struct. Div.*, Vol. 95, No. ST10, pp. 2305–2316.

Onat, E. T., and Prager, W. (1953), "Limit Analysis of Arches," *J. Mech. Phys. Solids*, Vol. 1, pp. 77–89.

Östlund, L. (1954), "Lateral Stability of Bridge Arches Braced with Transverse Bars," *Trans. R. Inst. Technol.*, Stockholm, No. 84.

Sakimoto, T. (1979), "Elasto-Plastic Finite Displacement Analysis of Three Dimensional Structures and Its Application to Design of Steel Arch Bridges," Dissertation, Osaka University, Osaka, Japan.

Sakimoto, T., and Komatsu, S. (1977a), "A Possibility of Total Breakdown of Bridge Arches due to Buckling of Lateral Bracings," *2nd Int. Collo. Stab. Steel Struct., Final Rep.*, Liege, April, pp. 299–301.

Sakimoto, T., and Komatsu, S. (1977b), "Ultimate Load Carrying Capacity of Steel Arches with Initial Imperfections," *2nd Int. Colloq. Stab. Steel Struct., Prelim. Rept.*, Liege, April, pp. 545–550.

Sakimoto, T., and Komatsu, S. (1979), "Ultimate Strength of Steel Arches under Lateral Loads," *Proc. Jpn. Soc. Civ. Eng.*, No. 292, pp. 83–94.

Sakimoto, T., and Komatsu, S. (1982), "Ultimate Strength of Arches with Bracing Systems," *ASCE J. Struct. Div.*, Vol. 108, No. ST5, pp. 1064–1076.

Sakimoto, T., and Komatsu, S. (1983a), "Ultimate Strength Formula for Steel Arches," *ASCE J. Struct. Div.*, Vol. 109, No. 3, pp. 613–627.

Sakimoto, T., and Komatsu, S. (1983b), "Ultimate Strength Formula for Central-Arch-Girder Bridges," *Proc. Jpn. Soc. Civ. Eng.*, No. 333, pp. 183–186.

Sakimoto, T., and Namita, Y. (1971), "Out-of-Plane Buckling of Solid Rib Arches Braced with Transverse Bars," *Proc. Jpn. Soc. Civ. Eng.*, No. 191, pp. 109–116.

Sakimoto, T., and Yamao, T. (1983), "Ultimate Strength of Deck-Type Steel Arch Bridges," *3rd Int. Colloq. Stab. Met. Struct., Prelim. Rep.*, Paris, November.

Sakimoto, T., Yamao, T., and Komatsu, S. (1979), "Experimental Study on the Ultimate Strength of Steel Arches," *Proc. Jpn. Soc. Civ. Eng.*, No. 286, pp. 139–149.

Shinke, T., Zui, H., and Namita, Y. (1975), "Analysis of In-Plane Elasto-Plastic Buckling and Load Carrying Capacity of Arches," *Proc. Jpn. Soc. Civ. Eng.*, No. 244, pp. 57–69 (in Japanese).

Shinke, T., Zui, H., and Namita, Y. (1977), "Analysis and Experiment on In-Plane Load Carrying Capacity of Arches," *Proc. Jpn. Soc. Civ. Eng.*, No. 263, pp. 11–23 (in Japanese).

Shinke, T., Zui, H., and Nakagawa, T. (1980), "In-Plane Load Carrying Capacity of Two-Hinged Arches with a Stiffening Girder," *Trans. Jpn. Soc. Civ. Eng.*, No. 301, pp. 47–59.

Shukla, S. N., and Ojalvo, M. (1971), "Lateral Buckling of Parabolic Arches with Tilting Loads," *ASCE J. Struct. Div.*, Vol. 97, No. ST6, pp. 1763–1773.

Stevens, L. K. (1957), "Carrying Capacity of Mild Steel Arches," *Proc. Inst. Civ. Eng.*, Vol. 6, pp. 493–514.

Stüssi, F. (1943), "Lateral-Buckling and Vibration of Arches," *Proc., Int. Assoc. Bridge Struct. Eng. Pub.*, Vol. 7, pp. 327–343. (in German).

Timoshenko, S. P., and Gere, J. M. (1961), *Theory of Elastic Stability*, 2nd ed., McGraw-Hill, New York.

Tokarz, F. J. (1971), "Experimental Study of Lateral Buckling of Arches," *ASCE J. Struct. Div.*, Vol. 97, No. ST2, pp. 545–559.

Tokarz, F. J., and Sandhu, R. S. (1972), "Lateral-Torsional Buckling of Parabolic Arches," *ASCE J. Struct. Div.*, Vol. 98, No. ST5, pp. 1161–1179.

Wästlund, G. (1960), "Stability Problems of Compressed Steel Members and Arch Bridges," *ASCE J. Struct. Div.*, Vol. 86, No. ST6, pp. 47–71.

Yabuki, T. (1981), "Study on Ultimate Strength Design of Steel Arch Bridge Structures," Dissertation, Tohoku University, Sendai, Japan (in Japanese).

Yabuki, T., and Kuranishi, S. (1973), "Out-of-Plane Behavior of Circular Arches under Side Loadings," *Proc. Jpn. Soc. Civ. Eng.*, No. 214, pp. 71–82.

Yabuki, T., and Vinnakota, S. (1984), "Stability of Steel Arch-Bridges, A State-of-the-Art Report," *Solid Mech. Arch.*, Vol. 9, Issue 2, Noordhoff International Publishers, Leyden, Netherlands.

Yabuki, T., Vinnakota, S., and Kuranishi, S. (1983), "Lateral Load Effect on Load Carrying Capacity of Steel Arch Bridge Structures," *ASCE J. Struct. Div.*, Vol. 109, No. 10, pp. 2434–2449.

Chapter Eighteen

Doubly Curved Shells and Shell-Like Structures

18.1 INTRODUCTION

A shell-like metal structure is one that resists loads in a manner similar to that of a thin shell. That is, the major mode of resistance is by membrane action by which forces are carried from point to point by biaxial tension or compression and by shear in the plane of the shell. In addition to the membrane resistance, the shell-like structure has bending resistance to help resist loads. Examples of shell-like structures are latticed shells, reticulated shells, stiffened shells, orthortropic shells, sandwich shells, framed shells, and other types where a modified shell theory may be used to determine their structural performance. Several types of spherical shell-like structures are illustrated in Fig. 18.1.

A latticed or reticulated shell is defined as the form resulting from approximating a solid shell surface by a framework of relatively short linear structural members. A stiffened shell is defined as a thin solid shell to which relatively closely spaced stiffeners have been added to greatly increase the bending resistance and consequently the stability of the structure. An orthotropic shell is defined as a shell-like structure that has different properties (area and bending stiffness) in different directions. A sandwich shell is defined as a shell-like structure that is made up of two thin shell layers separated by a center core that has shear-carrying capacity. Domes, barrel vaults, pressure vessels, and hypars have been constructed by using some of the methods described herein.

Three types of buckling may occur in shell-like structures. These are general buckling, local buckling, and member buckling. General buckling occurs when a relatively large area of the shell-like structure becomes unstable, and a relatively large number of joints or nodes are included in

(*a*)

(*b*)

Fig. 18.1 Spherical shell-like structures: (*a*) latticed shell dome; (*b*) framed shell dome; (*c*) stiffened shell dome.

(c)

Fig. 18.1 (*Continued*)

the buckle. Local buckling in a reticulated shell occurs when one node with its adjacent members is included in the buckle. Local buckling in a stiffened shell occurs when the shell buckles between stiffeners and the stiffeners do not buckle. Member buckling occurs when an individual member buckles as a column and the nodes do not buckle.

Many of the factors that affect the stability of thin-shell structures also affect the stability of shell-like structures. The effect of the edge conditions and residual stresses must be considered in the stability analysis. As edge-bending effects cause deflections from a perfect surface, lower buckling loads result. Deviations from a perfect surface in the fabricated and erected structure usually affect the buckling load. Such deviations also include those due to loads from zero to just prior to buckling which include the effects of the boundary conditions. The deviations from a perfect surface are considerably more serious in the buckling of a shell-like structure than in a column structure. For many domes, pressure vessels, and similar structures, plasticity reduction factors must be also considered.

Three methods are used to calculate buckling loads: (1) closed-form solutions, (2) computer solutions, and (3) code-type solutions. The results given in the remainder of this chapter apply only to spherical shells and shell-like structures.

18.2 GENERAL CONSIDERATIONS

The actual buckling stress of a shell, σ_c, can be expressed as the product of the theoretical elastic buckling stress, σ_e; a "capacity reduction factor," α, applied to account for edge effects, geometric imperfections (out-of-roundness), and deflections in general; and a "plasticity reduction factor," η, to take account of material nonlinearity due to the stress-strain curve and residual stresses. The buckling stress equation is

$$\sigma_c = \eta \alpha \sigma_e \tag{18.1}$$

In the case of stiffened shells the local buckling of the stiffener elements must also be considered.

Many formulations for the plasticity reduction factor η have been introduced for various load conditions and geometries. However, some of the more common expression are variations of the form given by Gerard and Becker (1957):

$$\eta = \frac{\sqrt{E_t E_s}}{E} \sqrt{\frac{1 - \nu_e^2}{1 - \nu^2}} \tag{18.2}$$

where E, E_t, and E_s are the elastic, the tangent, and the secant moduli, respectively, and ν_e and ν are, respectively, the elastic and the plastic Poisson's ratios. The following equation is simpler to use and is in close agreement with Eq. 18.2

$$\eta = \frac{E_t}{E} \tag{18.3}$$

Equations 18.2 and 18.3 are in good agreement for materials with a flat yield plateau. To find the tangent modulus the Ramberg–Osgood (1943) equation can be used to define the stress-strain curve:

$$\epsilon = \frac{\sigma}{E} + k\left(\frac{\sigma}{E}\right)^a \tag{18.4}$$

where

$$a = \frac{\ln (\bar{\epsilon}_y / \epsilon_1)}{(\sigma_y / \sigma_1)} \tag{18.5}$$

and

$\bar{\epsilon}_y$ = offset strain at yield stress, σ_y (0.2% offset)
ϵ_1 = offset strain at stress σ_1 (usually 0.1% offset is used)

$$k = \bar{\epsilon}_y / (\sigma_y / E)^a \tag{18.6}$$

The derivative $d\sigma / d\epsilon = E_t$, so from Eqs. 18.1, 18.3, 18.4, and 18.6, the value of the plasticity reduction factor η can be found by trial and error from the following equation:

$$\eta\left[1+\frac{a\bar{\epsilon}_y E}{\sigma_y}\left(\frac{\eta\alpha\sigma_e}{\sigma_y}\right)^{a-1}\right]=1 \tag{18.7}$$

Note should be made of the need to determine a (Eq. 18.5) so as to accurately represent the actual stress-strain curve.

In lieu of Eq. 18.7, the following simplified equations are suggested for steel with a well-defined yield stress plateau. For "as fabricated" shells:

$$\eta = 1.0 \qquad \text{for } \Delta \le 0.55 \tag{18.8a}$$

$$\eta = \frac{0.45}{\Delta} + 0.18 \qquad \text{for } 0.55 < \Delta \le 1.6 \tag{18.8b}$$

$$\eta = \frac{1.31}{1+1.15\Delta} \qquad \text{for } 1.6 < \Delta < 6.25 \tag{18.8c}$$

$$\eta = \frac{1}{\Delta} \qquad \text{for } \Delta \ge 6.25 \tag{18.8d}$$

where

$$\Delta = \frac{\alpha\sigma_e}{\sigma_y} \tag{18.9}$$

For stress-relieved or machined shells

$$\eta = 1.0 \qquad \text{for } \Delta \le 0.67 \tag{18.10a}$$

$$\eta = \frac{2.53}{1+2.29\Delta} \qquad \text{for } 0.67 < \Delta < 4.2 \tag{18.10b}$$

$$\eta = \frac{1}{\Delta} \qquad \text{for } \Delta \ge 4.2 \tag{18.10c}$$

These relations for η were obtained as the lower bounds of test data (Miller, 1983). When $\Delta \le 0.55$ for as-fabricated shells and $\Delta \le 0.67$ for stress-relieved and machined shells, the buckling stress is in the elastic range.

18.3 LOCAL BUCKLING OF SPHERICAL SHELLS

18.3.1 Theory

The theoretical buckling behavior of spherical shells subjeced to external pressure has been investigated extensively. A summary of these studies is presented by Timoshenko and Gere (1961) for solutions to the buckling of a complete sphere. However, a more likely case is the loading of a spherical cap, as a head, or a spherical segment located somewhere on the surface of the shell as in the transition from a cylindrical shell to a head.

The number of waves in the buckled region of the spherical shell, n, is determined by Timoshenko and Gere (1961) in Section 11.13. Using a combination of Eq. n on p. 515 and Eq. t on p. 517 of that reference gives

$$n \cong \tfrac{1}{2}(\sqrt{13.064R/t+1}-1) \tag{18.11}$$

where R/t is the ratio of the shell radius to its thickness. This formula applies for a Poisson's ratio of $\frac{1}{3}$. For the practical ranges of the values of R/t, the $+1$ and -1 term can be neglected, thus giving

$$n = 1.8\sqrt{R/t} \tag{18.12}$$

The corresponding theoretical wavelength of a complete sphere is $2\pi R/1.8\sqrt{R/t}$, or $3.49\sqrt{Rt}$. The theoretical elastic local buckling pressure of a spherical segment is the same as that for the complete sphere if the meridional length of the segment equals or exceeds $3.49\sqrt{Rt}$ (i.e., Eq. 11-31, Timoshenko and Gere, 1961).

$$p_e = \frac{E}{\sqrt{3(1-\nu^2)}}\left(\frac{t}{R}\right)^2 \tag{18.13}$$

For a value of $\nu = 0.3$, and introducing an adjustment for the difference between the outside radius R_o and the nominal centerline radius R,

$$\sigma_e = \frac{p_e R}{2t}\left(\frac{R_o}{R}\right)^2 = 0.605\left(\frac{Et}{R}\right)\left(\frac{R_o}{R}\right)^2 \tag{18.14}$$

Equation 18.14 has been verified by extensive studies made with the finite-difference program BOSOR 4. For arc lengths of less than $3.49\sqrt{Rt}$ a substantial increase in the theoretical local buckling stress was observed for spherical caps or for spherical segments located anywhere along a meridian of a sphere.

Substitution of Eq. 18.14 into Eq. 18.1 gives the following buckling stress equation for a fabricated sphere:

$$\sigma_{cr} = 0.605\eta\alpha\left(\frac{Et}{R}\right)\left(\frac{R_o}{R}\right)^2 \tag{18.15}$$

The critical pressure can be obtained from the equation

$$p_{cr} = \eta\alpha p_e \tag{18.16}$$

18.3.2 Effects of Imperfections

Most of the theoretical and experimental studies related to buckling behavior of spherical structures have indicated that initial imperfections, particularly out-of-roundness, residual stresses, and adverse boundary conditions, have a severe effect on the buckling strength. Much of this theoretical and experimental work is summarized by Fung and Sechler (1974). Various studies have attempted to improve the large discrepancy between the classical theory and the experimental results because the latter are much lower than these predicted by theory. This discrepancy is strongly dependent on the difference between the spherical shell as assumed in the theory and the spherical shell as produced by practical fabrication techniques. Allowable fabrication tolerances are specified to adjust for weight penalties. The actual buckling strength varies widely with different tolerance specifications.

Theoretical methods for dealing with the imperfection problem were proposed by Buchert (1964), Hutchinson (1967), Koiter (1970), Krenzke and Kiernan (1965), and many others. These studies all emphasize the severe effects of out-of-roundness on the buckling strength, and a need for a better understanding of the actual profile of the shell. Capacity reduction factors α (Eq. 18.1) for each of the above methods, as well as for a method presented below, are shown in Fig. 18.2. The ordinate is the ratio of the theoretical or test pressure of a spherical shell with imperfections to the theoretically predicted pressure with zero imperfection. The

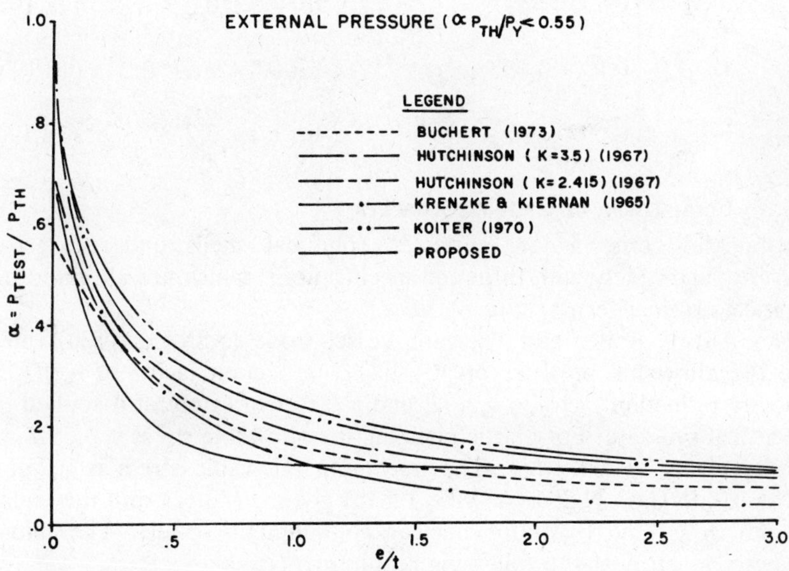

Fig. 18.2 Comparison of various theoretical and experimentally obtained methods for determining α versus e/t for elastic failures.

abscissa is the ratio of the imperfection amplitude e to the shell thickness. It can be concluded from Fig. 18.2 that the basic shapes of the curves are similar even though the definition of the imperfection amplitude is different for the various proposed methods.

Krenzke and Kiernan (1965) found that the reduction of strength depends primarily on the imperfections in the local geometry over a critical arc length L_c. From experiments it was determined that for $\nu = 0.3$ the value of L_c is

$$L_c = 2.42\sqrt{R_1 t} \tag{18.17}$$

where R_1 is the local radius. The relationship between R_1 and the maximum deviation e of the shell from the theoretical radius, when measured over an arc length L_c, can be determined from the approximate relationship

$$R_1 = (1 + 1.36e/t)R \tag{18.18}$$

The capacity reduction factor α accounts for the effects of initial imperfections, geometric nonlinearity, and boundary conditions. The following recommended values of α were determined as the lower bound of all available tests which failed by elastic buckling (Miller, 1983). The values of e were determined over the arc length L_c:

$$\alpha = \begin{cases} 0.7 - 1.75e/t & \text{for } e/t \leq 0.2 \quad (18.19a) \\ 0.09 + 0.0326(e/t)^{-1.29} & \text{for } 0.2 < e/t < 3.2 \quad (18.19b) \\ 0.097 & \text{for } e/t \geq 3.2 \quad (18.19c) \end{cases}$$

18.3.3 Comparison of Code Approaches

The buckling criteria for fabricated spherical shells under external pressure vary widely with different specifications, which are all based on empirical approaches.

The ASME Boiler and Pressure Vessel Code (ASME, 1983a) rules base the allowable pressure on a capacity reduction factor of 0.207, a plasticity reduction factor $\eta = E_t/E$ and a factor of safety of 4 applied to the critical pressure. For elastic buckling the allowable stress is $0.0518\sigma_e$, with σ_e given by Eq. 18.3. The maximum allowable stress is $0.25\sigma_y$. ASME Code Case N-284 (ASME, 1983b) gives $\alpha = 0.124$ and the equation for η is Eq. 18.10 for nuclear containment vessels. The initial geometry must meet the following requirements:

1. The overall out-of-roundness is limited to $D_{\max} - D_{\min} \leq 0.01D_{\text{nominal}}$, where D is the diameter of the shell.

2. The maximum deviation e from the true circular arc length L over a distance equal to one-half of the outside diameter is limited to a value between $0.3t$ and $1.0t$, where t is the shell thickness. The value of e/t is

$$\frac{e}{t} = 0.033(R/t)^{0.62} \tag{18.20}$$

and the value of L can be determined from

$$L = [4.59(e/t)^{0.323}]\sqrt{Rt} \tag{18.21}$$

The British Standard Rules for Unifined Pressure Vessels (BS 5500) (BSI, 1977) are based on an allowable pressure that can be determined from

$$\Delta = (0.1326k - 0.01262k^{1.953}) \qquad \text{for } k < 4.0 \tag{18.22a}$$

$$\Delta = 0.515k - 0.363k^{1.123} \qquad \text{for } 4.0 \leq k < 6.5 \tag{18.22b}$$

$$\Delta = 0.38 \qquad \text{for } k \geq 6.5 \tag{18.22c}$$

where Δ is the allowable stress divided by the yield stress and k is the theoretical buckling stress divided by the yield stress. The deviation from sphericity must be limited to 1.0 percent of the nominal radius, and the local radius over an arc length of $2.4\sqrt{Rt}$ should not exceed the nominal radius by more than 30%. For elastic buckling the allowable stress equals $0.12\sigma_e$. This value is 2.3 times higher than the ASME allowable stress. The allowable deviation from sphericity is also much less, and by limiting R_1/R to 1.3 a maximum value of $a/t = 0.22$ is determined from Eq. 18.18. The value of α from Eq. 18.19 is 0.32, and $\sigma_c = 0.32\sigma_e$. This indicates a factor of safety of 2.67.

The American Petroleum Institute rules (API 1982) require that the computed stress shall not exceed a value of S_{ca} (in psi) established for the applicable minimum thickness-to-radius ratios as follows:

$$S_{ca} = 1{,}000{,}000(t/R) \qquad \text{for } t/R < 0.00667 \tag{18.23a}$$

$$S_{ca} = 5650 + 154{,}200(t/R) \qquad \text{for } 0.0067 < t/R < 0.0175 \tag{18.23b}$$

$$S_{ca} = 8340 \qquad \text{for } t/R > 0.0175 \tag{18.23c}$$

The allowable stress for elastic buckling equals $0.057\sigma_e$. The shape tolerances are described as follows: The surface shall not deviate outside the design shape by more than 1.25% of D, and inside the specified shape by more than 0.625% of D, where D is the nominal inside diameter.

The collapse pressure given by the ECCS rules for steel construction

(ECCS, 1983) for elastic buckling is $p_c = 0.18p_e$. The shape tolerances are the same as for BS 5500.

18.4 LOCAL BUCKLING: SPHERICAL SHELL-LIKE STRUCTURES

Local buckling of a reticulated or framed shell-like structure is said to occur when one node is loaded and it deflects or snaps through and the local curvature reverses. The local-buckling load is a function of joint rigidity as well as member geometry. Wright (1965) gives the elastic local-buckling load of an equilateral triangular reticulated shell-like dome, with inextensible supports (support at adjacent nodes that do not move) and zero bending stiffness of the nodes, as

$$W_{cr} = \frac{2AEh^3}{\sqrt{3}L^3} \tag{18.24}$$

where W_{cr} is the concentrated load at buckling that is applied at a node and h is the difference in elevation between the node in question and the plane formed by adjacent nodes. If the supports are completely extensible (i.e., the remainder of the shell-like structures does not resist movement of adjacent nodes),

$$W_{cr} = \frac{AEh^3}{\sqrt{3}L^3} \tag{18.25}$$

For the case of continuous nodes (nodes with the same bending rigidity as the members) Wright gives the criteria for buckling as: If

$$\left(\frac{R}{L^2}\sqrt{\frac{I}{A}}\right) \leq 0.092 \tag{18.26}$$

buckling will occur for inextensible supports. If

$$\left(\frac{R}{L^2}\sqrt{\frac{I}{A}}\right) \leq 0.132 \tag{18.27}$$

buckling will occur for extensible supports.

Crooker and Buchert (1968) gave the local buckling criteria of reticulated shell-like structures as follows: If

$$\left(\frac{R}{L^2}\sqrt{\frac{I}{A}}\right) \leq 0.10 \tag{18.28}$$

local buckling can occur. Lind (1969) gives basically the same critieria and more details on methods of computing local buckling loads.

Buchert (1965) used a von Kármán–Tsien type of criterion to determine the local buckling load (buckling between stiffeners of a stiffened shell) as

$$p_{cr} \approx \frac{7.4Et^3}{Rd^2} \tag{18.29}$$

where d is the distance between stiffeners in a two-way grid. This is the pressure that will cause the shell to buckle or wrinkle between stiffeners and is based on the assumption that the stiffeners will not buckle.

ASME Code Case N-284 (ASME, 1983b) uses Eq. 18.1 to determine the local buckling stress of a stiffened sphere. Equation 18.10 is used to determine η, and α is given by the following equations:

$$\alpha = 0.826(L/\sqrt{Rt})^{-0.6} \qquad \text{for } 1.73 < L/\sqrt{Rt} < 23.6 \tag{18.30}$$

$$\alpha = 0.124 \qquad \text{for } L/\sqrt{Rt} > 23.6 \tag{18.31}$$

18.5 GENERAL BUCKLING: SPHERICAL SHELL-LIKE STRUCTURES

The so-called split-rigidity concept has been used by a number of investigators to determine the buckling pressure for a spherical shell-like structure. Klöppel and Jungbluth (1953), Buchert (1965a), Lakshmikantham and Gerard (1964), Crawford and Schwartz (1965), Wright (1965), Mitchell (1953), and others have used the concept to study shell-like structures such as reticulated, orthotropic, sandwich, and stiffened shells. When a shell buckles there is stretching of the middle surface as well as bending of the shell element. Both actions are important in resisting buckling. The shell-like structure may be treated conveniently by shell theory, by using, for membrane action, a rigidity based on a membrane thickness equal to the area per unit length of the shell-like structure, and for bending, a rigidity based on the effective moment of inertia per unit length of the shell-like structure. The critical pressure for elastic buckling of a spherical isotropic shell-like structure can be expressed as

$$p_{crE} = CE\left(\frac{t_m}{R}\right)^2\left(\frac{t_B}{t_m}\right)^{3/2} \tag{18.32}$$

where

p_{crE} = critical pressure applied uniformly and normal to the shell
C = a coefficient
E = modulus of elasticity
R = spherical radius
t_m = effective membrane thickness
t_B = effective bending thickness

This equation (with the proper values of t_m and t_B) may be used for latticed shells, stiffened shells, orthotropic shells, and sandwich shells.

Although Eq. 18.32 was derived using the split-rigidity concept, other methods of derivation (differential equation, energy, and continum) give results of similar form with the same basic assumptions. The differences in the results are mostly apparent in the coefficient C. Klöppel and Jungbluth (1953) used a "Zoelly" classical differential-equation theory and found that $C = 1.16$ with Poisson's ratio equal to zero. Buchert (1965a) used an approach similar to that used by von Kármán and Tsien (1939) and found that $C = 0.365$. Buchert also used an energy method and found that $C = 0.82$. Lakshmikantham and Gerard (1964) and Crawford and Schwartz (1965) used a differential-equation type of approach and found that $C = 1.16$. (Note that Gerard and Lakshmikantham also included shear and torsional stiffnesses, which are not considered in Eq. 18.32.) Wright (1965) used a modified continuum approach and recommended a value of $C = 0.38$. Sections III and VIII of the ASME Boiler and Pressure Vessel Code use a critical value of C of about 0.25 for thin metal shells when the specified values of permissible deviations from a perfect surface are not exceeded. All of the values above give the critical value of C in the form of Eq. 18.32 and thus do not include a factor of safety.

Many tests have been conducted to determine the value of C. The results show that values of C from zero to about 0.90 may be obtained depending on edge conditions, imperfections, plasticity effects, and such. This variable value of C has been the subject of many papers and has caused some designers to avoid the use of shell-like structures in the past. However, recent developments in the theory and partial understanding of the buckling process have resulted in a marked increase in the use of this type of structure. The following paragraphs give some of the quantitative results of recent developments.

Plasticity reduction factors have been derived by several investigators. The plastic buckling pressure is given by

$$p_{cr} = \eta p_{crE} \tag{18.33}$$

where η is the plasticity reduction factor. Lakshmikantham and Gerard (1964) and Bijlaard (1949) found that

$$\eta = \frac{(E_s E_t)^{1/2}}{E} \tag{18.34}$$

where E_s is the secant modulus and E_t is the tangent modulus associated with the maximum membrane stress. Buchert (1968) found that

$$\eta = \frac{3}{4E}\left(E_t + \frac{E_s}{3}\right) \tag{18.35}$$

Equations 18.31 and 18.32 give results that are within 5% of each other and experimental values are within about 10% of the theoretical values (Buchert, 1968).

Edge effects, imperfections, and deflections in general (caused by shell bending near the edges or by fabrication and erection procedures) are important in the buckling of shell-like structures. For example, a deflection-to-thickness ratio of 3 has been shown (both theoretically and experimentally) to reduce the value of C in the equation

$$p_{cr} = CE\left(\frac{t}{R}\right)^2 \tag{18.36}$$

from 0.84 to 0.06 (Buchert, 1964) for a thin shell. Similar large effects can be demonstrated for shell-like structures. Ratios of deflection to effective membrane thickness of the order of 20 to 30 are common in shell-like structures. Buchert (1968) derived an expression for the critical buckling pressure that included inelastic as well as deflection effects. The equation is

$$p_{cr} = \frac{8}{3}\eta E\left(\frac{t_m}{R}\right)^2 \left\{\left[0.210\left(\frac{\Delta}{t_m}\right)^2 + 0.0715\left(\frac{t_B}{t_m}\right)^3\right]^{1/2} - 0.459\frac{\Delta}{t_m}\right\} \tag{18.37}$$

where Δ is the maximum normal deflection from a perfect spherical surface due to edge deflections and also allows for fabrication and erection tolerances. The equation above agrees within ±15% with the results of tests made on thin shells at the David Taylor Model Basin (Krenzke and Kiernan, 1965) and at the University of Missouri, Columbia (Buchert, 1968).

For thin elastic spherical shells, Eqs. 18.32, 18.33, and 18.37 reduce to the classical Eq. 18.36.

The effective membrane thickness of the framed shell with equally spaced orthogonal rings is (Buchert, 1965a)

$$t_m = \frac{A}{d} \tag{18.38}$$

where A is the member area and d is the distance between members.

The effective bending stiffness of a shell per unit width is (assuming Poisson's ratio to be zero) $Et_B^3/12$. Therefore, the effective bending thickness of a framed shell with equally spaced orthogonal rings is

$$\frac{Et_n^3}{12} = \frac{EI}{d}$$

or

$$t_B = \left(\frac{12I}{d}\right)^{1/3} \tag{18.39}$$

where I represents member moment of inertia. These equations can often be used on a shell-like dome with radial members and circumferential rings where the members are nearly orthogonal.

In the case of a stiffened shell the effective membrane thickness is A/d plus the thickness of the shell. The effective bending thickness is computed by assuming that a shell width of $1.56\sqrt{Rt}$ acts with the stiffener in calculating the moment of inertia (Fig. 18.3).

The triangular-type pattern with an equilateral triangular grid with members of length L, area A, and moment of inertia I has an equivalent membrane thickness of (Wright, 1965)

$$t_m = \frac{2A}{\sqrt{3}L} \tag{18.40}$$

and an effective bending thickness of

$$t_B = \left(9\sqrt{3}\,\frac{I}{L}\right)^{1/3} \tag{18.41}$$

The double-layer grid, with each grid having an effective membrane thickness given by Eq. 18.40, has an effective membrane thickness of

$$t_m = \frac{4A}{\sqrt{3}L} \tag{18.42}$$

The effective bending thickness is given as

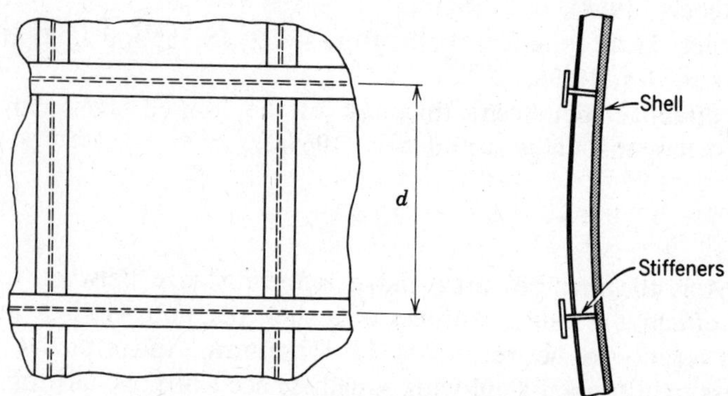

Fig. 18.3 Orthogonal ring-stiffened shell.

$$t_B = 2\left(2\sqrt{3}\,\frac{A}{L}\left(\frac{a}{2}\right)^2\right]^{1/3} \tag{18.43}$$

where α is the distance between the centers of the grids. Similar equations can be derived or experimentally determined for other types of latticed shells or for sandwich shells.

Currently, there is much effort directed toward applying the finite-element method of analysis to solving buckling and large-displacement problems of axisymmetric shells. A number of researchers have given solutions for such problems for the shell proper (Gallagher and Yang, 1968; Gallagher et al., 1967; Navaratna et al., 1968).

The same problem for stiffened shells has been treated by considering the stiffened shell to be made of an orthotropic material. Work is also in progress where the stiffeners are treated as discrete elements, and their bending and torsion rigidities are considered. As yet, this work is unavailable in technical literature.

The advantage of the finite-element method of analysis is that the geometry need not be restricted to spheres. The same procedure is applicable for shells of free form.

18.6 MEMBER BUCKLING

Member buckling of shell-like structures is basically a column-stability problem. The methods described in Chapters 3 and 4 may be used to calculate the critical loads.

18.7 DESIGN TRENDS AND RESEARCH NEEDS

Recent developments in the application of shell-like structures have resulted in a marked increase in their use. Formerly, this type of structure was rarely considered because of the high cost of the engineering design and analysis, fabrication, and erection. However, with the development of new joint systems, and new fabrication and erection techniques, costs of the finished structure have been reduced considerably. With the increased use of such structures, the method of analysis has been studied in greater detail with the result that the designer is now able to analyze the structure faster and with more accuracy than previously. Buchert (1973) has published *Buckling of Shell and Shell-Like Structures, for Engineers, Architects, Fabricators, Builders and Designers*, which contains formulas and procedures for the design of shell-like structures. Formulas for the effects of plasticity, imperfections and deflections, fabriation and erection tolerances, concentrated loads and moments, and interaction loads on the buckling of spherical, cylindrical, and hyperbolic paraboloids are given.

As is the case with beam- and column-type structures, additional research and development on shell-like structures is needed. Theoretical and experimental studies on stability need to be performed on barrel vaults and hypars. Such items as the effects of edges appear to be important in the buckling process.

Local buckling and effects of joint efficiencies need further study. Closed-form solutions for unsymmetrical loads should be developed.

A very fruitful area for investigation appears to be erection methods. Many structures have been erected without falsework. With proper development of applicable shell-like structural equations, larger structures have been erected in this inexpensive way.

Finally, the differences between continuous symmetric loads and the loads that actually occur on shell-like structures need further investigation. The stability of shell-like structures under nonsymmetric loads should be investigated in depth. Tests by Klöppel and Roos (1956) indicate that for most nonsymmetric pressures the nonsymmetric buckling pressure is greater than the same critical pressure applied symmetrically over the entire spherical shell.

References

American Petroleum Institute (1982), *Recommended Rules for Design and Construction of Large, Welded, Low-Pressure Storage Tanks*, API Standard 620, API.

American Society of Mechanical Engineers (1983a), *ASME Boiler and Pressure Vessel Code*, Sections III or VIII, ASME, New York.

American Society of Mechanical Engineers (1983b), "Metal Containment Shell Buckling Design Methods," *ASME Boiler and Pressure Vessel Code Case N-284*, ASME, New York.

Bijlaard, P. P. (1949), "Theory and Tests on the Plastic Stability of Plates and Shells," *J. Aeronaut. Sci.*, Vol. 16, No. 9.

British Standard Institute (1977), *British Standard Institute Specification for Unfired Fusion Welded Pressure Vessels*, BS 5500, BSI, London.

Buchert, K. P. (1964), "Stability of Doubly Curved Stiffened Shells," Dissertation, University of Missouri, January.

Buchert, K. P. (1965a), "Zur Stabilität grosser, doppelt, gekrümmter und versteifter Schalen," *Stahlbau*, February, pp. 55–62.

Buchert, K. P. (1965b), "Buckling of Doubly Curved Orthotropic Shells," Engineering Experiment Station, University of Missouri, Columbia, November.

Buchert, K. P. (1968), "Space Frame Buckling," *Eng. J. Am. Inst. Steel Constr.*, Vol. 5, No. 4, pp. 152–154.

Buchert, K. P. (1973), *Buckling of Shells and Shell-Like Structures for Engineers, Architects, Fabricators, Builders and Designers*, University of Missouri, Columbia.

Crawford, R., and Schwartz, D. (1965), "General Instability and Optimum Design of Grid-Stiffened Spherical Domes," *Am. Inst. Aeronaut. Astronaut.*, Vol. 3, No. 3.

Crooker, J., and Buchert, K. (1968), "Reticulated Space Structures," *ASCE Annu. Natl. Meet. Struct. Eng., Preprint 731*. Pittsburg, Pa. October.

European Convention for Constructional Steelwork (1983), *European Recommendations for Steel Construction: Buckling of Shells*, 2nd Ed., Technical Committee 8, Structural Stability, ECCS, Brussels.

Fung, Y. C., and Sechler, E. E., eds. (1974), *Thin Shell Structures*, Prentice-Hall, Englewood Cliffs, N.J.

Gallagher, R. H., and Yang, H. T. Y. (1968), "Elastic Instability Predictions for Doubly Curved Shells," *2nd AFIT Conf. Matrix Methods Struct. Mech.*, October.

Gallagher, R. H., Gellatly, R., Mallett, R., and Padlog, J. (1967), "A Discrete Element Procedure for Thin Shell Instability Analysis," *AIAA J.*, Vol. 5, No. 1.

Gerard, G., and Becker, H. (1957), "*Handbook of Structural Stability*: *Part III*. Buckling of Curved Plates and Shells," *NACA Tech. Note No. 3783*, August.

Hutchinson, J. W. (1967), "Imperfection Sensitivity of Externally Pressurized Spherical Shells," *Trans. ASME J. Appl. Mech.*, Vol. 34, pp. 49–55.

Kiernan, T. J., and Nishida, K. (1966), "The Buckling Strength of Fabricated HY-80 Steel Spherical Shells," *David Taylor Model Basin Rep. 1721*, July.

Klöppel, K., and Jungbluth, O. (1953), "Beitrag zum Durchschlag-problem dünnwändiger Kugelschalen," *Stahlbau*, Berlin, Vol. 22, pp. 121–130.

Klöppel, K., and Roos, E. (1956), "Contribution to the Buckling Problem of Thin-Walled Stiffened and Unstiffened Spherical Shells under Full and Half-Sided Loading," *Stahlbau*, Berlin, March, Vol. 25, pp. 49–60 (in German).

Koiter, W. T. (1970), "The Stability of Elastic Equilibrium," *Tech. Rep. No. AFFDL TR*-70-25, Air Force Flight Dynamics Laboratory, Wright-Patterson AFB, Ohio, February.

Krenzke, M. A., and Kiernan, T. J. (1965), "The Effect of Initial Imperfections on the Collapse Strength of Deep Spherical Shells," *David Taylor Model Basin Rep. No. 1757*, February.

Lakshmikantham, C., and Gerard, G. (1964), "Elastic and Plastic Stability of Geometrically Orthotropic Spherical Shells," *Tech. Rep. No. 235-3*, Allied Research, Inc., Concord, Mass., January 15.

Lind, N. (1969), "Local Instability Analysis of Triangulated Dome Frameworks," *Struct. Eng.*, Vol. 47, No. 8, pp. 317–324.

Miller, C. D. (1983), "Research Related to Buckling Design of Nuclear Containment," *7th Int. Conf. Struct. Mech. in Reactor Technol.*, Chicago, August 22–26.

Mitchell, L. (1953), "Shell Analogy for Framed Domes," *Struct. Mater, Note No. 208*, Aeronautical Research Laboratories, Department of Supply, Melbourne, Australia.

Navaratna, D., Pian, T., and Witmer, E. (1968), "Analysis of Elastic Stability of Shells of Revolution by the Finite Element Method," *AIAA J.*, Vol. 6, No. 2.

Ramberg, W., and Osgood, W. R. (1943), "Description of Stress-Strain Curves by Three Parameters," *NACA Tech. Note No. 902*, July.

Timoshenko, S. P., and Gere, J. M. (1961), *Theory of Elastic Stability*, 2nd ed., McGraw-Hill, New York.

von Kármán, T., and Tsien, H. S. (1939), "The Buckling of Spherical Shells by External Pressure," *J. Aeronaut. Sci.*, Vol. 7, No. 2, pp. 43–50.

Wright, D. T. (1965), "Membrane Forces and Buckling in Reticulated Shells," *ASCE J. Struct. Div.*, Vol. 91, pp. 173–202.

Chapter Nineteen

Selected Topics in Dynamic Stability

19.1 INTRODUCTION

The stability of structures subjected to dynamic loads encompasses a wide and diverse class of problems. In general, a system can be defined as stable with respect to a particular perturbation if a small change in that perturbation results in an arbitrarily small change of a particular aspect of structural response during a desired period of time. In other words, a structural system may be:

- Stable with respect to one system of parameters and unstable with respect to the other.
- Stable for a given time period and unstable thereafter.

For example, a column subjected to an axial compression in excess of the statical Euler force may not buckle if the duration of the force application is sufficiently short. Conversely, a periodic axial force having a magnitude being only a fraction of the Euler force may cause unstable lateral motion for a certain range of frequencies.

It is always difficult to summarize a discipline that is not only diverse but which is still growing at an increasing rate. In absence of similar texts it was decided to cover five distinctly different topics in hope to clarify at least some of the aspects of greatest practical interest, as follows:

1. Parametric resonance.
2. Stability of impulsively loaded columns.
3. Dynamic snap-through of shallow structures.
4. Flow induced instability.
5. Suddenly loaded structures.

19.2 PARAMETRIC RESONANCE

The phenomenon known as parametric excitation can be illustrated on an example of a simply supported column subjected to a harmonic axial force $P(t)$ (Fig. 19.1). Ordinarily, for a load intensity small in comparison with Euler's static buckling force, the expected response would consist of axial vibrations. However, for certain frequencies of the axial force $P(t)$, large and potentially dangerous transverse vibrations can occur. The occurrence of such a response characterized by motion which is typically orthogonal to the direction in which the external force is applied is referred to as parametric excitation. The exponential increase in the amplitude of these transverse vibrations is labeled as parametric resonance. For a comprehensive study of the phenomenon, see Bolotin (1964, 1968), Mettler (1962), Ibrahim et al. (1978), Nemat-Nasser (1972), and Schmidt (1975).

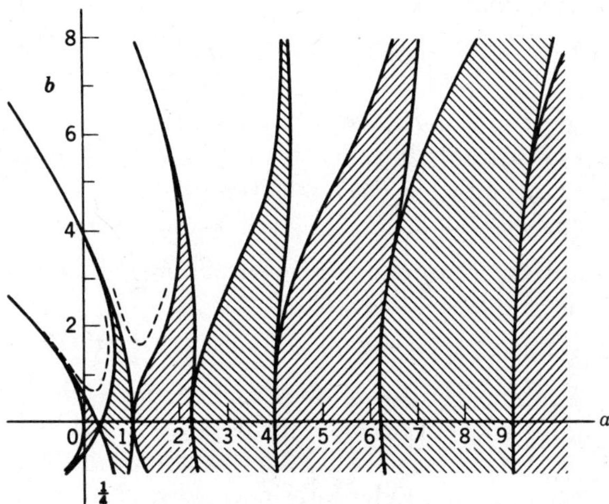

Fig. 19.1 Strutt's diagram.

19.2.1 Formulation of the Problem

Neglecting the axial component of the motion, the equation of the small, elastic transverse vibrations of a column subjected to a time-dependent axial force $P(t)$ is

$$EI \frac{\partial^4 w}{\partial x^4} + P(t) \frac{\partial^2 w}{\partial x^2} + m \frac{\partial^2 w}{\partial t^2} = 0 \tag{19.1}$$

where E, I, m, $w(x, t)$ are the elastic modulus, moment of inertia, mass per unit length, and the transverse displacement correspondingly. Seeking the solution in the form of standing waves (i.e., in separable form) gives us

$$w(x, t) = X(x)f(t) \tag{19.2}$$

Equation 19.1 implies that

$$\ddot{f} + \omega^2[1 - P(t)/P_e]f = 0 \tag{19.3}$$

where P_e is the static buckling force and ω the free vibration frequency. It is recognized that the standing-wave solution is, strictly speaking, possible only if $X(x)$ is simultaneously the static buckling mode and the free vibration mode. This occurs only in case of the simply supported column and uniform radial loading of a circular ring. In all other cases the solution (Eq. 19.2) is approximate (in sense of the Ritz or the Galerkin method).

The stability of the motion, as measured by the magnitude of the transverse displacement $w(x, t)$ in Eq. 19.2, depends on $f(t)$, which is the solution of the ordinary differential equation with periodic coefficients (Eq. 19.3). In case of deterministic loading, this governing equation is

a Mathieu equation if $$P(t) = P_0 + P_t \operatorname{cost} \theta t \tag{19.4}$$

a Hill equation if $$P(t) = P_0 + \sum_k P_{tk} \cos k\theta t \tag{19.5}$$

Both Mathieu and Hill equations were thoroughly investigated in the past (Bolotin, 1964, Schmidt, 1975; McLachlan, 1947). In addition, Eq. 19.3 admits a closed-form solution for the case of the periodic impact loading (Krajcinovic and Herrmann, 1968; Hsu, 1972).

19.2.2 Regions of Instability

The qualitative study of the small parametric vibrations of an elastic structure reduces to the determination of the combinations of the load intensity (P_0, P_t) and the load frequency (defined by θ) for which the response amplitude starts to grow exponentially.

For arbitrary boundary conditions the partial differential equation (Eq. 19.1) admits a solution in terms of a sum of standing waves:

$$w(x, t) = \sum_{k=1}^{\infty} f_k(t)X_k(x) \tag{19.6}$$

where $X_k(x)$ satisfy boundary conditions at $x = 0, L$. If X_k are chosen as the free vibration modes, Eq. 19.3 can be rewritten as (see Bolotin, 1964)

$$\ddot{f}_i + \omega_i^2\left[f_i - P(t)\sum_k a_{ik} f_k\right] = 0 \tag{19.7}$$

where

$$a_{ik} = -\frac{1}{m\omega_i^2}\int_0^L X_k X_i''\,dx \qquad \int_0^L X_i X_k\,dx = \begin{cases} 1 & \text{if } i = k \\ 0 & \text{if } i \neq k \end{cases} \tag{19.8}$$

with L being the column length. If the loading function is given by Eq. 19.4, then Eq. 19.7 can be recast into

$$\ddot{f}_i + \omega_i^2\left[f_i - (P_0 + P_t \cos\theta t)\sum_k a_{ik} f_k\right] = 0 \tag{19.9}$$

The system of Eqs. 19.9 of an infinite number of ordinary differential equations with periodic coefficients is coupled through the last term in the brackets. The solution of the system is greatly facilitated if the diagonal terms are much larger than the off-diagonal terms.

In case of a simply supported beam or a circular ring, $a_{ik} = 0$ if $i \neq k$. Then with

$$a_{ii} = \frac{1}{P_i} \tag{19.10}$$

it is possible to write, instead of Eq. 19.9,

$$\ddot{f}_i + \Omega_i^2(1 - \mu_i \cos\theta t) f_i = 0 \tag{19.11}$$

where

$$\Omega_i = \omega_i\sqrt{1 - \frac{P_0}{P_i}} \qquad \mu_i = \frac{P_t}{2(P_0 - P_i)} \tag{19.12}$$

Introducing the simple transformation

$$2z = \theta t \qquad a = (2\Omega/\theta)^2 \qquad q = \mu(2\Omega/\theta)^2 \tag{19.13}$$

Eq. 19.11 can be rewritten in a typical form of a Mathieu equation,

$$\frac{d^2 f}{dz^2} + (a - 2q\cos 2z) f = 0 \tag{19.14}$$

The solutions of Eq. 19.14 are either

- Stable; decreasing in time $f(t+T) < f(t)$, or
- Unstable; increasing in time $f(t+T) > f(t)$.

The regions of stable and unstable solutions in the (a, q) plane are separated by the loci of points representing harmonic solutions of (Eq. 19.14) $f(t+T) = f(t)$. These harmonic solutions are known as Mathieu functions $se_n(z)$ and $ce_n(z)$ of the integer order n. The graphical depiction of the stable and unstable (shaded in Fig. 19.1) regions is known as the Strutt diagram. The Mathieu functions defining the first two instability regions are listed in Table 19.1.

Table 19.1. Boundaries of First Two Instability Regions (Bolotin, 1968)

Region	Function	Relation between Ω_i and μ_i
1	$se_1(z)$	$(\theta_*/2\Omega)^2 = 1 + \mu + \frac{1}{8}\mu^2 + O(\mu^3)$
	$ce_2(z)$	$= 1 - \mu + \frac{1}{8}\mu^2 + O(\mu^3)$
2	$se_2(z)$	$= 1 + \frac{1}{3}\mu^2 + O(\mu^4)$
	$ce_2(z)$	$= 1 - \frac{5}{3}\mu^2 + O(\mu^4)$

Table 19.2. Boundary of the Principal Instability Region

Column	C_1	C_2	$P_c l^2/EI$
	7.04	2.66	2.47
	19.74	9.87	9.87
	30.84	21.3	20.2
	44.7	40.7	39.48
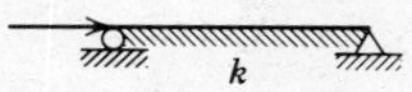 k	$19.74\sqrt{1+\beta}$	$9.87\sqrt{1+\beta}$	$9.87(n^2 + \beta/n^2)$, $\beta = \frac{kl^4}{\pi^4 EI}$ $n = 1, 0 \le \beta \le 4$; $n = 2, 4 \le \beta \le 36$; $n = 3, 36 \le \beta \le 144$

The analysis is much more complicated for columns other than simply supported. The boundaries of the principal instability region can be computed approximately using the information in Table 19.2, where k is the elastic foundation modulus,

$$C_1 = 2\omega_1 \left(\frac{mL^4}{EI} \right)^{1/2} \tag{19.15}$$

and

$$\left(\frac{\theta_*}{2\omega_1} \right)^2 = 1 - \frac{L^2}{EIC_2} (P_0 \pm \tfrac{1}{2} P_t) \tag{19.16}$$

A closed-form solution of the problem was derived only in two cases:

1. For the step load the harmonic solutions (separating the stable from unstable regions) are obtained as the roots of the transcendental equation (Bolotin, 1968)

$$\left| \cos \frac{\pi p_1}{\theta} \cos \frac{\pi p_2}{\theta} - \frac{p_1^2 + p_2^2}{2P_1 P_2} \sin \frac{\pi p_1}{\theta} \sin \frac{\pi p_2}{\theta} \right| = 0 \tag{19.17}$$

where

$$p_{1,2} = \Omega \sqrt{1 \pm \frac{P_t}{P_0 - P_{cr}}}$$

with Ω defined by Eq. 19.12a, while π/θ is the duration of each step.

2. For the loading consisting of impulses repeating after a period T,

$$P(t) = P_0 + P_t \sum_k \delta(\theta t - k\theta T)$$

where δ is the Dirac delta functional, the harmonic solutions are roots of the transcendental equation (see Krajcinovic and Herrmann, 1968)

$$\left| \frac{P_t}{2(P_{cr} - P_0)} \frac{\Omega_k T}{2\pi} \sin \Omega_k T + \cos \Omega_k T \right| = 1 \tag{19.18}$$

where Ω_k is again defined by Eq. 19.12a.

It is important to notice that depending on the loading frequency, in all of these cases:

- The instability may occur when the combined load magnitude $P_0 + P_t$ does not exceed the static buckling force P_e.
- The column can remain stable even though $P_0 + P_t$ may exceed P_e.

19.2.3 Summary of Results for Linear Systems

Most of the research conducted in the past was focused on the practically most important case of a harmonic load given by Eq. 19.4. The derived results can be summarized for the designer in the following way:

The loading frequencies leading to an unstable response are located in regions defined by boundaries $ce_n(z)$ and $se_n(z)$ arranged in Table 19.1. For a properly designed structure loaded dynamically ($P_t > P_0$, $P_e > P_0$) $|\mu| < 1$ from Eq. 19.12. Hence the first (principal) instability region is bounded by

$$1 - \mu < (\theta_{cr}/2\Omega) < 1 + \mu \tag{19.19}$$

The higher critical loading frequencies are located within very narrow bands surrounding lines

$$\theta_{cr} = \frac{2\Omega_i}{n} \qquad (n = 3, 4, 5, \ldots) \tag{19.20}$$

where Ω_i are frequencies of the natural vibrations of the structure subjected to a statical axial force P_0 (Eq. 19.12).

Linear or viscous damping. In case of viscous damping the governing equation (Eq. 19.11) must be amended by a term $2\eta f$, where 2η is the damping coefficient. The viscous damping has the effect of shrinking the instability regions without eliminating them. The critical loading frequencies belonging to the primary instability region satisfy the inequalities

$$1 - \sqrt{\mu^2 - 4\eta^2} < (\theta_{cr}/2\Omega) < 1 + \sqrt{\mu^2 - 4\eta^2}$$

Combination resonance. Investigation of multiple-degree-of-freedom systems (Cesari, 1973) revealed the existence of additional regions of unbounded response. The parametric combination resonance of order n occurs if

$$\theta_{cr} = \frac{1}{n} |\Omega_i \pm \Omega_j| \qquad (i \neq j; n = 1, 2, \ldots)$$

where Ω_i is the natural frequency of the ith mode (in the presence of the static force P_0).

Simultaneous parametric and forced excitation. If an elastic column is subjected to a harmonic axial load of frequency θ and a harmonic transverse load of frequency θ_f the instability will occur (Yamamoto and Saito, 1970) when

$$\Omega = \theta \pm \theta_f$$

Influence of Axial Motion. The governing equation (Eq. 19.1) was derived neglecting the displacements along the axis of the column (i.e., propagation of compressive and tensile waves). The system of equations governing the coupled axial and transverse (parametric) elastic vibrations were derived by Bolotin (1964). A solution of this problem reveals fairly significant modification of the instability region provided that the natural frequency of axial vibrations is close to the frequency 2Ω (given by Eq. 19.12). This is a natural consequence of the fact that near resonance the dynamic axial force in the column becomes very large.

19.2.4 Summary

Trying to summarize almost a century of intensive work, single out the most salient features of a complex phenomenon, and highlight the points of paramount practical importance is in many ways an impossible task. It is quite possible that the equations in Table 19.1 contain all that a designer should know in the beginning phase of a design. If the structure is designed such that the (a, Z) point is well within the stable region of the Strutt diagram, and if the frequency of the axial force is not close to the frequency of axial vibrations, the chances are that no problems will occur. Yet a complex and significant structure may warrant a much more comprehensive study, including a variety of nonlinear effects.

19.3 STABILITY OF IMPULSIVELY LOADED COLUMNS

An impulsively loaded column is defined to be a structural member subjected to a compressive stress that depends significantly on inertia properties and on time. The term "impulse" is used here to encompass all of the various loading conditions that can create time dependent excitation such as impact, which specifically refers to collisions between two bodies where the mass effect of both bodies is considered (Goldsmith, 1960), boundary velocity, such as that produced in a constant-speed testing machine (Hoff, 1951), and force pulse, such as that caused by overpressure distribution from air-blast conditions (Kornhauser, 1964).

The duration of the excitation is probably the most important consideration in the study of the stability of impulsively loaded columns. It determines whether or not the effects of wave propagation must be considered to describe the response accurately. This in turn influences the numerical approach required for the solution. If the duration is sufficiently long, the axial force can be considered constant along the length, being a function of time only. (This is equivalent to the assumption of an infinite axial wave speed.) If the duration is sufficiently short, the longitudinal wave must be considered in the formulation. If, in addition to the duration being short, the strain rate is sufficiently high, elastic-plastic wave propagation may be present, and the consideration of material strain rate sensitivity may be required.

19.3.1 Classification

Theoretical and experimental studies concerned with the stability of a single member subjected to impulsive loading are classified in Table 19.3 according to the following characteristics: mathematical model, material, imperfections, loading, and solution technique.

19.3.2 Low-Velocity Excitation

The distinction between the buckling characteristics of the lower- and higher-speed excitation is significant. The lower-speed (strain rates of the order 10^{-1} to 10^{2}/sec), longer-duration forcing function produces buckling forms that involve the entire column length. The usual assumption made—that the axial wave traverses the column length a sufficient number of times so that the axial force is not a function of position along the length but is a function of time only—has been generally confirmed in several studies (Davidson, 1953; Sevin, 1960; Lindberg, 1965; Abrahamson and Goodier, 1966). Justification has also been established for neglecting the axial inertia effects under the restrictive conditions of elastic material property and sufficiently long load duration (Sevin, 1960; Hayashi and Sano, 1972a). This conclusion was also reached for the restriction of a sufficiently large material dissipation (McIvor and Bernard, 1973). In addition, the axial-flexural coupling effect (Davidson, 1953; Huffington, 1963) and the rotary inertia and shear deformation effects (Housner and Tsu, 1962; Huffington, 1963; Hayashi and Sano, 1972) were found to be insignificant under these restrictions.

The classical conclusion of impulsive stability, discovered in the earliest work (Koning and Taub, 1934; Taub, 1934) and substantiated by many later analyses, is that the impulsive axial force can significantly exceed the static buckling load of the column without excessive lateral displacements due to the delaying effect of the column's inertia. The larger axial force overloads are associated with shorter duration, smaller initial imperfections, and larger slenderness values. However, the dynamic buckling load is never less than the static buckling load (Ari-Gur et al., 1980). Material damping can enhance the axial overload capacity (Suzuki, 1969).

The dynamic buckling mode shape, which grows with time for the axial overload condition, is strongly dependent on the magnitude of the axial force (Housner and Tsu, 1962; Lindberg, 1965) and the stress-strain function (Lee, 1978). For relatively slow loading rates and long durations, the fundamental mode, similar to the static configuration, dominates the shape, which is strongly influenced by the initial imperfections (Lee, 1981). For higher loading rates and shorter durations, the higher modes are excited (Housner and Tsu, 1962; McIvor and Bernard, 1973;

Table 19.3

	Model						Material			Imperfections			
Reference	BE	LI	AI	RI	SD	CS	LE	IE	VE	HS	GD	GV	RD
Meier (1945)	×	×					×			×			
Hoff (1951)	×	×				×	×			×			
Chawla (1951)	×	×				×	×	×		×			
Hoff et al. (1951)	×	×				×	×			×			
Gerard and Becker (1952)	×		×				×						
Davidson (1953)	×	×				×	×			×			
Schmitt (1956)	×	×					×			×			
Erickson et al. (1956)													
Hartz and Clough (1957)	×	×						×			×		
Sevin (1960)	×	×	×			×	×			×			
Hausner and Tsu (1962)	×	×		×	×		×				×		
Huffington (1963)		×	×	×	×	×	×			×	×		
McIvor (1964)	×	×	×			×	×					×	
Lindberg (1965)	×	×					×				×		×
Abrahamson and Goodier (1966)	×	×						×			×	×	
Malishev (1966)	×	×					×						
Kornev (1968)	×	×					×				×		
Holzer (1969)	×	×					×				×	×	
Suzuki (1969)	×	×				×			×	×			
Holzer (1970)	×	×					×				×	×	
Holzer (1971)	×	×					×				×		
Plaut (1971)	×	×					×				×		
Malyi and Efimov (1972)		×		×	×		×				×		
Hayashi and Sano (1972a)	×	×	×			×	×			×			
Hayashi and Sano (1972b)	×	×	×	×	×	×	×			×			
Malyi (1972)	×	×					×						
McIvor (1973)	×	×	×			×			×		×		
Malyi (1973)	×	×					×				×		
Kornev (1974)	×	×					×						
Kiryukhin and Malyi (1974)	×	×	×				×				×		
Grybos (1975)		×	×	×	×	×	×						
Lee (1978)	×	×	×			×		×					
Elishakoff (1978a)	×	×					×			×			×
Elishakoff (1978b)	×	×					×						×
Amiro (1979)	×	×					×				×	×	
Ari-Gur et al. (1980)	×	×	×			×	×				×		
Lee (1981)	×	×						×			×		

Table 19.3 (*Continued*)

Loading				Solution						
MI	BV	FP	IP	CF	AS	FE	FD	WP	ES	Comments
			×	×						Solution of equations of motion based on half-sine-wave shape
	×			×	×					Buckling of a column due to loading by a constant-speed testing machine
	×						×			Numerical solution to Hoff's problem (1951) extended to inelastic material
	×			×	×		×			Extension of Hoff's problem (1951) to include a wider range of parameters
	×			×				×	×	Buckling of a column confined to a fraction of total length due to stress wave
×					×				×	Study of mass impact through a spring connection to column
	×				×			×		Numerical method based on wave travel used to solve equations of motion
									×	Experimental study used to verify Hoff's results (1951)
		×			×				×	Numerical study of columns loaded beyond the linear material range
	×						×			Numerical solution to coupled equations of motion to study the axial inertia effects
		×	×	×						Detailed development of five column models
		×	×				×	×		Useful comparison of linear and nonlinear models
				×						Instability study of a column undergoing axial vibrations
			×	×				×	×	Study of the effects of imperfections expressed in a random form
			×	×				×	×	Linear strain hardening material used to study plastic buckling phenomenon
			×	×				×	×	Study of the stability of an infinitely long column with step load (in Russian)
			×	×						Russian publication
		×	×							Displacement bounds are determined
	×				×					Solution of Hoff's problem (1951) with damping and different loading velocities
		×								Displacement bounds are determined
		×								Displacement bounds are determined
		×								Displacement bounds are determined
	×			×				×		Exact solution to semi-infinite bar subjected to longitudinal impact (in Russian)
×					×		×		×	Study of column buckling due to mass impact at low velocity—first report
×							×		×	Study of column buckling due to mass impact at high velocity—second report
	×			×				×		Simplified solution to Malyi-Efimov problem (in Russian)
		×			×					Axial-lateral motion coupling with material dissipation
	×			×				×		Russian publication
			×	×						Russian publication
	×			×						Russian publication
×				×	×					Defines parameters associated with critical velocity of impacting mass
	×				×			×		Nonlinear axial strain and Ramberg–Osgood type of material function
			×	×						Random imperfections as the amplitude of half sine wave
			×		×					More general random imperfections than Elishakoff (1978a) utilizing Monte Carlo method
			×	×						Russian publication
×							×		×	Study of column buckling due to mass impact with several different materials
	×				×					Study of inelastic column buckling as an initial value-eigenvalue problem

Notes to Table 19.3

1. Model properties: equations of motion:

   BE = Bernoulli-Euler assumptions for bending
   LI = lateral inertia included
   AI = axial inertia included
   RI = rotary inertia included
   SD = shear deformation included
   CS = coupled strain-displacement relations

2. Material properties:

   LE = linear-elastic
   IE = inelastic
   VE = viscoelastic

3. Initial imperfections for flexural response:

   HS = half-sine-wave displacement function
   GD = general displacement function
   GV = general velocity function
   RD = random displacement variable

4. External loading condition:

   MI = mass impact, mass of impacting body considered
   BV = boundary velocity
   FP = finite pulse, finite duration
   IP = infinite pulse, including step load

5. Solution procedure:

   CF = closed-form solution to differential equations
   AS = approximate solution to differential equations
   FE = finite-element discretization
   FD = finite-difference discretization
   WP = wave propagation considered
   ES = experimental studies

Notes:

1. The order of papers in the classification is chronological.
2. Boundary conditions are not considered as a model property in the classification because they are frequently selected on the basis of convenience for a solution studied rather than a meaningful part of the model. For example, semi-infinite bars are frequently used in wave propagation studies to avoid the complexity of wave reflection and material unloading. End conditions are considered by Koning (1934).
3. All models considered have prismatic section properties: area and moment of inertia.

Lee, 1981; Lindberg, 1965; Abrahamson and Goodier, 1966) and the dominant shape falls between the highest and lowest modes associated with a critical load smaller than the applied axial load (Housner and Tsu, 1962). These modes are less influenced by imperfections (Lindberg, 1965; Hayashi and Sans, 1972).

The inelastic impulsive buckling phenomenon is more complex because it includes the nonlinear material property and the possibility of energy dissipation due to unloading. Some numerical studies of these effects have been made (Chawla, 1951; Hartz and Clough, 1957; Lee, 1981; Abrahamson and Goodier, 1966). It has been found that inelastic buckling is more abrupt than elastic buckling (Chawla, 1951).

19.3.3 High-Velocity Excitation

High-velocity excitation (strain rate of the order of 10^2 to 10^6/sec) is typically associated with impact, and its response is usually described by the propagation of stress waves. The axial stress function cannot be considered constant along the length since it is determined by the movement of the stress wave (i.e., the segment covered by the wave is fully stressed while the portion of the length head of the wave remains unstressed). The entire column length is no longer a parameter of stablity since the propagating wave may cause buckling in a shorter segment of the member (Gerard and Becker, 1952). The higher the strain rate, the closer the buckled segment is to the end where impact occurs and the shorter the buckled segment (Hayashi and Sano, 1972b). The higher buckling modes that are excited under these conditions influence the buckled shape and the initial imperfections are not a factor (Lindberg, 1965; Hayashi and Sano, 1972). Slender columns are more susceptible to high-velocity buckling (Abrahamson and Goodier, 1966).

Axial wave propagation caused by sufficiently high strain rates may involve both elastic and plastic wavefronts. Von Kármán and Duwez (1950) developed a solution to this problem for a uniaxial tensile wave, and although they did not address the stability problem, Lee (1978) used their solution procedure to study the stability of a column with an axial plastic compressive wave. Very few stability studies have been made for these conditions (Abrahamson and Goodier, 1966; Hayashi and Sano, 1972b; Lee, 1978). Some experimental testing has produced inelastically buckled shapes due to impact (Abrahamson and Goodier, 1966, Hayashi and Sans, 1972). Under these severe conditions, material properties may be affected. Indication of strain rate sensitivity has been noted (Von Kármán, 1950), while other studies have indicated that the effect is negligible (Lee, 1978).

19.3.4 Numerical Solutions

The numerical solutions to the differential equations of motion have been based on various approximating procedures (see Table 19.3, "Solution"). The low-velocity solution is relatively uncomplicated, but the high-velocity solution is more difficult, particularly for the elastic-plastic wave problem with special consideration required for solution discontinuities. Hyperbolic systems that model wave propagation effects must account for the discontinuities that may exist at the characteristic lines (Belytschko, 1976a). In either category, the solution must be developed in a detailed evolutionary manner that is classified as a nonlinear transient analysis problem. Recent surveys are available that discuss the computer methods for the numerical solution to this class of problems (Belytschko, 1976b; Hughes, 1980). Finite-difference and finite-elment discretization schemes are most commonly used (Belytschko, 1976a).

19.4 DYNAMIC SNAP-THROUGH OF SHALLOW STRUCTURES

Shallow structures or structural elements, such as arches or domes with low height-to-span ratios, can also be subjected to dynamic loads (e.g., from winds or earthquakes). Under extreme loading conditions, part or all of the structure may invert (snap-through). This part of the chapter considers the dynamic snap-through of shallow structures.

Critical values of the loads are usually defined by the Budiansky-Roth criterion (Holzer, 1976, 1979). First, a measure of the response is chosen, such as the area or volume displaced by the structure from its initial configuration, or the displacement of the crown. The maximum value of this response measure is then computed for increasing levels of load. If, at some load level, a small increase in load is accompanied by a large increase in the maximum response, this load is termed the critical load.

19.4.1 Types of Loads

Spatial Distribution. In most studies the load is assumed to be distributed uniformly over the structure. If the structure is symmetric, such as an arch symmetric about its center or a spherical cap symmetric about the vertical axis through its crown, it may have a particularly large resistance to snap-through under symmetric loads. The addition of small asymmetric components to the uniform load may cause a sharp drop in critical load (Gregory and Plaut, 1982). Therefore, results for a symmetric problem may be quite misleading if applied to an actual structure, which will not be perfectly symmetric in geometry, boundary conditions, and loading.

The effect of changes in the load distribution can be investigated by considering two or more independent loads. If two independent loads are applied in varying ratios, the critical load combinations can be plotted as interaction curves, and the effects of changes in the distribution of the combined loads can be studied. Such interaction curves have been determined for a shallow arch (Gregory and Plaut, 1982) and a lattice dome.

In general, as far as snap-through is concerned, the worst spatial distribution is probably a localized one, applied somewhere on one side of the structure. A snow load on one sector of a spherical roof, for example, may be much more dangerous than a fairly uniform load with the same pressure.

Temporal Variation. For blast loads, the duration is often assumed to be so short that the load is modeled as an impulse, which acts at an instant of time. An initial velocity is imparted to the structure, and the ensuing response is computed. Another common type of load is a step load with infinite duration. The load is applied at a certain time and then maintains the same magnitude during the motion. The determination of the structural response under these two types of loads is particularly simple, because the loads do not change with time after the instant of application (i.e., they are zero in the first case, and constant in the second).

Pulse loads, which act on the structure over some finite time interval, are considered in some investigations. They may be shaped like a half-sine wave, an N, or a triangle, for example. Right triangular pulses with varying time lengths of decay are treated by Kao (1978). The critical value of the initial load magnitude decreases as the time length of the pulse increases, with the step load of infinite duration providing a lower bound on these critical values.

A few authors consider snap-through of shallow structures subjected to stochastic loads (Ariaratnam and Sankar, 1968; Pi et al., 1971). The probability distribution for the time of snap-through is determined. Another type of load which has been analyzed is a pulsating load (Huang, 1972; Huang and Plaut, 1981). The critical value becomes lower when the forcing frequency is near a resonance condition, as one would expect.

19.4.2 Types of Structures

Shallow arches under dynamic loads have been studied extensively. This is partly due to the simplicity of the mathematical formulation. Usually, the ends are assumed to be simply supported or clamped, and the initial shape to be circular or sinusoidal. Spherical caps have also received much attention. In most cases the edge is taken to be clamped.

The dynamic response of shallow space trusses (lattice domes) also has been investigated. The bars remain straight, so that the configuration can be defined by the locations of the joints. Snap-through can be local, with inversion of one portion of the structure, or global, with complete inversion. Finally, dynamic snap-through of shallow space frames, in which the joints are rigid, has been considered in one study (Belytschko et al., 1977). The analysis is much more complicated than that for trusses, since bending and torsion are involved as well as axial forces in the members.

In almost all of the investigations, the structures are assumed to be made of homogeneous, isotropic, linearly elastic material. An orthotropic spherical cap is considered in one paper (Alwar and Reddy, 1979). Another paper treats a spherical cap having elastic-plastic material behavior (Kao, 1980), while viscoelastic behavior is assumed in a couple of studies of dynamic snap-through of an arch (Johnson, 1980; Huang and Nachbar, 1968). Plasticity reduces the critical load significantly, while the presence of internal damping (viscoelasticity), and also of external damping (Mescall and Tsui, 1971), tends to raise the critical load.

The effect of structural imperfections is examined in several papers (Kao and Perrone, 1978; Kao, 1980; Belytschko et al., 1977; Huang and Nachbar, 1968). The imperfection usually is chosen as an initial displacement of the structure, such as a dimple in the spherical cap. Such imperfections generally tend to decrease the critical load. The size of the decrease may be significant, even if the imperfection is small.

19.4.3 Numerical Techniques

There are two aspects to the numerical solution of a typical structural dynamics problem: spatial discretization and time integration (Ball, 1975). For a space truss, there are a finite number of degrees of freedom (the coordinates of the joints), so spatial discretization is not needed. For the other types of structures, various techniques are utilized, such as Galerkin's method, the finite-difference method, and the finite-element method.

After spatial discretization, the resulting equations of motion are usually integrated numerically over time in order to determine the structural response. The most popular methods are Newmark's beta method, Houbolt's technique, and the Runge-Kutta method. A Hamming predictor-corrector method and the finite-difference method are also used sometimes.

19.4.4 Comparison with Critical Static Loads

For step loads with infinite duration, the critical values for dynamic snap-through are often compared to the critical values of the static problem. For a clamped spherical cap with a uniform load and no damping, the critical step load is usually about one-half of the critical static load. For an undamped shallow arch (Gregory and Plaut, 1982) and for a lattice dome, the decrease is often around 25%. However, the percentage difference depends on many factors, including the type of structure, its shallowness, the spatial distribution of the load, and the amount of damping.

19.5 FLOW-INDUCED INSTABILITY

Flow-induced vibrations have been known to man since ancient times. But systematic studies of the flow effect on circular cylinders were not made until about a century ago when Strouhal established the vortex shedding frequency for a single circular cylinder across a flow (Goldstein, 1965). Since the collapse of the Tacoma Narrows Bridge in 1940, flow-induced instability has attracted much attention. Recently, due to the use of high-strength materials and the development of advanced structural analysis techniques, structures have become progressively lighter and more flexible and consequently more prone to vibration and instability. Significant progress has been made in understanding the complex interaction phenomena of flow and cylinder motion. The increasing study is evidenced by many conferences (Naudascher, 1972; Naudascher and Rockwell, 1980; IAEA, 1977; Chen and Bernstein, 1979) directed to this subject and numerous publications, including reviews and books (Scanlan and Wardlaw, 1973; Mulcahy and Chen, 1974; Mulcahy and Wambsganss, 1976; Shin and Wambsganss,, 1975; Chenoveth and Kistler, 1976; Chen, 1977; King, 1977; Junger and Feit, 1972; Blevins, 1977; Scanlan and Simiu, 1978).

19.5.1 Instability Mechanisms

The dynamics of circular cylinders subject to flow is described by the following equation of motion (Chen, 1977b).

$$[M]\{\ddot{q}\} + [C]\{\dot{q}\} + [K]\{q\} = \{Q\} \tag{19.21}$$

where

$[M] = [M_s] + [M_f]$
$[M_s]$ = structural mass
$[M_f]$ = added mass of fluid
$[C] = [C_s] + [C_f] + [C_v]$
$[C_s]$ = viscous damping of structures
$[C_f]$ = viscous damping of fluid
$[C_v]$ = velocity-dependent damping of fluid

$[K] = [K_s] + [K_f]$
$[K_s]$ = structural stiffness
$[K_f]$ = fluid elastic stiffness
$[Q]$ = excitation forces

Instability of the system can be studied by setting $\{Q\}$ equal to zero. It is obvious that the fluid elastic stiffness (K_f) and flow velocity-dependent damping (C_v) are the important parameters in a stability analysis. In general, C and K are not symmetric; therefore, the system may be subjected to different types of instability.

Divergence (buckling). Divergence of circular cylinders is caused by fluid elastic force, such as steady drag or lift force for flow across circular cylinders, and fluid centrifugal force for pipes conveying fluid. Divergence is a static phenomenon; it is associated with the characteristics of the matrix K only. Buckling of a pipe conveying steady flow is a typical example of divergence.

Fluid-damping-controlled instability. Fluid-damping-controlled instability is the instability caused by fluid damping force. When the flow velocity is increased, the modal damping of a mode becomes zero and the system loses stability. The instability of the system is attributed to the matrix C_v. Galloping of transmission lines is a typical example of instability controlled by fluid damping.

Fluidelastic-stiffness-controlled instability. When the matrix K is not symmetric, the system may become dynamically unstable. This type of instability is defined as fluid elastic-stiffness-controlled instability. Flutter of cantilevered pipes conveying fluid and whirling of tube arrays subject to cross flow are of flutter type instability. The instability is attributed to fluid elastic force.

Parametric instability. When the matrices M, C, or K are periodic functions of time, parametric resonance and/or combination resonance may occur. This type of instability has been observed in pipes conveying pulsating flow and two-phase flow.

19.5.2 Coupling Effect of Fluid

When circular cylinders vibrate in a fluid, the motion of a cylinder will affect other cylinders because of fluid coupling; therefore, all cylinders will respond as a group rather than as an individual structural element. This type of motion is called coupled vibration. Significant progress has been made in the characterization of fluid coupling and coupled modes; those include two cylinders, tube rows, and tube arrays (Chen, 1975; Chen and Chung, 1976; Chen and Jendrzejczyk, 1978; Lubin et al., 1977).

For a structural system oscillating in a still fluid, the fluid coupling effect may be taken into account using the added mass matrix $[M_f]$ and viscous damping matrix $[C_f]$. The added mass, in general, can be calculated by the potential flow theory. A summary of potential flow results, including formulas, graphs, and a computer program, is available (Chen and Chung, 1976). For small oscillations, viscous damping can be calculated based on the linearized viscous flow theory. Some of the available data are summarized in a recent review (Chen, 1981b).

19.5.3 Parallel-Flow-Induced Instability

Parallel flow may be either internal or external (Benjamin, 1961; Paidoussis, 1970, 1975; Paidoussis and Deksnis, 1970; Chen, 1971; Holmes, 1978). In some cases, both internal and external flows may exist at the same time. A review of this problem, including a list of over 100 references, has been published by S. S. Chen in 1974.

In parallel flow, M_f, C_v, and K_f are important in determining the critical flow velocity. C_v and K_f are in general not symmetric; therefore, the system may be subjected to divergence and dynamic instability.

Many intriguing phenomena of parallel flow-induced instability have been investigated in detail:

1. Destabilizing effect of velocity-dependent forces is an interesting phenomenon. The critical flow velocity at which large cylinder oscillations occur obtained from the equation of motion in the absence of a velocity-dependent force is higher than the critical flow velocity obtained by including the velocity-dependent force and letting it approach zero (Chen, 1971). Experimental verification of the analytical result is difficult because any practical system possesses velocity-dependent force.
2. Divergence is possible for a cantilevered, articulated pipe conveying fluid (Benjamin, 1961), but it is not possible for a cantilevered continuous pipe (Paidoussis, 1970; Paidoussis and Deksnis, 1970). This implies that a discrete model may not always be used as a model for a continuous system.
3. The system may be destabilized by increasing structural stiffness. For example, a cantilevered pipe may become unstable by divergence by placing an additional constraint at the free end; it is stable without the additional restraint (Chen, 1971).
4. A small change in system parameters may result in significant effects on system characteristics: for example, the sharp change in the critical flow velocity with mass ratio.

5. The linear theory predicts that dynamic instability is possible for a pipe supported at both ends (Paidoussis, 1975), while nonlinear theory shows that such a system can become unstable by divergence only (Holmes, 1978).

The general dynamic characteristics of cylinders subject to parallel flow are generally understood. Recent studies have been directed toward periodic flows, two-phase flow, and curved pipes.

19.5.4 Cross-Flow-Induced Instability

When a cylinder array in a cross flow oscillates, the flow field is disturbed. In return, cylinder motion may be enhanced and becomes unstable. Extensive studies have been published to establish the stability limit (Connors, 1970; Blevins, 1974; Chen, 1977a, 1980, 1981a,b). A brief review in this subject area is available (Chen, 1980).

Various forms of instability are possible, depending on whether C_v or K_f is dominant. When C_v is dominant, the instability of a cylinder is attributed to the fluid force associated with its own motion. In this case, the system possesses a classical mode. On the other hand, when K_f is dominant, the instability is attributed to the fluid elastic coupling among different cylinders. The instability mode is not a classical normal mode. In general, C_v is dominant for a heavy fluid while K_f is dominant for a light fluid (Chen, 1981a).

The threshold flow velocity is generally determined by the equation

$$\frac{V}{fD} = k\left(\frac{2\pi\zeta M}{\rho D}\right)^{\alpha} \tag{19.22}$$

where

M = cylinder mass
ρ = fluid density
f = cylinder natural frequency
D = cylinder diameter
ξ = modal damping ratio
V = flow velocity
α, k = constants

The constant α is found to be 0.5 in most cases; however, recent test results show that it varies from about 0.03 to 1.0. The constant k depends on tube arrangements and has been determined by many investigators; it varies from 1 to 10.

The mathematical model for a single tube in a tube row was developed initially by Connors (1970). Fluidelastic coupling of adjacent tubes for tube arrays was included later to improve the model (Blevins, 1974); critical flow velocity can be calculated if the instability mode is known. Further development of the model to include inertia, damping, and fluid elastic couplings has been proposed (Chen, 1981a). Based on the full coupled equation, instability modes as well as critical flow velocities can be calculated.

19.5.5 Design Guides for Stability of Cylinder Arrays in Fluid

Some of the general design guidelines that can be applied in design of system components consisting of a group of circular cylinders are discussed in this section.

Still Fluid. The effect of still fluid is to contribute to fluid added mass and fluid damping. A design guide for calculating the added mass of circular cylindrical structures is available (Chen and Chung, 1976). For a group of n cylinders, each cylinder can move in two directions, the order of added mass matrix is $2n$. Once the added mass is known, it can be incorporated in the equation of motion for calculating the natural frequencies and mode shapes of coupled cylinder/fluid vibration.

It has been shown that $[M_f]$ is a symmetric matrix (Chen, 1975); therefore, there are $2n$ eigenvalues. Let the eigenvalues be designated γ_i, $i = 1$ to $2n$. γ_i plays an important role for coupled vibration of cylinder arrays. For example, let the natural frequency of a group of identical cylinders in vacuum be f_v, and mass per unit length m. The natural frequencies for a single cylinder and multiple cylinders are as follows:

1. A single cylinder in infinite fluid

$$f_s = \frac{f_v}{(1 + m_d/m)^{1/2}} \qquad m_d = \text{displaced mass of fluid;}$$

2. Coupled vibration of multiple cylinders

$$f_c = \frac{f_v}{(1 + \gamma_i/m)^{1/2}} \qquad \gamma_i = \text{eigenvalue of the added mass matrix } M_f$$

For a group of n cylinders, there are $2n$ natural frequencies in a frequency band corresponding to a single natural frequency of a single cylinder; the mode shapes are the same as the eigenvectors of the added mass matrix.

Parallel Flow. Any type of instability in a structural system is considered to be unacceptable as it may lead to catastrophic failure. Fortunately, the critical flow velocity is usually very high and is unlikely to occur in practical system components. In general, it is of no concern except when the cylinders are extremely flexible.

Analysis of the critical flow velocity can be made using Eq. 19.21. The system may lose stability by divergence (static instability) or dynamic instability. The calculation procedure is straightforward if the matrices in Eq. 19.21 are known. The exact critical flow velocity is more of academic interest. An estimate of the critical flow velocity can be made rather easily. Let the critical load of the corresponding single cylinder subjected to a compressive force be P_{cr}. The critical flow velocity for cylinder arrays subjected to parallel flow is approximately given by

$$V = [(P_{cr} - pA)/\gamma_i]^{1/2} \tag{19.23}$$

where γ_i is the eigenvalue of the added mass matrix, A the cylinder cross section and p the fluid pressure. If the critical flow velocity calculated by this method is much higher than the required flow velocity, no further analysis is needed.

Cross Flow. Based on the available data, the lowest critical flow velocity is determined from the equation

$$\frac{U}{fD} = k\left(\frac{2\pi\zeta M}{\rho D^2}\right)^{\alpha} \tag{19.24}$$

The values of k and α depend on the tube arrangement; available experimental data are summarized in a recent report (Chen, 1981b).

19.6 SUDDENLY LOADED STRUCTURES

As already seen, the term "dynamic stability" encompasses many classes of problems and it has been used, by the various investigators, in connection with a particular study. Therefore, it is not surprising that there are various interpretations of the meaning of the term. The class of problems falling in the category of parametric excitation are the best defined, conceived, and understood problems of dynamic stability.

Moreover, many authors refer to problems of the "follower force" type as problems of dynamic stability (Bolotin, 1963; Herrmann, 1967). The primary reason for this is that critical conditions can be obtained (in many cases) only through the use of the "kinetic" or "dynamic" approach to static stability problems (flutter instead of divergence type of instability). In addition, problems of aeroelastic instability and flow-induced instability also fall under the general heading of dynamic stability.

A large class of structural problems that has received attention recently and does qualify as a category of dynamic stability is that of impulsively loaded configurations and configurations which are suddenly loaded with loads of constant magnitude and infinite duration. These configurations under static loading are subject to either limit-point instability or bifurcational instability with unstable post-buckling branch (violent buckling).

The two types of loads may be thought of as mathematical idealizations of blast loads of (1) large decay rates and small decay times, and (2) small decay rates and large decay times, respectively. For these loads, the concept of dynamic stability is related with the observation that for sufficiently small values of the loading the system simply oscillates about the near-static equilibrium point and the corresponding amplitudes of oscillation are sufficiently small. If the loading is increased, some systems will experience large-amplitude oscillations or, in general, divergent type of motion. For this phenomenon to happen the configuration must possess two or more static equilibrium positions and "tunneling through" (Hsu, 1966) occurs by having trajectories that can pass through an unstable static equilibrium point. Consequently, the methodologies developed by the various investigators are for structural configurations that exhibit snap-through buckling when loaded quasistatically.

Solutions to such problems started appearing in the open literature in the early 1950s. Hoff and Bruce (1954) considered the dynamic stability of a pinned half-sine arch under a half-sine distributed load. Budiansky and Roth (1962) in studying the axisymmetric behavior of a shallow spherical cap under suddenly applied loads defined the load to be critical, when the transient response increases suddenly with very little increase in the magnitude of the load. This concept was adopted by numerous investigators (Simitses, 1974) in the subsequent years because it is tractable to computer solutions. Conceptually, one of the best efforts in the area of dynamic buckling, under impulsive and suddenly applied loads, is the work of Hsu and his collaborators (1966, 1967). In his studies, he defined sufficiency conditions for stability and sufficiency conditions for instability, thus finding upper and lower bounds for the critical impulse or critical sudden load. Independently, Simitses (1965) in dealing with the dynamic buckling of shallow arches and spherical caps termed the lower bound as a minimum possible critical load (MPCL) and the upper bound as a minimum guaranteed critical load (MGCL). Finally, there exist a few reported investigations for the case of suddenly loaded systems with constant loads and finite duration (Zimcik and Tennyson, 1979; Simitses, 1980, 1982). Note that this entire class of problems falls in the category of dynamic analysis of conservative systems.

The concepts and methodologies used in estimating critical conditions are as follows:

1. *The equations of motion approach* (Budiansky-Roth, 1962). The equations of motion are (numerically) solved for various values of the load parameter (ideal impulse or sudden load), thus obtaining the system response. The load parameter, at which there exists a large (finite) change in the response, is called critical.

2. *The total energy–phase plane approach* [Hoff–Hsu (Hoff and Bruce, 1954; Hsu, 1966, 1967; Hsu et al., 1968)]. Critical conditions are related to characteristics of the system phase plane, and the emphasis is on establishing sufficient conditions for stability (lower bounds) and sufficient conditions for instability (upper bounds).
3. *The total potential energy approach* [Hoff–Simitses (Hoff and Bruce, 1954; Simitses, 1974, 1980, 1982)]. Critical conditions are related to characteristics of the system total potential. Through this approach, also, lower and upper bounds of critical conditions are established. This last approach is applicable to conservative systems only. The concepts and procedure related to the last approach are next explained, with some detail.

19.6.1 Concepts and General Procedure

The concept of dynamic stability is best explained through a single-degree-of-freedom system. First the case of ideal impulse is treated and then the case of constant load of infinite duration.

Ideal Impulse. Consider a single-degree-of-freedom system for which the total potential (under zero load) curve is plotted versus the generalized coordinate (independent variable) θ (see Fig. 19.2). Clearly, points A, B, C denote static equilibrium points and point B denotes the initial position ($\theta = 0$) of the system.

Since the system is conservative, the sum of the total potential, $\bar{U}_T^0$ (under "zero" load) and the kinetic energy, T^0 is the constant C, or

$$\bar{U}_T^0 + T^0 = C \tag{19.25}$$

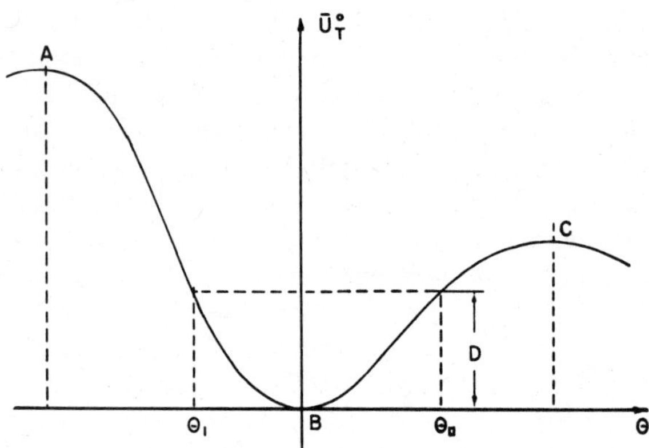

Fig. 19.2 "Zero-load" total potential curve.

Moreover (see Fig. 19.2), since $\bar{U}_T^0$ is zero at the initial position ($\theta = 0$), the constant C, can be related to some initial kinetic energy, T_i^0. Then

$$\bar{U}_T^0 + T^0 = T_i^0 \tag{19.26}$$

Next, consider an ideal impulse applied to the system. Through the impulse-momentum theorem, the impulse is related to the initial kinetic energy T_i^0. Clearly, if T_i^0 is equal to D (see Fig. 19.2), or $\bar{U}_T^0\,(\Theta_{\text{II}})$, the system will simply oscillate between Θ_1 and Θ_{II}. On the other hand, if the initial kinetic energy, T_i^0, is equal to the value of the total potential at the unstable static equilibrium point C, $U_T^0\,(C)$, then the system can reach point C with zero velocity ($T^0 = 0$), and there exists a possibility of motion escaping (passing position C) or becoming unbounded. Such a motion is termed "buckled motion" by Simitses (1965). In the case for which motion is bounded and the path may include the initial point (B), the motion is termed "unbuckled motion." Through this, both a concept of dynamic stability is presented, and the necessary steps for estimating critical impulses are suggested. Note that once the unstable static equilibrium positions (points A and C) are established, the critical initial kinetic energy is estimated by

$$T_{i_{cr}}^0 = \bar{U}_T^0(C) \tag{19.27}$$

Moreover, since T_i^0 is related to the ideal impulse, the critical impulse is estimated through Eq. 19.27. Observe that an instability of this type can occur only when the system under zero load possesses unstable static equilibrium points. Furthermore, if position C corresponds to a very large and thus unacceptable position θ (from physical considerations), one may still use this concept and estimate a maximum allowable (and therefore critical) ideal impulse. For example, if one restricts motion to the region between Θ_{I} and Θ_{II}, the maximum allowable ideal impulse is obtained from Eq. 19.27 but with D or $\bar{U}_T^0\,(\Theta_{\text{II}})$ replacing $\bar{U}_T^0\,(C)$. Because of this, a critical or an allowable ideal impulse can be obtained for all systems (including those that are not subject to buckling under static conditions such as beams, shafts, etc.).

For multi-degree-of-freedom systems, it is possible to use the same concept of dynamic stability and procedure for estimating critical conditions, but with one exception. For these systems, critical conditions can be bracketed between lower and upper bounds (see Hsu, 1966, 1967; Simitses, 1974, 1980). One final comment for the case of ideal impulse: Note from Fig. 19.2, in the absence of damping (as assumed), the direction of the ideal impulse is immaterial. If the system is loaded in one direction (say that the resulting motion corresponds to positive θ), a critical condition exists when the system reaches position C with zero kinetic energy. If the system is loaded in the opposite direction, some negative θ position will be reached with zero kinetic energy; after that the direction of the motion will reverse, and finally the system will reach position C with zero kinetic energy. Both of these phenomena will occur for the same value of the ideal impulse.

Constant Load of Infinite Duration. Consider again a single-degree-of-freedom system. Total potential curves are plotted versus the generalized coordinate θ on Fig. 19.3. Note that the various curves correspond to different load values, P_i. The index i varies from 1 to 5 and the magnitude of the load increases with increasing index value. These curves are typical of systems that for each load value, contain at least two static equilibrium points, A_i and B_i. This is the case when the system is subject to limit point instability and/or bifurcational buckling with unstable branching under static application of the load (shallow arches and spherical caps, perfect or imperfect cylindrical and spherical shells, two-bar frames, etc.).

Given such a system, one applies a given load suddenly with constant magnitude and infinite duration. For a conservative system,

$$\bar{U}_T^P + T^P = C \tag{19.28}$$

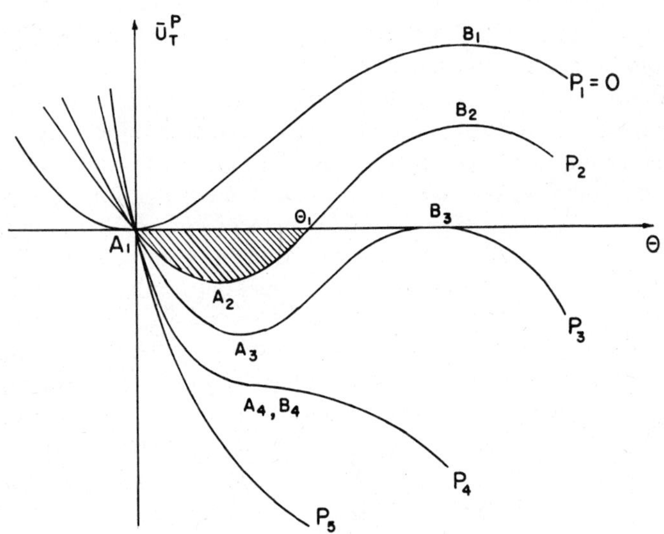

Fig. 19.3 Total potential curves for various loads.

The potential may be defined in such a way that it is zero at the initial position ($\theta = 0$). In such case, the constant is zero, or

$$\bar{U}_T^P + T^P = 0 \tag{19.29}$$

Since the kinetic energy is a positive-definite function of the generalized velocity, motion is possible when the total potential nonpositive (shaded area, on Fig. 19.3, for P_2). From this it is clear that for small values of the applied load, the system simply oscillates about the mean static equilibrium position. This is also an observed physical phenomenon. As the load increases, the total potential at the unstable point B_i decreases, it becomes zero (point B_3), and then it increases negatively until points A_i and B_i (A_4, B_4) coincide (the corresponding load, P_4, denotes the limit point under static loading). For loads higher than this (P_4), the stationary points (static equilibrium positions) disappear from the neighborhood. When the sudden load reaches the value corresponding to P_3, a critical condition exists because the system can reach position B_3 with zero kinetic energy and then move toward larger θ values ("buckled motion" can occur). Thus P_3 is a measure of the critical condition. Note that the value P_3 is smaller than the value of the limit point, P_4. This implies that the critical load under sudden application (infinite duration) is smaller than the corresponding static critical load.

In this case, also, one may wish to limit the dynamic response of the system to a value smaller than B_3 (see Fig. 19.3), say I. Then the maximum allowable (critical dynamic) load corresponds to P_2. Note that in multi-degree-of-freedom systems, one may easily establish upper and lower bounds for the critical dynamic load (see Hsu, 1966, 1967, 1968; Simitses, 1965, 1980, 1982).

The foregoing concept has been extended to the case of suddenly loaded systems with constant load and finite duration (Simitses, 1965, 1980). Moreover, the effect of static preloading on the critical dynamic conditions has been investigated (Simitses, 1982).

References

Abrahamson, G. R., and Goodier, J. N. (1966), "Dynamic Flexural Buckling of Rods, within an Axial Plastic Compression Wave," *Trans. ASME J. Appl. Mech.*, Vol. 33, No. 2, pp. 241–247.

Ari-Gur, J., Weller, T., and Singer, J. (1980), "Experimental and Theoretical Studies of Columns under Axial Impact," *15th Int. Cong. Theoretical and Appl. Mech.*, IUTAM, Toronto, August.

Alwar, R. S., and Reddy, B. S. (1979), "Dynamic Buckling of Isotropic and Orthotropic Shallow Spherical Cap with Circular Hole," *Int. J. Mech. Sci.*, Vol. 21, pp. 681–688.

Amiro, I. Y. (1979), "Critical Values of Compressive Forces Which Increase Rapidly with Time," *Sov. Appl. Mech.*, Vol. 15, No. 5, pp. 403–407 (translation).

Ariaratnam, S. T., and Sankar, T. S. (1968), "Dynamic Snap-Through of Shallow Arches under Stochastic Loads," *AIAA J.*, Vol. 6, No. 5, pp. 798–802.

Ball, R. E. (1975), "Dynamic Buckling of Structures," in *Shock and Vibration Computer Programs: Reviews and Summaries* (ed. W. Pilkey and B. Pilkey), Shock and Vibration Information Center, Washington, D.C., pp. 299–321.

Belytschko, T. (1976a), "A Survey of Numerical Methods and Computer Programs for Dynamic Structural Analysis," *Nucl. Eng. Des.*, Vol. 37, No. 1, pp. 23–34.

Belytschko, T. (1976b), "Computer Methods in Shock Wave Propagation Analysis," in *Computing in Applied Mechanics*, (ed., R. F. Hartung), Applied Mechanics Symposia Series, AMD-18, American Society of Mechanical Engineers, New York, pp. 139–161.

Belytschko, T., Schwer, L., and Klein, M. J. (1977), "Large Displacement, Transient Analysis of Space Frames," *Int. J. Numer. Methods Eng.*, Vol. 11, pp. 65–84.

Benjamin, T. B. (1961), "Dynamics of a System of Aritculated Pipes Conveying Fluid," *Proc. R. Soc.*, Vol. A261, pp. 457–499.

Blevins, R. D. (1974), "Fluid Elastic Whirling of a Tube Row," *Trans. ASME, J. Press. Vessel Tech.*, Vol. 96, No. 4, pp. 263–267.

Blevins, R. D. (1977), *Flow-Induced Vibration*, Van Nostrand Reinhold, New York.

Bolotin, V. V. (1963), *Nonconservative Problems of the Theory of Elastic Stability*, Moscow 1961; English translation Pergamon Press, Elmsford, N.Y.

Bolotin, V. V. (1964), *Dynamic Stability of Elastic Systems*, Holden-Day, San Francisco.

Bolotin, V. V. (1968), "Parametric Vibrations of Elastic Systems," *Prochost*, '*Ustoichivost*,' *Kolebania*, Vol. 3 (ed., I. A. Birger and L. G. M. Panovko), Mashinostroenie, Moscow, Chap. 6.

Budiansky, B., and Roth, R. S. (1962), "Axisymmetric Dynamic Buckling of Clamped Shallow Spherical Shells," Collected Papers on Instability of Shell Structures, *NASA Tech Note No. D-1510.*

Cesari, L. (1973), *Asymptotic Behavior and Stability Problems in Ordinary Differential Equations*, Academic Press, New York.

Chawla, J. P. (1951), "Numerical Analysis of the Process of Buckling of Elastic and Inelastic Columns," *Proc. 1st US Natl. Cong. Appl. Mech.*, June 11–16, pp. 435–441.

Chen, S. S. (1971), "Flow-Induced Instability of an Elastic Tube," *ASME Pap.* 71-Vibr-39.

Chen, S. S. (1974), "Parallel Flow-Induced Vibrations and Instabilities of Cylindrical Structures," *Shock Vib. Dig.*, Vol. 6, No. 10, pp. 1–11.

Chen, S. S. (1975), "Vibration of Nuclear Fuel Bundles," *Nucl. Eng. Des.*, Vol. 35, No. 3, pp. 399–422.

Chen, S. S. (1977a), "A Mathematical Model for Cross-Flow-Induced Vibrations of Tube Rows," *3rd Int. Conf. Pressure Vessel Tech.*, Part 1, pp. 415–426.

Chen, S. S. (1977b), "Flow-Induced Vibrations of Circular Cylindrical Structures," *Shock Vib. Dig.*, Vol. 9, No. 10, pp. 25–38; Vol. 9, No. 11, pp. 21–27.

Chen, S. S. (1980), "Cross-Flow-Induced Instabilities of Circular Cylinders," *Shock Vib. Dig.*, Vol. 12, No. 5, pp. 21–34.

Chen, S. S. (1981a), "Design Guide for Calculating the Instability Flow Velocity of Tube Arrays in Crossflow," *Tech. Memo. ANL-CT-81-40, Argonne National Laboratory*, December.

Chen, S. S. (1981b), "Fluid Damping for Circular Cylindrical Structures," *Nucl. Eng. Des.*, Vol. 63, No. 1, pp. 81–100.

Chen, S. S., and Bernstein, M. D., eds. (1979), "Flow Induced Vibrations," *3rd Natl. Cong. Press. Piping Technol.*, San Francisco, June.

Chen, S. S., and Chung, H. (1976), "Design Guide for Calculating Hydrodynamic-Mass; Part I. Circular Cylindrical Structures," *Tech. Memo ANL-CT-76-45*, Argonne National Laboratory.

Chen, S. S., and Jendrzejczyk, J. A. (1978), "Experiments on Fluidelastic Vibration of Cantilevered Tube Bundles," *Trans. ASME J. Mech. Des.*, Vol. 110, pp. 540–548.

Chenoweth, J. M., and Kistler, R. S. (1976), "Tube Vibrations in Shell-and-Tube Heat Exchanger," *Tech. Rep.* Heat Transfer Research, Inc.

Connors, H. J. (1970), "Fluidelastic Vibration of Tube Arrays Excited by Cross-Flow," *Symp. Flow-Induced Vib. Heat Exch. ASME Winter Annu. Meet.*, pp. 42–56.

Davidson, J. F. (1953), "Buckling of Struts under Dynamic Loading," *J. Mech. Phys. Solids*, Vol. 2, pp. 54–66.

Elishakoff, I. (1978a), "Axial Impact Buckling of a Column with Random Initial Imperfections," *Trans. ASME J. Appl. Mech.*, Vol. 45, No. 2, pp. 361–365.

Elishakoff, I. (1978b), "Impact Buckling of Thin Bar via Monte Carlo Method," *Trans. ASME Appl. Mech.*, Vol. 45, No. 3, pp. 586–590.

Erickson, B., Nardo, S. V., Patel, S. A., and Hoff, N. J. (1956), "An Experimental Investigation of the Maximum Loads Supported by Elastic Columns in Rapid Compression Tests," *Proc. Soc. Exp. Stress Anal.*, Vol. 14, No. 1, pp. 13–20.

Gerard, G., and Becker, H. (1952), "Column Behavior under Conditions of Impact," *J. Aeronaut. Sci.*, Vol. 19, No. 1, pp. 58–60.

Goldsmith, W. (1960), *Impact, The Theory and Physical Behavior of Colliding Solids*, Edward Arnold, London.

Goldstein, S. (1965), *Modern Developments in Fluid Dynamics*, Dover, New York.

Gregory, W. E., Jr., and Plaut, R. H. (1982), "Dynamic Stability Boundaries for Shallow Arches," *ASCE J. Eng. Mech. Div.*, Vol. 108, No. EM6, pp. 1036–1050.

Grybos, R. (1975), "Impact Stability of a Bar," *Int. J. Eng. Sci.*, Vol. 13, No. 5, pp. 463–477.

Hartz, B. J., and Clough, R. W. (1957), "Inelastic Response of Columns to Dynamic Loadings," *ASCE J. Eng. Mech. Div.*, Vol. 83, No. EM2, Pap. No. 1213.

Hayashi, T., and Sano, Y. (1972a), "Dynamic Buckling of Elastic Bars: First Report. The Case of Low Velocity Impact," *Bull. JSME*, Vol. 15, No. 88, pp. 1167–1175.

Hayashi, T., and Sano, Y. (1972b), "Dynamic Buckling of Elastic Bars: Second Report. The Case of High Velocity Impact" *Bull. JSME*, Vol. 15, No. 88, pp. 1176–1184.

Herrmann, G. (1967), "Stability of Equilibrium of Elastic Systems Subjected to Nonconservative Forces," *Appl. Mech. Rev.*, Vol. 20, pp. 103–108.

Hoff, N. J. (1951), "The Dynamics of the Buckling of Elastic Columns," *J. Appl. Mech. ASME*, Vol. 18, No. 1, pp. 68–74.

Hoff, N. J., and Bruce, V. C. (1954), "Dynamic Analysis of the Buckling of Laterally Loaded Flat Arches," *J. Math. Phys.*, Vol. 32, pp. 276–388.

Hoff, N. J., Nardo, S. V., and Erickson, B. (1951), "The Maximum Load Supported by an Elastic Column in a Rapid Compression Test., *Proc. 1st US Natl. Cong. Appl. Mech.*, June 11-16, pp. 419–423.

Holmes, P. J. (1978), "Pipes Supported at Both Ends Cannot Flutter," *Trans. ASME J. Appl. Mech.*, Vol. 45, No. 3, pp. 619–622.

Holzer, S. M. (1970), "Stability of Columns with Transient Loads," *ASCE J. Eng. Mech. Div.*, Vol. 96, No. EM6, pp. 913–930.

Holzer, S. M. (1971), "Response Bounds for Columns with Transient Loads," *Trans. ASME J. Appl. Mech.*, Vol. 38, No. 1, pp. 157–161.

Holzer, S. M. (1976), "Dynamic Stability of Elastic Imperfection-Sensitive Shells," *Shock Vib. Dig.*, Vol. 8, No. 4, pp. 3–10.

Holzer, S. M. (1979), "Dynamic Snap-Through of Shallow Arches and Spherical Caps," *Shock Vib. Dig.*, Vol. 11, No. 3, pp. 3–6.

Holzer, S. M., and Eubanks, R. A. (1969), "Stability of Columns Subject to Impulsive Loading," *ASCE J. Eng. Mech. Div.*, Vol. 95, No. EM4, pp. 897–420.

Housner, G. W., and Tso, W. K. (1962), "Dynamic Behavior of Supercritically Loaded Struts," *ASCE J. Eng. Mech. Div.*, Vol. 88, No. EM5, pp. 41–65.

Hsu, C. S. (1966), "On Dynamic Stability of Elastic Bodies with Prescribed Initial Conditions," *Int. J. Eng. Sci.*, Vol. 4, pp. 1–21.

Hsu, C. S. (1967), "The Effects of Various Parameters on the Dynamic Stability of a Shallow Arch," *J. Appl. Mech.*, Vol. 34, No. 2, pp. 349–356.

Hsu, C. S. (1972), "Impulsive Parametric Excitation: Theory," *J. Appl. Mech.*, Vol. 39, pp. 551–558.

Hsu, C. S., Kuo, C. T., and Lee, S. S. (1968), "On the Final States of Shallow Arches on Elastic Foundations Subjected to Dynamic Loads," *J. Appl. Mech.*, Vol. 35, No. 4, pp. 713–723.

Huang, N. C. (1972), "Dynamic Buckling of Some Elastic Shallow Structures Subject to Periodic Loading with High Frequency," *Int. J. Solids Struct.*, Vol. 8, No. 3, pp. 315–326.

Huang, N. C., and Nachbar, W. (1968), "Dynamic Snap-Through of Imperfect Viscoelastic Shallow Arches," *J. Appl. Mech.*, Vol. 35, No. 2, pp. 289–296.

Huang, K.-Y., and Plaut, R. H. (1981), "Snap-Through of a Shallow Arch under Pulsating Load," in *Stability in the Mechanics of Continua* (ed. F. H. Schroeder), Proceedings of an IUTAM Symposium, Numbrecht, West Germany, August 30–September 4, Springer-Verlag, Berlin.

Huffington, N.J., Jr. (1963), "Response of Elastic Columns to Axial Pulse Loading," *AIAA J.*, Vol. 1, No. 9, pp. 2099–2104.

Hughes, T. J. R. (1980), "Recent Developments in Computer Methods for Structrual Analysis," *Nucl. Eng. Des.*, Vol. 57, No. 2, pp. 427–439.

Ibrahim, R. A., Barr, A. D. S., Roberts, J. M. (1978), "Parametric Resonance Parts 1 to 5," *Shock Vib. Dig.*, Vol. 10, Nos. 1 to 5.

International Atomic Energy Agency (1977), "Summary Report—Specialists Meetings on LMFBR Flow Induced Vibrations," *IAEA IWGFR/21*, Argonne, Ill., September.

Johnson, E. R. (1980), "The Effect of Damping on Dynamic Snap-Through," *J. Appl. Mech.*, Vol. 47, No. 3, pp. 601–606.

Junger, M., and Feit, D. (1972), *Sound, Structures and Their Interaction*, MIT Press, Cambridge, Mass.

Kao, R. (1980), "Nonlinear Dynamic Buckling of Spherical Caps with Initial Imperfections," *Comput. Struct.*, Vol. 12, No. 1, pp. 49–63.

Kao, R., and Perrone, N. (1978), "Dynamic Buckling of Axisymmetric Spherical Caps with Initial Imperfections," *Comput. Struct.*, Vol. 9, pp. 463–473.

King, R. (1977), "A Review of Vortex Shedding Research and Its Applications," *Ocean Eng.*, Vol. 4, pp. 141–171.

Kiryukhin, L. V., and Malyi, V. I. (1974), "Buckling of an Elastic Bar under Longitudinal Impact," *Moscow Univ. Mech. Bull.*, Vol. 29, No. 3, pp. 24–28 (translation).

Koning, C., and Taub, J. (1934), "Impact Buckling of Thin Bars in the Elastic Range Hinged at Both Ends," NACA *Tech. Memo. No.* 748.

Kornev, V. M. (1968), "Modes of Stability Loss in an Elastic Rod under Impact," *J. Appl. Mech. Tech. Phys.*, Vol. 9, No. 3, pp. 275–277.

Kornev, V. M. (1974), "Asymmetric Analysis of the Behavior of an Elastic Bar under A-Periodic Intensive Loading," *J. Appl. Mech. Tech. Phys.*, Vol. 13, No. 3, pp. 398–406 (translation).

Kornhauser, M. (1964), *Structural Effects of Impact*, Spartan Books,.

Krajcinovic, D., and Herrmann, G. (1968), "Parametric Resonance of Straight Bars Subjected to Repeated Impulsive Compression," *AIAA J.*, Vol. T, No. 10.

Lee, L. H. N. (1978), "Quasi-Bifurcation of Rods within an Axial Plastic Compressive Wave," *Trans. ASME J. Appl. Mech.*, Vol. 45, No. 1, pp. 100–104.

Lee, L. H. N. (1981), "Dynamic Buckling of an Inelastic Column," Int. J. Solids Struct., Vol. 17, No. 3, pp. 271–279.

Lindberg, H. E. (1965), "Impact Buckling of a Thin Bar," *Trans. ASME J. Appl. Mech.*, Vol. 32, No. 2, pp. 315–322.

Lubin, B. T., Haslinger, K. H., Puri, A., and Goldberg, J. E. (1977), "Experimental Data on Natural Frequency of a Tube Array," *Am. Soc. Mech. Eng.* Pap. No. 77-FE-10.

Malyi, V. I. (1972), "Long-Wave Approximation in Problems of Stability Loss under Impact," *Mech. Solids*, Vol. 7, No. 4, pp. 120–125 (translation).

Malyi, V. I. (1973), "Buckling of a Rod under Longitudinal Impact: Small Deflections," *Mech. Solids*, Vol. 8, No. 4, pp. 162–166 (translation).

Malyi, V. I., and Efimov, A. B. (1972), "Stability Loss of a Rod in Longitudinal Impact," *Sov. Phys. Dokl.*, Vol. 17, No. 2, pp. 176–177.

Malyshev, B. M. (1966), "Stability of Columns under Impact Compression," *Mech. Solids*, Vol. 1, No. 4, pp. 86–89 (translation).

McIvor, I. K. (1964), "Dynamic Stability of Axially Vibrating Columns," *ASCE J. Eng. Mech. Div., ASCE*, Vol. 90, No. EM6, pp. 191–210.

McIvor, I. K., and Bernard, J. E. (1973), "The Dynamic Response of Columns under Short Duration Axial Loads," *Trans. ASME J. Appl. Mech.*, Vol. 40, No. 3, pp. 688–692.

McLachlan, N. W. (1947), *Theory and Application of Mathieu Functions*, Oxford, London.

Meier, J. H. (1945), "On the Dynamics of Elastic Buckling," *J. Aeronaut. Sci.*, Vol. 12, No. 4, pp. 433–440.

Mescall, J., and Tsui, T. (1971), "Influence of Damping on the Dynamic Stability of Spherical Caps under Step Pressure Loading," *AIAA J.*, Vol. 9, No. 7, pp. 1244–1248.

Mettler, E. (1962), "Dynamic Buckling," *Handbook of Engineering Mechanics* (ed. W. Flugge), McGraw-Hill, New York, Chap. 62.

Mulcahy, T. M., and Chen, S. S. (1974), "Annotated Bibliography on Flow-Induced Vibrations," *Tech. Memo. ANL-CT-74-05*, Argonne National Laboratory, January.

Mulcahy, T. M., and Wambsganss, M. W. (1976), "Flow-Induced Vibration of Nuclear Reactor System Components," *Shock Vib. Dig.*, Vol. 8, No. 7, pp. 33–45.

Naudascher, E., ed. (1972), "Flow-Induced Structural Vibrations," *IUTAM-IAHR Symp.*, Karlsruhe, 1972, Springer-Verlag, Berlin.

Naudascher, E., and Rockwell, D., eds. (1980), *Practical Experiences with Flow-Induced Vibrations*, Springer-Verlag, Berlin.

Nemat-Nasser, S. (1972), "On the Stability of Nonconservative Systems," in *Stability* (ed. H. H. E. Leipholz), Canada Study No. 6, University of Waterloo, Waterloo, Ontario.

Paidoussis, M. P. (1970), "Dynamics of Tubular Cantilevers Conveying Fluid," *J. Mech. Eng. Sci.*, Vol. 12, No. 2, pp. 85–103.

Paidoussis, M. P. (1975), "Flutter of Conservative System of Pipes Conveying Incompressible Fluid," *J. Mech. Eng. Sci.*, Vol. 17, No. 1, pp. 19–25.

Paidoussis, M. P., and Deksnis, E. B. (1970), "Articulated Models of Cantilevers Conveying Fluid: The Study of a Paradox," *J. Mech. Eng. Sci.*, Vol. 12, No. 4, pp. 288–300.

Pi, H. N., Ariaratnam, S. T., and Lennox, W. C. (1971), "First-Passage Time for the Snap-Through of a Shell-Type Structure," *J. Sound Vib.*, Vol. 14, No. 3, pp. 375–384.

Plaut, R. H. (1971), "Displacement Bounds for Beam-Columns with Initial Curvature Subjected to Transient Loads," *Int. J. Solids Struct.*, Vol. 7, No. 9, pp. 1229–1235.

Scanlan, R. H., and Simiu, E. (1978), *Wind Effects on Structures: An Introduction to Wind Engineering*, Wiley, New York.

Scanlan, R. H., and Wardlaw, R. L. (1973), "Reduction of Flow-Induced Structural Vibrations," *ASME Colloquim on Isolation of Mech. Vib., Impact, and Noise*, Cincinnati, Ohio, pp. 35–63.

Schmidt, G. (1975), *Parametererregte Schwingungen*, Deutscher Verlag der Wissenschaften, East Berlin.

Schmitt, A. F. (1956), "A Method of Stepwise Integration in Problems of Impact Buckling," *J. Appl. Mech. ASME*, Vol. 23, No. 2, pp. 291–294.

Sevin, E. (1960), "On the Elastic Bending of Columns due to Dynamic Axial Forces Including Effects of Axial Inertia," *Trans. ASME J. Appl. Mech.*, Vol. 27, No. 1, pp. 125–131.

Shin, Y. S., and Wambsganss, M. W. (1975), "Flow-Induced Vibration in LMFBR Steam Generators: A State-of-the-Art Review," *Tech. Memo. ANL-75-16*, Argonne National Laboratory.

Simitses, G. J. (1965), "Dynamic Snap-Through Buckling of Low Arches and Shallow Caps," Ph.D. dissertation, Department of Aernautics and Astronautics, Stanford University, Stanford, Calif.

Simitses, G. J. (1974), "On the Dynamic Buckling of Shallow Spherical Caps," *J. Appl. Mech.*, Vol. 41, No. 1, pp. 299–300.

Simitses, G. J. (1980), "Dynamic Stability of Structural Elements Subjected to Step-Loads," *Proc. Army Symp. Solid Mech.*, Designing for Extremes: Environment, Loading, and Structural Behavior, South Yarmouth, Cape Cod., Mass., September 30–October 2, pp. 87–107.

Simitses, G. J. (1982), "Effect of Static Preloading on the Dynamic Stability of Structures," *Proc. AIAA/ASME/ASCE/AHS 23rd Struct., Struct. Dyn. Mater. Conf.*, New Orleans, Part 2, pp. 299–307.

Suzuki, S. I. (1969), "Effects of Solid Viscosities, Loading Velocities, and Initial Deflections to Dynamic Buckling Loads of a Column," *Aeronaut. J.*, Vol. 73, No. 706, pp. 890–894.

Taub, J. (1934), "Impact Buckling of Thin Bars in the Elastic Range for Any End Condition," *NACA Tech. Memo. No. 749.*

Von Kármán, T., and Duwez, P. (1950), "The Propagation of Plastic Deformation in Solids," *J. Appl. Phys.*, Vol. 12, pp. 987–994.

Yamamoto, T., and Saito, A. (1970), "On the Vibrations of 'Summed and Differential Types' under Parametric Excitation," *Mem. Fac. Eng. Nagoya Univ.*, Japan, pp. 54–123.

Zimcik, D. G., and Tennyson, R. C. (1979), "Stability of Circular Cylindrical Shells under Transient Axial Impulsive Loading," *Proc. AIAA/ASME/ASCE/AHS 20th Struct. Struct. Dyn. Mater. Conf.*, St. Louis, April 4–6, pp. 275–281.

Chapter Twenty

Structural Safety

20.1 INTRODUCTION

Structural safety is an overriding consideration in the design of structures. In a proper design, safety, serviceability, and economy are in balance throughout the intended life of the structure. A "safe" structure serves usefully under the expected loads, with little or no damage and without injury or loss of life due to structural malfunctions. The chance of partial or total collapse due to catastrophic events such as earthquakes, severe windstorms, or extreme accidential overloads is also very small in a well-designed structure.

The discipline of "structural reliability" is concerned with developing methods for the design of safe structures. A reliable structure is one where the chance of exceeding a "limit state" is acceptably small. Because loads and resistances are "random" quantities (i.e., it is not possible to know their exact value a priori), structural reliability studies use probabilistic methodologies.

A "limit state" is a condition that defines the limit of structural usefulness. Many types of limit states exist, and these are customarily categorized as "serviceability" and "ultimate" limit states. One of the ultimate limit states is the condition when the whole structure or any of its elements becomes unstable. The entire contents of this guide deal with the determination of the instability limit states of columns, beams, plates, shells, arches, frames, and so on. In the broader context of structural reliability the subject matter of this guide thus provides the crucial input on one of the most important types of limit states (i.e., the loss of stability).

20.2 CONCEPTS OF STRUCTURAL SAFETY

Instability is characterized by large changes in displacement for small changes in load. The correct determination of this limit state is crucial since instability may often occur without prior warning and it can precipitate catastrophic collapse, as illustrated by many of the spectacular failures in the history of structural engineering.

The limit state of instability is affected by many factors, such as material properties, geometry, the magnitude and the nature of the loads, structural imperfections, the assumptions underlying the theory, the assessment of the boundary conditions, and so on, as demonstrated in the previous chapters of this guide (see also the flowcart in Fig. 20.1). Each of these effects are "random" quantities (Fig. 20.1). However, this randomness is not arbitrary, since design rules, controls, tolerances, and nature itself limit the extent of the variability. Each parameter affecting the limit state of instability can be characterized by a probability density curve. Knowledge about the probability density distributions of these parameters is variously incomplete, but at least the mean values and the standard deviations are known or can be estimated with fair accuracy.

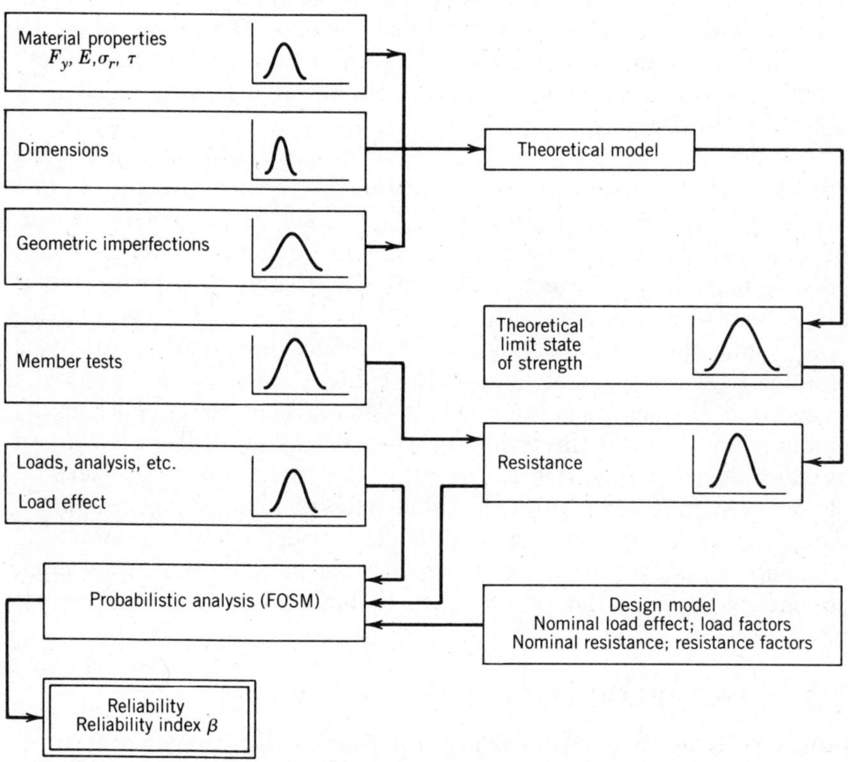

Fig. 20.1 Flowchart of reliability analysis.

The chapters in the guide contain a wealth of such information on the ultimate strength of steel structures and aluminum structures.

There are many approaches to the probability-based methods of evaluating the reliability of structures, and these are discussed in a number of textbooks (see, e.g., Bolotin, 1969; Benjamin and Cornell, 1970; Ang and Tang, 1975; Madsen et al., 1986; Thoft-Christensen and Baker, 1982; Hart, 1982; and others). The most useful of these methods are those which use only the mean and standard deviation of the random parameters of design (Ang and Cornell, 1974). These are called "first-order second-moment" (FOSM) methods. The FOSM methodology is very useful in structural applications because complete information on the distribution properties of the structural design variables is not usually available. Such methods are applied extensively to develop a new generation of structural design specifications (CEB, 1978; Ravindra and Galambos, 1978; Galambos, 1978; AISC, 1986; CSA, 1974). These new specifications retain the traditional format of limit states design (LSD) with resistance factors and load factors, but these factors are determined by probabilistic principles (Galambos et al., 1982; Ellingwood et al., 1982; ANSI, 1982).

The probabilistic method of reliability analysis will next be illustrated as it applies to axially loaded columns to demonstrate its use in the development of design criteria for instability. Similar approaches have been presented by Bjorhovde (1978) and Hall (1981) for columns, and by others (Yura et al., 1978; Bjorhovde et al., 1978; Cooper et al., 1978; Galambos, 1981) for beams, beam-columns, and plate-girders. The reliability of aluminum columns and beam-columns was examined by Chapuis and Galambos (1982).

The two parameters defining the reliability of an axially loaded column are the axial force due to applied loading, Q, and the resistance R (or ultimate load, or capacity) of the member. Both of these are derived quantities, depending on the statistical properties which characterize the various load types, the idealizations used in the analysis, the material properties, the variations of the sectional properties, and the strength model (Ravindra and Galambos, 1978). Probability density distributions for Q and for R are shown in Fig. 20.2 for the case of a 15-ft-long nominally pinned-end column supporting 1800 ft^2 of roof designed for a 15-psf dead-load and a 40-psf snow-load intensity. For illustration it was assumed that the distributions of both R and Q are normal. The nominal design load is 99 kips.

The allowable design capacity according to the 1978 AISC Specification is 131 kips for the W8 × 31 column, and it is 95 kips for the W8 × 28 column, if $F_y = 36$ ksi. The former is thus underutilized, whereas the latter is not acceptable.

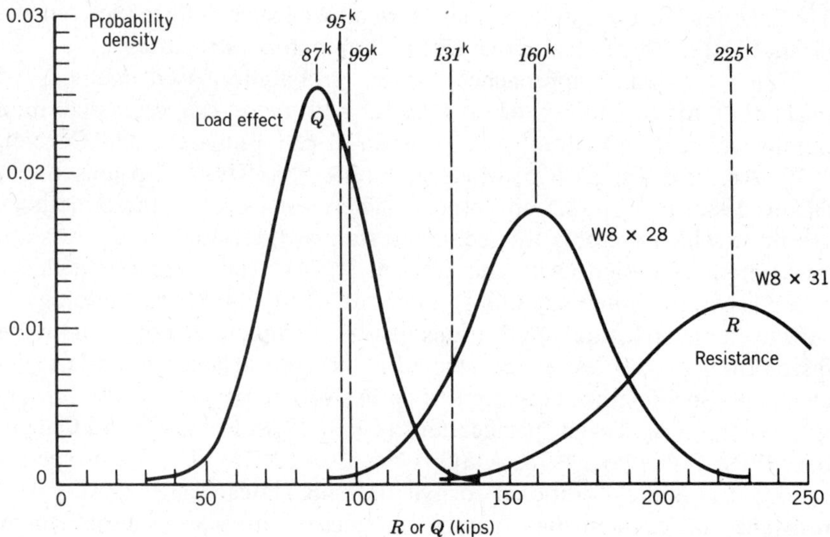

Fig. 20.2 Probability density curves for load effect and resistance.

The mean value and the standard deviation of the applied axial load are 87.4 kips and 15.6 kips, respectively (Galambos et al., 1982). The corresponding statistical properties for the two columns, as will be discussed subsequently, are: for the W8 × 31, 225 kips and 33.8 kips; for the W8 × 28, 160 kips and 22.5 kips, respectively.

The limit state is exceeded when $R \leq Q$ (i.e., when the distributions of R and Q overlap). As can be seen in Fig. 20.2, the region of overlap is larger for the smaller column, and thus the W8 × 28 column has a much greater chance of failing to meet the limit state than the W8 × 31 column. The central question here is whether or not the reliability of one or both of the columns is acceptable. This is a very difficult philosophical question and it has been answered by assuming that an individual design is acceptable if its reliability equals or exceeds the reliability of representative members, designed by traditional practice, which are considered by the profession to have performed with consistent satisfaction in the past. In the case of steel structures this means that the individual member reliability is "calibrated" against the approved AISC Specification (Ravindra and Galambos, 1978).

Two possible methods of evaluating the reliability of the two designs in Fig. 20.2 are illustrated in Fig. 20.3, where the natural log of R/Q is plotted against the probability density, assuming $\ln(R/Q)$ is normally distributed. The limit state is exceeded when $\ln(R/Q) \leq 1.0$, and the

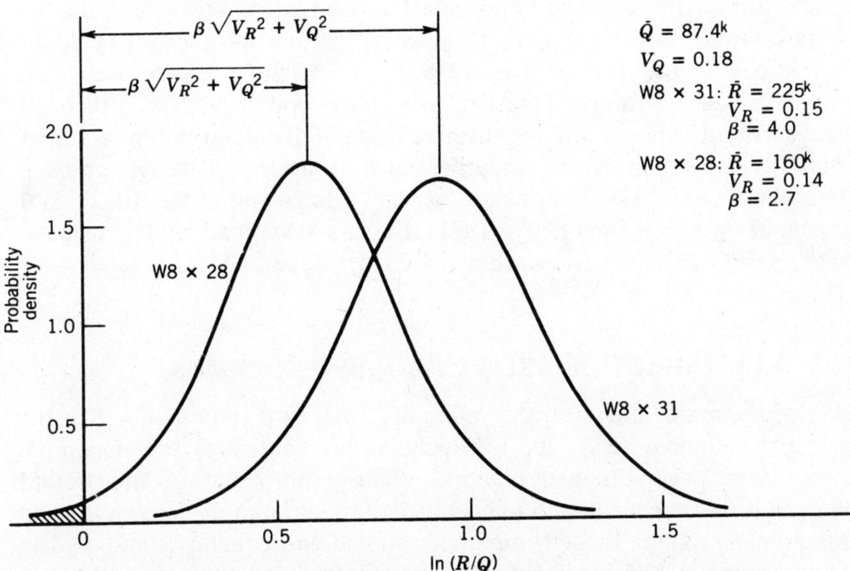

Fig. 20.3 Probability density curves for defining and reliability index β.

probability of exceeding the limit state is the shaded area to the left of $\ln(R/Q) = 1.0$ shown in Fig. 20.3. If the distribution is normal, these probabilities are approximately 1/500 and 1/10,000 for the W8 × 28 and the W8 × 31 columns, respectively. Such an evaluation is dependent on a knowledge of the type of distribution of R and Q, which in this case was assumed normal for convenience of illustration. Unfortunately, the statistical data at our disposal are rarely complete enough to estimate a reasonable distribution model, and therefore a more robust measure to comparative reliability in the form of the "reliability index, β" was proposed as the basis of probabilistic analysis (Ang and Cornell, 1974).

The reliability index β is defined as the number of standard deviations between the mean of $\ln(\bar{R}/\bar{Q})$, where $\bar{R}$ and $\bar{Q}$ are the mean values of R and Q, and the ordinate axis (Fig. 20.3), regardless of the type of distribution. Thus only mean values and standard deviations need to be known, hence the name "first-order, second-moment" (FOSM) probabilistic analysis. The formula for β is approximately given by (Ravindra and Galambos, 1978)

$$\beta = \frac{\ln(\bar{R}/\bar{Q})}{\sqrt{V_R^2 + V_Q^2}} \tag{20.1}$$

The use of this simplified "notional" reliability has been the basis for the development of the recently proposed structural specifications (AISC, 1986; CSA 1974). The original FOSM, as presented above, has been made considerably more sophisticated (Ellingwood et al., 1982) through the years, but the essential principles remain. By determining a target reliability β as that which underlies current successful design practice (Galambos et al., 1982), the load factors (Ellingwood et al., 1982), and resistance factors (Galambos, 1981) for the new load and resistance factor design (LRFD) criteria have been developed (AISC, 1986).

20.3 RELIABILITY OF AXIALLY LOADED COLUMNS

One of the major inputs to the reliability analysis is the resistance of the structural element (Fig. 20.1), which in this context is the maximum capacity of axially loaded columns. The requirement of the FOSM analysis is that the mean and the coefficient of variation be known. For an intentionally axially loaded nominally pinned-end column R and V_R can be expressed by the following relations (Galambos, 1983):

$$\bar{R} = \bar{B}_{\text{theory}} \times \bar{B}_{\text{test}} \times \bar{B}_{\text{material}} \times \bar{B}_{\text{cross section}} \times P_{cr} \tag{20.2}$$

$$V_R = (V^2_{\text{theory}} + V^2_{\text{test}} + V^2_{\text{material}} + V^2_{\text{cross section}})^{1/2} \tag{20.3}$$

B_{test} is the mean ratio of test to prediction, provided that the analysis is made with the measured actual properties of the test specimens. Such an analysis was performed by Bjorhovde (1972), and for axially loaded columns $\bar{B}_{\text{test}} = 1.03$ and $V_{\text{test}} = 0.05$, indicating that the theory, which is explained in Chapter 3, is excellent. The analysis of the variation of the yield stress and the cross-sectional properties (Ravindra and Galambos, 1978) gave the results $\bar{B}_{\text{material}} = 1.05$, $V_{\text{material}} = 0.10$, $\bar{B}_{\text{cross section}} = 1.00$, and $V_{\text{cross section}} = 0.05$. The critical load $\bar{B}_{\text{theory}} P_{cr}$ and V_{theory} was obtained and tabulated by Bjorhovde (1972) by considering initial crookedness and residual stress for many columns, as given in detail in Chapter 3 for perfectly pinned-end columns and for a sinusoidally varying initial crookedness having a maximum amplitude of $L/1000$. The resulting variation of V_R and $\bar{R}$ with the slenderness parameter λ is shown in Fig. 20.4 for Bjorhovde's Column Category 2 (see Chapter 3). By including a moderate end restraint ($G = 10$), Galambos (1983) determined R as shown also in Fig. 20.4. The latter curve will be used in the subsequent reliability analyses because practical column ends will always exhibit some end restraint (see Chapter 3).

The variations of the reliability index β with the slenderness parameter

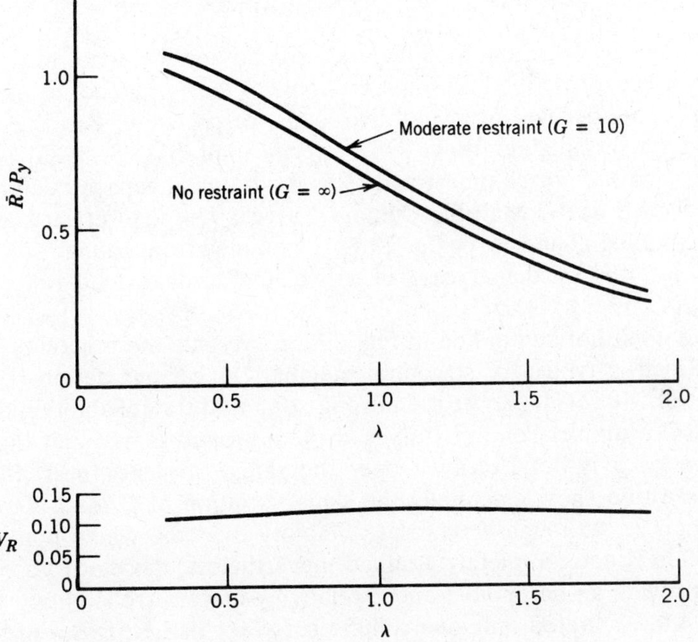

Fig. 20.4 Mean resistance and coefficient of variation for category 2 columns.

$$\lambda = \frac{L}{r} \frac{1}{\pi} \sqrt{\frac{F_y}{E}} \tag{20.4}$$

for the practical ranges of λ between 0.3 and 1.9 is shown in Fig. 20.5 for the SSRC Multiple Column Curves presented in Chapter 3. These curves are represented by Eqs. 3.13 through 3.15 (or, equivalently, by Eqs. 3.16 and 3.17) for the case of an initial crookedness of $L/1000$ (SSRC Column Curves 1, 2, and 3) and by Eqs. 3.18 through 3.20 for an initial crookedness of $L/1500$ (SSRC Column Curves 1P, 2P, and 3P). These equations in Chapter 3 were assumed to represent mean values, and the same coefficients of variation V_{theory} were used for both types of column curves [the data were taken from Bjorhovde (1972)]. Thus only one set of β's is obtained regardless of whether, respectively, curves 1, 2, and 3, or curves 1P, 2P, and 3P are used. The column resistance in this analysis included the effect of a small amount of end restraint (restraint coefficient $G = 10$, corresponding to an elastic effective length factor of 0.96).

The columns were designed by the LRFD limit state equation (AISC, 1986; ANSI, 1982)

$$\phi R_n = 1.2P_D + 1.6P_L \tag{20.5}$$

where 1.2 and 1.6 are the load factors for dead and live load, respectively, which are prescribed by ANSI (ANSI, 1982); P_D and P_L are the nominal code-specified dead and live loads, respectively; ϕ is the resistance factor specified in the LRFD Specification (AISC, 1986), which requires $\phi = 0.85$ for columns; and R_n is the nominal capacity defined by the respective SSRC Multiple Column Curves. The load effect statistics for determining $\bar{Q}$ and V_Q in Eq. 20 were obtained from Ellingwood et al. (1982). For the particular case of a live load-to-dead load ratio of 3, $\bar{Q} = P_D\ (1.05 + 3) = 4.05P_D$, and $V_Q = 0.19$.

The dashed horizontal line in Fig. 20.5 represents the reliability index $\beta = 3$ which is typical of structural members in current design (Ellingwood et al., 1982). It can be seen in Fig. 20.5 that the reliability index of the SSRC Multiple Column Curves clusters around $\beta = 3$, and that the variation of β is not excessive over the range of slenderness. Similar analyses would show essentially the same variation of β for L/D ratios other than 3. Although a uniform reliability over the whole domain of parameters is not completely attained, nevertheless, these SSRC column curves provide an acceptable and a relatively uniform reliability.

The AISC adopted only one column curve for its LRFD Specification

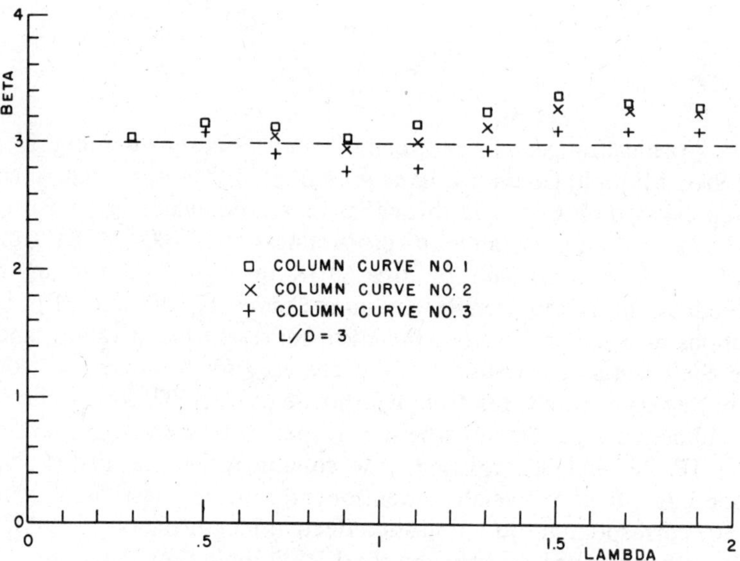

Fig. 20.5 Reliability of SSRC Column Curves.

(AISC, 1986). This curve is essentially the same as SSRC Curve 2P, but a different mathematical representation is used, as follows:

$$\begin{aligned} &\text{Resistance factor:} \qquad \phi = 0.85 \\ &\text{Nominal resistance:} \qquad R_n = AF_{cr} \end{aligned} \tag{20.6}$$

where A is the cross-sectional area of the column and

$$F_{cr} = 0.658^{\lambda^2} F_y \qquad \text{for } \lambda \leq 1.5 \tag{20.7}$$

$$F_{cr} = 0.877 F_y / \lambda^2 \qquad \text{for } \lambda \geq 1.5 \tag{20.8}$$

This column formula assumes an initial crookedness of $L/1500$. The variation of the reliability index β with the slenderness parameter λ for a live load-to-dead load ratio of 3 is shown in Fig. 20.6 by the squared marks. Also shown in this figure are the corresponding β's for the 1978 AISC Specification (AISC, 1978) column curve. This column curve is used with a variable factor of safety in an allowable stress design format, and it is given by the equations in Sec. 1.5.1.3.1 in the currently (1986) applicable criteria for the design of steel structures (AISC, 1978). The β's for the 1978 AISC column curve are marked by X's in Fig. 20.6. The horizontal dashed line in this figure represents $\beta = 2.6$, which is the basic reliability index adopted by the LRFD Specification (AISC, 1986) for

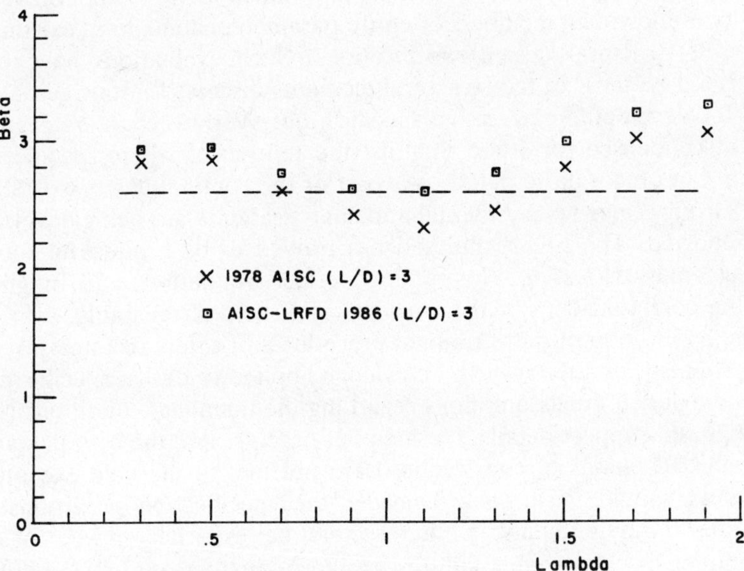

Fig. 20.6 Reliability of AISC Column Curves.

members. The analysis of the resistance, also includes the effect of small end restraints ($G = 10$).

Since only one column curve is used by AISC, the stratification of data into three levels, as was used in developing the SSRC Multiple Column Curves, is no longer appropriate. Thus the statistical properties (e.g., means and coefficients of variation) for all 120 of the column curves developed by Bjorhovde (1972) had to be used for the determination of the resistance parameters. Since the column curves now are spread over a larger bandwidth, the coefficients of variation are greater, and both the spread and the minimum values of β for the AISC column curves are different from the SSRC curves (compare Figs. 20.5 and 20.6).

From Fig. 20.6 it is evident that the LRFD column curve provides a somewhat higher reliability than that furnished by the current column design criteria. Since one column curve only is used, uniformity of reliability is sacrificed. Long and short columns have a higher reliability than columns of intermediate length.

20.4 CONCLUSIONS

The particular instances of column design development discussed above are not meant to be specific recommendations to specification writing groups; they are presented to illustrate the versatility of the first-order second-moment probabilistic method in evaluating the comparative reliability of individual members of entire parameters domains in existing or proposed structural design specifications. Such evaluations have been performed on most of the structural elements discussed in this guide as a basis for developing new design specifications. Various chapters gave the data in condensed form and they list the references where specification writers can obtain more details. As part of the continuing work of SSRC Task Groups, more experimental and analytical data are being developed and reported. The FOSM method has proved to be a powerful tool in placing analysis, testing, theory, load effect evaluation, and judgment into the common focus of the concepts of structural reliability.

The application of more rational procedures of safety risk analysis and more sophisticated theoretical knowledge applied to design specifications is meaningless if the assumptions regarding the boundary conditions (i.e., member end-support details, the member bracing, and the overall stability during all phases of construction) are not met in the field execution. Designers should particularly consider the specification provisions for adequate framing details, while erectors be especially aware of the possibilities of catastrophic failures caused by inadequate bracing during construction. This subject is too broad for further elaboration here, but it brought up as a general caveat to both designers and contractors: Structural safety is of concern to all the professionals involved with structures.

References

American Institute of Steel Construction (1978), *Specification for the Design, Fabrication and Erection of Structural Steel for Buildings*, AISC, Chicago.

American Institute of Steel Construction (1986), *Load and Resistance Factor Design Specification for Structural Steel Buildings*, AISC, Chicago.

American National Standards Institute (1982), *Building Code Requirements for Minimum Design Loads in Buildings and Other Structures*, ANSI A58.1-1982, ANSI, New York.

Ang, A. H.-S., and Cornell, C. A. (1974), "Reliability Bases of Structural Safety and Design," *ASCE J. Struct. Div.*, Vol. 100, No. ST9, pp. 1755–1769.

Ang, A. H.-S., and Tang, W. H. (1975), *Probability Concepts in Engineering Planning and Design*, Wiley, New York.

Benjamin, J. R., and Cornell, C. A. (1970), *Probability, Statistics and Decision for Civil Engineers*, McGraw-Hill, New York.

Bjorhovde, R. (1972), "Deterministic and Probabilistic Approaches to the Strength of Steel Columns," Ph.D. thesis, Lehigh University, Bethlehem, Pa.

Bjorhovde, R. (1978), "The Safety of Steel Columns," *ASCE J. Struct. Div.*, Vol. 104, No. ST3, pp. 463–478.

Bjorhovde, R., Galambos, T. V., and Ravindra, M. K. (1978), "LRFD Criteria for Steel Beam-Columns," *ASCE J. Struct. Div.*, Vol. 104, No. ST9, pp. 1371–1388.

Bolotin, V. V. (1969), *Statistical Methods in Structural Mechanics* (transl. Samuel Aroni), Holden-Day, San Francisco.

Canadian Standards Association (1974), *Steel Structures for Buildings-Limit States Design*, CSA Standard S16.1-1974, CSA, Rexdale, Ontario.

Chapuis, J., and Galambos, T. V. (1982), "Reliability of Aluminum Beam-Columns," *ASCE J. Struct. Div.*, Vol. 108, No. ST4, pp. 709–727.

Comite Euro-International du Beton (1978), "Common Unified Rules for Different Types of Construction and Material," *Bull. d'Inf. No. 124E*, CEB, Paris, April.

Cooper, P. B., Galambos, T. V., and Ravindra, M. K. (1978), "LRFD Criteria for Plate Girders," *ASCE J. Struct. Div.*, Vol. 104, No. ST9, pp. 1389–1408.

Ellingwood, B., MacGregor, J. G., Galambos, T. V., and Cornell, C. A. (1982), "Probability-Based Load Criteria: Load Factors and Load Combinations," *ASCE J. Struct. Div.*, Vol. 108, No. ST5, pp. 978–997.

Galambos, T. V. (1978), "Proposed Criteria for Load and Resistance Factor Design of Steel Building Structures," *AISI Bull.*, No. 27, January.

Galambos, T. V. (1981), "Load and Resistance Factor Design," *Eng. J. Am. Inst. Steel Constr.*, Vol. 18, No. 3, pp.

Galambos, T. V. (1983), "Reliability of Axially Loaded Columns," *Eng. Struct.*, Vol. 5, pp. 73–78.

Galambos, T. V., Ellingwood, B., MacGregor, J. G., and Cornell, C. A. (1982), "Probability-Based Load Criteria: Assessment of Current Design Practice," *ASCE J. Struct. Div.*, Vol. 108, No. ST5, pp. 959–977.

Hall, D. H. (1981), "Proposed Column Strength Criteria," *ASCE J. Struct. Div.*, Vol. 107, No. ST4, pp. 649–670.

Hart, G. C. (1982), *Uncertainty Analysis, Loads and Safety in Structural Engineering*, Prentice-Hall, Englewood Cliffs, N.J.

Madsen, H. O., Krenk, S., and Lind, N. C. (1986), *Methods of Structural Safety*, Prentice-Hall, Englewood Cliffs, N.J.

Ravindra, M. K., and Galambos, T. V. (1978), "Load and Resistance Factor Design for Steel," *ASCE J. Struct. Div.*, Vol. 104, No. ST9, pp. 1337–1354.

Thoft-Christensen, P. and Baker, M. J. (1982), *Structural Reliability Theory and Its Application*, Springer-Verlag, Berlin.

Yura, J. A., Galambos, T. V., and Ravindra, M. K. (1978), "The Bending Resistance of Steel Beams, *ASCE J. Struct. Div.*, Vol. 104, No. ST9, pp. 1355–1370.

Chapter Twenty-One

Finite-Element Analysis of Stability Problems

21.1 INTRODUCTION

The finite-element method is a procedure for obtaining numerical solutions to boundary value problems in engineering. It was developed for and has enjoyed its widest utilization in application to problems in structural mechanics. The basic concept of the method involves the subdivision of the total analytical model of the structure into subdomains of simple geometric form, these being the finite elements. The behavior of each element is described by approximating functions, resulting first in algebraic relationships for the elements and then, upon connection of the elements to form the representation of the complete structure, in a large-order system of algebraic equations. Solution of these algebraic equations results in an approximate solution for the response of the structure to the applied loads. Under conditions which are today well understood and which can be satisfied under most circumstances, the approximate solution approaches the exact solution as more and more finite elements are employed.

The single most important feature of the finite-element analysis is its amenability to computer programming in such a way that a single program can be written to deal with the widest variety of structural forms, materials, and loading conditions. In this era of "portable" software and of access through local and nationwide computer networks, virtually every practicing analyst can make use of the finite-element method in linear design analyses. The situation is less robust in connection with finite-element solutions of stability problems.

Even today only a minority of the programs that enjoy wide utilization incorporate the terms and procedures that are needed to deal with a full range of stability problems. This situation is in a state of change as new finite-element analysis capabilities continue to emerge. It is worth emphasizing, however, that although this chapter describes the state of the art in finite-element formulations and procedures for instability analysis, most practitioners can employ the method in stability analysis only within the limits prescribed by available software. The exceptions are those who are able to modify or develop the relevant software.

The finite-element method first gained widespread acceptance in the mid-1950s, for linear analysis problems. The extension of the method to stability problems took place a decade later (Gallagher and Padlog, 1963; Martin, 1965) and this aspect of the method was brought to a mature state during the 1970s. This chronology explains, in part, why the available software for finite-element analysis is limited in its ability to offer capabilities for stability analyses. Another reason is that the instability analysis capability adds substantial complexity to the program and computational process.

The initial developments of the finite-element method for instability phenomena were for linear elastic instability (bifurcation) analysis. The literature on this topic is not very extensive. The basic theory of a finite-element approach to linear stability analysis can be, and has been, written in fairly general terms (e.g., Gallagher, 1975a, Chap. 13; Cook, 1981, Chap. 12; Przemieniecki, 1968). As already made clear in this volume, many design situations require consideration of nonlinear effects. Developments pertaining to the finite-element method in nonlinear stability analysis are, in contrast to linear stability analysis, voluminous and of extremely wide scope. The nonlinear problem is essentially open-ended, extending on the theoretical side beyond mere large-deflection effects to inelastic behavior, finite strains, and very large rotations, and to the coupling of these effects. The computational side invites a wealth of alternative solution techniques. Entire conferences are devoted to nonlinear finite-element analysis (Wunderlich et al., 1981; Bergan et al., 1978a) and contributions to the literature show no sign of abating.

In this chapter we give details only for the topics in finite-element linear stability analysis. Key issues and developments for nonlinear analysis are discussed in outline. Many aspects of the linear analysis procedures are fundamental, however, to the nonlinear analysis procedures.

21.2 FORMULATION OF FINITE-ELEMENT ANALYSIS FOR ELASTIC INSTABILITY

The principle of stationary potential energy provides a convenient approach to the formulation of finite-element relationships and is chosen here as a basis for the description of finite-element elastic instability analysis. When nonlinear effects are significant it is more appropriate to base the development on the virtual work concept (Thomas, 1970), but this corresponds to the potential energy approach for the simplifications adopted below.

Potential energy is written in terms of strains, which are in turn expressed as derivatives of displacements. Thus the approximating functions of finite-element analysis, in the present context, are in the form of expressions for displacements. These expressions are basically polynomial functions of the spatial coordinates of the element. Other measures of behavior (e.g., the stress state or the exact solution of the governing differential equation) have been used and the approximating functions are by no means limited to polynomials. Nevertheless, the polynomial form of assumed displacement is so pervasive in its role in practical finite-element stability analysis that we emphasize it here.

The uniform section prismatic element, or "beam-column" element (Fig. 21.1), is the simplest medium for description of the formulation of finite-element relationships for stability analysis, and at the same time, illustrates key aspects of this type of formulation for all structural forms. The element lies in the x-z plane and remains in this plane as displacements take place under the application of loads.

The element is assumed to sustain only axially directed stresses. Transverse shear deformation is disregarded. Bending moments and axial loads produce the axially directed stresses, which vary linearly across the depth of the element, and the relationship between the corresponding strains and the displacements on the centroidal axis is given by

$$\varepsilon_x = \frac{du}{dx} - z\,\frac{d^2w}{dx^2} + \frac{1}{2}\left(\frac{dw}{dx}\right)^2 + \frac{1}{2}\left(\frac{du}{dx}\right)^2 \tag{21.1}$$

The first and second terms on the right side are the familiar components of axial and flexural strain for linear behavior in the absence of instability

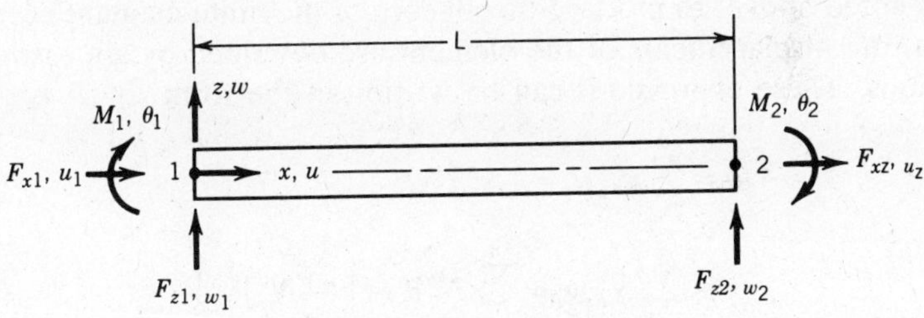

Fig. 21.1 Beam-column element.

effects. The third and fourth terms represent the axial strain due to the elongation of the neutral line in the deflected state. The fourth term can be viewed as proportional to the square of the axial strain and is negligible for metallic structures in the elastic range when the rotation of the member in the displaced state is moderate, say less than 10°. Assessments of the significance of this term are given in Thomas (1970) and Ebner and Uniffero (1972). It is neglected in the following.

The potential energy, denoted here by Π_p, is equal to the difference between the strain energy of deformation, U, and the potential of the applied loads,

$$\Pi_p = U - V \tag{21.2}$$

The strain energy for the element is given by

$$U^e = \frac{1}{2}\int_{\text{vol}} \varepsilon_x E \varepsilon_x \, d(\text{vol}) \tag{21.3}$$

and by substitution of Eq. 21.1 into Eq. 21.2 [with $d(\text{vol}) = dA\,dx$ and neglecting of $\frac{1}{2}(du/dx)^2$]

$$U^e = \frac{1}{2}\int_L \int_A \left[\left(\frac{du}{dx}\right)^2 + z^2\left(\frac{d^2w}{dx^2}\right)^2 + \frac{1}{4}\left(\frac{dw}{dx}\right)^4 - 2z\,\frac{du}{dx}\left(\frac{d^2w}{dx^2}\right) - z\,\frac{d^2w}{dx^2}\left(\frac{dw}{dx}\right)^2 + \frac{du}{dx}\left(\frac{dw}{dx}\right)^2\right] E\,dA\,dx \tag{21.4}$$

Next, we integrate across the depth of the member, noting that

$$\int_A dA = A \qquad \int_A z\,dA = 0 \qquad \int_A z^2\,dA = I \tag{21.5}$$

for z measured from the centroid. This yields

$$U^e = \frac{1}{2}\int_L \left[A\left(\frac{du}{dx}\right)^2 + I\left(\frac{d^2w}{dx^2}\right)^2 + A\left(\frac{du}{dx}\right)\left(\frac{dw}{dx}\right)^2 + \frac{A}{4}\left(\frac{dw}{dx}\right)^4\right] E\,dx \tag{21.6}$$

As noted above, to produce the algebra of the finite-element representation, the displacements of the element are described by approximating functions. These expressions can be written in the form

$$u = \sum_{i=1}^{n} N_i^u u_i = \lfloor N^u \rfloor \{u\} \tag{21.7}$$

$$w = \sum_{i=1}^{n} N_i^w w_i + \sum_{i=1}^{n} N_i^\theta \theta_i = \lfloor N^w N^\theta \rfloor \begin{Bmatrix} w \\ \theta \end{Bmatrix} \tag{21.8}$$

where N_i^u, N_i^w, and N_i^θ are functions of the coordinate x and are known as *shape functions*, and $\lfloor N^u \rfloor$, $\lfloor N^w N^\theta \rfloor$ are row matrices containing these functions. u_i and w_i are coordinate direction displacements at the element joints (*nodal point displacements*), θ_i are angular displacements at these joints, and $\{u\}$, $\{w\}$, and $\{\theta\}$ are columnar listings of the joint displacements.

Substitution of the expressions for u and w into the strain energy expression gives, after integration,

$$U^e = \frac{\lfloor \Delta \rfloor}{2}[k_0]\{\Delta\} + \frac{\lfloor \Delta \rfloor}{6}[k_1]\{\Delta\} + \frac{\lfloor \Delta \rfloor}{12}[k_2]\{\Delta\} = \frac{\lfloor \Delta \rfloor}{2}[\hat{k}]\{\Delta\} \tag{21.9}$$

where $\lfloor \Delta \rfloor = \lfloor \lfloor u \rfloor \lfloor w\theta \rfloor \rfloor$ and

$$[\hat{k}] = \left[[k_0] + \frac{1}{3}[k_1] + \frac{1}{6}[k_2]\right] \tag{21.10}$$

$[k_0]$ is the stiffness matrix for linear, stable behavior, $[k_1]$ is the "initial stress' or "geometric" stiffness matrix (a linear function of $\{\Delta\}$), and $[k_2]$ is the "initial displacement" stiffness matrix (a second-order function of $\{\Delta\}$). The alternative ways of arranging terms in $[k_1]$ and $[k_2]$ are reviewed by Murray and Rajasekaran (1975) and Wood and Schreffler (1978). Although Eq. 21.9 has been derived for the specific case of a beam-column element, its format holds true for all finite elements in geometrically nonlinear analysis.

The strain energy for the complete structure is obtained by summing all of the element strain energies

$$U = \sum U^e = \frac{\lfloor \Delta \rfloor}{2}[\hat{K}]\{\Delta\} \tag{21.11}$$

where now $\lfloor \Delta \rfloor$ includes the displacements at all the joints and $[\hat{K}]$ is the "global" stiffness matrix, obtained by appropriate summation of the element stiffness matrices. This development is for nonlinear behavior and in such cases $[\hat{K}]$ is known as the "secant" stiffness matrix for strain energy.

If the external loads are concentrated forces at the joints, whose direction remains unchanged throughout the displacement history, the potential of the applied loads is

$$V = \lfloor \Delta \rfloor \{P\} \tag{21.12}$$

where $\{P\}$ lists the joint forces. Distributed loads require special treatment but lead to the same algebraic form of V. Certain forces, especially distributed loads, might "follow" the directions of displaced structure; this is particularly important for problems of elastic instability (e.g., the buckling of pressure-loaded shells). In those cases the formulation of the load potential leads to an additional stiffness and complications in the analysis sequence. The finite-element approach to such problems is presented by Loganathan et al. (1979), Hibbitt (1979), and Jensen (1980).

Considering, for simplicity, only those cases for which the complications of follower forces are absent, the potential energy is, by insertion of Eqs. 21.11 and 21.12 into Eq. 21.2,

$$\Pi_p = \frac{\lfloor \Delta \rfloor}{2} [\hat{K}]\{\Delta\} - \lfloor \Delta \rfloor \{P\} \tag{21.2a}$$

For equilibrium, the first derivative of Π_p with respect to the displacements must be taken and set equal to zero. This gives

$$\left[[K_0] + \frac{1}{2}[K_1] + \frac{1}{3}[K_2]\right]\{\Delta\} - \{P\} = 0 \tag{21.13}$$

or

$$[K]\{\Delta\} - \{P\} = 0 \tag{21.14}$$

The difference between $[\hat{K}]$ (Eq. 21.10) and $[K]$ is in the multipliers of $[K_1]$ and $[K_2]$; this is the result of the differentiation process, which affects these matrices because they are functions of the displacement.

Most nonlinear analysis processes involve solutions for a series of increments on the load-displacement path. If Eq. 21.13 is differentiated, the incremental equations are obtained:

$$[K_T]\{\delta\Delta\} - \{\delta P\} = 0 \tag{21.15}$$

where $[K_T]$, the global "tangent stiffness," is

$$[K_T] = [K_0] + [K_1] + [K_2] \tag{21.16}$$

Equations 21.13 and 21.15 are the bases for stability analysis by the finite-element method. The next section describes the manner in which they are used for that purpose.

21.3 COMPUTATIONAL PROCEDURES FOR FINITE-ELEMENT STABILITY ANALYSIS

21.3.1 Types of Load-Displacement Response

It is pertinent to review, briefly, some of the elementary situations that can arise in structural instability. Figure 21.2 illustrates the load-displacement behavior for these situations, tracing in each case the response of a representative degree of freedom.

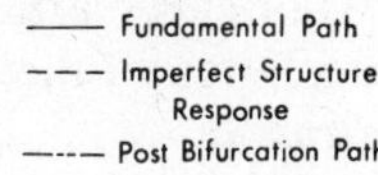

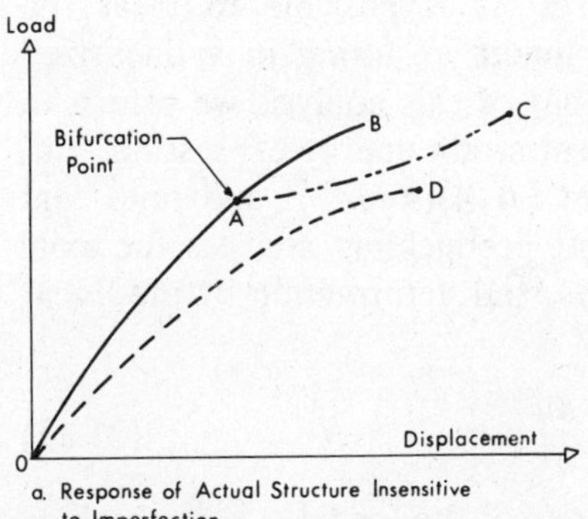

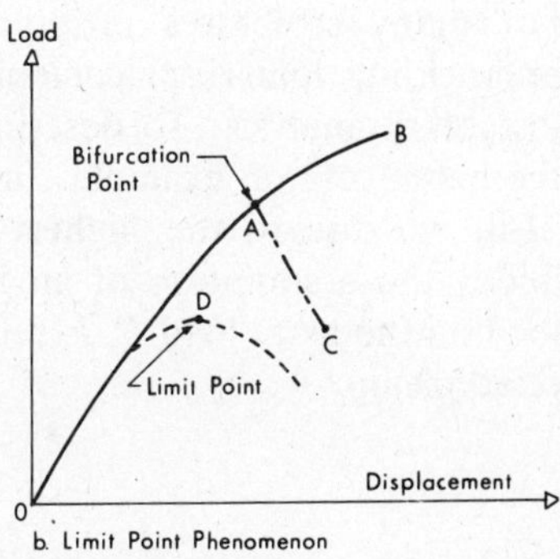

Fig. 21.2 Idealized load-displacement paths. Elementary situations in structural instability.

The solid line in both sketches (Fig. 21.2) applies to "perfect" structures. In these, the structure first displaces along the "fundamental path" (OAB), with bifurcation to a secondary path at point A. The postbuckling path AC may rise (Fig. 21.2*a*) or descend (Fig. 21.2*b*), depending on the characteristics of the structure and the loading.

For certain structural forms, or when fabricational imperfections are taken into account for the types of perfect structures that are portrayed with the solid lines, the load-displacement behavior follows the paths indicated by the dashed lines, OD. Structures with a rising postbuckling path (AC, Fig. 21.2*a*) will have a strength exceeding the bifurcation load. The load-carrying capacity of a real structure with a descending postbuckling path (Fig. 21.2*b*) will be less than the bifurcation load of the corresponding perfect structure unless the load-displacement path rises again at larger displacements. This maximum load-carrying capacity, or limit point, occurs at point D.

In view of the above, finite-element analysis procedures are needed for at least four different circumstances or load-displacement regimes: (1) general nonlinear analysis for tracing of the load-displacement path, (2) calculation of the bifurcation point, (3) determination of the load-displacement response along a postbuckling path, and (4) the calculation of limit points. There is extensive literature on finite element procedures for all of these, but space permits specific discussion of only (1) and (2) in this chapter; brief remarks are made about (3) and (4). We first consider (2), the calculation of the bifurcation load, because of the simplifications associated with it.

21.3.2 Instability (Bifurcation) Analysis

For many structures of practical interest it is possible to treat the prebuckling load-displacement path as linear, resulting in a linearized instability analysis. To describe the format of this analysis we return to the beam-column example. In the element strain energy expression, Eq. 21.6, we discard the highest-order term $(A/4)(dw/dx)^4$ and note that under the assumption of an independent prebuckling analysis for axial loading the axial load F_x is related to the axial deformation by the linear relationship

$$\lambda F_x = EA\,\frac{du}{dx} \tag{21.17}$$

where λ is a scalar that defines the intensity of the load F_x. F_x is obtained in an analysis for a reference intensity of the applied loads. Equation 21.6 then becomes

$$U^e = \frac{1}{2}\int_L \left[EA\left(\frac{du}{dx}\right)^2 + EI\left(\frac{d^2w}{dx^2}\right)^2 + F_x\left(\frac{dw}{dx}\right)^2\right] dx \tag{21.18}$$

The strain energy is thus reduced to a form in which the axial and flexural strain energies are uncoupled:

$$U^e = U^e_a + U^e_f \tag{21.19}$$

with

$$U^e_a = \frac{1}{2}\int_L EA\left(\frac{du}{dx}\right)^2 dx \tag{21.20}$$

$$U^e_f = \frac{1}{2}\int_L \left[EI\left(\frac{d^2w}{dx^2}\right)^2 + F_x\left(\frac{dw}{dx}\right)^2\right] dx \tag{21.21}$$

The strain energy U_a is the basis for the independent prebuckling analysis for axial behavior. Attention can now be restricted to the flexural behavior, it being assumed that the solution for axial load (or in-plane or membrane forces in the case of plates and shells) is accomplished through an independent analysis.

If the description of w (Eq. 21.8) is substituted into Eq. 21.21, one obtains

$$U^e_f = \frac{\lfloor w\theta \rfloor}{2}\,[k^f_0]\begin{Bmatrix} w \\ \theta \end{Bmatrix} + \lambda \lfloor w\theta \rfloor [\bar{k}_1]\begin{Bmatrix} w \\ \theta \end{Bmatrix} \tag{21.22}$$

where $[k_0^f]$ stems from the first term of the integral of Eq. 21.21 and $[\bar{k}_1]$ forms the second term. Globally, we have after summation of the element strain energies

$$U = \frac{\lfloor w\theta \rfloor}{2} [\bar{K}_0^f] \begin{Bmatrix} w \\ \theta \end{Bmatrix} + \lambda \lfloor w\theta \rfloor [\bar{K}_1] \begin{Bmatrix} w \\ \theta \end{Bmatrix} \tag{21.23}$$

For the more general case of other types of elements we write, analogous to Eq. 21.23,

$$U = \frac{\lfloor \Delta \rfloor}{2} [\bar{K}_0]\{\Delta\} + \lambda \lfloor \Delta \rfloor [\bar{K}_1]\{\Delta\} \tag{21.24}$$

Thus, for other elements the same type of simplifications are introduced into their strain energy expressions as were applied to the beam-column strain energy, Eq. 21.6, resulting in a $[\bar{K}_0^f]$ for flexure alone and $[\bar{K}_1]$ independent of $\{\Delta\}$.

The condition for elastic instability is that the second variation of the above strain energy is zero. This gives the alternative conditions

$$[\bar{K}_0^f + \lambda \bar{K}_1]\{\delta\Delta\} = 0 \tag{21.25}$$

or

$$\|[\bar{K}_0^f + \lambda \bar{K}_1]\| = 0 \tag{21.26}$$

The second condition states that the determinant of $[\bar{K}_0^f + \lambda \bar{K}_1]$ is zero.

The actual calculation of critical loads in finite-element analysis, based on the above, differs from classical instability analysis in that it operates on large-order systems of algebraic equations and therefore involves special techniques. The issues of reliability and computational efficiency are of particular concern.

As illustrated in Fig. 21.2*a*, at bifurcation (Eqs. 21.25 and 21.26) the applied load has reached an intensity such that alternative states of equilibrium are possible: the prebuckling state (OA) and the postbuckled state (AB). The applied load enters into the problem implicitly in the calculation of the internal load distribution, the intensity of which is denoted by the parameter λ. The loads $\{P\}$ play no explicit role in the bifurcation analysis itself, because the bifurcation analysis represents a study of small displacements $\{\delta\Delta\}$ about the bifurcation point.

Equations 21.25 and 21.26 are identical in form to the relationships that prevail in frequency and mode-shape calculation, where solution techniques are well established (Jensen, 1980; Bathe and Wilson, 1973). There are major differences in the desired results, however. In frequency calculations, many roots λ_i and associated displacement vectors $\{\Delta^1\}$ are usually required. In bifurcation analysis the value of one root, the lowest, is sought and on occasion the buckling-mode shape is only of incidental interest. For this reason a popular approach is to find the lowest root of the stability determinant (Eq. 21.26). One such approach, based on the very popular "frontal" method of equation solution, is given by Cedolin and Gallagher (1978). Another scheme, based on finding the lowest root of the determinant through extrapolation, was given by Gallagher (1973a) and has been used by Kiciman and Popov (1978), among others.

Figure 21.2 indicated that in general the bifurcation load is reached following a nonlinear load-displacement path. This path is linear only when flexural behavior is absent in the prebuckling regime. Although the actual behavior can be linearized through neglect of flexural behavior, as was done in the beam-column examples, this may be awkward to do in finite-element analysis because of the irregularity of procedure it introduces. Thus finite-element software (Section 21.8) frequently solves for a nonlinear prebuckling state. The latter is discussed next.

21.3.3 Nonlinear Analysis

Two broad options are open for tracing the nonlinear load-displacement behavior of structures. One option, the Lagrangian formulation, is based on referencing all displacements to the original axes of the structure; the other approach tracks the behavior through a frame of reference that moves with the element and is termed the Eulerian approach. Both have significant advantages. A clear description of these options, including elementary examples, is given in Chapter 13 of the text by Cook (1981).

It is important to recognize that nearly all schemes employed in practical nonlinear analysis are based on a linearization of the stiffness relationships within chosen increments of behavior. Thus, considering the load path *A-F* of Fig. 21.3, one divides this path into load intervals *A-B*, *B-C*, *C-D*, and so on, and beginning at the lower point, seeks to establish the solution at the upper point by one or more linear analyses.

An obvious approach to this task is to form the tangent stiffness at the start of each interval and solve Eq. 21.15 for the change in displacement $\{\delta P\}$. The displacements at the end of the interval are added to the previously accumulated displacements and used to construct a new tangent stiffness. Unfortunately, the numerical solution tends to drift away from the actual solution after a number of intervals and this approach, by itself, is not generally satisfactory.

The most sophisticated approach, within practical limits, is through Newton-Raphson iteration. If the solution obtained in the foregoing process is substituted into the equilibrium equations (Eq. 21.13) the error is represented by a vector of force imbalances $\{R^i\}$, where the superscript i denotes the cycle of the Newton-Raphson iterative process. Actually,

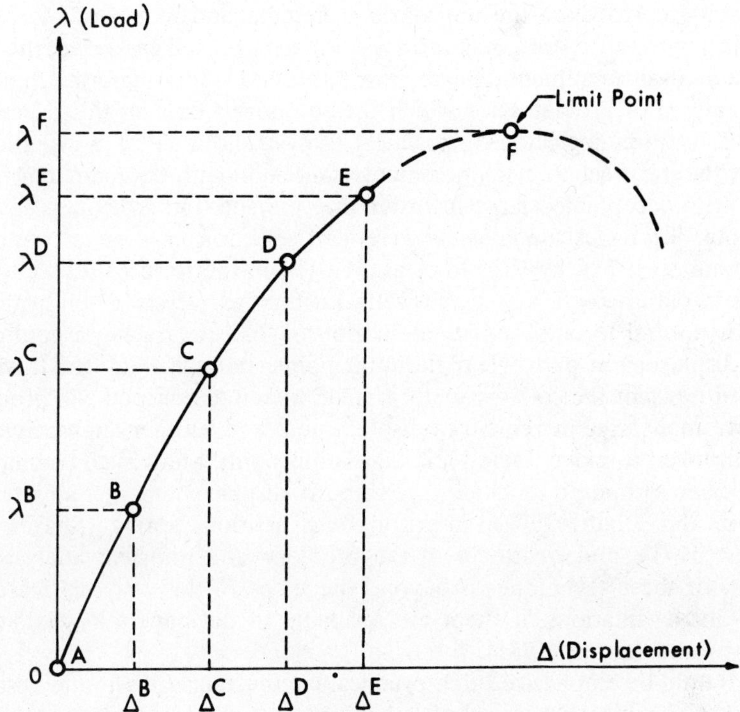

Fig. 21.3 Load-displacement path.

each term on the left side of Eq. 21.13 (e.g., $[K_0]\{\Delta\}$) represents a set of joint forces. Thus calculation of the force imbalance can be accomplished conveniently by evaluation of joint forces. If $[K_T^i]$ denotes the tangent stiffness matrix (Eq. 21.16) that has been updated to account for already calculated displacements, the change in displacement ($\{\delta\Delta^i\}$) due to the force imbalance is obtained by solving

$$[K_T^i]\{\delta\Delta^i\} = \{R^i\} \tag{21.27}$$

The displacement changes $\{\delta\Delta^i\}$ are added to the previously calculated displacements and a new force imbalance is determined. The process above is continued until convergence, to some criterion, is achieved. Such criteria, for nonlinear finite-element analysis, are discussed by Bergan and Clough (1972). There are many variants of the above, known as "modified" Newton-Raphson schemes, as well as extensions of the Newton-Raphson approach. Some of these alternatives are described and reviewed by Mondkar and Powell (1978), Bathe (1976), and Riks (1974).

Having decided on the appropriate element and global relationships and an appropriate solution strategy, it is then necessary to select the size of the analysis increments. Since most published studies describe numerical analyses of problems for which the solution is known, this aspect of analysis is often neglected. In practical analysis, however, it is essential to adopt the most economical increments consistent with obtaining convergence with acceptable error, in order that the solution cost be as low as possible. Strategies for rational increment selection have been proposed by Thomas (1973), by Bergan et al. (1978b) in the form of the "current stiffness parameter," and by Powell and Simons (1981) among others.

It is natural to employ incrementation of loads in tracing a nonlinear load-displacement path. Near the limit point (point F, Fig. 21.3), which might represent the collapse state, a small increment of load will produce an extremely large increment of displacement and another approach must be employed to move toward and across this point. Many such techniques have been proposed, including a shift to displacement incrementation (Batoz and Dhatt, 1979), insertion of a fictitious spring (Sharifi and Popov, 1971), and treatment of the problem as a transient analysis. A review of these techniques is beyond the scope of this chapter. Descriptions and evaluations of them are available in the papers by Gallagher (1973b) and Ramm (1981).

It should be emphasized that practical circumstances do not necessarily evidence the idealized form of behavior pictured in Fig. 21.3 or which underlies many reported analysis. Thus, in practice, the performance of nonlinear analysis must be approached with care and significant numerical experimentation might be required to obtain acceptable analysis performance and solution accuracy.

21.4 PRISMATIC MEMBERS

21.4.1 Flexural Instability

The simplest form of prismatic (one-dimensional) element is one that sustains flexural instability under the action of axial loading (i.e., Euler-column buckling). The governing differential equation for this case can be solved to obtain and expression, in terms of transcendental functions, for the transverse displacement. These "stability functions" can be used as a basis for element formulation for elastic instability analysis. This was the approach taken (Livesley, 1956) in the earliest finite-element analysis of beam and frame flexural instability analysis.

Gallagher and Padlog (1963) formulated the element stiffness equations for the uniform-section element for flexural instability based on a polynomial function for transverse displacement. This expression is the simple cubic which is applicable to simple flexure (see Fig. 21.1 for a description of the element and coordinates).

$$w = \left[1 - 3\left(\frac{x}{L}\right)^2 + 2\left(\frac{x}{L}\right)^3\right]w_1 + \left[3\left(\frac{x}{L}\right)^2 - 2\left(\frac{x}{L}\right)^3\right]w_2$$
$$- x\left(\frac{x}{L} - 1\right)^2 \theta_1 - x\left[\left(\frac{x}{L}\right)^2 - \frac{x}{L}\right]\theta_2 \qquad (21.28)$$

Subsequently many formulations of the same element stiffness matrices have been published (e.g., Martin, 1965). It is the basis for most later studies of special cases, such as tapered elements (Gallagher and Lee, 1970), elements with significant transverse shear deformations (Chugh, 1977), and the phenomena described in the following.

21.4.2 Torsional and Lateral Instability Formulations

Even for straight beams a general capability for elastic instability analysis requires the inclusion of lateral and torsional instability phenomena. This general capability is obviously essential to the analysis of two- and three-dimensional frameworks. The relevant finite-element formulation, based on polynomial displacement functions, was given by Barsoum and Gallagher (1970) and Powell and Klingner (1970). Applications and extensions of the formulation of Barsoum and Gallagher (1970) have appeared in Krajcinovic (1969), Szymczak (1980), Medland (1973), Meek and Swannell (1976), Nethercot and Rockey (1973), Rajasekaran and Murray (1973), and Wang et al. (1977). A formulation based on the transcendental functions that result from the classical solution to the member equations has been presented by Krajcinovic (1969).

Rajasekaran and Murray (1973) extended the work of Barsoum and Gallagher (1970) to include coupled local buckling of wide-flange beam columns by including the deformational modes associated with the distortion of the cross section. Szymczak (1980) considers the behavior of thin-walled I-columns and specializes the developments of Barsoum and Gallagher (1970) to account for a large angle of twist; postbuckling behavior is also studied. Wang et al. (1977) introduce effective-width concepts into a computational procedure for beam buckling which is otherwise based on the formulations of Barsoum and Gallagher (1970). A summary of the finite-element method as applied to elastic-plastic beam-columns, in plane and in space, is given by Chen and Atsuta (1977).

21.5 FRAMEWORK INSTABILITY ANALYSIS

There is a wealth of reported finite-element analyses of the stability of frameworks. This work is concerned predominantly with planar structures (e.g., Halldorsson and Wang, 1968; Remeth, 1979; Bagci, 1980; Beskos, 1977) although a significant number of studies of three-dimensional frameworks have also been reported (Oran, 1973; Jagannathan et al., 1975; Epstein and Murray, 1976; Bathe and Bolourchi, 1979). The references cited differ principally in the paricular structural forms and loadings that are studied and in the algorithms employed for either bifurcation or general nonlinear analysis. They generally use the stiffness relationships for one-dimensional elements based on polynomial functions, discussed in the preceding section, although a few use stiffness matrices based on the stability functions (e.g., Halldorsson and Wang, (1968). Beskos (1977) has performed numerical studies of a variety of structures to assess the difference between these two approaches. His results show that the computer times are always less for frameworks represented by polynomial-based one-element members, but that there can be problems of solution accuracy for highly braced structures. Epstein and Murray (1976) employ a polynomial-based formulation that includes consideration of the distortion of the cross sections of thin-walled members.

21.6 FLAT PLATES

21.6.1 Unstiffened Flat Plates

The formulation of terms to account for elastic instability and the finite displacement of thin flat-plate elements follows the same lines as for one-dimensional elements and presents no special difficulties. One major problem arises, however, due to the absence of general agreement about the most effective form of the element stiffness relationships.

Flat-plate buckling and the more general problem of the nonlinear analysis of flat plates by the finite-element method are governed by expressions which are simple extensions of the relationships for one-dimensional members. Thus the potential energy expression for linearized elastic instability is now written in the form

$$\Pi_p = \frac{1}{2}\int_A \lfloor k_f \rfloor [E_f] \{\kappa_f\}\, dA + \frac{1}{2}\int_A N_x \left(\frac{\partial w}{\partial x}\right)^2 dA + \frac{1}{2}\int_A N_y \left(\frac{\partial w}{\partial y}\right)^2 dA + \int_A N_{xy}\left(\frac{\partial w}{\partial x}\right)\frac{\partial w}{\partial y}\, dA \qquad (21.29)$$

Here N_x, N_y, and N_{xy} are the known or otherwise calculated forces per unit length in the plate middle surface,

$$\lfloor \kappa_f \rfloor = \left[\frac{\partial^2 w}{\partial x^2} \quad \frac{\partial^2 w}{\partial y^2} \quad 2\frac{\partial^2 w}{\partial x \partial y}\right]$$

and $[E_f]$ contains the plate flexural stiffness and material constants.

The first integral in Eq. 21.29 is the usual expression for the flexural strain energy of thin plates and is the basis for the linear, stable stiffness matrix $[K_0]$. The presence of second derivatives introduces the requirement that certain first derivatives, in particular the normal slope, be continuous across element interfaces. The next three integrals represent the instability effects and are the basis for the initial stress stiffness matrix $[\bar{K}_1]$ (see Eq. 21.22). The extension of this formulation to geometrically nonlinear behavior leads to expressions that correspond to von Kármán's equations (Timoshenko and Woinowsky-Krieger, 1959, pp. 415–420) and to the inclusion of the terms for the initial displacement stiffness matrix.

The difficulty in the finite-element instability analysis of flat plates is in the choice of expressions that describe the transverse displacement (w) of the individual elements. This is in contrast to one-dimensional elements, where the cubic polynomial is the customary choice for flexural behavior and leads to satisfactory results, and where equally satisfactory choices can be made in the presence of torsion and other phenomena. In two dimensions the definition of functions that satisfy continuity of not only the transverse displacement itself, but also of slope continuity, has been the central problem.

This problem has not been difficult to solve for rectangles. The full product of the cubic function in the two perpendicular directions gives a 16-term "bicubic" function in x and y. Considering a typical rectangle (Fig. 21.4), we see that at the four vertices there are three obvious displacement (w, $\partial w/\partial x$, $\partial w/\partial y$), for a total of 12. To make up the total of 16 terms the usual approach taken is to introduce the twist derivative ($\partial^2 w/\partial x\, \partial y$) as a degree of freedom at each vertex. This gives accurate results for test cases of buckling and nonlinear analysis (see Carson and Newton, 1969; Yang, 1972; Anderson et al., 1968), but the twist derivative is awkward to handle in many practical situations.

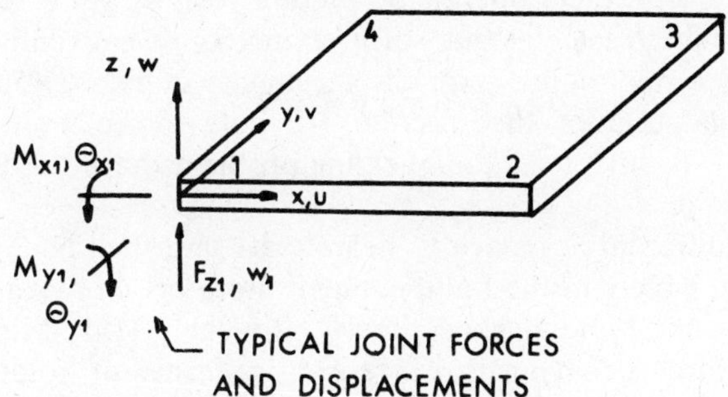

Fig. 21.4 Rectangular plate bending finite element.

The rectangle does not give the generality in geometric representation that one expects of the finite-element method. The triangle is the desired shape of element for geometric representation of practical structures, but the identification of displacement descriptions that are continuous across element interfaces, especially in the slope, has not been easy. A quintic (fifth-order) polynomial serves this purpose, but it contains 21 terms and the degrees of freedom of the corresponding element include all of the second derivatives. This element is rarely used in practical computations. Fourth-order complete polynomials lack continuity of slope across element interfaces and also require many (15) degrees of freedom.

Representation by means of third-order polynomials appears desirable because of the correspondence with beam flexure. It would also be desirable to represent the element in terms of a transverse displacement and two rotations at each vertex—9 degrees of freedom. A complete polynomial has 10 terms, which seemingly presents no difficulties because the tenth term can be eliminated through algebraic operations, but it is unsatisfactory due to a severe lack of slope continuity across element interfaces. Many special schemes have been proposed to define approximating functions of third order, and the most popular of these have been due to Bazeley, Cheung, Irons, and Zienkiewicz [for details, see Zienkiewicz (1977)], a modification of it proposed by Razzague (1973), and a formulation by Clough and Tucker (1965). Plate-buckling applications for these formulations are given by Vos and Vann (1973).

More recently, the entire issue of formulations for triangular bending elements was given an extensive assessment by Batoz et al. (1980). They found that the elements of Zienkiewicz (1977) and Clough and Tucker (1965) sustain a severe loss of accuracy, compared to equilateral triangles or 45° right triangles, when they are elongated. They also demonstrated that simple, reliable formulations for the triangle in bending can be established by use of "discrete Kirchoff" concepts. A description of these concepts is beyond the scope of the present review, but it suffices to say that they result in an element stiffness matrix of conventional format, expressed in terms of the joint displacements of Fig. 21.5. Indeed, an explicit formulation of the discrete Kirchoff formulation is available (Batoz 1982). Results for plate-buckling problems have not yet appeared for this scheme.

A rather different approach to beam plate and shell bending behavior has appeared widely in the finite-element literature in recent years. This invokes Reissner/Mindlin plate theory (Reissner, 1945; Mindlin, 1951), wherein flexural behavior is expressed in terms of rotations of the normals to the midplane and terms are included which account for transverse shear deformation. The latter are written in terms of the

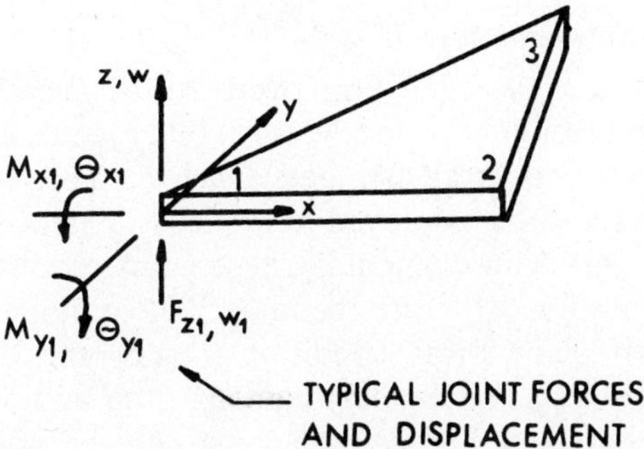

Fig. 21.5 Triangular plate bending finite element.

transverse displacement. A detailed account and background of this approach, encompassing buckling and large-displacement solutions, has been published by Pica and Wood (1980). In general, such formulations do not converge to the correct solutions in the case of extreme thinness; the source of this difficulty is explained by Prathap and Bhashyam (1982). A resolution of the problem has been established for thin rectangular elements by Hughes and Cohen (1979). It appears that most Reissner-Mindlin formulations, for practical width-to-thickness ratios, are not adversely affected.

21.6.2 Stiffened Flat Plates

The most efficient form of construction of flat plates to resist buckling loads often involves stiffeners. The finite-element method would appear to be ideally suited to this class of problem, because of its facility in the combination of different types of elements. Yet the direct combination of plate-bending elements and beam elements (for the stiffeners) is not always convenient. The beam element is based on a cubic polynomial, but as explained in the preceding section, not all of the plate-bending elements are based on cubics. When the stiffener and plate element displacements are consistent, care must be exercised in accounting for the eccentricity of the axis of the stiffeners and the middle surface of the plate (Miller, 1980).

Finite-element stiffened-plate buckling analyses are reported by Crisfield (1976), Allman (1975), and Hagedoorn (1973). Hagedoorn (1973) gives a thorough review of the topic, details of representation of stiffener eccentricity, and numerical comparisons of the finite-element approach, clasical solutions for discretely stiffened plates, and solutions for "smeared" representation of the stiffeners.

21.6.3 Flat-Plate Assemblies

Flat-plate elements have been employed as the basic components of finite-element representations in the instability analysis of rolled and formed thin-walled structural members, such as I-beams and channels. Plate elements can also be employed to model stiffeners, an alternative to modeling them with beam elements. These cases require that the finite-element representation of plate bending be combined with element representations of plane-stress conditions. Since the overall structural forms usually consist of rectangular segments, the finite elements can be rectangular and there is no difficulty in constructing a satisfactory analytical model with available element formulations. Akay et al. (1977) describe analyses of thin-walled beams and frames modeled entirely with triangular flat-plate elements. Some background on finite-element representations of similar problems is found in their work.

An approach related to the finite-element method known, as the "finite strip method," has also been employed in this class of problem (Kleiber and Zacharski, 1982; Haucock, 1978; Plank and Wittrick, 1974). This approach involves the use of a series representation or classical solution in one direction and a finite-element discretization in the other direction. The former reduces the number of required elements and therefore the cost of solution.

21.7 SHELL BUCKLING

Three distinct popular approaches to finite-element analysis of curved thin-shell structures can be identified: (1) by means of three-dimensional (solid) elements with curved boundaries, (2) in "faceted" form with flat elements, and (3) via elements formulated on the basis of curved shell theory.

An appeal of solid elements is that they stem directly from three-dimensional theory of elasticity. The curved boundaries are easily defined through use of "isoparametric" curvilinear coordinates, which employ the same form of description for both geometry and displacement. Conditions of the Love-Kirchoff type are imposed, but this is insufficient to achieve economical solutions; it is necessary either to introduce certain supplementary displacement modes to describe bending action (Wood, 1973) or to deemphasize the representation of shear deformation by approximation (reduced numerical integration) of the shear strain energy (Hughes et al., 1981). There has not been sufficient development of the nonlinear analysis side of this approach to measure its effectiveness in thin-shell stability analysis.

The use of flat-plate elements is appealing in its simplicity. Among the objections to this approach, however, are the following: (1) it excludes coupling of stretching and bending within the element; (2) only triangular elements can be employed when general shells are to be handled; (3) difficulty of treating junctions where all elements meeting at the point are coplanar; (4) the presence of "discontinuity" bending moments at element juncture lines, which do not appear in the continuously curved actual structure; and for stability analysis, (5) the influence of the geometric approximation upon the solution for imperfection-sensitive structures. We might add the problems of definition of displacement fields which meet interelement continuity conditions, discussed already for flat elements, but these are even more serious for curved elements.

Flat finite elements represent the behavior of curved structures in the limit and the errors due to (1) can be made small by use of a refined element network. Nearly all finite-element programs suitable for stability analysis will contain triangles, so that (2) is not a serious objection. In item (3), coplanarity, a null stiffness corresponding to rotation about the axis normal to the plane will be present, but steps can always be taken to identify and resolve the problem. Use of flat elements was one of the earliest approaches in finite-element thin-shell instability analysis (Gallagher et al., 1967) and it has been the basis of more recent investigations (Horrigmoe and Bergan, 1978; Argyris et al., 1977).

The formulation and practical utilization of curved thin-shell finite elements raises a number of significant issues. Adequate means must be found for input of the structural and element geometry in curvilinear coordinates. It is difficult to establish a general procedure, although advances in interactive computer graphics give a comprehensive solution when the necessary equipment is available (Haber and Abel, 1982). A satisfactory shell theory must be chosen; this has posed unresolved questions until relatively recently. Proper accounting for nonlinear terms is of particlar importance (Batoz, 1979). Moreover, there are conditions that are basic to finite-element modeling to be satisfied in the choice of approximating functions that have proved difficult to meet. One challenging problem, for example, is to devise assumed displacement functions which give zero strain under the rigid body motion of the element. Also, in the presence of nonlinear strain-displacement relationships, the assumed displacements of linear flat-plate elements will generally not meet the conditions for convergence of finite-element solutions (Dawe, 1971; Allman, 1982). Such functions are widely used as the basis for curved elements.

Curved thin-shell element formulations up through 1976 are reviewed by Gallagher (1975b). New developments have appeared since then and others can be expected to appear in the years to come.

Many of the published thin-shell elements have been formulated to account for elastic instability and geometrically nonlinear behavior and applied to a wide variety of shell buckling problems. Gallagher (1973a) reviews work through 1973. Subsequent contributions include works by Kanodia et al. (1977) and Matsui and Matsuoka (1976). A number of widely distributed finite-element programs also contain curved thin-shell elements which have been applied extensively to practical problems of shell instability.

A special type of thin-shell finite element is the axisymmetric element, which is useful for the analysis of shells of revolution. Application of this type of element to buckling problems is described by Venkateswara Rao et al. (1974).

21.8 FINITE-ELEMENT SOFTWARE

Practitioners today depend, for the most part, on "packaged" finite-element programs for linear design analysis, rather than on programs that have been developed in-house. Unquestionably, there is even greater dependence on the former when the design analysis problem involves elastic instability and/or geometrically nonlinear analysis. Existing software represents an extensive capability for the types of analyses discussed in this chapter.

Noor (1981) has conducted a comprehensive study of existing software for nonlinear analysis. It emphasizes programs that are well documented and maintained, extensively tested, and widely available. Detailed information is given regarding program capabilities. Another source of information is the Structural Mechanics Software Series, a series of books published by the University of Virginia Press (e.g., Perrone and Pilkey, 1981). These contain state-of-the-art reviews of software in many different aspects of structural mechanics, of format similar to that of Noor (1981).

References

Akay, H., Johnson, C. P., and Will, K. M. (1977), "Local and Lateral Buckling of Beams and Frames," *ASCE J. Struct. Div.*, Vol. 103, No. ST9, pp. 1821–1832.

Allman, D. (1975), "Calculation of the Elastic Buckling Loads of Thin Flat Reinforced Plates Using Triangular Finite Elements," *Int. J. Numer. Methods Eng.*, Vol. 9, pp. 415–432.

Allman, D. J. (1982), "Improved Finite Element Model for Large Displacement Bending and Postbuckling Analysis of Thin Plates," *Int. J. Solids Struct.*, Vol. 18, No. 9, pp. 737–762.

Anderson, R. G., Irons, B., and Zienkiewicz, O. C. (1968), "Vibration and Stability of Plates Using Finite Elements," *Int. J. Solids Struct.*, Vol. 4, pp. 1031–1055.

Argyris, J., Dunne, P. C., Malejannakis, G. A., and Schelke, E. (1977), "A Simple Triangular Facet Shell Element with Applications to Linear and Non-linear Equilibrium and Elasstic Stability Problems," *Comput. Methods Appl. Mech. Eng.*, Vol. 11, pp. 97–131.

Bagci, C. (1980), "Elastic Stability and Buckling Loads of Multispan Nonuniform Beams, Shafts and Frames on Rigid or Elastic Support Using Planar Uniform Line Elements," *Comput. Struct.*, Vol. 12, pp. 233–243.

Barsoum, R. S., and Gallagher, R. H. (1970), "Finite Element Analysis of Torsional and Lateral Stability Problems," *Int. J. Numer. Methods Eng.*, Vol. 2, No. 3, pp. 335–352.

Bathe, K.-J. (1976), "An Assessment of Current Finite Element Analysis of Nonlinear Problems in Structural Mechanics," in *Numerical Solution of Partial Differential Equations—III* (Ed. B. Hubbard), Academic Press, New York.

Bathe, K.-J., and Bolourchi, S. (1979), "Large Displacement Analysis of Three-Dimensional Beam Structures," *Int. J. Numer. Methods Eng.*, Vol. 14, pp. 961–986.

Bathe, K.-J., and Wilson, E. L. (1973), "Solution Methods for Eigenvalue Problems in Structural Mechanics," *Int. J. Numer. Methods Eng.*, Vol. 6, pp. 213–226.

Batoz, J. L. (1979), "Curved Finite Elements and Shell Theories with Particular Reference to the Buckling of a Circular Arch," *Int. J. Numer. Methods Eng.*, Vol. 14, pp. 774–779.

Batoz, J. L. (1982), "An Explicit Formulation for an Efficient Triangular Plate-Bending Element," *Int. J. Numer. Methods Eng.*, Vol. 18, pp. 1077–1090.

Batoz, J. L., and Dhatt, G. (1979), "Incremental Displacement Algorithms for Nonlinear Problems," *Int. J. Numer. Methods Eng.*, Vol. 14, pp. 1262–1267.

Batoz, J. L., Bathe, K.-J., and Ho, L. W. (1980), "A Study of Three-Node Triangular Plate Bending Elements," *Int. J. Numer. Methods Eng.*, Vol. 15, pp. 1771–1812.

Bergan, P., and Clough, R. (1972), "Convergence Criteria for Iterative Processes," *AIAA J.*, Vol. 10, pp. 1107–1108.

Bergan, P., et al., (eds.) (1978a), *Finite Elements in Nonlinear Mechanics*, Tapir, Trondheim, Norway.

Bergan, P. G., Horrigmoe, G., Krakelan, B., and Soriede, T. (1978b), "Solution Techniques for Nonlinear Finite Element Problems," *Int. J. Numer. Methods Eng.*, Vol. 12, pp. 1677–1696.

Beskos, D. (1977), "Framework Stability by Finite Element Methods," *ASCE J. Struct. Div.*, Vol. 103, No. ST11, pp. 2273–2278.

Carson, W., and Newton, R. E. (1969), "Plate Buckling Analysis Using a Fully Compatible Finite Element," *AIAA J.*, Vol. 7, No. 3, pp. 527–529.

Cedolin, L., and Gallagher, R. H. (1978), "Frontal-Based Solver for Frequency Analysis," *Int. J. Numer. Methods Eng.*, Vol. 12, pp. 1659–1666.

Chen, W. F., and Atsuta, A. (1977), *Theory of Beam-Columns*, Vol. 2, *Space Behavior and Design*, McGraw-Hill, New York.

Chugh, A. K. (1977), Stiffness Matrix for a Beam Element Including Transverse Shear and Axial Force Effects," *Int. J. Numer. Methods Eng.*, Vol. 11, pp. 1681–1697.

Clough, R. W., and Tucker, J. L. (1965), "Finite Element Stiffness Matrices for Analysis of Plates in Bending," *Proc. Conf. Matrix Methods Struct. Mech.*, AFFDL TR 66-80, Wright-Patterson AFB, Ohio.

Cook, R. D. (1981), *Concepts and Applications of Finite Element Analysis*, 2nd ed., Wiley, New York.

Crisfield, M. A. (1976), "Large Deflection Elasto-Plastic Buckling Analysis of Eccentrically-Stiffened Plates Using Finite Elements," *Transp. Road R & S Lab. Rep. 725*.

Dawe, D. J. (1971), "A Finite Deflection Analysis of Shallow Arches by the Discrete Element Method," *Int. J. Numer. Methods Eng.*, Vol. 3, pp. 529–552.

Ebner, A., and Uciffero, J. (1972), "Theoretical Numerical Comparison of Elastic Nonlinear Finite Element Methods," *Comput. Struct.*, Vol. 2, pp. 1043–1061.

Epstein, M., and Murray, D. W. (1976), "Three-Dimensional Large Deformation Analysis of Thin-Walled Beams," *Int. J. Solids Struct.*, Vol. 12, pp. 867–876.

Gallagher, R. H. (1973a), "The Finite Element Method in Shell Stability Analysis," *Comput. Struct.*, Vol. 3, pp. 543–547.

Gallagher, R. H. (1973b), "Finite Element Analysis of Geometrically Nonlinear Problems," in *Theory and Practice in Finite Element Structural Analysis* (eds. Y. Yamada and R. Gallagher), University of Tokyo Press, Tokyo, pp. 109–124.

Gallagher, R. H. (1975a), *Finite Element Analysis; Fundamentals*, Prentice-Hall, Englewood Cliffs, N.J.

Gallagher, R. H. (1975b), "Shell Elements," *Proc. 1st World Conf. Finite Elements*, Bournemouth, England.

Gallagher, R. H., and Lee, B. (1970), "Matrix Dynamic and Instability Analysis with Nonuniform Elements," *Int. J. Numer. Methods Eng.*, Vol. 2, pp. 265–275.

Gallagher, R. H., and Padlog, J. (1963), "Discrete Element Approach to Structural Stability Analysis," *AIAA J.*, Vol. 1, No. 6, pp. 1437–1439.

Gallagher, R. H., Gellatly, R. A., Mallett, R. H., and Padlog, J. (1967), "A Discrete-Element Procedure for Thin Shell Instability Analysis," *AIAA J.*, Vol. 5, No. 1, pp. 138–144.

Haber, R., and Abel, J. F. (1982), "Discrete Transfinite Mappings for the Description and Meshing of Three-Dimensional Surfaces Using Interactive Computer Graphics," *Int. J. Numer. Methods Eng.*, Vol. 18, pp. 41–66.

Hagedoorn, A. H. (1973), "An Elastic Finite Element Analysis of the Buckling of Eccentrically Stiffened Plates and Shells," Ph.D. dissertation, Department of Structural Engineering, Cornell University, Ithaca, N.Y.

Halldorsson, O. P., and Wang, C. K. (1968), "Stability Analysis of Frameworks by Matrix Methods," *ASCE J. Struct. Div.*, Vol. 94, ST7, pp. 1745–1760.

Hancock, G. J. (1978), "Local, Distortional and Lateral Buckling of I-Beams," *ASCE J. Struct. Div.*, Vol. 104, No. ST11, pp. 1787–1798.

Hibbitt, H. D. (1979), "Some Follower Forces and Load Stiffness," *Int. J. Numer. Methods Eng.*, Vol. 14, pp. 937–941.

Horrigmoe, G., and Bergan, P. (1978), "Nonlinear Analysis of Free Form Flat Shells by Flat Finite Elements," *Comput. Methods Appl. Mech. Eng.*, Vol. 16, pp. 11–35.

Hughes, T. J. R., and Cohen, M. (1979), "The 'Heterosis' Finite Element for Plate Bending," *Comput. Struct.*, Vol. 10, pp. 445–450.

Hughes, T. J. R., Kanok-Nukulchai, W., and Taylor, R. L. (1981), "A Large Deformation Formulation for Shell Analysis by the Finite Element Method," *Comput. Struct.*, Vol. 13, pp. 19–30.

Jagannathan, D. S., Epstein, H. K., and Christiano, P. (1975), "Nonlinear Analysis of Reticulated Spaced Trusses," *ASCE J. Struct. Div.*, Vol. 101, No. ST12, pp. 2041–2658.

Jensen, P. S. (1980), "Eigenvector Algorithms for Structural Analysis," in *Pressure Vessels and Piping: Design Technology, 1980. A Decade of Progress* (eds. S. Zamrik and D. Dietrich), American Society of Mechanical Engineers, New York.

Kanodia, V., Gallagher, R. H., and Mang, H. (1977), "Instability Analysis of Torispherical Pressure Vessel Heads with Triangular Thin-Shell Finite Elements," *Trans. ASME J. Press. Vess. Tech.*, Vol. 99, pp. 64–74.

Kiciman, O., and Popov, E. (1978), "Post-Buckling Analysis of Cylindrical Shells," *ASCE J. of the Eng. Mech. Div.*, Vol. 104, No. EM4, pp. 751–762.

Kleiber, M., and Zacharski, A. (1982), "Numerical Analysis of Local Instabilities in Elastic and Elasto-Plastic Prismatic Plate Assemblies," *Comput. Methods Appl. Mech. Eng.*, Vol. 31, No. 2.

Krajcinvoic, D. (1969), "A Consistent Discrete Element Technique for Thin-Walled Assemblages," *Int. J. Solids Struct.*, Vol. 5, pp. 639–662.

Livesley, R. K. (1956), "The Application of an Electronic Digital Computer to Some Problems of Structural Analysis," *Struct. Eng.*, Vol. 34, No. 1, pp. 1–12.

Loganathan, K., Chang, S., Gallagher, R. H., and Abel, J. F. (1979), "Finite Element Representation and Pressure Stiffness in Shell Stability Analysis," *Int. J. Numer. Methods Eng.*, Vol. 14, pp. 1413–1420.

Martin, H. C. (1965), "On the Derivation of Stiffness Matrices for the Analysis of Large Deflection and Stability Problems," *Proc. Conf. Matrix Methods Struct. Mech.*, AFFDL TR 66-80, Wright-Patterson AFB, Ohio.

Matsui, T., and Matsuoka, O. (1976), "A New Finite Element Scheme for Instability Analysis of Thin Shells," *Int. J. Numer. Methods Eng.*, Vol. 10, pp. 145–170.

Medland, I. C. (1973), "The Buckling, Bracing and Strength of I-Section Continuous Members," *Proc. 4th Australasian Conf. Mech. Struct. Mater.*, Brisbane, pp. 166–173.

Meek, J. L., and Swannell, P. (1976), "Stiffness Matrices for Beam Members Including Warping Torsion Effects," *ASCE J. Eng. Mech. Div.*, Vol. 102, No. EM1, pp. 193–197.

Miller, R. E. (1980), "Reduction of Error in Eccentric Beam Modelling," *Int. J. Numer. Methods Eng.*, Vol. 15, pp. 575–582.

Mindlin, R. D. (1951), "Influence of Rotatory Inertia and Shear on a Flexural Motion of Isotropic Elastic Plates," *J. Appl. Mech.*, Vol. 18, pp. 31–38.

Mondkar, D. P., and Powell, G. H. (1978), "Evaluation of Solution Schemes for Nonlinear Structures," *Comput. Struct.*, Vol. 9, pp. 223–236.

Murray, D. W., and Rajasekaran, S. (1975), "Technique for Formulating Beam Equations," *ASCE J. Eng. Mech. Div.*, Vol. 101, No. EM5, pp. 561–573.

Nethercot, D., and Rockey, K. C. (1973), "Lateral Buckling with Mixed End Conditions," *Struct. Eng.*, Vol. 51, No. 4.

Noor, A. K. (1981), "Survey of Computer Programs for Solution of Nonlinear Structural and Solid Mechanics Problems," *Comput. Struct.*, Vol. 13, pp. 425–465.

Oran, C. (1973), "Tangent Stiffness in Space Frames," *ASCE J. Struct. Div.*, Vol. 99, No. ST6, pp. 981–1001.

Perrone, N., and Pilkey, W., eds. (1981), *Structural Mechanics Software Series*, Vol. IV, University Press of Virginia, Charlottesville, Va.

Pica, A., and Wood, R. D. (1980), "Postbuckling Behavior of Plates and Shells Using a Mindlin Shallow Shell Formulation," *Comput. Struct.*, Vol. 12, pp. 759–768.

Plank R., and Wittrick, W. (1974), "Buckling under Combined Loading of Thin-Flat-Walled Structures by a Complex Finite Strip Method," *Int. J. Numer. Methods Eng.*, Vol. 8, pp. 323–334.

Powell, G., and Klingner, R. (1970), "Elastic Lateral Buckling of Steel Beams," *ASCE J. Struct. Div.*, Vol. 96, No. ST9, pp. 1919–1929.

Powell, G. H., and Simons, J. (1981), "Improved Iteration Strategy for Nonlinear Structures," *Int. J. Numer. Methods Eng.*, Vol. 17, pp. 1455–1467.

Prathap, G., and Bhashyam, G. R. (1982), "Reduced Integration and the Shear-Flexible Element," *Int. J. Numer. Methods Eng.*, Vol. 18, No. 2, pp. 195–210.

Przemieniecki, J. S. (1968), "Discrete-Element Methods for Stability Analysis," *Aeronaut. J.*, Vol. 72, No. 12, pp. 1077–1086.

Rajasekaran, S., and Murray, D. W. (1973), "Coupled Local Buckling in Wide-Flange Beam Columns," *ASCE J. Strct. Div.*, Vol. 99, pp. 1003–1023.

Ramm, E. (1981), "Strategies for Tracing the Nonlinear Response Near Limit Points," in *Nonlinear Finite Element Analysis in Structural Mechanics* (eds. W. Wunderlich, E. Stein, and K.-J. Bathe), Springer-Verlag, New York, pp. 63–89.

Razzague, A. (1973), "Program for Triangular Bending Element with Derivative Smoothing," *Int. J. Numer. Methods Eng.*, Vol. 5, pp. 588–589.

Reissner, E. (1945), "Effect of Transverse Shear Deformation on the Bending of Elastic Plates," *J. Appl. Mech. ASME*, Vol. 67, pp. A-69–A-77.

Remeth, S. (1979), "Nonlinear Static and Dynamic Analysis of Framed Structures," *Comput. Struct.*, Vol. 10, pp. 879–897.

Riks, E. (1979), "An Incremental Approach to the Solution of Snapping and Buckling Problems," *Int. J. Solids Struct.*, Vol. 15, pp. 529–551.

Sharifi, P., and Popov, E. (1971), "Nonlinear Buckling Analysis of Sandwich Arches," *ASCE J. Eng. Mech. Div.*, Vol. 97, No. EM5, pp. 1397–1412.

Szymczak, C. (1980), "Buckling and Initial Post-Buckling Behavior of Thin-Walled I-Columns," *Comput. Struct.* Vol. 11, pp. 481–487.

Thomas, J. M. (1970), "A Finite Element Approach to the Structural Instability of Beam Columns, Frames, and Arches," *NASA Tech. Note No. D-5782.*

Thomas, G. R. (1973), "A Variable Step Incremental Procedure," *Int. J. Numer. Methods Eng.*, Vol. 7, pp. 563–566.

Timoshenko, S., and Woinowsky-Krieger, S. (1959), *Theory of Plates and Shells*, 2nd ed., McGraw-Hill, New York.

Venkateswara Rao, G., Radhamohan, S., and Raju, I. S. (1974), "Reinvestigation of Buckling of Shells of Revolution by a Refined Finite Element," *AIAA J.*, Vol. 12, No. 1, pp. 100–101.

Vos, R., and Vann, W. (1973), "A Finite Element Tensor Approach to Plate Buckling and Postbuckling," *Int. J. Numer. Methods Eng.*, Vol. 5, pp. 351–366.

Wang, S. T., Yost, M., and Tien, Y. (1977), "Lateral Buckling of Locally Buckled Beams Using Finite Element Techniques," *Comput. Struct.*, Vol. 7, pp. 469–473.

Wood, R. D. (1973), "The Application of Finite Element Methods to Geometrically Nonlinear Finite Element Analysis," Ph.D. thesis, University of Wales, Swansea.

Wood, R. D., and Schreffler, B. (1978), "Geometrically Nonlinear Analysis—A Correlation of Finite Element Notations," *Int. J. Numer. Methods Eng.*, Vol. 12, No. 4, pp. 635–642.

Wunderlich, W., Stein, E., and Bathe, K.-J., eds. (1981), *Nonlinear Finite Element Analysis in Structural Mechanics*, Springer-Verlag, Berlin.

Yang, H. T. Y. (1972), "Finite Displacement Plate Flexure by Use of Matrix Incremental Approach," *Int. J. Numer. Methods Eng.*, Vol. 4, pp. 415–432.

Zienkiewicz, O. C. (1977), *The Finite Element Method*, 3rd ed., McGraw-Hill, New York, pp. 241–244.

Appendix A

General References on Structural Stability

Allen, H. G., and Bulson, P. S., (1980), *Background to Buckling*, McGraw-Hill, Maidenhead, Berkshire, England.

American Society of Civil Engineers (1971), "Plastic Design in Steel—A Guide and a Commentary," *ASCE Man. Rep. Eng. Pract.*

American Society for Testing and Materials (1967), "Test Methods for Compression Members," *ASTM Spec. Publ. No.* 419, ASTM, Philadelphia.

Baker, J. F., Horne, M. R., and Heyman, J. (1956), *The Steel Skeleton*, Vol. II, *Plastic Behavior and Design*, Cambridge University Press, Cambridge.

Ballio, G., and Mazzolani, F. M. (1983), *Theory and Design of Steel Structures*, Chapman & Hall, London.

Bleich, F. (1952), *Buckling Strength of Metal Structures*, Engineering Societies Monograph, McGraw-Hill, New York, (prepared in collaboration with Column Research Council).

Brockenbrough, R. L., and Johnston, B. G. (1968), *USS Steel Design Manual*, U.S. Steel Corporation, Pittsburgh, Pa.

Brush, D. O., and Almroth, B. O. (1975), *Buckling of Bars, Plates, and Shells*, McGraw-Hill, New York.

Burgermeister, G., and Steup, H. (1957), *Stabilitätstheorie*, Akademie-Verlag, East Berlin, (in German).

Chajes, A. (1974), *Principles of Structural Stability Theory*, Prentice-Hall, Englewood Cliffs, N.J.

Chen, W. F., and Atsuta, T. (1977), *Theory of Beam-Columns*, Vols. 1 and 2, McGraw-Hill, New York.

Chen, W. F., and Han, D. J. (1985), *Tubular Members in Offshore Structures*, Pitman Advanced Publishing Program, Boston.

Chilver, A. H., ed. (1967), *Thin-Walled Structures*, Wiley, New York.

Column Research Committee of Japan (1971), *Handbook of Structural Stability*, Corona Publishing Co., Tokyo (in English).

Column Research Council (1962), "Column Research Council Symposium on Metal Compression Members," *ASCE Trans.*, Vol. 127.

Eckhaus, W. (1965), *Studies in Nonlinear Stability Theory*, Springer-Verlag, New York.

Galambos, T. V. (1968), *Structural Members and Frames*, Prentice-Hall, Englewood Cliffs, N.J.

Gallagher, R. H. (1975), *Finite Element Analysis Fundamentals*, Prentice-Hall, Englewood Cliffs, N.J.

Gerard, G. (1962), *Introduction to Structural Stability Theory*, McGraw-Hill, New York.

Hetenyi, M. (1946), *Beams on Elastic Foundations*, University of Michigan Press, Ann Arbor, Mich.

Hoff, N. J. (1956), *The Analysis of Structures*, Wiley, New York.

Johnston, B. G., (1981), "Selected Papers," FERS and SSRC, Bethlehem Pa.

Johnston, B. G., ed. (1976), *Guide to Stability Design Criteria for Metal Structures*, 3rd ed., Wiley, New York.

Kollbrunner, C. F., and Meister, M. (1961), *Knicken, Biegedrillknicken, Kippen*, Springer-Verlag, Berlin.

Langhaar, H. L. (1962), *Energy Methods in Applied Mechanics*, Wiley, New York.

Massonnet, C. E. (1976), "Forty Years of Research on Beam-Columns in Steel," *SM Arch.*, Vol. 1, Issue 1, Noordhoff International Publishing, Leyden, The Netherlands.

Mazzolani, F. M. (1985), *Aluminum Alloy Structures*, Pitman Advanced Publishing Program, Boston.

McGuire, W. (1968), *Steel Structures*, Prentice-Hall, Englewood Cliffs, N.J.

Narayanan, R., ed. (1982), *Axially Compressed Structures*, Applied Science Publishers, Barking, Essex, England.

Narayanan, R., ed. (1983), *Beams and Beam-Columns*, Applied Science Publishers, Barking, Essex, England.

Narayanan, R., ed. (1985), *Shell Structures*, Elsevier Applied Science Publishers, Barking, Essex, England.

Roorda, J. (1980), *Buckling of Elastic Structures*, University of Waterloo, Waterloo, Ontario.

Salmon, C. G., and Johnson, J. E. (1980), *Steel Structures, Design and Behavior*, Harper & Row, New York.

Simitses, G. J. (1976), *An Introduction to the Elastic Stability of Structures*, Prentice-Hall, Englewood Cliffs, N.J.

Structural Stability Research Council (1982), *Stability of Metal Structures—A World View*, American Institute of Steel Construction, Chicago.

Supple, W. J. (1973), *Structural Instability—Fundamentals of Post-buckling Behavior of Structures*, IPC Science and Technology Press Ltd., Guilford, England.

Tall, L., ed. (1974), *Structural Steel Design*, Ronald Press, New York.

Timoshenko, S. P. (1983/1953), *History of Strength of Materials*, Dover, New York, 1983; McGraw-Hill, New York. 1953.

Timoshenko, S. P. and Gere, J. M. (1961), *Theory of Elastic Stability*, Engineering Societies Monograph, McGraw Hill, New York.

Trahair, N. S. (1977), *The Behavior and Design of Steel Structures*, Chapman & Hall, London.

Vlasov, V. Z. (1961), *Thin-Walled Elastic Beams*, Israel Program for Scientific Translation, for NSF, Jerusalem, (in English; original Russian edition, 1959).

Winter, G. (1975), *The Collected Papers of George Winter*, Cornell University, Ithaca, N.Y.

Yu, W. W. (1985), *Cold-Formed Steel Design*, Wiley, New York.

Ziegler, H. (1968), *Principles of Structural Stability*, Blaisdell, Waltham, Mass.

Appendix B

Technical Memorandums of Structural Stability Research Council

B.1 Technical Memorandum No. 1: The Basic Column Formula*

The Column Research Council has brought out that it would be desirable to reach agreement among engineers as to the best method for predicting the ultimate load-capacity in compression of straight, prismatic, axially loaded, compact members of structural metals. It was proposed that Research Committee A of the Council be assigned the problem of reporting on the correctness and desirability of the tangent-modulus column formula. This formula involves simply the substitution of the tangent modulus, E_t, for E in the Euler formula. This formula may be written

$$\frac{P}{A} = \frac{\pi^2 E_t}{(KL/r)^2}$$

where

P = the ultimate load (lb)
A = the cross-sectional area (in.^{2})
E_t = the compressive tangent modulus (slope of the compressive stress-strain curve) of the material in the column at the stress P/A (lb/in.^{2})
r = least radius-of-gyration of cross section (in.)
L = the length of the column (in.),
K = a constant depending on end conditions:
$K = 2$ for one end fixed and the other end free

*Issued May 19, 1952. Technical Memorandum No. 5 reflects current (1987) position of SSRC and replaces Technical Memorandum No. 1.

$K = 1$ for both ends simply supported
$K = 0.7$ for one end fixed and the other end simply supported
$K = 0.5$ for both ends fixed.

For materials which exhibit upper and lower yield points in compression, the lower yield point is to be considered as the limiting value of P/A.

Information and reference to literature supporting the foregoing statement will be made available on request to the Secretary of the Column Research Council.

It is the considered opinion of the Column Research Council that the tangent-modulus formula for the buckling strength affords a proper basis for the establishment of working-load formulas.

The column formula presented here differs in form from the familiar Euler formula only in that the tangent modulus-of-elasticity is substituted for the ordinary modulus-of-elasticity. There is, however, a great practical difference between the two formulas, for, whereas the Euler formula can be solved directly for the average stress corresponding to any given slenderness ratio, the tangent-modulus formula cannot. It is not the intention to advocate the use of the tangent-modulus formula in design, but rather to propose it as the basis for relating the compressive stress-strain properties of the material to the column strength of the material. The formula furnishes the information for approximating to the average stress in terms of the ratio of slenderness, for any type of centrally loaded column under consideration, by making suitable assumptions with respect to such items as accidental eccentricity, initial curvature of member, residual stresses, and variation in properties of the material.

Advisory Preface to Technical Memoranda Nos. 2, 3, and 4

The reader is advised that Technical Memoranda Nos. 2, 3, and 4, although accurate, need care and interpretation when used in conjunction with modern testing and data acquisition equipment. Thus it may be inappropriate to adhere to a specific sensitivity of measured strain in Technical Memorandum No. 2 when the strain-measuring device referred to is no longer in general use. Similarly, for Technical memoranda Nos. 3 and 4, it may not be appropriate to specify crosshead speeds in the testing machine when the testing machine has stroke or strain controlled by a servomechanism. In Technical Memorandum No. 4, the first alignment method of preparing the column for testing is no longer in general use—rather, most tests use the second alignment method, by which columns are tested "as is" with exact measurements of initial out-of-straightness which are used with analytical strength predictions. Although these Technical Memoranda remain correct, it has not been possible to update them to reflect today's practice in this edition of the guide, since the methods in use today have not yet been standardized. These Technical Memoranda have been retained in this edition so as to provide a basis of comparison for the researcher.

B.2 Technical Memorandum No. 2: Notes on the Compression Testing of Metals

For predicting column strength it is necessary to have compressive and tensile stress-strain curves of the column material. In order that the tension and compression specimens be as nearly as possible equally representative of the material, the thickness of compression specimens should approximate that of the tension specimens. Preferably each pair of tensile and compressive specimens should be cut from the same coupon (conveniently, the section or slice remaining from the residual strain measurements).

Specimens taken from a flange or web of a rolled shape, or from a plate, should be rectangular in cross section. They should be machined on all four sides with grinding as the final machining operation (on a magnetic grinder for steel). The ends of the specimens should be ground plane and normal to the longitudinal axis of the specimen. The ends should be parallel within close limits.

In general, compression specimens (Fig. B1) should be no longer than necessary of accomodate a compressometer or resistance strain gages and have, between each end of the specimen and the adjacent end of the gage length, a distance, $\frac{1}{2}(L - G)$, at least equal to the greatest cross-sectional dimension, b. The compressometer should meet the specifications for

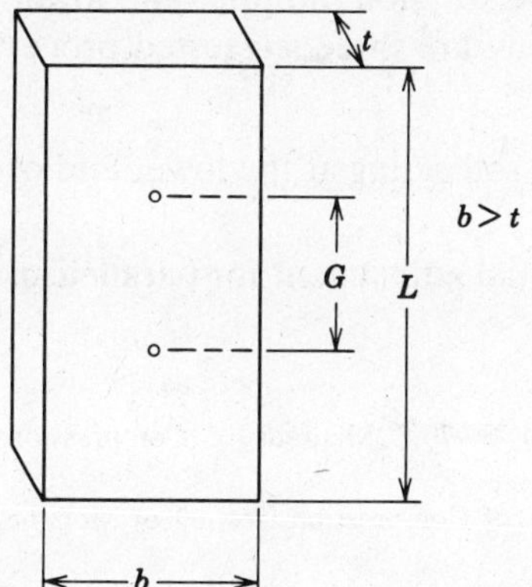

$b > t$

Fig. B.1 Compression test specimen.

Class A Extensometers (ASTM Standard E83–67, "Verification and Classification of Extensometers") which limit the error of indicated strain to 0.00001. The length should not be less than the greatest cross-sectional dimension; and in order to keep the specimen short, the gage length should not exceed twice this dimension. If the length of a rectangular specimen is more than about 4.5 times the length of the shorter side of the rectangle or, for a circular specimen, more than four times the diameter, difficulty may be expected in avoiding premature bending (column action), the special precautions must be taken to prevent excessive bending.*

Referring to Fig. B.1, the following relationships summarize the preceding requirements for a member of rectangular cross section:

$$G \gtreqless b$$

$$G \geqq t$$

$$4.5t \gtreqless L \gtreqless (G + 2b)$$

in which G = gage length; b = specimen width; L = specimen length; and t = specimen thickness. Specimens should be measured with a micrometer reading to 0.001 in. (0.025 mm) and the dimensions recorded. Nominal dimensions should not be used in computations.

Both ends of a compressive specimen should bear on smoothly finished plane surfaces. The bearing blocks should be made of, or faced with, a suitably hard material so that the faces of the blocks will not suffer permanent deformation during the test. The blocks should be at least as thick as the smallest cross-sectional dimension of the specimen and should project beyond the contact area a distance at least half as great as the smallest cross-sectional dimension.

Precautions should be taken to ensure uniform distribution of strain over the cross section and to prevent relative rotation of upper and lower bearing surfaces during testing. The following are three suggested procedures for attaining these goals:

1. Use of a subpress loaded through a push rod acting at the lower end of the hollow plunger.
2. Use of bearing blocks that will permit initial adjustment for parallelism of bearing surfaces.**

*For rectangular specimens see ASTM Standard E9-70, "Methods of Compression Testing of Metallic Materials."

**See Fig. 1 in ASTM Standard E9-70, "Methods of Compression Testing of Metallic Materials."

3. Use of a thin capping layer of Hydrostone between the upper bearing block and the testing machine crosshead. While the Hydrostone is setting, a small load should be maintained.

If tilting or lateral displacement of one testing machine crosshead relative to the other during loading is a suspected possibility, it is suggested that a subpress be used to load the compressive specimen so as to reduce the probability of bending during compression. The hollow plunger should fit closely within the annulus of the subpress frame (but vertical motion of the plunger should not be restricted).

Adjustable bearing blocks cannot be depended upon to compensate for tilting of the testing machine heads during loading and should be used only if appreciable relative tilting of the heads does not occur. If a spherical bearing block is used, it should be at the upper end of the specimen (for specimens tested with the longitudinal axis vertical). It is desirable that the center of the spherical surface lie within the flat surface on which the specimen bears. Also, it is essential that the longitudinal axis of the specimen pass, closely, the center of the spherical surface, so that the eccentricity of loading may not be great enough to overcome the friction necessary to rotate the block.

The compressive specimen should be aligned so that the deviation in strain indicated by any gage is less than 5% of the average of all gages when the specimen is subjected to a stress of about one-half the yield strength of the material. At least three strain gages need to be mounted on the specimen and monitored during the aligning operation to meet this requirement.

If the length of the specimen does not exceed the maximum recommended length (4 or 4.5 times a cross-sectional dimension for a circular or rectangular specimen, respectively), strains should be measured during the test with an averaging compressometer or with two strain gages mounted opposite each other. In the case of longer specimens tested without lateral support, reasonable certainty of uniform strain distribution can be obtained only with the use of not fewer than two strain gages on the wide sides of thin rectangular specimens, or three gages on thick rectangular or square specimens (one at the center of each of three sides), or three gages on circular specimens. Strain measurement with only one gage is unreliable. The stress-strain curve should extend from zero stress, and strain, to values for which the ratio of total stress to total strain is less than $0.7\,E$, or to a total strain of at least 0.01, whichever results in the larger strain.

Material properties will be a function of the loading rate and therefore the rate should be recorded. Dynamic loading should be interrupted in the yielding region so as to obtain at least three values of the static yield stress (yield stress at essentially zero strain rate) for metals with a yield plateau.

Compressive stress-strain curves should be plotted with stress as ordinate and strain as abscissa to as large a scale as the quality of the data justifies. The individual values of stress and strain should also be reported. When applying the procedures above to material that is suspected of showing a considerable variation in properties, the specimens should be taken from a sufficient number of locations to define the extent of the variation in properties.

Determination of Typical Stress-Strain Curve from a Number of Stress-Strain Curves

It is assumed that the compressive stress-strain relationships of enough specimens will be determined so that all variations of the material likely to be submitted under a given specification will be represented. The yield strength values determined from the individual tests should be presented in the form of a distribution plot in which the percentage of the total number of tests for which the yield strength is within a certain range is plotted against the average of the range. The standard deviation should be indicated if the number of specimens is significant.

Several methods for constructing a typical stress-strain curve from a number of individual curves have been proposed. One simple method, which has had considerable use, is described as follows:

1. Record the strain departures from the modulus line for various fixed percentages of the particular individual yield strength value. These percentages should cover stresses from the proportional limit to above the yield strength.
2. Average all offset values for each of the fixed percentages. (For steel shapes it is recommended that the offsets be weighted in proportion to relative flange and web areas.) A curve may be plotted in which the ordinate is the percent of yield strength and the abscissa is the average strain offset from the initial modulus line.
3. For any appropriate yield strength value, a typical curve can then be plotted by adding the offset values to the strain consistent with the elastic-modulus values.

Figure B.2 shows a typical compressive stress-strain curve of a high-strength aluminum alloy. Lines have been drawn tangent to this curve at different values of stress, P/A. The slopes of these lines define the

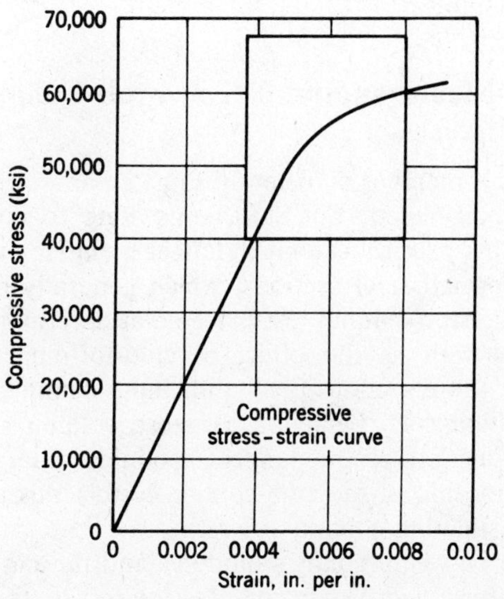

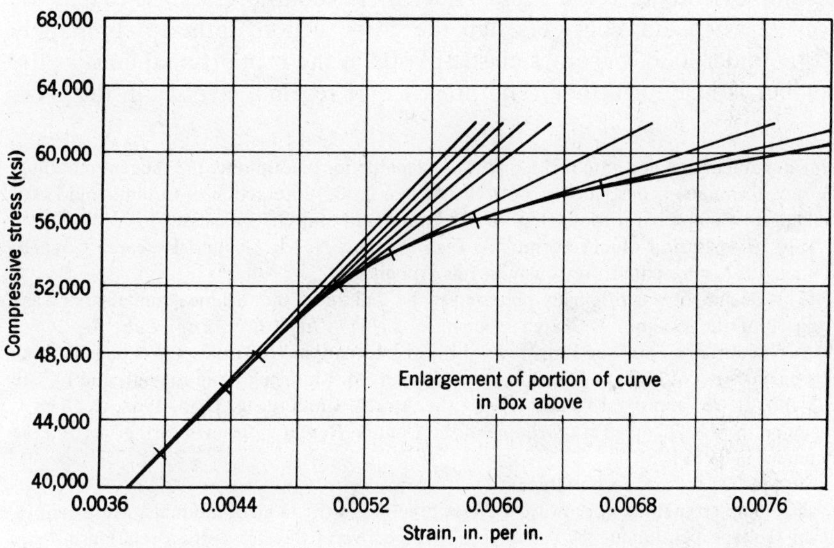

Fig. B.2 Typical compression stress-strain curve of a high-strength aluminum alloy. 1 ksi = 6.9 MPa.

corresponding tangent modulus, E_t, essential to the determination of the basic column strength.*†

B.3 Technical Memorandum No. 3: Stub-Column Test Procedure‡§

A stub column is a member sufficiently short so as to preclude member buckling when compressed, but sufficiently long to contain the same initial residual stress pattern as a much longer member cut from the same stock. For cold-formed steel sections, which generally have thin-walled plate elements, the stub-column test is aimed at determining the effect of local buckling as well as the effect of cold-forming on the column performance. For these sections the stub-column length should be sufficiently long to exhibit such behavior. Because column strength may be expressed as a function of the tangent modulus determined from the stress-strain relationship of the stub-column test,§§ this test is an important tool for investigating column strength.

The difference between Young's modulus and the tangent modulus at any load level, determined from a compression test on the complete cross section, essentially reflects the effect of residual stresses. This may be realized when one considers that the cross section, hitherto completely elastic under load, becomes elastic-plastic at the proportional limit as the member is loaded further. The presence of residual stresses in the cross

*Convenient and accurate techniques are available for determining the tangent modulus; one such technique is described in *NACA TN*, No. 2640, "Interaction of Column and Local Buckling in Compression Members," by P. P. Bijlaard and G. P. Fisher.

†See "The Basic Column Formula," *Tech. Memo.*, No. 1, Column Research Council, May 19, 1952. (Presented previously in this Appendix.)

‡This document was originally prepared by L. Tall under the technical guidance of Task Group 1 of the Column Research Council as *Lehigh Univ. Fritz Eng. Lab. Rep.*, No. 220A.36 (February 1961), and was revised by an International Institute of Welding Working Group consisting of H. Louis (Belgium), Chairman, M. Marincek (Yugoslavia), and L. Tall (U.S.A.). It was approved by the IIW at the Annual Conference, Oslo, 1962, as Class C Document No. X-282-61. Task Group 6 of the Column Research Council further revised the document in 1974.

§See p. 702 for advisory preface.

§§Column strength is not always a direct function of the tangent modulus. For example, for an H-shape bent about the strong axis, irrespective of the stress-strain relationship and the pattern of residual stress, the function is direct. However, for an H-shape bent about the weak axis (only for rolled or welded built-up shapes with universal mill plates whose stress-strain curve can be considered as elastic-perfectly plastic) the strength is approximately a function of the cube of the tangent modulus. Because there is no direct or simple relationship for other cases, care must be taken in applying the stub-column test results to the prediction of column strength.

section implies that some fibers are in a state of residual compression. The fibers in a state of residual compression are the first to reach the yield point under load.

The difference between the behavior of a column free of residual stresses and one containing residual stresses lies in the fact that, beyond the proportional limit for the latter, both the tangent modulus and the effective moment-of-inertia are greater for the column free of residual stresses. (The behavior of a stub column, however, because of its shorter length, reflects only the effect of residual stresses on the tangent modulus; the reduction in the effective moment-of-inertia due to plastification has no effect on its behavior.) Under load, some parts of the column cross section will yield before others, leading to a decrease of the effective moment of inertia and hence in the strength of the column, as those proportions of the cross section which have yielded support no additional load if strain hardening is neglected. The residual stress distribution over the cross section, through its influence on the effective moment-of-inertia, is the connecting link between column strength and the tangent modulus of the stress-strain relationship of the stub column. That residual stresses are, indeed, a major factor affecting the strength of axially loaded, initially straight columns, and that a conservative value for this strength may be specified in terms of the tangent modulus determined from the results of a stub-column test, have been documented extensively.

Stub-Column Test Procedure

1. *Object.* To determine the average stress-strain relationship over the complete cross section by means of a compressive test of a stub column.

2. *Specimen.*

a. The stub column should be cold-sawed from the stock at a distance at least equal to the shape depth away from a flame-cut end.

b. The length of hot-rolled stub columns should not be less than $2d + 10$ in. ($d + 250$ mm) or $3d$, whichever is smaller, and not greater than $20r_y$ or $5d$, whichever is larger, in which d = depth of the shape and r_y = radius-of-gyration about the weak axis. For cold-formed shapes the length of the stub column should not be less than three times the largest dimension of the cross section and no more than 20 times the least radius-of-gyration.

c. The ends of the columns should be milled plane and perpendicular to the longitudinal axis of the column.* This operation may be omitted

*A tolerance across the milled surface of ± 0.001 in. (± 0.025 mm) is usually satisfactory.

for light gage members which are difficult to mill, if their ends are welded to base plates.

d. The thickness of flanges and webs and the length and cross-sectional area of the stub column should be measured and recorded.

3. *Instrumentation.* Mechanical dial indicator gages or electrical resistance gages may be used to determine the strains during testing. The use of dial gages over a comparatively large gage length is to be preferred as they provide a better average strain indication. The dial gages should read to 0.0001 in. (0.0025 mm) when read over a 10-in. (250 mm) gage length, or to 0.001 in. (0.025 mm) when installed between base plates over the whole length of the stub column. Where it can be demonstrated that electrical resistance gages give the same or better results, they may be used instead of dial gages.

The gage length should be placed symmetrically about the mid-height of the stub column. At least two gages in opposite positions should be used and the average of the readings taken. Corner gages over the complete column length are used for alignment; mid-height gages are used for determining the stress-strain relationship. When four mid-height gages are used instead of two, the corner gages may be omitted. (This is possible with the flange tips of an H-shape.) Figure B.3 depicts typical gage arrangements for structural shapes.

For uniformity in stub-column testing, the following instrumentation is recommended for H-shapes:

a. Four 0.001-in. (0.025-mm) dial gages over the complete length of the stub column, at the four corners; to be used during alignment.

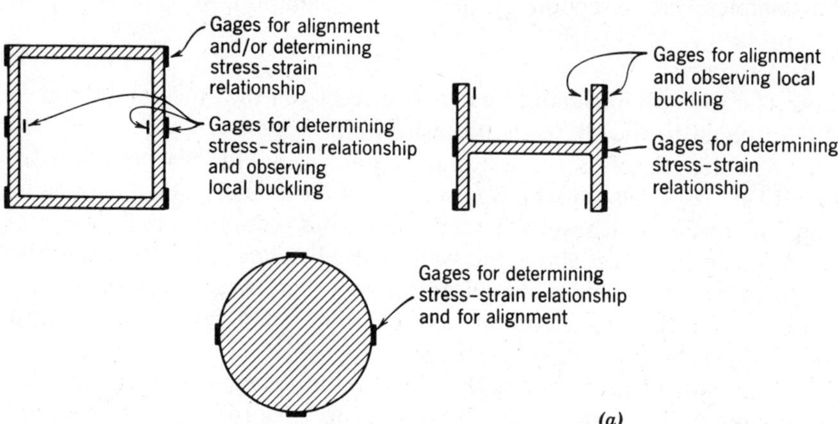

Fig. B.3 Position of gages for alignment and testing: (*a*) location of electrical resistance gages; (*b*), (*c*), (*d*) location of dial indicators;

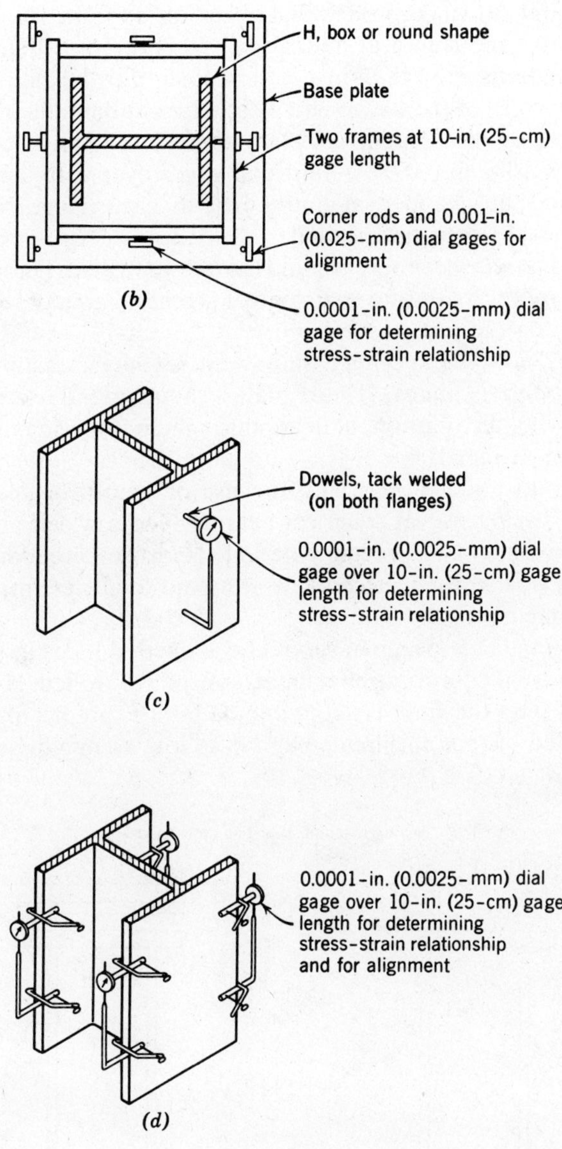

Fig. B.3 (*continued*)

b. Two 0.0001-in. (0.0025-mm) dial gages on opposite flanges over a 10-in. (250-mm) gage length at mid-height; to be used to determine the stress-strain relationship. The points of attachment for the gage length are to be at the junction of the flange and web, to avoid the influence of local flange crippling on the readings. When early local flange crippling is unlikely, four 0.0001-in. (0.0025-mm) dial gages over a 10-in. (250-mm) gage length may be clamped at mid-height to each flange tip. Corner alignment gages are then not needed.

c. As-rolled steel specimens should be whitewashed before testing. Flaking of the mill scale during testing gives a general area of the progress of yielding during the test.

4. *Test Set-Up.* The specimen should be set in the testing machine between flat bearing plates. These plates should be thick enough to ensure a uniform distribution of load through the specimen. The test set-up is shown in Fig. B.4.

Alignment may be achieved with the use of special beveled bearing plates, or else by the use of spherical bearing blocks which are fixed by wedges after alignment to prevent rotation. Hydrostone bedding for the column ends has been used successfully as an aid to alignment, especially for light-gage members.

5. *Alignment.* The specimen should be aligned at loads less than that corresponding to the proportional limit stress. For rolled H-shapes of mild structural steel this limit is about one-half of the predicted yield level load; for welded shapes the limit may be as low as one-quarter of the yield level load.

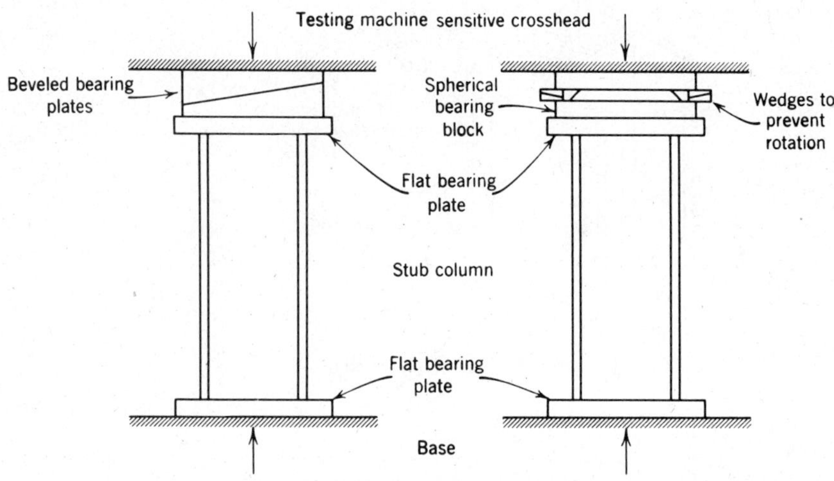

Fig. B.4 Subcolumn set-up for testing.

The alignment is preformed by noting the variation of strain at the four corners of the specimen. The variation between the strains at any corner from the average strain should be less than 5% at the maximum alignment load. Alignment at low loads is unsatisfactory. The alignment loading should consist of several increments up to the maximum alignment load.

To check that the load is below the proportional limit, the stress-strain relationship may be plotted during the test and its linearity observed. It is inadvisable to exercise this control by observing the whitewash for flaking of the mill scale, because flaking begins at a load greater than that corresponding to the proportional limit stress as indicated by the plotted stress-strain relationship.

6. *Testing*. The stress-strain curve should be constructed from as many experimental data points as possible. To this end, the load increments in the elastic region should be less than 10% of the expected yield load. After the proportional limit has been reached, the load increments should be reduced so that there are sufficient data points to delineate the "knee" of the stress-strain curve. Strain increments may be more convenient than load increments to delineate the "knee" in the inelastic region. The proportional limit* will be marked by the beginning of the deviation of the stress-strain relationship from linear behavior. Yield lines (made clearly visible by the whitewash as the mill scale flakes off) will indicate the progress of yielding. This matter is covered further in Item 10.

After the onset of yielding, readings should be recorded when both load and strain have stabilized. The criteria used to specify when data may be recorded depend on the type of machine used for testing. This is explained further in Item 7.

To ensure correct evaluation of the yield level and other material properties, the test should be continued until one of the following conditions is satisfied:

a. After an immediate drop in load due to local plate buckling, the test should be continued until the load has dropped to about half the predicted yield level load.

b. For a specimen that exhibits a plastic region of considerable extent, the test should be continued until the load had dropped to about 80% the predicted yield level load.

*It is assumed that the residual stresses are symmetrical with respect to the principal axes of the cross section and constant in the longitudinal direction, so that the proportional limit does not indicate localized yielding.

c. For a specimen that strain hardens without apparent buckling, or which strain hardens without a plastic range, the test should be continued until the load is about 25% above that corresponding to the load computed from the yield strength based on the 0.2% strain offset criterion.

The load and strain at all initial load levels should be recorded. This is further outlined in Item 9. It may be necessary to remove some of the dial gages before the test is completed to avoid damage due to local buckling.

7. *Criteria for Stabilization of Load.* Standard criteria should be followed for recording of test data when the load is greater than that at the proportional limit. The criterion depends on the type of testing machine used, whether hydraulic or mechanical.

For mechanical testing machines (screw type) the criterion is as follows: no relative crosshead movement with both the loading and bypass valves closed. For hydraulic systems that leak the criterion is a simulation of that for the screw-type machine. This is accomplished by balancing the load and bypass flows so that no relative motion of the crossheads occurs, and then by waiting for the load to stabilize.

These criteria are best applied by plotting the load change, or crosshead movement versus time, and noting the value corresponding to the asymptote (see Fig. B.5). The test data are recorded when:

a. The asymptotic load is approached, when using the load criterion.

b. The asymptotic crosshead movement is approached when using the crosshead movement criterion.

Readings should not be recorded until the asymptote is definite. Experience will indicate the time intervals required, but 3-min intervals are usually satisfactory. The crosshead movement should be measured with a 0.001-in. (0.0025-mm) dial gage.

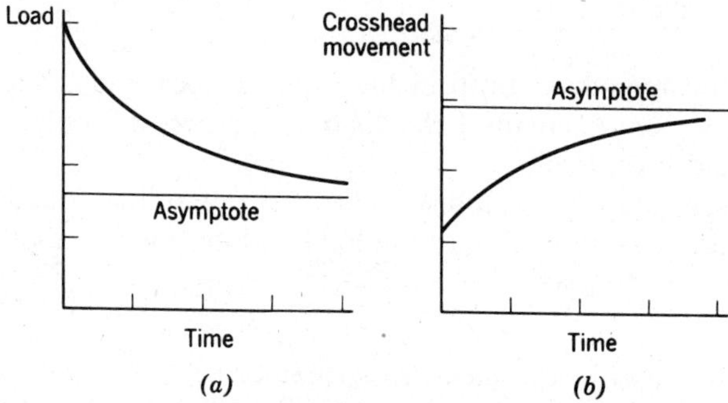

Fig. B.5 Criteria for load stabilization.

8. *Evaluation of Data.* The test data should be evaluated in the following manner:

a. Plot the test data during the test to detect any inconsistencies.

b. Translate the test data to stress versus strain (based on the actual cross-sectional area) and plot the stress-strain relationship to a large scale. A typical stress-strain diagram is shown in Fig. B.6. A Ramberg Osgood type curve is often fitted to the test data.

c. Determine the tangent-modulus curve from the stress-strain relationship. This may be done by using a strip of mirror. The mirror is held normal to the curve, the normal being determined from the continuity of the stress-strain curve and its mirror image at the tangent point considered. Then a line is drawn along the mirror edge.

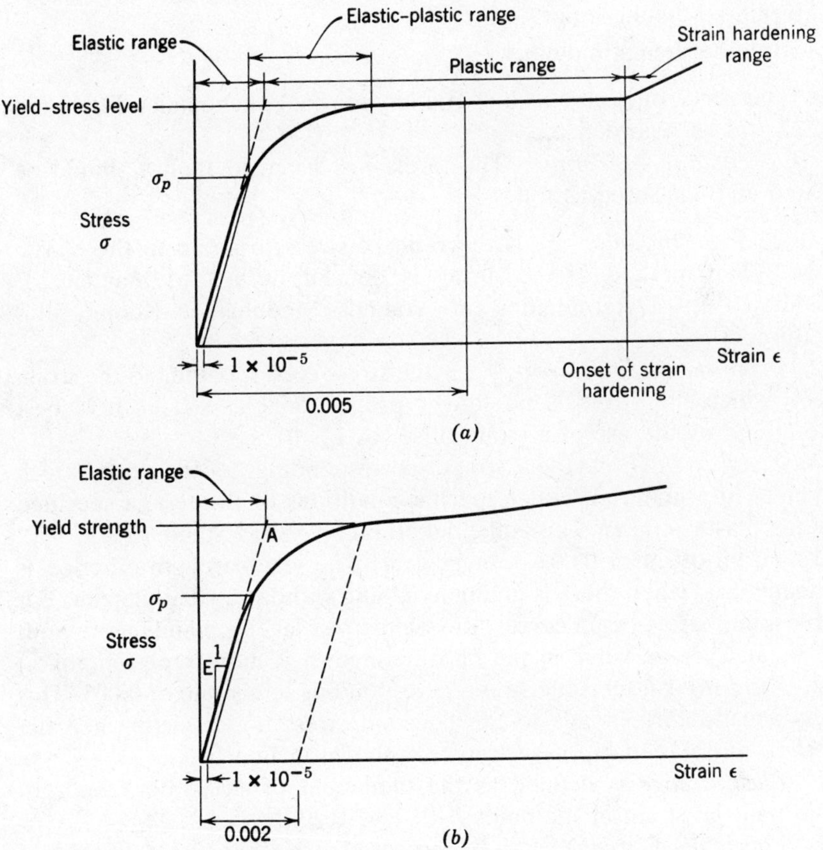

Fig. B.6 The stress-strain diagram.

9. *Data To Be Reported.* In addition to presenting the stress-strain and stress-(tangent modulus) curves the following information, obtained from the stress-strain relationship given by the stub-column test, should be reported:

1. Young's modulus of elasticity.
2. Proportional limit stress.
3. Yield strength.
4. Yield stress level.
5. Elastic range.
6. Elastic-plastic range.
7. Plastic range.
8. Onset of strain hardening.
9. Strain-hardening range.
10. Strain-hardening modulus.

The occurrence of local buckling and any unusual phenomena during the test should be recorded.

10. *Definition of Terms.* The terms just listed in Item 9 should be defined and measured as follows:

1. *Young's modulus, E,* is the ratio of stress to strain in the elastic range. (The method of measuring is defined by ASTM Standard E 111-61, 1972, "Determination of Young's Modulus at Room Temperature.")

2. *Proportional limit stress,* σ_p, is the stress corresponding to the strain above which the stress is no longer proportional to strain. It is best determined by the use of a strain offset of 1×10^{-5}.

3. *Yield strength* is the "stress, corresponding to the load which produces in a material, under specific conditions of the test, a specified limiting plastic strain." This is the definition of ASTM Standard A370-72, and a strain offset of 0.002 is suggested. (The yield-strength criterion is normally used when there is gradual yielding without a yield plateau. For stub-column stress-strain curves, the yield stress level is mainly used, and, as it is an average value in the plastic range, it is more representative.)

4. *Yield stress level* is the stress corresponding to a strain of 0.005. This stress usually corresponds to the constant stress under yield when the stress-strain relationship is such as that shown in Fig. B.6*a*.

5. *Elastic range* is defined as the increment of strain between zero strain and the strain at the point *A* in Fig. B.6*b*.

6. *Elastic-plastic range* is the increment of strain between the strain at the proportional limit stress and the strain at which the stress first reaches the yield stress level.

7. *Plastic range* is defined as the increment of strain betwen the elastic range and the onset of strain hardening.

8. *Onset of strain hardening* may be defined as the strain corresponding to the intersection on the stress-strain curve of the yield stress level in the plastic range with the tangent to the curve in the strain-hardening range. This tangent is drawn as the average value in the strain increment of 0.002 after the apparent onset of strain hardening.

9. *Strain-hardening range* is the range of strain after the plastic range in which the material no longer strains at a constant or near-constant stress.

10. *Strain-hardening modulus* is the ratio of stress to strain in the initial strain-hardening range. It is taken as the average value in the strain increment of 0.005 after the onset of strain hardening.

B.4 Technical Memorandum No. 4: Procedure for Testing Centrally Loaded Columns*†

A column may be defined as a member whose length is considerably larger than any of its cross-sectional dimensions and which is subjected to compression in the longitudinal direction. If the resultant compressive force is approximately coincident with the longitudinal centroidal axis of the member, the column is said to be centrally loaded. Although columns have been extensively studied for more than two centuries, both analytically and experimentally, technological developments may necessitate further testing of centrally loaded columns. The purpose of this memorandum is to set forth a suggested procedure for conducting such experiments.

The experimentally determined values of column strength from a wide scatter band when plotted versus the effective slenderness ratio, KL/r, in which KL denotes the effective column length and r the appropriate radius-of-gyration of the cross section. The scatter is due to geometrical imperfections of the column specimens, eccentric application of load, nonhomogeneity of the column material, residual stresses from the rolling and fabricating processes, variations in the action of loading machines, imperfections in the end fixtures, and other factors. The major sources of scatter are briefly discussed subsequently.

*This document was prepared by Task Group 6 of the Column Research Council based on *Lehigh Univ. Fritz Eng. Lab. Rep.* No. 351.6, authored by N. Tebedge and L. Tall (1970).

†See page 702 for advisory preface.

A geometrically perfect centrally loaded column would not deflect laterally at loads less than the critical load. However, all column specimens deflect from the beginning of loading because of bending that results from the initial curvature and twist of the specimen and the unavoidable eccentricity of load application. Nonhomogeneity of the column material results in bending of columns which are stressed above the proportional limit because the pattern of yielded zones of the cross section is not perfectly symmetrical with respect to the principal axes of the cross section.

Residual stresses from the rolling and fabricating processes, present in the column specimen prior to testing, cause scatter in the observed column strength because of patterns of the residual stresses among different specimens of the same size and shape cause variations in the load at the onset of yielding and, as residual stress patterns are generally not symmetric about the principal axes of the cross section, cause variations in the amount of column bending and twisting which, in turn, affects the column strength.

In column tests, as in stability tests of other structural elements, the response of the column is influenced by the action of the loading device. Loading devices may be categorized as gravity, deformation, and pressure types. The force-deflection characteristics of these types differ. The oldest form of testing device used for columns is the gravity-type. For such a system, the load-deflection characteristic is simple and can be graphically represented by a series of straight lines parallel to the deflection axis. Later, the screw-type testing machine became a common laboratory apparatus. These machines have the advantage of a well-defined load-deflection characteristic, the slope of which depends, essentially, on the elastic response of the loading system. As higher capacity loading machines were needed, the hydraulic testing machine was developed. This system, however, does not have an easily defined load-deflection characteristic as it depends on the properties of the hydraulic system, temperature, and other factors. Loading of a column in a testing machine is always conducted under some finite loading rate and the experimental results are influenced by this.

Centrally loaded columns may have different end conditions ranging, theoretically, from full restraint (fixed) to zero restraint (pinned), with respect to end rotation and warping. Most investigators have used the pinned-end condition for column testing for a number of reasons. Under the pinned-end conditions the critical cross section is located near the mid-height of the column, thus making the cross section of interest remote from the boundary and, therefore, little influenced by end effects. For the same effective slenderness ratio, the pinned-end condition re-

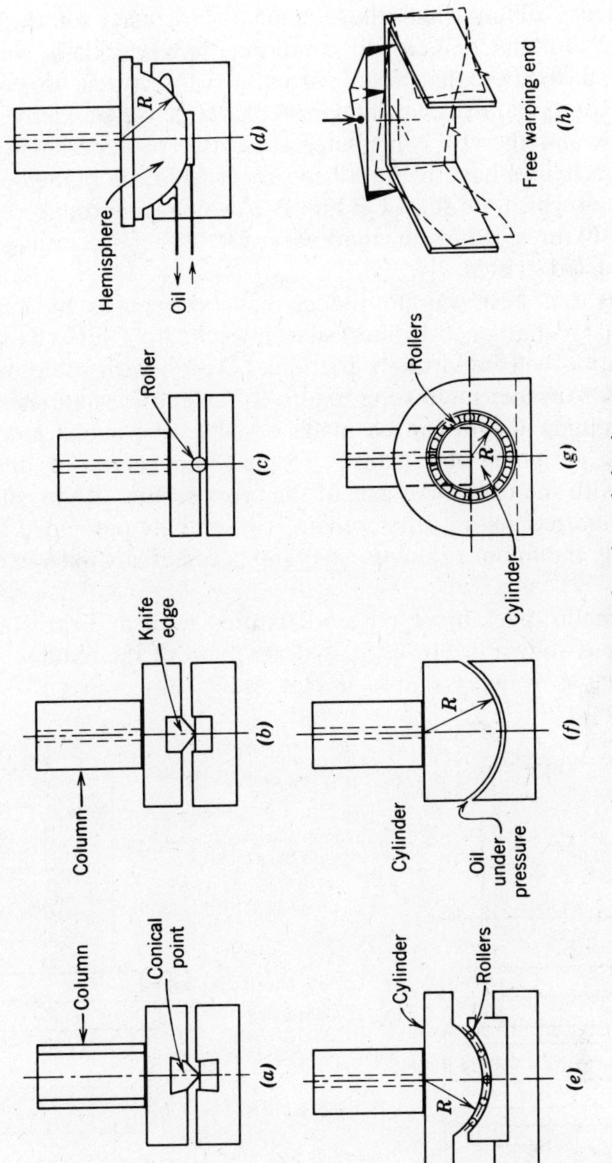

Fig. B.7 End fixtures for pin-ended columns.

quires the use of only half the column length used for the fixed-end condition. With the pinned-end condition, however, it is necessary to provide end fixtures that offer virtually no restraint to column-end rotation. Rotational restraint influences the effective slenderness ratio of the column and thereby contributes to scatter of experimental results.

Several schemes have been utilized to provide the pinned-end condition, some of which are shown in Fig. B.7, which is reproduced from Ref. B.1. The fixtures differ in that they are either "position-fixed" or "direction-fixed" (B.2).

Probably the best way to reduce end restraint is by means of a relatively large hardened cylindrical surface bearing on a hard flat surface. Rotation will be virtually frictionless, even with some indentation under load. Another interesting feature of cylindrical fixtures is that the effective column length can be made equal to the actual length of the column by designing the fixtures so that the center of the cylinder coincides with the centroidal axis of the cross section at the column end. With a cylindrical fixture, the column is essentially pin-ended about one axis (usually the minor principal axis) and is essentially fixed-ended about the other.

A schematic diagram of the end fixtures used at Fritz Engineering Laboratory is shown in Fig. B.8. A description of the fixtures, and their performance as "pins", is given in Ref. B.3.

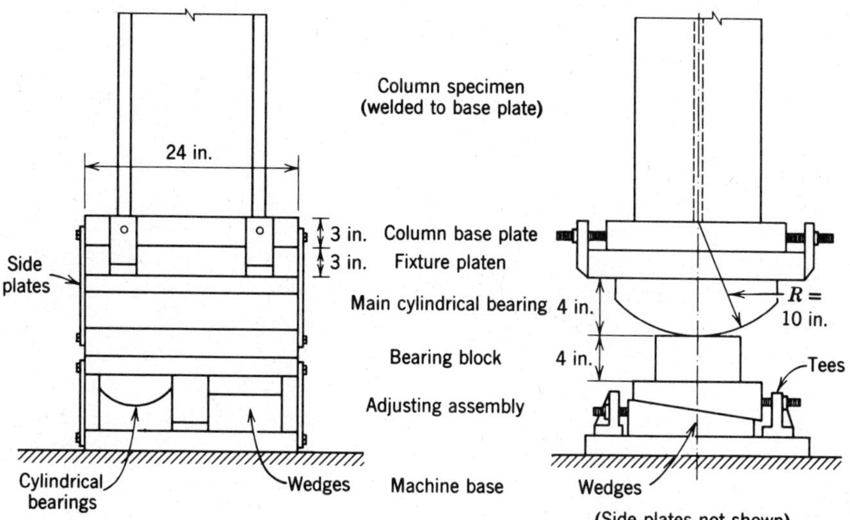

Fig. B.8 Standard column end fixture at Fritz Engineering Laboratory (capacity = 2.5) million lb).

In testing columns under the fixed-end condition, the full restraint may not be provided in the entire range of the test loads; thus the effective length of the column is not a constant but a function of the applied load. This may be due partly to the fact that the rigidity of the testing machine varies with the applied load and partly to the indeterminate nature of the stress distribution at the column end, particularly in the load range in which the material yields. These problems are eliminated by using pinned-end conditions because the critical conditions exists at about the mid-height cross section.

(1) Column Test Procedure

Preparation of Specimens. To minimize initial geometrical imperfections of the specimen, the column specimen is cut from a straight portion of the stock. Both ends of the specimen are milled. Columns may be tested with the ends bearing directly on the loading fixtures, provided the material of which the loading fixtures are made is sufficiently harder than that of the column to avoid damaging the fixtures. Otherwise, base plates should be welded to the specimen ends, matching the geometric center of the specimen to the center of the base plate. The welding procedure should be such that compressive residual stresses at the flange tips caused by the welding are minimized. For columns initially curved, the milled surfaces may not be parallel to each other, but will be perpendicular to the centerline at the ends because milling is usually performed with reference to the end portions of the columns. For relatively small column specimens, it is possible to machine the ends flat and parallel to each other by mounting the specimens on an arbor in a lathe. For small deviations in parallelism, the leveling plates at the sensitive crosshead of the testing machine may be adjusted to improve alignment. The tolerance in deviation must not exceed the range of adjustment of the leveling plates of the particular testing machine.

Initial Dimensions. The variation in cross-sectional area and shape, and the initial curvature and twist, will affect the column strength. Therefore, initial measurement of the specimen is an important step in column testing.

The cross section is measured to determine the variation between the actual dimensions of the section and the nominal catalog dimensions and to enable the computation of the required geometrical cross-sectional parameters. The dimensions shown in Fig. B.9 are measured at a number of stations along the column (the quarter points of the specimen are recommended, as a minimum).

The initial camber, sweep, and twist of the specimen are measured at intervals. Nine stations, spaced at one-eighth of the column lengths, are

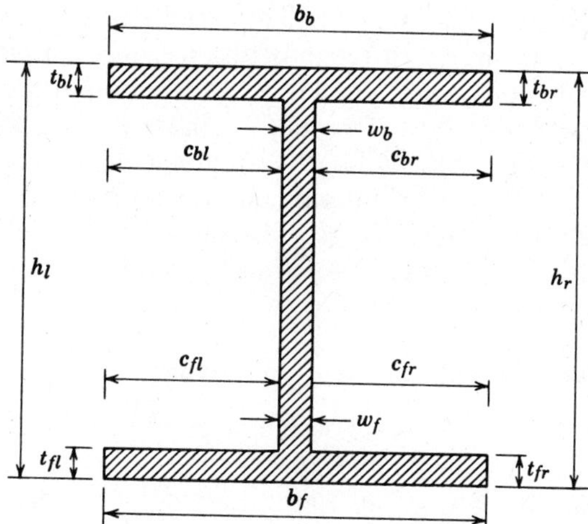

Fig. B.9 Measurements required to determine cross-sectional properties of H-shaped column.

suggested. A method of determining the initial out-of-straightness and twist is shown in Fig. B.10. Readings are taken with a theodolite (stationed in line with the column and near one of the ends) on a strip scale mounted onto a movable carpenters square. The sweep offset is determined from four readings—each referenced to a flange tip. The average of the four readings is considered as the sweep offset. The camber offset is obtained from two readings, one referenced to each flange surface at the intersection of web and flange. The initial angle of twist is computed from the sweep offset readings and the cross-sectional dimensions. Values of initial out-of-straightness and twist are used in evaluating test results.

Aligning the Column Specimen. Aligning the specimen within the testing machine is the most important step in the column testing procedure, prior to loading. Two approaches have been used to align centrally loaded columns. In the first approach the column is aligned under load such that the axial stresses are essentially uniform over the mid-height and the quarter-point cross sections. The objective in this alignment method is to maximize the column load by minimizing the bending stresses caused by geometrical imperfections of the specimen.

In the second alignment method, the column is carefully aligned geometrically, but no special effort is made to secure a uniform stress distribution over the critical cross section. Geometric alignment is performed with respect to a specific reference point on the cross section (the

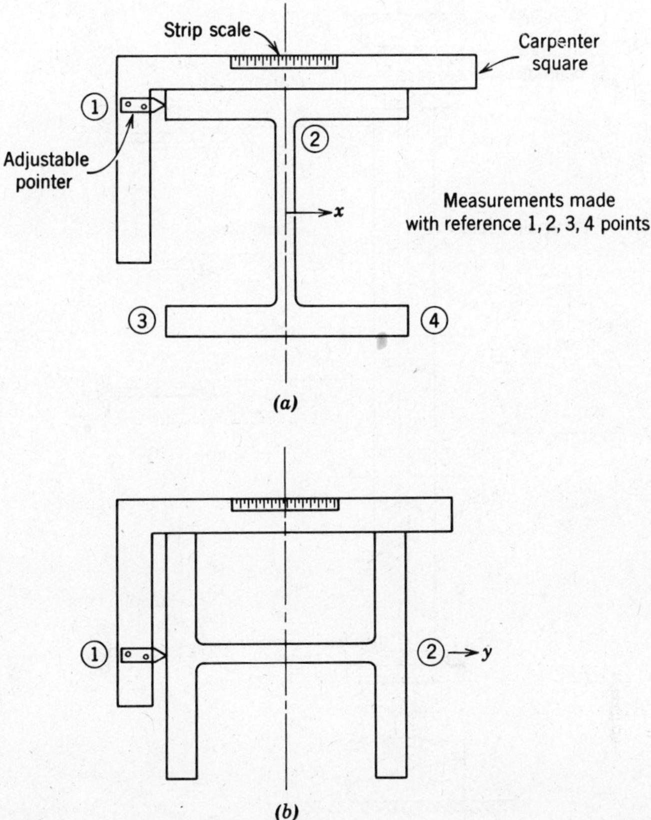

Fig. B.10 Method for measuring initial camber and sweep: (*a*) measurement normal to minor axis (sweep); (*b*) measurement normal to major axis (camber).

specific reference point will be defined later). The method of geometric alignment is recommended as it is, generally, simpler and quicker. The end plates can easily be centered with reference to the centerline of the testing machine (B.4).

The specific reference point on the cross section utilized in geometrical alignment depends on the cross-sectional shape. For H-shaped cross sections the best centering point is the center of flanges because the web has little effect on buckling about the minor axis. This reference point may be located at the midpoint of the line connecting the two centers of the flanges (B.4).

Instrumentation. In some column investigations, only the ultimate load is measured during the test. However, it is usually desirable to

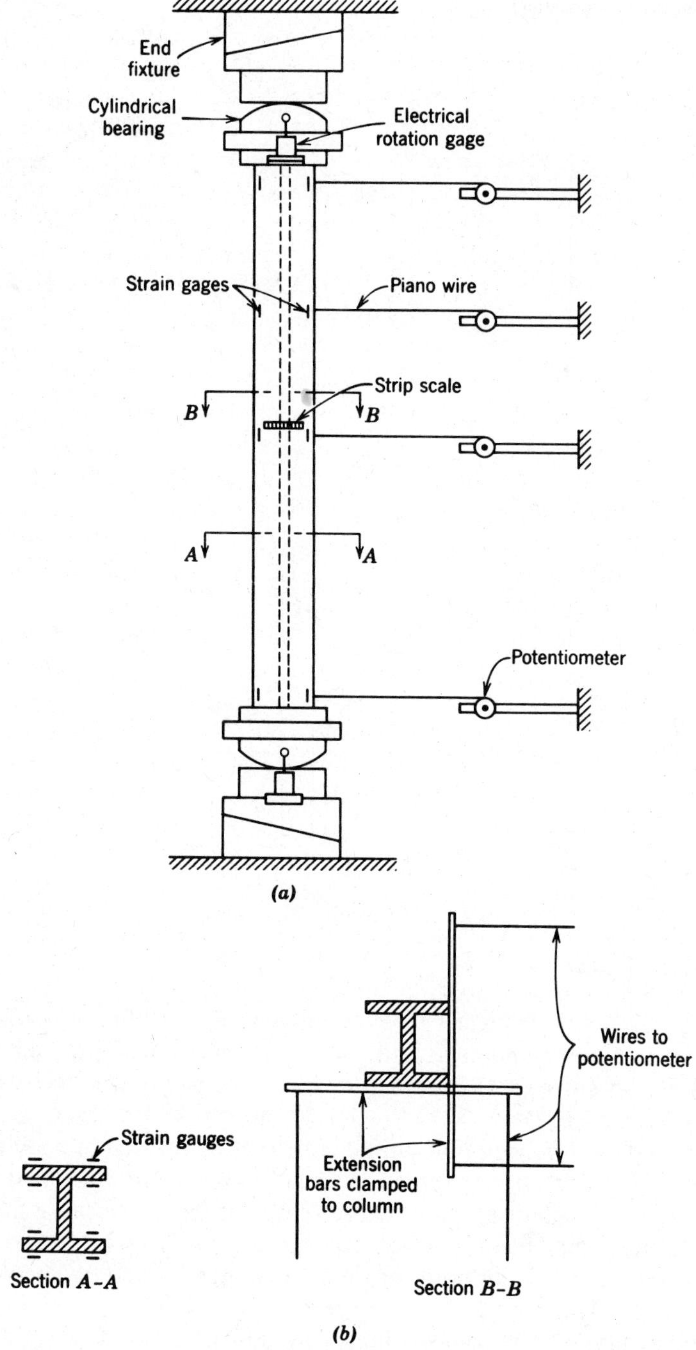

Fig. B.11 Column test set-up.

measure the more important deflections and twists to compare the behavior of the column specimen under load with theoretical predictions of behavior. The instrumentation for column tests has changed markedly in the past few years due to progress made in measuring techniques and data acquisition systems, and it is now possible to obtain automatic recordings and plotting of the measurements. Such recordings have been found to be more convenient and more precise than manual readings.

The most important records needed in column testing are the applied load and the corresponding lateral displacements, twist, and overall column shortening. A typical column set-up and instrumentation are shown in Fig. B.11.

Lateral deflections normal to both principal cross-sectional axes may be automatically recorded by means of potentiometers attached at quarter points of the column (more points may be used for larger columns). Lateral deflections may also be measured from strip scales attached to the column and read with the aid of a theodolite.

Strains are measured using electric-resistance strain gages. For ordinary pinned-end column tests, it is sufficient to mount eight strain gages at each end and at the midheight level. For long columns, it may be necessary to mount eight more strain gages at the quarter- and three-quarter points. As shown in Section *A-A* of Fig. B.11, the gages should be mounted in pairs "back-to-back" to enable the local flange bending effects to be cancelled by averaging the readings of each pair of "back-to-back" gages.

In the fixed-end test condition more strain gages are mounted below and above the quarter- and three-quarter levels. This is done to determine the actual effective length of the column by locating the inflection points using the strain gage measurements.

End rotations are measured by mechanical or electrical rotation gages. Mechanical rotation gages (B.5) are assembled by mounting levels bars on support brackets welded to the base plate and the top plate of the column as shown in Fig., B.12*a*. Angle changes due to column-end rotation are measured by centering the level bubble with the micrometer screw adjustment. A dial gage attached to the end of the level bar gives an indication of the rotation of the bar over a gage length of 20 in. (508 mm). In the electrical rotation gage, rotations are determined from bending strains induced in a thin metallic strip from which a heavy pendulum is suspended as depicted in Fig. B.12*b*. It has been shown that the strain at any location of the strip is proportional to the end rotation (B.6).

The angles of twist are determined at mid-height and at the two ends by measuring at each level the differences in lateral deflections of the two flanges. For better accuracy, the measurements may be taken at points

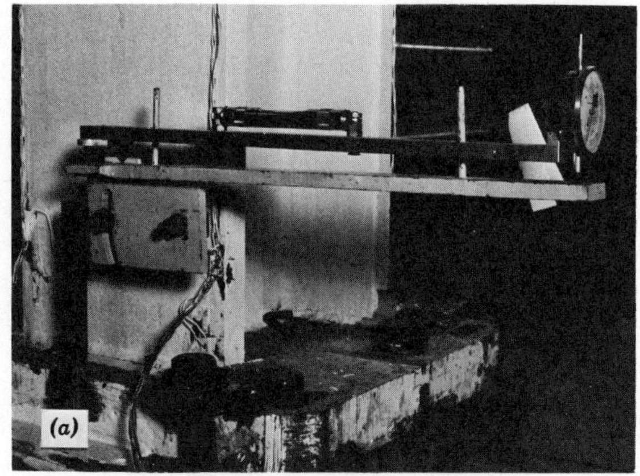

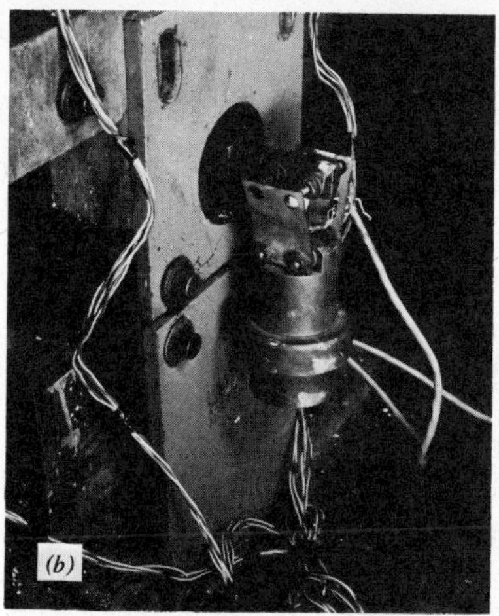

Fig. B.12 Rotation gages: (*a*) mechanical; (*b*) electrical.

located at the ends of two rods attached transversely on the adjacent sides of the column, as shown in Section *B-B* of Fig. B.11.

The overall shortening is determined by measuring the movement of the sensitive crosshead relative to the fixed crosshead using the dial gage or potentiometer.

Steel column specimens are whitewashed with hydrated lime. During testing, the whitewash cracking pattern indicates the progression of yielding in the column (the cracking reflects the flaking of the mill-scale at yielding zones).

Testing Procedure. After the specimen is aligned in the testing machine, the test is started with an initial load of 1/20 to 1/15 of the estimated ultimate load capacity of the column. This is done to preserve the alignment established at the beginning of the test. At this load all measuring devices are adjusted for initial readings.

Further load is applied at a rate of 1 ksi/min (6.9 MPa/min), and the corresponding deflections are recorded instantly. This rate is established when the column is still elastic. The dynamic curve is plotted until the ultimate load is reached, immediately after which the "maximum static" load is recorded. (The procedure for determining a "static" load is described subsequently.) After the maximum static load is recorded, compression of the specimen is resumed at the "strain rate" which was utilized for the elastic range. In hydraulic testing machines this may be accomplished, approximately, by using the same bypass valve and load valve settings as had been used in the elastic range. The specimen is compressed in the "unloading range" until the desired load-displacement curve is attained. An example of such a curve is shown in Fig. B.13.

A static condition, as is needed to obtain the "maximum static" load, is when the column shape is unchanged under a constant load for a period of time. This means that the chord length of the column must remain constant, or practically, the distance between the crossheads must remain constant during the period. For screw-type testing machines the criteria can normally be satisfied by maintaining the crossheads in a stationary position. However, it is difficult to maintain the distance between crossheads in hydraulic machines because of oil leakage and changes in oil properties due to the temperature changes that accompany pumping and throttling. To attain the static condition in the hydraulic machine, from

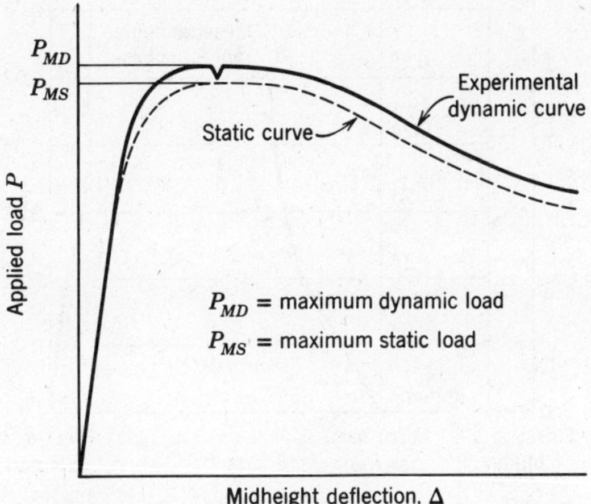

Fig. B.13 Typical load-deflection curve of column.

the dynamic state, the bypass valve is further opened slowly until further lateral deflection of the column at mid-height ceases. The cessation of lateral deflection amounts to the condition of constant chord length. Alternatively, the relative crosshead displacements may be monitored, but this parameter is usually not as sensitive as lateral displacement.

(2) Test Results

Presentation of the Data. The behavior of the test specimen under load is determined with the assistance of measurements of lateral deflections at various levels along the two principal directions, rotations at the ends, strains at selected cross sections, angles of twist, and the column shortening. These measurements are compared to theoretical predictions. The results of the test are best presented in diagrammatic form. Such plots are shown in Figs. B.14 through B.19.

In Fig. B.14 typical plots of initial camber and sweep for a column are shown. They are used to determine the reduction in column strength due to initial out-of-straightness.

Figure B.15*a* shows the midheight load-deflection curve of the column

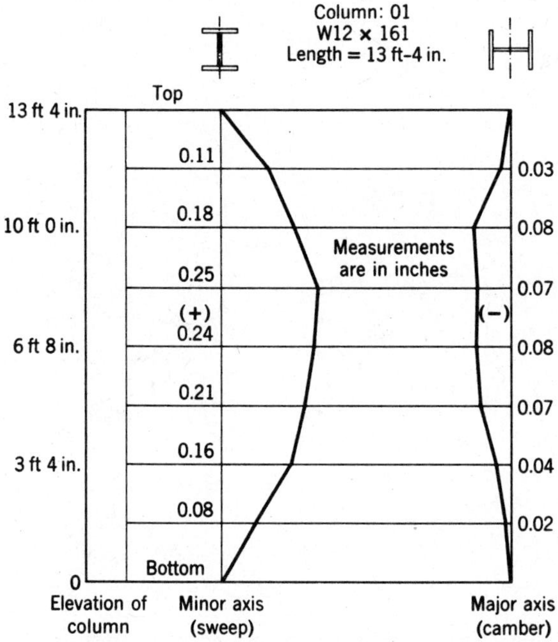

Fig. B.14 Initial camber and sweep of a column specimen.

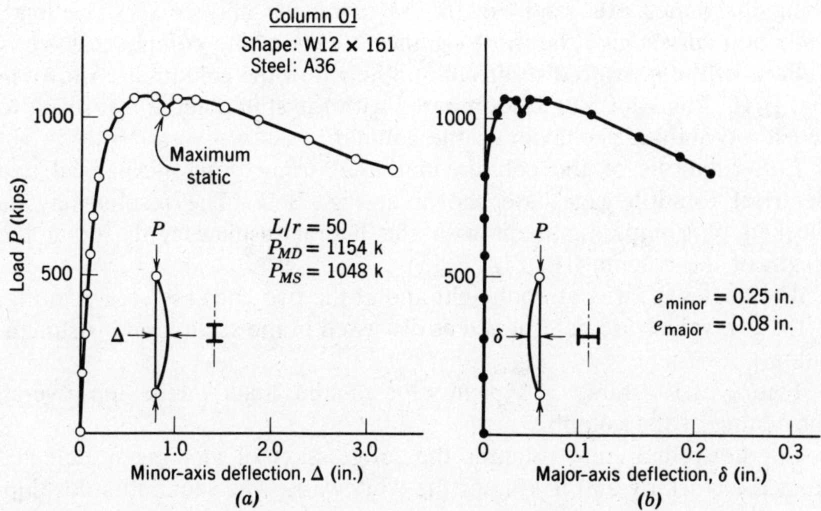

Fig. B.15 Load-deflection curves.

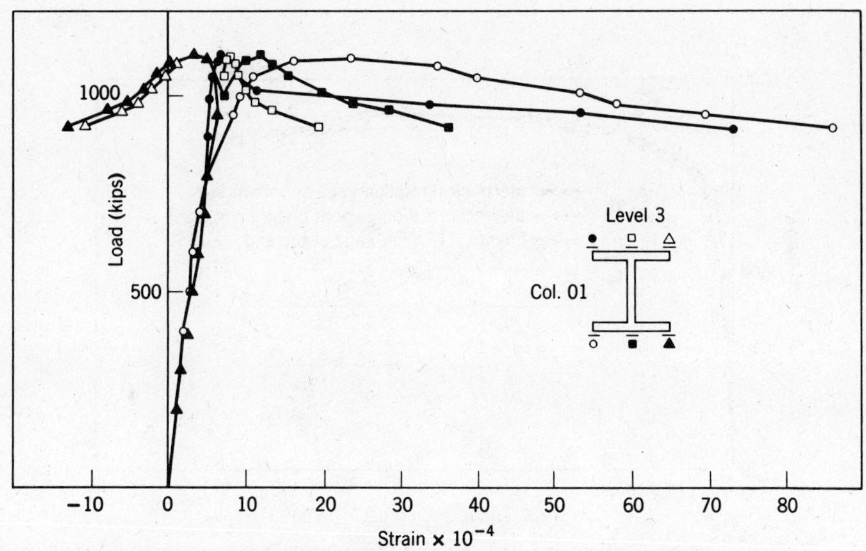

Fig. B.16 Strain measurements at mid-height section using strain gages.

along the minor axis, and Fig. B.15*b* along the major axis. The load-deflection curves give the most significant data of the column test.

Plots of the measured strains at mid-height of the column are shown in Fig. B.16. This plot may be compared with the stub column test result to detect any unusual behavior of the column.

End rotations of the column measured using both mechanical and electrical rotation gages are shown in Fig. B.17. The results may be checked by comparing them with the lateral displacements along the length of the column (see Fig. B.15).

The angles of twist at midheight and at the two ends are shown in Fig. B.18. The values are determined as discussed in the section on instrumentation.

Figure B.19 shows a typical plot of the load versus the overall shortening of the column.

For hot-rolled steel columns the progression of yielding is detected from the cracking and flaking of the whitewash. The sequential development of the whitewash cracks may be recorded to indicate the yielding pattern during loading. Whenever local buckling or any other phenomena occur during the test they should be recorded.

Evaluation of Test Results. The test results may be evaluated by comparing the experimental load-deflection behavior and the theoretical prediction. A preliminary theoretical prediction can be made based on simplified assumptions of material properties, residual stresses, and mea-

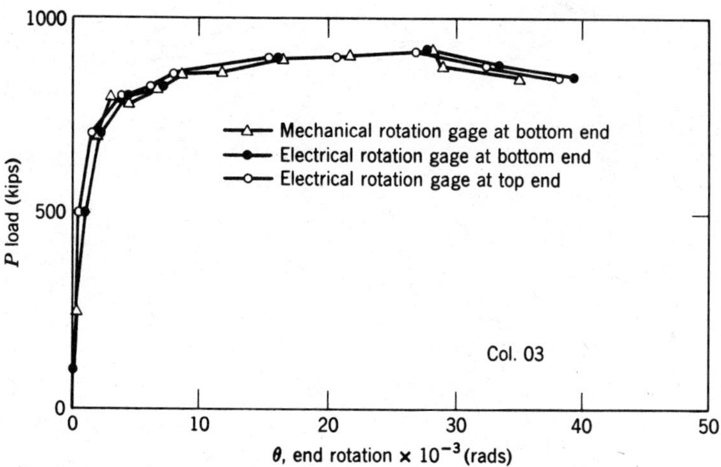

Fig. B.17 End rotations of column 03 determined from mechanical and electrical rotation gages.

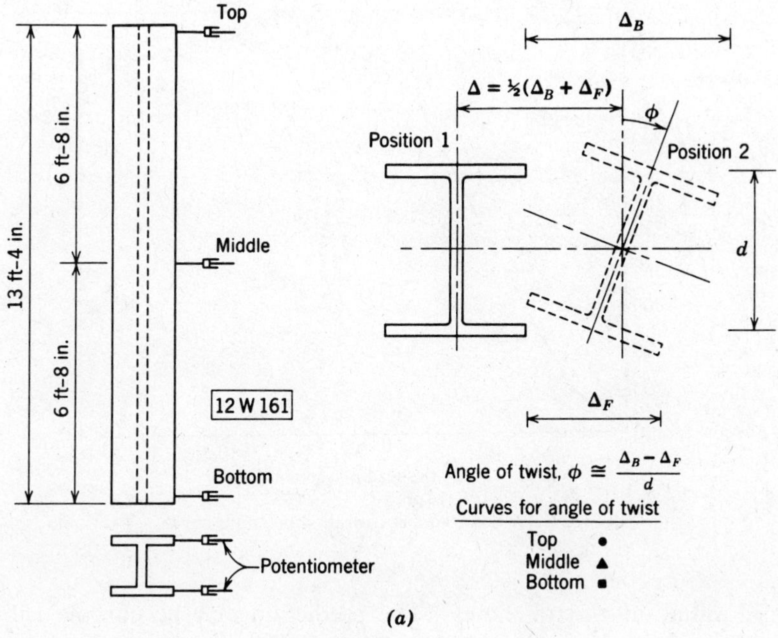

(a)

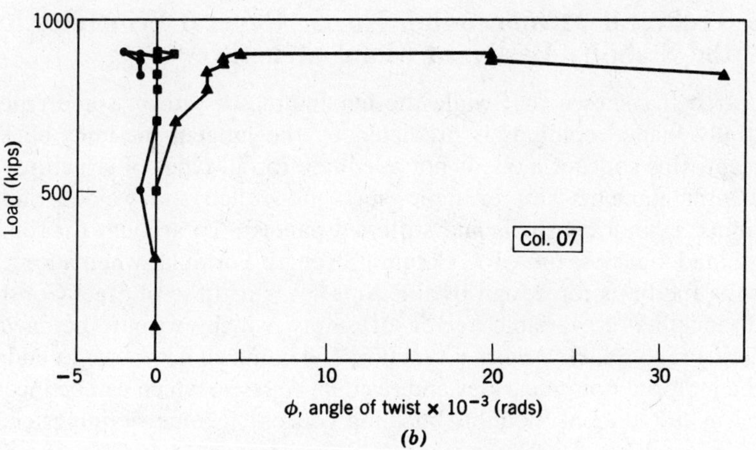

(b)

Fig. B.18 Angles of twist at three levels.

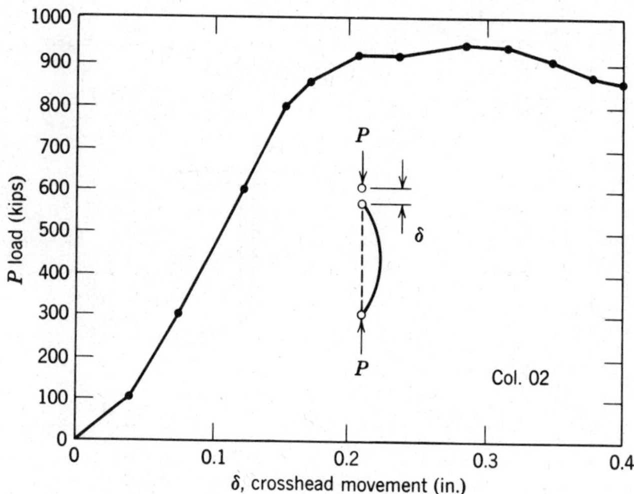

Fig. B.19 Load versus overall shortening curve.

sured initial out-of-straightness. The prediction may be improved if the actual residual stresses and the variations in material properties are used in the analysis. These properties should be determined from preliminary tests of specimens obtained from the original source stock. The preliminary test specimens should be cut immediately adjacent to that portion of the stock utilized for the column specimen.

B.5 Technical Memorandum No. 5: General Principles for the Stability Design of Metal Structures*

Research has shown that while the maximum strength of some types of centrally loaded columns is predicted by the tangent modulus buckling concept, this concept alone is not adequate for all types of structures and structural elements (for example, most hot-rolled and welded built-up columns, cylindrical shells and stiffened panels). To account for some of these inadequacies, the CRC Column Strength Formula, when adopted in 1960 as the basis for design by the American Institute of Steel Construction, included a variable factor of safety which, within the inelastic buckling range, increased as a function of column slenderness. In addition to the material non-linearities and residual stresses, which can be incorporated in the tangent modulus buckling concept, geometric imperfections

*Published in February 1981, *Civil Engineering* (ASCE).

(such as out-of-straightness), loading history, large deflections, post-buckling strength and behavior and connection response may affect significantly the limit of structural usefulness.

The maximum load resisting capacity of a member or frame determined inelastically by the inclusion of the effects mentioned above, has been termed its "maximum strength". Although the strength of an element may be thought of as a uniquely defined quantity, "maximum strength" is definitive in contrast with other concepts of strength such as tangent modulus, elastic modulus or first yield strength. For members, as distinct from frames, the dominant factor in the determination of the maximum strength, in addition to residual stress and material non-linearity, is member out-of-straightness.

In elements for which the effect of imperfections must be assessed in the determination of strength, end restraint may be significant, as it is in the tangent-modulus approach, and likewise, should be considered in formulation of design criteria.

In accordance with the foregoing discussion, the following principle represents the currently (1979) held position of Structural Stability Research Council with respect to stability in the design of metal structures:

> Maximum strength, determined by evaluation of those effects that influence significantly the maximum load-resisting capacity of a frame, member or element, is the proper basis for the establishment of strength design criteria.

This philosophy underlies the research effort and the *Guide* of SSRC. Also, it is the stated philosophy of the *Manual of Stability of Steel Structures* of the European Convention for Constructional Steelwork. It incorporates the tangent modulus buckling approach insofar as it provides the proper basis for defining the maximum strength of certain types of structures and elements, but it encompasses the members of frames in which initial geometric imperfections, large deflections, post-buckling strength and behavior, residual stresses, material non-linearities, load eccentricities and end restraint must be considered.

Implicit in the tangent modulus approach recommended in TM No. 1 was a procedure to establish the column buckling load/slenderness curve, determined theoretically from the tangent-modulus curve for the specific column section under consideration. Whenever possible, the procedure for the establishment of the load carrying capacity of frames, members or elements on the basis of maximum strength should be based on a mathematical model which incorporates:

1. Experimentally determined physical characteristics, such as residual stresses, material non-linearities, and cross-sectional variations in yield strength, rationalized as may be appropriate.

2. A statistically appropriate combination of acceptance characteristics that are specified in supply, fabrication and erection standards, such as out-of-straightness, underrun of cross-section, cross-section dimensional variations, material properties and erection tolerances.
3. Effect of boundary conditions, such as restraint applied to the end of members.

When it is not possible to determine maximum strength theoretically, experimentally determined values of maximum strength may be accepted provided that the tests have been controlled and the results adjusted to compensate for the inclusion of the most adverse combination of unfavorable factors, contributing to a reduction in strength below the experimentally determined values, which has an acceptable probability of occurrence.

Although the maximum strength of frames and the maximum strength of the component members are interdependent (but not necessarily coexistent), it is recognized that in many structures it is not practical to take this interdependence into account rigorously. At the same time, it is known that difficulties are encountered in complex frameworks when attempting to compensate automatically in column design for the instability of the entire frame (for example, by adjustment of column effective length). Therefore, SSRC recommends that, in design practice, the two aspects, stability of individual members and elements of the structure and stability of the structure as a whole, be considered independently. The proper basis for member design is the maximum strength of the restrained imperfect member. Where appropriate, second order effects (such as P-Δ effects in frames) determined with due regard for non-linear and non-coexistent response, should be included with the first order effects among the actions for which the member is to be designed.

B.6 Technical Memorandum No. 6: Determination of Residual Stresses*

Abstract: The paper explains the origin of residual stresses as well as the techniques and procedures for measurement. Resisdual stresses may be induced during manufacture as a result of non-uniform cooling of the metal and/or from cold working. The magnitude and pattern of residual stress can have a pronouned influence on the behavior of structural

*Prepared by an SSRC Task Group. Chairman: T. Peköz; Members: R. Bjorhovde, S. J. Errera, B. G. Johnston, D. R. Sherman, and L. Tall. Published in *Exp. Tech.*, Vol. 5, No. 3, September 1981.

members. The most widely used technique of determining residual stresses is the method of sectioning and is fully described in detail in this paper.

List of Symbols

d = diameter of the contact edge in the gage hole
t = thickness of strip
L_g = gage length
N_i = initial middle ordinate
N_f = final middle ordinate
S_i = correction for initial middle ordinate
S_f = correction for final middle ordinate
α = internal angle of the extensometer gage point
δ = difference between initial and final middle ordinates
λ_h, λ_n, λ_s, λ_t = correction factor

Introduction

Most metal products contain residual stresses induced during manufacture. One source of these stresses in hot-rolled steel shapes is the nonuniform cooling of the metal after it leaves the rolls. Steel shapes fabricated from hot-rolled products by means of welding have additional residual stresses due to nonuniform heating of the base metal as the weld metal is deposited, and the subsequent nonuniform cooling of the weldment.

In addition to these thermal residual stresses, steel products may possess residual stresses as a result of cold working. For example, rolled structural shapes and welded built-up shapes are often bent by gagging to remove camber or sweep, or to introduce camber or sweep. In some mills hot-rolled shapes are rotary straightened, a process in which the shape is successively bent in opposite curvature several times as it passes through rolls at ambient temperature, rendering the shape essentially straight. The metal undergoes plastic deformation during rotary straightening as evidenced by flaking of the mill scale on the product.

The magnitude of the residual thermal and cold-work stresses resulting from the processes described above is far greater in the longitudinal direction of the shape than in any transverse direction except for surface effects. Longitudinal residual stresses can have a pronounced influence on the behavior of structural members, especially in the case of columns and plate structures built up by welding, and it is with the determination of such stresses that this memorandum is primarily concerned. Longitudinal residual stresses vary with respect to the width and thickness of each plate-element comprising the shape. However, except for very thick elements, and for walls of cold-formed tubular members, the variation of residual stresses through the thickness is usually not important.

Residual stresses due to cold work are present in all products whose cross section is cold-formed from sheets, either by press-braking or roll-forming. The magnitude of residual stresses is greatest in the direction transverse to the bent line, and the variation through the thickness is pronounced. However, because of Poisson's ratio, longitudinal stresses are created by the transverse stresses and these longitudinal stresses affect the behavior of members subjected to longitudinal compression, bending, and twist.

Aluminum structural shapes are usually extruded. Extruded shapes are straightened by stretching so that the finished product is virtually free of residual stresses. Aluminum sheets and plates may contain residual stresses from rolling, or from quenching if subjected to heat treatment after rolling. The flattening operation, which may consist of roller leveling or stretching, or both, does not always produce a stress-free product. These stresses may cause inconvenience in machining operations but are rarely of structural significance. Shapes built-up by welding contain residual stresses due to nonuniform heating and cooling during the welding process. The magnitude of the longitudinal stresses from this source may reach the yield strength of the aluminum in its annealed state.

Techniques for Determining Residual Stress Magnitude

The techniques for determining the magnitude and distribution of residual stresses may be classified as nondestructive, semidestructive, and destructive. X-ray and ultrasonic methods are termed as nondestructive. Unfortunately, at the present time these methods are not practical for determining residual stresses in structural members. However, recent research such as that reported in Ref. B.7, shows significant potential benefits. The ultrasonic technique provides information on the difference between the principal (in the geometric sense) stresses only, and it is not practicable to interpret the residual stress in a specific direction.

In the semidestructive and destructive techniques the residual stresses are determined from distortions caused by the removal of material. Since the residual stresses in the body are in equilibrium, removal of stressed material by cutting, planing, drilling, grooving, or etching causes a relaxation in stress and a corresponding strain. The strain is measured, and the relaxation of stress is obtained by using Hooke's Law. The testing technique is said to be semi-destructive if the amount of material removed is small compared to the initial volume of the specimen and if the specimen can be made whole again, as by welding. The technique is termed destructive if so much material is removed that the specimen is virtually destroyed.

Semidestructive techniques usually involve drilling holes in the specimen. The action of drilling alters the internal stress distribution resulting in deformation at the surface of the specimen. This deformation is interpreted as caused by residual stress. The hole-drilling method of residual stress determination as developed by Mathar, and that of Soete and Vancrombrugge were studied by Tebedge, Alpsten, and Tall (B.8, B.9). In Mathar's method two gage points are installed diametrically opposite and equidistant from the center of the hole to be drilled, to suit a sensitive mechanical extensometer. The axis of the gage points and hole should be in the direction of the stress to be determined. The initial distance between the gage points is measured, the hole drilled, and the final spacing of the gage points measured. From the displacement of the gage points, caused by the deformation of the specimen during drilling, the relaxation in stress may be determined theoretically or by utilizing a calibration test.

In the semidestructive technique developed by Soete and Vancrombrugge (B.10, B.11) the strains induced in the specimen by drilling are measured with the aid of electrical strain gages. These strain gages are more convenient and reliable for one not experienced in the use of mechanical extensometers. They are available in rosette form and in small size. The advantage of small rosettes is that the drilled hole can be of correspondingly small diameter, thus minimizing damage to the specimen and allowing the work to be done with smaller tools. The direction of the principal strains can then be determined. Although Soete and Vancrombrugge's technique seems to possess many advantages compared with destructive techniques, it has not been utilized extensively in the United States. The foregoing semidestructive techniques are described in more detail in the papers by Tebedge, Alpsten, and Tall, which also includes an extensive bibliography.

Most studies of residual stresses in structural shapes have been performed using the destructive technique. A portion of the specimen (the test piece), located at a suitable distance from the ends of the specimen, is marked into strips as shown in Figure B.20*a*. Gage holes are drilled near each end at the midwidth of each strip, and their longitudinal spacing is measured. The test piece is then cold-sawed from the specimen, appearing as in Figure B.20*b* after this operation, and the strips are cut from the test piece often by means of a thin milling cutter or band-saw. They have come to be termed *sections* and such a section is depicted in Figure B.20*c*. The distance between gage holes of each section is

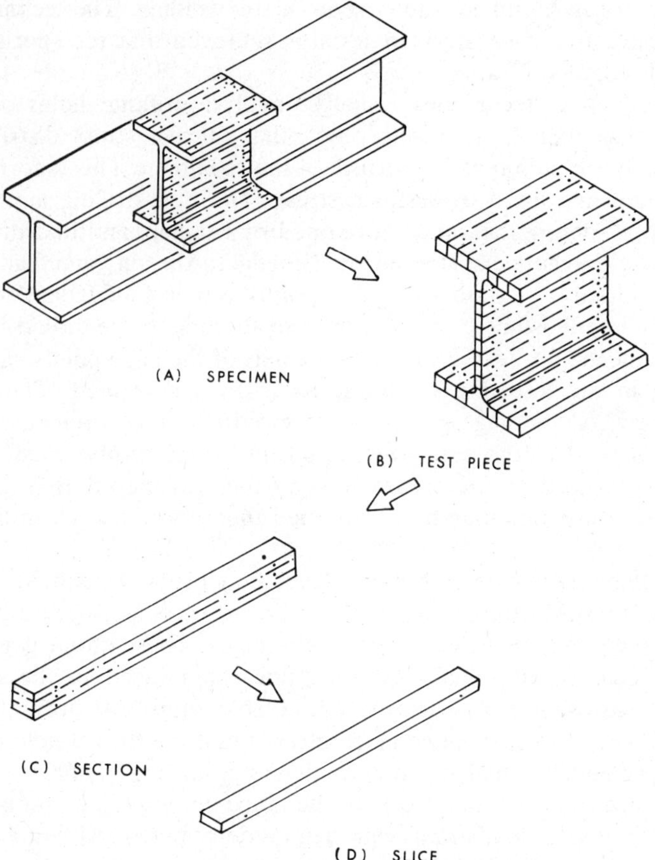

Fig. B.20 Method of sectioning.

measured and the change in length of a section is interpreted with Hooke's Law as the average value of the residual stress present in that section prior to its removal from the specimen. It is believed the method of sectioning was first used in Fritz Laboratory in the late 1940's by Johnston and Luxion (B.12).

If it is desired to determine the variation of residual stress through the thickness of the section, the section is marked into strips as shown in Figure B.20*c*, gage holes are drilled at the ends of each strip, and the longitudinal distance between these is measured. The strips are then cut from the section to obtain what are termed *slices*. The distance between the gage holes of each slice is measured again, and the change in length of a slice is a measure of the average residual stress in the section from which it was cut. This destructive technique has been named *the method of sectioning* and will be described in more detail subsequently.

Electrical strain gages can be used to measure the strains in the method of sectioning. They have the distinct advantages that smaller sections may be used and fewer corrections are necessary. However, very careful and time-consuming techniques are required to protect the gages from damage during the cutting operation and to ensure the same zero base if the gage wires have to be re-attached to the instrumentation for the final readings.

Method of Sectioning

Preparation of Test Piece. The stock should be carefully examined for evidence of cold work prior to selecting the portion to be used as a specimen. If it is desired to determine the residual stresses due to thermal effects only, then the specimen should be cut from the portion of the stock that does not have transverse cracks in the mill scale if any are present. Such cracks are usually evidence that the member has been cold-straightened, either by gagging or rotary straightening. On the other hand, if it is desired to evaluate the residual stresses caused by both cold work and thermal effects (the usual situation), then the specimen should have a representative pattern of mill scale cracks. In either case, the specimen should be cut to a minimum length equal to three times the largest transverse dimension, plus the gage length, plus 2 in. (50 mm), if it is to be cold-sawed from the stock. If the specimen is removed by flame cutting, it should be longer by an amount equal to the largest transverse dimension in order to minimize the possibility that this operation will disturb the residual stress pattern in the central portion of the specimen. The specimen should be wire brushed and washed with a solvent to remove traces of cutting oil and its ends should be deburred. The specimen is then ready for marking of the test piece.

The central portion of the specimen is usually coated with machinist's blueing or a similar compound for a distance equal to at least the gage length plus 2 in. (50 mm). Next, the lines defining the sections are scribed on both sides of all elements of the cross section, where possible. The width of the strips should be based on the expected residual stress gradient and the needs of the investigation.

When the sections have been scribed on the test piece, the gage holes for the extensometer are laid out. A gage length of 10 in. (254 mm) is recommended, partly because this is the length of one of the standard

Whittemore Gages. The centers of the holes are punched on the centerline of each strip. These punch marks serve to guide the drill bit during drilling. It is convenient to use the punch fixture in the Whittemore kit, as the gage hole centers will then be located within the measuring range of the gage. If the relaxation of strain is to be measured on one side of the section only (either as a matter of convenience or because the second side is not accessible) the gage holes need be located only on the first side. If strains are to be measured on both sides, but only one side is accessible for drilling and the thickness of the section is less than about 1 in. (25 mm), gage hole centers need only be punched on the side accessible to the drill because those on the opposite side can be located by drilling through the section. In the case of thick sections the strains should be measured on both sides. It is preferable to punch the gage hole centers on both sides provided that drilling on both sides is possible.

The importance of careful preparation of the gage holes to receive the gage points of the Whittemore Gage cannot be overemphasized. The diameter of the gage holes should be as small as practicable. Where shallow holes are drilled into the sections, or through thin sections, drill bits of about 0.03 in. (0.75 mm) diameter are suitable. For holes through thick material it is advisable to double the drill bit diameter so as to avoid excessive "wander" as the bit advances. It is desirable to utilize a drill press or, for large shapes, a magnetic base drill stand, to ensure that the holes will be as close to normal to the surface as possible. The holes must be chamfered to remove burrs from the drilling operation and to bring the contact surfaces between the test piece and the gage points of the instrument below the surface of the test piece, in order to protect them from damage during subsequent handling and machining. The depth of chamfer should be on the order of 0.02 in. (0.50 mm), and it is conventional practice to accomplish this with the reamer provided in the Whittemore kit. However, since the reamer is hand-held some researchers have found it desirable to use a guide block to aid in maintaining the reamer normal to the test piece. Such a guide block can be readily made by drilling an appropriately sized hole in a steel bar of about 1 in. (25 mm) thickness. Some investigators have reported success with a special bit with which the holes can be drilled and chamfered in an single operation (B.8, B.9).

Measuring Technique. For measuring the relaxation of strain resulting from removal of material the Whittemore Gage, or other comparable quality mechanical gage, with a nominal gage length of 10 in. (254 mm) is recommended. This is a portable mechanical extensometer with a dial gage graduated to 0.0001 in. (0.0025 mm) which is clamped to one of two steel coaxial tubes and bears on the other. To minimize temperature changes of the tubes due to handling, they are attached (internally) to a three-sided metal housing by means of flexible links. The observer holds the housing to operate the gage.

Measurements should be made with respect to the steel reference bar which is part of the Whittemore kit so that, if the temperatures of the bar and the coaxial tubes are essentially equal, no correction need be made to account for instrument errors due to ambient temperature variations during the test. Furthermore, if the test piece is steel and precautions are taken to maintain the reference bar, Whittemore Gage, and test piece at the same temperature, then all the errors due to temperature change will be minimal and no temperature corrections are necessary.

In addition to measuring the longitudinal gage lengths it is also necessary to measure the curvature in planes normal and parallel to the strip surface for both the "initial" and "final" states when through the thickness residual stresses exist. Because the equilibrium of the strip is disturbed during the cutting operation, these curvatures change. This phenomenon is especially evident in strips cut from structural tubes (B.13). It is necessary to determine these curvature changes in order to correct for the following errors:

1. Error due to "final" gage length measurement along chord between gage holes, rather than along the arc lying in the surface plane. Correction denoted as λ_s.
2. Error due to "final" gage length measurement along chord, rather than along arc in plane normal to strip surface. Correction denoted as λ_n.
3. Error due to misalignment of instrument gage point and gage hole axes due to change of curvature in plane normal to strip surface. Correction denoted as λ_h.

In order to compute corrections for the first of these errors (due to the change in curvature in the surface planes), the middle ordinates in the "initial" and "final" states, respectively S_i and S_f, should be measured. Normally, $S_i = 0$. The corrections for the last two errors are functions of the change in curvature in a plane normal to the strip and can be computed from the middle ordinates N_i and N_f.

The suggested procedures for taking the "initial" and "final" measurements are essentially alike and may be outlined as follows:

1. Thoroughly clean all gage holes utilizing cotton swabs, solvent, and compressed air blasts. If the holes were covered with adhesive tape for protection during machining, they may be coated with a gummy residue because some adhesives are soluble in some cutting oils. Such deposits must be removed in order to secure repeatability of measurements.

2. Place the test piece, sections, or slices upon a sturdy table or other support so as to bring the work to a convenient height. The table should be located away from windows, radiators, and fans, and out of the path of drafts. The material should preferably not rest directly upon the table but rather on strips of wood, to reduce heat conduction. The parts should be arranged so that minimal handling is required after testing is commenced. Place the Whittemore gage and the reference bar directly upon the material to be measured, cover with a clean cloth, and wait (usually overnight) for the specimens, reference bar, and instrument to reach the same temperature.
3. When testing is to begin, remove the cloth cover and proceed with the measurements as expeditiously as possible. The data required are indicated on the suggested data sheet shown as Figure B.21. For each side of a strip take a reference bar reading before (columns 2 and 8) and after (columns 4 and 10) the three readings for the gage length (record average in columns 3 and 9). If the reference bar readings

DATA SHEET—RESIDUAL STRAIN MEASUREMENTS

Date ______________ Test piece indent. ______________

Time start ________ Temp. start ________ Initial/final measure. (indicate which).

Time finish ________ Temp. finish ________

Observer ______________ Recorder ______________

	Side A					
(1) Section/Slice Identification	(2) Ref. Bar Reading	(3) Gage Length Avg. of 3	(4) Ref. Bar Reading	(5) Uncorrected Length with Respect to Ref. Bar	(6) Middle Ordinates S	(7) N

	Side B					
	(8) Ref. Bar Reading	(9) Gage Length Avg. of 3	(10) Ref. Bar Reading	(11) Uncorrected Length with Respect to Ref. Bar	(12) Middle Ordinates S	(13) N

Fig. B.21 Heading for residual strain data sheet.

differ by more than one or two dial indicator units, discard the set of measurements. Such a difference indicates excessive temperature change, slipping of the dial indicator in its clamp, other instrument malfunction, or improper operation of the gages. If the largest difference between the three "initial" gage length readings exceeds three units, discard the set and redress the gage holes with the reamer. (After the "initial" readings have been completed, the gage holes must not be touched, except with a cotton swab.) The uncorrected gage lengths (columns 5 and 11) are determined by subtracting the average reference bar readings from the average gage-length readings.

4. The values of the middle ordinates, S (columns 6 and 12) and N. (columns 7 and 13), are often measured with a dial indicator gage and fixture of the type described by Sherman (B.13). If the strips are cut from the test piece by a milling machine it is usually to take $S_i = 0$. When using the fixture, precautions should be taken to avoid contact with the gage holes.

Calculation of Curvature Corrections

The bending deformations accompanying removal of a section from the test piece, or a slice from a section, may have to be accounted for in determining the final gage distances. The three errors are listed previously. The first two errors, due to "final" measurement along the chord instead of the arc, may be corrected adequately by adding the following quantity to the final measured values:

$$\lambda_n = \lambda_s = \frac{8\delta^2}{3L_g}$$

where L_g is the gage length, and $\delta = (S_f - S_i)$ or $(N_f - N_i)$ depending on the change of curvature under consideration. To account for misalignment of the conical extensometer point and gage hole axes, the correction to the observed final gage length is

$$\lambda_h = \frac{4d\delta}{L_g} \tan\frac{\alpha}{2}$$

where d is the diameter of the contact edge in the gage hole and α is the internal angle of the extensometer gage point.

If measurements are made on both surfaces of a section or slice, the middle thickness relaxation of strain is usually obtained from the average of the gage length changes on both surfaces, neglecting the corrections λ_n and λ_h. On the other hand if measurements are made only on one side, then all corrections λ_n, λ_s, and λ_h may be required. Furthermore, because curvature corrections λ_n, λ_s, and λ_h refer to the surface of the strip it may be necessary to apply a correction for curvature to obtain the final length at mid-thickness, if measurements are made on one surface only. This correction is

$$\lambda_t = \frac{4t\delta}{L_g}$$

where t is the thickness of the strip.

A further discussion of the above corrections is given in Ref. B.13.

Acknowledgments

The Council wishes to thank the former members of Task Group 6 who were active during the time of preparation of earlier drafts of the report. They were: L. S. Beedle, C. Birnstiel, J. W. Clark, E. W. Gradt, R. A. Hechtman, T. R. Higgins, and B. M. McNamee.

B.7 SSRC Technical Memorandum No. 7: Tension Testing*

Introduction

The tension yield strength is the key mechanical property required by most material specifications and design practice. Because of its standard usage, it is the most accepted value for analysing and comparing test data. Usually, the comparison is performed by "normalization," that is, the test results are non-dimensionalized with respect to the yield strength (stress). Thus, the tension test becomes a most important aspect of a "test of stability" in which all or some portion of a structural shape is tested in compression. Since the tensile yield is sensitive to the rate of straining, normalized stability test data can easily be shifted by more than 20% if care is not exercised in conducting the tension test and in reporting the test method employed and its results.

Yield strength is not the only parameter that is important in evaluating tests of stability and theory, as it is often desirable to know other material properties such as the proportional limit and the strain hardening characteristics which can be obtained from a tension test. At the same time, the tension test method must conform as closely as possible to those of standard quality control tests so that the stability test results can be interpreted with respect to design standards.

*Prepared by an SSRC Task Group; Chairman: L. Tall; Members: P. C. Birkemoe, R. Bjorhovde, S. J. Errera, K. H. Klippstein, R. A. LaBoube, T. Peköz, D. R. Sherman, and R. B. Testa. Approved by SSRC Executive Committee: October 23, 1986.

Ideally, the strain rate in the tension test and the stability test should be the same. Due to the difficulty in conducting a stability test at a constant, known strain rate, the SSRC advocates the use of the static yield strength in stub column tests and the static load in tests of stability. Static values are obtained by loading the specimen with a load or deflection increment and then holding a constant distortion until the load is stabilized. This stable load is the static value. A static tensile yield strength is, therefore, the most appropriate value to be used in normalizing test data.

Purpose

This Technical Memorandum is intended to provide guidelines for obtaining the static yield strength level in a tension test so that consistent, uniform values are obtained and reported.

Equipment

Tension tests are performed on different types of testing machines in different laboratories. These machines can be grouped as screw-driven machines, manually controlled hydraulic machines and servo-controlled closed loop systems of the screw or hydraulic type. One common aspect of all these machines is that the specimen is loaded by the motion of a cross head, although a feedback mechanism can be used to relate the cross head motion to load or strain in the specimen. In manually operated machines, the feedback mechanism is the operator who watches a load dial and the output of a strain extensometer, whereas a closed loop machine uses a servo controller which provides the appropriate feedback to drive the cross head and thus maintain the desired rate of load, stroke, or strain. These systems control strain, load, or stroke more precisely and with faster response than a manually operated machine.

Another difference in testing machines is the manner of gripping the specimen. Threads, button heads, wedge grips or hydraulic grips are the most common methods. The characteristics of the grips have a distinct influence on the relationship between cross head motion and strain in the specimen. Due to inelastic creep in the gripping, the strain in the specimen may change even though the cross head motion is completely stopped.

A Class 2B extensometer (strain error <0.0002) will normally be satisfactory for monitoring strain. If accurate values of the modulus of elasticity are required in addition to the yield information, a more accurate extensometer or strain gage should be used.

Specimen

The tension test specimen should be prepared in accordance with ASTM E8 and any applicable product specification for the specimen in the test for stability. The end section for gripping and in some cases the size of the tension test specimen will be dictated by the testing machine. If the test for stability involves a rolled or standard shape, the tension test specimen should be taken from the piece as required by ASTM or other product specification. For fabricated test pieces, two tension test specimens are desirable. One should be taken from the plate or sheet material prior to fabrication. This provides a correlation with the material requirements. The other tension test specimen should be taken from the fabricated piece at a location that represents average properties resulting from strain hardening or work hardening and residual stresses. The tensile properties of this specimen are required for comparison with other data and for correlation with theory.

Procedure

Although it is desirable to operate the testing machine in a strain control mode, this may not always be practical. Even servo controlled machines may go out of control if the extensometer slips, or it may be difficult to switch to a cross head or displacement control mode during loading. Consequently, operation of the testing machine in a cross head or stroke control mode is acceptable. In this event, it will be necessary to determine the rate of cross head motion that will produce the desired strain rate by loading the specimen to less than 50% of the anticipated yield and making adjustments in the rate of cross head motion as required. When yielding occurs, it is usually necessary to reduce the rate of cross head motion to obtain approximately the same rate of straining.

The rate of strain should be approximately the same as that obtained during loading in the stability test, but still within ASTM limits. A graphic plot of stress (or load) vs. strain is desirable but the data may be taken manually in sufficient increments to produce a well defined curve. When the strain reaches a value corresponding to approximately 0.2% offset $(0.002 + \sigma_y E)$,* the test should be interrupted by holding a constant strain or stopping the cross head motion. This condition should be maintained for at most five minutes or until the load stabilizes. The lowest value of the load and the corresponding strain should be recorded. Straining is then resumed at the original strain rate. The test should be

*If the applicable material or product specification defines yield as 0.005 total strain, that value should be used in place of 0.2% offset throughout this technical memorandum.

interrupted with static load values recorded at least two more times before strain hardening begins or at 0.005 increments of strain. Straining at the original rate should continue until the initial strain hardening characteristics of the material are evident, at which time the rate can be increased according to ASTM procedures until failure occurs.

Results

The stress-strain information should form one of the two curves shown in Fig. B.22. If the resulting curve is the flat yielding Type A, the static yield should be reported as the average of the three low values obtained. For rounded Type B stress-strain curves, a curve should be drawn through the low points of the three interruptions and the static yield determined by the 0.2% offset intercept.

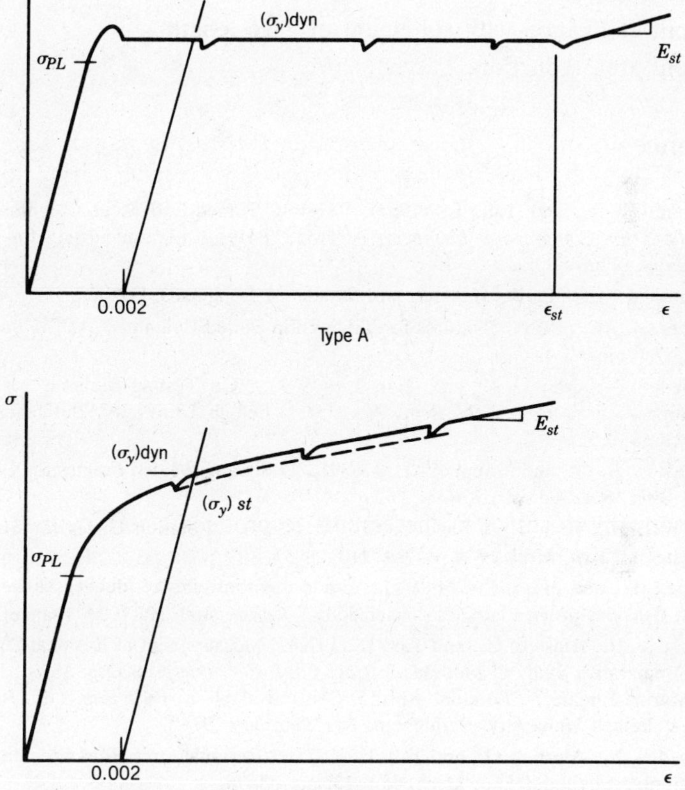

Fig. B.22 Typical tension test stress-strain curves.

Since the yield strength is not the only parameter important in stability theory, the following information obtained in the tension test should be reported:

1. Static yield strength (indicate if obtained by 0.2% offset).
2. Dynamic yield strength and strain rate.
3. Proportional limit.
4. Strain at initiation of strain hardening.
5. Strain hardening modulus.
6. Modulus of elasticity if other than the normally accepted value for the material has been obtained.

Three additional values that are normally part of a tension test should be reported for quality control purposes, although they are not a factor in stability.

7. Ultimate strength.
8. Percent elongation with statement of gage length.
9. Percent area reduction.

References

B.1 Estuar, F. R., and Tall, L. (1967), "Testing Pinned-End Steel Columns," *Test Methods for Compression Members*, ASTM STP419, American Society for Testing Materials, Philadelphia.

B.2 Salmon, E. H., (1931), *Materials and Structures*, Longmans, London.

B.3 Huber, A. W., (1958), "Fixtures for Testing Pin-Ended Columns," *ASTM Bull.*, No. 234, December.

B.4 Tebedge, N., Marek, P., and Tall, L. (1971), "On Testing Methods for Heavy Columns," *Fritz Eng. Lab. Rep., No. 351.4,* Lehigh University, Bethlehem, Pa., March.

B.5 Johnston, B. G., and Mount, E. H. (1939), "Designing Welded Frames for Continuity," *Weld. Res. J.*, Vol. 18.

B.6 Yarimci, E., Yura, J. A., and Lu, L. W. (1968), "Rotation Gages for Structural Research," *Exp. Mech.*, Vol. 8, No. 11.

B.7 Castex, L., and Maso, J. C. (1975), "Étude des contraites résiduelles, diffractometriex, dans des profilés laminés d'acier doux," *Constr. Met.* (CTICM, France), No. 2.

B.8 Tebedge, N., Alpsten, G., and Tall, L. (1972), "Measurement of Residual Stresses—A Comparative Study of Methods," *Proc. Conf. Rec. Interpret. Eng. Meas.*, Institute of Marine Engineers, London, April 5. (Also available as *Fritz Eng. Lab. Rep. No. 337.8,* Lehigh University, Bethlehem, Pa., February 1971.)

B.9 Tebedge, N., Alpsten, G., and Tall, L. (1973), "Residual-Stress Measurement by the Sectioning Method," *Exp. Mech.* Vol. 13, pp. 88–96.

B.10 Soete, W., Vancrombrugge, A. (1955), "An Industrial Method for the Determination of Residual Stresses," *Proc. Soc. Exp. Stress Anal.*, Vol. 8, No. 1.

B.11 Randle, N. J., and Vigness, F. (1966), "Hole-Drilling Strain Gage Method of Measuring Residual Stresses," *Exp. Mech.* Vol. 6.

B.12 Luxion, W., and Johnston, B. G. (1948), "Plastic Behavior of Wide Flange Beams," *Weld. J.*, Vol 27, p. 538s.

B.13 Sherman, D. R., (1969), "Residual Stress Measurement in Tubular Members," *ASCE J. Struct. Div.*, Vol. 95, No. ST4, pp. 635–648.

Appendix C

Structural Stability Research Council

The following is a summary of pertinent information about the SSRC.

Purposes

The general purposes of the Structural Stability Research Council shall be:

1. To maintain a forum where the structural stability aspects of metal and composite metal-and-concrete structures and their components can be presented for evaluation, and pertinent structural research problems proposed for investigation.
2. To review the world's literature on structural stability of metal and composite metal-and-concrete structures and study the properties of materials available for their construction, and to make the results widely available to the engineering profession.
3. To organize, administer and guide cooperative research projects in the field of structural stability, and to solicit financial support for such projects.
4. To promote publication and dissemination of research information in the field of structural stability.
5. To study the application of the results of research to stability design of metal and composite metal-and-concrete structures, and to develop comprehensive and consistent strength and performance criteria and encourage consideration thereof by specification-writing bodies.

Membership

Organizations or Firms concerned with investigation and design of metal and composite structures may be invited by the Council to become Sponsors, Participating Organizations or Participating Firms.

Sponsors. Minimum fee of $1,000 per year. May appoint up to 5 representatives.

Participating organization. Minimum fee of $250 per year. May appoint up to 3 representatives.

Participating firm. Fee of $125 per year. May appoint up to 2 representatives.

The voting membership of the Council consists of Representatives of Sponsors, Participating Organizations and Participating Firms; Members-at-Large; Corresponding Members; and Life Members.

Representatives. Individuals appointed by Organizations or Firms subject to the approval of the Executive Committee.

Members-at-large. Individuals who have expressed interest in the work of the Council, and who have done or are doing work germane to its interest, may be elected by the Council, following nomination by the Executive Committee. Fee—$45 for a 3-year period.

Corresponding members. Individuals appointed by the Executive Committee to maintain contact with organizations in other countries that are active in areas of interest to the Council. Fee—voluntary.

Life members. Active Council members of appropriate age and service to SSRC may be elected to the Council, following nomination by the Executive Committee. Fee—voluntary.

Sponsors (1987)

American Institute of Steel Construction
American Iron and Steel Institute
American Petroleum Institute
Canadian Institute of Steel Construction
Chevron USA, Inc.
Exxon Production Research Company
Federal Highway Administration
Metal Building Manufacturers Association
National Science Foundation
Nippon Steel U.S.A., Inc.
Shell Oil Company
Sumitomo Metal Industries, Ltd.

Participating Organizations (1987)

Aluminum Association
American Institute of Architects
American Society of Civil Engineers
American Society of Mechanical Engineers
Canadian Society for Civil Engineering
Corps of Engineers, U.S. Army
Earthquake Engineering Research Institute
EMC Steelal Limited
European Convention for Constructional Steelwork
Federal Highway Administration
General Services Administration
Institute of The Ironworking Industry
KEC International
Langley Research Center, NASA
National Bureau of Standards
Naval Ship Research & Development Center, U.S. Navy
Steel Joist Institute
Structural Engineers Association of California
The Steel Institute of New York
Welding Reseach Council

Participating Firms (1987)

Allison, McCormac & Nickolaus, P.A.
Amirikian Engineering Company
Amoco Production Company
Arabian American Oil Company
Michael Baker, Jr. of New York, Inc.
Balke Engineers
Basil Engineering Corporation
Alfred Benesch & Company
N.H. Bettigole, P.A.
Black & Veatch
Bakke, Kopp, Ballou, Mc Farlin
Brandow & Johnston Associates
Buckland and Taylor Ltd.
Butler Manufacturing Company
Carruthers & Wallace Limited
CBI Industries, Inc.
CH2M Hill

Computerized Structural Design, Inc.
Conoco Inc.
Copperweld Tubing Group
James Madison Cutts
DRC Consultants, Inc.
H.J. Degenkolb Associates
Earl & Wright
Edwards and Kelcey, Inc.
Envirodyne Engineers
Feld, Kaminetzky & Cohen, P.C.
Fluor Engineers, Inc.
Gannett Fleming Corddry and Carpenter, Inc.
Gilbert Associates, Inc.
Graef, Anhalt, Schloemer & Associates, Inc.
Greiner Engineering Sciences, Inc.
Hardesty & Hanover
Hazelet & Erdal, Inc.
Howard Needles Tammen & Bergendoff
Iffland Kavanagh Waterbury, P.C.
J.P. Kenny & Partners Ltd.
LeMessurier Consultants Inc.
Lev Zetlin Associates, Inc.
A. G. Lichtenstein & Associates, Inc.
Stanley D. Lindsay & Associates, Ltd.
Loomis and Loomis, Inc.
Chas. T. Main, Inc.
Marathon Oil Company
McDermott Incorported
Midgley-Clauer Associates, Inc.
Mobil Research & Development Corporation
Modjeski and Masters
MTS Systems Corporation
Walter P. Moore & Associates, Inc.
Morrison Hershfield Limited
Parsons Brinckerhoff Quade and Douglas, Inc.
PRC Engineering
Proctor & Redfern Limited
Leslie E. Robertson Associates
Sargent & Lundy
Science, Engineering, Management, Inc.
Seelye, Stevenson, Value & Knecht, Inc.
Skidmore, Owings & Merrill

Steinman Boynton Gronquist & Birdsall
Stone & Webster Engineering Corporation
Sverdrup & Parcel and Associates, Inc.
Tippetts-Abbett-McCarthy-Stratton
URS/John A. Bulme & Associates
URS Company, Inc.
Veritec
Vollmer Associates
Weidlinger Associates, Inc.
Weiskopf & Pickworth
Whitney, Bailey, Cox & Magnani
Wiss, Janney, Elstner Associates, Inc.

Activities (1987)

No.	Task Group	Chairman
1	Centrally Loaded Columns	R. Bjorhovde
3	Beam-Columns	Z. Razzaq
4	Frame Stability and Columns as Frame Members	M. H. Ackroyd
6	Test Methods	L. Tall
11	International Cooperation on Stabilty Studies	D. Sfintesco J. S. B. Iffland (V. Ch.)
12	Mechanical Properties of Metals	R. B. Testa
13	Thin-Walled Metal Construction	W. W. Yu
14	Horizontally Curved Girders	C. H. Yoo
15	Laterally Unsupported Beams	S. Vinnakota
17	Doubly Curved Shells and Shell-like Structures	K. P. Buchert
18	Unstiffened Tubular Members	P. C. Birkemoe
20	Composite Members and Systems	R. W. Furlong
22	Stiffened Cylindrical Members	C. C. Capanoglu
24	Stability Under Seismic Loading	F. Y. Cheng
25	Connection Restraint Characteristics	S. A. Ioannides
26	Stability of Angle Struts	L. A. Lutz
27	Plate and Box Girders	M. Elgaaly

No.	Subject	Task Report
11	Stability of Aluminum Structural Members	M. L. Sharp
14	Fire Effects on Structural Stability	K. H. Klippstein
15	Curved Compression Members	S. Kuranishi
16	Stiffened Plate Structures	A. Mansour
18	Application of Finite Element Methods to Stability Problems	R. H. Gallagher
19	Creep Buckling	
20	Large Deflection Buckling	A. Chajes
21	Tapered Members	G. C. Lee
22	Aerospace Structures	M. Stein

Officers (1987)

Chairman: Samuel J. Errera
Vice Chairman: Gerard F. Fox
Director: Lynn S. Beedle
Treasurer: Jackson L. Durkee

Executive Committee (1987)

L. S. Beedle—Lehigh University
W. F. Chen—Purdue University
J. L. Durkee—Consulting Structural Engineer
S. J. Errera—Bethlehem Steel Corporation
G. F. Fox—Howard Needles Tammen & Bergendoff
T. V. Galambos—University of Minnesota
J. S. B. Iffland—Iffland Kavanagh Waterbury, PC
N. Iwankiw—American Institute of Steel Construction
B. G. Johnston—Consultant
K. H. Klippstein—University of Pittsburgh
R. M. Meith—Chevron U.S.A., Inc.
C. D. Miller—CBI Industries, Inc.
D. R. Sherman—University of Wisconsin-Milwaukee
J. Springfield—Carruthers & Wallace Limited

Headquarters Staff (1987)
(215) 758-3522

Lynn S. Beedle—Director
Graham S. Stewart—Technical Secretary
Lesleigh G. Federinic—Administrative Secretary

Name Index

Abdel-Sayed, G.:
(1969), 99, 147, 261, 280
(1973), 302, 322
Abel, J. F.:
(1979), 695
(1982), 691, 694
Abrahamson, G. R., (1966), 635, 636
Ackroyd, M. H., 755
(1979), 48, 83, 547, 565, 571
(1981), 51, 83
(1983a), 565, 566, 571
(1983b), 565, 566, 571
Adams, P. F.:
(1965), 325
(1966a), 571
(1966b), 571
(1968), 571
(1970), 295, 322, 327, 420
(1971), 538, 539, 573
(1974), 295, 322
(1977), 552, 553
Ades, C. F., (1957), 475, 508
Aglan, A. A., (1973), 302, 322
Ajmani, J. L., (1971), 420, 421
Akay, H. V., (1977), 170, 184, 690, 692
Alexander, J. M.:
(1963a), 465, 509
(1963b), 465, 509
Allen, H. G., (1980), 93, 107, 147, 156, 184, 699
Allman, D. J.:
(1975), 689, 692
(1982), 691, 692
Allnutt, R. B., (1963), 148
Almeida, P. N., (1970), 597, 605
Almroth, B. O.:
(1966), 469, 508
(1975), 23, 25, 156, 184, 699
Alpsten, G. A., (737)
(1969), 86
(1970), 38, 43, 45, 83, 84
(1970a), 84
(1970b), 45, 85
(1972), 84, 748
(1972a), 38, 83
(1972b), 84
(1973), 748
Althouse, W. S., (1962), 499, 511
Alvarez, R. J., (1969), 547, 548, 571
Alwar, R. S., (1979), 642, 653
Amazigo, J. C., (1967), 496, 510
Amirikian, A., (1952), 330, 331, 357
Amiro, I. Y., (1979), 636, 654
Anderson, D., (1968), 547, 573
Anderson, G. B., (1959), 358
Anderson, G. C., (1963), 330, 357
Anderson, J. M., (1972), 167, 184
Anderson, R. G.:
(1968), 687, 693
(1969), 152
Ang, A. H. S.:
(1974), 663, 665, 671
(1975), 663, 671
Anslijn, R., (1983), 314, 315, 323
Apparao, T. V. S. R.:
(1969), 448
(1976a), 449, 455
(1976b), 448, 455
Appl, F. M., (1968), 357
Arai, H., (1970), 149

Argyris, J., (1977), 691, 693
Ariaratnam, S. T.:
(1968), 641, 654
(1971), 658
Aribert, J. M., (1981), 230, 234
Ari-Gur, J., (1980), 636, 653, 685
Arnold, P.:
(1966a), 559, 571
(1966b), 559, 571
(1968), 537, 571
Atrek, E., (1980), 446, 454
Atsuta, T.:
(1972a), 293, 310, 323
(1972b), 290, 323
(1974), 293, 309, 323
(1976), 58, 85
(1977), 284, 293, 294, 297, 298, 301, 305–310, 313, 324, 685, 693, 699
Attenbury, T. J., (1976), 511
Austin, W. J., 13
(1957), 165, 184
(1961), 305, 323
(1971), 577, 580, 605
(1976), 579, 581, 605

Back, G., (1930), 95, 152
Bagchi, D. K.:
(1970), 223, 238
(1972), 238, 458
(1975), 234
Bagci, C., (1980), 686, 693
Baker, E. H., (1968), 507, 508
Baker, J. F., (1948), 84
Baker, M. J., (1982), 663, 672
Baldry, J. A. S., (1974), 491, 512
Ball, R. E., (1975), 642, 654
Ball, W. E., (1962), 491, 508
Ballio, G.:
(1973), 290, 323
(1981), 290, 294, 305, 323
(1983), 699
Banavalkar, P. V., (1971), 456
Barnoff, R. M., (1957), 517
Baron, M. L., (1970–1971), 484, 510
Barr, A. D. S., (1971), 559, 571, 656
Barsch, W., (1973), 290, 292, 326
Barsoum, R. S., (1970), 169, 184, 685, 693
Bart, R., (1957), 488, 510
Basdekas, N. L.:
(1960), 484, 509
(1966), 491, 508
Basler, K.:
(1960), 190, 192
(1960a), 190, 234
(1960b), 190, 234
(1961), 223, 234, 264, 280
(1962), 190, 239
(1963), 194, 205, 206
(1963a), 190, 193, 219, 234
(1963b), 190, 203, 234
Basu, A. K.:
(1967), 363, 385
(1968), 368, 385
(1969), 368, 385
Batdorf, S. B.:
(1946), 109, 147
(1947), 109, 147, 508
(1947a), 469, 509
(1947b), 478, 509
(1948), 143, 151
Bathe, K. J.:
(1973), 681, 693
(1976), 683, 693
(1979), 686, 693
(1980), 693
(1981), 697
Batterman, R. H., (1967), 36, 37, 45, 68, 69, 72, 75, 83, 84
Batoz, J. L.:
(1979), 684, 691, 693
(1980), 693
(1982), 688, 693
Beaulieu, D.:
(1977), 552, 553, 571
(1985), 457
Becker, H.:
(1952), 636, 639, 655
(1957–1958), 91, 112, 127, 131, 132, 149, 468, 502, 510, 612, 625
(1958), 491, 509
(1963), 143, 147, 148
Beedle, L. S., 12, 744, 756
(1952), 87, 325
(1955), 326
(1960), 33, 34, 84
(1968), 231
(1969), 542, 571
Beer, G., (1970), 84

Beer, H., (1970), 33, 34, 46, 84
Belytschko, T.:
(1976a), 640, 654
(1976b), 640, 654
(1977), 642, 654
Benito, R., (1984), 436, 458
Benjamin, J. R., (1970), 663, 671
Benjamin, T. B., (1961), 645, 654
Benthem, J. P., (1959), 122, 147
Benton, M. D., (1980), 393, 406
Bergan, P.:
(1972), 683, 693
(1978), 691, 694
(1978a), 674, 693
(1978b), 684, 693
Bergfelt, A.:
(1968), 226, 230, 234
(1971), 230, 234
(1976), 227, 234
Bergmann, R.:
(1977), 327
(1984), 372, 384
Bergmann, S., (1932), 105, 147
Berks, W., (1952), 488, 509
Bernard, A., (1973), 76, 84, 635, 657
Bernard, J. E., (1973), 67, 84
Bernstein, M. D., (1979), 643, 655
Beskos, D., (1977), 686, 693
Bez, R., (1983), 310, 323
Bhashyam, G. R., (1982), 689, 696
Bijlaard, P. P., 708
(1949), 93, 148, 493, 509, 520, 620, 624
(1950), 93, 148
(1952), 122, 148
(1952a), 436, 454
(1952b), 436, 455
(1955), 292, 323
(1957), 103, 148
Birkemoe, P. C., 755
(1977a), 57, 84
(1977b), 57, 64, 84
(1979), 46, 58, 64, 84
(1983), 472, 509
Birnstiel, C., 744
(1968), 314, 315, 323
(1969), 546, 547, 548, 557, 571, 573
(1971), 573
(1980), 542, 553, 572
(1982), 552, 556, 573
Biswas, M., (1983), 568, 572
Bjørhovde, R.:
(1971), 34, 40, 41, 45, 57, 59, 84
(1972), 34, 37, 38, 40, 41, 45, 46, 47, 50, 58–62, 64, 83, 84, 666, 667, 670, 671
(1977), 46, 58, 64, 84
(1978), 58, 84, 663, 671
(1979), 46, 58, 64, 84
(1980), 64, 84
(1981), 51, 83, 84, 734
(1984), 51, 84
Bleich, F.:
(1951), 124, 148
(1952), 91, 92, 148, 156, 157, 169, 184, 318, 323, 388, 391, 398, 406, 516, 529, 699
Blevins, R. D.:
(1974), 646, 654
(1977), 643, 654
Blumenberg, W. F.:
(1959), 491, 513
(1965), 493, 496, 509
Bodner, S. R.:
(1952), 409, 488
(1957), 491, 509
Bogdanoff, J. L., (1964), 185
Boichot, L., (1964), 484, 509
Boley, B. A., (1963), 329, 357
Bolotin, V. V.:
(1956), 559, 560, 562, 572
(1963), 648, 654
(1964), 628, 629, 654
(1968), 654
(1969), 663, 671
Bolourchi, S., (1979), 686, 693
Borovikov, V., (1970), 238
Bossert, T. W., (1967), 223, 234
Braham, M., (1980), 223, 234
Bresler, B., (1968), 446, 455
Brockenbrough, R. L.:
(1968), 699
(1974), 99, 103, 112, 448
Brolin, C. A., (1972), 393, 406
Brown, J. E., (1968), 559, 572
Brown, P. T., (1968), 169, 184
Brozzetti, J.:
(1970a), 34, 84
(1970b), 38, 40, 43, 85
(1972), 84
Bruce, V. C., (1954), 649, 650, 655
Brungraber, R. J., (1962), 78, 79, 85

Brush, D. O., (1975), 23, 25, 156, 699, 842
Bryan, E. R., (1981), 446, 455
Bryan, G. H., (1891), 91, 148
Bryant, A. R., (1954), 482, 491, 493, 509
Buchert, K. P., 13, 755
(1964), 615, 621, 624
(1965a), 619, 620, 621, 624
(1965b), 624
(1968), 618, 620, 621, 624, 625
(1973), 615, 623, 624
Budiansky, B.:
(1962), 649, 654
(1964), 16, 25, 84
Bulson, P. S.:
(1970), 107
(1980), 93, 107, 147, 148, 156, 184, 699
Burgan, B. A., (1985), 261, 280
Butler, D. J.:
(1959), 358
(1963), 330, 357
(1966), 330, 357

Campanini, C., (1981), 290, 294, 305, 323
Campanoglu, C. C., 755
Campus, F., (1956), 297, 301, 303–305, 323
Carskaddan, P. S., (1968), 205
Carson, W., (1969), 693, 687
Carter, W. O., (1962), 357, 358
Castex, L., (1975), 748
Cedolin, L., (1978), 682, 693
Celebi, N.:
(1969), 457
(1971), 449, 455
Celigoj, C., (1979), 171, 184
Cesari, L., (1973), 633, 654
Cescotto, S.:
(1981), 193, 234
(1982), 290, 325, 292
Chagneau, A., (1973), 303, 323
Chajes, A., 12, 756
(1965), 440, 455
(1966), 440, 445, 455
(1974), 156, 184, 699
Chambers, R. S., (1977), 161, 186
Chang, C. K.:
(1973), 583, 605
(1982), 340, 358
Chang, J. G., (1971), 48, 51, 87
Chang, S., (1979), 695
Chapuis, J., (1982), 48, 70, 85, 663, 671
Chatterjee, S.:
(1977), 248, 280
(1978), 248, 280
(1980), 228, 234
Chawla, J. P., (1951), 636, 639, 654
Chen, S. S., 13
(1971), 645, 654
(1974), 643, 645, 654, 657
(1975), 644, 647, 654
(1976), 644, 645, 647, 655
(1977), 643
(1977a), 646, 654
(1977b), 643, 654
(1978), 644, 655
(1979), 643, 655
(1980), 646, 654
(1981a), 646, 647, 654
(1981b), 645, 646, 648, 655
Chen, W. F., 12, 756
(1968), 284, 310, 324
(1970), 290, 307, 323, 328
(1971a), 284, 290, 323
(1971b), 290, 323
(1972a), 293, 310, 323
(1972b), 290, 323
(1974), 293, 310, 323, 328
(1976), 58, 85, 328
(1977), 284, 293, 294, 297, 298, 301, 305–310, 313, 324, 685, 693, 699
(1977a), 306, 310, 323
(1977b), 311, 315, 323
(1978), 464, 465, 501
(1980), 48, 51, 85, 295, 324
(1981), 284, 310, 314, 317, 323
(1982), 51, 87, 314, 323
(1983), 314, 317, 324
(1983a), 50, 86, 317, 325
(1983b), 50, 86, 317, 325
(1984), 62, 86
(1985), 58, 85, 565, 572, 699
Chen, Y. C., (1979), 358
Cheney, L., (1942), 292, 325
Cheng, F. Y., 755
(1973), 559, 560, 561–564, 572
(1976a), 561, 572
(1976b), 564, 572
(1979), 565, 572
Cheng, J., (1985), 170, 188
Chenoweth, J. M., (1976), 643, 655

Cheong-Siat-Moy, F.:
(1974a), 324
(1974b), 290, 324
(1979a), 290
(1980), 295, 316, 324
Chern, C.:
(1969), 194, 197, 201, 205, 234
(1970), 217, 219, 231, 237
(1970a), 207, 234
(1970b), 202, 210, 216, 234
(1971), 213, 215, 218, 234
Cherry, S., (1960), 122, 148
Cheung, Y. K., (1969), 152
Childs, S. B., (1967), 585, 605
Chilver, A. H., (1967), 440, 455, 699
Chong, C. K., (1981), 228, 237
Christiano, P., (1975), 695
Chubkin, G. M., (1959), 314, 315, 324
Chugh, A. K., (1977), 685, 693
Chung, H., (1976), 644, 645, 647, 655
Chwalla, E.:
(1936a), 112, 148
(1936b), 148
Clark, J. W., 12, 744
(1951a), 297, 325
(1951b), 297, 325
(1955), 292, 324
(1956), 325
(1960), 165, 167, 168, 184
(1962), 78, 79, 85, 100, 150
(1963), 86
(1964), 471, 472, 477, 479, 509
(1966), 72, 73, 74, 85, 101, 148
(1968), 101, 105, 150, 433, 451, 452, 456
(1971), 192, 194, 200, 234, 238
Clough, R. W.:
(1957), 636, 639, 655
(1958), 516, 529
(1965), 688, 693
(1972), 683, 693
Cohen, M., (1979), 689, 694
Conley, W. F., (1963), 97, 148
Connors, H. J., (1970), 446, 655
Cook, I. T., (1963), 107, 148
Cook, N. E., Jr., (1983), 547, 572
Cook, R. D., (1981), 674, 682, 694
Cooper, P. B.:
(1965), 216, 234
(1966), 235
(1967), 212, 213, 217, 220, 234, 218, 280
(1971), 206, 234
(1978), 663, 671
Cornell, C. A.:
(1970), 663, 671
(1974), 663, 665, 671
(1982), 671, 672
Coronforth, R. C., (1967), 585, 605
Corrado, J. A., (1970), 237
Cottingham, W., (1962), 239
Couch, W. P., (1961), 481, 483, 484, 510
Crate, H., (1951), 511
Crawford, R. F.:
(1965), 619, 620, 624
(1980), 393, 406
Crisfield, M. A.:
(1975), 133, 148
(1976), 689, 694
(1982), 280
Crooker, J., (1968), 618, 625
Cuk, P., (1981), 304, 306, 324
Culver, C. G.:
(1966), 440, 455
(1966a), 315, 324
(1966b), 315, 324
(1968), 357, 358
(1969), 186
(1971), 181, 182, 184

Dabrowski, R.:
(1961), 440, 455
(1964), 181, 184
Daniels, J. H.:
(1967), 547, 572
(1973), 446, 455
D'Apice, M. A., (1966), 218, 235
Dat, D. T., (1980), 435, 455
Davidson, J. F., (1953), 635, 636, 655
Davies, G., (1979), 204, 235
Davies, J. M., (1981), 446, 455
Davis, C. S., (1973), 113, 114, 154
Dawe, D. J., (1971), 691, 694
Dawson, R. G., (1972), 97, 148
DeFalco, F., (1966), 48, 85
DeHeart, R. C., (1960), 484, 509
Deksnis, E. B., (1970), 645, 658
Demuts, E., (1969), 596, 605, 606
Denton, A. A.:
(1963a), 465, 509
(1963b), 465, 509
Der-Avanessian, N. G. V., (1985), 233, 237

Desmond, T. P.:
(1981a), 118, 122, 134, 148, 458
(1981b), 118, 122, 134, 148, 458
Deutsch, E., (1940), 583, 605
DeWolf, J. T.:
(1974), 122, 148, 426, 436, 437, 455
(1976), 436, 455
(1983), 187
Dhatt, G., (1979), 684, 693
Disque, R. O.:
(1973), 51, 85, 544
(1975), 48, 85
Djalaly, H.:
(1971), 324
(1973), 324
(1975), 295, 303, 304, 306, 313, 314, 324
Djubek, J., (1976), 228, 235
Dogaki, M., (1978), 239
Donald, P. T. A.:
(1961a), 604, 605
(1961b), 604, 605
Donnell, L. H.:
(1932), 153
(1934), 475, 509
(1950), 468, 469
(1956), 468, 469
Dow, N. F., (1954), 149
Dowling, P., 12
(1975), 259, 280, 281
(1976), 259, 281
(1977), 248
(1977a), 248, 269, 280
(1977b), 244, 280
(1979), 261, 280
(1980), 153, 261, 280, 281
(1981), 264, 280
(1982), 280
(1985), 261, 280
Downs, T., (1980), 295, 316, 324
Dubas, C., (1948), 222, 235
Dubas, P.:
(1971), 216, 222, 235
(1986), 258, 280
Duberg, J. C., (1950), 68, 85
Dunne, P. C., (1977), 693
Durkee, J. L., 756
Durscher, H. E., (1972), 406
Duschnitsky, V. M., (1970), 238
Duwez, P., (1950), 639, 659
Dux, P. F., (1985), 167, 185
Dwight, J. B.:
(1968), 97, 149
(1969), 102, 149
(1971), 102, 149
(1976), 149, 248, 280
Dwyer, T. U., (1965), 292, 324

Easley, J. T., (1977), 446, 455
Ebner, A., (1972), 676, 694
Eckhaus, W., (1965), 699
Eder, M. F., (1984), 509
Edwards, S. H., (1939), 504, 509
Efimov, A. B., (1972), 636, 657
Einarsson, B., (1982), 280
El Dakhakhni, W. M., (1973), 446, 455
Elgaaly, M. A., 12, 13, 755
(1970), 516, 529
(1972), 238, 458
(1973), 204, 223, 225, 235
(1975), 223, 225, 235
(1977), 223, 226, 235
(1983), 230, 235
El-Hakim, N., (1979), 456
Elishakoff, I.:
(1978a), 636, 637, 655
(1978b), 636, 655
Ellingwood, B., (1982), 663, 666, 668, 671, 672
Ellis, J. S., (1970), 314, 326
Elnawawy, O., (1981), 234
Elsharkawi, K., (1980), 134, 149
Emkin, L. Z., (1979), 546, 547, 573
Emperger, F. V., (1907), 363, 385
Engesser, F.:
(1884), 516, 529
(1891), 388, 406
(1893), 516, 529
Epstein, M.:
(1975), 686, 695
(1976), 694
Erickson, B.:
(1951), 655
(1956), 636, 655
Errera, S. J., 12, 13, 734, 744, 756
(1967), 448, 655
(1969), 454
(1971), 450, 459
(1975), 153
(1976a), 449, 455
(1976b), 448, 455
Estuar, F. R., (1967), 748

Esztergar, E. P., (1970), 484, 488, 510
Eubanks, R. A., (1969), 656
Euler, L., (1744), 1, 543, 572
Evans, H. R.:
(1973), 238
(1974), 238
(1975), 237, 281
(1980), 247, 281

Falby, W. W., (1976), 204, 236, 357, 358
Fang, P. J., (1966), 455
Faravelli, L., (1980), 322, 324
Farmer, L. E., (1966), 496, 510
Fazio, P. P., (1974), 456
Federinic, L. G., 756
Feit, D., (1972), 643, 656
Fenves, S. J., (1969), 310, 327
Fielding, D. J., (1966), 235
Fisher, G. P., 708
(1943), 150
(1952), 122, 148
(1952a), 436, 454
(1952b), 436, 455
(1955), 323
(1958), 326
(1966), 447, 458
(1967), 455
(1969), 454
Flügge, W.:
(1932), 467, 475, 499, 510
(1973), 510
Foehl, F. P., (1948), 320, 324
Fogel, C. M., (1962), 357, 358
Fok, W. C., (1977), 129, 149
Fox, G. F., 756
Fralich, R. W., (1950), 220, 238
Franciosi, G., (1955), 585, 605
Freeman, B. G., (1973), 406
Frey, F.:
(1969), 43, 85
(1973), 84
(1975), 85
(1977), 67, 86
(1980), 76, 86
(1983), 292, 295, 304, 327
Frieze, P. A.:
(1975), 280
(1977a), 280
(1977b), 280
(1979), 261, 280
Frye, M. J., (1975), 48, 85
Fujii, T.:
(1968a), 194, 195, 196, 207, 236
(1968b), 196, 236
(1971), 196, 201, 207, 236
Fujita, Y.:
(1970), 112, 149
(1981), 255, 280
Fukasawa, Y., (1983), 596, 605
Fukumoto, Y., 12
(1962), 295, 327
(1965), 325
(1966), 302, 324
(1971), 236
(1977), 129, 133, 149, 171, 181, 184
(1980), 171, 181, 184
(1981), 171, 181, 183, 184
(1982), 122, 149, 153
(1983), 32, 34, 46, 64, 85
Fung, Y. C.:
(1957), 507, 510
(1974), 615, 625
Furlong, R. W., 13, 755
(1967), 363, 383, 385
(1968), 364, 385
(1974), 368, 385
(1983), 372, 386

Galambos, T. V., 756
(1961), 288, 290, 292, 294, 324
(1962), 295, 325
(1964), 290, 292, 294, 303, 324, 328
(1965), 171, 185, 302, 304, 324, 325, 326
(1966), 302, 324
(1967), 171, 185
(1968), 156, 157, 166, 169, 184, 318, 325, 439, 456, 700
(1970), 169, 184
(1971), 320, 328
(1972), 326
(1978), 188, 663, 665, 666, 671, 672
(1981), 255, 280, 294, 304, 325, 663, 664, 666, 671
(1982), 48, 70, 85, 663, 666, 671, 672
(1983), 85, 181, 184, 666
Gallagher, R. H., 13, 756
(1963), 674, 684, 694
(1967), 623, 625, 691, 694
(1968), 623, 625
(1969), 337, 358
(1970), 169, 184, 685, 693, 694
(1973a), 682, 692, 694

Gallagher, R. H. (*Continued*)
(1973b), 684, 694
(1975), 700
(1975a), 674, 694
(1975b), 691, 694
(1976), 149, 236
(1977), 695
(1978), 682, 693
(1979), 695
Galletly, G. D., (1957), 488, 510
Galli, A., (1955), 585, 605
Gatto, F., (1979), 78
Gaylord, C. N., (1972), 191, 236
Gaylord, E. H.:
(1963), 195, 236
(1972), 191
Gedies, R. W., (1983), 172, 185, 236
Gehri, E., (1986), 258, 280
Gellatly, R. A., (1967), 625, 694
Gellin, S., (1980), 475, 510
Gent, A. R., (1977), 303, 325
Gerard, G., 25
(1952), 655
(1956), 471, 510
(1957), 22, 58, 91, 112, 127, 131, 132, 149, 468, 470, 479, 484, 502, 507, 510, 612
(1962), 700
(1963), 143, 147
(1964), 619, 620, 625
Gere, J. M.:
(1961), 91, 124–128, 153, 156, 157, 168, 169, 318, 322, 328, 388, 407, 439, 458, 475, 478, 485, 491, 492, 497, 513, 577, 579, 594, 595, 607, 613, 614, 625, 636, 639, 700
(1962), 357, 358
Gergley, P., (1971), 446, 456
Gerstle, K. H.:
(1979), 565, 572
(1983a), 565, 566, 571
(1983b), 565, 566, 571
Gietzelt, R.:
(1983), 161, 185, 326
(1984), 46, 86
Gilson, S., (1982), 290, 292, 325
Girijavallabhan, C. V., (1969), 357, 358 358
Girkmann, K., (1936), 223, 236
Glantz, W. D., (1964), 185
Glauser, E. C., (1970), 406
Godden, W. G.:
(1954), 597, 605
(1961a), 604, 605
(1961b), 604, 605
Goder, W., (1957), 363, 386
Goldberg, J., (1977), 657
Goldberg, J. E.:
(1954), 52, 85
(1964), 185
Goldsmith, W., (1960), 634, 655
Goldstein, S., (1965), 643, 655
Goodier, J. N.:
(1942), 439, 456
(1966), 635, 636, 639, 653
Goodman, L. E., (1953), 514
Goverdhan, A., (1983), 565, 572
Gradt, E. W., 744
Graham, R. R., (1965), 462, 510
Graham, T. C., (1957), 410, 421
Granholm, C. A., (1960), 226, 236
Graves-Smith, T. R.:
(1968), 122, 149
(1969), 122, 149, 436, 455
(1978), 122, 149
(1980), 149
(1981), 122, 152
Greenspan, M., (1948), 403, 407
Gregory, W. E., Jr., (1982), 640, 641, 643, 655
Grimm, D. F., (1980), 472, 474, 512
Grosskurth, J. F., (1976), 115, 149, 233, 236
Grove, R. B.:
(1983), 512
(1984), 509
(1985), 511
Grybos, R. B., (1975), 636, 655
Guiaux, P., (1974), 435, 456
Gunzelman, S.X.:
(1976), 472, 512
(1978), 472, 512

Ha, K. H., (1979), 446, 456
Haaijer, G., (1981), 321, 325
Haber, R., (1982), 691, 694
Hagedoorn, A. H., (1973), 689, 694
Hall, D. H., (1981), 32, 85, 663, 672
Hall, J. B., Jr., (1960), 112, 164
Halldorsson, O. P., (1968), 686, 694
Han, D. J.:
(1983), 324

(1983a), 317, 325
(1983b), 317
(1985), 58, 85, 699
Hancock, G. J.:
(1977), 313
(1978), 325
(1980), 170, 185, 690, 694
(1981), 122, 149
Harari, A., (1970–1971), 484, 510
Harding, J. E.:
(1977a), 280
(1977b), 280
(1979), 232, 236
(1981), 264, 280
Hariri, R., (1967), 68–71, 85
Harries, H., (1970), 583, 605
Harris, H. G., (1969), 136, 137, 149
Harris, L. A., (1957), 471, 510
Harrison, H. B., (1982), 584, 585, 605
Harstead, G. A., (1971), 531, 573
Hart, G. C., (1982), 663, 672
Hartmann, A. J., (1967), 162, 185
Hartmann, E. C.:
(1938), 392, 406
(1956), 325
(1963), 72, 86
Hartz, B. J., (1957), 636, 639
Harvey, J. M.:
(1971), 104, 122, 152
(1976), 104, 122, 152
(1977), 104, 152
Haslinger, K. H., (1977), 657, 667
Haussler, R. W., (1964), 430, 456
Hayashi, T.:
(1972a), 635, 636, 639, 655
(1972b), 636, 639, 655
Hayman, B., (1977), 237
He, B. K., (1981), 427, 457
Hechtman, R. A., 744
Hedgepeth, J. M., (1979), 393, 406
Heimerl, G. J.:
(1943), 150
(1952), 153
Heins, C. P.:
(1975), 181, 185
(1979), 161, 185
Heise, O., (1970), 484, 488, 510
Herrmann, G.:
(1967), 559, 573, 648, 655
(1968), 559, 573, 629, 632, 657
Herzog, M.:
(1973), 208, 236
(1974a), 194, 200, 210, 236
(1974b), 194, 200, 208, 230, 236
Hetenyi, M., (1946), 700
Hetrakul, N., (1978), 433, 456
Heyman, J., (1956), 699
Hibbit, H. D., (1979), 678, 694
Higgins, T. R., 744
Hill, H. N.:
(1951a), 297, 303, 325
(1951b), 297, 303, 325
(1954), 428, 456
(1956), 292, 325
(1960), 165, 167, 168, 184
Hill, W. F., (1968), 368, 385
Hlavacek, V., (1972), 433, 456
Ho, L. W., (1980), 693
Hobbs, R. E.:
(1975), 280
(1979), 232, 236
Hoff, N. J.:
(1951), 634, 636, 655
(1954), 649, 650, 655
(1956), 655, 700
(1966), 16, 25, 468, 510
Höglund, T.:
(1971a), 198, 236
(1971b), 194, 198, 205, 236
(1973), 194, 198, 201, 202, 205, 206, 236
Holmes, P. J., (1978), 645, 646, 656
Holmquist, J. L., (1939), 484, 503, 504, 510
Holt, M.:
(1938), 406
(1951), 516, 529
(1952), 488, 510, 516, 520, 523, 529
(1956), 516, 520, 529
(1957), 516, 517, 524, 529
Holzer, S. M.:
(1969), 636
(1970), 636, 656
(1971), 636, 656
(1976), 640, 656
(1979), 640, 656
Hom, K.:
(1961), 481, 483, 484, 510
(1962), 495, 510
Horne, M. R.:
(1954), 157, 185
(1956), 297, 305, 325, 699

Horne, M. R. (*Continued*)
(1960), 169, 185
(1963), 554, 555
(1971), 420, 421
(1975), 134, 149
(1977), 134, 149, 244, 281
(1980), 221, 236
Horrigmoe, G.:
(1978), 691, 694
(1978b), 693
Horsington, R. W., (1984), 145, 154
Houbolt, J. C., (1946), 109, 147
Housner, G. W., (1962), 635, 636, 639, 656
Hovik, J., (1968), 226, 230, 234
Howard, J. E., (1908), 33, 86
Hribar, J. A., (1968), 169
Hsu, C. S.:
(1966), 649, 650, 651, 653, 656
(1967), 650, 651, 653, 656
(1968), 650, 653, 656
(1972), 629, 656
Hsu, T. L.:
(1979), 358
(1981), 353, 357, 358
Hu, L. S., (1952), 516, 522, 529
Hu, X. R., (1983), 326
Huang, H. C.:
(1968a), 411, 421
(1968b), 412, 421
Huang, J. S., (1970), 237
Huang, K. Y., (1981), 641, 656
Huang, N. C.:
(1968), 642, 656
(1972), 641, 656
Huber, A. W.:
(1956), 86
(1958), 748
Hubka, R. E.:
(1951), 145, 149, 151
(1954), 149
Huffington, N. J., Jr., (1963), 636, 656
Hughes, T. J. R.:
(1979), 689, 694
(1980), 640, 656
(1981), 690, 694
Hutchinson, J. W.:
(1964), 16
(1967), 496, 510, 615, 625
(1970), 23, 25
Hutt, J. M., (1968), 572

Ibrahim, R. A., (1978), 628
Iffland, J. S. B., 756
(1980), 542, 553, 572
(1982), 536, 552, 573
Iguchi, S., (1938), 107, 109, 150
Ingvaarsson, L.:
(1975), 435, 456
(1977), 435, 456
Irons, B., (1968), 693
Itoh, Y.:
(1980), 184
(1981), 181, 184
Iványi, G., (1966), 594, 596
Iwankiw, N., 756

Jagannathan, D. S., (1975), 686, 695
Jaquet, J., (1970), 57, 86
Jendrzejczyk, J. A., (1978), 644, 655
Jensen, P. S., (1980), 678, 681, 695
Jetteur, P.:
(1983), 237, 247, 258, 281
(1984), 204, 236, 261, 281
Jirsa, J. O., (1972), 476, 477, 510
Johns, D. J., (1971), 143, 144, 150
Johns, K., C., (1977), 237
Johns, T. G.:
(1975), 505, 507, 511
(1976), 505, 507, 511
Johnson, A. E., Jr., (1957), 135, 150
Johnson, A. L.:
(1966), 96, 150, 450, 456
(1969), 450, 456
Johnson, C. P., (1977), 170, 184, 692
Johnson, E. R., (1980), 642, 656
Johnson, J. E., (1980), 700
Johnston, B. G., 734, 738, 756
(1939), 745
(1942), 292, 325
(1948), 749
(1952), 325
(1963), 68, 86
(1964), 68, 70, 86
(1967), 36, 37, 45, 68, 69, 72, 75, 84
(1968), 699
(1970), 406
(1971), 400, 406
(1974), 99, 102, 103, 112, 148
(1976), 156, 162, 185, 284, 287, 288, 310, 325, 388, 391, 406, 545, 562, 573, 700

(1981), 1, 13, 700
(1983), 1, 13
Jombock, J. R.:
(1962), 100, 150
(1968), 101, 105, 150, 433, 451, 452, 456
Jonatowski, J. J., (1969), 546, 556, 573
Jones, R. M., (1975), 145, 150
Jones, S. W.:
(1980), 48, 51, 86
(1982), 48, 49, 50, 51, 86
Jungbluth, O., (1953), 619, 620, 625
Junger, M., (1972), 643, 656

Kahl, T. L., (1979), 572
Kalyanaraman, V.:
(1977), 100, 150, 426, 436, 456
(1978), 100, 122, 150, 426, 456
Kamalvand, H., (1968), 147, 151, 326
Kaminsky, E. L., (1955), 326
Kanchanalai, T.:
(1977), 551, 554, 573
(1979), 291, 294, 325, 557, 573
Kanodia, V., (1977), 692, 695
Kanok-Nukulchai, W., (1981), 694
Karren, K. W., (1967), 423, 435, 456
Kao, R.:
(1978), 641, 642, 656
(1980), 642, 656
Kato, B.:
(1977), 58, 86
(1978), 383, 386
Kawai, T., (1968), 112, 113, 150
Kawana, K., (1965), 223, 237
Kee, C. F., (1961), 597, 605
Kelsen, G. A., (1969), 450, 456
Kemp, A. R., (1984), 295, 297, 325
Kendrick, S.:
(1953a), 491, 493, 495, 511
(1953b), 493, 495, 511
Kennard, E. H., (1956), 495, 514
Kennedy, J. B.:
(1972), 319, 325
(1982), 317, 325
Ketter, R. L.:
(1952), 292, 325
(1955), 292, 326
(1961), 288, 290, 292, 294, 324
(1962), 147, 150, 290, 294, 325, 357, 358
(1972), 358
(1981), 358
Khalifa, M. K., Jr., (1970), 516, 529
Khan, M. Z.:
(1972), 237
(1977), 223, 237
Kiciman, O., (1978), 682, 695
Kiernan, T. J.:
(1963), 484, 491, 511
(1965), 615, 616, 621, 625
(1966), 625
Kim, S. W., (1971), 547, 573
Kindmann, R., (1983), 327
King, R., (1977), 643, 657
Kinra, R. K., (1982), 512
Kirby, P. A.:
(1979), 185
(1980), 86
(1982), 86
Kiryukin, L., (1974), 636, 657
Kishima, Y., (1969), 40, 86
Kistler, R. S., (1976), 643, 655
Kitipitayangkul, P., (1979), 565, 572
Kitipornchai, S., 12
(1972), 185, 330, 358
(1975), 330, 358
(1975a), 185
(1975b), 180, 185
(1980), 167, 185, 321, 328
(1983), 319, 326
(1985), 167, 185
(1986), 321, 328
Kiziltug, A. Y., (1985), 503, 511
Kleiber, M., (1982), 690, 695
Klein, M. J., (1977), 654
Klingner, R., (1970), 162, 169, 187, 685, 696
Klippstein, K. H., 744, 756
Klitchieff, J. M., (1949), 125, 150
Klöppel, K.:
(1953), 619, 620, 625
(1956), 624, 625
(1957), 363, 386
(1958), 440, 456
(1960), 237
(1961), 595, 606
(1962), 314, 315, 326

Klöppel, K. (*Continued*)
(1964), 223, 237
(1969), 122, 150
(1973), 290, 292, 326
Knoll, W. D., (1947), 151
Knowles, R. B., (1969), 353, 363, 386
Koiter, W. T.:
(1963), 131, 150
(1970), 17, 23, 25, 615, 625
Kollár, L. U., (1966), 594, 596, 606
Kollbrunner, C. F.:
(1936), 579, 606
(1942), 579, 606
(1961), 700
Komatsu, S.:
(1971), 194, 196, 202, 237
(1977), 586, 599, 606
(1977a), 598, 603, 607
(1977b), 589, 599, 607
(1979), 604, 607
(1982), 598, 600, 607
(1983a), 601, 602, 603, 607
(1983b), 603, 607
Konig, J., (1980), 436, 457
Koning, C., (1934), 635, 638, 657
Konishi, I., (1965), 196, 237
Korn, A., (1967), 547, 548, 573
Kornev, V. M.:
(1968), 636, 657
(1974), 636, 657
Kornhauser, M., (1964), 634, 657
Korol, R. M., (1978), 476, 477, 511
Kotoguchi, H., (1983), 181, 186
Krajcinovic, D.:
(1968), 559, 573, 629, 632, 657
(1969), 685, 695
Krakelan, B., (1978b), 693
Krefeld, W. J., (1959), 358
Krenk, H. O., (1986), 672
Krenzke, M. A.:
(1959), 495, 511
(1963), 484, 491, 511
(1965), 615, 616, 621, 625
Kroll, W. D., (1943), 119, 121, 150
Kubo, M.:
(1977), 171, 184
(1980), 184
(1982), 122, 149
Kuhn, P., (1956), 190, 237
Kulak, G.:
(1982), 513
(1983), 513
Kumai, T., (1952), 114, 150
Kuo, C. T., (1968), 656
Kuper, E. J., (1969), 465, 511
Kuranishi, S.:
(1961a), 597, 606
(1961b), 604, 606
(1972), 583, 587, 606
(1973), 587, 592, 604, 606, 608
(1977), 592, 604, 606
(1979), 588, 606
(1980), 588, 606
(1981), 598, 605
(1983), 608
(1984a), 606
(1984b), 592, 605, 606
Kurth, W., (1982a), 326
Kusuda, T., (1959), 129, 150

Laboube, R. A., 744
(1982), 150, 432, 457
(1983), 427, 457
Lachal, A., (1981), 234
Lakshmikantham, C., (1964), 619, 620, 625
Lamas, A. R. G., (1980), 261, 281
Langhaar, H. L., (1962), 700
Lansing, W.:
(1949), 459
(1950), 154
Larson, M. A., (1960), 447, 457
Lau, J. H., (1981), 145, 150
Laughlin, W. P., (1968), 169, 185
Lay, M. G.:
(1965), 120, 150, 171, 185, 292, 326
(1967), 171, 185
(1973), 328
(1974), 297, 326
Layrangues, P., (1959), 604, 606
Lee, B., (1970), 685, 694
Lee, G. C., 12, 756
(1960), 156, 188
(1964), 440, 458
(1967), 339, 358
(1972), 331, 342, 344, 351, 356, 358
(1974), 353, 357, 358
(1975), 357, 358
(1976), 204, 236, 357, 358

(1979), 348, 358
(1981), 330, 337, 342, 348, 353, 357, 358
(1982), 340, 358
Lee, L. H. N.:
(1959), 329, 358
(1978), 635, 636, 639, 657
(1981), 635, 636, 639, 657
Lee, S. L., (1958), 516, 529
Lee, S. S., (1968), 656
Leggett, D. M. A.:
(1941), 107, 151
(1962), 212, 238
Leigh, J. M.:
(1972), 320, 326
(1973), 328
Leissa, A. W., (1969), 145, 154
LeMessurier, W. J., (1977), 557, 569, 573
Lennox, W. C., (1971), 658
Leu, K. C., (1971), 573
Levy, S.:
(1939), 151
(1947), 112, 114, 151
Lew, H. S., (1968), 196, 237
Lian, C. Y., (1971), 152
Libove, C.:
(1948), 143, 151
(1951), 145, 151
(1952), 153
(1954), 149
(1978), 145, 151
Lim, L. C.:
(1969), 571
(1970), 302, 326
Lin, F. J., (1970), 390, 395, 406
Lin, T. Y., (1968), 455
Lind, N. C.:
(1969), 618, 625
(1973), 129, 151
(1976), 100, 151
(1986), 672
Lindberg, H. E., (1965), 635, 636, 639, 657
Lindner, J., 12
(1972), 313, 326
(1974), 171, 185
(1978), 302, 317, 326
(1981), 317, 328
(1982a), 303, 317
(1982b), 313, 316, 326
(1983), 161, 185, 314, 316, 326
(1984), 46, 86
(1985), 170, 185
Little, G. H.:
(1976), 133, 134, 149, 151, 248, 280
(1979), 122, 151
Livesley, R. K., (1956), 684, 695
Lo, H., (1951), 507, 511
Logan, J. L., (1962), 499, 511
Loganathan, K., (1979), 678, 695
Loh, T. S., (1985), 445, 457
Lord, A. R., (1912), 363, 386
Lorenz, R., (1908), 467, 511
Loughlan, J., (1980), 122, 152
Louis, H., 708
Louw, J. M., (1957), 53, 87
Lu, L. W.:
(1966a), 571
(1966b), 571
(1968), 147, 290, 294, 305, 326, 371, 748
(1969), 571
(1970), 326
(1972), 583, 587, 606
(1979), 291, 294, 325, 557, 573
(1983), 48, 87, 290, 326, 557, 573
Lubin, B. T., (1977), 644, 657
Lubinski, A.:
(1951), 499, 511
(1962), 499, 511
Lui, E. M.:
(1983), 314, 324
(1983a), 50, 86
(1983b), 50, 86
(1984), 62, 86
(1985), 565, 572
Lunchick, M. E.:
(1957), 488, 511
(1959), 494, 495, 511
(1961a), 484, 511
(1961b), 484, 511
(1963), 484, 511
Lundquist, E. E.:
(1933), 469, 512
(1952), 153
Luttrell, L. C., (1967), 446, 457
Lutz, L. A., 755
Luxion, W., (1948), 738, 749

Macadam, J. N., (1969), 465, 511

McCalley, R. B., Jr.:
(1949), 459
(1950), 154
McDermott, R. J., (1981), 434, 457
McDonald, W. S., Jr., (1966), 322, 327
MacGregor, J. G., (1982), 671, 672
McGuire, P. J., (1971), 186, 188
McGuire, W., (1968), 55, 86, 172, 700
McIvor, I. K.:
(1964), 636, 657
(1973), 635, 636, 657
McKenzie, K. I., (1964), 112, 151
McLachlan, N. W., (1947), 629, 657
McLellan, E. R., (1970), 295, 327
McManus, P. F.:
(1969), 181, 186
(1971), 181, 182, 184
McNamee, B. M., (1967), 540, 547, 573, 744
McPherson, A. E., (1939), 151
McVinie, W. W., (1982), 313, 327
McWhannell, D. C., (1971), 559, 571
Madsen, H. O., (1986), 663, 672
Madsen, I. E., (1941), 33, 86
Madugala, M. K. S., (1982), 317, 325
Maeda, Y., (1971), 206, 230, 237
Maegawa, K.:
(1983), 182, 183, 187
(1984), 182, 183, 188
Majid, K. I., (1968), 547, 573
Majumdar, S. N. G., (1971), 538, 539, 573
Malejannakis, G. A., (1977), 693
Mallett, R., (1967), 625, 694
Malyi, V. I.:
(1972), 636, 657
(1973), 636, 657
(1974), 636, 657
Malyshev, B. M., (1966), 636
Mandal, S. N., (1979), 204, 235
Mang, H., (1977), 695
Mansour, A. E., 756
Maquoi, R.:
(1971), 247, 258, 281
(1979), 62, 63, 87
(1981), 234
(1982), 314, 317, 326
(1983), 219, 222, 237
Marek, P., (1971), 748
Marincek, M., 708
Marino, F. J., (1966), 48, 85
Marlow, R. S., (1982), 512
Marsh, C.:
(1969), 319, 326, 457, 454
(1971), 152
(1972), 321, 326
(1977), 79, 81, 86
Marshall, P. J., (1970), 314, 326
Marshall, P. W., (1971), 474, 498, 512
Martin, C. H., (1965), 674, 685, 695
Marzullo, M. A., (1978), 472, 512
Maso, J. C., (1975), 748
Mason, R. G., (1958), 292, 326
Massey, C., (1971), 168, 186
Massicotte, B., (1985), 448, 457
Massonnet, C. E.:
(1947), 157, 186
(1956), 297, 301, 303, 304, 305, 323
(1962), 216, 222, 237
(1965), 289, 290, 294, 327
(1971), 247, 258, 281
(1976), 284, 306, 310, 326, 700
(1980), 455
(1981), 234
(1983), 237
Masur, E. F.:
(1954), 86
(1957), 170
Matsui, C.:
(1970), 574
(1973), 386
(1976), 692, 695
Matsuoka, O., (1976), 692, 695
Matthey, P. A.:
(1982), 327
(1983), 314
Matthiesen, R. B., (1970), 433, 457
Maugh, L. C., (1963), 411, 421
Mayrbourl, R. M., (1980), 248, 254, 264, 266, 281
Mazzonlani, F. M.:
(1974), 78, 86, 699
(1977), 67, 86
(1979), 85
(1980), 76, 83, 86, 295, 304, 327
(1983), 295, 327, 304
(1985), 700
Meck, H. R., (1965), 125, 491

Medland, I. C.:
(1973), 685, 695
(1979), 57, 86, 517, 529
(1980), 57, 87, 172, 186
Meek, J. L., (1976), 685, 695
Meier, J. H., (1945), 636, 657
Meister, M., (1961), 700
Meith, R. M., 756
Mensch, J. L., (1917), 363, 386
Mertz, K. L., (1970), 433, 457
Mescall, J., (1971), 642, 657
Mesloh, R. E.:
(1975), 511
(1976), 511
Mettler, E., (1962), 628, 657
Miakami, I., (1978), 239
Michael, M. E., (1960), 115, 116, 151
Michalos, J., (1957), 53, 87
Milbrandt, K. P., (1957), 170, 186
Miller, C. D., 13, 756
(1982), 503, 512
(1983), 466, 512, 514, 613, 616, 625
(1984), 467, 507, 509, 512
(1985), 511
Miller, C. J., (1972), 446, 457
Miller, C. P., (1939), 504, 509
Miller, R. E., (1980), 689, 695
Miller, R. K., (1979), 393, 406
Minami, K.:
(1970), 574
(1978), 385, 386
Mindlin, R. D., (1951), 688, 695
Miranda, C., (1965), 302, 327
Mitani, I., (1970), 574
Mitchell, L., (1953), 619, 625
Moffatt, K. R.:
(1975), 259, 281
(1976), 259, 281
Moheit, W., (1939), 106, 151
Mondkar, D. P., (1978), 683, 695
Montgomery, C. J.:
(1982), 513
(1983), 513
Mooney, W. G., (1957), 517
Moore, R. L., (1938), 406
Morgan, P. R., (1974), 80, 87
Morrell, M. L.:
(1972), 358
(1974), 353, 357, 358
(1975), 357, 358
Morri, (1979), 85
Morris, G. A.:
(1969), 310, 327
(1975), 48, 85
Mount, E. H., (1939), 748
Moxham, K. E., (1969), 149
Mueller, J. A.:
(1960), 234
(1966), 213, 239
Mulcahy, T. M.:
(1974), 643, 657
(1976), 643, 657
Mulligan, G. P., (1983), 122, 426, 436, 445, 457
Mungan, I., (1974), 507, 512
Munshi, R. K., (1971), 358
Murray, D. W.:
(1973), 187, 685, 696
(1975), 133, 151, 170, 677, 695
(1976), 686, 694
Murray, J. J., (1957), 410, 421
Murray, J. M., (1946), 95, 151
Murray, N. W., (1973), 133, 151
Murty, M. K. S., (1972), 319, 325
Mutton, B. R., (1973), 172, 186

Nachbar, W., (1968), 642, 656
Nadai, A., (1939), 484, 504, 510
Nair, R. S., (1975), 570, 573
Nakagawa, T., (1980), 607
Nakai, H., (1983), 181, 186
Namita, Y.:
(1968), 596, 606
(1971), 597, 598, 607
(1975), 607
(1977), 607
Narayanan, R.:
(1975), 134, 149
(1977), 134, 149
(1982), 700
(1983), 700
(1985), 233, 237
Nardo, S. V.:
(1951), 655
(1956), 655
Narmoka, M., (1971), 103, 154
Nash, W. A., (1957), 512, 478

Nasir, G., (1969), 186
Naudascher, E.:
  (1972), 643, 657
  (1980), 643, 658
Navaratna, D., (1968), 623, 625
Neal, B. G., (1950), 171, 186
Nemat-Nasser, S., (1972), 628, 658
Nethercot, D. A.:
  (1972), 158, 160, 168, 186
  (1973), 330, 358, 685, 695
  (1973a), 162, 186
  (1973b), 167, 186
  (1973c), 168, 186
  (1975), 172, 449, 457
  (1979), 185
  (1980), 86
  (1982), 86, 316, 327
  (1983), 85, 156, 158, 162, 167–170, 186, 314, 316, 317, 327
  (1984), 172, 186
Newell, J. S., (1930), 95, 151
Newmark, N. M.:
  (1933), 472, 473, 514
  (1953), 514
Newton, R. E., (1969), 687, 692
Nguyen, R. P.:
  (1978), 430, 457
  (1982), 134, 151
Nilson, A. H.:
  (1960), 446, 457
  (1980), 446, 454
Nishida, K., (1966), 625
Nishida, S., (1981), 181–184
Nishino, F., (1971), 236
Nolke, H., (1976), 232, 238
Noor, A. K., (1981), 692, 695
Novak, P., (1972), 230, 238
Novotny, R., (1965), 259, 281

Ohtsubo, H., (1968), 112, 113, 150
Ojalvo, M.:
  (1962), 295, 327
  (1965), 302, 327
  (1969), 595
  (1971), 597, 600, 607
  (1977), 155, 161, 186
  (1981), 170, 186
Okumura, T., (1971), 236
Oliveto, G., (1980), 517, 529
Olson, F. D., (1941), 475, 514
Onat, E. T., (1953), 585, 606
Opperman, H. P., (1983), 313, 327
Oran, C., (1973), 695
Osgood, W. R.:
  (1943), 612, 625
  (1951), 33, 87
Ostapenko, A.:
  (1967), 147, 153, 223, 234, 248, 281
  (1968), 223, 237
  (1969), 194, 197, 201, 205, 235
  (1970), 216, 217, 219, 220, 230, 237, 238
  (1970a), 207, 235
  (1970b), 210, 216, 235
  (1971), 207, 213, 215, 218, 235, 237
  (1976), 472, 512
  (1978), 472, 512
  (1980), 472, 474, 512
Oster, K. B.:
  (1976a), 561, 572
  (1976b), 564, 572
Osterrieder, P., (1983), 313, 327
Östlund, L., (1954), 597, 598, 604, 607
Otsuki, M., (1980), 606

Padlog, J.:
  (1963), 674, 684, 694
  (1967), 625, 694
Paidoussis, M. P.:
  (1970), 645, 658
  (1975), 645, 646, 658
Palmer, A. C., (1974), 499, 512
Pao, H. Y.:
  (1977), 122
  (1981), 153
Parikh, B. P., (1966), 547, 573
Park, R., (1969), 363, 386
Parker, J. E., (1971), 456
Parsanejad, S.:
  (1970), 230, 237
  (1971), 237
Patel, S. A., (1956), 655
Patterson, P. J., (1970), 230
Pavolovic, M. N., (1981), 45, 87
Pekoz, T., 12, 13, 734, 744
  (1969), 440, 442–445, 457, 458
  (1971), 455
  (1974), 148, 455
  (1975), 458
  (1976), 448, 455, 458
  (1977), 150, 456

(1978), 100, 150, 426, 456
(1980), 427, 435, 455, 457, 459
(1981), 427, 457
(1981a), 148, 426, 458
(1981b), 148, 426, 458
(1982), 448, 449, 457, 459
(1983), 172, 186, 426, 436, 445, 457
(1986), 437, 445, 457
Perel, D., (1978), 145, 151
Perrone, N.:
(1978), 656
(1981), 692, 695
Peters, R. W., (1954), 110, 151
Peters, S. W., (1984), 509, 511
Petrini, V., (1973), 323
Pfeiffer, P. A., (1983), 181, 187
Pfluger, A., (1961), 440, 458
Pi, H. N., (1971), 641, 658
Pian, R. H., (1946), 433, 459
Pica, A., (1980), 689, 696
Picard, A., (1985), 487
Pifko, A. B., (1969), 136, 137, 149
Pilkey, W., (1981), 692, 695
Pillai, U. S.:
(1970), 311, 327
(1974), 293, 308, 327
(1980), 314, 327
(1981), 314, 327
Pincus, G. H., (1966), 447, 458
Plank, R. J., (1974), 170, 186, 690, 696
Plantema, F. J., (1946), 472, 474, 475, 512
Plaut, R. H.:
(1971), 636, 658
(1981), 656
(1982), 640, 641, 643, 655
Popov, E. P.:
(1971), 684, 696
(1978), 682, 695
Porter, D. M.:
(1973), 238
(1974), 238
(1975), 197, 198, 202, 211, 213, 217, 237, 264, 281
Potocko, R. A., (1979), 185
Powell, G.:
(1970), 162, 169, 187, 683, 685, 696
(1978), 695
(1981), 684, 696
Prager, W., (1953), 585, 606
Prasad, J., (1962), 295, 325
Prathap, G., (1982), 689
Prawel, S. P., Jr.:
(1964), 440, 458
(1974), 330, 358
Preg, S. M., Jr., (1968), 357, 358
Prescott, J., (1946), 499, 512
Prion, H. G. L., (1983), 509
Protte, W., (1961), 595, 606
Przemieniecki, J. S., (1968), 674
Pulos, J. G., (1961), 494, 495, 512
Pulos, J. O., (1963), 484, 512
Puri, A., (1977), 657

Radhamohan, S., (1974), 696
Rajasekaran, S.:
(1973), 170, 187, 685
(1975), 677, 695
(1977), 313, 327
Raju, I. S., (1974), 696
Ramberg, W.:
(1939), 96, 151
(1943), 612, 625
Ramm, E.:
(1981), 684
(1983), 313, 327
Ramsey, L. B., (1951), 124, 148
Randle, N. J., (1966), 749
Ratcliffe, A. T., (1968), 99, 149
Ravindra, M. K.:
(1976), 151
(1978), 663–666, 671, 672
Razzague, A., (1973), 688
Razzaq, Z., 755
(1981), 48, 51, 87
(1982), 313, 327
Reck, H., (1975), 427, 458
Reddy, B. S., (1979), 642, 653
Redwood, R. G., (1979), 115, 152, 233, 237
Reissner, H., (1932), 105, 147
Remeth, S., (1979), 686
Reynolds, T. E.:
(1957), 491, 493, 513
(1959), 491, 513
(1960), 484, 513
(1962), 480, 481, 483, 484, 513
(1964), 484, 509
(1971), 513
Rhodes, J.:
(1971), 104, 122, 152

Rhodes, J. (*Continued*)
(1976), 104, 122, 152
(1977), 104, 149, 152
(1979), 423, 458
(1980), 122, 152, 457
Riks, E., (1979), 683
Roberts, J. M., (1978), 656
Roberts, T. M.:
(1979), 228, 237
(1981), 228, 237
Rockey, K. C.:
(1962), 212, 238
(1963), 107, 148
(1968), 200, 238
(1969), 114, 115, 152
(1970), 238
(1971a), 211, 222, 238
(1971b), 211, 222, 238
(1971c), 238
(1972), 158, 160, 168, 186, 197, 211, 223, 238, 433, 458
(1973), 223, 225, 235, 238, 685, 695
(1974), 194, 214, 218, 238
(1975), 223, 234, 237, 281
(1979), 225, 228, 237
(1980), 114, 152, 247, 281
Rockwell, D., (1980), 643, 658
Roderick, J. W., (1948), 84
Roeder, C. W., (1984), 384, 386
Roik, K.:
(1976), 295, 327
(1977), 295, 327
(1983), 295, 316, 327
(1984), 372, 384, 386
Rolf, R. L.:
(1964), 471, 472, 477, 479, 509
(1966), 72, 73, 74, 85, 101
Romstad, K. M., (1970), 48, 87
Rondal, J.:
(1979), 62, 63, 87
(1980), 455
(1982), 317, 326
(1965), 19, 25
(1979), 100, 152
(1980), 700
Roos, E., (1956), 624, 625
Ross, D. A.:
(1976), 312, 328, 579, 581
(1978), 464, 465, 509
Roth, R. S., (1962), 649, 654
Rutenberg, A.:
(1981), 568, 569, 573
(1982), 568, 569, 573

Saito, A., (1970), 659
Sakimoto, T.:
(1971), 597, 598, 607
(1973a), 607
(1973b), 607
(1977), 599, 606
(1977a), 598, 603, 607
(1977b), 598, 599, 607
(1979), 600, 603, 604, 607
(1979a), 598
(1982), 598, 600, 607
(1983), 603, 607
(1983a), 601, 602, 603, 607
(1983b), 603, 607
Sakino, K., (1973), 386
Salama, A. E., (1968), 572
Salerno, V. L., (1961), 494, 495, 512
Salmon, C. G., (1980), 700
Salmon, E. H.:
(1921), 33, 87
(1931), 748
Salvadori, M. G.:
(1955), 157, 162, 169, 187
(1956), 157, 187, 297, 305, 328
Sandhu, R. S., (1972), 597, 600, 607
Sankar, T. S., (1968), 641, 654
Sano, Y.:
(1972a), 635, 636, 639, 655
(1972b), 636, 639, 655
Santathadaporn, S.:
(1968), 284, 310, 324
(1970), 307, 328
Sato, T.:
(1980), 606
(1983), 509
Save, M., (1965), 289, 290, 294, 327
Scalzi, J. B., (1968), 455
Scanlan, R. H.:
(1973), 643, 658
(1978), 643, 658
Schardt, R., (1958), 440, 456
Scheer, J. S.:
(1960), 237
(1976), 232, 238
Schelke, E., (1977), 693
Schibler, W., (1946), 523, 529

Schildcrout, M.:
(1947a), 509
(1947b), 509
Schilling, C. G., (1965), 462, 463, 476, 478, 480, 502, 513
Schlack, A. L., (1964), 112, 114, 152
Schmidt, G., (1975), 628, 629, 658
Schmidt, L. C., (1974), 80, 87
Schmied, R., (1969), 150
Schmitt, A. F., (1956), 636, 658
Schorn, G., (1976), 151
Schreffler, B., (1978), 677, 697
Schroter, W., (1971), 192, 194, 200, 239
Schubert, J., (1969), 150
Schueller, W., (1970), 220, 238
Schultz, G., (1970), 33, 34, 46, 56, 57, 84
Schuman, L., (1930), 95, 152
Schwartz, D., (1965), 619, 620, 624
Schwartz, E. B., (1951), 511
Schwer, L., (1977), 654
Sechler, E. E., 95
(1932), 153
(1957), 507, 510
(1974), 615, 625
Segedin, C. M., (1979), 517, 529
Seide, P.:
(1949), 124, 152
(1960), 499, 507, 513
(1961), 475, 514
Selberg, A., (1973), 195, 238
Sen, T. K., (1977), 303, 325
Senne, J. H., (1984), 423, 459
Sevin, E., (1960), 635, 636, 658
Seydel, E., (1933), 105, 152
Sfintesco, D., 755
(1970), 33, 34, 57, 85
(1976), 57, 87
Shanley, F. R.:
(1947), 30, 87
(1968), 433, 458
Sharp, M. L., 756
(1966), 124, 152
(1970), 122, 152, 438, 458
(1971), 192, 194, 200, 235, 238
Shelestanko, L. P., (1970), 230, 238
Shen, Z. Y., (1983), 48, 87, 326
Sherbourne, A. N., (1971), 125, 131, 152
Sherman, D. R., 13, 734, 744, 756
(1969), 465, 513, 749
(1971), 433, 458
(1976), 58, 87, 476, 477, 513
(1978), 513
(1984), 477
Shin, Y. S., (1975), 643, 658
Shinke, T.:
(1975), 588, 607
(1977), 586, 592, 606, 607
(1980), 588, 607
Short, R. D.:
(1957), 488, 511
(1959), 495, 511
Shukla, S. N., (1971), 597, 607
Silva, N. F., (1977), 80, 81, 86
Simaan, A.:
(1973), 458
(1976), 458
Simitses, G. J.:
(1965), 649, 651, 653, 658
(1974), 649, 650, 651, 658
(1976), 700
(1980), 649, 650, 651, 653, 658
(1982), 649, 650, 653, 658
Simiu, E., (1978), 643, 658
Simons, J., (1981), 684
Singer, J.:
(1967), 496, 513
(1980), 653
Skaloud, M.:
(1962), 219, 238
(1965), 259, 281
(1968), 200, 238
(1970), 122, 152
(1971), 217, 238
(1972), 197, 211, 230, 238
(1976), 228, 235
Skan, S., (1924), 106, 152
Smith, E. A., (1983), 187, 322, 328
Smith, J. O., (1968), 357
Smith, R. E., (1955), 454, 458
Soete, W., (1955), 737, 749
Sommers, A. E., 13
Sommerville, W., (1969), 368, 385
Sorenson, J. E.:
(1975), 511
(1976), 511
Soriede, T., (1978b), 693
Soroushian, P., (1982), 449, 457
Southwell, R. V.:
(1914), 467, 513

Southwell, R. V. (*Continued*)
(1915), 480, 513
(1924), 106, 152
Springfield, J., 528, 756
Sridharan, S.:
(1978), 122, 149
(1980), 12, 13, 122, 149
(1981), 122, 152
(1982), 152
(1984), 458
Stang, A. H., (1948), 403, 407
Stein, E., (1981), 697
Stein, M.:
(1947), 110
(1947a), 509
(1947b), 509
(1949), 124, 152
(1950), 220, 238
(1959), 131, 152
(1968), 513
Stein, O., (1936), 111, 152
Steinhardt, O., (1971), 87, 192, 194, 200, 239
Stephens, M. J.:
(1982), 476, 513
(1983), 513
Steup, H., (1957), 699
Stevens, L. K.:
(1957), 585, 607
(1981), 45, 87
Stevens, R. F., (1965), 364, 386
Stewart, G. S., 756
Stowell, E. Z.:
(1948), 93, 153
(1949), 110, 153
(1952), 91, 153
Sturm, R. G., (1941), 488, 513
Stüssi, F., (1943), 597, 604, 607
Subramaniam, C. V., (1970), 48, 87
Sugimoto, H., (1982), 48, 87
Supple, W. J.:
(1973), 700
(1978), 80, 87
(1980), 146, 153
Suzuki, S. I., (1969), 635, 636, 658
Swanger, M. H., (1979), 546, 547, 573
Swannell, P., (1976), 685, 695
Szabo, B. A., (1967), 340, 358
Szewczak, R. M., (1983), 161, 187
Szymczak, C., (1980), 685
Takeuchi, T., (1964), 194, 196, 239
Talbot, A. N., (1912), 363
Tall, L., 12, 708, 734, 737, 755
(1960), 33, 34, 84
(1964), 34, 87
(1966), 38, 87
(1967), 748
(1970), 38, 40, 43, 84, 717
(1970a), 84
(1970b), 85
(1971), 34, 40, 41, 45, 57, 59, 84, 290, 292, 328, 748
(1972), 84, 748
(1973), 748
(1974), 700
Tanavde, A. S.:
(1978), 513
(1984), 477
Tang, W. H., (1975), 663, 671
Taub, J., (1934), 635, 657, 658
Taylor, R. L., (1981), 694
Tebedge, N.:
(1970), 717
(1971), 748
(1972), 748
(1973), 748
(1974), 308, 312, 328
Tennyson, R. C.:
(1964), 23, 25
(1979), 649, 659
Testa, R. B., 744
Thoft-Christensen, P., (1982), 663, 672
Thomas, G. R., (1973), 320, 328, 684
Thomas, J. M., (1970), 675, 676
Thomasson, P. O.:
(1978), 122, 153, 436, 458
(1980), 436, 457
Thürlimann, B.:
(1953), 440, 458
(1956), 403, 407
(1960a), 190, 234
(1960b), 190, 234
(1963), 190, 205, 206, 234
Tien, Y. L.:
(1977), 153, 696
(1979), 122, 153
Timoshenko, S. P.:
(1910), 105, 153, 467, 474, 513
(1934), 111, 153
(1953–1983), 700

(1959), 687
(1961), 91, 124–128, 153, 156, 157, 168, 169, 187, 322, 328, 388, 406, 439, 458, 475, 478, 485, 491, 492, 497, 513, 579, 594, 595, 607, 613, 614, 625, 700
Tokarz, F. J.:
(1969), 606
(1971), 597, 607
(1972), 597, 600, 607
Tokugawa, T., (1929), 474, 491, 514
Tolke, F., (1949), 514
Tomii, M., 364
Tomonaga, K., (1971), 46, 87
Toprac, A. A., (1968), 196, 237
Trahair, N. S., 12
(1968), 169, 184
(1969), 162, 187, 320, 328
(1972), 167, 184, 185, 330, 358
(1973), 170, 172, 186, 187
(1974), 161, 170, 187
(1975), 162, 172, 186, 187, 330, 358, 449, 457
(1975a), 167, 185
(1975b), 180, 185
(1976), 170, 187
(1977), 32, 87, 156, 700
(1977a), 162, 167, 171, 172, 187
(1977b), 156, 162, 171, 187
(1980), 167, 185
(1981), 304, 306, 324
(1983), 171, 181, 186, 187
(1984), 172, 186
Trestain, T. W. J., (1982), 437, 458
Trilling, C., (1934), 480, 481, 484, 514
Troitsky, D. S. C., (1976), 122, 153
Tseng, W. H., (1973), 559, 560, 562, 563, 564, 572
Tsien, H. S.:
(1939), 620, 625
(1941), 514
(1942), 131, 153
Tso, W. K., (1962), 635, 636, 656
Tsui, T., (1971), 642, 657
Tucker, J. L., (1965), 688, 692, 693
Tuma, J. J., (1971), 358
Tung, T. P., (1957), 184
Tvergaard, V., (1973), 129, 153

Uciffero, J., (1972), 676, 694
Uenoya, M., (1979), 115, 152
Uno, H., (1978), 239
Urbano, C., (1973), 323
Uribe, J., (1970), 423, 459
Usami, T.:
(1970), 320, 328
(1971), 320, 328
(1982), 104, 149, 153

Vacharajittiphan, P.:
(1974), 161, 187
(1975), 162, 187
Vanaskie, W., (1979), 572
Vancrombrugge, A., (1955), 737, 749
Van den Broek, J. A., (1948), 79, 87
Van Kuren, R. C., (1964), 292, 303, 328
Van Manen, S. E., (1982), 328
Vann, W. P., (1973), 688, 696
Varghese, P. C., (1961), 364, 386
Venkataramaiah, K. R., (1979), 100, 152
Venkateswara Rao, G., (1974), 692
Venoya, M., (1979), 233, 237
Vigness, F., (1966), 749
Vinnakota, S., 12, 13, 755
(1967), 547, 548, 574
(1971), 547, 548, 574
(1972), 548, 574
(1977), 180, 301, 302, 313, 328
(1977a), 170, 171, 187
(1977b), 169, 187
(1982), 48, 88
(1983), 48, 88, 608
(1984), 48, 88, 585, 588, 608
Virdi, K. S., (1981), 313, 328
Vlasov, V. Z.:
(1959), 440, 459
(1961), 156, 169, 181, 187, 318, 328, 700
Vogel, V., (1981), 328
Vojta, J. F.:
(1967), 147, 153
(1983), 503, 512, 514
(1984), 507, 512
von Kármán, T.:
(1932), 95, 153
(1939), 620, 625
(1941), 514
(1950), 639, 659
von Mises, R.:
(1931), 480, 514
(1933), 480, 514
Von Sanden, K., (1949), 480, 514

Vos, R., (1973), 688
Vroonland, E. J., (1971), 393, 407

Wagemann, C. H., (1964), 223, 237
Wagner, H., (1931), 190, 193, 239
Wagonknecht, R., (1976), 327
Wah, T., (1967), 496, 514
Wakabayashi, M.:
(1970), 538, 559, 574
(1978), 385, 386
(1984), 383, 386
Walker, A. C.:
(1964), 122, 153
(1972), 99, 148, 223, 237
(1977), 149
(1979), 423, 458
(1980), 134, 149, 457
Wambsganss, M. W.:
(1975), 643, 658
(1976), 643, 657
Wan, C. C., (1950), 468, 469, 509
Wang, C. K., (1968), 686, 694
Wang, S. T.:
(1969), 96, 153, 450, 459
(1971), 450, 459
(1974), 427, 459
(1975), 96, 153
(1977), 122, 153, 685
(1979), 122, 153
(1981), 122, 153
Wardlaw, R. L., (1973), 643, 658
Warkenthin, W., (1965), 223, 239
Wästlund, G., (1960), 598, 604
Way, S., (1936), 111, 153
Webb, S. E., (1980), 153
Weingarten, V. I.:
(1961), 475, 514
(1965), 470, 514
(1968), 470, 471, 514
Weller, T., (1980), 653
Wenk, E., (1956), 495, 514
White, M. W., (1956), 403, 407
White, R. N.:
(1962), 239
(1976), 149, 236
Whitney, J. M., (1969), 154
Wiechart, G., (1978), 326
Wilder, T. W., (1950), 85
Wilkesmann, F. W., (1960), 223, 239
Will, K. M., (1977), 692
Wilson, E. L., (1973), 681, 693
Wilson, J. M., (1886), 189, 239
Wilson, W. M.:
(1933), 472, 473, 514
(1937), 472, 473, 514
(1941), 475, 514
Windenburg, D. F.:
(1934), 480–484, 514
(1960), 488, 574
Winegardner, R., (1975), 511
Wing, B. A., (1981), 434, 459
Winkelmann, E., (1962), 314, 315, 326
Winter, G.:
(1943), 427, 459
(1946), 433, 459
(1947), 96, 154, 427, 459
(1949), 449, 459
(1950), 96, 154
(1955), 323, 449, 459
(1958), 326, 446, 447, 459
(1960), 55, 88, 172, 187
(1965), 440, 455
(1966), 96, 150, 450, 455, 456
(1967), 423, 435, 456
(1969), 440, 443, 458
(1970), 423, 427, 429, 459
(1971), 455
(1974), 148, 455
(1975), 153, 458, 700
(1976), 455
(1977), 150, 456
(1981a), 148, 458
(1981b), 148, 458
(1983), 95, 154
Witmer, E., (1968), 625
Wittrick, W. H.:
(1952), 141, 142, 154
(1968), 133, 154
(1974), 170, 186, 690
(1984), 145, 154
Woinowsky-Krieger, S., (1959), 687, 696
Wolchuk, R., (1980), 248, 254, 264, 266, 281
Wolley, R. M., (1947), 151
Wood, R. D.:
(1973), 690
(1978), 677
(1980), 689, 696

Wood, R. H., (1974), 556, 574
Woolcock, S. T.:
(1973), 170, 187
(1974), 170, 187
(1976), 170, 187
(1980), 321, 328
(1986), 321, 328
Wright, D. T., (1965), 618, 619, 620, 622, 625
Wright, E. W., (1966), 546, 547, 548
Wu, T. S., (1953), 488, 514
Wunderlich, W., (1981), 674
Wyly, L. T., (1940), 387, 407

Yabuki, T.:
(1973), 604
(1977), 604, 606
(1979), 588, 606
(1981), 588, 598, 605, 606
(1983), 605
(1984), 585, 588
(1984a), 606
(1984b), 592, 605
Yamaguchi, K., (1977), 149
Yamaki, N., (1959–1960), 154
Yamakoshi, M., (1965), 237
Yamamoto, T., (1970), 659
Yamao, T.:
(1979), 607
(1983), 603, 607
Yang, H. T. Y.:
(1952), 33, 88
(1968), 623, 625
(1969), 114, 154
(1972), 687, 697
Yao, J. C., (1962), 475, 514
Yarimci, E.:
(1966), 547
(1968), 574
Yegian, S., (1957), 184
Yeh, S. S., (1974), 427, 459
Yen, B. T.:
(1960), 234
(1962), 190, 213, 239
(1966), 239
(1968), 237
Yener, M., (1980), 427, 459
Yonezawa, H., (1978), 203, 239
Yoo, C., 755
(1983), 181, 187
Yoshida, H.:
(1983), 182, 183, 187
(1984), 182, 183, 187
Yoshida, K., (1970), 149
Yoshizuka, J., (1971), 103, 154
Yost, M. I., (1977), 153, 696
Young, B. W., (1973), 290, 294, 328
Yu, C. K., (1971), 290, 292, 328
Yu, W. W., 12, 13, 755
(1973), 113, 114, 154
(1978), 430, 433, 456, 457
(1980), 434, 459
(1982), 104, 134, 150, 432, 457
(1984), 459
(1985), 424, 459, 700
Yura, J. A.:
(1971), 33, 51, 88, 544, 574
(1978), 180, 663, 672
(1985), 170

Zacharski, A., (1982), 690, 695
Zanon, P., (1980), 322, 324
Zender, W., (1960), 112, 154
Zetlin, L., (1955), 223, 239, 449, 459
Zhang, Y., (1982), 448, 459
Ziegler, H.:
(1968), 700
(1982), 388, 407
Zienkiewicz, O. C.:
(1968), 693
(1977), 688, 697
Zimcik, D. G., (1979), 649, 659
Zornerova, M., (1970), 122, 152
Zui, H.:
(1975), 607
(1977), 607
(1980), 607

Subject Index

AASHTO:
perforated cover plate requirements, 403
plate girder stiffener design, 220
plate girder web design, 191
pony truss design requirements, 524
shear allowance for columns, 389
web-thickness ratios, 191
Acknowledgments, chapter contributors, 12
AISC Specification, beam-column interaction formula, 315
column design formula, 669
plate girder stiffeners, 221
tapered members, 349
AISE Standard No. 13, 409
AISI specification, angle struts:
cold-formed steel members, 425
effective plate width, 425
laterally unsupported, 425
thin-walled metal construction, 425
unstiffened compression elements, 425
width-thickness ratios for stiffened plates, 425
Z-section beams, 429
Aluminum alloys, 4, 66
stress-strain curves, 4
Arches:
critical load parameters, 578
design for stability, 592
out-of-plane buckling, 593
bracing systems to prevent, 597
circular arches, 594
in plane stability, 575
linear, 577
nonlinear, 581
parabolic arches, 597
references, 605
ultimate load, 585
ultimate strength of steel arch bridges, 603
ultimate strength of steel arches, 599
AREA specification:
plate girder webs, 191
shear allowance for columns, 389
ASME Boiler and Pressure Vessel Code, circular tubes and shells, 485

Battened columns, 388
design example, 396
effective length, 398
references, 406
shear flexibility, 390
tie plates, 399
Beam-columns, 283
angle struts, 317
approximate interaction equation, 296
biaxial bending, interaction curves, intermediate and slender columns, 310
reference summary, 313–314
short (zero length) members, 307
cyclic loading, 317
equivalent uniform-moment factor, 305
in-plane bending strength, 287
reference summary, 294–295
lateral-torsional buckling, 297
reference summary, 302–304
references, 322

Beams, *see also* Box girders; Plate girders
  bracing requirements, 171
  design methods, 177
  horizontally curved beams, 181
  inelastic lateral-torsional buckling, 170
  lateral-torsional buckling, 157
    cantilever beams, 168
    doubly symmetric beams, 157
      end restraint, 160
    monosymmetric beams, 165
  references, 184
Bifurcation, definition, 6
Box girders, 241
  bending strength, 263
  combined bending and shear strength, 266
  design basis, 241
  diaphragms, 269
    stiffened, 277
    unstiffened, 272
  failures, 242, 244
  longitudinal stiffeners, 265, 268. *See also* Stiffened flat plates
  orthotropic plate approach, 258
  references, 280
  research needs, 279
  shear lag, 259
  shear strength, 263
  wide flange buckling, 246
Braced frame, definition, 7
Buckling load, definition, 7
Building frames, mill buildings, 409
  multistory frames, 531

Centrally loaded columns, 27. *See also* Column strength; Column-strength curves; Composite columns
  references, 83
Channels, lateral buckling strength, 165
Circular tubes and shells, 461
  axial stress combined with pressure, 502
  under axial compression, column type buckling, 498
    interaction between column and local buckling, 498
  general buckling behavior, 465
  material properties, 462
  production practice, 461
  references, 508
  residual stresses, 462
  stringer or ring and stringer stiffened cylinders, 496
  unstiffened or heavy ring-stiffened shells in bending, 475
    combined column and pressure loads, 502
    elastic shell buckling between ring stiffeners, 467
    under external pressure, 480, 485, 491, 495
    imperfection effects, 469
    inelastic shell buckling, 471
    shear, 479
    test result summary, 474
    in torsion, 478
Column strength, aluminum-alloy members, 68, 72
  cold-straightening effect, 43
  CRC basic column formula, 32, 34
  critical load as a basis, 29
  design of aluminum alloy columns, 72
  design procedure alternatives, 63
  design specifications, AISC, 65
    Canadian (CSA) standards, 64
    European (ECCS) recommendations, 65
    load factor and limit-states (LRFD) design, 661
  effective length, 51
  end restraint, 48, 75
  Euler load theory, 27
  initially curved members, 37
  initial yield basis, 32
  multiple column curves, 66, 67, 68
  residual stress influence, 33
  T-sections, 71
Column strength curves, aluminum alloys, 73
  ECCS multiple column curves, 65
  multiple column curves, 47, 57
  scatter band, steel columns, 59
  steel shapes, 37
  steel W-shapes, initially curved, 37
  tangent-modulus curve construction, 30
  welded steel shapes, 38
Column tests, alignment, 722
  end fixtures, 719, 720
  instrumentation, 723

presentation of data, 728
procedure, 721
Column types, battened, 394. *See also* Beam-columns; Circular tubes and shells; Composite columns; Mill-building columns
centrally loaded, 27
laced, 390
mill building, 409
perforated cover plated, 403
pipe, 498
spaced, 399
tapered, 329
T-section, 71
tubular, 498
wide-flange, 34
Composite columns, 359
column strength, 363, 377
column types, 359–363
concrete-filled steel tubes, 372
connections, 383
encased structural shapes, 373
moment-thrust interactions, 366, 367
references, 385
shear strength, 383
Compression testing of metals, 703
Compression tests, specimen preparation, 704
Critical load, definition, 7
Critical load column curves, compared with maximum strength curves, 37
Critical load theory, Euler load, 29
residual stress effect on steel columns, 33
Shanley concept, 30
tangent-modulus load, 30
Critical stress, *see* Critical load; Critical load column curves; Critical load theory
CSA Standard, thin-walled metal construction, 425
Curved members, *see* Arches
Cylinders, *see* Circular tubes and shells

Definitions, 6
Doubly curved shells and shell-like structures, 609
buckling behavior, 613
design trends, 623
general buckling, elastic, 619
references, 624
research needs, 623
Dynamic load effects, *see* Selected topics in dynamic stability

Effective column length, 7, 27, 51
battened columns, 398
chart of idealized end conditions, 52
laced columns, 398
truss members, 53
Effective width, defined, 7. *See also* Plates, post-buckling strength
Euler formula, 27
European Convention of Constructional Steelworks (ECCS), multiple column curves, 65

Factor-of-safety, *see* Structural safety
Finite-element analysis of stability, problems:
computational procedures, 678
bifurcation analysis, 680
nonlinear analysis, 682
development, 674
flat plates, 686
flexural instability, 684
formulation for elastic instability, 674
framework instability analysis, 685
references, 692
shell buckling, 690
software, 692
stiffened plates, 689
Flat plates, *see* Plates
Flexural members, *see* Beams; Beam-columns; Box girders; Plate Girders; Thin-walled flexural members
Frame stability, 531
analytical models, 536
factors influencing stability, 540
flexibly connected steel frames, 565
frame behavior, 532
references, 571
stability design procedure, 542
critical load concept, 542
Merchant–Rankine formula, 554
Moment amplification method, 557
P-delta method, 548
second order inelastic analysis, 545
stability under dynamic loads, 558

Gag-straightening effects, 43
Girders, *see* Beams; Box girders; Plate girders
Guyed towers, 528

Imperfections:
  effect on buckling behavior, 19
  effect on circular cylinders, 469
  effect on shells, 614
Instability, definitions, 8. *See also* Stability theory

Laced columns, 390
  equivalent length, 391
  references, 406
Lateral bracing requirements, centrally loaded columns, 55
Lateral support of beams, 171
Load and resistance factor (LRFD) design, 669

Members with elastic lateral restraint, 515. *See also* Pony trusses
  guyed towers, 528
  references, 529
Mill-building columns, 409
  design example, 415
  design procedure, 415
  references, 420
  research needs, 420
  stepped columns, effective length, 411
  types, 410
Multiple column curves, 57
Multistory frames, *see* Frame stability

Parametric resonance, 628
Perforated cover plates, 403
Pipe columns, *see* Circular tubes and Shells; Composite columns
Plate girders, 189. *See also* Beams; Box girders; Plates, local buckling
  bending-shear-strength interaction, 208
  bending strength, hybrid girders, 207
    post web buckling, 206
    webs stiffened longitudinally, 208
  compression flange buckling, vertical, 205
  design principles, 231
  edge-loaded panels, 223
  elastically braced compression flange, 557
  end panel of web, 200
  fatigue from web buckling, 205
  lateral buckling strength, 177
  post-buckling concepts, 189
  references, 234
  research needs, 233
  shear strength, of girder panels, 192
    end panels, 199, 219
    tabular comparison of ten theories, 194
    tabular comparison of test results, 201, 202, 205, 216, 217, 218
  stiffeners, longitudinal, 211, 222
  tension field strength, 192
  transverse, 220
  transverse and longitudinal, combined, 221
  web buckling, 190
Plates, local buckling of, 90. *See also* Thin-walled metal construction
  critical stress, channels and Z-sections, 121
    combined compression and bending, 102, 103
    inelastic buckling in compression, 92
    interaction between elements, 117
    rolled shapes and box sections, 119, 120
    shear, 105, 106
    shear combined with bending, 110, 111
    shear combined with compression, 109
    short plates in compression, 92
    uniform edge compression, 91
    width-thickness ratios for design, 96
  perforation effects, 112
  post-buckling strength, 93, 104, 107
  references, 147
  stiffened plates, 122
    compression and shear combined, 134
    laterally loaded plates, 146
    orthotropic plates, 138
    post-buckling strength, 130
    uniaxial compression, 123
Pony trusses, 515
  AASHTO design requirements, 524
  compression chord buckling, 517
  design procedures, 520, 524
    example, 525
  effective-length factor table, 518
  Engesser theory, 518
  secondary factors, 523

top chord stresses, 523
transverse spring constant, 522
Properties of metals, 3
Proportional limit, definition, 8

Quebec Bridge failure, 387

References, general, 699
Research needs, *see chapter summaries*
Residual stress:
effect on steel columns, 33
rolled steel shapes, 34
welded shapes, 38
Resistance probability functions, 664

Safety, *see* Structural safety
Selected topics in dynamic stability, 627
columns impulsively loaded, 634
flow-induced instability, 643
parametric resonance, 628
references, 653
snap through of shallow structures, 640
suddenly loaded structures, 648
Shanley concept, 30
Shear effect, battened columns, 390
Shells and shell-like structures, *see* Doubly curved shells and shell-like structures
Specifications, *see specific designation,* e.g. AASHTO, AREA
Stability, definition, 8
Stability theory, 15
bifurcation buckling, 15
initially imperfect systems, 19
initially perfect systems, 15
limit load buckling, 23
references, 25
Steels, stress-strain curves, 4
Stepped columns, 409
Stiffened flat plates, 122. *See also* Box girders; Plates; Plate girders
Strain hardening modulus, 9
Stress-strain curve determination, 703, 744
Stress-strain curves, 4
Structural safety, concepts, 661
probability functions, 664
references, 671
reliability index, 665
Structural Stability Research Council, purposes, membership, sponsors, participating organizations, firms and other pertinent information, (Appendix C), 751
Structural steels, stress-strain relationships, 4
Stub-column tests, 9, 708

Tangent modulus, 9. *See also* critical load theory
formula, 30
Shanley interpretation, 30
Tapered members, 329
beam-column interaction, 356
buckling strength, 341
carry-over factors, 335
column strength formulas, 343
frame analysis, 330
lateral-torsional buckling strength, 349
references, 357
research review, 330
slope-deflection equations, 332
stiffness factors, 333
stress computation, 337
Technical memorandums of the Structural Stability Research Council:
No. 1: The Basic Column Formula, 701, (replaced by No. 5)
No. 2: Notes on the Compression Testing of Metals, 703
No. 3: Stub-Column Test Procedure, 708
No. 4: Procedure for Testing Centrally Loaded Columns, 717
No. 5: General Principles for the Stability Design of Metal Structures, 732
No. 6: Determination of Residual Stresses, 734
No. 7: Tension Testing, 744
Tests and test methods, centrally loaded columns, 717. *See also specific types of structures or structural members*
compression tests, 703
Stub-columns, 708
tension tests, 744
Thin-walled compression members, bracing requirements, 446
circular tubes and shells, 461
diaphragm action of panels, 446
flexural buckling, 434
general behavior, 434
local buckling interaction, 436

Thin-walled compression members, bracing requirements (*Continued*)
  singly-symmetric sections, 440
  torsional-flexural buckling theory, 438
    boundary condition coefficients, 441
    design considerations, 442, 445
  wall studs, 438
Thin-walled flexural members, 424. *See also* Plates, critical stress
  box members, 429
  cold-forming effects, 434
  hat-shaped sections, 430
  lateral buckling, 427
  moment capacity, 424
  shear buckling, 431
  unbraced compression flanges, 430
  unstiffened compression elements, 425
  web crippling, 433
  web design, 430
Thin-walled metal construction, references, 454
  specification coverage, 423
Tie plates in battened columns, 399
Torsional-flexural buckling, *see* Thin-walled compression members
Towers, guyed, 528
Truss frameworks, 53

Web buckling, plate girders, 190
Welded columns, 38
Wide-flange (W) rolled shapes, columns, 34

Yield point, 10
Yield strength, structural metals, 10
Yield stress, 10
Yield stress level, 10

Zero strain rate, 747